Gauge theory of elementa

Gauge theory of elementary particle physics

TA-PEI CHENG
University of Missouri–St. Louis

and

LING-FONG LI
Carnegie-Mellon University

CLARENDON PRESS · OXFORD

Oxford University Press, Walton Street, Oxford OX2 6DP
Oxford New York Toronto
Delhi Bombay Calcutta Madras Karachi
Petaling Jaya Singapore Hong Kong Tokyo
Nairobi Dar es Salaam Cape Town
Melbourne Auckland
and associated companies in
Berlin Ibadan

Oxford is a trademark of Oxford University Press

Published in the United States
by Oxford University Press, New York

First published 1984
Reprinted 1985
Reprinted (with corrections) 1988, 1989

British Library Cataloguing in Publication Data
Cheng, Ta-Pei
Gauge theory of elementary particle physics.
1. Gauge fields (Physics)
I. Title. II. Li, Ling-Fong
539.'21 QC793.3.F5
ISBN 0-19-851956-7
ISBN 0-19-851961-3 Pbk

Library of Congress Cataloging in Publication Data
Cheng, Ta-Pei
Gauge theory of elementary particle physics.
Bibliography: p.
Includes index.
1. Gauge fields (Physics) 2. Particles (Nuclear Physics)
I. Li, Ling-Fong. II. Title
QC793.3.F5C48 1983 539.7'21'01 83–7623
ISBN 0-19-851956-7
ISBN 0-19-851961-3 (pbk)

Printed in Great Britain by St. Edmundsbury Press
Bury St Edmunds, Suffolk

Preface

Elementary particle physics has made remarkable progress in the past ten years. We now have, for the first time, a comprehensive theory of particle interactions. One can argue that it gives a complete and correct description of all non-gravitational physics. This theory is based on the principle of gauge symmetry. Strong, weak, and electromagnetic interactions are all gauge interactions. The importance of a knowledge of gauge theory to anyone interested in modern high energy physics can scarcely be overstated. Regardless of the ultimate correctness of every detail of this theory, it is the framework within which new theoretical and experimental advances will be interpreted in the foreseeable future.

The aim of this book is to provide student and researcher with a practical introduction to some of the principal ideas in gauge theories and their applications to elementary particle physics. Wherever possible we avoid intricate mathematical proofs and rely on heuristic arguments and illustrative examples. We have also taken particular care to include in the derivations intermediate steps which are usually omitted in more specialized communications. Some well-known results are derived anew, in a way more accessible to a non-expert.

The book is not intended as an exhaustive survey. However, it should adequately provide the general background necessary for a serious student who wishes to specialize in the field of elementary particle theory. We also hope that experimental physicists with interest in some general aspects of gauge theory will find parts of the book useful.

The material is based primarily on a set of notes for the graduate courses taught by one of us (L.F.L.) over the past six years at the Carnegie–Mellon University and on lectures delivered at the 1981 Hefei (China) Summer School on Particle Physics (Li 1981). It is augmented by material covered in seminars given by the other author (T.P.C.) at the University of Minnesota and elsewhere. These notes have been considerably amplified, reorganized, and their scope expanded. In this text we shall assume that the reader has had some exposure to quantum field theory. She or he should also be moderately familiar with the phenomenology of high energy physics. In practical terms we have in mind as a typical reader an advanced graduate student in theoretical physics; it is also our hope that some researchers will use the book as a convenient guide to topics that they wish to look up.

Modern gauge theory may be described as being a 'radically conservative theory' in the sense used by Wheeler (see Wilczek 1982*b*). Thus, one extrapolates a few fundamental principles as far as one can, accepting some 'paradoxes' that fall short of contradiction. Here we take as axioms the principles of *locality*, *causality*, and *renormalizability*. We discover that a certain class of relativistic quantum field theory, i.e. the gauge theory, contains unexpected richness (Higgs phenomena, asymptotic freedom, confinement, anomalies, etc.), which is necessary for an understanding of elementary particle interactions. And yet, this does not occasion any revision of the basic principles of relativity and quantum mechanics. Thus the prerequisite for the study of gauge theory is just the traditional preparation in advanced quantum mechanics and quantum field theory, especially the prototype gauge theory of quantum electrodynamics (QED).

The book is organized in two parts. Part I contains material that can be characterized as being 'pre-gauge theory'. In Chapters 1, 2, and 3 the basics of relativisitic quantum field theory (quantization and renormalization) are reviewed, using the simple $\lambda\phi^4$ theory as an illustrative example. In Chapters 4 and 5 we present the elements of group theory, the quark model, and chiral symmetry. The interrelationship of the above main topics—renormalization and symmetry—is then studied in Chapter 6. The argument that quarks are the basic constituents of hadrons is further strengthened by the discovery of Bjorken scaling. Scaling and the quark–parton model are described in Chapter 7. These results paved the way for the great synthesis of particle interaction theories in the framework of the non-Abelian gauge theories, which is treated in Part II. After the classical and quantized versions of gauge theories are discussed in Chapters 8 and 9, we are then ready for the core chapters of this book—Chapters 10–14—where gauge theories of quantum chromodynamics (QCD), quantum flavourdynamics (QFD), and grand unification (GUT) are presented. As a further illustration of the richness of the gauge theory structure we exhibit its nonperturbative solutions in the form of magnetic monopoles and instantons in Chapters 15 and 16.

We have also included at the end of the book two appendices. In Appendix A one can find the conventions and normalizations used in this book. Appendix B contains a practical guide to the derivation of Feynman rules as well as a summary of the propagators and vertices for the most commonly used theories—the $\lambda\phi^4$, Yukawa, QCD, and the (R_ξ gauge) standard model of the electroweak interaction.

In the table of contents we have marked sections and chapters to indicate whether they are an essential part (unmarked), or details that may be omitted upon a first reading (marked by an asterisk), or introductions to advanced topics that are somewhat outside the book's main line of development (marked by a dagger). From our experience the material covered in the unmarked sections is sufficient for a one-semester course on the gauge theory of particle physics. Without omitting the marked sections, the book as a whole is adequate for a two-semester course. It should also be pointed out that although we have organized the sections according to their logical interconnection there is no need (it is in fact unproductive!) for the reader to strictly follow the order of our presentation. For example, §1.2 on path integral quantization can be postponed until Chapter 9 where it will be used for the first time when we quantize the gauge theories. As we anticipate a readership of rather diverse background and interests, we urge each reader to study the table of contents carefully before launching into a study programme. A certain amount of repetition is deliberately built into the book so that the reader can pick and choose different sections without any serious problems. An experimentally inclined reader, who is not particularly interested in the formal aspects of relativistic quantum field theory, can skip Chapters 1, 2, 3, and 6 on quantization and renormalization. After an introductory study of group theory and the quark model in Chapters 4, 5, and 7 she or he should proceed directly to the parts of Chapters 8, 10, 11, 12, 14, etc. where a general introduction to and applications of gauge theory can be found.

The sections on references and bibliography at the end of the book represent some of the commonly cited references that we ourselves are familiar with. They are not a comprehensive listing. We apologize to our colleagues who have been inadequately referenced. Our hope is that we have provided a sufficient set so that an interested reader can use it to go on to find further reviews and research articles.

It is a pleasure to acknowledge the aid we have received from our colleagues and students; many have made helpful comments about the preliminary version of the book. We are very grateful to Professor Mahiko Suzuki who undertook a critical reading of the manuscript, and also to Professors James Bjorken, Sidney Drell, Jonathan Rosner, and Lincoln Wolfenstein for having encouraged us to begin the conversion of the lecture notes into a book. One of us (T.P.C.) would like to thank the National Science Foundation, UMSL Summer Research Fellowship Committee, and the Weldon Spring Endowment for support. During various stages of working on this project he has enjoyed the hospitality of the theoretical physics groups at the Lawrence Berkeley Laboratory, the Stanford Linear Accelerator Center and the University of Minnesota. L.F.L. would like to thank the Institute for Theoretical Physics at the University of California–Santa Barbara for hospitality and the Department of Energy and the Alfred P. Sloan Foundation for support. Finally, we also gratefully acknowledge the encouragement and help given by our wives throughout this project. And, we are much indebted to Ms Susan Swyers for the painstaking task of typing this manuscript. Other technical assistance by Ms Tina Ramey and Mr Jerry McClure is also much appreciated.

Note added in proof. As this manuscript was being readied for publication we received the news that the CERN UA1 and UA2 groups have observed events in $p\bar{p}$ collisions which may be interpreted as the production of an intermediate vector boson W with a mass approximately 80 GeV. Also, the Irvine–Michigan–Brookhaven collaboration reported a preliminary result setting a lower bound for the lifetime $\tau(p \to e^+ \pi^0) > 6.5 \times 10^{31}$ years.

St. Louis and Pittsburgh T.P.C.
September 1982 L.F.L.

Contents

Unmarked sections are the basic part of the book; those labelled with an asterisk contain details that may be omitted upon a first reading. Sections and chapters marked with a dagger are elementary introductions to advanced topics that are somewhat outside the book's main line of development.

PART I

Part I

1 Basics in field quantization

The dynamics of a classical field $\phi(x)$ are determined by the Lagrangian density $\mathscr{L}(\phi, \partial_\mu\phi)$ through the action principle

$$\delta S = 0 \tag{1.1}$$

where S is the action

$$S = \int \mathrm{d}^4x \mathscr{L}(\phi, \partial_\mu\phi).$$

This extremization leads to the Euler–Lagrange equation of motion

$$\partial_\mu \frac{\delta\mathscr{L}}{\delta(\partial_\mu\phi)} - \frac{\delta\mathscr{L}}{\delta\phi} = 0. \tag{1.2}$$

To quantize a system we can adopt either of two equivalent approaches. The canonical formalism involves the identification of the true dynamical variables of the system. They are taken to be operators and are postulated to satisfy the canonical commutation relations. The Hamiltonian of the system is constructed and used to find the time evolution of the system. This allows us to compute the transition amplitude from the state at an initial time to the state at final time. Alternatively, we can use the Feynman path-integral formalism to describe the quantum system. Here the transition amplitude is expressed directly as the sum (a functional integral) over all possible paths between the initial and final states, weighted by the exponential of i times the action (in units of the Planck's constant $\hbar$) for the particular path. Thus in the classical limit ($\hbar \to 0$) the integrand oscillates greatly, making a negligible contribution to the integral except along the stationary path selected by the action principle of eqn (1.1).

In this chapter we present an elementary study of field quantization. First we review the more familiar canonical quantization procedure and its perturbative solutions in the form of Feynman rules. Since we will find that gauge field theories are most easily quantized using the path-integral formalism we will present an introduction to this technique (and its connection to Feynman rules) in §1.2. For the most part the simplest case of the self-interacting scalar particle will be used as the illustrative example; path-integral formalism for fermions will be presented in §1.3.

Since the path-integral formalism will not be used until Chapter 9 when we quantize the gauge fields, the reader may wish to postpone the study of §§1.2 and 1.3 until then. It should also be pointed out that even for gauge theories we shall use these two quantization formalisms in an intermixed fashion. By this we mean that we will use whatever language is most convenient for the task at hand, regardless of whether it implies path-integral or canonical quantization. For example, in the discussion of the short-distance phenomena in Chapter 10, we continue to use the language of 'operator product expansion' even though strictly speaking this implies canonical quantization. The reader is also referred to Appendix B at the end of the book where one can find a practical guide to derivation of Feynman rules via path-integral formalism.

1.1 Review of canonical quantization formalism

We assume familiarity with the transition from a classical nonrelativistic particle system to the corresponding quantum system. The Schrödinger equation is obtained after we replace the canonical variables by operators and the Poisson brackets by commutators. These operators act on the Hilbert space of square integrable functions (the wavefunctions), and they satisfy equations of motion which are formally identical to the classical equations of motion.

A relativistic field may be quantized by a similar procedure. For a system described by the Lagrangian density $\mathscr{L}(\phi, \partial_\mu\phi)$, the field $\phi(x)$ satisfies the classical equation of motion given in eqn (1.2). We obtain the corresponding quantum system by imposing the canonical commutation relations at equal time

$$[\pi(\mathbf{x}, t), \phi(\mathbf{x}', t)] = -\mathrm{i}\delta^3(\mathbf{x} - \mathbf{x}')$$
$$[\pi(\mathbf{x}, t), \pi(\mathbf{x}', t)] = [\phi(\mathbf{x}, t), \phi(\mathbf{x}', t)] = 0 \tag{1.3}$$

where the conjugate momentum is defined by

$$\pi(x) = \frac{\delta\mathscr{L}}{\delta(\partial_0\phi)}. \tag{1.4}$$

The Hamiltonian

$$H = \int \mathrm{d}^3x[\pi(x)\,\partial_0\phi(x) - \mathscr{L}(x)] \tag{1.5}$$

governs the dynamics of the system

$$\partial_0\phi(\mathbf{x}, t) = \mathrm{i}[H, \phi(\mathbf{x}, t)]$$
$$\partial_0\pi(\mathbf{x}, t) = \mathrm{i}[H, \pi(\mathbf{x}, t)]. \tag{1.6}$$

Example 1.1. Free scalar field. Given the Lagrangian density

$$\mathscr{L} = \tfrac{1}{2}[(\partial_\lambda\phi)(\partial^\lambda\phi) - \mu^2\phi^2],$$

eqn (1.2) yields the Klein–Gordon equation

$$(\partial^2 + \mu^2)\phi(x) = 0. \tag{1.7}$$

In quantum field theory the field $\phi(x)$ and its conjugate momentum operators given by eqn (1.4), $\pi(x) = \partial_0\phi(x)$, satisfy the canonical commutation relations

$$[\partial_0\phi(\mathbf{x}, t), \phi(\mathbf{x}', t)] = -\mathrm{i}\delta^3(\mathbf{x} - \mathbf{x}')$$
$$[\partial_0\phi(\mathbf{x}, t), \partial_0\phi(\mathbf{x}', t)] = [\phi(\mathbf{x}, t), \phi(\mathbf{x}', t)] = 0. \tag{1.8}$$

The Hamiltonian is given by

$$H_0 = \int \mathrm{d}^3x \tfrac{1}{2}[(\partial_0\phi)^2 + (\nabla\phi)^2 + \mu^2\phi^2]. \tag{1.9}$$

The time evolution equation (1.6), which is basically Hamilton's equation of motion, can be cast in the form of (1.7). Thus the field operator $\phi(x)$ formally satisfies the Klein–Gordon equation. This simple non-interacting case can be solved and we have

$$\phi(\mathbf{x}, t) = \int \frac{\mathrm{d}^3k}{[(2\pi)^3 2\omega_k]^{1/2}} [a(\mathbf{k})\, \mathrm{e}^{\mathrm{i}(\mathbf{k}\cdot\mathbf{x} - \omega_k t)} + a^\dagger(\mathbf{k})\, \mathrm{e}^{-\mathrm{i}(\mathbf{k}\cdot\mathbf{x} - \omega_k t)}] \tag{1.10}$$

where $\omega_k = (\mathbf{k}^2 + \mu^2)^{1/2}$. The coefficients of expansion $a(\mathbf{k})$ and $a^\dagger(\mathbf{k})$ are operators. The canonical commutation relations of eqn (1.8) are transcribed into

$$[a(\mathbf{k}), a^\dagger(\mathbf{k}')] = \delta^3(\mathbf{k} - \mathbf{k}')$$
$$[a(\mathbf{k}), a(\mathbf{k}')] = [a^\dagger(\mathbf{k}), a^\dagger(\mathbf{k}')] = 0 \tag{1.11}$$

and the Hamiltonian of eqn (1.9) can be expressed as

$$H_0 = \int \mathrm{d}^3k \omega_k a^\dagger(\mathbf{k}) a(\mathbf{k}) \tag{1.12}$$

where we have discarded an irrelevant constant. Remembering the situation of the harmonic oscillator, we see immediately that $a(\mathbf{k})$ and $a^\dagger(\mathbf{k})$ can be interpreted as destruction and creation operators. Thus the one-particle state with momentum $\mathbf{k}$ is given by the creation operator acting on the vacuum state

$$|\mathbf{k}\rangle = [(2\pi)^3 2\omega_k]^{1/2} a^\dagger(\mathbf{k})|0\rangle \tag{1.13}$$

where the normalization is

$$\langle \mathbf{k}'|\mathbf{k}\rangle = (2\pi)^3 2\omega_k\, \delta^3(\mathbf{k} - \mathbf{k}').$$

The product $a^\dagger a$ has the usual interpretation as a number operator and eqn (1.12) shows that H_0 is the Hamiltonian for a system of non-interacting particles.

Given the solution, (1.10), and (1.11), we can easily calculate the Feynman propagator function, which is the vacuum expectation value for a time-ordered product of two fields,

$$\begin{aligned}\mathrm{i}\Delta(x_1 - x_2) &\equiv \langle 0|T(\phi(x_1)\phi(x_2))|0\rangle \\ &= \theta(t_1 - t_2)\langle 0|\phi(x_1)\phi(x_2)|0\rangle + \theta(t_2 - t_1)\langle 0|\phi(x_2)\phi(x_1)|0\rangle \\ &= \int \frac{\mathrm{d}^4k}{(2\pi)^4} \frac{\mathrm{i}}{k^2 - \mu^2 + \mathrm{i}\varepsilon} \exp\{\mathrm{i}k\cdot(x_1 - x_2)\}.\end{aligned} \tag{1.14}$$

Example 1.2. Scalar field with $\lambda\phi^4$ interaction. The Lagrangian density is given by

$$\mathscr{L} = \tfrac{1}{2}[(\partial_\lambda\phi)(\partial^\lambda\phi) - \mu^2\phi^2] - \frac{\lambda}{4!}\phi^4,$$

the equation of motion is

$$(\partial^2 + \mu^2)\phi(x) = -\frac{\lambda}{3!}\phi^3(x). \tag{1.15}$$

The conjugate momentum and canonical commutation relations are the same as those for the free-field case in Example 1.1. The Hamiltonian is of the form

$$H = H_0 + H' \tag{1.16}$$

where H_0 is given by eqn (1.9) and

$$H' = \int d^3x\mathscr{H}'$$

where

$$\mathscr{H}' = \frac{\lambda}{4!}\phi^4 \tag{1.17}$$

is the interaction Hamiltonian density. Since the free-field theory is soluble we can obtain transition amplitudes and matrix elements of physical interest by a systematic expansion in λ. This approximation scheme of perturbation theory will be briefly outlined below.

In the usual *Heisenberg picture* the operators are time-dependent and the time evolutions of the dynamical variables of the system are governed by the Hamiltonian

$$\phi(\mathbf{x}, t) = e^{iHt}\phi(\mathbf{x}, 0)\, e^{-iHt}$$

$$\pi(\mathbf{x}, t) = e^{iHt}\pi(\mathbf{x}, 0)\, e^{-iHt}.$$

The state vector $|a\rangle$ is time-independent. But, in the *Schrödinger picture*, the operators are time-independent and state vectors carry time dependence. They are related to those in the Heisenberg picture by

$$\phi^S(\mathbf{x}) = e^{-iHt}\phi(\mathbf{x}, t)\, e^{iHt}$$

$$\pi^S(\mathbf{x}) = e^{-iHt}\pi(\mathbf{x}, t)\, e^{iHt}$$

$$|a, t\rangle^S = e^{-iHt}|a\rangle.$$

In perturbation theory we introduce another picture—the *interaction picture*—with operators and states defined by

$$\begin{aligned}\phi^I(\mathbf{x}, t) &\equiv e^{iH_0t}\phi^S(\mathbf{x})\, e^{-iH_0t} \\ &= e^{iH_0t}\, e^{-iHt}\phi(\mathbf{x}, t)\, e^{iHt}\, e^{-iH_0t} \\ &= U(t, 0)\phi(\mathbf{x}, t)U^{-1}(t, 0)\end{aligned} \tag{1.18}$$

Similarly, for $\pi^{\mathrm{I}}(\mathbf{x}, t)$,

$$|a, t\rangle^{\mathrm{I}} \equiv \mathrm{e}^{\mathrm{i}H_0 t}|a, t\rangle^{\mathrm{S}}$$

$$= U(t, 0)|a\rangle \tag{1.19}$$

where

$$U(t, 0) = \mathrm{e}^{\mathrm{i}H_0 t}\,\mathrm{e}^{-\mathrm{i}Ht} \tag{1.20}$$

is the unitary time-evolution operator. Since the operators in the interaction picture (1.18) satisfy the (soluble) free-field equation

$$\partial_0 \phi^{\mathrm{I}}(\mathbf{x}, t) = \mathrm{i}[H_0, \phi^{\mathrm{I}}(\mathbf{x}, t)]$$

$$\partial_0 \pi^{\mathrm{I}}(\mathbf{x}, t) = \mathrm{i}[H_0, \pi^{\mathrm{I}}(\mathbf{x}, t)], \tag{1.21}$$

the dynamic problem in this language becomes that of finding the solution for the U-matrix.

The time-evolution operator U can be defined more generally than in (1.20),

$$|a, t\rangle^{\mathrm{I}} \equiv U(t, t_0)|a, t_0\rangle^{\mathrm{I}} \tag{1.22}$$

where $U(t_0, t_0) = 1$ and satisfies the multiplication rule,

$$U(t, t')U(t', t_0) = U(t, t_0)$$

$$U(t, 0)U^{-1}(t_0, 0) = U(t, t_0). \tag{1.23}$$

The equation of motion for the U-operator can be deduced from eqns (1.19) and (1.20)

$$\mathrm{i}\frac{\partial}{\partial t} U(t, t_0) = H'^{\mathrm{I}}(t)U(t, t_0) \tag{1.24}$$

where

$$H'^{\mathrm{I}} = \mathrm{e}^{\mathrm{i}H_0 t}H'\,\mathrm{e}^{-\mathrm{i}H_0 t}$$

is the interaction Hamiltonian in the interaction picture, i.e.

$$H'^{\mathrm{I}} = H'(\phi^{\mathrm{I}}). \tag{1.25}$$

Eqn (1.24) has the solution

$$U(t, t_0) = T\exp\left[-\mathrm{i}\int_{t_0}^{t} \mathrm{d}t_1 H'^{\mathrm{I}}(t_1)\right]$$

$$= T\exp\left[-\mathrm{i}\int_{t_0}^{t} \mathrm{d}t_1 \int \mathrm{d}^3x_1 \mathscr{H}'^{\mathrm{I}}(\mathbf{x}_1, t_1)\right] \tag{1.26}$$

which can be expanded in a power series

$$U(t, t_0) = 1 + \sum_{p=1}^{\infty} \frac{(-\mathrm{i})^p}{p!} \int_{t_0}^{t} \mathrm{d}^4x_1 \int_{t_0}^{t} \mathrm{d}^4x_2 \ldots$$

$$\times \int_{t_0}^{t} \mathrm{d}^4x_p T(\mathscr{H}'^{\mathrm{I}}(x_1)\mathscr{H}'^{\mathrm{I}}(x_2)\cdots\mathscr{H}'^{\mathrm{I}}(x_p)). \tag{1.27}$$

Green's functions in $\lambda\phi^4$ theory. Next we need to translate this formal perturbative solution into quantities that will have more direct physical meaning.

In field theory we are interested in calculating Green's function defined by

$$G^{(n)}(x_1, \ldots, x_n) = \langle 0|T(\phi(x_1), \ldots, \phi(x_n))|0\rangle \tag{1.28}$$

from which the S-matrix elements can be readily obtained. By a straightforward application of eqns (1.18), (1.23), and (1.26) we have

$$G^{(n)}(x, \ldots, x_n) = \langle 0|U^{-1}(t, 0)T(\phi^{\mathrm{I}}(x_1) \ldots \phi^{\mathrm{I}}(x_n) \times \exp\left[-\mathrm{i}\int_{-t}^{t} \mathrm{d}t' H'^{\mathrm{I}}(t')\right])U(-t, 0)|0\rangle \tag{1.29}$$

where t is some reference time which we shall eventually let approach ∞. In this limit the vacuum state becomes an eigenstate of the U-operator, and the eigenvalue product of the two Us in (1.29) becomes

$$\frac{1}{\langle 0|T\left(\exp\left[-\mathrm{i}\int_{-\infty}^{\infty} \mathrm{d}t' H'^{\mathrm{I}}(t')\right]\right)|0\rangle}. \tag{1.30}$$

The effect of these two Us acting on the vacuum states is to take out 'the disconnected part' of the vacuum expectation value (see, for example, Bjorken and Drell 1965). Also, after we substitute the power series expansion of (1.27), the n-point Green's function with the notation of (1.25) takes on the form

$$G^{(n)}(x_1, \ldots, x_n) = \sum_{p=0}^{\infty} \frac{(-\mathrm{i})^p}{p!} \int_{-\infty}^{\infty} \mathrm{d}^4 y_1, \ldots, \mathrm{d}^4 y_p \ \langle 0|T(\phi^{\mathrm{I}}(x_1), \ldots, \phi^{\mathrm{I}}(x_n) \times \mathscr{H}'(\phi^{\mathrm{I}}(y_1)) \ldots \mathscr{H}'(\phi^{\mathrm{I}}(y_p))|0\rangle_c. \tag{1.31}$$

The subscript c denotes 'the connected part'. The terminology clearly reflects features in the graphic representation of the Green's function.

Consider the simplest example of a first-order ($p = 1$) term for a four-point ($n = 4$) Green's function in the theory with $\mathscr{H}'(\phi) = \lambda/4!\ \phi^4$ as in (1.17)

$$G_1^{(4)}(x_1, \ldots, x_4) = -\frac{\mathrm{i}\lambda}{4!}\int \mathrm{d}^4 y \langle 0|T(\phi^{\mathrm{I}}(x_1), \ldots, \phi^{\mathrm{I}}(x_4)[\phi^{\mathrm{I}}(y)]^4)|0\rangle. \tag{1.32}$$

We then normal-order the entire time-ordered product by moving the creation operators to the left of the annihilation operators and eliminating those terms which end up with the annihilation operator on the right and/or the creation operator on the left. After this application of Wick's theorem (1950) the connected part of the expression in (1.32) decomposes into a product of two-point functions

$$G_1^{(4)}(x_1, \ldots, x_4) = (-\mathrm{i}\lambda)\int \mathrm{d}^4 y \langle 0|T(\phi^{\mathrm{I}}(x_1)\phi^{\mathrm{I}}(y))|0\rangle\langle 0|T(\phi^{\mathrm{I}}(x_2)\phi^{\mathrm{I}}(y))|0\rangle \langle 0|T(\phi^{\mathrm{I}}(x_3)\phi^{\mathrm{I}}(y))|0\rangle\langle 0|T(\phi^{\mathrm{I}}(x_4)\phi^{\mathrm{I}}(y))|0\rangle. \tag{1.33}$$

The original 4! factor in the denominator of (1.32) is cancelled because there are a corresponding number of ways to contract the $\phi^{\mathrm{I}}(x_i)$s with each field in $[\phi^{\mathrm{I}}(y)]^4$. The interaction picture field obeying the free-field equation, the propagator $\mathrm{i}\Delta(x, y) \equiv \langle 0|T(\phi^{\mathrm{I}}(x)\phi^{\mathrm{I}}(y))|0\rangle$, is a known quantity and is given by eqn (1.14). A graphic representation of the expression in (1.33) is shown in Fig. 1.1.

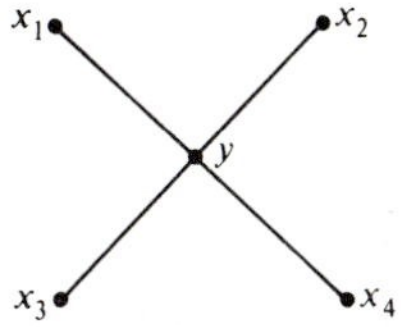

FIG. 1.1. Graphic representation of eqn (1.33).

We next consider the example of the second-order term for the four-point function, i.e. the $p = 2$, $n = 4$ term in eqn (1.31)

$$G_2^{(4)}(x_1 \ldots x_4) = \frac{1}{2!}\left(\frac{-\mathrm{i}\lambda}{4!}\right)^2 \int \mathrm{d}^4y_1\, \mathrm{d}^4y_2 \langle 0|T(\phi^{\mathrm{I}}(x_1) \ldots \phi^{\mathrm{I}}(x_4)[\phi^{\mathrm{I}}(y_1)]^4 \times [\phi^{\mathrm{I}}(y_2)]^4)|0\rangle. \tag{1.34}$$

We then use Wick's theorem to reduce it and keep only the connected parts,

$$\begin{aligned} G_2^{(4)}(x_1 \ldots x_4) &= \frac{1}{2!}(-\mathrm{i}\lambda)^2 \int \mathrm{d}^4y_1\, \mathrm{d}^4y_2 [\mathrm{i}\Delta(y_1, y_2)]^2 \{[\Delta(x_1, y_1)\, \Delta(x_2, y_1)] \\ &\quad \times [\Delta(x_3, y_2)\, \Delta(x_4, y_2)] + [\Delta(x_1, y_1)\, \Delta(x_3, y_1)][\Delta(x_2, y_2)\, \Delta(x_4, y_2)] \\ &\quad + [\Delta(x_1, y_1)\, \Delta(x_4, y_1)][\Delta(x_2, y_2)\, \Delta(x_3, y_2)]\} + \frac{1}{2!}(-\mathrm{i}\lambda)^2 \int \mathrm{d}^4y_1\, \mathrm{d}^4y_2 \\ &\quad \times [\mathrm{i}\Delta(y_1, y_1)][\mathrm{i}\Delta(y_1, y_2)]\{\Delta(x_1, y_1)\, \Delta(x_2, y_2)\, \Delta(x_3, y_2)\, \Delta(x_4, y_2) \\ &\quad + \Delta(x_1, y_2)\, \Delta(x_2, y_1)\, \Delta(x_3, y_2)\, \Delta(x_4, y_2) + \Delta(x_1, y_2)\, \Delta(x_2, y_2) \\ &\quad \times \Delta(x_3, y_1)\, \Delta(x_4, y_2) + \Delta(x_1, y_2)\, \Delta(x_2, y_2)\, \Delta(x_3, y_2)\, \Delta(x_4, y_1)\}. \end{aligned} \tag{1.35}$$

The symmetry factor 2! in the first term on the right-hand side of (1.35) can be understood as follows. The original factors of $(1/2!)(1/4!)^2$ in eqn (1.34) are cancelled by the permutation of y_1 and y_2 and by the multiple ways in which we can attach the fields emanating out of each vertex. However, this is an over-counting. Since there are two identical internal lines connecting a pair of vertices, we need to divide out a factor of 2!. The factor 1/2 in the second term of (1.35) has a different origin. It comes from the fact that an internal line starts and terminates in the same vertex. Eqn (1.35) is shown graphically in Fig. 1.2.

Similarly, for a first-order ($p = 1$) two-point ($n = 2$) function, we have

$$G_1^{(2)}(x_1, x_2) = \frac{-\mathrm{i}\lambda}{2!} \int \mathrm{d}^4y [\mathrm{i}\Delta(x_1, y)][\mathrm{i}\Delta(y, y)][\mathrm{i}\Delta(x_2, y)] \tag{1.36}$$

as shown in Fig. 1.3.

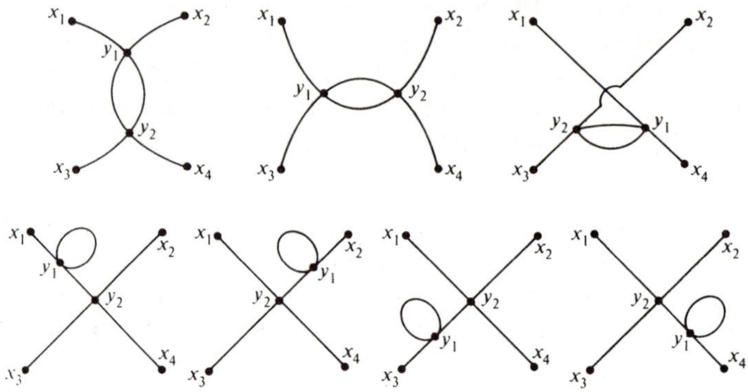

FIG. 1.2. Graphic representation of eqn (1.35).

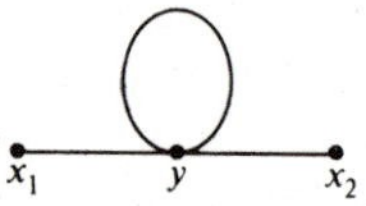

FIG. 1.3. Graphic representation of eqn (1.36).

Usually it is more convenient to work with Green's function in momentum space

$$(2\pi)^4\,\delta^4(p_1+\ldots+p_n)G^{(n)}(p_1\ldots p_n)\equiv\int\prod_{i=1}^{n}\mathrm{d}^4x_i\,\mathrm{e}^{-\mathrm{i}p_ix_i}G^{(n)}(x_1\ldots x_n) \tag{1.37}$$

and with the *amputated Green's function*, which is related to $G^{(n)}(p_1\ldots p_n)$ by removing the propagators on external lines

$$G^{(n)}_{\mathrm{amp}}(p_1\ldots p_n)=\left[\prod_{j=1}^{n}\frac{-\mathrm{i}}{\Delta(p_j)}\right]G^{(n)}(p_1\ldots p_n) \tag{1.38}$$

where $p_1+p_2+\ldots p_n=0$. In fact for spin-0 particles the amputated Green's function is just the usual transition amplitude (the *T-matrix element*) from which the cross-section can be directly computed.

Feynman rules of $\lambda\phi^4$ theory. The result of perturbation theory may be conveniently summarized in terms of the *Feynman rules* for the transition amplitude. With the $\mathscr{H}_1=(\lambda/4!)\phi^4(x)$ interaction, we have the following prescription for calculating the N-point amputated Green's function.

1. Draw all possible connected, topologically distinct, graphs with N external lines;
2. For each internal line, put in the propagator factor

(i) $\mathrm{i}\Delta_{\mathrm{F}}(p)=\dfrac{\mathrm{i}}{p^2-\mu^2+\mathrm{i}\varepsilon}$;

For each vertex,

(ii)  $-i\lambda$;

3. For each internal momentum l not fixed by momentum conservation at each vertex, perform an integration $\int d^4l/(2\pi)^4$;
4. Each graph has to be divided by a symmetry factor S corresponding to the number of permutations of internal lines one can make for fixed vertices.

.2 Introduction to path integral formalism

In the first section the canonical quantization procedure via operator formalism was briefly reviewed. We have outlined the steps through which the perturbative solution of an interacting field theory may be obtained in the form of Feynman rules. In this section the same set of rules will be recovered using the path-integral (PI) formalism (Dirac 1933; Feynman 1948*a*; Schwinger 1951*b*). This alternative quantization approach has the advantage of exhibiting a closer relationship to the classical dynamical description and the manipulation involves only ordinary functions. This allows us to see more clearly the effect of any nonlinear transformations on the fundamental variables. Thus the PI formalism is particularly suited for handling constrained systems such as gauge theories.

Quantum mechanics in one dimension

We first introduce the PI formalism in the simplest quantum-mechanical system in one dimension. Generalization to field theory with infinite degrees of freedom, together with its perturbative solution, will be presented in a later part of this section.

In quantum mechanics a fundamental quantity is the transition matrix element corresponding to the overlap between initial and final stages

$$\langle q'; t'|q; t\rangle = \langle q'|\, e^{-iH(t'-t)}|q\rangle \tag{1.39}$$

where the $|q\rangle$s are eigenstates of the position operator Q in the Schrödinger picture with eigenvalue q

$$Q|q\rangle = q|q\rangle \tag{1.40}$$

and the $|q; t\rangle$s on the left-hand side of (1.39) denote the states in the Heisenberg picture, $|q; t\rangle = e^{iHt}|q\rangle$. It should be remembered that the Heisenberg-picture states do not carry time-dependance. The notation used here means that the Heisenberg-picture states $|q; t\rangle$ and $|q'; t'\rangle$ in eqn (1.39) coincide with two distinctive Schrödinger-picture states $|q(t)\rangle$ and $|q'(t)\rangle$ at time t and t' respectively. In the PI formalism the transformation matrix element of eqn (1.39) is written as a functional integral

$$\langle q'; t' | q; t\rangle = N \int [\mathrm{d}q] \exp\left\{ \mathrm{i} \int_t^{t'} L(q, \dot{q})\, \mathrm{d}\tau \right\} \tag{1.41}$$

where N is the normalization factor and $L(q, \dot{q})$ is the Lagrangian. The integration is performed in the function space $q(t)$. It represents the sum of contributions over all paths that connect (q, t) and (q', t'), weighted by the exponential of i times the action. In the following we will derive (1.41) using the familiar canonical-operator formalism. The definition of the integration measure $[\mathrm{d}q]$ will be given in eqns (1.50) and (1.51) and this should clarify the meaning of the functional integral.

We first divide the interval (t', t) into n segments with space $\delta t = (t' - t)/n$. Then the transition amplitude in eqn (1.39) may be written

$$\langle q' | \mathrm{e}^{-\mathrm{i}H(t'-t)} | q \rangle = \int \mathrm{d}q_1 \ldots \mathrm{d}q_{n-1} \langle q' | \mathrm{e}^{-\mathrm{i}H\delta t} | q_{n-1} \rangle \langle q_{n-1} | \mathrm{e}^{-\mathrm{i}H\delta t} | q_{n-2} \rangle \ldots$$
$$\times \ldots \langle q_1 | \mathrm{e}^{-\mathrm{i}H\delta t} | q \rangle \tag{1.42}$$

where we have inserted complete sets of eigenstates of the Schrödinger picture operator Q^{S}. For sufficiently small δt,

$$\langle q' | \mathrm{e}^{-\mathrm{i}H\delta t} | q \rangle = \langle q' | [1 - \mathrm{i}H(P, Q)\, \delta t] | q \rangle + O(\delta t)^2. \tag{1.43}$$

If the Hamiltonian has the form

$$H(P, Q) = \frac{P^2}{2m} + V(Q), \tag{1.44}$$

then

$$\begin{aligned}
\langle q' | H(P, Q) | q \rangle &= \langle q' | \frac{P^2}{2m} | q \rangle + V\left(\frac{q + q'}{2}\right) \delta(q - q') \\
&= \int \frac{\mathrm{d}p}{2\pi} \langle q' | p \rangle \langle p | \frac{P^2}{2m} | q \rangle + V\left(\frac{q + q'}{2}\right) \int \frac{\mathrm{d}p}{2\pi} \mathrm{e}^{\mathrm{i}p(q'-q)} \\
&= \int \frac{\mathrm{d}p}{2\pi} \mathrm{e}^{\mathrm{i}p(q'-q)} \left[\frac{p^2}{2m} + V\left(\frac{q + q'}{2}\right) \right].
\end{aligned}$$

We have used $\langle q' | q \rangle = \delta(q' - q)$ and $\langle q | p \rangle = \mathrm{e}^{\mathrm{i}pq}$. Also, symmetric ordering of operators in $V(Q)$ is assumed. Then

$$\begin{aligned}
\langle q' | \mathrm{e}^{-\mathrm{i}H\delta t} | q \rangle &\simeq \int \frac{\mathrm{d}p}{2\pi} \mathrm{e}^{\mathrm{i}p(q'-q)} \left\{ 1 - \mathrm{i}\, \delta t \left[\frac{p^2}{2m} + V\left(\frac{q + q'}{2}\right) \right] \right\} \\
&\simeq \int \frac{\mathrm{d}p}{2\pi} \mathrm{e}^{\mathrm{i}p(q'-q)}\, \mathrm{e}^{-\mathrm{i}\,\delta t H(p, q + q'/2)}.
\end{aligned} \tag{1.45}$$

Thus, $H(p, q)$ is the classical Hamiltonian. Substituting into (1.42), we have

$$\langle q'|\, e^{-iH(t'-t)}|q\rangle \simeq \int\left(\frac{dp_1}{2\pi}\right)\cdots\left(\frac{dp_n}{2\pi}\right)\int dq_1\ldots dq_{n-1}$$

$$\times \exp\left\{i\sum_{i=1}^{n}\left[p_i(q_i - q_{i-1}) - \delta t H\left(p_i, \frac{q_i + q_{i-1}}{2}\right)\right]\right\}. \tag{1.46}$$

The transition amplitude can then be written symbolically as

$$\langle q'|\, e^{-iH(t'-t)}|q\rangle = \int\left[\frac{dp\, dq}{2\pi}\right]\exp\left\{i\int_t^{t'} dt[p\dot{q} - H(p,q)]\right\} \tag{1.47}$$

$$\equiv \lim_{n\to\infty}\int\left(\frac{dp_1}{2\pi}\right)\cdots\left(\frac{dp_n}{2\pi}\right)\int dq_1\ldots dq_{n-1}$$

$$\times \exp\left\{i\sum_{i=1}^{n}\delta t\left[p_i\left(\frac{q_i - q_{i-1}}{\delta t}\right) - H\left(p_i, \frac{q_i + q_{i-1}}{2}\right)\right]\right\}. \tag{1.48}$$

The second line defines the path integral. We almost have the promised result of eqn (1.41) if we can perform the momentum-space $[dp/2\pi] = \Pi_i^n\, dp_i/2\pi$ part of the path integral. The integrand being oscillatory, we analytically continue it to Euclidean space by formally treating $(i\,\delta t)$ as real. The Gaussian integral formula

$$\int_{-\infty}^{\infty}\frac{dx}{2\pi}\, e^{-ax^2+bx} = \frac{1}{\sqrt{4\pi a}}\, e^{b^2/4a} \tag{1.49}$$

can then be used to obtain

$$\int\frac{dp_i}{2\pi}\exp\left[\frac{-i\,\delta t}{2m}p_i^2 + ip_i(q_i - q_{i-1})\right] = \left(\frac{m}{2\pi i\,\delta t}\right)^{1/2}\exp\left[\frac{im(q_i - q_{i-1})^2}{2\delta t}\right].$$

In this way we have for eqn (1.48)

$$\langle q'|\, e^{-iH(t'-t)}|q\rangle = \lim_{n\to\infty}\left(\frac{m}{2\pi i\,\delta t}\right)^{n/2}\int\prod_i^{n-1} dq_i$$

$$\times \exp\left\{i\sum_{i=1}^{n}\delta t\left[\frac{m}{2}\left(\frac{q_i - q_{i-1}}{\delta t}\right)^2 - V\right]\right\} \tag{1.50}$$

or

$$\langle q; t|q'; t'\rangle = \langle q'|\, e^{-iH(t'-t)}|q\rangle = N\int[dq]\exp\left\{i\int_t^{t'} d\tau\left[\frac{m}{2}\dot{q}^2 - V(q)\right]\right\} \tag{1.51}$$

which is the stated result of eqn (1.41), where $L = (m\dot{q}^2/2) - V(q)$.

Green's functions of one-dimensional quantum mechanics. We will next translate the basic result of eqns (1.41) and (1.47) into forms that can be easily generalized to PI formulae for Green's functions in field theory.

Let us start with the simplest two-point function: the matrix element of a time-ordered product between ground states,

$$G(t_1, t_2) = \langle 0|T(Q^{\mathrm{H}}(t_1)Q^{\mathrm{H}}(t_2))|0\rangle$$

where $|0\rangle$ denotes the ground state. Inserting complete sets of states,

$$G(t_1, t_2) = \int \mathrm{d}q\, \mathrm{d}q' \langle 0|q'; t'\rangle\langle q'; t'|T(Q^{\mathrm{H}}(t_1)Q^{\mathrm{H}}(t_2))|q; t\rangle\langle q; t|0\rangle. \tag{1.52}$$

The matrix element

$$\langle 0|q; t\rangle = \phi_0(q)\, \mathrm{e}^{-\mathrm{i}E_0 t} = \phi_0(q, t) \tag{1.53}$$

is the wavefunction for the ground state. We next concentrate on the PI formulation of $\langle q'; t'|T(Q^{\mathrm{H}}(t_1)Q^{\mathrm{H}}(t_2))|q; t\rangle$. For $t_1 > t_2$ (i.e., $t' > t_1 > t_2 > t$), we have

$$\begin{aligned}\langle q'; t'|T(Q^{\mathrm{H}}(t_1)Q^{\mathrm{H}}(t_2))|q; t\rangle &= \langle q'|\, \mathrm{e}^{-\mathrm{i}H(t'-t_1)}Q^{\mathrm{S}}\, \mathrm{e}^{-\mathrm{i}H(t_1-t_2)}Q^{\mathrm{S}}\, \mathrm{e}^{-\mathrm{i}H(t_1-t_2)}|q_2\rangle \\ &= \int \langle q'|\, \mathrm{e}^{-\mathrm{i}H(t'-t_1)}|q_1\rangle\langle q_1|Q^{\mathrm{S}}\, \mathrm{e}^{-\mathrm{i}H(t_1-t_2)}|q_2\rangle \\ &\quad \times \langle q_2|Q^{\mathrm{S}}\, \mathrm{e}^{-\mathrm{i}H(t_2-t)}|q\rangle\, \mathrm{d}q_1\, \mathrm{d}q_2.\end{aligned}$$

Taking eigenvalues in the Schrödinger picture and applying the basic PI result of eqn (1.47), it follows that

$$\begin{aligned}&\langle q'; t'|T(Q^{\mathrm{H}}(t_1)Q^{\mathrm{H}}(t_2))|q; t\rangle \\ &\quad = \int \left[\frac{\mathrm{d}p\, \mathrm{d}q}{2\pi}\right] q_1(t_1)q_2(t_2) \exp\left\{\mathrm{i}\int_t^{t'} \mathrm{d}\tau[p\dot{q} - H(p, q)]\right\}.\end{aligned} \tag{1.54}$$

A minute of thought will convince us that exactly the same PI formula holds for the time sequence $t_2 > t_1$ (i.e., $t' > t_2 > t_1 > t$). Thus eqn (1.54) is a general result. Substituting eqns (1.54) and (1.53) into eqn (1.52) we have

$$\begin{aligned}G(t_1, t_2) = \int \mathrm{d}q\, \mathrm{d}q' \phi_0(q', t')\phi_0^*(q, t) \int \left[\frac{\mathrm{d}q\, \mathrm{d}p}{2\pi}\right] q_1(t_1)q_2(t_2) \\ \times \exp\left\{\mathrm{i}\int_t^{t'} \mathrm{d}\tau[p\dot{q} - H(p,q)]\right\}\end{aligned} \tag{1.55}$$

or

$$\begin{aligned}G(t_1, t_2) = \int \left[\frac{\mathrm{d}q\, \mathrm{d}p}{2\pi}\right] \phi_0(q', t')\phi_0^*(q, t)q_1(t_1)q_2(t_2) \\ \times \exp\left\{\mathrm{i}\int_t^{t'} \mathrm{d}\tau[p\dot{q} - H(p, q)]\right\}.\end{aligned} \tag{1.56}$$

The presence of the ground-state wavefunctions $\phi_0(q', t')$ and $\phi_0^*(q, t)$ in eqn (1.55) makes it clumsy to do practical calculations. To remove them, consider the matrix element

$$\langle q'; t'|\mathcal{O}(t_1, t_2)|q; t\rangle = \int dQ\, dQ' \langle q'; t'|Q'; T'\rangle \times \langle Q'; T'|\mathcal{O}(t_1, t_2)|Q; T\rangle\langle Q; T|q; t\rangle \quad (1.57)$$

where $\mathcal{O}(t_1, t_2) = T(Q^{\rm H}(t_1)Q^{\rm H}(t_2))$ and $t' \geq T' \geq (t_1, t_2) \geq T \geq t$. Let $|n\rangle$ be the energy eigenstate with energy E_n and wavefunction $\phi_n(q)$,

$$H|n\rangle = E_n|n\rangle$$

$$\langle q|n\rangle = \phi_n^*(q).$$

Then we have

$$\langle q'; t'|Q'; \mathrm{T}'\rangle = \langle q'|\, e^{-iH(t'-T')}|Q'\rangle = \sum_n \langle q'|n\rangle\langle n|\, e^{-iH(t'-T')}|Q'\rangle$$

$$= \sum_n \phi_n^*(q')\phi_n(Q')\, e^{-iE_n(t'-T')}. \quad (1.58)$$

To isolate the ground-state wavefunction in this equation, we use the fact that $E_n > E_0$ for all $n \neq 0$, and take the limit $t' \to -i\infty$, which yields

$$\lim_{t' \to -i\infty} \langle q'; t'|Q'; T'\rangle = \phi_0^*(q')\phi_0(Q')\, e^{-E_0|t'|}\, e^{iE_0T'}. \quad (1.59)$$

Similarly,

$$\lim_{t \to i\infty} \langle Q; T|q; \mathrm{t}\rangle = \phi_0(q)\phi_0^*(Q)\, e^{-E_0|t|}\, e^{-iE_0T}. \quad (1.60)$$

Then eqn (1.57) becomes

$$\lim_{\substack{t' \to -i\infty \\ t \to i\infty}} \langle q'; t'|\mathcal{O}(t_1t_2)|q; \mathrm{t}\rangle = \int dQ\, dQ' \phi_0^*(q')\phi_0(Q')\langle Q'; T'|\mathcal{O}(t_1t_2)|Q, T\rangle \times \phi_0^*(Q)\phi_0(q)\, e^{-E_0|t'|}\, e^{+iE_0T'}\, e^{-iE_0T}\, e^{-E_0|t|}$$

$$= \phi_0^*(q')\phi_0(q)\, e^{-E_0|t'|}\, e^{-E_0|t|} G(t_1, t_2) \quad (1.61)$$

where we have used eqn (1.52). From eqns (1.59) and (1.60) it is clear that

$$\lim_{\substack{t' \to -i\infty \\ t \to i\infty}} \langle q'; t'|q; t\rangle = \phi_0^*(q')\phi_0(q)\, e^{-E_0|t'|}\, e^{-E_0|t|}. \quad (1.62)$$

Combining eqns (1.61) and (1.62), we obtain for Green's function

$$G(t_1, t_2) = \lim_{\substack{t' \to -i\infty \\ t \to i\infty}} \left[\frac{\langle q'; t'|T(Q^{\rm H}(t_1)Q^{\rm H}(t_2))|q; t\rangle}{\langle q'; t'|q; t\rangle}\right]$$

$$= \lim_{\substack{t' \to -i\infty \\ t \to i\infty}} \frac{1}{\langle q'; t'|q; t\rangle} \int \left[\frac{dq\, dp}{2\pi}\right] q(t_1)q(t_2) \times \exp\left\{i \int_t^{t'} d\tau [p\dot{q} - H(p, q)\right\} \quad (1.63)$$

where we have used eqn (1.54) and the path-integral representation for the factor $\langle q'; t' | q; t\rangle$ in the denominator is given in eqn (1.51). This clearly can be generalized to the n-point Green's function

$$G(t_1 \dots t_n) = \langle 0 | T(q(t_1)q(t_2) \dots q(t_n)) | 0\rangle$$

$$= \lim_{\substack{t' \to -i\infty \\ t \to i\infty}} \frac{1}{\langle q'; t' | q; t\rangle} \int \left[\frac{dq\, dp}{2\pi}\right] q(t_1) \dots q(t_n)$$

$$\times \exp\left\{ i \int_t^{t'} d\tau [p\dot{q} - H(p, q)] \right\}. \qquad (1.64)$$

This entire set of Green's functions can be generated as follows.

$$G(t_1 \dots t_n) = \left.\frac{(-i)^n \, \delta^n W[J]}{\delta J(t_1) \dots \delta J(t_n)}\right|_{J=0} \qquad (1.65)$$

with

$$W[J] = \lim_{\substack{t' \to -i\infty \\ t \to i\infty}} \frac{1}{\langle q'; t' | q; t\rangle} \int \left[\frac{dq\, dp}{2\pi}\right]$$

$$\times \exp\left\{ i \int_t^{t'} d\tau [p\dot{q} - H(p, q) + J(\tau)q(\tau)] \right\}. \qquad (1.66)$$

Comparing this expression for $W[J]$ with the Green's function in eqn (1.64), we see that the generating functional $W[J]$ corresponds to the transition amplitude from the ground state at t to the ground state at t' in the presence of an external source $J(\tau)$,

$$W[J] = \langle 0|0\rangle_J \qquad (1.67)$$

with the normalization $W[0] = 1$. Thus the computation of Green's functions is now reduced to the computation of $W[J]$. We will see later in the case of quantum field theory that the $J(t)$-independent factor $\langle q; t | q'; t'\rangle$ in eqn (1.66) is irrelevant for generating the connected Green's functions and can be neglected.

Euclidean Green's function. In the formulae for Green's function (eqns (1.63) and (1.64)) the unphysical boundary condition $t' \to -i\infty, t \to i\infty$ should be interpreted in terms of the 'Euclidean' Green's functions which are defined by

$$S^{(n)}(\tau_1, \dots, \tau_n) = i^n G^{(n)}(-i\tau_1, \dots, -i\tau_n). \qquad (1.68)$$

The generating functional for the S-function is then given by

$$W_E[J] = \lim_{\substack{\tau' \to \infty \\ \tau \to -\infty}} \int [dq] \exp\left\{ \int_\tau^{\tau'} d\tau'' \left[-\frac{m}{2}\left(\frac{dq}{d\tau''}\right)^2 - V(q) + J(\tau'')q(\tau'') \right] \right\}$$

$$(1.69)$$

with

$$S^{(n)}(\tau_1 \ldots \tau_n) = \left.\frac{\delta^n W_{\mathrm{E}}[J]}{\delta J(\tau_1) \ldots \delta J(\tau_n)}\right|_{J=0}$$

The unphysical limits $t' \to -\mathrm{i}\infty$, $t \to \mathrm{i}\infty$ make sense in 'Euclidean space' where t is replaced by $-\mathrm{i}\tau$. Furthermore, the path integral in Euclidean space (eqn (1.69)), is well-defined for those potentials $V(q)$ which are bounded below. This is because we can always readjust the zero point of $V(q)$ such that

$$\frac{m}{2}\left(\frac{\mathrm{d}q}{\mathrm{d}\tau}\right)^2 + V(q) > 0 \tag{1.70}$$

and the exponential in (1.69) will always give a damping factor so that the path integral converges. Note that (1.70) is satisfied for a physically stable system.

Thus, the path-integral formalism has well-defined meaning only in Euclidean (or imaginary-time) space. To obtain physical quantities in real space, we have to do an analytic continuation. In practice, we will just do the manipulations in real space with the understanding that they can be justified in Euclidean space.

Let us summarize the discussion of the PI formulation of the quantum-mechanical description of a one-dimensional system. The basic results are the functional-integral formulae for the transition amplitude of eqns (1.41) and (1.47). In preparation for generalizing the formalism to field theory we have derived from these results the n-point Green's functions in (1.64). All these $G^{(n)}(t_1 \ldots t_n)$s can be generated from $W[J]$, the ground-state transition amplitude in the presence of an external source J. This central quantity can be computed according to (1.66) with an obvious generalization to systems with N degrees of freedom as

$$W[J_1, \ldots, J_N] \sim \lim_{\substack{t' \to -\mathrm{i}\infty \\ t \to \mathrm{i}\infty}} \int \prod_i^N [\mathrm{d}q_i\, \mathrm{d}p_i] \times \exp\left\{\mathrm{i}\int_t^{t'} \mathrm{d}\tau \left[\sum_i^N p_i \dot{q}_i - H(p_i, q_i) + \sum_i^N J_i q_i\right]\right\}$$

or

$$W[J_1, \ldots, J_N] \sim \lim_{\substack{t' \to -\mathrm{i}\infty \\ t \to \mathrm{i}\infty}} \int \prod_i^N [\mathrm{d}q_i] \times \exp\left\{\mathrm{i}\int_t^{t'} \mathrm{d}\tau \left[L(q_i, \dot{q}_i) + \sum_i^N J_i q_i\right]\right\}. \tag{1.71}$$

Field theory

We consider a field theory as a quantum-mechanical system with infinite degrees of freedom and make the following identifications for the results presented above

$$\prod_{i=1}^{N} [\mathrm{d}q_i\, \mathrm{d}p_i] \to [\mathrm{d}\phi(x)\, \mathrm{d}\pi(x)]$$

$$L(q_i, \dot{q}_i),\ H(q_i, p_i) \to \int \mathrm{d}^3x \mathscr{L}(\phi, \partial_\mu \phi),\ \int \mathrm{d}^3x \mathscr{H}(\phi, \pi) \tag{1.72}$$

with $\pi(x)$, $\mathscr{L}(x)$, and $\mathscr{H}(x)$ being the conjugate momentum field, the Lagrangian density, and the Hamiltonian density, respectively. The ground state in field theory is generally referred to as the vacuum state. Thus the generating functional $W[J]$ is the vacuum-to-vacuum transition amplitude in the presence of an external source $J(x)$. The generalization of eqn (1.71) takes the form

$$W[J] \sim \int [\mathrm{d}\phi\, \mathrm{d}\pi] \exp\left\{\mathrm{i} \int \mathrm{d}^4x[\pi(x)\, \partial_0\phi(x) - \mathscr{H}(\pi, \phi) + J(x)\phi(x)]\right\} \tag{1.73}$$

or

$$W[J] \sim \int [\mathrm{d}\phi] \exp\left\{\mathrm{i} \int \mathrm{d}^4x[\mathscr{L}(\phi(x)) + J(x)\phi(x)]\right\}. \tag{1.74}$$

Furthermore, the limit $t \to \mathrm{i}\infty$ in (1.71) suggests that we first calculate the Euclidean-space quantity $W_{\mathrm{E}}[J]$, which is the analytic continuation of $W[J]$ with $\bar{x}_\mu \equiv (\tau = \mathrm{i}t, \mathbf{x})$ replacing $x_\mu \equiv (t, \mathbf{x})$.

$$W_{\mathrm{E}}[J] \sim \int [\mathrm{d}\phi] \exp\left\{\int \mathrm{d}^4\bar{x}[\mathscr{L}(\phi(\bar{x})) + J(\bar{x})\phi(\bar{x})]\right\}. \tag{1.75}$$

For field theory what we are interested in is the *connected* Green's function which is related to the generating functional by

$$G^{(n)}(\bar{x}_1, \ldots, \bar{x}_n) = \left[\frac{1}{W_{\mathrm{E}}[J]}\, \frac{\delta^n W_{\mathrm{E}}[J]}{\delta J(\bar{x}_1) \ldots \delta J(\bar{x}_n)}\right]\Bigg|_{J=0} \tag{1.76}$$

Thus in order to remove the disconnected part of the Green's function, an extra factor of $W[J]$ has been inserted in the denominator of the definition (1.69). We recall that the same division was involved in our previous discussion of Green's function (eqns (1.30) and (1.31)). The important practical consequence of this division is that the J-independent absolute normalization of $W[J]$ is immaterial for any subsequent calculation of the Green's function.

We now return to our illustrative example of $\lambda\phi^4$ theory

$$\mathscr{L}(\phi) = \mathscr{L}_0(\phi) + \mathscr{L}_{\mathrm{I}}(\phi)$$

with

$$\mathscr{L}_0(\phi) = \tfrac{1}{2}(\partial_\lambda \phi)(\partial^\lambda \phi) - \tfrac{1}{2}\mu^2\phi^2$$

$$\mathscr{L}_{\mathrm{I}}(\phi) = \frac{-\lambda}{4!}\phi^4.$$

The Euclidean generating functional

$$W[J] = \int [\mathrm{d}\phi] \exp\left\{ - \int \mathrm{d}^4 x \left[\frac{1}{2}\left(\frac{\partial\phi}{\partial\tau}\right)^2 + \frac{1}{2}(\nabla\phi)^2 + \frac{1}{2}\mu^2\phi^2 + \frac{\lambda}{4!}\phi^4 + J\phi \right]\right. \tag{1.77}$$

may be written as

$$W[J] = \left[\exp \int \mathrm{d}^4 x \mathscr{L}_{\mathrm{I}}\left(\frac{\delta}{\delta J}\right)\right] W_0[J] \tag{1.78}$$

where

$$W_0[J] = \int [\mathrm{d}\phi] \exp\left[\int \mathrm{d}^4 x (\mathscr{L}_0 + J\phi)\right]$$

is the free-field generating functional. (For simplicity of notation we drop the subscript E and the bar over x indicating Euclidean space.) The factors $-(\partial\phi/\partial\tau)^2 - (\nabla\phi)^2$ in eqn (1.77) can be replaced by $\phi(\partial^2/\partial\tau^2 + \nabla^2)\phi$ because the difference is a total four-divergence and we have

$$W_0[J] = \int [\mathrm{d}\phi] \exp\left[-\frac{1}{2}\int \mathrm{d}^4 x\, \mathrm{d}^4 y \phi(x) K(x, y) \phi(y) + \int \mathrm{d}^4 z J(z)\phi(z)\right] \tag{1.79}$$

where

$$K(x, y) = \delta^4(x - y)\left(-\frac{\partial^2}{\partial\tau^2} - \nabla^2 + \mu^2\right). \tag{1.80}$$

As x and y may be taken as 'continuous indices', $W_0[J]$ of eqn (1.79) can be considered an infinite-dimensional ($N \to \infty$) Gaussian integral of form

$$\int \mathrm{d}\phi_1 \ldots \mathrm{d}\phi_N \exp\left[-\frac{1}{2}\sum_{i,j} \phi_i \mathrm{K}_{ij}\phi_j + \sum_k J_k\phi_k \right]$$
$$\sim \frac{1}{\sqrt{\det K}} \exp\left[\frac{1}{2}\sum_{i,j} J_i (K^{-1})_{ij} J_j\right]. \tag{1.81}$$

The right-hand side is obtained by a generalization of the result cited in eqn (1.49). In this way the ϕ functional integral in (1.79) can be performed and we obtain, up to an inessential multiplicative factor,

$$W_0[J] = \exp\left[\frac{1}{2}\int \mathrm{d}^4 x\, \mathrm{d}^4 y J(x)\, \Delta(x, y) J(y)\right] \tag{1.82}$$

where $\Delta(x, y)$ should be the inverse of $K(x, y)$ in (1.80). Thus,

$$\int \mathrm{d}^4 y K(x, y)\, \Delta(y, z) = \delta^4(x - z). \tag{1.83}$$

It is not difficult to see that

$$\Delta(x, y) = \int \frac{\mathrm{d}^4\kappa}{(2\pi)^4} \frac{\mathrm{e}^{\mathrm{i}\kappa\cdot(x-y)}}{\kappa^2 + \mu^2} \tag{1.84}$$

where $\kappa = (\mathrm{i}k_0, \mathbf{k})$ forms a Euclidean momentum four-vector. The perturbative expansion in powers of $\mathscr{L}_{\rm I}$ of the exponential in (1.78) gives

$$W[J] = W_0[J]\{1 + \lambda\omega_1[J] + \lambda^2\omega_2[J] + \ldots\}, \tag{1.85}$$

where

$$\omega_1[J] = -\frac{1}{4!} W_0^{-1}[J]\left\{\int \mathrm{d}^4x \left[\frac{\delta}{\delta J(x)}\right]^4\right\} W_0[J]$$

$$\omega_2[J] = -\frac{1}{2(4!)^2} W_0^{-1}[J]\left\{\int \mathrm{d}^4x \left[\frac{\delta}{\delta J(x)}\right]^4\right\}^2 W_0[J]$$

$$= -\frac{1}{2(4!)} W_0^{-1}[J]\left\{\int \mathrm{d}^4x \left[\frac{\delta}{\delta J(x)}\right]^4\right\} \omega_1[J]. \tag{1.86}$$

When we plug in the explicit form (eqn (1.82)) for $W_0[J]$, we obtain

$$\omega_1[J] = -\frac{1}{4!}[\Delta(x, y_1)\,\Delta(x, y_2)\,\Delta(x, y_3)\,\Delta(x, y_4) J(y_1)J(y_2)J(y_3)J(y_4)$$

$$+ 3!\,\Delta(x, y_1)\,\Delta(x, y_2)\,\Delta(x, x) J(y_1)J(y_2)], \tag{1.87}$$

and

$$\omega_2[J] = \frac{1}{2}\omega_1^2[J]$$

$$+ \frac{1}{2(3!)^2}\Delta(x_1, y_1)\,\Delta(x_1, y_2)\,\Delta(x_1, y_3)\,\Delta(x_1, x_2)\,\Delta(x_2, y_4)$$

$$\times\,\Delta(x_2, y_5)\,\Delta(x_2, y_6) J(y_1)J(y_2)J(y_3)J(y_4)J(y_5)J(y_6)$$

$$+ \frac{3}{2(4!)}\Delta(x_1, y_1)\,\Delta(x_1, y_2)\,\Delta^2(x_1, x_2)\,\Delta(x_2, y_3)\,\Delta(x_2, y_4)$$

$$\times\, J(y_1)J(y_2)J(y_3)J(y_4) + \frac{2}{24!}\Delta(x_1, y_1)\,\Delta(x_1, x_1)\,\Delta(x_1, x_2)$$

$$\times\,\Delta(x_2, y_2)\,\Delta(x_2, y_3)\,\Delta(x_2, y_4) J(y_1)J(y_2)J(y_3)J(y_4)$$

$$+ \frac{1}{8}\Delta(x_1, y_1)\,\Delta(x_1, x_1)\,\Delta(x_1, x_2)\,\Delta(x_2, x_2)\,\Delta(x_2, y_2) J(y_1)J(y_2)$$

$$+ \frac{1}{8}\Delta(x_1, y_1)\,\Delta^2(x_1, x_2)\,\Delta(x_2, x_2)\,\Delta(x_1, y_2) J(y_1)J(y_2)$$

$$+ \frac{1}{12}\Delta(x_1, y_1)\,\Delta^3(x_1, x_2)\,\Delta(x_2, y_2) J(y_1)J(y_2) \tag{1.88}$$

where we have dropped all J-independent terms (see Figs. 1.4 and 1.5). It is understood that all arguments (x_i, y_i) are integrated over.

It is clear that the first factor on the right-hand side of (1.88), $\frac{1}{2}\omega_1^2$, corresponds to a disconnected contribution. For the connected Green's function defined by (1.76)

$$G^{(n)}(x_1 \ldots x_n) = \left.\frac{\delta^n \ln W[J]}{\delta J(x_1) \ldots \delta J(x_n)}\right|_{J=0}, \tag{1.89}$$

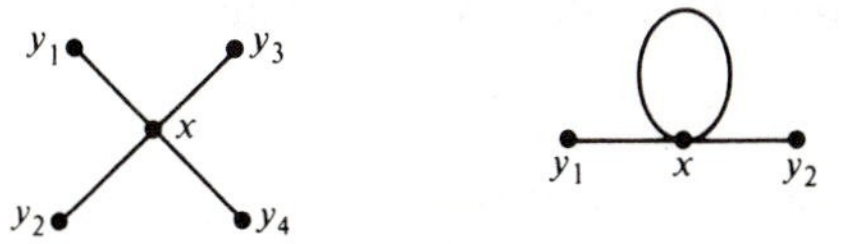

FIG. 1.4. Graphic representation of ω_1 in eqn (1.87).

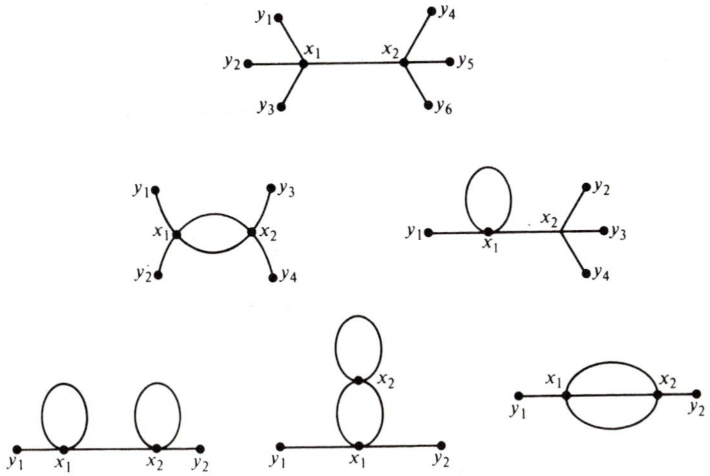

FIG. 1.5. Connected parts of ω_2 in eqn (1.88).

such terms would not contribute. To see this explicitly,

$$\begin{aligned}\ln W[J] &= \ln W_0[J] + \ln\{1 + W_0^{-1}[J](W[J] - W_0[J])\}\\ &= \ln W_0[J] + \ln\{1 + W_0^{-1}[J](e^{\langle \mathcal{L}_I \rangle} - 1)W_0[J]\}, \qquad (1.90)\end{aligned}$$

where we have used (1.78). Since $W_0^{-1}(e^{\langle \mathcal{L}_I \rangle} - 1)W_0$ is also small, we can expand the exponential as well as the logarithm. Thus from eqn (1.85)

$$\begin{aligned}\ln W[J] &= \ln W_0[J] + (\lambda\omega_1 + \lambda^2\omega_2 + \ldots) - \tfrac{1}{2}(\lambda\omega_1 + \lambda^2\omega_2 + \ldots)^2 + \ldots\\ &= \ln W_0[J] + \lambda\omega_1 + \lambda^2(\omega_2 - \tfrac{1}{2}\omega_1^2) + \ldots. \qquad (1.91)\end{aligned}$$

Thus the disconnected $\frac{1}{2}\omega_1^2$ in ω_2 is in fact cancelled. It is not difficult to generalize this, to prove that all disconnected contributions disappear in $\ln W[J]$.

We note that what corresponds to Wick's theorem is simply the rule for functional differentiation

$$\begin{aligned}\frac{\delta J(y)}{\delta J(x)} &= \frac{\delta}{\delta J(x)} \int J(x)\,\delta(x - y)\,dx\\ &= \delta(x - y). \qquad (1.92)\end{aligned}$$

Differentiation according to (1.89) finally yields the (Euclidean) Green's function. For example, the terms with four Js in (1.87) and (1.88) give rise to the first- and second-order four-point functions. These results are the same as those in (1.33) and (1.35) (with the propagator given in eqn (1.14)) except that they are valid in Euclidean space.

The analytic continuation $\bar{x}_\mu \to x_\mu$ and $\kappa_\mu \to k_\mu$ of $\Delta(\bar{x} - \bar{y})$ in (1.84) yields the familiar Feynman propagator (1.14)

$$-\Delta(\bar{x} - \bar{y}) \to i\Delta_F(x - y) = i \int \frac{d^4k}{(2\pi)^4} \frac{e^{-ik\cdot(x-y)}}{k^2 - \mu^2 + i\varepsilon}.$$

The iε factor in the denominator indicates how the boundary condition on the propagator is to be imposed. It corresponds to the addition of a $\frac{1}{2}i\varepsilon\phi^2$ term in the Lagrangian and hence provides a suitable damping factor for the path integral (1.77) in Minkowski space.

Clearly the same set of Feynman rules, which we briefly reviewed in §1.1, follow from PI quantization formalism. The reader is referred to Appendix B where a practical guide to the derivation of the Feynman rules is given.

1.3 Fermion field quantization

Here we discuss the quantization procedure for systems with fermions. After briefly reviewing the canonical formalism, we indicate how the corresponding path-integral quantization can be formulated (see, for example, Berezin 1966). This involves the subject of Grassmann algebra.

Canonical quantization for fermions

In §1.1 we reviewed the canonical quantization procedure for a scalar field. Bose–Einstein statistics follow naturally from the commutation relations of the particle creation and annihilation operators (1.11), i.e. from the commutation of scalar field operators (1.3). For a many-fermion system, in order to arrive at an exclusion principle the field operations must satisfy a set of anticommutation relations. Consider the case of free Dirac field,

$$\mathscr{L}(x) = \bar{\psi}(x)(i\gamma^\mu \partial_\mu - m)\psi(x). \tag{1.93}$$

Eqn (1.2) yields the Dirac equation

$$(i\gamma^\mu \partial_\mu - m)\psi(x) = 0. \tag{1.94}$$

In quantum theory, the field $\psi(x)$ and its conjugate momentum $\pi(x) = i\psi^\dagger(x)$ are postulated to be operators; they satisfy the canonical anticommutation relations

$$\{\psi(\mathbf{x}, t), \psi^\dagger(\mathbf{x}', t)\} = \delta^3(\mathbf{x} - \mathbf{x}')$$
$$\{\psi(\mathbf{x}, t), \psi(\mathbf{x}', t)\} = \{\psi^\dagger(\mathbf{x}, t), \psi^\dagger(\mathbf{x}', t)\} = 0$$

where $\{A, B\} \equiv AB + BA$. Following the same steps as in the scalar case of §1.1 we formally solve the Dirac equation and calculate the Feynman propagator function

$$\mathrm{i}S_{\mathrm{F}}(x_1 - x_2)_{\alpha\beta} \equiv \langle 0|T(\psi_\alpha(x_1)\bar{\psi}_\beta(x_2))|0\rangle$$
$$= \int \frac{\mathrm{d}^4k}{(2\pi)^4}\left(\frac{\mathrm{i}}{\not{k} - m + \mathrm{i}\varepsilon}\right)_{\alpha\beta} \mathrm{e}^{-\mathrm{i}k\cdot(x_1-x_2)}. \tag{1.95}$$

For interacting systems the perturbative solution in the form of Feynman rules can be developed, again much in the same manner as the scalar case of §1.1. We shall not repeat the steps here except noting that a consequence of the anticommutation relation is that there will be a minus sign for each closed fermion loop in a Feynman diagram.

Path-integral quantization for fermions

Quantization of the fermion system can also be carried out by expressing the transition amplitude directly as the sum over all possible world lines connecting the initial and final states. The generating functional is then

$$W[\eta, \bar{\eta}] = \int [\mathrm{d}\psi(x)][\mathrm{d}\bar{\psi}(x)] \exp\left\{\mathrm{i}\int \mathrm{d}^4x[\mathscr{L}(\psi, \bar{\psi}) + \bar{\psi}\eta + \bar{\eta}\psi]\right\} \tag{1.96}$$

where $\psi(x)$, $\bar{\psi}(x)$, $\eta(x)$, and $\bar{\eta}(x)$ are (classical) fermion fields and sources, respectively. While the sum over the path for a boson system is a functional integral over ordinary c-number functions (classical scalar fields), the functional integral in (1.96) must be taken over anticommuting c-number functions ('classical' fermion fields)

$$\{\psi(x), \psi(x')\} = \{\psi(x), \bar{\psi}(x')\} = \{\bar{\psi}(x), \bar{\psi}(x')\} = 0$$
$$\{\eta(x), \eta(x')\} = \{\eta(x), \bar{\eta}(x')\} = \{\bar{\eta}(x), \bar{\eta}(x')\} = 0.$$

Thus they are elements of Grassmann algebra. In the following section we shall provide a brief introduction to this subject.

Grassmann algebra

In an n-dimensional Grassmann algebra, the n generators $\theta_1, \theta_2, \ldots, \theta_n$ satisfy

$$\{\theta_i, \theta_j\} = 0 \qquad i, j = 1, 2, \ldots, n \tag{1.97}$$

and every element can be expanded in a finite series

$$p(\theta) = P_0 + P^{(1)}_{i_1}\theta_{i_1} + P^{(2)}_{i_1 i_2}\theta_{i_1}\theta_{i_2} + \ldots + P^{(n)}_{i_1\ldots i_n}\theta_{i_1}\ldots\theta_{i_n}$$

where each of the summed-over indices $i_1, i_2, \ldots, i_n$ ranges from 1 to n. The expansion terminates because of (1.97). We shall now discuss the subject of differentiation and integration in such an algebra. Before stating the general n-dimensional results, we first motivate them with the simplest case of one Grassmann variable,

$$\{\theta, \theta\} = 0 \quad \text{or} \quad \theta^2 = 0. \tag{1.98}$$

Thus any element of the algebra has the simple expansion

$$p(\theta) = P_0 + \theta P_1. \tag{1.99}$$

If we take $p(\theta)$ to be an ordinary number, then P_0 and P_1 are ordinary and Grassmann numbers, respectively. (We can imagine embedding this one-dimensional Grassmann algebra into a higher-dimensional one so that we would have more than one anticommuting element.)

The operation of differentiation may be taken from left or right with the basic definition

$$\frac{\mathrm{d}}{\mathrm{d}\theta}\theta = \theta\frac{\overleftarrow{\mathrm{d}}}{\mathrm{d}\theta} = 1. \tag{1.100}$$

We have the 'left derivative'

$$\frac{\mathrm{d}}{\mathrm{d}\theta}p(\theta) = P_1 \tag{1.101}$$

and the 'right derivative'

$$p(\theta)\frac{\overleftarrow{\mathrm{d}}}{\mathrm{d}\theta} = -P_1 \tag{1.102}$$

because $(\mathrm{d}P_0/\mathrm{d}\theta) = 0$ and $(\mathrm{d}/\mathrm{d}\theta)$ anticommutes with P_1.

We next introduce the integration operation, which ordinarily is taken to be the inverse of differentiation. However such an inverse is ill defined in a Grassmann algebra, as can be seen by the fact that, for either type of derivative,

$$\frac{\mathrm{d}^2}{\mathrm{d}\theta^2}p(\theta) = 0. \tag{1.103}$$

Thus we must be content with a formal definition of the integration operation which preserves some general properties of our intuitive notion of integration. We require it to be invariant under a translation of the integration variable by a constant. Thus

$$\int \mathrm{d}\theta p(\theta) = \int \mathrm{d}\theta p(\theta + \alpha). \tag{1.104}$$

From (1.99) we must have

$$\int \mathrm{d}\theta P_1\alpha = 0 \quad \text{or} \quad \int \mathrm{d}\theta = 0 \tag{1.105}$$

where α is another element in the Grassmann algebra which is independent of θ and anticommutes with θ. We can normalize the remaining integral using

$$\int \mathrm{d}\theta\theta = 1. \tag{1.106}$$

From (1.105) and (1.106) it follows that

$$\int \mathrm{d}\theta p(\theta) = P_1 \tag{1.107}$$

which is the same as left differentiation in (1.101). Thus our definitions of integration and (left) differentiation lead to the same result

$$\int \mathrm{d}\theta p(\theta) = \frac{\mathrm{d}}{\mathrm{d}\theta} p(\theta) = P_1. \tag{1.108}$$

We next consider the problem of change of integration variable $\theta \to \tilde{\theta} = a + b\theta$, where a and b are anticommuting and ordinary numbers, respectively. For an ordinary c-number we have the familiar relation

$$\int \mathrm{d}\tilde{x} f(\tilde{x}) = \int \mathrm{d}x \left(\frac{\mathrm{d}\tilde{x}}{\mathrm{d}x}\right) f(\tilde{x}(x)). \tag{1.109}$$

What will be the corresponding result for Grassmann numbers? Since, by (1.108),

$$\int \mathrm{d}\tilde{\theta} p(\tilde{\theta}) = \frac{\mathrm{d}}{\mathrm{d}\tilde{\theta}} p(\tilde{\theta}) = P_1 \tag{1.110}$$

and

$$\int \mathrm{d}\theta p(\tilde{\theta}) = \int \mathrm{d}\theta b\theta P_1 = bP_1, \tag{1.111}$$

we have

$$\int \mathrm{d}\tilde{\theta} p(\tilde{\theta}) = \int \mathrm{d}\theta \left(\frac{\mathrm{d}\tilde{\theta}}{\mathrm{d}\theta}\right)^{-1} p(\tilde{\theta}(\theta)). \tag{1.112}$$

Thus for anticommuting numbers the 'Jacobian' is the inverse of what we would ordinarily expect.

We now proceed to generalize our one-variable results of (1.101), (1.102), (1.105), and (1.106) to the n-dimensional Grassmann algebra. We have the 'left derivative'

$$\frac{\mathrm{d}}{\mathrm{d}\theta_i}(\theta_1\theta_2 \ldots \theta_n) = \delta_{i1}\theta_2 \ldots \theta_n - \delta_{i2}\theta_1\theta_3 \ldots \theta_n + \ldots (-1)^{n-1} \delta_{in}\theta_1 \ldots \theta_{n-1}$$

and the 'right derivative'

$$(\theta_1\theta_2 \ldots \theta_n)\frac{\overleftarrow{\mathrm{d}}}{\mathrm{d}\theta_i} = \delta_{in}\theta_1 \ldots \theta_{n-1} - \ldots + (-1)^{n-1} \delta_{1i}\theta_2 \ldots \theta_n.$$

Thus, to calculate the left (right) derivative $(\mathrm{d}/\mathrm{d}\theta_i)$ of $\theta_1\theta_2, \ldots, \theta_n$, commute θ_i all the way to the left (right) in the product; then drop that θ_i. The symbol $\mathrm{d}\theta_1, \mathrm{d}\theta_2, \ldots, \mathrm{d}\theta_n$ is introduced with the conditions

$$\{\mathrm{d}\theta_i, \mathrm{d}\theta_j\} = 0$$

and

$$\int \mathrm{d}\theta_i = 0$$

$$\int \mathrm{d}\theta_i\theta_j = \delta_{ij} \tag{1.113}$$

which defines the integration operation. For a change of integration variable

$$\tilde{\theta}_i = b_{ij}\theta_j, \tag{1.114}$$

we have the generalization of (1.112)

$$\int \mathrm{d}\tilde{\theta}_n \ldots \mathrm{d}\tilde{\theta}_1 p(\tilde{\theta}) = \int \mathrm{d}\theta_n \ldots \mathrm{d}\theta_1 \left[\det \frac{\mathrm{d}\tilde{\theta}}{\mathrm{d}\theta}\right]^{-1} p(\tilde{\theta}(\theta)). \tag{1.115}$$

To show this result we follow the same steps as in the one-θ case. Just as in (1.110) and (1.111), we need to compare $\int \mathrm{d}\tilde{\theta}_n \ldots \mathrm{d}\tilde{\theta}_1 p(\tilde{\theta})$ and $\int \mathrm{d}\theta_n \ldots \mathrm{d}\theta_1 p(\tilde{\theta}(\theta))$. The only terms in $p(\theta)$ which can contribute to these integrals are terms with n $\tilde{\theta}$s,

$$\tilde{\theta}_1 \ldots \tilde{\theta}_n = b_{1i_1} \ldots b_{ni_n}\theta_{i_1} \ldots \theta_{i_n}. \tag{1.116}$$

The right-hand side is non-zero only if $i_1, \ldots, i_n$ are all different and we can write

$$\begin{aligned}\tilde{\theta}_1 \ldots \tilde{\theta}_n &= b_{1i_1} \ldots b_{ni_n}\varepsilon_{i_1 \ldots i_n}\theta_1 \ldots \theta_n \\ &= (\det b)\theta_1 \ldots \theta_n.\end{aligned} \tag{1.117}$$

However, in order to maintain the normalization conditions (1.113), we must have

$$\mathrm{d}\tilde{\theta}_1 \ldots \mathrm{d}\tilde{\theta}_n = (\det b)^{-1}\, \mathrm{d}\theta_1 \ldots \mathrm{d}\theta_n; \tag{1.118}$$

hence the result of (1.115). To repeat, for anticommuting variables integration is equivalent to differentiation and we get $[\det(\mathrm{d}\tilde{\theta}/\mathrm{d}\theta)]^{-1}$ rather than $[\det(\mathrm{d}\tilde{\theta}/\mathrm{d}\theta)]$.

As we have seen in §1.2 the Gaussian integral plays an important role in the PI formalism. Thus we need to evaluate

$$G(A) = \int \mathrm{d}\theta_n \ldots \mathrm{d}\theta_1 \exp(\tfrac{1}{2}(\theta, A\theta)) \tag{1.119}$$

where A is an antisymmetric matrix and $(\theta, A\theta) = \theta_i A_{ij}\theta_j$. First consider the simple case of $n = 2$

$$A = \begin{pmatrix} 0 & A_{12} \\ -A_{12} & 0 \end{pmatrix}$$

and

$$\begin{aligned}G(A) &= \int \mathrm{d}\theta_2\, \mathrm{d}\theta_1 \exp(\theta_1\theta_2 A_{12}) \\ &= \int \mathrm{d}\theta_2\, \mathrm{d}\theta_1 (1 + \theta_1\theta_2 A_{12}) \\ &= A_{12} = \sqrt{\det A}.\end{aligned} \tag{1.120}$$

For the general case where A is an $n \times n$ antisymmetric matrix, we can first put A in the standard form by a unitary transformation. (Here n is taken to be even as the integral vanishes for odd n.)

$$UAU^{\dagger} = A_s \tag{1.121}$$

with

$$A_s = \begin{bmatrix} a\begin{pmatrix} 0 & 1 \\ -1 & 0 \end{pmatrix} & & \\ & b\begin{pmatrix} 0 & 1 \\ -1 & 0 \end{pmatrix} & \\ & & \ddots \end{bmatrix}. \tag{1.122}$$

This can be seen as follows. Since $\mathrm{i}A$ is hermitian, it can be diagonalized by a unitary transformation V

$$V(\mathrm{i}A)V^{\dagger} = A_{\mathrm{d}} \tag{1.123}$$

where A_{d} is real and diagonal with diagonal elements which are solutions to the secular equation

$$\det|\mathrm{i}A - \lambda I| = 0. \tag{1.124}$$

Since $A^T = -A$, we have $\det|\mathrm{i}A - \lambda I|^T = \det|-\mathrm{i}A - \lambda I| = 0$. Thus, if λ is a solution, so is $(-\lambda)$, and A_{d} is of the form

$$A_{\mathrm{d}} = \begin{bmatrix} a & & & & \\ & -a & & & \\ & & b & & \\ & & & -b & \\ & & & & \ddots \end{bmatrix}. \tag{1.125}$$

To put A_{d} into the standard form of (1.122), we use the 2×2 unitary matrix

$$S_2 = \frac{1}{\sqrt{2}}\begin{pmatrix} \mathrm{i} & 1 \\ 1 & \mathrm{i} \end{pmatrix} \tag{1.126}$$

which has the property

$$S_2(-\mathrm{i})\begin{pmatrix} 1 & 0 \\ 0 & -1 \end{pmatrix} S_2^{\dagger} = \begin{pmatrix} 0 & 1 \\ -1 & 0 \end{pmatrix}. \tag{1.127}$$

Thus $S(-\mathrm{i}A_{\mathrm{d}})S^{\dagger} = A_s$ for

$$S = \begin{pmatrix} S_2 & & \\ & S_2 & \\ & & \ddots \end{pmatrix} \tag{1.128}$$

and the unitary matrix in (1.122) must be the product $U = SV$ because

$(SV)A(SV)^\dagger = S(-iA_d)S^\dagger = A_s$. Furthermore, let

$$T = \begin{bmatrix} a^{-1/2} & & & & \\ & a^{-1/2} & & & \\ & & b^{-1/2} & & \\ & & & b^{-1/2} & \\ & & & & \ddots \end{bmatrix}. \tag{1.129}$$

Thus,

$$\det(T^{-1}) = \sqrt{\det A}. \tag{1.130}$$

We can then write

$$T(UAU^\dagger)T = TA_sT \equiv \overline{A_s}$$
$$= \begin{bmatrix} 0 & 1 & & & \\ -1 & 0 & & & \\ & & 0 & 1 & \\ & & -1 & 0 & \\ & & & & \ddots \end{bmatrix}. \tag{1.131}$$

The Gaussian integral (1.119) can then be written as

$$G(A) = \int d\theta_n \ldots d\theta_1 \exp(\tfrac{1}{2}(\theta, U^\dagger T^{-1}\overline{A_s}T^{-1}U\theta). \tag{1.132}$$

Change the integration variable

$$\tilde{\theta} = (T^{-1}U)\theta \tag{1.133}$$

and use (1.115)

$$G(A) = \int d\tilde{\theta}_n \ldots d\tilde{\theta}_1 \exp(\tfrac{1}{2}(\tilde{\theta}, \overline{A_s}\tilde{\theta}))\left[\det\left(\frac{d\tilde{\theta}}{d\theta}\right)\right]$$
$$= \det\left(\frac{d\tilde{\theta}}{d\theta}\right). \tag{1.134}$$

Since, by (1.133) and (1.130)

$$\det\left(\frac{d\tilde{\theta}}{d\theta}\right) = \det(T^{-1}U) = \det(T^{-1}) = \sqrt{\det A}, \tag{1.135}$$

we obtain the result

$$G(A) = \int d\theta_n \ldots d\theta_1 \exp(\tfrac{1}{2}(\theta, A\theta)) = \sqrt{\det A} \tag{1.136}$$

which should be contrasted with the Gaussian integral with ordinary commuting real variables

$$\int \frac{dx_1}{\sqrt{2\pi}} \cdots \frac{dx_n}{\sqrt{2\pi}} \exp(-\tfrac{1}{2}(x, Ax)) = \frac{1}{\sqrt{\det A}} \tag{1.137}$$

or with ordinary commuting complex variables ($z = x + iy$)

$$\int \frac{dz_1}{\sqrt{\pi}} \cdots \frac{dz_n}{\sqrt{\pi}} \frac{dz_1^*}{\sqrt{\pi}} \cdots \frac{dz_n^*}{\sqrt{\pi}} \exp(-(z^*, Az)) = \frac{1}{\det A} \tag{1.138}$$

where $\int dz\, dz^* = \int dx\, dy$. The Gaussian integral for complex Grassmann variables can be shown to have the value

$$\int d\theta_1\, d\bar{\theta}_1 \ldots d\theta_n\, d\bar{\theta}_n \exp(\bar{\theta}, A\theta) = \det A \tag{1.139}$$

where θ_i and $\bar{\theta}_i$ are independent generators of the algebra.

The classical fermion fields $\psi(x)$ and $\bar{\psi}(x)$ are then taken to be elements of an infinite-dimensional Grassmann algebra. All the above results for the general n-dimensional case can be naïvely extended.

Since the fermion fields always enter the Lagrangian quadratically $\mathscr{L} = (\bar{\psi} A \psi)$, the functional integral of (1.96) will be a generalized Gaussian integral. The result in (1.139) can then be applied

$$\begin{aligned} W &= \int [d\psi(x)][d\bar{\psi}(x)] \exp\left\{\int d^4x \bar{\psi} A \psi\right\} \\ &= \det A \end{aligned} \tag{1.140}$$

where we have not bothered to display the source fields. W is the vacuum-to-vacuum amplitude and the (connected) Feynman diagram representation, as generated by $\ln W$, will be a set of single-closed-fermion-loop graphs (Fig. 1.6). The change of going from the ordinary functional integral (1.138) to the anticommuting variable functional integral (1.140), with the replacement of $(\det A)^{-1}$ by $(\det A)$, corresponds to changing the overall sign of $\ln W$. This is the familiar Feynman rule of an extra minus sign for each closed fermion loop.

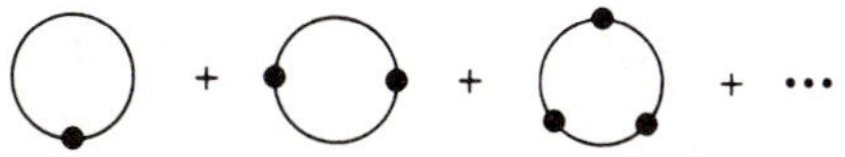

FIG. 1.6. Vacuum-to-vacuum amplitude as represented by single closed loops.

2 Introduction to renormalization theory

Given any quantum field theory one can construct the Feynman rules for calculating the Green's functions and S-matrix elements in perturbation theory as described in Chapter 1. But in relativistic field theory one often encounters infinities in the calculation of diagrams containing loops. This is because the momentum variable in the loop integration ranges all the way from zero to infinity. In other words, for a relativistic theory, there is no intrinsic cut-off in momenta. These divergences will render the calculation meaningless. The theory of renormalization is a prescription which allows us to consistently isolate and remove all these infinities from the physically measurable quantities. It has been of utmost importance to the development of relativistic quantum field theory.

It should be emphasized however that the need for renormalization is rather general and is not unique to the relativistic field theories. Renormalization has its own intrinsic physical basis and is not brought about solely by the necessity to expurgate infinities. Even in a totally finite theory we would still have to renormalize physical quantities. The following example should illustrate this point. Consider an electron moving inside a solid. Due to the interaction of the electron with the lattice, the effective mass of the electron m^*, which determines its response to an externally applied force, is certainly different from the mass of the electron m measured outside the solid. The electron mass is changed (renormalized) from m to m^* by the interaction of the electron with the lattice in the solid. In this simple case one can in principle measure both m^* and m by switching on and off the interaction (i.e. by placing the electron inside or outside of the solid). Clearly the difference is finite since both m and m^* are finite and measurable. For the relativistic field theory, the situation is the same except for two important distinctions. First, renormalization due to the interaction is generally infinite (corresponding to the divergent loop diagrams). Second, there is no way to switch off the interaction; hence quantities in the absence of interaction, called the *unrenormalized* or the *bare* quantities, are not measurable. For example, in quantum electrodynamics the difference between the bare electron mass m and the renormalized mass m^* is infinite, and the bare mass cannot be measured because the electron interacts with the virtual photon field constantly and there is no way to turn off this interaction.

The programme of removing infinities from physically measurable quantities in a relativistic theory, the *renormalization programme*, involves shuffling all divergences into the bare quantities. In other words, the unrenormalized quantities are assumed to be appropriately divergent to begin with and the infinite renormalization due to interaction then cancels these divergences to produce finite renormalized quantities. We should recall that in a relativistic quantum field theory the renormalized quantities are physically measurable while the bare ones are not. This difficult programme, as originally formulated for quantum electrodynamics by Feynman (1948*b*), Schwinger (1948, 1949), Tomonaga (1948), and Dyson (1949), has been quite successful and in the case of QED the agreement between theory and experiment has been spectacular.

Technically the theory of renormalization is rather complicated. A detailed and thorough discussion of this subject is beyond the scope of this book. In this chapter we shall explain the principal ideas behind it and give examples to illustrate how it works.

2.1 Conventional renormalization in $\lambda\phi^4$ theory

We shall first use the simple $\lambda\phi^4$ theory as an example to illustrate the renormalization procedure. The Lagrangian density is separated into free and interacting parts

$$\mathscr{L} = \mathscr{L}_0 + \mathscr{L}_1 \tag{2.1}$$

with

$$\mathscr{L}_0 = \tfrac{1}{2}[(\partial_\mu\phi_0)^2 - \mu_0^2\phi_0^2] \tag{2.2}$$

and

$$\mathscr{L}_\mathrm{I} = -\frac{\lambda_0}{4!}\phi_0^4. \tag{2.3}$$

The propagator and the vertex of this theory are displayed in Fig. 2.1.

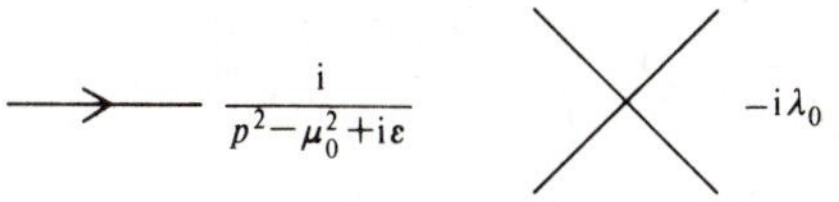

FIG. 2.1.

We will concentrate on the *one-particle-irreducible* (1PI) diagrams. They are the Feynman diagrams which cannot be disconnected by cutting any one internal line. Correspondingly, we define the one-particle-irreducible (1PI) Green's functions, denoted by $\Gamma^{(n)}(p_1 \ldots p_n)$, which have contributions coming from 1PI diagrams only. For example the graph in Fig. 2.2(a) is a 1PI diagram while the one in Fig. 2.2(b) is not. The reason for selecting 1PI diagrams is that any one-particle-reducible diagram can be decomposed into 1PI diagrams without further loop integration, and if we know how to take care of the divergences of 1PI diagrams we will also be able to handle the

FIG. 2.2.

reducible diagrams. For example, the two-point Green's function (propagator)

$$i\Delta(p) = \int d^4x\, e^{-ip\cdot x}\langle 0|T(\phi_0(x)\phi_0(0))|0\rangle \tag{2.4}$$

can be decomposed in terms of the 1PI self-energy parts $\Sigma(p)$ as in Fig. 2.3.

FIG. 2.3. The propagator as a sum of 1PI self-energy insertions.

Then we can write the propagator as

$$\begin{aligned} i\Delta(p) &= \frac{i}{p^2-\mu_0^2+i\varepsilon} + \frac{i}{p^2-\mu_0^2+i\varepsilon}(-i\Sigma(p^2))\frac{i}{p^2-\mu_0^2+i\varepsilon} + \dots \\ &= \frac{i}{p^2-\mu_0^2+i\varepsilon}\left[\frac{1}{1+i\Sigma(p^2)\dfrac{i}{p^2-\mu_0^2+i\varepsilon}}\right] \\ &= \frac{i}{p^2-\mu_0^2-\Sigma(p^2)+i\varepsilon}. \end{aligned} \tag{2.5}$$

Clearly if we can make the proper self-energy part $\Sigma(p^2)$ finite, the propagator $\Delta(p)$ will also be finite.

Since there is no divergence in the tree (zero-loop) diagrams, we begin our calculation with the one-loop 1PI graphs. It is not difficult to see that Figs. 2.4 and 2.5 represent an exhaustive listing of all the one-loop divergent 1PI diagrams in this $\lambda\phi^4$ theory. Fig. 2.4 is the self-energy graph

$$-i\Sigma(p^2) = -\frac{i\lambda_0}{2}\int\frac{d^4l}{(2\pi)^4}\frac{i}{l^2-\mu_0^2+i\varepsilon}. \tag{2.6}$$

The factor of 1/2 is the symmetry factor, of which some examples were given in §1.1. Or we can deduce it directly from the fact that there are $4\cdot 3 = 12$

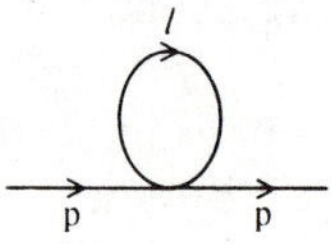

FIG. 2.4.

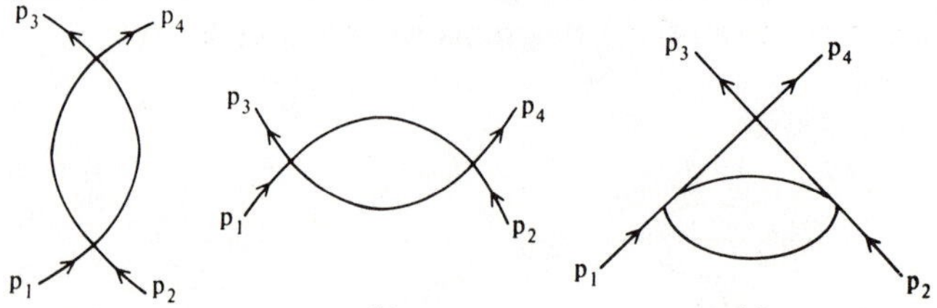

FIG. 2.5.

ways to 'contract' $\phi(x_1)\phi(x_2)$ into the interaction term $\phi^4(x)$ and this does not completely compensate for the 1/4! factor in (2.3). The integral in (2.6) is quadratically divergent. Fig. 2.5 shows the vertex corrections with contributions given by

$$\Gamma_a = \Gamma(p^2) = \Gamma(s) = \frac{(-i\lambda_0)^2}{2}\int \frac{d^4 l}{(2\pi)^4}\frac{i}{(l-p)^2-\mu_0^2+i\varepsilon}\frac{i}{l^2-\mu_0^2+i\varepsilon} \tag{2.7}$$

$$\Gamma_b = \Gamma(t), \qquad \Gamma_c = \Gamma(u) \tag{2.8}$$

where

$$s = p^2 = (p_1+p_2)^2, \qquad t = (p_1-p_3)^2, \qquad u = (p_1-p_4)^2 \tag{2.9}$$

are the Mandelstam variables. The contributions in (2.7) and (2.8) diverge logarithmically.

In the renormalization programme one first introduces some appropriate regularization schemes so that all divergent integrals are made finite. We are then free to manipulate (formally) these quantities, which are divergent only when the regularization is removed (e.g. by letting the cutoff approach infinity) at the end of the calculation. The commonly used regularization schemes will be discussed in §2.3. In the meantime it should be understood that by divergences we mean the regulated divergent quantities which are finite and cutoff-dependent.

For any divergent diagram we will first separate the divergent part from the finite part, then absorb the divergences in some appropriate redefinitions of mass, coupling, and field operators. To make the separation one uses an important property of the Feynman integrals given in (2.6) and (2.7): if one differentiates the divergent integral with respect to the external momenta, this increases the power of the internal momenta in the denominator and makes the integral less divergent. (These are examples of the 'primitively divergent' diagrams—for further discussion, see §2.2.) Therefore, when differentiated a sufficient number of times, the result is completely convergent. For example, if one differentiates $\Gamma(p^2)$ with respect to p^2, one finds

$$\begin{aligned}\frac{\partial}{\partial p^2}\Gamma(p^2) &= \frac{1}{2p^2}p_\mu\frac{\partial}{\partial p_\mu}\Gamma(p^2)\\ &= \frac{\lambda_0^2}{p^2}\int\frac{d^4 l}{(2\pi)^4}\frac{(l-p)\cdot p}{[(l-p)^2-\mu_0^2+i\varepsilon]^2}\frac{1}{l^2-\mu_0^2+i\varepsilon}\end{aligned} \tag{2.10}$$

which is finite. This means that the divergences will reside only in the first few terms of a Taylor series expansion in external momenta of the Feynman diagrams. For example, the Taylor expansion $\Gamma(p^2)$ around $p^2 = 0$ is of the form

$$\Gamma(p^2) = a_0 + a_1 p^2 + \ldots \frac{1}{n!}a_n(p^2)^n + \ldots$$

where

$$a_n = \left.\frac{\partial^n}{\partial^n p^2}\Gamma(p^2)\right|_{p^2=0}. \tag{2.11}$$

The a_ns are finite for $n \geq 1$ and only a_0 contains the logarithmic divergence. We can sum up all the finite terms and write

$$\Gamma(s) = \Gamma(0) + \tilde{\Gamma}(s) \tag{2.12}$$

where $\Gamma(0) = a_0$ is divergent and $\tilde{\Gamma}(s)$ is finite with the property

$$\tilde{\Gamma}(0) = 0. \tag{2.13}$$

These functions $\Gamma(0)$ and $\tilde{\Gamma}(s)$ will be calculated explicitly in §2.3. Note that the finite part $\tilde{\Gamma}(s)$ is just the original $\Gamma(s)$ with its value at $s = 0$ subtracted out. Hence this procedure is sometimes referred to as the *subtraction*.

In the following discussion we shall use the Taylor expansion of (2.6) and (2.7) to separate the divergent part from the finite part and absorb the divergent parts into redefinitions of the bare quantities.

Mass and wavefunction renormalization

The self-energy contribution given in eqn (2.6) is quadratically divergent. But this one-loop contribution has the peculiar property of being independent of the external momentum p. Hence the Taylor expansion is trivial; i.e. $\Sigma(p^2) = \Sigma(0)$. This is true only for the one-loop approximation in $\lambda\phi^4$ theory. For example the two-loop self-energy diagram in Fig. 2.2(a) is quadratically divergent and has a non-trivial dependence on p^2. Thus in general the Taylor expansion in external momenta around some arbitrary value μ^2 will have two divergent terms

$$\Sigma(p^2) = \Sigma(\mu^2) + (p^2 - \mu^2)\Sigma'(\mu^2) + \tilde{\Sigma}(p^2) \tag{2.14}$$

where $\Sigma(\mu^2)$ is quadratically and $\Sigma'(\mu^2)$ logarithmically divergent, as each differentiation with respect to the external momentum $\partial/\partial p_\mu$ decreases the degree of divergence by one unit and $\Sigma'(\mu^2)$ can be written in the form $\frac{1}{8}(\partial/\partial p_\mu)(\partial/\partial p^\mu)\Sigma(p^2)|_{p^2=\mu^2}$. Note that in general a quadratically divergent diagram will have three divergent terms with quadratic, linear, and logarithmic divergences. But in $\Sigma(p^2)$ there is no linearly divergent term because a term proportional to p_μ is not Lorentz invariant. The last term in (2.14) is finite and has the properties

$$\tilde{\Sigma}(\mu^2) = 0, \tag{2.15}$$

$$\tilde{\Sigma}'(\mu^2) = 0. \tag{2.16}$$

Of course in the one-loop approximation $\Sigma'(p^2) = \tilde{\Sigma}(p^2) = 0$ for all values of p^2. But in general the self-energies do not vanish. Substituting (2.14) into the expression for the full propagator in (2.5), we have

$$\mathrm{i}\Delta(p) = \frac{\mathrm{i}}{p^2 - \mu_0^2 - \Sigma(\mu^2) - (p^2 - \mu^2)\Sigma'(\mu^2) - \tilde{\Sigma}(p^2) + \mathrm{i}\varepsilon}. \tag{2.17}$$

The physical mass is defined as the position of the pole of the propagator. Since up to this point μ^2 is arbitrary, we can choose it to satisfy the equation

$$\mu_0^2 + \Sigma(\mu^2) = \mu^2. \tag{2.18}$$

Then

$$i\Delta(p^2) = \frac{i}{(p^2 - \mu^2)[1 - \Sigma'(\mu^2)] - \tilde{\Sigma}(p^2) + i\varepsilon}. \tag{2.19}$$

Using (2.15) one sees that $\Delta(p^2)$ has a pole at $p^2 = \mu^2$. Thus μ^2 is the physical mass and is related to the bare mass through eqn (2.18). This is the *mass renormalization*. Since $\Sigma(\mu^2)$ is divergent, the bare mass μ_0^2 must also be divergent so that the physical mass μ^2 is finite. To remove the divergent term $\Sigma'(\mu^2)$ we note that both $\Sigma'(\mu^2)$ and $\tilde{\Sigma}(p^2)$ are of order λ_0 (again keep in mind that all divergent quantities are regulated to be finite); we have

$$\tilde{\tilde{\Sigma}}(p^2) \simeq [1 - \Sigma'(\mu^2)]\tilde{\Sigma}(p^2) \tag{2.20}$$

and the propagator function can be written as

$$i\Delta(p^2) = \frac{iZ_\phi}{p^2 - \mu^2 - \tilde{\tilde{\Sigma}}(p^2) + i\varepsilon} \tag{2.21}$$

where

$$Z_\phi = [1 - \Sigma'(\mu^2)]^{-1} = 1 + \Sigma'(\mu^2) + 0(\lambda_0^2). \tag{2.22}$$

In this form the divergence is a multiplicative factor and can be removed by rescaling the field operator ϕ_0. More specifically, if we define the *renormalized field* ϕ by

$$\phi = Z_\phi^{-1/2}\phi_0, \tag{2.23}$$

then the renormalized propagator function given by

$$\begin{aligned} i\Delta_R(p) &= \int d^4x\, e^{-ip\cdot x}\langle 0|T(\phi(x)\phi(0))|0\rangle \\ &= Z_\phi^{-1}\int d^4x\, e^{-ip\cdot x}\langle 0|T(\phi_0(x)\phi_0(0))|0\rangle \\ &= \frac{i}{p^2 - \mu^2 - \tilde{\tilde{\Sigma}}(p^2) + i\varepsilon} = iZ_\phi^{-1}\,\Delta(p) \end{aligned} \tag{2.24}$$

is completely finite. Z_ϕ is usually referred to as the *wavefunction renormalization constant*. In this way, the divergences in self-energy are removed by mass renormalization (2.18) and wavefunction renormalization (2.23).

The renormalized field ϕ given in eqn (2.23) defines the renormalized Green's functions $G_R^{(n)}$ which are related to the unrenormalized ones by

$$\begin{aligned} G_R^{(n)}(x_1 \ldots x_n) &= \langle 0|T(\phi(x_1)\ldots\phi(x_n))|0\rangle \\ &= Z_\phi^{-n/2}\langle 0|T(\phi_0(x_1)\ldots\phi_0(x_n))|0\rangle \\ &= Z_\phi^{-n/2}G_0^{(n)}(x_1 \ldots x_n). \end{aligned} \tag{2.25}$$

Or, in momentum space,

$$G_R^{(n)}(p_1 \ldots p_n) = Z_\phi^{-n/2}G_0^{(n)}(p_1 \ldots p_n) \tag{2.26}$$

where

$$(2\pi)^4\,\delta^4(p_1 + \ldots + p_n)G_{\mathrm{R}}^{(n)}(p_1 \ldots p_n) = \int \left(\prod_{i=1}^{n} \mathrm{d}x_i\, \mathrm{e}^{-\mathrm{i}p_i \cdot x_i} \right) \times G_{\mathrm{R}}^{(n)}(x_1 \ldots x_n) \tag{2.27}$$

$$(2\pi)^4\,\delta^4(p_1 + \ldots + p_n)G_{0}^{(n)}(p_1 \ldots p_n) = \int \left(\prod_{i=1}^{n} \mathrm{d}x_i\, \mathrm{e}^{-\mathrm{i}p_i \cdot x_i} \right) \times G_{\mathrm{R}}^{(n)}(x_1 \ldots x_n) \tag{2.28}$$

To go from the connected Green's function given in (2.26) to the 1PI (amputated) Green's function, we have to eliminate the one-particle reducible diagrams, and also to remove the propagators for the external lines in 1PI Green's functions, i.e. remove the $\Delta_{\mathrm{R}}(p_i)$s from $\Gamma_{\mathrm{R}}^{(n)}(p_1 \ldots p_n)$ and the $\Delta(p_i)$s from $\Gamma_0^{(n)}(p_1 \ldots p_n)$. Since $\Delta_{\mathrm{R}}(p)$ and $\Delta(p)$ are related by

$$\Delta_{\mathrm{R}}(p_i) = Z_\phi^{-1}\,\Delta(p_i), \tag{2.29}$$

the renormalized and unrenormalized 1PI Green's functions are related by

$$\Gamma_{\mathrm{R}}^{(n)}(p_1 \ldots p_n) = Z_\phi^{n/2}\Gamma_0^{(n)}(p_1 \ldots p_n). \tag{2.30}$$

Coupling constant renormalization

We now proceed to renormalize the 1PI four-point function of Fig. 2.5. From eqns (2.7) and (2.8), this unrenormalized Green's function is given, to order λ_0^2, by

$$\Gamma_0^{(4)}(s, t, u) = -\mathrm{i}\lambda_0 + \Gamma(s) + \Gamma(t) + \Gamma(u) \tag{2.31}$$

where on the right-hand side the first term is the tree-graph contribution and the last three terms are the one-loop contributions which are divergent. We want to absorb these divergences by a redefinition of the coupling constant.

How is the coupling constant measured in $\lambda\phi^4$ theory? Since the basic vertex involves four particles, it would be natural to define the coupling constant in terms of the two-particle scattering amplitude, which is physically measurable. But for the discussion of the renormalization, it is more convenient to define the coupling constant in terms of the closely related renormalized 1PI (amputated) four-point function $\Gamma_{\mathrm{R}}^{(4)}(p_1, \ldots, p_4)$. Since $\Gamma_{\mathrm{R}}^{(4)}$ is a function of the kinematical variables s, t, and u (i.e. it is not a constant), some particular point in the kinematical region has to be chosen to define the physical coupling constant. Remembering that for particles on the shell $p_i^2 = \mu^2$ these variables satisfy the relation $s + t + u = 4\mu^2$, one may choose, as a convention, the *symmetric point*,

$$s_0 = t_0 = u_0 = \frac{4\mu^2}{3} \tag{2.32}$$

to define the coupling constant. Thus,

$$\Gamma_{\mathrm{R}}^{(4)}(s_0, t_0, u_0) = -\mathrm{i}\lambda \tag{2.33}$$

where λ is the physical coupling constant.

We will now separate the divergent and finite parts in the unrenormalized vertex function of (2.31) by making a Taylor series expansion around the symmetric point given in (2.32)

$$\Gamma_0^{(4)}(s, t, u) = -i\lambda_0 + 3\Gamma(s_0) + \tilde{\Gamma}(s) + \tilde{\Gamma}(t) + \tilde{\Gamma}(u) \tag{2.34}$$

where $\tilde{\Gamma}(s) = \Gamma(s) - \Gamma(s_0)$ is finite and has the property

$$\tilde{\Gamma}(s_0) = 0. \tag{2.35}$$

One defines the *vertex renormalization constant* Z_λ by

$$-iZ_\lambda^{-1}\lambda_0 = -i\lambda_0 + 3\Gamma(s_0). \tag{2.36}$$

Eqn (2.34) becomes

$$\Gamma_0^{(4)}(s, t, u) = -iZ_\lambda^{-1}\lambda_0 + \tilde{\Gamma}(s) + \tilde{\Gamma}(t) + \tilde{\Gamma}(u) \tag{2.37}$$

which at the symmetric point gives

$$\Gamma_0^{(4)}(s_0, t_0, u_0) = -iZ_\lambda^{-1}\lambda_0. \tag{2.38}$$

From the relation between the unrenormalized and the renormalized 1PI Green's functions eqn (2.30), we have

$$\Gamma_R^{(4)}(s, t, u) = Z_\phi^2\Gamma_0^{(4)}(s, t, u). \tag{2.39}$$

Then using eqns (2.33), (2.38), and (2.39), we see that the renormalized (physical) coupling constant λ defined in (2.33) is related to the unrenormalized coupling constant λ_0 by

$$\lambda = Z_\phi^2 Z_\lambda^{-1}\lambda_0. \tag{2.40}$$

It is now easy to demonstrate the finiteness of the renormalized 1PI four-point function. From eqns (2.37), (2.39), and (2.40), one has

$$\begin{aligned}\Gamma_R^{(4)}(p_1, \ldots, p_4) &= Z_\phi^2\Gamma_0^{(4)}(p_1, \ldots, p_4)\\ &= -iZ_\lambda^{-1}Z_\phi^2\lambda_0 + Z_\phi^2[\tilde{\Gamma}(s) + \tilde{\Gamma}(t) + \tilde{\Gamma}(u)]\\ &= -i\lambda + Z_\phi^2[\tilde{\Gamma}(s) + \tilde{\Gamma}(t) + \tilde{\Gamma}(u)].\end{aligned} \tag{2.41}$$

Since $Z_\phi = 1 + O(\lambda_0)$, $\tilde{\Gamma} = O(\lambda_0^2)$, and $\lambda = \lambda_0 + O(\lambda_0^2)$, we write to order λ^2

$$\Gamma_R^{(4)}(p_1, \ldots, p_4) = -i\lambda + \tilde{\Gamma}(s) + \tilde{\Gamma}(t) + \tilde{\Gamma}(u) + O(\lambda^3) \tag{2.42}$$

which is completely finite.

For the renormalization of the connected four-point Green's function to one loop, we have to add the one-particle reducible one-loop diagram (Fig. 2.6) and attach propagators for the external lines. Thus the unrenormalized

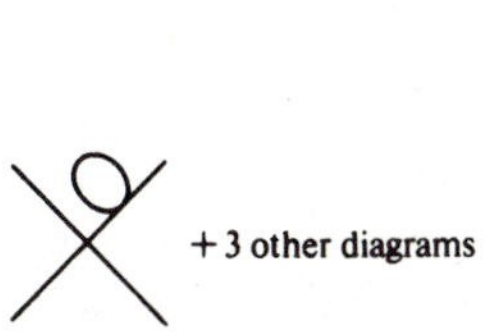

FIG. 2.6.

Green's function $G_0^{(4)}(p_1 \ldots p_n)$ is given by

$$G_0^{(4)}(p_1 \ldots p_4) = \left[\prod_{j=1}^{4}\left(\frac{1}{p_j^2 - \mu_0^2 + \mathrm{i}\varepsilon}\right)\right]\left\{-\mathrm{i}\lambda_0 + 3\Gamma(s_0) + \tilde{\Gamma}(s) + \tilde{\Gamma}(t) + \tilde{\Gamma}(u)\right.$$
$$\left. + (-\mathrm{i}\lambda_0)\sum_{k=1}^{4}[-\mathrm{i}\Sigma(p_k^2)]\left(\frac{\mathrm{i}}{p_k^2 - \mu_0^2 + \mathrm{i}\varepsilon}\right)\right\}. \qquad (2.43)$$

The first and last terms in (2.43) can be combined to give

$$(-\mathrm{i}\lambda_0)\left[\prod_{j=1}^{4}\left(\frac{1}{p_j^2 - \mu_0^2 + \mathrm{i}\varepsilon}\right)\right]\left[1 + \sum_{k=1}^{4}\Sigma(p_k^2)\left(\frac{1}{p_k^2 - \mu_0^2 + \mathrm{i}\varepsilon}\right)\right]$$
$$= (-\mathrm{i}\lambda_0)\left[\prod_{j=1}^{4}\left(\frac{1}{p_j^2 - \mu_0^2 - \Sigma(p_j^2) + \mathrm{i}\varepsilon}\right)\right] + O(\lambda_0^3). \qquad (2.44)$$

Since $\Gamma \sim O(\lambda_0^2)$, $\tilde{\Gamma} \sim O(\lambda_0^2)$, we can also write

$$\left[\prod_{j=1}^{4}\left(\frac{1}{p_j^2 - \mu_0^2 + \mathrm{i}\varepsilon}\right)\right][3\Gamma(s_0) + \tilde{\Gamma}(s) + \tilde{\Gamma}(t) + \tilde{\Gamma}(u)]$$
$$= \left[\prod_{j=1}^{4}\left(\frac{1}{p_j^2 - \mu_0^2 - \Sigma(p_j^2) + \mathrm{i}\varepsilon}\right)\right]$$
$$\times [3\Gamma(s_0) + \tilde{\Gamma}(s) + \tilde{\Gamma}(t) + \tilde{\Gamma}(u)] + O(\lambda_0^3). \qquad (2.45)$$

Using eqns (2.44) and (2.45), we can write eqn (2.43) as

$$G_0^{(4)}(p_1 \ldots p_4) = \prod_{j=1}^{4}\left[\frac{1}{p_j^2 - \mu_0^2 - \Sigma(p_j^2)}\right]$$
$$\times [-\mathrm{i}\lambda_0 + 3\Gamma(s_0) + \tilde{\Gamma}(s) + \tilde{\Gamma}(t) + \tilde{\Gamma}(u)]$$
$$= \left[\prod_{j=1}^{4}\mathrm{i}\Delta(p_j)\right]\Gamma_0^{(4)}(p_1 \ldots p_4) \qquad (2.46)$$

where we have used eqns (2.5) and (2.31). The renormalized four-point Green's function is defined by (2.26) as

$$G_R^{(4)}(p \,..\, p_4) = Z_\phi^{-2} G_0^{(4)}(p_1 \ldots p_4). \qquad (2.47)$$

Then from eqn (2.46) and the relations between the renormalized and the unrenormalized quantities (2.29) and (2.39), we get

$$G_R^{(4)}(p_1 \ldots p_4) = Z_\phi^{-2}\left[Z_\phi^4 \prod_{j=1}^{4}\mathrm{i}\Delta_R(p_j)\right] Z_\phi^{-2}\Gamma_R^{(4)}(p_1 \ldots p_4)$$
$$= \prod_{j=1}^{4}[\mathrm{i}\Delta_R(p_j)]\Gamma_R^{(4)}(p_1 \ldots p_4) \qquad (2.48)$$

which is also finite because $\Delta_R(p)$ and $\Gamma_R^{(4)}(p_1 \ldots p_4)$ have been shown to be finite.

We see that the mass, wavefunction, and vertex renormalizations remove all the divergences in the two- and four-point Green's functions in the one-loop approximation. There is no divergence in the other 1PI diagrams although the one-particle reducible graphs for the higher-point functions have divergent one-loop graphs. For example, the six-point function in Fig. 2.7 is divergent. However, it is clear that the divergence is brought about by that of the four-point vertex function and it is removed once we renormalize the four-point vertex function.

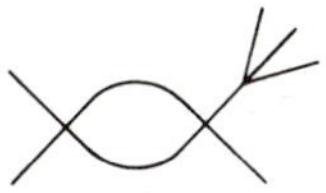

FIG. 2.7.

In summary, Green's function can be made finite if we express the bare quantities in terms of the renormalized ones through relations (2.18), (2.23), and (2.40)

$$\phi = Z_\phi^{-1/2}\phi_0 \tag{2.49}$$

$$\lambda = Z_\lambda^{-1} Z_\phi^2 \lambda_0 \tag{2.50}$$

$$\mu^2 = \mu_0^2 + \delta\mu^2 \tag{2.51}$$

where $\delta\mu^2 = \Sigma(\mu^2)$. More specifically, for an n-point Green's function when we express the bare mass μ_0 and bare coupling constant λ_0 in terms of the renormalized mass μ and coupling λ, and multiply by $Z_\phi^{-1/2}$ for each external field as in (2.26), then the result (the renormalized n-point Green's function) is completely finite

$$G_{\mathrm{R}}^{(n)}(p_1, \ldots, p_n; \lambda, \mu) = Z_\phi^{-n/2} G_0^{(n)}(p_1, \ldots, p_n; \lambda_0, \mu_0, \Lambda) \tag{2.52}$$

where Λ is the cut-off needed to define the divergent integrals. This feature, in which all the divergences, after rewriting λ_0 and μ_0 in terms of λ and μ, are aggregated into some multiplicative constants [$Z_\phi^{-n/2}$ in eqn (2.52)], is called being *multiplicatively renormalizable*. Equivalently, the 1PI Green's functions are made finite as in (2.30) by multiplying by $Z_\phi^{n/2}$ and expressing the bare quantities λ_0, μ_0 in terms of the physical quantities λ, μ,

$$\Gamma_{\mathrm{R}}^{(n)}(p_1, \ldots, p_n; \lambda, \mu) = Z_\phi^{n/2} \Gamma_0^{(n)}(p_1, \ldots, p_n; \lambda_0, \mu_0, \Lambda). \tag{2.53}$$

The programme of removing divergences as outlined in this section is closely related to the one originally developed and we shall refer to this as the *conventional renormalization scheme*.

2.2 BPH renormalization in $\lambda\phi^4$ theory

BPH renormalization (Bogoliubov and Parasiuk 1957; Hepp 1966; Zimmermann 1970) is completely equivalent to conventional renormalization. This alternative formulation of the programme is often more convenient for many applications of the renormalization theory. In this section we shall simply illustrate the connection between these two renormalization schemes. For a concise and lucid presentation see Coleman (1971*b*).

For the original (unrenormalized) Lagrangian (2.1)

$$\mathscr{L}_0 = \tfrac{1}{2}[(\partial_\mu \phi_0)^2 - \mu_0^2 \phi_0^2] - \frac{\lambda_0}{4!}\phi_0^4 \tag{2.54}$$

we can replace the bare quantities by renormalized quantities using eqns (2.49), (2.50), and (2.51) to obtain

$$\mathscr{L}_0 = \mathscr{L} + \Delta\mathscr{L}$$

where

$$\mathscr{L} = \tfrac{1}{2}[(\partial_\mu \phi)^2 - \mu^2 \phi^2] - \frac{\lambda}{4!}\phi^4 \tag{2.55}$$

and

$$\Delta\mathscr{L} = \frac{(Z_\phi - 1)}{2}[(\partial_\mu \phi)^2 - \mu^2\phi^2] + \frac{\delta\mu^2}{2} Z_\phi \phi^2 - \frac{\lambda(Z_\lambda - 1)}{4!}\phi^4. \tag{2.56}$$

$\mathscr{L}$, which has exactly the same form as $\mathscr{L}_0$ but with all the unrenormalized quantities replaced by renormalized ones, is called the *renormalized Lagrangian density*. $\Delta\mathscr{L}$ contains the divergent renormalization constants. $(Z_\phi - 1)$, $(Z_\lambda - 1)$, and $\delta\mu^2$ are all of order λ and this makes $\Delta\mathscr{L}$ of order $\lambda\mathscr{L}$. We call $\Delta\mathscr{L}$ the *counterterm Lagrangian*.

The BPH renormalization prescription consists of the following sequence of steps

(1) One starts with the renormalized Lagrangian of eqn (2.55) to construct propagators and vertices.

(2) The divergent part of the one-loop 1PI diagrams is isolated by the Taylor expansion. One then constructs a set of counterterms $\Delta\mathscr{L}^{(1)}$ which is designed to cancel these one-loop divergences.

(3) A new Lagrangian $\mathscr{L}^{(1)} = \mathscr{L} + \Delta\mathscr{L}^{(1)}$ is used to generate two-loop diagrams and to construct the counterterm $\Delta\mathscr{L}^{(2)}$ which cancels the divergences up to this order and so on, as this sequence of operations is iteratively applied.

The resulting Lagrangian is of the form

$$\mathscr{L}^{(\infty)} = \mathscr{L} + \Delta\mathscr{L}$$

where the counterterm Lagrangian $\Delta\mathscr{L}$ is given by

$$\Delta\mathscr{L} = \Delta\mathscr{L}^{(1)} + \Delta\mathscr{L}^{(2)} + \ldots + \Delta\mathscr{L}^{(n)} + \ldots. \tag{2.57}$$

In order to show that this renormalization scheme is equivalent to the conventional one which develops the unrenormalized perturbation theory directly we need to show that the counterterm Lagrangian (2.57) has the same structure as that of eqn (2.56). To demonstrate this we shall use the power-counting method to study the counterterms.

Power-counting method

To analyse the divergent structure of any Feynman diagram we introduce the term *superficial degree of divergence D*, which is the number of loop momenta in the numerator minus the number of loop momenta in the denominator. For example the graph shown in Fig. 2.8 has $D = 4 - 4 = 0$. Hence it is expected to be logarithmically divergent. To calculate D for any graph in the $\lambda\phi^4$ theory we define the following numbers.

B = number of external lines;
IB = number of internal lines;
n = number of vertices.

Since each vertex has four lines and both ends of an internal line must

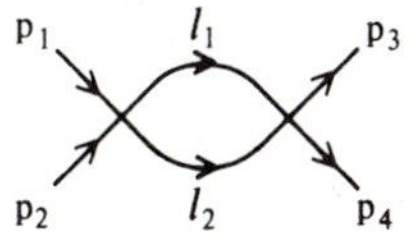

FIG. 2.8.

terminate on vertices while only one end of an external line is connected to a vertex, we have the relation

$$4n = 2(IB) + B. \tag{2.58}$$

We need to convert some of these to the number of loop momenta. The usual Feynman rule requires us to integrate over internal momenta which are not fixed by momentum conservation at each vertex. Thus we expect the number of loop momenta (L) to be the number of internal lines (IB) minus the number of vertices (n). But one of the combinations of momentum conservation δ-functions just expresses the overall momentum conservation and it does not depend on the internal momenta. For example the graph in Fig. 2.8 has two vertices and hence two δ-functions: $\delta^4(p_1 + p_2 - l_1 - l_2)$ $\delta^4(l_1 + l_2 - p_3 - p_4)$. But this can be written as $\delta^4(p_1 + p_2 - p_3 - p_4)$ $\delta^4(p_1 + p_2 - l_1 - l_2)$. The two vertices eliminate only one, rather than two, internal momenta. Therefore, we have

$$L = IB - n + 1. \tag{2.59}$$

For each internal line the propagator contributes two powers of loop momenta in the denominator and each loop integration contributes four powers of loop momenta in the numerator. For $\lambda\phi^4$ theory the vertices do not contribute any momentum factors and the superficial degree of divergence is given by

$$D = 4L - 2(IB). \tag{2.60}$$

We can eliminate L and IB in favour of B and n by using eqns (2.59) and (2.58),

$$D = 4 - B. \tag{2.61}$$

Since $\lambda\phi^4$ theory has reflection symmetry $\phi \to -\phi$, B must be an even number and eqn (2.61) implies that only the two-point function ($B = 2$) and four-point function ($B = 4$) are superficially divergent.

From this power counting, which is valid to all orders of perturbation theory, we can now study the structure of the counterterms. For the two-point function we have, according to (2.61), $D = 2$. Being quadratically divergent, the necessary Taylor expansion is taken to be

$$\Sigma(p^2) = \Sigma(0) + p^2\Sigma'(0) + \tilde{\Sigma}(p^2)$$

where $\Sigma(0)$ and $\Sigma'(0)$ are divergent while $\tilde{\Sigma}(p^2)$ is finite. There is no term linear in p_μ as $\Sigma(p^2)$ is a Lorentz scalar. We need to add two counterterms $\frac{1}{2}\Sigma(0)\phi^2 + \frac{1}{2}\Sigma'(0)(\partial_\mu\phi)^2$ to cancel the divergences. They correspond to the Feynman-rule vertices shown in Fig. 2.9(a), (b). The four-point function has $D = 0$ and the Taylor expansion

$$\Gamma^{(4)}(p_i) = \Gamma^{(4)}(0) + \tilde{\Gamma}^{(4)}(p_i)$$

where $\Gamma^{(4)}(0)$ is a logarithmically divergent term which is to be cancelled by a counterterm of the form $(\mathrm{i}\Gamma^{(4)}(0)/4!)\phi^4$. This has the graphic representation shown in Fig. 2.9(c).

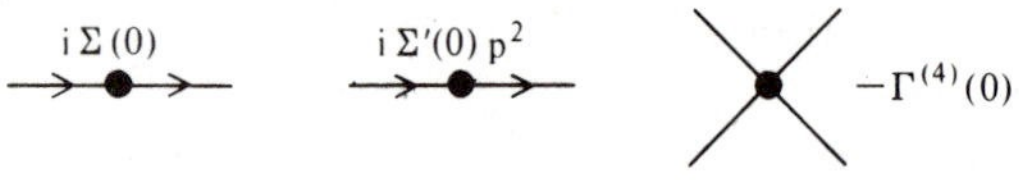

FIG. 2.9. Feynman-rule vertices corresponding to the counterterm Lagrangian (2.11).

The general counterterm Lagrangian is then of the form

$$\Delta\mathscr{L} = \frac{\Sigma(0)}{2}\phi^2 + \frac{\Sigma'(0)}{2}(\partial_\mu\phi)^2 + \frac{\mathrm{i}\Gamma^{(4)}(0)}{4!}\phi^4 \tag{2.62}$$

which is clearly the same as eqn (2.56) with the correspondences

$$\begin{aligned} \Sigma'(0) &= Z_\phi - 1 \\ \Sigma(0) &= -(Z_\phi - 1)\mu^2 + \delta\mu^2 = -\Sigma'(0)\mu^2 + \delta\mu^2 \\ \Gamma^{(4)}(0) &= -\mathrm{i}\lambda(1 - Z_\lambda). \end{aligned} \tag{2.63}$$

They are consistent with eqns (2.22), (2.51), and (2.36) as the renormalized coupling λ here is defined at the zero momentum point, thus $\Gamma^{(4)}(0) = 3\Gamma(0)$. This demonstrates the equivalence of BPH renormalization and conventional renormalization.

Comments on subgraph divergences

We shall not present any proof that, to all orders in the perturbation theory, this renormalization programme removes all divergences in the Green's functions. We merely illustrate some general features of the renormalization procedure for higher-order diagrams and the convergence properties of Feynman integrals with the following remarks.

(1) We state without proof the following convergence theorem (Weinberg

1960). *The general Feynman integral converges if the superficial degree of divergence of the graph together with the superficial degree of divergence of all subgraphs are negative.* To be more explicit, consider a Feynman graph with n external lines and l loops. Put a cut-off Λ in the momentum integration to estimate the order of divergence

$$\Gamma^{(n)}(p_1 \dots p_{n-1}) = \int_0^\Lambda d^4 q_1 \dots d^4 q_l I(p_1 \dots p_{n-1}; q_1 \dots q_l) \qquad (2.64)$$

where I is the product of vertices and propagators depending on p_i (external momenta) and q_i (internal momenta). Take a subset $S = \{q'_1 \dots q'_m\}$ of the loop momenta $\{q_1 \dots q_l\}$ and scale them to infinity (all $q'_i \to \Lambda$ with $\Lambda \to \infty$), all other momenta being fixed. Let $D(S)$ be the superficial degree of divergence associated with the integration over this set, namely

$$\left| \int_0^\Lambda d^4 q'_1 \dots d^4 q'_m I \right| \leq \Lambda^{D(S)}\{\ln \Lambda\} \qquad (2.65)$$

where $\{\ln \Lambda\}$ is some function of $\ln \Lambda$. Then the above theorem states that the integral over $\{q_1 \dots q_l\}$ converges if the $D(S)$s for all possible choices of S are negative. For example the graph in Fig. 2.10 being a six-point function has $D = -2$. But the integration inside the box having $D = 0$ is logarithmically divergent. Thus a successful renormalization programme must systematically remove all divergences including those associated with the subintegrations. In the BPH procedure these subdiagram divergences are in fact renormalized by low-order counterterms. For example, the graph in Fig. 2.11 with its counterterm vertex will cancel the subgraph divergence of Fig. 2.10.

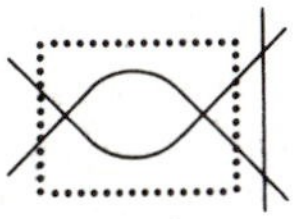

FIG. 2.10.

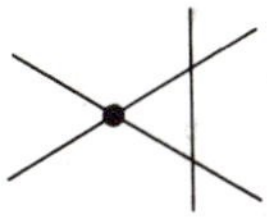

FIG. 2.11.

(2) There is another aspect of the renormalization programme related to these graphs with divergent subintegrations: not all divergences in a multi-loop diagram can be removed by subtracting out the first few terms in the Taylor expansion around the external momenta. This can be illustrated by the following example. Consider the two-loop graph of Fig. 2.12(a) which has the Feynman integral

$$\Gamma_a^{(4)}(p) \propto \lambda^3 [\Gamma(p)]^2 \qquad (2.66)$$

where

$$\Gamma(p) = \frac{1}{2} \int d^4 l \frac{1}{l^2 - \mu^2 + i\varepsilon} \frac{1}{(l-p)^2 - \mu^2 + i\varepsilon} \qquad (2.67)$$

with $p = p_1 + p_2$. With each of the $\Gamma(p)$ factors being logarithmically divergent, $\Gamma_a^{(4)}$ cannot be made convergent no matter how many derivatives operate on it, even though the overall superficial degree of divergence is zero. However we have the lower-order counterterm $-\lambda^2\Gamma(0)$ corresponding to the subtraction introduced at the one-loop level. This generates the additional λ^3 contributions of Fig. 2.12(b), (c) with $\Gamma_b^{(4)} \propto -\lambda^3\Gamma(p)\Gamma(0)$ and $\Gamma_c^{(4)} \propto -\lambda^3\Gamma(0)\Gamma(p)$, respectively.

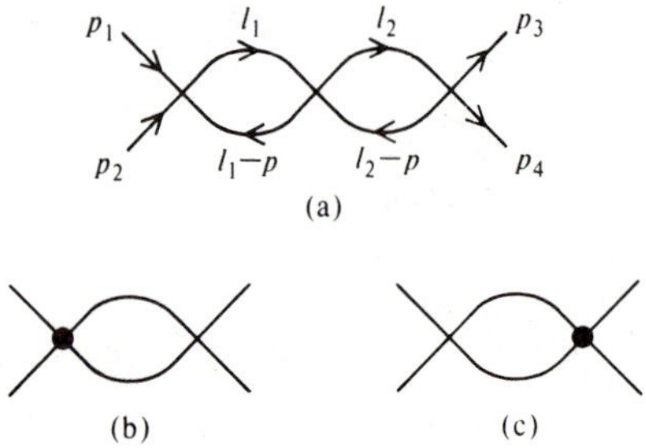

FIG. 2.12. s-channel λ^3 four-point functions. The black spots represent the counterterm $-\lambda^2\Gamma(0)$.

Adding the three graphs, Fig. 2.12(a), (b), (c), we have

$$\begin{aligned}\Gamma^{(4)}(p) &= \Gamma_a^{(4)} + \Gamma_b^{(4)} + \Gamma_c^{(4)} \\ &= -\lambda^3[\Gamma(0)]^2 + \lambda^3[\Gamma(p) - \Gamma(0)]^2 \\ &= \Gamma^{(4)}(0) + \tilde{\Gamma}^{(4)}(p).\end{aligned} \tag{2.68}$$

Only the first term on the right-hand side is divergent and can be removed by a λ^3 counterterm of the form $\mathrm{i}\Gamma^{(4)}(0)\phi^4/4!$. We see how, with the inclusion of the lower-order counterterms, divergences take on the form of polynomials in the external momenta. Thus for diagrams with more than one loop it is useful to characterize a divergent contribution as being primitively divergent or not. A *primitively divergent* graph has a non-negative overall superficial degree of divergence but is convergent for all subintegrations. Thus, they are diagrams in which the *only* divergence is caused by *all* of the loop momenta growing large together. In general only primitively divergent graphs such as Fig. 2.13 can have their divergences isolated by direct Taylor-series expansion. For other cases, diagrams with lower-order counterterm insertions must be included in order to aggregate the divergences into the form of polynomials in the external momenta.

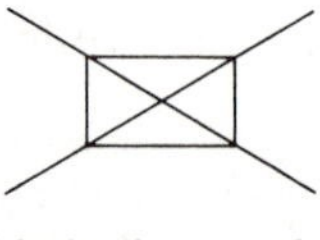

FIG. 2.13. A primitively divergent four-point function.

(3) In the above example of a two-loop, four-point function we have seen how the overall divergence can be isolated when diagrams with lower-order counterterms are included. For such cases where the divergent subinteg-

rations are *disjoint* this can be accomplished in a fairly direct manner. Similarly, it is also relatively easy for cases with *nested divergences*, i.e. for cases where one of each pair of divergent 1PI subgraphs is entirely contained within the other (see the example in Fig. 2.14). After the subgraph divergence

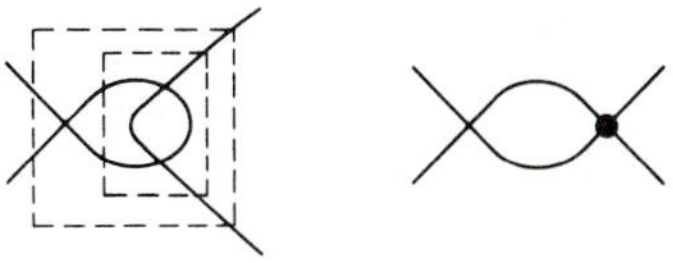

FIG. 2.14. Nested divergences and a diagram with a lower-order counterterm which cancels the subintegration divergence.

is removed by diagrams with lower-order counterterms (Fig. 2.14(b)), the overall divergence is then renormalized by a λ^3 counterterm. Thus for both disjoint and nested divergences the renormalization procedure is rather straightforward. The difficult step in the proof of the convergence (to all orders) involves disentangling the *overlapping divergences*, which are neither disjoint nor nested divergent 1PI diagrams. Fig. 2.2(a) is an example of overlapping divergence. Here it is difficult to see in a simple way how the subintegration divergences can be removed in a systematic fashion because they do not factorize in a simple manner. Nevertheless, this problem has been overcome and we refer the interested reader to the literature (Hepp 1966; Zimmermann 1970; Itzykson and Zuber 1980). The purpose of these comments is to indicate how the proof of renormalizability generally involves complicated graph classifications and combinatorial analysis.

2.3 Regularization schemes

In this section we will give detailed calculations of the various renormalization constants in the renormalized perturbation theory described in the previous sections. To make any meaningful mathematical manipulations on the divergent integrals we must cut off, or regularize, the momentum integration to make the integral finite. The divergent part will then be a function of the cut-off Λ while the finite part will be cut-off-independent in the limit $\Lambda \to \infty$. The cut-off procedure must be chosen in such a way that it maintains the Lorentz invariance and symmetry of the problem. There are two commonly used regularization schemes: the covariant cut-off and dimensional regularization. We shall illustrate them in turn.

Covariant regularization

In this procedure (Pauli and Villars 1949) the propagator will be modified as

$$\frac{1}{l^2 - \mu^2 + i\varepsilon} \quad \to \quad \frac{1}{l^2 - \mu^2 + i\varepsilon} + \sum_i \frac{a_i}{l^2 - \Lambda_i^2 + i\varepsilon} \tag{2.69}$$

where $\Lambda_i^2 \gg \mu^2$ and the a_is are chosen in such a way that in the asymptotic region the modified propagator will have a sufficient number of internal momenta in the denominator so that the integral is convergent.

Let us start with the four-point function. The graph in Fig. 2.5(a) yields a contribution (2.7)

$$\Gamma_a = \Gamma(p^2) = \frac{(-i\lambda)^2}{2} \int \frac{d^4 l}{(2\pi)^4} \frac{i}{(l-p)^2 - \mu^2 + i\varepsilon} \frac{i}{l^2 - \mu^2 + i\varepsilon}. \quad (2.70)$$

Clearly the replacement

$$\frac{1}{l^2 - \mu^2 + i\varepsilon} \to \frac{1}{l^2 - \mu^2 + i\varepsilon} - \frac{1}{l^2 - \Lambda^2 + i\varepsilon} = \frac{\mu^2 - \Lambda^2}{(l^2 - \mu^2 + i\varepsilon)(l^2 - \Lambda^2 + i\varepsilon)}$$

will be sufficient to render the integral finite. Eqn (2.70) then becomes

$$\Gamma(p^2) = \frac{-\lambda^2 \Lambda^2}{2} \int \frac{d^4 l}{(2\pi)^4} \frac{1}{((l-p)^2 - \mu^2 + i\varepsilon)(l^2 - \mu^2 + i\varepsilon)(l^2 - \Lambda^2 + i\varepsilon)}. \quad (2.71)$$

We choose to make the Taylor expansion around $p^2 = 0$ (or to make subtraction at $p^2 = 0$),

$$\Gamma(p^2) = \Gamma(0) + \tilde{\Gamma}(p^2) \quad (2.72)$$

with

$$\Gamma(0) = \frac{-\lambda^2 \Lambda^2}{2} \int \frac{d^4 l}{(2\pi)^4} \frac{1}{(l^2 - \mu^2 + i\varepsilon)^2 (l^2 - \Lambda^2 + i\varepsilon)} \quad (2.73)$$

$$\begin{aligned} \tilde{\Gamma}(p^2) &= \frac{-\lambda^2 \Lambda^2}{2} \int \frac{d^4 l}{(2\pi)^4} \frac{1}{(l^2 - \mu^2 + i\varepsilon)(l^2 - \Lambda^2 + i\varepsilon)} \\ &\quad \times \left[\frac{1}{(l-p)^2 - \mu^2 + i\varepsilon} - \frac{1}{l^2 - \mu^2 + i\varepsilon} \right] \\ &= \frac{\lambda^2}{2} \int \frac{d^4 l}{(2\pi)^4} \frac{2l \cdot p - p^2}{(l^2 - \mu^2 + i\varepsilon)^2 ((l-p)^2 - \mu^2 + i\varepsilon)} \end{aligned} \quad (2.74)$$

where in the last line we have taken the limit $\Lambda \to \infty$ inside the integral because $\tilde{\Gamma}(p^2)$ is convergent. The standard method to evaluate these integrals is to first use the identity to combine the denominator factors

$$\frac{1}{a_1 a_2 \ldots a_n} = (n-1)! \int_0^1 \frac{dz_1\, dz_2 \ldots dz_n}{(a_1 z_1 + a_2 z_2 + \ldots a_n z_n)^n} \delta\left(1 - \sum_{i=1}^{n} z_i\right) \quad (2.75)$$

where the z_is are called the *Feynman parameters*. We can also differentiate with respect to a_1 to get

$$\frac{1}{a_1^2 a_2 \ldots a_n} = n! \int_0^1 \frac{z_1\, dz_1\, dz_2 \ldots dz_n}{(a_1 z_1 + a_2 z_2 + \ldots a_n z_n)^{n+1}} \delta\left(1 - \sum_{i}^{n} z_i\right). \quad (2.76)$$

This formula has the advantage that one less Feynman parameter is needed for the case where there are two identical factors in the denominator. Using

(2.76), we can combine the denominators in (2.74) to give

$$\frac{1}{(l^2-\mu^2+i\varepsilon)^2}\frac{1}{(l-p)^2-\mu^2+i\varepsilon}=2\int_0^1\frac{(1-\alpha)\,d\alpha}{A^3} \tag{2.77}$$

where

$$A=(1-\alpha)(l^2-\mu^2)+\alpha[(l-p)^2-\mu^2]+i\varepsilon$$
$$=(l-\alpha p)^2-a^2+i\varepsilon$$

with

$$a^2=\mu^2-\alpha(1-\alpha)p^2.$$

Thus,

$$\tilde{\Gamma}(p^2)=\lambda^2\int_0^1(1-\alpha)\,d\alpha\int\frac{d^4l}{(2\pi)^4}\frac{2l\cdot p-p^2}{[(l-\alpha p)^2-a^2+i\varepsilon]^3}$$

$$=\lambda^2\int_0^1(1-\alpha)\,d\alpha\int\frac{d^4l}{(2\pi)^4}\frac{(2\alpha-1)p^2}{(l^2-a^2+i\varepsilon)^3} \tag{2.78}$$

where we have changed the variable l to $l+\alpha p$ and have dropped the term linear in l which will vanish upon symmetric integration. It is more convenient to do the integration by the *Wick rotation*, which transforms the Minkowski momentum to the Euclidean momentum. First we note that $d^4l=dl_0\,dl_1\,dl_2\,dl_3$ and

$$l^2-a^2+i\varepsilon=l_0^2-\mathbf{l}^2-a^2+i\varepsilon$$
$$=l_0^2-[(\mathbf{l}^2+a^2)^{1/2}-i\varepsilon]^2.$$

This shows that the integral (2.78) has poles in the complex l_0-plane as shown in Fig. 2.15.

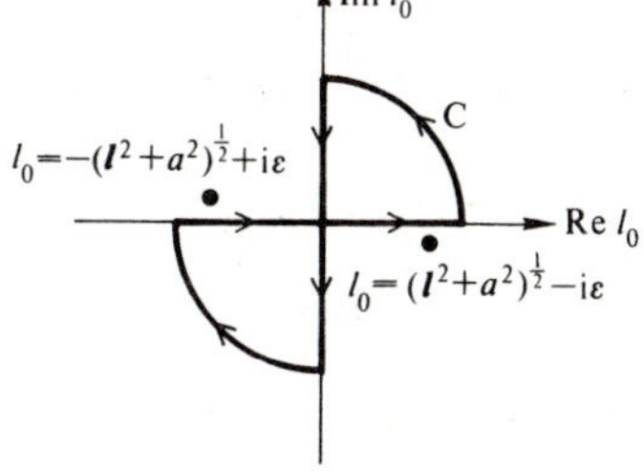

FIG. 2.15.

Using Cauchy's theorem we then have

$$\oint_C dl_0\,f(l_0)=0 \tag{2.79}$$

where

$$f(l_0) = \frac{1}{[l_0^2 - ((\mathbf{l}^2 + a^2)^{1/2} - i\varepsilon)^2]^3}.$$

Since $f(l_0) \to l_0^{-6}$ as $l_0 \to \infty$, the contribution from the circular part of contour C vanishes. Eqn. (2.79) implies that

$$\int_{-\infty}^{\infty} dl_0 f(l_0) = \int_{-i\infty}^{+i\infty} dl_0 f(l_0).$$

Thus, the integration along the real axis has been rotated to that along the imaginary axis. Change the variable $l_0 = il_4$ so that l_4 is real and

$$\int_{-i\infty}^{+i\infty} dl_0 f(l_0) = i \int_{-\infty}^{\infty} dl_4 f(il_4)$$

$$= -i \int_{-\infty}^{\infty} \frac{dl_4}{(l_1^2 + l_2^2 + l_3^2 + l_4^2 + a^2 - i\varepsilon)^3}. \tag{2.80}$$

If we define Euclidean momentum $k_i = (l_1, l_2, l_3, l_4)$ with $k^2 = l_1^2 + l_2^2 + l_3^2 + l_4^2$, then the results in eqns (2.79) and (2.80) may be written

$$\int \frac{d^4 l}{(2\pi)^4} \frac{1}{(l^2 - a^2 + i\varepsilon)^3} = -i \int \frac{d^4 k}{(2\pi)^4} \frac{1}{(k^2 + a^2 - i\varepsilon)^3} \tag{2.81}$$

where $d^4k = dl_1\, dl_2\, dl_3\, dl_4$. Using polar coordinates in four-dimensional Euclidean space, we have

$$\int d^4 k = \int_0^{\infty} k^3\, dk \int_0^{2\pi} d\phi \int_0^{\pi} \sin\theta\, d\theta \int_0^{\pi} \sin^2\chi\, d\chi \tag{2.82}$$

and

$$\int \frac{d^4 k}{(2\pi)^4} \frac{1}{(k^2 + a^2 - i\varepsilon)^3} = 2\pi^2 \int_0^{\infty} \frac{k^3\, dk}{(2\pi)^4} \frac{1}{(k^2 + a^2 - i\varepsilon)^3}$$

$$= \frac{1}{16\pi^2} \int_0^{\infty} \frac{k^2\, dk^2}{(k^2 + a^2 - i\varepsilon)^3}. \tag{2.83}$$

Using the formula for beta functions

$$\int_0^{\infty} \frac{t^{m-1}\, dt}{(t + a^2)^n} = \frac{1}{(a^2)^{n-m}} \frac{\Gamma(m)\Gamma(n-m)}{\Gamma(n)}, \tag{2.84}$$

we obtain

$$\int \frac{d^4 k}{(2\pi)^4} \frac{1}{(k^2 + a^2 - i\varepsilon)^3} = \frac{1}{32\pi^2(a^2 - i\varepsilon)} \tag{2.85}$$

or the vertex function in eqn (2.78) becomes

$$\tilde{\Gamma}(p^2) = \frac{-i\lambda^2}{32\pi^2}\int_0^1 \frac{d\alpha(1-\alpha)(2\alpha-1)p^2}{[\mu^2 - \alpha(1-\alpha)p^2 - i\varepsilon]}. \tag{2.86}$$

Since $0 < \alpha < 1$ we get $\mu^2 - \alpha(1-\alpha)p^2 > 0$ for $p^2 < 4\mu^2$ and we can drop $i\varepsilon$ in the denominator. It is straightforward to evaluate the integral to give

$$\begin{aligned}\tilde{\Gamma}(p^2) = \tilde{\Gamma}(s) &= \frac{i\lambda^2}{32\pi^2}\left\{2 + \left(\frac{4\mu^2 - s}{|s|}\right)^{\frac{1}{2}} \ln[\{(4\mu^2 - s)^{\frac{1}{2}} - (|s|)^{\frac{1}{2}}\}/\{(4\mu^2 - s)^{\frac{1}{2}} + (|s|)^{\frac{1}{2}}\}]\right\} \quad \text{for} \quad s < 0 \\ &= \frac{i\lambda^2}{32\pi^2}\left\{2 - 2\left(\frac{4\mu^2 - s}{s}\right)^{\frac{1}{2}} \tan^{-1}\left(\frac{s}{4\mu^2 - s}\right)^{\frac{1}{2}}\right\} \quad \text{for} \quad 0 < s < 4\mu^2 \\ &= \frac{i\lambda^2}{32\pi^2}\left\{2 + \left(\frac{s - 4\mu^2}{s}\right)^{\frac{1}{2}} \ln\left[\frac{s^{\frac{1}{2}} - (s - 4\mu^2)^{\frac{1}{2}}}{s^{\frac{1}{2}} + (s - 4\mu^2)^{\frac{1}{2}}}\right] + i\pi\right\} \\ &\qquad \text{for} \quad s > 4\mu^2. \end{aligned} \tag{2.87}$$

With the same procedure, the divergent term $\Gamma(0)$ given in eqn (2.73) can be calculated

$$\Gamma(0) = \frac{i\lambda^2\Lambda^2}{32\pi^2}\int_0^1 \frac{\alpha\, d\alpha}{\alpha(\mu^2 - \Lambda^2) + \Lambda^2}. \tag{2.88}$$

For large Λ^2, this gives

$$\Gamma(0) \simeq \frac{i\lambda^2}{32\pi^2}\ln\frac{\Lambda^2}{\mu^2}. \tag{2.89}$$

Thus the one-loop contribution to the four-point function is

$$\Gamma^{(4)}_{\text{1-loop}}(s, t, u) = 3\Gamma(0) + \tilde{\Gamma}(s) + \tilde{\Gamma}(t) + \tilde{\Gamma}(u) \tag{2.90}$$

where the cut-off-dependent $\Gamma(0)$ is given by eqn (2.89) and the finite $\tilde{\Gamma}(s)$ is given by eqn (2.87). We have to add the counterterm $(3i\Gamma(0)/4!)\phi^4$ to cancel these divergences. By (2.36) this corresponds to the renormalization constant

$$Z_\lambda^{-1} = 1 + \frac{3i\Gamma(0)}{\lambda} = 1 - \frac{3\lambda}{32\pi^2}\ln\frac{\Lambda^2}{\mu^2}. \tag{2.91}$$

Having cancelled the divergences, the total four-point function up to this order is then given by (2.42)

$$\Gamma^{(4)}_{R}(s, t, u) = -i\lambda + \tilde{\Gamma}(s) + \tilde{\Gamma}(t) + \tilde{\Gamma}(u). \tag{2.92}$$

For the two-point function of eqn (2.6), corresponding to the graph in Fig. 2.4, we have

$$-i\Sigma(p^2) = \frac{-i\lambda}{2}\int \frac{d^4 l}{(2\pi)^4}\frac{i}{l^2 - \mu^2 + i\varepsilon}. \tag{2.93}$$

This is a quadratically divergent integral and it can be regularized by choosing a_1 and a_2 in eqn (2.69) such that

$$\frac{1}{l^2-\mu^2+\mathrm{i}\varepsilon}+\frac{a_1}{l^2-\Lambda_1^2+\mathrm{i}\varepsilon}+\frac{a_2}{l^2-\Lambda_2^2+\mathrm{i}\varepsilon}\to\frac{1}{l^6}\quad\text{as}\quad l^2\to\infty.$$

It is not difficult to see that we need

$$a_1=\frac{\mu^2-\Lambda_2^2}{\Lambda_2^2-\Lambda_1^2}\quad\text{and}\quad a_2=\frac{\Lambda_1^2-\mu^2}{\Lambda_2^2-\Lambda_1^2}.$$

Then the modified propagator becomes

$$\frac{1}{l^2-\mu^2+\mathrm{i}\varepsilon}+\frac{a_1}{l^2-\Lambda_1^2+\mathrm{i}\varepsilon}+\frac{a_2}{l^2-\Lambda_2^2+\mathrm{i}\varepsilon}$$
$$=\frac{(\Lambda_1^2-\mu^2)(\Lambda_2^2-\mu^2)}{(l^2-\mu^2)(l^2-\Lambda_1^2)(l^2-\Lambda_2^2)}\to\frac{\Lambda^4}{(l^2-\mu^2)(l^2-\Lambda^2)^2}$$

for Λ_1 and Λ_2 both approach a large Λ. The regularized self-energy is

$$\begin{aligned}-\mathrm{i}\Sigma(p^2)&=\frac{\lambda}{2}\int\frac{\mathrm{d}^4l}{(2\pi)^4}\frac{\Lambda^4}{(l^2-\mu^2+\mathrm{i}\varepsilon)(l^2-\Lambda^2+\mathrm{i}\varepsilon)^2}\\&=\frac{-\mathrm{i}\lambda\Lambda^4}{32\pi^2}\int_0^1\frac{\alpha\,\mathrm{d}\alpha}{\alpha\Lambda^2+(1-\alpha)\mu^2}\\&=\frac{-\mathrm{i}\lambda}{32\pi^2}\left[\Lambda^2-\mu^2\ln\frac{\Lambda^2}{\mu^2}\right].\end{aligned}\tag{2.94}$$

Since it is independent of the external momentum p, the Taylor expansion is trivial,

$$\Sigma(p^2)=\Sigma(0)\simeq\frac{\lambda}{32\pi^2}\Lambda^2.\tag{2.95}$$

As we have mentioned before, this p-independence is a special property of the one-loop approximation in $\lambda\phi^4$ theory. For a more general self-energy graph, $\Sigma(p)$ will have a nontrivial dependence on p and the Taylor series around $p^2=0$ will be

$$\Sigma(p^2)=\Sigma(0)+p^2\Sigma'(0)+\tilde{\Sigma}(p^2)\tag{2.96}$$

where $\Sigma(0)$ and $\Sigma'(0)$ are cut-off-dependent and $\tilde{\Sigma}(p^2)$ is finite. And we have to add $\frac{1}{2}\Sigma(0)\phi^2$ and $\frac{1}{2}\Sigma'(0)(\partial_\mu\phi)^2$ counterterms to cancel these divergences.

To summarize, the total Lagrangian up to one loop has the form

$$\mathscr{L}^{(1)}=\mathscr{L}^{(0)}+\Delta\mathscr{L}^{(1)}\tag{2.97}$$

where

$$\mathscr{L}^{(0)}=\frac{1}{2}[(\partial_\mu\phi)^2-\mu^2\phi^2]-\frac{\lambda}{4!}\phi^4$$

$$\Delta\mathscr{L}^{(1)}=\frac{3\mathrm{i}\Gamma(0)}{4!}\phi^4+\frac{1}{2}\Sigma(0)\phi^2+\frac{1}{2}\Sigma'(0)(\partial_\mu\phi)^2.$$

Combining terms of the same structure, we can write (2.97) as

$$\mathscr{L}^{(1)} = \frac{Z_\phi}{2}(\partial_\mu\phi)^2 - \frac{(\mu^2 - \delta\mu^2)\phi^2}{2} - \frac{\lambda Z_\lambda}{4!}\phi^4 \tag{2.98}$$

with

$$\begin{aligned} Z_\phi &= 1 + \Sigma'(0), \\ \lambda Z_\lambda^{-1} &= \lambda + 3\mathrm{i}\Gamma(0), \\ \delta\mu^2 &= \Sigma(0). \end{aligned} \tag{2.99}$$

The values of these renormalization constants in the one-loop approximation are

$$\begin{aligned} Z_\phi &= 1 \quad \text{since} \quad \Sigma'(0) = 0, \\ Z_\lambda &= 1 + \frac{3\lambda}{32\pi^2}\ln\frac{\Lambda^2}{\mu^2}, \\ \delta\mu^2 &= \frac{\lambda}{32\pi^2}\Lambda^2. \end{aligned} \tag{2.100}$$

If we express everything in terms of the bare quantities through eqns (2.49), (2.50), and (2.51), we find

$$\mathscr{L}^{(1)} = \frac{1}{2}[(\partial_\mu\phi_0)^2 - \mu_0^2\phi_0^2] - \frac{\lambda_0}{4!}\phi_0^4 \tag{2.101}$$

which is exactly the same as the unrenormalized Lagrangian (2.1) as it should be.

Finally we comment on the convention used in making the Taylor series expansions (2.72) and (2.96) around $p_i = 0$ to fix the finite part of the Green's function. An equivalent way to state the same convention is to specify the normalization conditions of Green's function. From (2.96), the finite part of the self-energy has the properties

$$\tilde{\Sigma}(p^2)|_{p^2=0} = 0 \tag{2.102}$$

and

$$\left.\frac{\partial\tilde{\Sigma}(p^2)}{\partial p^2}\right|_{p^2=0} = 0. \tag{2.103}$$

These properties imply that the full propagator

$$\mathrm{i}\Delta_\mathrm{R}(p^2) = \frac{\mathrm{i}}{p^2 - \mu^2 - \tilde{\Sigma}(p^2) + \mathrm{i}\varepsilon} \tag{2.104}$$

will satisfy the normalization conditions

$$\Delta_\mathrm{R}^{-1}(p^2)|_{p^2=0} = -\mu^2 \tag{2.105}$$

and

$$\left.\frac{\partial\Delta_\mathrm{R}^{-1}}{\partial p^2}\right|_{p^2=0} = 1. \tag{2.106}$$

Similarly from (2.72) and thus from $\tilde{\Gamma}(0) = 0$, we have from (2.92) the normalization condition for the vertex function

$$\Gamma_R^{(4)}(0, 0, 0) = -i\lambda. \tag{2.107}$$

(*Remark:* Although (2.104) was originally derived with a Taylor expansion of $\Sigma(p^2)$ around $p^2 = \mu^2$ it also holds for the present $p^2 = 0$ expansion as a derivation entirely similar to eqns (2.14)–(2.22) will show.)

In short, one can use conditions (2.105), (2.106), and (2.107) to replace the prescription 'Taylor expansion around $p_i = 0$' to fix the finite part of Green's function.

In this connection we observe that the renormalized coupling constant defined by (2.107) differs from that defined by eqn (2.41) where a Taylor expansion has been made around the symmetric point $s_0 = t_0 = u_0 = 4\mu^2/3$. It implies condition (2.33)

$$\Gamma_R^{(4)}(s_0, t_0, u_0) = -i\lambda \tag{2.108}$$

to be contrasted with (2.107). Thus, different Taylor expansions or subtraction points yield different definitions of the coupling constant. This leads to the concept of a running coupling constant (see Chapter 3). Clearly the physics should not depend on the choice of subtraction point which is purely a convention. In practice how is this apparent difference taken care of? Consider the two-body scattering cross-sections calculated using two different definitions of the coupling constant. The calculated cross-sections may appear to be different by an overall constant (the angular distributions are identical). But this is immaterial because we need to define the coupling constant operationally as the value of the cross-section at some kinematical point. Thus the difference is only apparent and the two seemingly different calculations really yield the same result.

Dimensional regularization

The basic idea of this scheme ('t Hooft and Veltman 1972; Bollini and Giambiagi 1972; Ashmore 1972; Cicuta and Montaldi 1972) is that, since the ultraviolet divergences in Feynman diagrams come from the integration of internal momenta in four-dimensional space, the integrals can be made finite by lowering the dimensionalities of the space–time. Then the Feynman integrals can be defined as analytic functions of the space–time dimension n. The ultraviolet divergences will manifest themselves as singularities as $n \to 4$. As before, the finite part can be obtained by subtracting out the first few terms in the Taylor expansion. This regularization scheme has the important advantage that it will not destroy any algebraic relations among Green's functions that do not depend on space–time dimensions. In particular, the Ward identities, which are relations among Green's functions resulting from the symmetries of the theory, can be maintained in this dimensional regularization scheme. For a review see Leibrandt (1975).

We will illustrate this method with an example. Consider the one-loop four-point Green's function in eqn (2.7) corresponding to the diagram in Fig.

2.5(a). It is proportional to the integral

$$I = \int d^4 l \frac{1}{(l-p)^2 - \mu^2 + i\varepsilon} \frac{1}{l^2 - \mu^2 + i\varepsilon} \tag{2.109}$$

which is logarithmically divergent. To define the integral in n-dimensional space, we take the internal momentum to have n components: $l_\mu = (l_0, l_1, \ldots, l_{n-1})$, while the external momentum has four nonvanishing components: $p_\mu = (p_0, p_1, p_2, p_3, 0 \ldots 0)$. The integral in n-dimensional space is then defined as

$$I(n) = \int d^n l \frac{1}{(l-p)^2 - \mu^2 + i\varepsilon} \frac{1}{l^2 - \mu^2 + i\varepsilon} \tag{2.110}$$

which is convergent for $n < 4$. To define this integral for non-integer values of n, we first combine the denominators using Feynman parameters and make the Wick rotation (eqn (2.75)),

$$\begin{aligned} I(n) &= \int_0^1 d\alpha \int \frac{d^n l}{[(l - \alpha p)^2 - a^2 + i\varepsilon]^2} \\ &= i \int_0^1 d\alpha \int \frac{d^n l}{[l^2 + a^2 - i\varepsilon]^2} \end{aligned} \tag{2.111}$$

with $a^2 = \mu^2 - \alpha(1-\alpha)p^2$.

The integrand is now independent of the angles of the integration momentum, which can then be integrated out

$$\begin{aligned} \int d^n l &= \int_0^\infty l^{n-1}\, dl \int_0^{2\pi} d\theta_1 \int_0^\pi \sin\theta_2\, d\theta_2 \int_0^\pi \sin^2\theta_3\, d\theta_3 \ldots \\ &\quad \times \int_0^\pi \sin^{n-2}\theta_{n-1}\, d\theta_{n-1} \\ &= \frac{2\pi^{n/2}}{\Gamma\left(\frac{n}{2}\right)} \int_0^\infty l^{n-1}\, dl \end{aligned} \tag{2.112}$$

where we have used the formula

$$\int_0^\pi \sin^m\theta\, d\theta = \frac{\pi^{1/2}\Gamma\left(\frac{m+1}{2}\right)}{\Gamma\left(\frac{m+2}{2}\right)}. \tag{2.113}$$

Thus eqn (2.111) may be written

$$I(n) = \frac{2i\pi^{n/2}}{\Gamma\left(\frac{n}{2}\right)} \int_0^1 d\alpha \int_0^\infty \frac{l^{n-1}\, dl}{[l^2 + a^2 - i\varepsilon]^2}. \tag{2.114}$$

The dependence on n is now explicit. For complex n, the integral is well-defined as long as $0 < \mathrm{Re}(n) < 4$; the lower bound results from the apparent divergence of the integral at the $l = 0$ limit. This infrared divergence is actually an artefact of our procedure as it is cancelled by the singularity in $\Gamma(\frac{1}{2}n)$ as $n \to 0$. We can extend this domain of analyticity by integration by parts

$$\frac{1}{\Gamma\left(\frac{n}{2}\right)}\int_0^\infty \frac{l^{n-1}\,\mathrm{d}l}{[l^2 + a^2 - \mathrm{i}\varepsilon]^2} = \frac{-2}{\Gamma\left(\frac{n}{2}+1\right)}\int_0^\infty \mathrm{d}l\,\frac{\mathrm{d}}{\mathrm{d}l}\left(\frac{1}{[l^2 + a^2 - \mathrm{i}\varepsilon]^2}\right) \tag{2.115}$$

where we have used

$$z\Gamma(z) = \Gamma(z+1). \tag{2.116}$$

The integral is now well defined for $-2 < \mathrm{Re}(n) < 4$. If we repeat this procedure v times, the analyticity domain is extended to $-2v < \mathrm{Re}(n) < 4$ and eventually to $\mathrm{Re}(n) \to -\infty$. Thus the integral given in eqn (2.114) can be taken as an analytic function for $\mathrm{Re}(n) < 4$. To see what happens as $n \to 4$, we use (2.84) to evaluate the integral for $n < 4$,

$$I(n) = \mathrm{i}\pi^{n/2}\Gamma\left(2 - \frac{n}{2}\right)\int_0^1 \frac{\mathrm{d}\alpha}{[a^2 - \mathrm{i}\varepsilon]^{2-n/2}}. \tag{2.117}$$

Using formula (2.116)

$$\Gamma\left(2 - \frac{n}{2}\right) = \frac{\Gamma\left(3 - \frac{n}{2}\right)}{2 - \frac{n}{2}} \to \frac{2}{4-n} \quad \text{as} \quad n \to 4,$$

we see that the singularity at $n = 4$ is a simple pole. If we now expand everything around $n = 4$

$$\Gamma\left(2 - \frac{n}{2}\right) = \frac{2}{4-n} + A + (n-4)B + \dots \tag{2.118}$$

$$a^{n-4} = 1 + (n-4)\ln a + \dots, \tag{2.119}$$

where A and B are some constants, we obtain the limit

$$I(n) \underset{n\to 4}{\to} \frac{2\mathrm{i}\pi^2}{4-n} - \mathrm{i}\pi^2\int_0^1 \mathrm{d}\alpha \ln[\mu^2 - \alpha(1-\alpha)p^2] + \mathrm{i}\pi^2 A. \tag{2.120}$$

With the one-loop contribution of (2.7) ($\Gamma = \lambda^2 I/32\pi^4$), we have

$$\Gamma(p^2) = \frac{\lambda^2}{32\pi^2}\left\{\frac{2\mathrm{i}}{4-n} - \mathrm{i}\int_0^1 \mathrm{d}\alpha \ln[\mu^2 - \alpha(1-\alpha)p^2] + \mathrm{i}A\right\}. \tag{2.121}$$

The Taylor expansion around $p^2 = 0$ gives

$$\Gamma(p^2) = \Gamma(0) + \tilde{\Gamma}(p^2) \tag{2.122}$$

where

$$\Gamma(0) = \frac{\lambda^2}{32\pi^2}\left(\frac{2\mathrm{i}}{4-n} - \mathrm{i}\ln\mu^2 + \mathrm{i}A\right)$$

$$\simeq \frac{\mathrm{i}\lambda^2}{16\pi^2(4-n)} \tag{2.123}$$

and

$$\tilde{\Gamma}(p^2) = \frac{-\mathrm{i}\lambda^2}{32\pi^2}\int_0^1 \mathrm{d}\alpha \ln\left[\frac{\mu^2 - \alpha(1-\alpha)p^2}{\mu^2}\right]$$

$$= \frac{-\mathrm{i}\lambda^2}{32\pi^2}\int_0^1 \frac{\mathrm{d}\alpha(1-\alpha)(2\alpha-1)p^2}{[\mu^2 - \alpha(1-\alpha)p^2]} \tag{2.124}$$

where we have performed an integration by parts. Clearly the finite part is exactly the same as that given by the method of covariant regularization in eqn (2.86). Thus the finite part of Green's function is independent of the regularization schemes as it should be and only depends on the subtraction point. The $\Gamma(0)$ term diverges as a simple pole at $n = 4$ corresponding to the $\ln\Lambda$ term (2.89) in the covariant regularization calculation.

The one-loop self-energy (Fig. 2.4) is given by eqn (2.6) which in the dimensional-regularization scheme becomes

$$-\mathrm{i}\Sigma(p^2) = \frac{\lambda}{2}\int \frac{\mathrm{d}^n l}{(2\pi)^4}\frac{1}{l^2 - \mu^2 + \mathrm{i}\varepsilon}$$

$$= \frac{-\mathrm{i}\lambda\pi^{n/2}\Gamma\left(1-\dfrac{n}{2}\right)}{32\pi^4(\mu^2)^{1-n/2}}. \tag{2.125}$$

Since, from eqn (2.116),

$$\Gamma\left(1-\frac{n}{2}\right) = \frac{\Gamma\left(3-\dfrac{n}{2}\right)}{\left(1-\dfrac{n}{2}\right)\left(2-\dfrac{n}{2}\right)}, \tag{2.126}$$

the quadratic divergent term (2.95) has poles at $n = 4$ and also at $n = 2$. For $n \to 4$ we have

$$-\mathrm{i}\Sigma(0) = \frac{\mathrm{i}\lambda\mu^2}{16\pi^2}\left(\frac{1}{4-n}\right). \tag{2.127}$$

To compare the two regularization methods we list the results for the divergences in Table 2.1. Thus divergent Feynman integrals when evaluated in n-dimensional space appear as poles of the resulting Γ function at

$n = 4, \ldots$ etc., keeping in mind that the quadratic divergence also has a pole at $n = 2$, see eqn (2.126).

TABLE 2.1

	Covariant regularization	Dimensional regularization
$\Gamma(0)$	$\frac{i\lambda^2}{32\pi^2} \ln \frac{\Lambda^2}{\mu^2}$	$\frac{i\lambda^2}{32\pi^2}\left(\frac{2}{4-n}\right)$
$\Sigma(0)$	$\frac{\lambda}{32\pi^2} \Lambda^2$	$\frac{\lambda}{32\pi^2}\left(\frac{-2\mu^2}{4-n}\right)$

2.4 Power counting and renormalizability

In the previous sections the renormalization procedure in $\lambda\phi^4$ theory has been illustrated in some detail. Here we will discuss the problem of renormalization for the more general types of interaction. The BPH renormalization procedure will be followed in this discussion. In a later part of this section, renormalization of composite operators will also be examined.

Theories with fermion and scalar particles

For simplicity we shall first concentrate on theories with spin-1/2 and spin-0 particles only. For the Lagrangian density, $\mathscr{L} = \mathscr{L}_0 + \Sigma_i \mathscr{L}_i$, where $\mathscr{L}_0$ is the free Lagrangian quadratic in the fields and the $\mathscr{L}_i$s are the interaction terms (for example, $\mathscr{L}_i = g_1\bar{\psi}\gamma_\mu\psi\, \partial^\mu\phi, g_2(\bar{\psi}\psi)^2, g_3\bar{\psi}\psi\phi, g_4\phi^3, g_5\phi^4, \ldots$), for a given graph we can define the quantities

n_i = number of ith type vertices;
b_i = number of scalar lines in the ith type vertex;
f_i = number of fermion lines in the ith type vertex;
d_i = number of derivatives in the ith type vertex;
B = number of external scalar lines;
F = number of external fermion lines;
IB = number of internal scalar lines;
IF = number of internal fermion lines.

Thus for $\mathscr{L}_1 = g_1\bar{\psi}\gamma_\mu\psi\, \partial^\mu\phi$ we have $b_1 = 1$, $f_1 = 2$, $d_1 = 1$. From the structure of the graph we have relations like that of (2.58)

$$B + 2(IB) = \sum_i n_i b_i \tag{2.129a}$$

$$F + 2(IF) = \sum_i n_i f_i. \tag{2.129b}$$

Just as in (2.59), the number of loop integrations L can be calculated

$$L = (IB) + (IF) - n + 1 \tag{2.130}$$

where

$$n = \sum_i n_i. \tag{2.131}$$

The superficial degree of divergence D is then given by

$$\begin{aligned} D &= 4L - 2(IB) - (IF) + \sum_i n_i d_i \\ &= 4 + 2(IB) + 3(IF) + \sum_i n_i(d_i - 4). \end{aligned} \tag{2.132}$$

Using (2.129) we can eliminate IB and IF,

$$D = 4 - B - \tfrac{3}{2}F + \sum_i n_i \, \delta_i \tag{2.133}$$

where

$$\delta_i = b_i + \tfrac{3}{2}f_i + d_i - 4 \tag{2.134}$$

is called the *index of divergence* of the interaction $\mathscr{L}_i$. For $\lambda\phi^4$ theory, $\delta = 0$ and (2.133) reduces to (2.61). In general δ_i can be related to the dimension of the coupling constant in units of mass. Knowing that the Lagrangian density has dimension four and that the scalar field, the fermion field, and the derivative have dimensions 1, 3/2, and 1, respectively, the dimension of the coupling constant is given by

$$\dim(g_i) = 4 - b_i - \tfrac{3}{2}f_i - d_i = -\delta_i. \tag{2.135}$$

From (2.133) we see that, for a fixed number of external lines, the superficial degree of divergence will have different behaviour for the following three cases.

(1) *g_i has positive dimension* (*or* $\delta_i < 0$). Then D decreases with the number of ith type vertices. In this case $\mathscr{L}_i$ is called a *super-renormalizable* interaction and the divergences are restricted to a finite number of graphs. For example, consider the graphs for the two-point Green's functions in the super-renormalizable $\lambda\phi^3$ theory. The one-loop diagram in Fig. 2.16(a) is divergent while the two-loop one in Fig. 2.16(b) is not.

(a) (b)

FIG. 2.16.

(2) *g_i is dimensionless* (*or* $\delta_i = 0$). Here D is independent of the number of ith type vertices. The divergences are present in all higher-order diagrams of a finite number of Green's functions. $\mathscr{L}_i = g_1\phi^4, g_2\bar{\psi}\psi\phi$ are such examples, and they are called *renormalizable* interactions.

(3) *g_i has negative dimension* (*or* $\delta_i > 0$). In this case, D increases with the number of ith type vertices and all Green's functions are divergent for sufficiently large n_i. These types of interactions are *non-renormalizable*, and are exemplified by $\mathscr{L}_i = g_1\bar{\psi}\gamma_\mu\psi\,\partial^\mu\phi,\ g_2(\bar{\psi}\psi)^2,\ g_3\phi^5, \ldots$ etc.

The index of divergence δ_i is also related to the *canonical dimension* of the field operator. The latter is defined in terms of the high-energy behaviour of the free-field propagator, which is clearly relevant for power counting. Write the propagator for the free-field operator as

$$D_A(p^2) = \int d^4x\, e^{-ip\cdot x}\langle 0|T(A(x)A(0))|0\rangle. \tag{2.136}$$

If the asymptotic behaviour is of form

$$D_A(p^2) \underset{p^2\to\infty}{\to} (p^2)^{-\omega_A/2}, \tag{2.137}$$

then the canonical dimension for the field operator is defined as

$$d(A) = (4 - \omega_A)/2. \tag{2.138}$$

Thus for the scalar and fermion fields and their derivatives, we have

$$\begin{aligned} d(\phi) &= 1, \qquad d(\partial^n\phi) = 1 + n, \\ d(\psi) &= \tfrac{3}{2}, \qquad d(\partial^n\psi) = \tfrac{3}{2} + n. \end{aligned} \tag{2.139}$$

For composite operators that are polynomials in the fields the canonical dimension is the algebraic sum of the constituent fields: for example, $d(\phi^2) = 2d(\phi) = 2$, $d(\bar\psi\psi\phi) = 2d(\psi) + d(\phi) = 4$. In the case of theories with fermions and scalars only, the canonical dimension of an operator is the same as that of the naïve dimension in units of mass. But as we shall see later, these dimensions are different for massive vector fields. With these definitions and those in (2.128), the canonical dimension for each term in the interaction Lagrangian density becomes

$$d(\mathscr{L}_i) = b_i + \tfrac{3}{2}f_i + d_i. \tag{2.140}$$

With the index of divergence $\delta_i = d(\mathscr{L}_i) - 4$, we see that a dimension-four term corresponds to a renormalizable interaction, that less than four is super renormalizable, and that greater than four is nonrenormalizable.

Counterterms

Since the counterterms are constructed to cancel the divergences in the n-point Green's function, their structures are closely related to that of the superficially divergent Green's function. For example, we have seen that in $\lambda\phi^4$ theory to cancel the quadratically divergent parts in the two-point function, we need counterterms $(\partial_\mu\phi)(\partial^\mu\phi)$ with dimension 4 and ϕ^2 terms with dimension 2, while the logarithmically divergent four-point function needs the dimension-4 counterterm ϕ^4. In general we have to add counterterms to cancel all divergences in Green's functions with superficial degrees of divergence $D \geq 0$ as determined by (2.133). For convenience we will use the Taylor expansion around zero external momenta $p_i = 0$ to isolate the divergent terms. The structure of the counterterms depends on the number of divergent terms in the Taylor expansion. For example, if a Green's function is quadratically divergent, the first three terms in the expansion will be divergent

$$\Gamma^{(n)}(p_i) = a + b_i^{\mu} p_{i\mu} + c_{ij}^{\mu\nu} p_{i\mu} p_{j\nu} + \tilde{\Gamma}^{(n)}(p_i). \tag{2.141}$$

The counterterms designed to cancel the a-term will have no derivative, the terms designed to cancel the b-term will have one derivative, etc. In the notation of (2.128) the counterterm will have the form $O_{\text{ct}} = (\partial_\mu)^\alpha(\psi)^{\text{F}}(\phi)^{\text{B}}$ with $\alpha = 0, 1, \ldots, D$. For $\lambda\phi^4$ theory, for example, we have terms corresponding to $F = 0$, $B = 2$ with $\alpha = 0$ and 2, $B = 4$ with $\alpha = 0$. The canonical dimension of O_{ct} is given by

$$d_{\text{ct}} = \tfrac{3}{2}F + B + \alpha. \tag{2.142}$$

The index of divergence of the counterterm can then be written through (2.133) as

$$\begin{aligned}\delta_{\text{ct}} &= d_{\text{ct}} - 4 \\ &= (\alpha - D) + \sum_i n_i \, \delta_i.\end{aligned} \tag{2.143}$$

Since $\alpha \leq D$, we have the result

$$\delta_{\text{ct}} \leq \sum_i n_i \, \delta_i. \tag{2.144}$$

Thus, the counterterms induced by a Feynman diagram have indices of divergence δ_{ct} less or equal to the sum of the indices of divergence of all interactions δ_i in the diagram.

The renormalizable interactions which have $\delta_i = 0$ will generate counterterms with $\delta_{\text{ct}} \leq 0$. If all the $\delta_i \leq 0$ terms are present in the original Lagrangian, so that here the counterterms have the same structure as the terms in the original Lagrangian, they may be considered as redefining parameters like masses and coupling constants in the theory. These renormalized parameters are inputs of the theory and we need measurements of some physical processes to determine them. With these inputs, we can then predict the outcome of all other physical processes. For example, in $\lambda\phi^4$ theory we have two free parameters, the coupling constant λ and mass μ. We can use the two-particle elastic scattering cross-section at two different scattering angles to determine the values of λ and μ. The cross-sections for all other angles and/or all other energies (and also all other inelastic cross-sections) can then be predicted. Much the same holds for super-renormalizable theories. On the other hand, non-renormalizable interactions which have $\delta_i > 0$ will generate counterterms with arbitrary large δ_{ct} in sufficiently high orders and clearly they cannot be absorbed into the original Lagrangian by a redefinition of parameters. For example, the non-renormalizable interaction $\lambda\phi^6$ which has $\delta = 2$, will produce counterterms consisting of all even powers of ϕ and their derivatives: ϕ^{2n} and $\partial^{2m}\phi^{2n}$ with $n, m = 1, 2, \ldots, \infty$. We need an infinite number of measurements to fix the coefficients of these terms. Thus non-renormalizable theories will not necessarily be infinite; however the infinite number of counterterms associated with a non-renormalizable interaction will make it lack in predictive power and hence be unattractive, in the framework of perturbation theory.

We will adopt a more restricted definition of renormalizability. A Lagrangian is said to be *renormalizable by power counting* if all the counterterms induced by the renormalization procedure can be absorbed by redefinitions of the parameters in the Lagrangian. With this definition, the theory with a single-fermion interaction with a single scalar through the Yukawa coupling $\bar{\psi}\gamma_5\psi\phi$ is not renormalizable even though the coupling constant is dimensionless. This is because the one-loop diagram of Fig. 2.17 is logarithmically divergent and we need a ϕ^4 counterterm. But such a term is not present in the original Lagrangian. The same theory with a ϕ^4 interaction

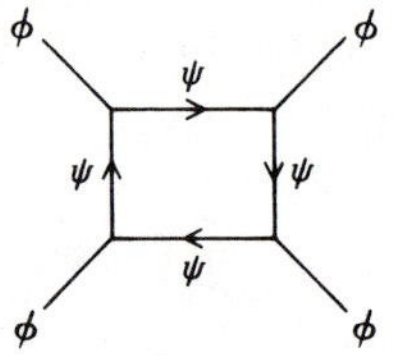

FIG. 2.17.

is renormalizable. On the other hand, if a term can be excluded on symmetry grounds, then the renormalizability of the theory is not disturbed because higher-order terms will not generate such a term. For example, in a theory with only one scalar field,

$$\mathscr{L} = \tfrac{1}{2}(\partial_\mu\phi)^2 - \tfrac{1}{2}\mu^2\phi^2 - \frac{\lambda}{4!}\phi^4$$

is renormalizable because it contains *all* terms with $\delta \leq 0$ (equivalently with dimension less than or equal to 4) which are consistent with the symmetry $\phi \to -\phi$. The ϕ^3 counterterm will be forbidden by such a reflection symmetry. Also, in this context we can understand result (2.133), or the more restricted $\lambda\phi^4$ result (2.61). The higher-order contributions to, say, a six-point function should be finite in (renormalizable) $\lambda\phi^4$ theory. This must be the case because, if they were not, one would need a ϕ^6 counterterm to absorb the divergences. Such a counterterm having $\delta = 2$ would ruin the renormalizability of the theory.

Theories with vector fields

Since the asymptotic behaviour of free vector-field propagators is very different for the massless and massive cases, we will discuss them separately.

Massless vector field. In a theory with local gauge invariance such as QED, the vector field is massless. The asymptotic behaviour of the free propagator is mild. For example the Feynman-gauge photon propagator in QED is given by

$$\Delta_{\mu\nu}(k) = \frac{-\mathrm{i}g_{\mu\nu}}{k^2 + \mathrm{i}\varepsilon} \underset{k\to\infty}{\rightarrow} O(k^{-2}). \tag{2.145}$$

This implies that the photon field will have unit canonical dimension: $d(A) = 1$, like that for the scalar ϕ. Consequently power counting for a massless vector field is the same as that for the scalar field. Theories with a massless vector field will be renormalizable if they contain all interactions with dimension less than or equal to four and consistent with local gauge invariance. Denoting the massless vector field by A_μ, we have, for example, the dimension-4 operators

$$\bar{\psi}\gamma_\mu\psi A^\mu,\ \phi^2 A_\mu A^\mu,\ (\partial_\mu\phi)\phi A^\mu.$$

This in fact represents an exhaustive listing of all possible renormalizable interactions (i.e. dimension-4 or -3) of spin-0 and -1/2 fields with massless vector fields. The only possible dimension-3 operator $(\partial_\mu\phi)A^\mu$, which is bilinear in fields, is part of the free Lagrangian.

Massive vector field. Generally the free Lagrangian for a massive vector field V_μ has the form

$$\mathscr{L}_0 = -\tfrac{1}{4}(\partial_\mu V_\nu - \partial_\nu V_\mu)(\partial^\mu V^\nu - \partial^\nu V^\mu) + \tfrac{1}{2}M_V^2 V_\mu V^\mu. \tag{2.146}$$

The vector propagator in momentum space

$$D_{\mu\nu}(k) = \frac{-\mathrm{i}(g_{\mu\nu} - k_\mu k_\nu/M_V^2)}{k^2 - M_V^2 + \mathrm{i}\varepsilon}, \tag{2.147}$$

has the asymptotic behaviour

$$D_{\mu\nu}(k) \underset{k\to\infty}{\rightarrow} O(1). \tag{2.148}$$

This means that the canonical dimension for the vector field is two which differs from its (naïve) dimension by a mass unit of one. The power counting is now modified with the superficial degree of divergence given by

$$D = 4 - B - \tfrac{3}{2}F - 2V + \sum_i n_i(\Delta_i - 4) \tag{2.149}$$

and

$$\Delta_i = b_i + \tfrac{3}{2}f_i + 2v_i + d_i \tag{2.150}$$

where V is the number of external vector lines, v_i is the number of vector fields in the ith type of vertex, and Δ_i is the canonical dimension of the interaction term $\mathscr{L}_i$. To have a renormalizable interaction we need $\Delta_i \leq 4$ but, from (2.150), the only such term trilinear in the fields is $\phi^2 A_\mu$, which is not Lorentz-invariant. There is no nontrivial interaction of the massive vector field which is renormalizable. However, two important exceptions to this statement should be noted.

(A) In a gauge theory with spontaneous symmetry breakdown, the vector (gauge) boson will acquire mass in such a way as to preserve the renormalizability of the theory. We will discuss this in detail in Chapter 8.

(B) A theory with a neutral massive vector boson coupled to conserved current is also renormalizable. Heuristically we can understand this as follows. The propagator $D_{\mu\nu}(k)$ given in (2.147) always appears between the conserved currents $J^{\mu}(k)$ and $J^{\nu}(k)$ and the $k_{\mu}k_{\nu}/M_{\mathrm{V}}^2$ term will not contribute because of current conservation. $k^{\mu}J_{\mu}(k) = 0$ or, in the coordinate space, $\partial^{\mu}J_{\mu}(x) = 0$. Then power counting is essentially the same as for the massless vector field case.

Renormalization of composite operators

So far we have only considered Green's functions involving elementary field operators. In many practical applications we are interested also in functions of *composite operators*, i.e. local monomials of fields and their derivatives, e.g. $\bar{\psi}\gamma_{\mu}\psi$, ϕ^2, $\phi\,\partial^2\phi$, etc.

Again we will illustrate the renormalization of such composite operators in $\lambda\phi^4$ theory. Consider the composite operator $\frac{1}{2}\phi^2(x)$. The Green's function with one insertion of $\frac{1}{2}\phi^2(x)$ has the form

$$G_{\phi^2}^{(n)}(x; x_1 \ldots x_n) = \langle 0|T(\tfrac{1}{2}\phi^2(x)\phi(x_1)\ldots\phi(x_n))|0\rangle \tag{2.151}$$

or, in momentum space,

$$(2\pi)^4\,\delta^4(p + p_1 + \ldots p_n)G_{\phi^2}^{(n)}(p; p_1, \ldots, p_n)$$
$$= \int \mathrm{d}^4x\,\mathrm{e}^{-\mathrm{i}p\cdot x}\int\prod_{i=1}^{n}\mathrm{d}^4x_i\,\mathrm{e}^{-\mathrm{i}p_i\cdot x_i}G_{\phi^2}^{(n)}(x; x_1, \ldots, x_n). \tag{2.152}$$

In perturbation theory we can use Wick's theorem to work out Green's function in terms of Feynman diagrams. For example, for $G_{\phi^2}^{(2)}(x; x_1, x_2)$ to the zeroth order in λ we have

$$G_{\phi^2}^{(2)}(x; x_1, x_2) = \langle 0|T(\tfrac{1}{2}\phi^2(x)\phi(x_1)\phi(x_2))|0\rangle$$
$$= \mathrm{i}\Delta(x - x_1)\,\mathrm{i}\Delta(x - x_2) \tag{2.153}$$

or, in momentum space,

$$G_{\phi^2}^{(2)}(p; p_1, -p - p_1) = \mathrm{i}\Delta(p_1)\,\mathrm{i}\Delta(p + p_1). \tag{2.154}$$

If we truncate the propagators on the external lines, we have

$$\Gamma_{\phi^2}^{(2)}(p; p_1, -p - p_1) = 1 \tag{2.155}$$

as represented in Fig. 2.18(a). The same Green's function to first order in λ is

$$G_{\phi^2}^{(2)}(x; x_1, x_2) = \int \mathrm{d}^4y\langle 0|T(\tfrac{1}{2}\phi^2(x)\phi(x_1)\phi(x_2)\frac{(-\mathrm{i}\lambda)}{4!}\phi^4(y))|0\rangle$$
$$= \int \mathrm{d}^4y\frac{(-\mathrm{i}\lambda)}{2}\,[\mathrm{i}\Delta(x - y)]^2\,\mathrm{i}\Delta(x_1 - y)\,\mathrm{i}\Delta(x_2 - y)$$

with (amputated) 1PI momentum-space Green's function given below (see Fig. 2.18(b))

$$\Gamma_{\phi^2}^{(2)}(p; p_1, -p - p_1) = \frac{-\mathrm{i}\lambda}{2}\int\frac{\mathrm{d}^4l}{(2\pi)^4}\frac{\mathrm{i}}{l^2 - \mu^2 + \mathrm{i}\varepsilon}\frac{\mathrm{i}}{(l - p)^2 - \mu^2 + \mathrm{i}\varepsilon}. \tag{2.156}$$

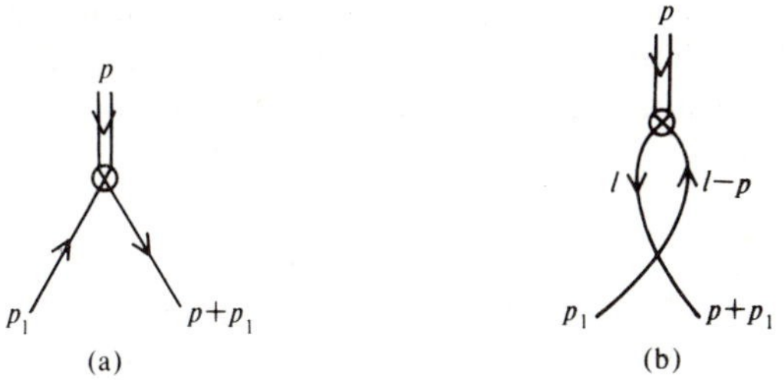

FIG. 2.18. Zeroth- and first-order diagrams of $\Gamma^{(2)}_{\phi^2}(p; p_1, -p-p_1)$.

We see that the composite operator generates a vertex very much like a term in the Lagrangian except that the composite operator can carry off momenta. This suggests the following method of systematically calculating Green's functions with composite operators. As we have seen in §1.2, we can generate Green's functions of elementary fields $\phi(x)$ with the insertion of $J(x)\phi(x)$ in the Lagrangian density, $J(x)$ being an arbitrary c-number function. For a composite operator $\Omega(x)$ we can similarly insert $\chi(x)\Omega(x)$ in the Lagrangian density where $\chi(x)$ is the c-number source function

$$\mathscr{L}[\chi] = \mathscr{L}[0] + \chi\Omega. \tag{2.157}$$

Following exactly the same procedure of constructing the generating functional $W[\chi]$, which is the vacuum-to-vacuum transition amplitude in the presence of this external source $\chi(x)$, we obtain the connected Green's functions by first differentiating $\ln W[\chi]$ with respect to χ and then setting the source χ to zero. With $\Omega(x) = \frac{1}{2}\phi^2(x)$ we have the vertex shown in Fig. 2.19(a) which may appear for example in the one-loop four-point ϕ function in Fig. 2.19(b),

$$\Gamma^{(4)}_{\Omega}(p; p_1 \dots p_4) = \frac{(-\mathrm{i}\lambda)^2}{2} \int \frac{\mathrm{d}^4 l}{(2\pi)^4} \frac{\mathrm{i}}{l^2 - \mu^2 + \mathrm{i}\varepsilon} \frac{\mathrm{i}}{(l+p)^2 - \mu^2 + \mathrm{i}\varepsilon} \times \frac{\mathrm{i}}{(l - p_1 - p_2)^2 - \mu^2 + \mathrm{i}\varepsilon}. \tag{2.158}$$

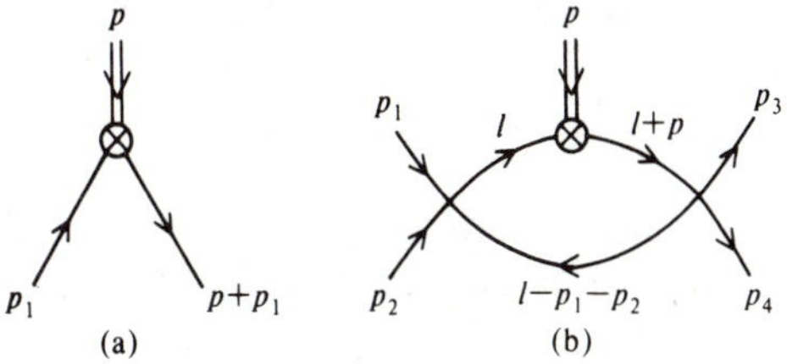

FIG. 2.19.

We are now ready to discuss the renormalization of this new set of Green's functions $\Gamma^{(n)}_{\Omega}(p; p_1 \dots p_n)$. The procedure is exactly the same as that for Green's functions without $\Omega(x)$, $\Gamma^{(n)}(p_1 \dots p_n)$. Since an insertion of a composite operator is like an additional vertex, a straightforward application of (2.133) will show that the superficial degree of divergence D_Ω of $\Gamma_\Omega^{(n)}$ differs from D of $\Gamma^{(n)}$ by the index of divergence δ_Ω of the composite operator

$$D_\Omega = D + \delta_\Omega = D + (d_\Omega - 4), \tag{2.159}$$

where d_Ω is the canonical dimension of Ω. Thus, for $d_\Omega \leq 4$, the insertion of (renormalizable or superrenormalizable) composite operators will not worsen the convergence property of the Green's function; the insertion of a $d_\Omega > 4$ operator worsens the divergence of the diagram. For the case of $\Omega = \frac{1}{2}\phi^2$, we have $d(\phi^2) = 2$ and, for an n-point function, $D_{\phi^2} = 2 - n$. Thus only $\Gamma_{\phi^2}^{(2)}$ is logarithmically divergent and needs to be renormalized. The relevant one-loop diagram shown in Fig. 2.18(b) has the Taylor expansion

$$\bar{\Gamma}_{\phi^2}^{(2)}(p; p_1, -p-p_1) = \bar{\Gamma}_{\phi^2}^{(2)}(0; 0, 0) + \bar{\Gamma}_{\phi^2\mathrm{R}}^{(2)}(p; p_1, -p-p_1) \tag{2.160}$$

where $\bar{\Gamma}_{\phi^2\mathrm{R}}^{(2)}$ is finite and has the normalization

$$\bar{\Gamma}_{\phi^2\mathrm{R}}^{(2)}(0; 0, 0) = 0. \tag{2.161}$$

We can combine the counterterm

$$\frac{-\mathrm{i}}{2}\bar{\Gamma}_{\phi^2}^{(2)}(0; 0, 0)\chi(x)\phi^2(x)$$

with the original term to write

$$\frac{-\mathrm{i}}{2}\chi\phi^2 - \frac{\mathrm{i}}{2}\bar{\Gamma}_{\phi^2}^{(2)}(0; 0, 0)\chi\phi^2 = \frac{-\mathrm{i}}{2}Z_{\phi^2}\chi\phi^2$$

with

$$Z_{\phi^2} = 1 + \bar{\Gamma}_{\phi^2}^{(2)}(0; 0, 0). \tag{2.162}$$

Thus, the total contribution to $\Gamma_{\phi^2\mathrm{R}}^{(2)}$ up to one loop is

$$\Gamma_{\phi^2\mathrm{R}}^{(2)}(p; p_1, -p-p_1) = 1 + \bar{\Gamma}_{\phi^2\mathrm{R}}^{(2)}(p; p_1, -p-p_1) \tag{2.163}$$

with the normalization

$$\Gamma_{\phi^2\mathrm{R}}^{(2)}(0; 0, 0) = 1. \tag{2.164}$$

In general we need to insert the counterterm $\Delta\Omega$ into the original addition of (2.157).

$$\mathscr{L} \to \mathscr{L} + \chi(\Omega + \Delta\Omega). \tag{2.165}$$

In particular, for the counterterm proportional to the original composite operator itself, $\Delta\Omega = C\Omega$, as is the case with $\frac{1}{2}\phi^2$, we have

$$\begin{aligned}\mathscr{L}[\chi] &= \mathscr{L}[0] + \chi Z_\Omega \Omega \\ &= \mathscr{L}[0] + \chi\Omega_0\end{aligned} \tag{2.166}$$

with

$$\Omega_0 = Z_\Omega \Omega = (1 + C)\Omega.$$

Such composite operators are said to be *multiplicatively renormalizable*. This means that the Green's function of the unrenormalized operator Ω_0 is related to that of the renormalized operator Ω by

$$G^{(n)}_{\Omega_0}(x; x_1, \ldots x_n) = \langle 0|T(\Omega_0(x)\phi_0(x_1), \ldots \phi_0(x_n))|0\rangle$$
$$= Z_\Omega Z_\phi^{n/2} G^{(n)}_{\Omega R}(x; x_1, \ldots x_n). \tag{2.167}$$

The composite operator $\Omega = \frac{1}{2}\phi^2$ is multiplicatively renormalizable because it is the only operator with dimension-two. For more general cases, $\Delta\Omega \neq C\Omega$, the renormalization of a composite operator may require counterterms proportional to other composite operators. In this way renormalization may introduce mixings among composite operators. For example, for $\Omega = \phi^4$, the counterterms $\Delta\Omega = \phi^2$, $(\partial_\mu\phi)^2$, and ϕ^4 will be needed. To be definite we will restrict our illustration to the case of two composite operators A and B which can mix under renormalization

$$\mathscr{L}[\chi] = \mathscr{L}[0] + \chi_A(A + \Delta A) + \chi_B(B + \Delta B). \tag{2.168}$$

The counterterms ΔA and ΔB are some linear combinations of A and B

$$\Delta A = C_{AA}A + C_{AB}B$$
$$\Delta B = C_{BA}A + C_{BB}B. \tag{2.169}$$

We can write $\mathscr{L}[\chi]$ as

$$\mathscr{L}[\chi] = \mathscr{L}[0] + (\chi_A, \chi_B)C\begin{pmatrix} A \\ B \end{pmatrix} \tag{2.170}$$

where

$$C = \begin{pmatrix} 1 + C_{AA} & C_{AB} \\ C_{BA} & 1 + C_{BB} \end{pmatrix}. \tag{2.171}$$

Such a matrix C can be diagonalized with a bi-unitary transformation (see §11.3). Thus,

$$UCV^\dagger = \begin{pmatrix} Z_{A'} & 0 \\ 0 & Z_{B'} \end{pmatrix} \tag{2.172}$$

where U and V are unitary matrices. The Lagrangian can then be written

$$\mathscr{L}[\chi] = \mathscr{L}[0] + Z_{A'}\chi_{A'}A' + Z_{B'}\chi_{B'}B'$$

where

$$\begin{pmatrix} A' \\ B' \end{pmatrix} = V\begin{pmatrix} A \\ B \end{pmatrix} \tag{2.173}$$

$$(\chi_{A'}, \chi_{B'}) = (\chi_A, \chi_B)U^\dagger.$$

This means that the linear combinators A' and B' as defined by (2.173) are multiplicatively renormalizable

$$\langle 0|T(A'(x)B'(y)\phi(x_1)\ldots\phi(x_n))|0\rangle$$
$$= Z_{A'}^{-1}Z_{B'}^{-1}Z_\phi^{-n/2}\langle 0|T(A'_0(x)B'_0(y)\phi_0(x_1)\ldots\phi_0(x_n))|0\rangle. \tag{2.174}$$

An example of such simple mixing involving only two composite operators is the theory defined by

$$\mathscr{L} = \bar{\psi}(i\gamma^\mu \partial_\mu - m)\psi - \frac{1}{2}(\partial_\mu \phi)^2 - \frac{1}{2}\mu^2\phi^2 - \frac{\lambda}{4!}\phi^4 - g\bar{\psi}\psi\phi - \frac{\lambda'\phi^3}{3!} \tag{2.175}$$

with $A = \phi^3$ and $B = \bar{\psi}\psi$. These two composite operators can mix under renormalization because of the divergences in the diagrams shown in Fig. 2.20.

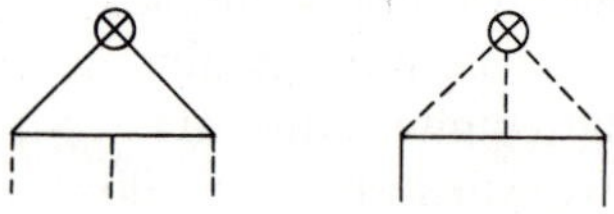

FIG. 2.20. One-loop divergent diagrams involving composite operators $A = \phi^3$ and $B = \bar{\psi}\psi$. The dashed lines represent ϕ-fields; solid lines ψ-fields.

3 Renormalization group

The renormalization theory discussed in the last chapter has some arbitrariness related to our choice of kinematic points in defining physical parameters such as the mass and the coupling constants. For example, the BPH renormalization prescription only requires that the divergent part of the 1PI graph be cancelled by counterterms constructed from Taylor expansions. However the reference points for the expansions are arbitrary. Different choices of the reference points, i.e. different subtraction points, lead to different definitions of the physical parameters of the theory. But any choice is as good as any other; the physics should not depend on the choices of the normalization conditions. This is the *renormalization group*: the physical content of the theory should be invariant under the transformations which merely change the normalization conditions. This seemingly empty statement actually provides us with highly nontrivial constraints on the asymptotic behaviour of the theory. In systems with infinite degrees of freedom (such as quantum field theory), renormalization can be defined in such a way that it involves a series of redefinitions of physical parameters on the relevant length or energy scales. There must be relations between the physical quantities so defined. Hence the renormalization group equation expresses the effect of a scale change in the theory or, more accurately, expresses the connection of renormalizability to scale transformations.

Gell-Mann and Low (1954) were the first ones to use renormalization group techniques to study the asymptotic behaviour of Green's functions in quantum electrodynamics. The renormalization group was discovered by Stueckelberg and Peterman (1953); its role in the Gell-Mann–Low analysis was discussed by Bogoliubov and Shirkov (1959). The recent interest in the applications of renormalization group has largely been brought about by the work of Wilson (1969). Our presentation is patterned after the lecture by Coleman (1971*a*). There are a number of ways to set up the renormalization group equation. In §3.1 we study this in the form of the Callan–Symanzik equation (Callan 1970; Symanzik 1970*b*) which is associated with momentum subtraction schemes. In §3.2 we briefly discuss the mass-independent renormalization or minimal subtraction scheme ('t Hooft 1973; Weinberg 1973*a*) and its associated renormalization group equation. The solutions to these equations in the asymptotic region are found in terms of the 'effective coupling constants' which are studied in more detail in §3.3.

3.1 Momentum subtraction schemes and the Callan–Symanzik equation

As stated above the existence of a renormalization group is related to the freedom one has in the choice of the reference points for Taylor expansions leading to different definitions of the physical parameters of the theory. These choices may be expressed as different normalization conditions on certain 1PI amplitudes. The physical parameters should then be regarded as dependent on the choices of normalization conditions. We shall first illustrate this in $\lambda\phi^4$ theory by giving two specific examples of mass-dependent normalization conditions (or momentum-subtraction schemes).

Intermediate renormalization

This corresponds to a Taylor expansion around zero external momenta. For the self-energy we have

$$\Sigma(p^2) = \Sigma(0) + \Sigma'(0)p^2 + \tilde{\Sigma}(p^2). \tag{3.1}$$

The finite part $\tilde{\Sigma}(p^2)$ will have the properties

$$\tilde{\Sigma}(0) = 0 \tag{3.2}$$

$$\left.\frac{\partial\tilde{\Sigma}(p^2)}{\partial p^2}\right|_{p^2=0} = 0. \tag{3.3}$$

The full propagator $\Delta_R(p^2)$ is related to the self-energy $\tilde{\Sigma}(p^2)$ by

$$\mathrm{i}\Delta_R(p^2) = \frac{\mathrm{i}}{p^2 - \mu^2 - \tilde{\Sigma}(p^2)} \tag{3.4}$$

and the 1PI two-point function $\Gamma_R^{(2)}(p^2)$ is given by

$$\begin{aligned}\mathrm{i}\Gamma_R^{(2)}(p^2) &= \mathrm{i}\Delta_R(p^2)[\mathrm{i}\Delta_R(p^2)]^{-2}\\ &= -\mathrm{i}[\Delta_R(p^2)]^{-1}\\ &= -\mathrm{i}[p^2 - \mu^2 - \tilde{\Sigma}(p^2)].\end{aligned} \tag{3.5}$$

The normalization conditions on $\tilde{\Sigma}(p^2)$ (eqns (3.2) and (3.3)) can be translated in terms of $\Gamma_R^{(2)}(p^2)$ as

$$\Gamma_R^{(2)}(0) = \mu^2 \tag{3.6}$$

$$\left.\frac{\partial\Gamma_R^{(2)}(p^2)}{\partial p^2}\right|_{p^2=0} = -1. \tag{3.7}$$

For the four-point function, the finite part of the higher-order contribution is defined by

$$\bar{\Gamma}_R^{(4)}(p_1, p_2, p_3) = \bar{\Gamma}^{(4)}(p_1, p_2, p_3) - \bar{\Gamma}^{(4)}(0, 0, 0). \tag{3.8}$$

Thus we have

$$\bar{\Gamma}_R^{(4)}(p_1, p_2, p_3) = 0 \quad \text{at} \quad p_1 = p_2 = p_3 = 0. \tag{3.9}$$

Including the tree-level contribution

$$\Gamma_R^{(4)}(p_1, p_2, p_3) = -\mathrm{i}\lambda + \bar{\Gamma}_R^{(4)}(p_1, p_2, p_3), \tag{3.10}$$

the normalization condition on the total four-point function reads

$$\Gamma_R^{(4)}(p_1, p_2, p_3) = -\mathrm{i}\lambda \quad \text{at} \quad p_1 = p_2 = p_3 = 0. \tag{3.11}$$

We note that μ^2 in this subtraction scheme is not the physical mass and that λ is not the physical coupling constant because the points $p_i = 0$ are not in the physically allowed region. But we can express all physically measurable quantities in terms of these two parameters. In this sense they are physical parameters.

On-shell renormalization

This corresponds to a Taylor expansion for external momenta on the mass shell, i.e. $p_i^2 = \mu^2$. For the self-energy, this gives

$$\Sigma(p^2) = \Sigma(\mu^2) + (p^2 - \mu^2)\Sigma'(\mu^2) + \tilde{\Sigma}(p^2). \tag{3.12}$$

Thus,

$$\tilde{\Sigma}(\mu^2) = 0 \tag{3.13}$$

$$\left.\frac{\partial\tilde{\Sigma}(p^2)}{\partial p^2}\right|_{p^2=\mu^2} = 0. \tag{3.14}$$

Or, in terms of $\Gamma_{\mathrm{R}}^{(2)}(p^2)$ of (3.5),

$$\Gamma_{\mathrm{R}}^{(2)}(\mu^2) = 0 \tag{3.15}$$

$$\left.\frac{\partial\Gamma_{\mathrm{R}}^{(2)}(p^2)}{\partial p^2}\right|_{p^2=\mu^2} = -1. \tag{3.16}$$

For the four-point function, a convenient choice of the reference point for the Taylor expansion will be the symmetric momentum point

$$\Gamma_{\mathrm{R}}^{(4)}(p_1, p_2, p_3) = -\mathrm{i}\lambda \quad \text{at} \quad p_i^2 = \mu^2, \qquad s = t = u = 4\mu^2/3 \tag{3.17}$$

where s, t, and u are the Mandelstam variables. In this case the parameters μ^2 and λ are the physical mass and, up to some kinematical factors, the physical differential cross-section at $s = t = u = 4\mu^2/3$, respectively.

These two examples are specific realizations of a general renormalization scheme where the normalization conditions R can be a function of several fixed 'reference momenta', $\xi_1, \xi_2 \ldots$ such that

$$\Gamma_R^{(2)}(\xi_1^2) = \mu^2 \tag{3.18a}$$

$$\left.\frac{\partial\Gamma_R^{(2)}(p^2)}{\partial p^2}\right|_{p^2=\xi_2^2} = -1 \tag{3.18b}$$

and

$$\Gamma_R^{(4)}(\xi_3, \xi_4, \xi_5) = -\mathrm{i}\lambda. \tag{3.18c}$$

Renormalization group. Consider two different renormalization procedures, R and R'. Since both start from the same bare Lagrangian

$$\begin{aligned}\mathscr{L} &= \mathscr{L}_R(R\text{-quantities}) \\ &= \mathscr{L}_{R'}(R'\text{-quantities}),\end{aligned} \tag{3.19}$$

in terms of the unrenormalized fields (see eqn (2.23)), we must have

$$\phi_R = Z_\phi^{-1/2}(R)\phi_0; \qquad \phi_{R'} = Z_\phi^{-1/2}(R')\phi_0. \tag{3.20}$$

Thus,

$$\phi_{R'} = Z_\phi^{-1/2}(R', R)\phi_R$$

where

$$Z_\phi(R', R) = Z_\phi(R')/Z_\phi(R). \tag{3.21}$$

This means that the renormalized fields in different subtraction schemes are related by a multiplicative constant. Since both ϕ_R and $\phi_{R'}$ are finite, $Z_\phi(R', R)$ must also be finite even though it is a ratio of two divergent quantities. Similar relations between the coupling constants, masses, and Green's functions can be worked out

$$\lambda_{R'} = Z_\lambda^{-1}(R', R)Z_\phi^2(R', R)\lambda_R \tag{3.22}$$

$$\mu_{R'}^2 = \mu_R^2 + \delta\mu^2(R', R) \tag{3.23}$$

where

$$Z_\lambda(R', R) = Z_\lambda(R')/Z_\lambda(R) \tag{3.24}$$

$$\delta\mu^2(R', R) = \delta\mu^2(R') - \delta\mu^2(R). \tag{3.25}$$

are all finite. The operation which takes the quantities in one renormalization scheme R to quantities in another scheme R' can be viewed as a transformation from R to R'. The set of all such transformations is said to form the *renormalization group*. We now translate this renormalization group invariance into the analytic form.

Callan–Symanzik equation

First we note that differentiation of the unrenormalized Green's function with respect to the bare mass is equivalent to an insertion of the composite operator $\Omega_0 = \frac{1}{2}\phi_0^2$ carrying zero momentum

$$\frac{\partial\Gamma^{(n)}(p_i)}{\partial\mu_0^2} = -\mathrm{i}\Gamma_{\phi^2}^{(n)}(0; p_i) \tag{3.26}$$

because $\Gamma^{(n)}(p_i)$ depends on μ_0^2 only through the bare propagator

$$\mathrm{i}\Delta_0(p) = \frac{\mathrm{i}}{p^2 - \mu_0^2 + \mathrm{i}\varepsilon} \tag{3.27}$$

and because

$$\frac{\partial}{\partial\mu_0^2}\left(\frac{\mathrm{i}}{p^2 - \mu_0^2 + \mathrm{i}\varepsilon}\right) = \frac{\mathrm{i}}{p^2 - \mu_0^2 + \mathrm{i}\varepsilon}(-\mathrm{i})\frac{\mathrm{i}}{p^2 - \mu_0^2 + \mathrm{i}\varepsilon}. \tag{3.28}$$

In terms of the renormalized (1PI) Green's functions, we can write

$$\Gamma_R^{(n)}(p_i; \lambda, \mu) = Z_\phi^{n/2}\Gamma^{(n)}(p_i; \lambda_0, \mu_0) \tag{3.29a}$$

$$\Gamma_{\phi^2 R}^{(n)}(p, p_i; \lambda, \mu) = Z_{\phi^2}^{-1} Z_\phi^{n/2}\Gamma_{\phi^2}^{(n)}(p, p_i; \lambda_0, \mu_0). \tag{3.29b}$$

After substituting (3.29) into (3.26) and using the following relation

$$\frac{\partial}{\partial \mu_0^2}\Gamma_R^{(n)}(p_i;\lambda,\mu)=\left[\frac{\partial \mu^2}{\partial \mu_0^2}\frac{\partial}{\partial \mu^2}+\frac{\partial \lambda}{\partial \mu_0^2}\frac{\partial}{\partial \lambda}\right]\Gamma_R^{(n)}(p_i;\lambda,\mu), \tag{3.30}$$

we have the Callan–Symanzik equation in $\lambda\phi^4$ theory

$$\left[\mu\frac{\partial}{\partial\mu}+\beta\frac{\partial}{\partial\lambda}-n\gamma\right]\Gamma_R^{(n)}(p_i;\lambda,\mu)=-\,\mathrm{i}\mu^2\alpha\Gamma_{\phi^2 R}^{(n)}(0,p_i;\lambda,\mu) \tag{3.31}$$

where α, β, and γ are dimensionless functions

$$\beta=2\mu^2\frac{\partial\lambda/\partial\mu_0^2}{\partial\mu^2/\partial\mu_0^2} \tag{3.32}$$

$$\gamma=\mu^2\frac{\partial\ln Z_\phi/\partial\mu_0^2}{\partial\mu^2/\partial\mu_0^2} \tag{3.33}$$

$$\alpha=\frac{\partial Z_{\phi^2}/\partial\mu_0^2}{\partial\mu^2/\partial\mu_0^2}. \tag{3.34}$$

The function α is related to γ: for $n=2$ we have the normalization conditions (3.6) and (2.164)

$$\Gamma_R^{(2)}(0;\lambda,\mu)=\mathrm{i}\mu^2 \quad \text{and} \quad \Gamma_{\phi^2 R}^{(2)}(0,0;\lambda,\mu)=1. \tag{3.35}$$

Hence, from eqn (3.31),

$$\alpha=2(\gamma-1). \tag{3.36}$$

Since the renormalized quantities $\Gamma_R^{(n)}$ and $\Gamma_{\phi^2 R}^{(n)}$ are both cut-off independent to all orders in λ, we expect that the functions α, β, and γ are also cut-off independent. To see this explicitly we set $n=2$ in (3.31) and differentiate with respect to p^2

$$\left[\mu\frac{\partial}{\partial\mu}+\beta\frac{\partial}{\partial\lambda}-2\gamma\right]\frac{\partial}{\partial p^2}\Gamma_R^{(2)}(p;\lambda,\mu)=-\mathrm{i}\mu^2\alpha\frac{\partial}{\partial p^2}\Gamma_{\phi^2 R}^{(2)}(0,p;\lambda,\mu). \tag{3.37}$$

Set $p^2=0$ and use the normalization condition (3.7)

$$\left.\frac{\partial\Gamma_R^{(2)}(p^2;\lambda,\mu)}{\partial p^2}\right|_{p^2=0}=-1. \tag{3.38}$$

Then (3.37) turns into

$$\gamma=\mu^2(1-\gamma)\left[\frac{\partial}{\partial p^2}\Gamma_{\phi^2 R}^{(2)}(0,p^2,\lambda,\mu)\right]_{p^2=0}. \tag{3.39}$$

This demonstrates that γ is cut-off independent. Every function except β in (3.31) is now independent of the cut-off; hence β is also cut-off independent. Since, α, β, and γ are dimensionless, the cut-off independence implies that they are functions of the dimensionless coupling constant only, i.e. $\alpha=\alpha(\lambda)$, $\beta=\beta(\lambda)$, and $\gamma=\gamma(\lambda)$.

In practical calculations of α, β, and γ it is convenient to use the cut-off (Λ)

dependence of the renormalization constants Z_λ, Z_ϕ as follows. In unrenormalized perturbation theory with unrenormalized λ_0 and μ_0, the renormalized parameters μ and λ, defined in (2.51) and (2.50),

$$\mu^2 = \mu_0^2 + \delta\mu^2 \tag{3.40}$$

and

$$\lambda = \bar{Z}\lambda_0 \tag{3.41}$$

with

$$\bar{Z} = Z_\lambda^{-1} Z_\phi^2 \tag{3.42}$$

are functions of λ_0, μ_0, and Λ. From dimensional argument, λ and the Z_is can depend only on dimensionless quantities like λ_0 and Λ/μ_0. If we further replace μ_0 by $\mu = \mu(\lambda_0, \mu_0, \Lambda)$, we have $\lambda = \lambda(\lambda_0, \Lambda/\mu)$ and $Z_i = Z_i(\lambda_0, \Lambda/\mu)$. Using the chain rule of differentiation

$$\frac{\partial}{\partial\mu_0^2}\lambda(\lambda_0, \Lambda/\mu)\bigg|_{\Lambda,\lambda_0} = \frac{\partial\mu^2}{\partial\mu_0^2}\frac{\partial}{\partial\mu^2}\lambda(\lambda_0, \Lambda/\mu)\bigg|_{\Lambda,\lambda_0}, \tag{3.43}$$

we have

$$\begin{aligned}\beta &= \mu\frac{\partial}{\partial\mu}[\lambda(\lambda_0, \Lambda/\mu)]_{\Lambda,\lambda_0} \\ &= \mu\frac{\partial}{\partial\mu}[\bar{Z}(\lambda_0, \Lambda/\mu)\lambda_0]_{\Lambda,\lambda_0} \\ &= -\lambda_0\Lambda\frac{\partial}{\partial\Lambda}[\bar{Z}(\lambda_0, \Lambda/\mu)]_{\mu,\lambda_0}\end{aligned} \tag{3.44}$$

or

$$\beta = -\lambda\frac{\partial}{\partial\ln\Lambda}[\ln\bar{Z}(\lambda_0, \Lambda/\mu)]. \tag{3.45}$$

Similarly, we obtain

$$\gamma = -\frac{1}{2}\frac{\partial}{\partial\ln\Lambda}[\ln Z_\phi(\lambda_0, \Lambda/\mu)]. \tag{3.46}$$

This means that to calculate the Callan–Symanzik β and γ functions we only need to know the $\ln\Lambda$ term in the Z_is. At the one-loop level we have (eqn (2.100))

$$\begin{aligned}Z_\lambda &= 1 + \frac{3\lambda_0}{32\pi^2}\ln\frac{\Lambda^2}{\mu^2} + O(\lambda_0^2) \\ Z_\phi &= 1 + O(\lambda_0^2).\end{aligned}$$

Hence

$$\beta(\lambda) = \frac{3\lambda^2}{16\pi^2} + O(\lambda^3) \tag{3.47}$$

$$\gamma(\lambda) = O(\lambda^2). \tag{3.48}$$

The generalization of the Callan–Symanzik equation to Green's functions involving several composite operators A, B, $C\ldots$ can be carried through in a straightforward manner. First we choose the appropriate linear combination of operators such that they are multiplicatively renormalizable (see §2.4).

$$\{G^{(n)}_{AB\ldots}\}_R = Z_A^{-1}Z_B^{-1}\ldots Z_\phi^{-n/2}\{G^{(n)}_{AB\ldots}\}_0 \tag{3.49}$$

or

$$\{\Gamma^{(n)}_{AB\ldots}\}_R = Z_A^{-1}Z_B^{-1}\ldots Z_\phi^{n/2}\{\Gamma^{(n)}_{AB\ldots}\}_0. \tag{3.50}$$

The Callan–Symanzik equation can be readily shown to be

$$\left[\mu\frac{\partial}{\partial\mu} + \beta\frac{\partial}{\partial\lambda} - n\gamma + \gamma_{AB\ldots}\right]\{\Gamma^{(n)}_{AB\ldots}\}_R = -\mathrm{i}\mu^2\alpha\{\Gamma^{(n)}_{\phi^2,AB\ldots}\}_R \tag{3.51}$$

with

$$\gamma_{AB\ldots} = -\frac{1}{2}\frac{\partial}{\partial \ln \Lambda}\ln[Z_A Z_B\ldots].$$

Weinberg's theorem on the asymptotic behaviour of Green's function

The large-momentum or short-distance behaviour of Green's function is clearly of great interest. It is related to the renormalizability properties of the theory. An important theorem here is the one due to Weinberg (1960). It concerns this behaviour for nonexceptional values of momenta in the Euclidean region. In the Euclidean region all momenta are space-like, $p_i^2 < 0$, which can be realized most easily by having real space and imaginary time components. A momentum configuration $p_1, p_2, \ldots, p_n$ is said to be nonexceptional if *no* nontrivial partial sum vanishes, $p_{i_1} + p_{i_2} + \ldots p_{i_k} = 0$ for $i_1, i_2, \ldots, i_k$ take on any of the labels, $1, 2, \ldots, n$. (A trivial partial sum which vanishes would be $p_1 + p_2 + \ldots p_n = 0$ because of the overall energy–momentum conservation.)

Again we state without proof *Weinberg's theorem. If the momenta are non exceptional and parametrized as* $p_i = \sigma k_i$, *the* 1*PI Green's function* $\Gamma_R^{(n)}$ *grows in the deep Euclidean region (corresponding to* $\sigma \to \infty$ *with* k_i *fixed) as* σ^{4-n} *times polynomials in* $\ln\sigma$ *to any finite order in the coupling* λ. *Similarly* $\Gamma^{(n)}_{\phi^2 R}$ *grows as* σ^{2-n} times polynomials in $\ln\sigma$.

We note that the powers of σ for $\Gamma_R^{(n)}$ and $\Gamma^{(n)}_{\phi^2 R}$ are just their superficial degrees of divergence (see Chapter 2), which are also their (naïve) dimensions in unit of the mass.

For convergent diagrams it is not difficult to understand this result intuitively. For a nonexceptional external momentum configuration, the hard momenta must flow through the internal loops and set the scale for the loop integration momenta as well. (For an exceptional momentum configuration this need not be true.) This explains why the same degrees of divergence appear in our study of the large internal momentum limit and of the large external momentum limit. For divergent diagrams the result stated by Weinberg's theorem may not be so obvious. One would expect that the ultraviolet portion of the integration would be controlled by the cut-off Λ even for hard external momenta. However, the cut-off-dependent part is cancelled when the necessary counterterms are included. The surviving leading contribution again corresponds to the portion of the loop integration with momentum of the same order of magnitude as the hard external momenta. To illustrate this remark, consider the one-loop four-point function in Fig. 2.5,

$$\Gamma \sim \int \frac{d^4 l}{[(l-p)^2 - \mu^2][l^2 - \mu^2]}. \tag{3.52}$$

In the three integration regions, we have $\Gamma \sim \ln \Lambda$ for $l \gg p$; $\Gamma \sim \ln p$ for $l \sim p$; $\Gamma \sim p^{-2}$ for $l \ll p$. After the inclusion of the counterterm of Fig. 2.9(c), the $\ln \Lambda$ term is cancelled and replaced by some term constant in p. Thus the dominant asymptotic behaviour comes from the region of integration where $l \sim p$. This is why the power of σ in the asymptotic behaviour is the same as the superficial degree of divergence. In this particular case, we have $D = 4 - n = 0$; we expect from Weinberg's theorem the asymptotic form $\Gamma = (\sigma^0) \times$ (polynomial in $\ln \sigma$). This agrees with the estimate given above.

We note that, in the deep Euclidean region, particles are very off their mass shell $p_i^2 \gg \mu^2$. Nevertheless, as we shall see later in Chapters 7 and 10, in cases such as deep inelastic lepton scatterings we can still extract useful information with the help of the operator product expansion.

Weinberg's theorem tells us that Green's function in perturbation theory takes on the asymptotic form

$$\Gamma^{(n)}(\sigma p_i, \lambda, \mu) \underset{\sigma \to \infty}{\to} \sigma^{4-n}[a_0(\ln \sigma)^{b_0} + a_1(\ln \sigma)^{b_1}\lambda + \ldots] \tag{3.53}$$

with the constants a_i and b_i unspecified. Thus, it leaves open the question as to what the power series in polynomials of $\ln \sigma$ sums up to. If this sums up to some power of σ, say σ^γ, then γ will be called the *anomalous dimension* as it modifies the canonical behaviour σ^{4-n} to $\sigma^{4-n-\gamma}$. Clearly, we would like to learn all we can about the anomalous dimension γ.

The asymptotic solution of the renormalization-group equation

If we can ignore the inhomogeneous term on the right-hand side of eqn (3.31) involving mass insertions, the Callan–Symanzik equation can provide information on the asymptotic behaviour of Green's function. As it relates quantities of different orders in the coupling ($\mu(\partial/\partial\mu) \sim O(1)$, $\beta(\partial/\partial\lambda) \sim O(\lambda)$, and γ is of even higher order), the equation can be viewed as some kind of recurrence relation among the a_is and b_is of (3.53). Thus the asymptotic solution of the Callan–Symanzik equation should be relevant to the study of the true large-momentum or short-distance behaviour of Green's function. In other words, this renormalization-group equation sums up all the leading logarithmic terms to all orders of the perturbation series.

From Weinberg's theorem we have $\Gamma_R^{(n)} \gg \Gamma_{\phi^2 R}^{(n)}$ to any finite order of λ in the deep Euclidean region ($\sigma \to \infty$). If we assume that this is true even when the perturbation series is summed to all orders, we can then drop the right-hand side of the Callan–Symanzik equation (3.31) and obtain a homogeneous differential equation

$$\left[\mu\frac{\partial}{\partial\mu} + \beta(\lambda)\frac{\partial}{\partial\lambda} - n\gamma(\lambda)\right]\Gamma^{(n)}_{\text{as}}(p_i, \lambda, \mu) = 0 \tag{3.54}$$

where $\Gamma^{(n)}_{\text{as}}$ is the asymptotic form of $\Gamma^{(n)}_R$. Thus in the deep Euclidean region, a small change in the mass parameter (the $\mu(\partial/\partial\mu)$ term) can always be compensated for by an appropriate small change in the coupling (the $\beta(\partial/\partial\lambda)$ term) and an appropriate small rescaling of the fields (the $-n\gamma$ term).

First we replace the change in mass parameter trivially by the corresponding change in the scale parameter. From dimensional analysis we can write

$$\Gamma^{(n)}_{\text{as}}(p_i, \lambda, \mu) = \mu^{4-n}\bar{\Gamma}^{(n)}_R(p_i/\mu, \lambda) \tag{3.55}$$

where $\bar{\Gamma}^{(n)}_R$ is dimensionless and satisfies

$$\left(\mu\frac{\partial}{\partial\mu} + \sigma\frac{\partial}{\partial\sigma}\right)\bar{\Gamma}^{(n)}_R(\sigma p_i/\mu, \lambda) = 0. \tag{3.56}$$

We have from (3.55) and (3.56)

$$\left[\mu\frac{\partial}{\partial\mu} + \sigma\frac{\partial}{\partial\sigma} + (n-4)\right]\Gamma^{(n)}_{\text{as}}(\sigma p_i/\mu, \lambda) = 0. \tag{3.57}$$

The asymptotic form of the Callan–Symanzik equation can be written

$$\left[\sigma\frac{\partial}{\partial\sigma} - \beta(\lambda)\frac{\partial}{\partial\lambda} + n\gamma(\lambda) + (n-4)\right]\Gamma^{(n)}_{\text{as}}(\sigma p_i, \lambda, \mu) = 0. \tag{3.58}$$

To solve this equation we first remove the nonderivative terms with the transformation

$$\Gamma^{(n)}_{\text{as}}(\sigma p_i, \lambda, \mu) = \sigma^{4-n}\exp\left[n\int_0^\lambda \frac{\gamma(x)}{\beta(x)}\,\mathrm{d}x\right]F^{(n)}(\sigma p_i, \lambda, \mu). \tag{3.59}$$

Thus,

$$\left[\sigma\frac{\partial}{\partial\sigma} - \beta(\lambda)\frac{\partial}{\partial\lambda}\right]F^{(n)}(\sigma p, \lambda, \mu) = 0. \tag{3.60}$$

For convenience, define $t = \ln\sigma$. We need to solve

$$\left[\frac{\partial}{\partial t} - \beta(\lambda)\frac{\partial}{\partial\lambda}\right]F^{(n)}(\mathrm{e}^t p_i, \lambda, \mu) = 0. \tag{3.61}$$

In order to do this we introduce the *effective*, or *running, coupling constant* $\bar{\lambda}$ as the solution to the equation

$$\frac{\mathrm{d}\bar{\lambda}(t, \lambda)}{\mathrm{d}t} = \beta(\bar{\lambda}) \tag{3.62}$$

with the boundary condition $\bar{\lambda}(t = 0, \lambda) = \lambda$. To obtain another form of (3.62) we first integrate it with respect to t

$$t = \int_\lambda^{\bar{\lambda}(t,\lambda)} \mathrm{d}x/\beta(x), \tag{3.63}$$

then differentiate both sides with respect to λ

$$0 = \frac{1}{\beta(\bar{\lambda})}\frac{\mathrm{d}\bar{\lambda}}{\mathrm{d}\lambda} - \frac{1}{\beta(\lambda)}$$

or

$$\left[\frac{\partial}{\partial t} - \beta(\lambda)\frac{\partial}{\partial \lambda}\right]\bar{\lambda}(t, \lambda) = 0. \tag{3.64}$$

Thus if $F^{(n)}$ depends on t and λ through the combination of $\bar{\lambda}(t, \lambda)$ it will satisfy (3.61). $\Gamma^{(n)}_{\mathrm{as}}$ must have the form

$$\Gamma^{(n)}_{\mathrm{as}}(\sigma p_i, \lambda, \mu) = \sigma^{4-n} \exp\left[n \int_0^{\lambda} \frac{\gamma(x)}{\beta(x)}\,\mathrm{d}x\right] F^{(n)}(p_i, \bar{\lambda}(t, \lambda), \mu). \tag{3.65}$$

We can write

$$\begin{aligned}
\exp\left[n \int_0^{\lambda} \frac{\gamma(x)}{\beta(x)}\,\mathrm{d}x\right] &\sim \exp\left[n \int_0^{\bar{\lambda}} \frac{\gamma(x)}{\beta(x)}\,\mathrm{d}x + n \int_{\bar{\lambda}}^{\lambda} \frac{\gamma(x)}{\beta(x)}\,\mathrm{d}x\right] \\
&= H(\bar{\lambda}) \exp\left[-n \int_{\lambda}^{\bar{\lambda}} \frac{\gamma(x)}{\beta(x)}\,\mathrm{d}x\right] \\
&= H(\bar{\lambda}) \exp\left[-n \int_0^{t} \gamma(\bar{\lambda}(t', \lambda))\,\mathrm{d}t'\right]
\end{aligned} \tag{3.66}$$

where

$$H(\bar{\lambda}) = \exp\left[n \int_0^{\bar{\lambda}} \frac{\gamma(x)}{\beta(x)}\,\mathrm{d}x\right].$$

Thus, we have

$$\Gamma^{(n)}_{\mathrm{as}}(\sigma p_i, \lambda, \mu) = \sigma^{4-n} \exp\left[-n \int_0^{t} \gamma(\bar{\lambda}(x', \lambda))\,\mathrm{d}x'\right] H(\bar{\lambda}(t, \lambda)) F^{(n)}(p_i, \bar{\lambda}(t, \lambda), \mu). \tag{3.67}$$

If we set $t = 0$ ($\sigma = 1$) in eqn (3.67), we see that the combination $H(\bar{\lambda})F^{(n)}(\bar{\lambda})$ is just $\Gamma^{(n)}_{\mathrm{as}}$. Therefore,

$$\Gamma^{(n)}_{\mathrm{as}}(\sigma p_i, \lambda, \mu) = \sigma^{4-n} \exp\left[-n \int_0^{t} \gamma(\bar{\lambda}(x', \lambda))\,\mathrm{d}x'\right] \Gamma^{(n)}_{\mathrm{as}}(p_i, \bar{\lambda}(t, \lambda), \mu). \tag{3.68}$$

In this form the asymptotic solution $\Gamma^{(n)}_{\mathrm{as}}$ has a simple interpretation. The effect of rescaling the momenta p_i in the Green's function $\Gamma^{(n)}_R$ is equivalent to replacing the coupling constant λ by the effective coupling constant $\bar{\lambda}$, apart from some multiplicative factors. The first factor σ^{4-n} in (3.68) is the canonical dimension coming from the fact that $\Gamma_R^{(n)}$ has dimension $4-n$ in units of mass. The exponential factor in (3.68) is the anomalous dimension term which is the result of summing up the leading logarithms in perturbation theory. This factor is controlled by the γ-function. Thus γ is often called the *anomalous dimension* (Wilson 1971).

The result in this section may be viewed as follows. The expectation that in the large-momentum limit masses become negligible and theory should be scale-invariant is too simple. Even without physical masses the renormalizable theory still has an energy scale as we must always impose normalization conditions at some mass scale. Thus naïve dimension analysis is generally inadequate and scale invariance is broken. However the dependence of the theory on this normalization mass scale is given by the renormalization-group equation which expresses the effect of a small change of scale. In favourable cases when the inhomogeneous term in the Callan–Symanzik equation may be dropped the solution indicates that the asymptotic behaviour displays a certain universal character with operators being assigned anomalous dimensions.

.2 The minimal subtraction scheme and its renormalization-group equation

In this section we will illustrate other forms of the renormalization-group equation. Again let us examine the multiplicative renormalizability statement (3.29a) which may be written as

$$\Gamma^{(n)}(p_i; \lambda_0, \mu_0) = Z_\phi^{-n/2}\Gamma_R^{(n)}(p_i; \lambda, \mu, \kappa). \tag{3.69}$$

If we regard the bare parameters λ_0, μ_0, ϕ_0 as independent variables, then the renormalized quantities are functions of these bare parameters and the normalization scale parameter κ. In this form, the right-hand side of (3.69) depends on κ explicitly as well as implicitly through the definitions of λ and μ. However the left-hand side is independent of κ; we then have

$$\left[\kappa\frac{\partial}{\partial\kappa} + \beta\frac{\partial}{\partial\lambda} + \gamma_m\mu\frac{\partial}{\partial\mu} - n\gamma\right]\Gamma_R^{(n)} = 0 \tag{3.70}$$

with

$$\beta\left(\lambda, \frac{\mu}{\kappa}\right) = \kappa\frac{\partial\lambda}{\partial\kappa} \tag{3.71}$$

$$\gamma_m\left(\lambda, \frac{\mu}{\kappa}\right) = \kappa\frac{\partial \ln\mu}{\partial\kappa} \tag{3.72}$$

$$\gamma\left(\lambda, \frac{\mu}{\kappa}\right) = \frac{1}{2}\kappa\frac{\partial \ln Z_\phi}{\partial\kappa}. \tag{3.73}$$

Compared to the Callan–Symanzik equation (3.31), this renormalization-group equation has no inhomogeneous term to begin with. We will try to approach it with a procedure similar to that used in solving the asymptotic form (3.54) of the Callan–Symanzik equation. But the coefficients β, γ_m, and γ are now dimensionless functions of *two* variables λ and μ/κ which makes the solution difficult. However, in contrast to the momentum subtraction schemes discussed in §3.1, there exists a mass-independent renormalization procedure in which the mass dependences of these renormalization group equation coefficients disappear. We now give an outline of this subtraction scheme due to 't Hooft (1973); see also the discussion in Ramond (1981).

Minimal subtraction scheme

This renormalization procedure is particularly suitable for dimensional regularization. Here the divergences show up as poles when the dimension $n \to 4$. The minimal subtraction scheme consists of adding counterterms to cancel these poles. In other words, the counterterms have no finite parts.

As an example, consider the one-loop self-energy (Fig. 2.4) in $\lambda\phi^4$ theory. In momentum subtraction schemes of §3.1 the presence of the arbitrary mass scale κ is obvious (e.g. as the normalization point). In the dimensional regularization one also needs to introduce a mass scale κ to compensate for the naïve dimensions of coupling constants and masses: $\lambda \to (\kappa)^{4-n}\lambda$ and $\mu \to \kappa\mu$. We have

$$-\mathrm{i}\Sigma(p^2) = \frac{-\mathrm{i}\lambda\kappa^{\varepsilon}}{2}\int\frac{\mathrm{d}^n l}{(2\pi)^n}\frac{\mathrm{i}}{l^2-\mu^2}$$

$$= \frac{-\mathrm{i}\lambda\kappa^{\varepsilon}}{2(2\pi)^n}\frac{\pi^{n/2}\Gamma(1-n/2)}{\mu^{2-n}} = \frac{-\mathrm{i}\lambda\mu^2}{32\pi^2}\Gamma(-1+\varepsilon/2)\left(\frac{\kappa}{\mu}\right)^{\varepsilon}2^{\varepsilon}\pi^{\varepsilon/2} \tag{3.74}$$

where

$$\varepsilon = 4 - n. \tag{3.75}$$

To make an expansion around $\varepsilon = 0$, we use the formulae

$$\Gamma(-n+\varepsilon) = \frac{(-1)^n}{n!}\left[\frac{1}{\varepsilon} + \psi(n+1) + O(\varepsilon)\right] \tag{3.76}$$

$$a^{\varepsilon} = \mathrm{e}^{\varepsilon\ln a} = 1 + \varepsilon\ln a + O(\varepsilon^2) \tag{3.77}$$

where

$$\psi(n+1) = 1 + \frac{1}{2} + \ldots\frac{1}{n} - \gamma$$

and $\gamma = 0.5772\ldots$ is the Euler constant. Thus as $\varepsilon \to 0$, eqn (3.74) becomes

$$-\mathrm{i}\Sigma(p^2) \to \frac{-\mathrm{i}\lambda\mu^2}{32\pi^2}\left[\frac{2}{\varepsilon} + \psi(2) + 2\ln(\kappa/\mu) + 2\ln 2\sqrt{\pi} + O(\varepsilon)\right]. \tag{3.78}$$

Thus the counterterm $\Sigma(0)\phi^2/2$ to be added, as in eqn (2.62), in the minimal subtraction scheme is

$$\Delta\mathscr{L}_{\phi^2} = \frac{\lambda\mu^2}{32\pi^2}\frac{1}{\varepsilon}\phi^2. \tag{3.79}$$

This is to be contrasted with the counterterm $(\lambda\mu^2/32\pi^2)[1/\varepsilon + \frac{1}{2}\psi(2) +$

$\ln(\kappa/\mu) + \ln 2\sqrt{\pi}]\phi^2$ that one would have added in the momentum subtraction schemes. Thus the minimal subtraction counterterm Lagrangian when expanded in a Laurent series in ε will contain only divergent terms. The relations between the physical λ, μ, and ϕ and the bare parameters are

$$\lambda_0 = \kappa^\varepsilon\left[\lambda + \sum_{r=1}^{\infty} a_r(\lambda)/\varepsilon^r\right] \tag{3.80}$$

$$\mu_0 = \kappa\left[1 + \sum_r b_r(\lambda)/\varepsilon^r\right] \tag{3.81}$$

$$\phi_0 = \phi\left[1 + \sum_r c_r(\lambda)/\varepsilon^r\right] \equiv \phi Z_\phi^{-1/2}. \tag{3.82}$$

Thus the coefficients, reflecting the same property of the counterterms, are independent of the arbitrary parameter κ and (since they are all dimensionless) the particle mass μ. Hence this minimal subtraction scheme is also called the *mass-independent renormalization* procedure. One can easily understand this feature as the counterterms have no finite part; they just have the 'barebone' structure needed to cancel the infinities at very large momenta where the theory is not sensitive to its masses (provided the amplitude are well-behaved as $p \to \infty$).

To calculate the renormalization-group parameters of eqns (3.71)–(3.73), we use the fact that the bare quantities are independent of κ. Thus, from (3.80),

$$\varepsilon\lambda + \left(a_1 + \kappa\frac{\partial\lambda}{\partial\kappa}\right) + \sum_{r=1}^{\infty}\frac{1}{\varepsilon^r}\left[\frac{\partial a_r}{\partial\lambda}\kappa\frac{\partial\lambda}{\partial\kappa} + a_{r+1}\right] = 0. \tag{3.83}$$

Since $\kappa(\partial\lambda/\partial\kappa)$ is analytic at $\varepsilon = 0$, we can write

$$\kappa\frac{\partial\lambda}{\partial\kappa} = d_0 + d_1\varepsilon + d_2\varepsilon^2 + \dots. \tag{3.84}$$

From (3.83), it is clear that $d_r = 0$ for $r > 1$ and

$$\varepsilon(\lambda + d_1) + \left(a_1 + d_0 + d_1\frac{\mathrm{d}a_1}{\mathrm{d}\lambda}\right) + \sum_r\frac{1}{\varepsilon^r}\left[a_{r+1} + d_0\frac{\mathrm{d}a_r}{\mathrm{d}\lambda} + d_1\frac{\mathrm{d}a_{r+1}}{\mathrm{d}\lambda}\right] = 0 \tag{3.85}$$

which implies that

$$(\lambda + d_1) = 0$$

$$a_1 + d_1\frac{\mathrm{d}a_1}{\mathrm{d}\lambda} = -d_0$$

$$\left(1 + d_1\frac{\mathrm{d}}{\mathrm{d}\lambda}\right)a_{r+1} = -\mathrm{d}_0\frac{\mathrm{d}a_r}{\mathrm{d}\lambda}. \tag{3.86}$$

Thus,

$$\kappa\frac{\partial\lambda}{\partial\kappa} = -a_1 + \lambda\frac{\mathrm{d}a_1}{\mathrm{d}\lambda} - \lambda\varepsilon$$

or

$$\beta(\lambda) = -a_1 + \lambda \frac{da_1}{d\lambda}. \tag{3.87}$$

We also have

$$\left(1 - \lambda \frac{d}{d\lambda}\right)[a_{r+1}(\lambda) - a_1(\lambda)] = \frac{d}{d\lambda} a_r(\lambda). \tag{3.88}$$

Similarly, we have, from (3.81),

$$\gamma_m = \kappa \frac{\partial \ln \mu}{\partial \kappa} = \lambda \frac{db_1}{d\lambda}. \tag{3.89}$$

$$\lambda \frac{db_{r+1}}{d\lambda} = b_r \lambda \frac{db_1}{d\lambda} - \frac{db_r}{d\lambda}\left(1 - \lambda \frac{d}{d\lambda}\right) a_1(\lambda) \tag{3.90}$$

and, from (3.82),

$$\gamma = \frac{1}{2} \kappa \frac{\partial \ln Z_\phi}{\partial \kappa} = \lambda \frac{dc_1}{d\lambda} \tag{3.91}$$

$$\lambda \frac{dc_{r+1}}{d\lambda} = c_r \lambda \frac{dc_1}{d\lambda} - \frac{dc_r}{d\lambda}\left(1 - \lambda \frac{d}{d\lambda}\right) a_1(\lambda). \tag{3.92}$$

Thus eqns (3.87), (3.89), and (3.91) enable us to calculate β, γ_m, and γ directly from the residues of the simple poles a_1, b_1, and c_1. The recursion relations (3.88) and (3.90) are useful in computing the residues of the higher-order pole terms in terms of the simple pole. (This is the same reason why the leading logarithms, the next-to-leading logarithms, etc. can be calculated to all orders by using the renormalization-group equation with the computation of just a few low-order terms.) Here we will just make a simple check that the β-function result agrees with previous calculation. From eqns (2.50), (2.63), and (2.123) we have

$$\lambda_0 = Z_\phi^{-2} Z_\lambda \lambda$$

with

$$Z_\phi = 1$$

$$\lambda Z_\phi = \lambda - i\Gamma(0) = \lambda + \frac{\lambda^2}{16\pi^2}\frac{1}{\varepsilon}. \tag{3.93}$$

Thus, a_1 being quadratic in λ, $\lambda(da_1/d\lambda) = 2a_1$ and

$$\beta = a_1 = \frac{3\lambda^2}{16\pi^2} \tag{3.94}$$

which agrees with (3.47).

The fact that β, γ, and γ_m in this subtraction scheme are functions of λ only will simplify the solution to the renormalization-group equation (3.70). The procedure will be similar to the steps of eqns (3.55)–(3.68). From dimensional analysis, eqn (3.70) can be written as

$$\left[\sigma \frac{\partial}{\partial \sigma} - \beta(\lambda)\frac{\partial}{\partial \lambda} - (\gamma_m - 1)\frac{\partial}{\partial \mu} + n\gamma(\lambda) + (n-4)\right]\Gamma_R^{(n)}(\sigma p_i, \mu, \lambda, \kappa) = 0. \tag{3.95}$$

To solve (3.95) we now introduce not only an effective coupling constant $\bar{\lambda}(t)$, but also an *effective mass* $\bar{\mu}(t)$ (with $t \equiv \ln \sigma$ as before)

$$\frac{d\bar{\lambda}(t)}{dt} = \beta(\bar{\lambda}) \tag{3.96}$$

$$\frac{d\bar{\mu}(t)}{dt} = [\gamma_m(\bar{\lambda}) - 1]\bar{\mu}(t) \tag{3.97}$$

with boundary conditions

$$\bar{\lambda}(t=0) = \lambda \tag{3.98}$$

$$\bar{\mu}(t=0) = \mu. \tag{3.99}$$

The solution to eqn (3.95) may be written down: just as eqn (3.68),

$$\Gamma_R^{(n)}(\sigma p_i, \mu, \lambda, \kappa) = \sigma^{4-n} \exp\left[-n \int_0^t \gamma(\bar{\lambda}(t'))\, dt'\right] \Gamma_R^{(n)}(\delta p_i, \bar{\mu}(t), \bar{\lambda}(t), \kappa).$$

In this formulation the large-momentum limit (3.54) and the validity of the asymptotic solution (3.68) hinge on whether or not the effective mass $\bar{\mu}(t)$ vanishes in the deep Euclidean limit $t \to \infty$.

We should remark that in the momentum-subtraction schemes of §3.1, one can also introduce the arbitrary mass scale μ as the subtraction point to derive a homogeneous renormalization-group equation (Weinberg 1973*a*). But then the functions γ, β, γ_m will depend on (m/μ) in addition to depending on the coupling constant λ. This will cause some difficulty in solving the renormalization-group equation. In practice one can get around this by choosing the subtraction point μ large enough so that the dependence on (m/μ) of γ, β, and γ_m can be neglected.

3.3 Effective coupling constants

Apart from the trivial dimensional factor σ^{4-n}, the Green's function $\Gamma_R^{(n)}(\sigma p_i, \lambda, \mu)$ in the deep Euclidean region with $\sigma \to \infty$ (or $e^t \to \infty$) depends on σ only through the effective coupling constant $\bar{\lambda}(t, \lambda)$, which we will concentrate on in our study of the asymptotic behaviour of Green's function.

As we discussed in §3.1, the definition of the coupling constant λ depends on the subtraction point. For example in the intermediate renormalization scheme, the four-point function in the $\lambda\phi^4$ theory is given by eqn (2.42),

$$\Gamma_I^{(4)}(s, t, u) = -i\lambda_I + \tilde{\Gamma}(s, \lambda_I) + \tilde{\Gamma}(t, \lambda_I) + \tilde{\Gamma}(u, \lambda_I) \tag{3.100}$$

where $\tilde{\Gamma}(s, \lambda)$ is given by eqn (2.86), and λ_I is the coupling constant defined in the intermediate renormalization scheme. The cross-section for two-particle elastic scattering is related to the four-point function by

$$\frac{d\sigma}{d\Omega} = \frac{1}{64\pi^2}\frac{1}{s}|\Gamma^{(4)}|^2. \tag{3.101}$$

But in the on-shell renormalization scheme the coupling is defined differently with the four-point function expanded around the symmetric point $s = t = u = 4\mu^2/3$.

$$\Gamma_S^{(4)}(s, t, u) = -i\lambda_S + \tilde{\Gamma}_S(s, \lambda_S) + \tilde{\Gamma}_S(t, \lambda_S) + \tilde{\Gamma}_S(u, \lambda_S) \tag{3.102}$$

where

$$\tilde{\Gamma}_S(s, \lambda_S)|_{s=4\mu^2/3} = 0 \quad \text{and} \quad \Gamma_S^{(4)}(s = t = u = 4\mu^2/3) = -i\lambda_S.$$

Since the cross-section (3.101) should be the same in these two schemes of renormalization, we have

$$\Gamma_I^{(4)}(s, t, u) = \Gamma_S^{(4)}(s, t, u). \tag{3.103}$$

Evaluating both sides at the symmetric point, $s_0 = t_0 = u_0 = 4\mu^2/3$, eqn (3.103) implies

$$-i\lambda_S = -i\lambda_I + \tilde{\Gamma}(s_0, \lambda_I) + \tilde{\Gamma}(t_0, \lambda_I) + \tilde{\Gamma}(u_0, \lambda_I).$$

This gives the relation between coupling constants defined by different subtraction schemes. Clearly the subtraction point can be taken at any point in the physical or unphysical region. And the coupling constant in any renormalization scheme should be regarded as a function of the subtraction point. In this sense the coupling constant is energy-dependent and is called the *effective*, or the *running coupling constant*.

There is another way to look at the running coupling constant. It simply reflects the effect of the leading radiative-correction terms. In perturbation theory the effective expansion parameter is actually the coupling constant multiplied by some logarithmical factors. Normally one picks the normalization point to be of the same order of magnitude as the typical momentum scale of the problem. The argument of the logarithm, which is typically the ratio of these scale factors, is then generally of order one. However, for a problem involving a large range of energy scale, the radiative correction through these large log factors can be substantial. The solution to the renormalization-group equation simply represents the summation of these logarithmic factors to all orders of perturbation theory.

The running coupling constant $\bar{\lambda}$ satisfies the differential equation (3.62)

$$\frac{d\bar{\lambda}(t, \lambda)}{dt} = \beta(\bar{\lambda}) \tag{3.104}$$

or more explicitly as a renormalization-group equation (3.64)

$$\left(\frac{\partial}{\partial t} - \beta\frac{\partial}{\partial \lambda}\right)\bar{\lambda}(t, \lambda) = 0. \tag{3.105}$$

Thus the change in the effective coupling $\bar{\lambda}$ induced by the change in energy scale is governed by the renormalization-group β-function. To study the asymptotic behaviour of $\bar{\lambda}$ let us suppose that $\beta(\lambda)$ has the form shown in Fig. 3.1. The points 0, λ_1, and λ_2 where β vanishes are called *fixed points*. If the coupling constant λ is at any one of these points at $t = 0$, it will remain there for all values of momenta. Furthermore, we can distinguish two types of fixed points. Consider the neighbourhood of λ_1. Because $\beta(\lambda) > 0$ for

$0 < \lambda < \lambda_1$, the effective coupling constant $\bar{\lambda}$ in (3.104) increases with increasing momenta (i.e. $d\bar{\lambda}/dt = \beta(\bar{t}) > 0$) and is driven to λ_1, as $t \to \infty$. As $\beta(\lambda) < 0$ for $\lambda_1 < \lambda < \lambda_2$, it decreases with increasing momenta and is again driven to λ_1. Thus in the interval $0 < \lambda < \lambda_2$, the coupling constant λ is always driven to λ_1 for large momenta. λ_1 is called an *ultraviolet stable fixed point*. By similar argument it is straightforward to see that in the neighbourhood of 0 and λ_2 the coupling will be driven to these points for small momenta, i.e. as $t \to 0$. Hence the origin and λ_2 are examples of *infrared stable fixed points*.

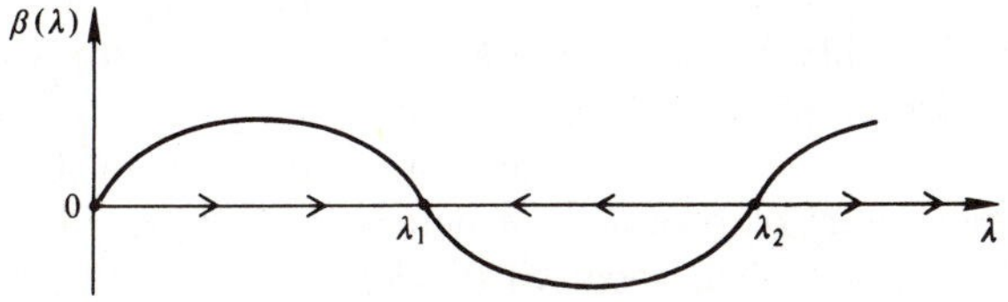

FIG. 3.1. An example of the Callan–Symanzik β-function exhibiting an ultraviolet stable fixed point at λ_1 and infrared stable fixed points at 0 and λ_2. The direction arrows indicate how the coupling constant will move for increasing momenta.

Now we can study the asymptotic solution of the Callan–Symanzik equation. Suppose $0 < \lambda < \lambda_2$. Then

$$\lim_{t \to \infty} \bar{\lambda}(t, \lambda) = \lambda_1 \tag{3.106}$$

and

$$\Gamma^{(n)}_{\text{as}}(p_i, \bar{\lambda}(t, \lambda), \mu) \underset{t \to \infty}{\to} \Gamma^{(n)}_{\text{as}}(p_i, \lambda_1, \mu). \tag{3.107}$$

For purposes of illustration let us assume that $\beta(\lambda)$ has a simple zero at λ_1 and that $\gamma(\lambda_1)$ does not vanish; then we have in the neighbourhood of λ_1

$$\beta(\lambda) \simeq a(\lambda_1 - \lambda) \quad \text{with} \quad a > 0. \tag{3.108}$$

From

$$\frac{d\bar{\lambda}}{dt} = a(\lambda_1 - \lambda), \tag{3.109}$$

we obtain

$$\bar{\lambda} = \lambda_1 + (\lambda - \lambda_1)\, e^{-at}. \tag{3.110}$$

Thus for (3.108) the approach of $\bar{\lambda}$ to λ_1 is exponential in the variable t. In the same approximation, we have

$$\begin{aligned}\int_0^t \gamma(\bar{\lambda}(x, \lambda))\, dx &= \int_\lambda^{\bar{\lambda}} \frac{\gamma(y)\, dy}{\beta(y)} \\ &\simeq \frac{-\gamma(\lambda_1)}{a} \int_\lambda^{\bar{\lambda}} \frac{d\lambda'}{\lambda' - \lambda_1} \\ &= \frac{-\gamma(\lambda_1)}{a} \ln\left(\frac{\bar{\lambda} - \lambda_1}{\lambda - \lambda_1}\right) \\ &= \gamma(\lambda_1)t = \gamma(\lambda_1) \ln \sigma.\end{aligned} \tag{3.111}$$

Thus the particular realization of (3.68) takes on the form

$$\lim_{\sigma\to\infty} \Gamma^{(n)}_{\mathrm{as}}(\sigma p_i, \lambda, \mu) = \sigma^{4-n[1+\gamma(\lambda_1)]}\Gamma^{(n)}_{\mathrm{as}}(p_i, \lambda_1, \mu). \tag{3.112}$$

This means that in the deep Euclidean region, the field scales with anomalous dimension $\gamma(\lambda_1)$ and Green's function takes on a value with λ replaced by λ_1.

In general it is difficult to calculate the zeros of the β-function since this requires results beyond perturbation theory. However $\beta(\lambda)$ has a trivial zero at the origin $\lambda = 0$, where the anomalous dimension $\gamma(\lambda)$ also vanishes. Besides the practicality of calculating $\beta(\lambda)$ for small λ it turns out that this may have particular phenomenological relevance. As we shall discuss in Chapter 7 deep inelastic lepton–hadron scattering probes the large-momentum behaviour of products of hadronic electromagnetic (or weak) currents. The observed phenomena of Bjorken scaling can be interpreted as indicating that the product of these currents has the free-field singularity structure. Hence, if we can find a field theory which has an ultraviolet stable fixed point at the origin $\lambda = 0$, it may be taken as a candidate theory for the hadron constituent (quark) interactions. In other words, the Bjorken scaling phenomena in deep inelastic lepton–hadron scattering may be explained if the effective interaction among quarks vanishes in the short-distance limit. This suggests that a theory of quark interactions should have the feature that it become a free-field theory in the ultraviolet asymptotic limit (*asymptotic freedom*) and one needs to calculate the β-function and to see whether $\beta(\lambda) < 0$ for $\lambda \gtrsim 0$.

For $\lambda\phi^4$ theory, from (3.47) we see that it is not ultraviolet asymptotically free. More explicitly we can integrate (3.104)

$$\frac{\mathrm{d}\bar{\lambda}}{\mathrm{d}t} = \frac{3\lambda^2}{16\pi^2} \tag{3.113}$$

to obtain

$$\bar{\lambda} = \frac{\lambda}{1 - \dfrac{3\lambda}{16\pi^2}t} \tag{3.114}$$

where $\lambda = \bar{\lambda}(t = 0, \lambda)$. Of course (3.113) and (3.114) are valid only for small $\bar{\lambda}$. We have dropped higher-order terms in $\bar{\lambda}$. Had it been applicable for large couplings also, eqn (3.114) would predict that interaction strength would blow up at the 'Landau singularity' of $t = 16\pi^2/3\lambda$.

The β-functions for other theories will be discussed in Chapter 10. It will be shown in particular that no theory without a non-Abelian gauge field can be asymptotically free.

We can summarize this introduction of the renormalization group and its effective couplings as follows. The aim of the renormalization-group approach is to describe how the dynamics of a system evolves as one changes the scale of the phenomena being observed. Generally one is particularly interested in the behaviour of the system at extremely small (ultraviolet) or extremely large (infrared) limits of the scale. These renormalization-group transformations (of the effective theories at different scales) after some iterations often have the property of approaching a fixed point in these limits. The attractive feature is that the behaviour of effective theory at the fixed point is relatively insensitive to details of the theory at ordinary length scales and in some cases these fixed-point effective theories are particularly simple to calculate.

4 Group theory and the quark model

Ever since Einstein, symmetry has played a fundamental role in theoretical physics. In this chapter and the next one, we shall discuss the more familiar subject of global symmetry. The notion of local gauge symmetry with its space–time-dependent transformation will be introduced in Chapter 8. Such gauge symmetries can be used to generate dynamics, the gauge interactions. The natural mathematical language of symmetry is group theory. After the development of quark models and non-Abelian gauge theories of strong and electroweak interactions, some knowledge of Lie groups has become indispensable for anyone interested in the study of elementary particle theory. Here we shall present a practical introduction to the subject. It begins with a mathematical preliminary section composed mostly of definitions and illustrative examples. Our approach is informal. The basic notions introduced here are for group theory as applied in practice in particle physics. The groups SU(2) and SU(3) are studied with elementary techniques and supplemented with graphic methods in §4.2. The tensor method which is appropriate for the general SU(n) groups is presented in §4.3. The physical realization of the flavour symmetry SU(3) of strong interactions is the quark model which is briefly studied in §4.4.

4.1 Elements of group theory

A *group* G is a set of elements $(a, b, c, \ldots)$ with a multiplication law having the following properties.

(i) *Closure*. If a and b are in G, $c = ab$ is also in G;
(ii) *Associative*. $a(bc) = (ab)c$;
(iii) *Identity*. There exists an element e such that $ea = ae = a$ for every a in G;
(iv) *Inverse*. For every a in G, there exists an element a^{-1} such that $aa^{-1} = a^{-1}a = e$.

Also, if the multiplication is commutative—$ab = ba$ for all a and b in G, G is an *Abelian group*. If the number of elements in G is finite, it is a *finite group*. A *subgroup* is a subset of G, which also forms a group.

Here are some examples. The *cyclic group of order n*, Z_n, consists of a, a^2, $a^3, \ldots, a^n = e$ (identity). It is a finite Abelian group. The *symmetric group* (or *permutation group*), S_n, being the set of all permutations of n objects is a finite non-Abelian group. The *unitary group*, $U(n)$, is the set of $n \times n$ unitary matrices: $UU^\dagger = U^\dagger U = 1$. It is non-Abelian for $n > 1$. The Abelian group $U(1)$ consists of 1×1 unitary matrices, i.e. they are phase transformations $e^{i\delta}$. The group of $n \times n$ unitary matrices with a unit determinant is called the *special unitary group*, SU(n). Similarly, SO(n) is the group of $n \times n$ orthogonal matrices: $AA^{\mathrm{T}} = A^{\mathrm{T}}A = 1$ with unit determinant. Thus SO(3) is just the familiar *rotational group*.

Given any two groups $G = \{g_1, g_2, \ldots\}$ and $H = \{h_1, h_2, \ldots\}$, if the g_is commute with the h_js, we can define a *direct-product group* $G \times H = \{g_i h_j\}$ with the multiplication law

$$g_k h_l \cdot g_m h_n = g_k g_m \cdot h_l h_n. \tag{4.1}$$

Examples of direct-product groups are SU(2) × U(1) (the group consists of elements which are direct products of SU(2) matrices and the U(1) phase factor) and SU(3) × SU(3) (the group consists of elements which are direct products of matrices of two different SU(3)s). These groups will play an important role in the application of group theory in particle physics (see Chapters 5 and 11). If we can write a group as a direct product of smaller groups, the study of group structure will be greatly simplified. To see whether this decomposition is possible, it is useful to introduce the notion of an *invariant subgroup*, which is the subgroup N such that for any element t in N then rtr^{-1} is still in N for all r in G. Thus each component of a direct-product group is an invariant subgroup. If the group does not contain any non-trivial invariant subgroup, i.e. it cannot be written as a direct-product group, it is *a simple group*. SU(n) is such an example, but U(n) is not because it can be written as SU(n) × U(1). The groups which are a direct product of simple groups without any Abelian factors are called *semi-simple groups*.

A *representation* is a specific realization of the multiplication of the group elements by matrices. Thus, it is a mapping of the abstract group elements to a set of matrices $a \to D(a)$ such that, if $ab = c$, then $D(a)D(b) = D(c)$, i.e. the group multiplications are preserved. Thus properly speaking the above definitions of the groups U(n) and SU(n) are given in terms of their *defining representations*. Also note that the permutation operations of S_n may be represented by a finite number of $n \times n$ matrices. If a representation $D(a)$ can be put in block-diagonal form, i.e. if there exists a non-singular matrix M, independent of the group elements, such that

$$MD(a)M^{-1} = \begin{bmatrix} D_1(a) & & 0 \\ & D_2(a) & \\ 0 & & \ddots \end{bmatrix} \quad \text{for all } a \text{ in } G, \tag{4.2}$$

$D(a)$ is called a *reducible representation*. It is denoted by a direct sum $D_1(a) \oplus D_2 \oplus \ldots$. If this cannot be done, $D(a)$ is said to be *irreducible*. We can consider the matrices $D(a)$ as linear transformations on a set of basis (or state) vectors. The *dimension* of a representation is just the dimension of the vector space on which it acts. The reducible representation means that a subset of states is never connected to other states and in irreducible representations all states are connected with each other through group transformations.

Of particular relevance to physical applications are the *Lie groups*, which we shall first define narrowly as continuous groups (having elements labelled by continuous parameters such as the Euler angles for the rotation group SO(3)) with representations by unitary operators. Let $a(\boldsymbol{\theta}) = a(\theta_1, \theta_2, \ldots, \theta_n)$ be the group elements labelled by n continuous real parameters. The identity element is taken to be $e = a(0)$. The group multiplication $a(\boldsymbol{\theta})a(\boldsymbol{\phi}) = a(\boldsymbol{\xi})$ corresponds to the mapping of the parameter space on to itself

$$\mathbf{f}(\boldsymbol{\theta}, \boldsymbol{\phi}) = \boldsymbol{\xi} \tag{4.3}$$

which satisfies the requirements of

$$\mathbf{f}(0, \boldsymbol{\theta}) = \mathbf{f}(\boldsymbol{\theta}, 0) = \boldsymbol{\theta}, \qquad \mathbf{f}(\boldsymbol{\theta}, \mathbf{f}(\boldsymbol{\phi}, \boldsymbol{\xi})) = \mathbf{f}(\mathbf{f}(\boldsymbol{\theta}, \boldsymbol{\phi}), \boldsymbol{\xi}), \tag{4.4}$$

and $\mathbf{f}(\boldsymbol{\theta}, \boldsymbol{\theta}') = 0$ if $a(\boldsymbol{\theta})^{-1}$ is parametrized as $a(\boldsymbol{\theta}')$. This is a Lie group if the function $\mathbf{f}$ in (4.3) is an analytic function (or continuously differentiable) with respect to its variables. Thus we can use the usual analytic methods in abstract group space when dealing with Lie groups. Also, since transformations in quantum mechanics are unitary operators in Hilbert space we are particularly interested in those Lie groups with unitary representations

$$a(\boldsymbol{\theta}) = \exp\{\mathrm{i}\boldsymbol{\theta}\cdot\mathbf{X}\} = a(0) + \mathrm{i}\theta_k \mathrm{X}_k + \ldots \tag{4.5}$$

where

$$X_k = -\mathrm{i}\left.\frac{\partial a}{\partial \theta_k}\right|_{\boldsymbol{\theta}=0} \tag{4.6}$$

are called the (infinitesimal) group *generators*. For unitary $a(\boldsymbol{\theta})$, the X_k are a set of linearly independent hermitian operators. For example, when $a(\theta)$ is an element of the SO(2) group, the group of two-dimensional rotations, the generator is simply the Pauli matrix

$$X = \sigma_2 = \begin{pmatrix} 0 & -\mathrm{i} \\ \mathrm{i} & 0 \end{pmatrix}. \tag{4.7}$$

Define the commutator of two group elements $a(\boldsymbol{\phi})$ and $a(\boldsymbol{\theta})$, lying near the identity, as $a(\boldsymbol{\phi})a(\boldsymbol{\theta})a(\boldsymbol{\phi})^{-1}a(\boldsymbol{\theta})^{-1}$. This product should also be a group element, call it $a(\boldsymbol{\xi})$. $\boldsymbol{\xi}$ must be a function of $\boldsymbol{\theta}$ and $\boldsymbol{\phi}$,

$$\xi_i = g_i(\boldsymbol{\theta}, \boldsymbol{\phi}) \quad \text{with} \quad \mathbf{g}(0, \boldsymbol{\phi}) = \mathbf{g}(\boldsymbol{\theta}, 0) = 0. \tag{4.8}$$

For small $\boldsymbol{\theta}$ and $\boldsymbol{\phi}$ we can expand $g_i(\boldsymbol{\theta}, \boldsymbol{\phi})$ in powers of θ_i and ϕ_i,

$$\xi^l = A^l + B^l_j\theta_j + B'^l_k\phi_k + C^l_{jk}\theta_j\phi_k + C'^l_{jk}\theta_j\theta_k + C''^l_{jk}\phi_j\phi_k + \ldots.$$

The boundary conditions in (4.8) imply that

$$A^l = B^l_j = B'^l_j = C'^l_{jk} = C''^l_{jk} = 0$$

or

$$\xi^l = C^l_{jk}\theta_j\phi_k + \ldots \tag{4.9}$$

When we equate

$$a(\boldsymbol{\xi}) = e + \mathrm{i}\xi_l X_l + \ldots \tag{4.10a}$$

to

$$a(\boldsymbol{\phi})a(\boldsymbol{\theta})a(\boldsymbol{\phi})^{-1}a(\boldsymbol{\theta})^{-1} = e + \theta_j\phi_k[X_j, X_k] + \ldots, \tag{4.10b}$$

we have the *Lie algebra*

$$[X_j, X_k] = \mathrm{i}C^l_{jk}X_l. \tag{4.11}$$

The C^l_{jk}s, called the *structure constants* of the group, are a set of real numbers with

$$C^l_{jk} = -C^l_{kj}. \tag{4.12}$$

For example, the generators of the rotation group in three dimensions SO(3) are just the angular momentum operators J_1, J_2, and J_3. They satisfy the commutation relation

$$[J_j, J_k] = \mathrm{i}\varepsilon_{jkl}J_l \tag{4.13}$$

where ε_{jkl} is the totally antisymmetric Levi–Civita symbol with $\varepsilon_{123} = 1$.

If the $D(a)$s form a representation of the group, the $D^*(a)$s form the *complex conjugate representation*, since $D(a_1)D(a_2) = D(a_1a_2)$ implies $D^*(a_1)D^*(a_2) = D^*(a_1a_2)$. From (4.6) we have the representation matrix of the generators $T(X_j) \equiv T_j$,

$$D(a(\boldsymbol{\theta})) = \exp\{\mathrm{i}\boldsymbol{\theta}\cdot\mathbf{T}\} \tag{4.14}$$

with

$$[T_j, T_k] = \mathrm{i}C^l_{jk}T_l. \tag{4.15}$$

Clearly the $-T^*_j$s also form a representation of the generators. If T_j and $-T^*_j$ are equivalent, i.e. if there exists a nonsingular matrix S such that

$$ST_jS^{-1} = -T^*_j \quad \text{for all } j, \tag{4.16}$$

then the T_j is called a *real representation*. As we shall see below in §4.2, all irreducible representations of SU(2) are real; some properties of real representations will also be discussed in §4.2.

From the Jacobi identity

$$[X_j, [X_k, X_l]] + [X_l, [X_j, X_k]] + [X_k, [X_l, X_j]] = 0 \tag{4.17}$$

and (4.11), we have the relation among structure constants

$$C^m_{jk}C^n_{lm} + C^m_{lj}C^n_{km} + C^m_{kl}C^n_{jm} = 0. \tag{4.18}$$

We can define a set of matrices

$$C^m_{jk} = \mathrm{i}(T_j)^m_k \tag{4.19}$$

which satisfies the commutation relation of (4.15). Thus the structure constants also generate a representation of the algebra, the *adjoint representation*. It has dimension equal to the number of real parameters necessary to specify a group element.

For the semi-simple group (i.e. one having no U(1) invariant subgroup) a normalization convention of the T_js that is compatible with the nonlinear commutation relation (4.15) is

$$\mathrm{tr}(T_iT_j) = \lambda\delta_{ij} \tag{4.20}$$

because $\mathrm{tr}(T_iT_j)$ is a real symmetric matrix and can be diagonalized by taking an appropriately chosen real linear combination of the generators. The diagonal coefficients have been set to a constant λ. With this basis in the vector space of the generators, the structure constants may be written

$$C^m_{jk} = \frac{-\mathrm{i}}{\lambda}\,\mathrm{tr}(T_m[T_j, T_k]) \tag{4.21}$$

which implies that C^m_{jk} is totally antisymmetric in all three indices.

Because the representation matrices of the group elements and their generators are related by exponentiation (4.14), many of their properties can be directly translated into one another. Trivially, they have the same dimension, etc. In the following, unless the ambiguity makes a difference, the term 'representation' will mean either that of the group elements or their generators. Also, the set of basis states of the representation is sometimes referred to, for brevity, as the representation.

4.2 SU(2) and SU(3)

The special unitary groups SU(n) are encountered repeatedly in particle physics theories. It is SU(2) in isospin invariance; SU(3) in 'the eightfold way'; the standard gauge model of strong and electroweak interactions uses SU(3) × SU(2) × U(1); the simplest grand unification group is SU(5). In this section we shall concentrate on groups SU(2) and SU(3). The subject of the tensor method in SU(n) is presented in §4.3.

SU(n) is the group of $n \times n$ unitary matrices with unit determinant: $U^\dagger U = UU^\dagger = 1$ and $\det U = 1$. Any unitary matrix U can be written in terms of a hermitian matrix H as $U = \mathrm{e}^{\mathrm{i}H}$. From the identity $\det(\mathrm{e}^A) = \mathrm{e}^{\mathrm{tr}\,A}$ and $\det U = 1$, it follows that $\mathrm{tr}\,H = 0$. Since there are $n^2 - 1$ traceless hermitian $n \times n$ matrices, an element of SU(n) can be written as

$$U = \exp\left\{\mathrm{i}\sum_{a=1}^{n^2-1} \varepsilon_a J_a\right\} \tag{4.22}$$

where the ε_as are (real) group parameters. The J_as are group generators represented by traceless hermitian matrices. Only $n - 1$ of $n^2 - 1$ generators are diagonal. We say SU(n) is a group of *rank* $n - 1$.

The SU(2) group

There are three group parameters. We write the 2 × 2 unitary unimodular matrices as

$$U(\varepsilon_1, \varepsilon_2, \varepsilon_3) = \exp\{\mathrm{i}\varepsilon_a \sigma_a\} \tag{4.23}$$

where the σ_as are 2 × 2 traceless hermitian matrices. We choose the basis to be the standard Pauli matrices.

$$\sigma_1 = \begin{pmatrix} 0 & 1 \\ 1 & 0 \end{pmatrix}, \quad \sigma_2 = \begin{pmatrix} 0 & -\mathrm{i} \\ \mathrm{i} & 0 \end{pmatrix}, \quad \sigma_3 = \begin{pmatrix} 1 & 0 \\ 0 & -1 \end{pmatrix}.$$

The generators defined by $J_i = \sigma_i/2$ will give the commutation relation

$$[J_a, J_b] = \mathrm{i}\varepsilon_{abc} J_c \tag{4.24}$$

where ε_{abc} is the totally antisymmetric Levi–Civita symbol and $\varepsilon_{123} = 1$. We then abstract this as the general Lie algebra of SU(2) and all representations of the generators satisfy this set of commutation relations.

SU(2) representations. The algebra (4.24) is the same as that in (4.13). We say SU(2) is isomorphic to the rotation group SO(3). The standard method of setting up angular momentum eigenstates will be followed here to get all the irreducible representations of SU(2).

First define

$$J^2 \equiv J_1^2 + J_2^2 + J_3^2 \tag{4.25}$$

which is an invariant operator, a *Casimir operator*, commuting with all the generators of the group

$$[J^2, J_a] = 0, \qquad a = 1, 2, 3. \tag{4.26}$$

Also define the raising and lowering operators

$$J_\pm \equiv J_1 \pm \mathrm{i} J_2 \tag{4.27}$$

then

$$J^2 = \tfrac{1}{2}(J_+ J_- + J_- J_+) + J_3^2. \tag{4.28}$$

We have from (4.24)

$$[J_+, J_-] = 2J_3 \tag{4.29}$$

$$[J_\pm, J_3] = \mp J_\pm. \tag{4.30}$$

Consider an eigenstate of J^2 and J_3 with eigenvalues λ and m

$$J^2|\lambda, m\rangle = \lambda|\lambda, m\rangle$$

$$J_3|\lambda, m\rangle = m|\lambda, m\rangle. \tag{4.31}$$

Because of (4.30) the states $J_\pm|\lambda, m\rangle$ are also eigenstates of J_3 with eigenvalues $m \pm 1$, and, because of (4.26), the same eigenvalue λ

$$J_\pm|\lambda, m\rangle = C_\pm(\lambda, m)|\lambda, m \pm 1\rangle \tag{4.32}$$

where the $C_\pm(\lambda, m)$s are constants to be determined later. For a given λ, values of m are bounded

$$\lambda - m^2 \geq 0 \tag{4.33}$$

because $J^2 - J_3^2 = J_1^2 + J_2^2 \geq 0$. Let j be the largest value of m

$$J_+|\lambda, j\rangle = 0. \tag{4.34}$$

Eqns (4.34), (4.28), and (4.29) then imply

$$\begin{aligned} 0 &= J_- J_+|\lambda, j\rangle \\ &= (J^2 - J_3^2 - J_3)|\lambda, j\rangle = (\lambda - j^2 - j)|\lambda, j\rangle \end{aligned} \tag{4.35}$$

or

$$\lambda = j(j+1). \tag{4.36}$$

Similarly, let j' be the smallest value of m

$$J_-|\lambda, j'\rangle = 0. \tag{4.37}$$

We obtain

$$\lambda = j'(j'-1). \tag{4.38}$$

Thus $j(j+1) = j'(j'-1)$ which has the solutions $j' = -j$ and $j' = j+1$. Since second solution violates the assumption that j is the largest value of m, we have

$$j' = -j. \tag{4.39}$$

Since J_- lowers the value of m by one unit, $j - j' = 2j$ must be an interger. This means that j can be either an integer or half-integer. To determine $C_\pm(\lambda, m)$ in (4.32) we use

$$\langle \lambda, m|J_- J_+|\lambda, m\rangle = |C_+(\lambda, m)|^2 \tag{4.40}$$

because $J_- = J_+^\dagger$ implies that $\langle \lambda, m|J_- = C_+^*(\lambda, m)\langle \lambda, m+1|$. We also have, from (4.35)

$$\begin{aligned}\langle \lambda, m|J_- J_+|\lambda, m\rangle &= \langle \lambda, m|(J^2 - J_3^2 - J_3)|\lambda, m\rangle \\ &= j(j+1) - m^2 - m.\end{aligned} \tag{4.41}$$

Hence,

$$C_+(\lambda, m) = [(j-m)(j+m+1)]^{1/2}. \tag{4.42}$$

Similarly,

$$C_-(\lambda, m) = [(j+m)(j-m+1)]^{1/2}. \tag{4.43}$$

These states $|j, m\rangle$ with $m = j, j-1, \ldots, -j$ form the basis of an SU(2) irreducible representation, characterized by j which is either an integer or half-integer. Thus the dimension of the representation is $2j+1$. We can use the relations

$$\begin{aligned}J_3|j, m\rangle &= m|j, m\rangle \\ J_\pm|j, m\rangle &= [(j \mp m)(j \pm m + 1)]^{1/2}|j, m \pm 1\rangle\end{aligned} \tag{4.44}$$

to work out the representation matrices.

Example 1. $J = \frac{1}{2}$, $m = \pm\frac{1}{2}$

$$J_3|\tfrac{1}{2}, \pm\tfrac{1}{2}\rangle = \pm\tfrac{1}{2}|\tfrac{1}{2}, \pm\tfrac{1}{2}\rangle. \tag{4.45}$$

If we denote

$$|\tfrac{1}{2}, \tfrac{1}{2}\rangle = \begin{pmatrix}1\\0\end{pmatrix} \qquad |\tfrac{1}{2}, -\tfrac{1}{2}\rangle = \begin{pmatrix}0\\1\end{pmatrix}, \tag{4.46}$$

then

$$J_3 = \tfrac{1}{2}\begin{pmatrix}1 & 0\\ 0 & -1\end{pmatrix}. \tag{4.47}$$

From $J_+|\frac{1}{2}, \frac{1}{2}\rangle = 0$ and $J_+|\frac{1}{2}, -\frac{1}{2}\rangle = |\frac{1}{2}, \frac{1}{2}\rangle$ we have

$$J_+ = \begin{pmatrix} 0 & 1 \\ 0 & 0 \end{pmatrix}. \tag{4.48}$$

Also,

$$J_- = J_+^\dagger = \begin{pmatrix} 0 & 0 \\ 1 & 0 \end{pmatrix} \tag{4.49}$$

$$J_1 = (J_+ + J_-)/2 = \tfrac{1}{2}\begin{pmatrix} 0 & 1 \\ 1 & 0 \end{pmatrix}, \qquad J_2 = (J_+ - J_-)/2\mathrm{i} = \tfrac{1}{2}\begin{pmatrix} 0 & -\mathrm{i} \\ \mathrm{i} & 0 \end{pmatrix}. \tag{4.50}$$

Example 2. $J = 1$, $m = 1, 0, -1$.
Denote

$$|1, 1\rangle = \begin{pmatrix} 1 \\ 0 \\ 0 \end{pmatrix}, \quad |1, 0\rangle = \begin{pmatrix} 0 \\ 1 \\ 0 \end{pmatrix}, \quad |1, -1\rangle = \begin{pmatrix} 0 \\ 0 \\ 1 \end{pmatrix}. \tag{4.51}$$

Then

$$J_3 = \begin{pmatrix} 1 & 0 & 0 \\ 0 & 0 & 0 \\ 0 & 0 & -1 \end{pmatrix}. \tag{4.52}$$

From $J_+|1, 1\rangle = 0$, $J_+|1, 0\rangle = \sqrt{2}|1, 1\rangle$, and $J_+|1, -1\rangle = \sqrt{2}|1, 0\rangle$, we have

$$J_+ = \begin{pmatrix} 0 & \sqrt{2} & 0 \\ 0 & 0 & \sqrt{2} \\ 0 & 0 & 0 \end{pmatrix}. \tag{4.53}$$

Then

$$J_- = \begin{pmatrix} 0 & 0 & 0 \\ \sqrt{2} & 0 & 0 \\ 0 & \sqrt{2} & 0 \end{pmatrix} \tag{4.54}$$

$$J_1 = \frac{1}{\sqrt{2}}\begin{pmatrix} 0 & 1 & 0 \\ 1 & 0 & 1 \\ 0 & 1 & 0 \end{pmatrix} \qquad J_2 = \frac{1}{\sqrt{2}}\begin{pmatrix} 0 & -\mathrm{i} & 0 \\ \mathrm{i} & 0 & -\mathrm{i} \\ 0 & \mathrm{i} & 0 \end{pmatrix}. \tag{4.55}$$

It is straightforward to check that the J_is satisfy the Lie algebra of (4.24).

SU(2) product representations. In applications, we often need to deal with *product representations*. For example, if we have two spin 1/2 particles, we want to know the total spin J of the product of the two wavefunctions. In this simple case, the answer is $J = 0$ or 1. Let us study this case in terms of group theory. Denote the spin-up and spin-down wavefunctions of the first particle by r_1 and r_2. Similarly denote those of the second particle by s_1 and s_2. Under SU(2) transformation

$$r_i' = U(\boldsymbol{\varepsilon})_{ij} r_j, \qquad s_k' = U(\boldsymbol{\varepsilon})_{kl} s_l \tag{4.56}$$

where $U(\boldsymbol{\varepsilon}) = \exp\{\mathrm{i}\varepsilon_a J_a\}$ and $J_a = \sigma_a/2$. Then the product will transform as

$$(r_i' s_k') = U(\boldsymbol{\varepsilon})_{ij} U(\boldsymbol{\varepsilon})_{kl} (r_j s_l) \equiv D(\boldsymbol{\varepsilon})_{ik,\,jl} (r_j s_l). \tag{4.57}$$

Generally $D(\boldsymbol{\varepsilon})$ is reducible. To see what irreducible representation it decomposes into, it is easier to work with the generators directly by taking $\varepsilon_i \ll 1$.

$$\begin{aligned} r_i' &= (1 + \mathrm{i}\varepsilon_a J_a)_{ij} r_j \equiv (1 + \mathrm{i}\varepsilon_a J_a^{(1)})_{ij} r_j \\ s_k' &= (1 + \mathrm{i}\varepsilon_a J_a)_{kl} s_l \equiv (1 + \mathrm{i}\varepsilon_a J_a^{(2)})_{kl} s_l \end{aligned} \tag{4.58}$$

where $J_a^{(1)}$ operates only on r_i and does not affect s_i; $J_a^{(2)}$ operates only on s_i and not on r_i. Define the total angular momentum operator as

$$\mathbf{J} = \mathbf{J}^{(1)} + \mathbf{J}^{(2)}. \tag{4.59}$$

We now change to the more familiar notation. Let α_i denote the spin-up wavefunction of the ith particle and β_i the spin-down wavefunction. There are four combinations of two-particle wavefunctions: $\alpha_1\alpha_2, \alpha_1\beta_2, \beta_1\alpha_2, \beta_1\beta_2$. Take the one with the largest value of J_3

$$\begin{aligned} J_3(\alpha_1\alpha_2) &= (J_3^{(1)}\alpha_1)\alpha_2 + \alpha_1(J_3^{(2)}\alpha_2) \\ &= (\alpha_1\alpha_2). \end{aligned} \tag{4.60}$$

Clearly it is a state with $J_3 = 1$. To find its J value, we use

$$\begin{aligned} \mathbf{J}^2 &= (\mathbf{J}^{(1)})^2 + (\mathbf{J}^{(2)})^2 + 2\mathbf{J}^{(1)}\cdot\mathbf{J}^{(2)} \\ &= \mathbf{J}(^{(1)})^2 + (\mathbf{J}^{(2)})^2 + 2[\tfrac{1}{2}(J_+^{(1)}J_-^{(2)} + J_-^{(1)}J_+^{(2)}) + J_3^{(1)}J_3^{(2)}] \end{aligned} \tag{4.61}$$

to find that

$$\mathbf{J}^2(\alpha_1\alpha_2) = 2(\alpha_1\alpha_2). \tag{4.62}$$

This means $J = 1$ and we can make the identification

$$|1, 1\rangle = (\alpha_1\alpha_2).$$

We use the lowering operator $J_- = J_-^{(1)} + J_-^{(2)}$ to reach all other states of the $J = 1$ irreducible representation

$$\begin{aligned} J_-(\alpha_1\alpha_2) &= (J_-^{(1)}\alpha_1)\alpha_2 + \alpha_1(J_-^{(1)}\alpha_2) \\ &= (\beta_1\alpha_2 + \alpha_1\beta_2). \end{aligned} \tag{4.63}$$

On the other hand, using eqn (4.44), we get

$$J_-|1, 1\rangle = \sqrt{2}|1, 0\rangle. \tag{4.64}$$

Thus,

$$|1, 0\rangle = \frac{1}{\sqrt{2}}(\alpha_1\beta_2 + \beta_1\alpha_2). \tag{4.65}$$

Obviously,

$$|1, -1\rangle = \beta_1\beta_2. \tag{4.66}$$

The remaining independent state must be identified as

$$|0, 0\rangle = \frac{1}{\sqrt{2}}(\alpha_1\beta_2 - \beta_1\alpha_2). \tag{4.67}$$

Again, we can check this assignment by applying eqn (4.61). In short, the two-particle wavefunctions can be organized as

$$\begin{aligned} |1, 1\rangle &= \alpha_1\alpha_2 \\ |1, 0\rangle &= \frac{1}{\sqrt{2}}(\alpha_1\beta_2 + \beta_1\alpha_2) \\ |1, -1\rangle &= \beta_1\beta_2 \end{aligned} \tag{4.68}$$

which is symmetric under the interchange of particles $1 \leftrightarrow 2$, and

$$|0, 0\rangle = \frac{1}{\sqrt{2}}(\alpha_1\beta_2 - \beta_1\alpha_2) \tag{4.69}$$

which is antisymmetric under $1 \leftrightarrow 2$.

More generally the product representations $|j_1, m_1\rangle \times |j_2, m_2\rangle$ can be combined into eigenstates $|J, M\rangle$ of total $\mathbf{J} = \mathbf{J}^{(1)} + \mathbf{J}^{(2)}$

$$|J, M\rangle = \sum_{m_1, m_2} \langle j_1 m_1 j_2 m_2 | JM\rangle |j_1 m_1\rangle |j_2 m_2\rangle. \tag{4.70}$$

The coefficients $\langle j_1 m_1 j_2 m_2 | JM\rangle$ are called the *Clebsch–Gordon coefficients*. Thus for the above case (eqns (4.68) and (4.69)) we have

$$\langle \tfrac{1}{2}\,\tfrac{1}{2}\,\tfrac{1}{2}\,\tfrac{1}{2}\,|1\;\;1\rangle = 1, \qquad \langle \tfrac{1}{2}\,-\tfrac{1}{2}\,\tfrac{1}{2}\,-\tfrac{1}{2}\,|1\;\;0\rangle = \frac{1}{\sqrt{2}}, \text{ etc.}$$

The procedure of working out the irreducible representations of the product representations can be summarized as follows.

(1) Start with the combination of states with the largest J_3. This is also the state with the largest total J.

(2) Use the lowering operator $J_- = J_-^{(1)} + J_-^{(2)}$ to get to all the other states in the same irreducible representation.

(3) Find the orthogonal combination to $|J_m, J_m - 1\rangle$ where J_m is the maximum value of J in the product. This should be the state $|J_m - 1, J_m - 1\rangle$. Then use the lowering operator to reach the other $J = (J_m - 1)$ states.

(4) Repeat these steps until $J = |j_1 - j_2|$.

We can also graphically represent SU(2) representations. The group is rank 1, i.e. it has one diagonal generator; each irreducible representation j can be characterized by a straight-line segment with points on it denoting values of m (see Fig. 4.1). In a product representation the eigenvalues of the diagonal generators $J_3^{(1)}$ and $J_3^{(2)}$ are additive. We can represent this addition graphically by repeatedly placing the centre ($m = 0$) of one representation, say j_1, over on every point of the other, j_2, representation (see examples in Fig. 4.2). As we shall see, this graphical method can be generalized to the rank-2 group of SU(3), where the results are less trivial and difficult to see without such a diagramatical aid.

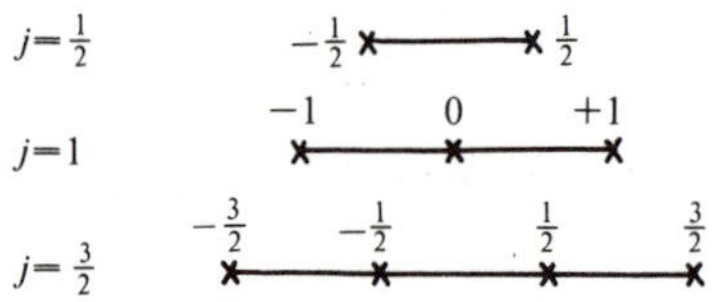

FIG. 4.1. Graphical representation of SU(2) multiplets. The raising (lowering) operator $J_+(J_-)$ moves a state to the right (left).

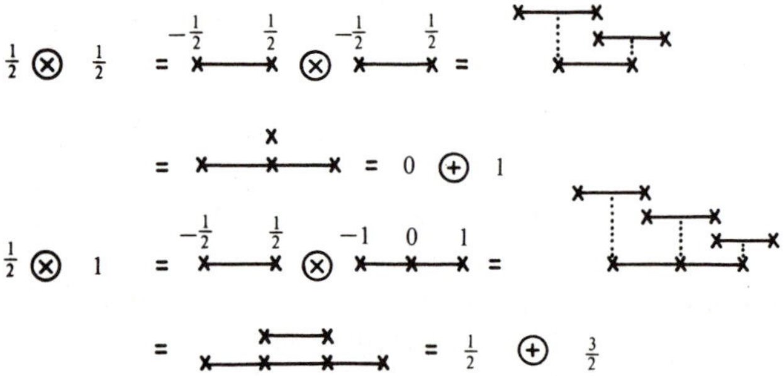

FIG. 4.2.

The Reality property of SU(2) representations. We shall denote the representation matrices of the generator by $T(J_a) \equiv T_a$. As we already mentioned in §4.1, SU(2) has the property that all its representations are real, i.e. there is a (fixed) matrix S such that

$$ST_aS^{-1} = -T_a^*. \tag{4.71}$$

For example, in the defining representation $T_a = \sigma_a/2$ we have $-\sigma_1^* = -\sigma_1$, $-\sigma_2^* = \sigma_2$, and $-\sigma_3^* = -\sigma_3$. The reality condition (4.71) can be satisfied with $S = \sigma_2$. In general the eigenvalues of diagonal generators change sign as we go from T_a to $-T_a^*$ because the T_as are hermitian with real eigenvalues. The eigenvalues of $-T_a^*$ are precisely the negatives of those for T_a. In SU(2) all the irreducible representations have the property that their J_3 eigenvalues occur in pairs, i.e. $m = \pm j, \ \pm(j-1), \ldots$. This is why they are real representations; $-T_a^*$ can be obtained from T_a by changing the basis from $|j, m\rangle$ to $|j, -m\rangle$. For example, in the $j = 1$ representation, the $|1, 1\rangle$ and $|1, -1\rangle$ states of (4.51) are interchanged, leaving $|1, 0\rangle$ invariant, by the transformation

$$S = \begin{pmatrix} 0 & 0 & 1 \\ 0 & -1 & 0 \\ 1 & 0 & 0 \end{pmatrix} \tag{4.72}$$

and we can easily check that (4.71) is satisfied. Clearly, it is a general property of any group representation that, if one of the diagonal generators does not have eigenvalues in pairs of opposite signs, then that representation is not real.

We state once again some of the SU(2) properties which the reader should keep in mind when studying the less familiar SU(3) group.

(1) Only the representation matrix of J_3 is diagonal; SU(2) is a rank-1 group.

(2) The irreducible representation labelled by j (dimension $2j + 1$) has basis states $|j, m\rangle$

$$J^2|j, m\rangle = j(j+1)|j, m\rangle$$

$$J_3|j, m\rangle = m|j, m\rangle .$$

(3) States with different values of m are connected through the raising and lowering operators

$$J_\pm|j, m\rangle = [(j \mp m)(j \pm m + 1)]^{1/2}|j, m \pm 1\rangle .$$

(4) Each irreducible representation can be pictured by a one-dimensional graph because of (1), with equally spaced points representing the $2j + 1$ states. The $T_\pm$ operator moves these points along the line.

Products of two representations j_1 and j_2 can be obtained simply by placing the first representation line $2j_2 + 1$ times over the second representation line, with the $m_1 = 0$ centres coinciding with each state of the j_2-representation.

The SU(3) group

There are eight group parameters. For the defining representation we write the 3×3 unitary unimodular matrices

$$U(\varepsilon_1, \ldots, \varepsilon_8) = \exp\{i\varepsilon_a\lambda_a\} \qquad a = 1, \ldots, 8. \tag{4.73}$$

The λ_as are 3×3 traceless hermitian matrices, which may be chosen to have the form (Gell-Mann 1962*a*)

$$\lambda_1 = \begin{pmatrix} 0 & 1 & 0 \\ 1 & 0 & 0 \\ 0 & 0 & 0 \end{pmatrix} \quad \lambda_2 = \begin{pmatrix} 0 & -i & 0 \\ i & 0 & 0 \\ 0 & 0 & 0 \end{pmatrix} \quad \lambda_3 = \begin{pmatrix} 1 & 0 & 0 \\ 0 & -1 & 0 \\ 0 & 0 & 0 \end{pmatrix}$$

$$\lambda_4 = \begin{pmatrix} 0 & 0 & 1 \\ 0 & 0 & 0 \\ 1 & 0 & 0 \end{pmatrix} \quad \lambda_5 = \begin{pmatrix} 0 & 0 & -i \\ 0 & 0 & 0 \\ i & 0 & 0 \end{pmatrix}$$

$$\lambda_6 = \begin{pmatrix} 0 & 0 & 0 \\ 0 & 0 & 1 \\ 0 & 1 & 0 \end{pmatrix} \quad \lambda_7 = \begin{pmatrix} 0 & 0 & 0 \\ 0 & 0 & -i \\ 0 & i & 0 \end{pmatrix} \quad \lambda_8 = \frac{1}{\sqrt{3}}\begin{pmatrix} 1 & 0 & 0 \\ 0 & 1 & 0 \\ 0 & 0 & -2 \end{pmatrix}. \tag{4.74}$$

They have the normalization

$$\mathrm{tr}(\lambda_a \lambda_b) = 2\delta_{ab} \tag{4.75}$$

and satisfy the commutation relation

$$\left[\frac{\lambda_a}{2}, \frac{\lambda_b}{2}\right] = \mathrm{i} f_{abc} \frac{\lambda_c}{2}. \tag{4.76}$$

f_{abc} is totally antisymmetric with nonvanishing members,

$$\begin{gathered} f_{123} = 1, f_{147} = 1/2, f_{156} = -1/2, f_{246} = 1/2, f_{257} = 1/2, \\ f_{345} = 1/2, f_{367} = -1/2, f_{458} = \surd(3/2), f_{678} = \surd(3/2), \end{gathered} \tag{4.77}$$

The generators F_a of SU(3) satisfy the Lie algebra

$$[F_a, F_b] = \mathrm{i} f_{abc} F_c. \tag{4.78}$$

We can follow the pattern of the SU(2) procedure to obtain irreducible representations of SU(3). Here we follow the presentation of Gasiorowicz (1966).

SU(3) is a rank-2 group; since λ_3 and λ_8 are both diagonal,

$$[F_3, F_8] = 0. \tag{4.79}$$

F_3 and F_8 can be diagonalized simultaneously. We define the following raising and lowering operators

$$T_\pm = F_1 \pm \mathrm{i}F_2, \quad U_\pm = F_6 \pm \mathrm{i}F_7, \quad V_\pm = F_4 \pm \mathrm{i}F_5.$$

We also define

$$T_3 = F_3, \qquad Y = \frac{2}{\sqrt{3}} F_8. \tag{4.80}$$

In terms of these operators, the communication relations of (4.78) can be written as

$$\begin{aligned} &[T_3, T_\pm] = \pm T_\pm && [Y, T_\pm] = 0 \\ &[T_3, U_\pm] = \mp 1/2 U_\pm && [Y, U_\pm] = \pm U_\pm \\ &[T_3, V_\pm] = \pm 1/2 V_\pm && [Y, V_\pm] = \pm V_\pm \end{aligned} \tag{4.81}$$

$$\begin{aligned} &[T_+, T_-] = 2T_3 \\ &[U_+, U_-] = 3/2Y - T_3 \equiv 2U_3 \\ &[V_+, V_-] = 3/2Y + T_3 \equiv 2V_3 \end{aligned} \tag{4.82}$$

$$\begin{aligned} &[T_+, V_+] = [T_+, U_-] = [U_+, V_+] = 0 \\ &[T_+, V_-] = -U_- \qquad [T_+, U_+] = V_+ \\ &[U_+, V_-] = T_-. \qquad [T_3, Y] = 0. \end{aligned} \tag{4.83}$$

SU(3) representations. Since T_3 and Y can be diagonalized simultaneously, the states in an SU(3) irreducible representation must be labelled by two eigenvalues: t_3 and y. A representation is then pictured as a two-dimensional figure on the t_3–y plane, just as an SU(2) representation is an one-dimensional line (Fig. 4.2). From the commutation relations in eqns (4.81)–(4.83), it is not difficult to see the results of raising and lowering operators acting on the states:

T_+ raises t_3 by 1 unit and leaves y unchanged;

U_+ lowers t_3 by 1/2 unit and raises y by 1 unit;

V_+ raises t_3 by 1/2 unit and raises y by 1 unit, etc. (4.84)

If the units of t_3 and y are appropriately scaled in the graph, these raising and lowering operators connect points along lines that are multiples of 60° with each other (Fig. 4.3).

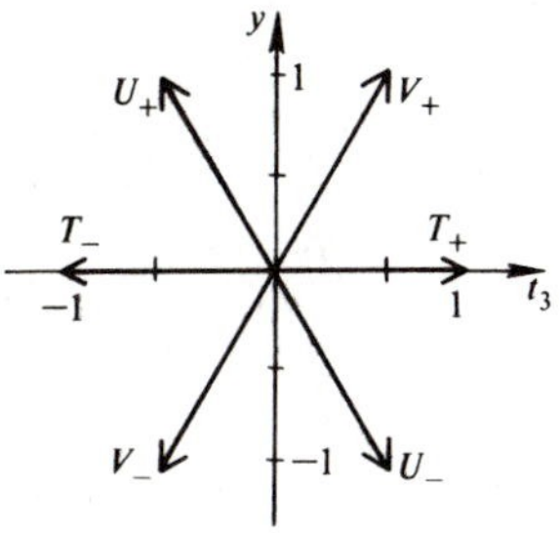

FIG. 4.3.

Each irreducible representation of SU(3) is characterized by a set of two integers (p, q). Graphically it shows up as a figure with a hexagonal boundary on the t_3–y plane: three sides having p units of length and the other three sides having q units (see Fig. 4.4(a)); the hexagon collapses into a equilateral triangle when either p or q vanishes (Fig. 4.4(b)). The boundary is symmetric under reflections in the y-axis. We recall that an SU(2) irreducible representation is characterized by one integer j; graphically it is a straight line of $2j$ units of length. There are $2j + 1$ sites, each of them singly occupied by one state. For the SU(3) representation (p, q) the multiplicity of states on each site in the t_3–y plane form the following pattern: the sites in the boundary are singly occupied, on the next layer they are doubly occupied, on the third layer triply occupied, etc., until a triangle layer is reached beyond

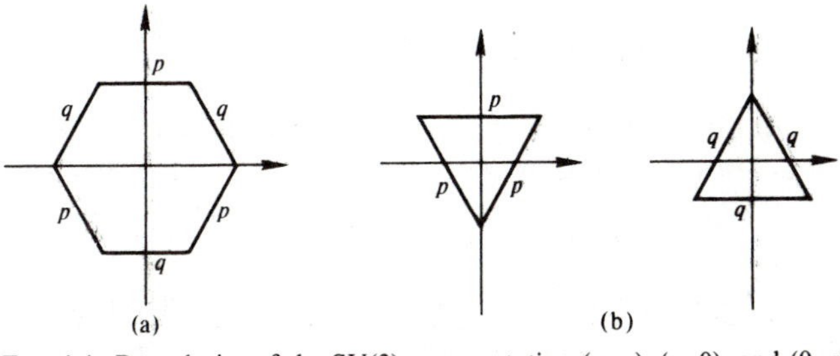

FIG. 4.4. Boundaries of the SU(3) representation (p, q), $(p, 0)$, and $(0, q)$.

which the multiplicity ceases to increase and remains $q+1$ for $p > q$ (or $p+1$ for $q > p$).

The procedures used to deduce these properties of the irreducible representation from the commutation relations are all similar. We shall present one such proof to illustrate the general algebraic technique. To show that the boundary layer is singly occupied, take two neighbouring states $|A\rangle$ and $|B\rangle$ on the boundary shown in Fig. 4.5. Thus

$$U_-|A\rangle = |B\rangle. \tag{4.85}$$

We need to show that, given $|A\rangle$, the state $|B\rangle$ is unique regardless of the path taken to go from $|A\rangle$ to $|B\rangle$. Consider an alternative path ACB; we have

$$V_-T_+|A\rangle = ([V_-, T_+] + T_+V_-)|A\rangle = U_-|A\rangle = |B\rangle \tag{4.86}$$

where we have used $V_-|A\rangle = 0$ and (4.83).

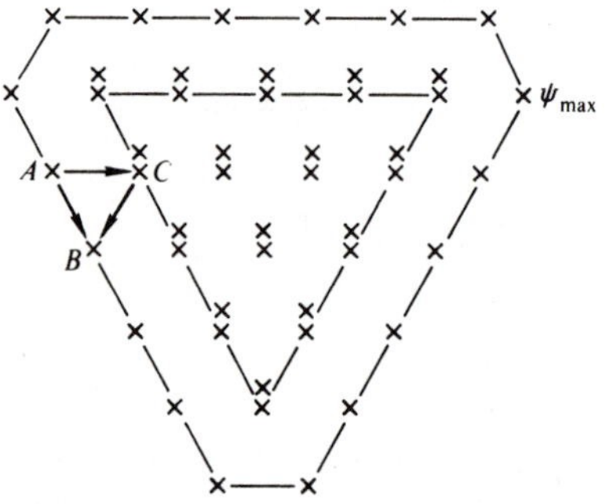

FIG. 4.5. A typical representation with $(p, q) = (5, 1)$. Multiplicity of states at each site is indicated by the crosses. ψ_{max} is the $t = t_3 = (p+q)/2 = 3$ state.

It is not difficult to convince oneself that the result holds independently of the path taken to go from $|A\rangle$ to $|B\rangle$; hence, given $|A\rangle$, the state $|B\rangle$ is unique. Since the state of maximum eigenvalue of T_3 is unique and resides on the boundary, all boundary sites are singly occupied.

Once the multiplicity of states at each site is given, we can add them up. This sum is the dimension of the irreducible representation. To do this we start with counting the number of sites in the inner triangle which has sites $p - q$

$$\sum_{l=1}^{p-q+1} l = \frac{1}{2}(p-q+1)(p-q+2). \tag{4.87}$$

Here the multiplicity is $(q+1)$. On the next outer layer there are $3(p-q+2)$ sites with multiplicity q; on the next one, $3(p-q+4)$ sites each with $(q-1)$ states, etc. Thus the dimension is equal to

$$\frac{1}{2}(q+1)(p-q+1)(p-q+2) + \sum_{v=0}^{q} 3(q-v)(p-q+2v+2) \tag{4.88}$$

or

$$d(p, q) = (p + 1)(q + 1)(p + q + 2)/2. \tag{4.89}$$

Instead of labelling an irreducible representation by (p, q), another common practice is to denote it by its dimensionality. Thus an m-dimensional irreducible representation is labelled by $\mathbf{m}$ and its complex conjugate by $\mathbf{m}^*$. Some of the more important representations are shown in Fig. 4.6.

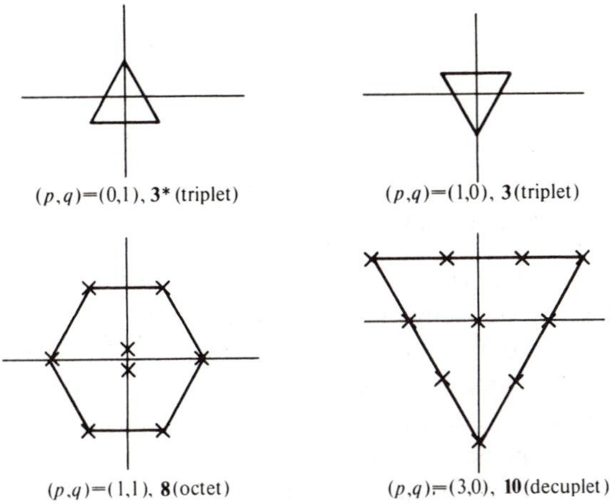

FIG. 4.6. Examples of SU(3) representations with states labelled by (t_3, y). Here all sites are singly occupied except the centre of **8**: one is a $t = 0$ state; another is the $t_3 = 0$ member of a $t = 1$ triplet.

One more remark about the graphical representation (p, q). Since there are generally several states for a given value of (t_3, y), at a given site we need further labelling to distinguish the different states. For this we can specify the SU(2) subgroup to which they belong. A convenient choice will be the T-spin value t. There are $p + 1$ sites each singly occupied on the top line, corresponding to $t = p/2$. The next line has two T-spin multiplets: $t_1 = (p + 1)/2$ and $t_2 = (p - 1)/2$. etc. Also since the widest portion of the hexagon has width $(p + q)$ we conclude that

$$t_{\max} = (p + q)/2. \tag{4.90}$$

For the product representation we can follow a procedure similar to that for the SU(2) group. The method of using the raising and lowering operators gives not only the decomposition of the product representations but also the Clebsch–Gordon coefficients. But this method is rather tedious as there are quite a few raising and lowering operators in SU(3). If we are interested only in the decomposition, we can use the simple graphical method. Again we will place one representation figure on top of *each* member state of the second representation: the centre ($t_3 = y = 0$) of the first one coinciding with the site of each state of the second representation. The simplest case of $\mathbf{3} \times \mathbf{3}^* = \mathbf{8} + \mathbf{1}$ is illustrated in Fig. 4.7. The more systematic approach of the tensor method will be presented in the next section.

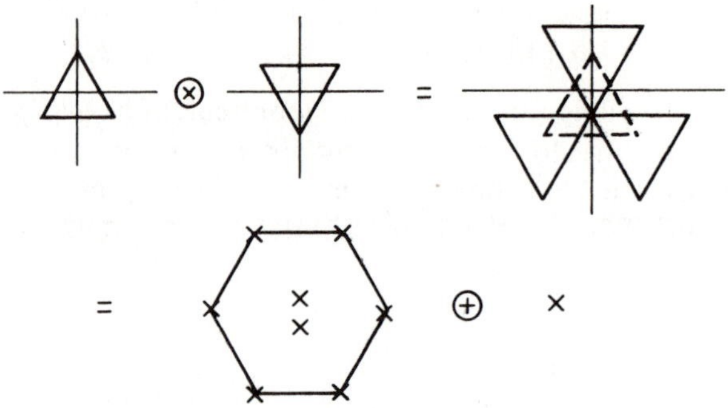

FIG. 4.7.

4.3 The tensor method in SU(n)

The analysis of SU(2) and SU(3) in the last section shows that, as the group gets larger, the elementary techniques used to dissect the representation structure and product become inadequate. For the SU(4) group, which is rank-3, the irreducible representations have to be pictured in a three-dimensional plot and one would need a keen spatial sense to work out the decomposition of the product representation. This approach becomes rather hopelessly complicated for groups of rank-4 or higher. Clearly one needs a more efficient approach. The tensor method turns out to be particularly appropriate for the study of irreducible representations and the decomposition of the product representations in the general SU(n) group.

Transformation law of tensors

The SU(n) group consists of $n \times n$ unitary matrices with unit determinant. We can regard these matrices as linear transformations in an n-dimensional complex vector space C_n. Thus any vector $\psi_i = (\psi_1, \psi_2, \ldots, \psi_n)$ in C_n is mapped by an SU(n) transformation U_{ij} as

$$\psi_i \to \psi_i' = U_{ij}\psi_j. \tag{4.91}$$

The ψ_i's also belong to C_n, with $UU^\dagger = U^\dagger U = 1$ and $\det U = 1$. Clearly for any two vectors we can define a scalar product

$$(\psi, \phi) = \psi_i^* \phi_i \tag{4.92}$$

which is an SU(n) invariant. The transformation law for the conjugate vector is given by

$$\psi_i^* \to \psi_i'^* = U_{ij}^* \psi_j^* = \psi_j^* U_{ji}^\dagger. \tag{4.93}$$

It is convenient to introduce upper and lower indices

$$\psi^i \equiv \psi_i^*, \; U_i^{\,j} \equiv U_{ij} \quad \text{and} \quad U^i_{\;j} \equiv U_{ij}^*. \tag{4.94}$$

Thus complex conjugation just changes the lower indices to upper ones, and vice versa. In this notation, eqns (4.91) and (4.93) read

$$\psi_i \to \psi'_i = U_i^{\ j}\psi_j$$
$$\psi^i \to \psi^{i\prime} = U^i_j\psi^j. \tag{4.95}$$

The SU(n) invariant scalar product is

$$(\psi, \phi) = \psi^i\phi_i. \tag{4.96}$$

and the unitarity condition becomes

$$U^i_{\ k}U^k_{\ j} = \delta^i_{\ j} \tag{4.97}$$

where the Kronecker delta is defined as

$$\delta^i_j \equiv \delta_{ij} = \begin{cases} 1 & \text{if } i = j \\ 0 & \text{otherwise} \end{cases} \tag{4.98}$$

Note that in this notation the summation is always over a pair of upper and lower indices. We call this a *contraction of indices*. The ψ_is are the basis for the SU(n) *defining representation* (also called the *fundamental* or *vector representation* and denoted as **n**), while the ψ^i's are the basis for the conjugate representation, **n***.

Higher-rank tensors are defined as those quantities which have the same transformation properties as the direct products of vectors. Thus tensors generally have both upper and lower indices with the transformation law

$$\psi'^{i_1 i_2 \ldots i_p}_{j_1 j_2 \ldots j_q} = (U^{i_1}_{k_1}U^{i_2}_{k_2}\ldots U^{i_p}_{k_p})(U^{l_1}_{j_1}U^{l_2}_{j_2}\ldots U^{l_q}_{j_q})\psi^{k_1 k_2 \ldots k_p}_{l_1 l_2 \ldots l_q}. \tag{4.99}$$

They correspond to the basis for higher-dimensional representations.

The Kronecker delta and Levi–Civita symbol are invariant tensors under SU(n) transformations. They play important role in the study of irreducible tensors.

(1) From the unitarity condition of (4.97) we immediately have

$$\delta^i_j = U^i_k U^l_j\,\delta^k_l. \tag{4.100}$$

Hence δ^i_j is an invariant tensor. Generally, contracting indices with the Kronecker delta will produce a tensor of low rank. For example,

$$\delta^{j_1}_{i_1}\psi^{i_1 i_2 \ldots i_p}_{j_1 j_2 \ldots j_q} = \psi^{i_2 \ldots i_p}_{j_2 \ldots j_q}. \tag{4.101}$$

We can regard the right-hand side as the *trace* between the pair of indices, in this case i_1 and j_1. Also, a tensor with all its indices contracted $\psi^{i_1 i_2 \cdots i_p}_{i_1 i_2 \ldots i_p}$ is an SU(n) invariant scalar.

(2) The Levi–Civita symbol is defined as the totally antisymmetric quantity

$$\varepsilon^{i_1 i_2 \ldots i_n} = \varepsilon_{i_1 i_2 \ldots i_n} = \begin{cases} 1 \text{ if } (i_1, \ldots, i_n) \text{ is an even permutation of } (1, \ldots, n) \\ -1 \text{ if } (i_1, \ldots, i_n) \text{ is an odd permutation of } (1, \ldots, n) \\ 0 \text{ otherwise.} \end{cases} \tag{4.102}$$

It is an invariant tensor

$$\begin{aligned}\varepsilon'_{i_1 i_2 \ldots i_n} &= U^{j_1}_{i_1}U^{j_2}_{i_2}\ldots U^{j_n}_{i_n}\varepsilon_{j_1 j_2 \ldots j_n} \\ &= (\det U)\varepsilon_{i_1 i_2 \ldots i_n} = \varepsilon_{i_1 i_2 \ldots i_n}\end{aligned} \tag{4.103}$$

where we have used the definition of determinant and det $U = 1$ for SU(n). Similar to the contraction by δ_i^j we can sum over indices by using the Levi–Civita symbol. For example,

$$\varepsilon^{i_1 i_2 \ldots i_n} \psi_{i_2 \ldots i_n} = \psi^{i_1}. \tag{4.104}$$

Thus tensors with upper indices can be constructed from those with lower indices, and vice versa

$$\psi^{i_1 j_1} = (\varepsilon^{i_1 i_2 \ldots i_n} \psi_{i_2 \ldots i_n})(\varepsilon^{j_1 j_2 \ldots j_n} \psi_{j_2 \ldots j_n}). \tag{4.105}$$

So, in principle, to study the transformation properties of the tensor we need work with a tensor having only upper (or lower) indices.

In this connection we also note that

$$\psi = \varepsilon_{i_1 i_2 \ldots i_n} \psi^{i_1 i_2 \ldots i_n} \tag{4.106}$$

is an SU(n)-invariant scalar. And eqns (4.103) and (4.106) imply that a totally antisymmetric tensor of rank n is invariant under SU(n) transformations.

Irreducible representations and the Young tableaux

Generally the tensors we have just defined are bases for reducible representations of SU(n). To decompose them into irreducible representations we use the following property of these tensors. The permutation of upper (or lower) indices commutes with the group transformations, as the latter consist of identical U_{ij}s (or $U_{ij}^\dagger$s). We will illustrate this with the following example. Consider the second-rank tensor ψ_{ij} whose transformation is given by

$$\psi'^{ij} = U_k^i U_l^j \psi^{kl}. \tag{4.107}$$

Since the Us are the same, we can relabel the indices

$$\psi'^{ji} = U_l^j U_k^i \psi^{lk} = U_k^i U_l^j \psi^{lk}. \tag{4.108}$$

Thus the permutation of the indices does not change the transformation law. If P_{12} is the permutation operator which interchanges the first two indices $P_{12}\psi^{ij} = \psi^{ji}$, then P_{12} commutes with the group transformation

$$P_{12}\psi'^{ij} = U_k^i U_l^j P_{12} \psi^{kl}. \tag{4.109}$$

This property can be used to decompose ψ^{ij} as follows. First we form eigenstates of the permutation operator P_{12} by symmetrization or antisymmetrization,

$$S^{ij} = \frac{1}{2}(\psi^{ij} + \psi^{ji}), \qquad A^{ij} = \frac{1}{2}(\psi^{ij} - \psi^{ji}).$$

Thus,

$$P_{12} S^{ij} = S^{ij}, \qquad P_{12} A^{ij} = -A^{ij}. \tag{4.110}$$

It is clear that S^{ij} and A^{ij} will not mix under the group transformation

$$S^{ij} = U_k^i U_l^j S^{kl}, \qquad A^{ij} = U_k^i U_l^j A^{kl}. \tag{4.111}$$

This shows that the second-rank tensor ψ^{ij} decomposes into S^{ij} and A^{ij} in such a way that group transformations never mix parts with different symmetries. It turns out S^{ij} and A^{ij} cannot be decomposed any further and they thus form the basics for irreducible representations of SU(n). This can be generalized to tensors of higher rank (hence the possibility of mixed symmetries) with the result that the basis for irreducible representations of SU(n) correspond to tensors with definite permutation symmetry among (the positions of) its indices. The task of finding irreducible tensors of an arbitrary rank f (i.e. number of upper indices) involves forming a complete set of permutation operations on these indices. The problem of finding the irreducible representation of the permutation group has a complete solution in terms of the *Young tableaux*. They are pictorial representations of the permutation operations of f objects as a set of f-boxes each with an index number in it. For example, for the second-rank tensors, the symmetrization of indices i and j in S_{ij} is represented by $\begin{array}{|c|c|}\hline i & j \\ \hline\end{array}$; the antisymmetrization operation in A_{ij} is represented by $\begin{array}{|c|}\hline i \\ \hline j \\ \hline\end{array}$. For the third-rank tensors, we have $\begin{array}{|c|c|c|}\hline i & j & k \\ \hline\end{array}$ in the case of the completely symmetric S_{ijk}, $\begin{array}{|c|}\hline i \\ \hline j \\ \hline k \\ \hline\end{array}$ in the totally antisymmetric A_{ijk}, and $\begin{array}{|c|c|}\hline i & j \\ \hline k \\ \cline{1-1}\end{array}$ for the tensor with mixed symmetry

$$\psi_{ij;k} \equiv \psi_{ijk} + \psi_{jik} - \psi_{jki} - \psi_{kji}.$$

A general Young tableau is shown in Fig. 4.8. It is an arrangement of f boxes in rows and columns such that the length of rows should not increase from top to bottom: $f_1 \geq f_2 \geq \ldots$ and $f_1 + f_2 + \ldots = f$. Each box has an index $i_k = 1, 2, \ldots, n$. To this tableau we associate the tensor

$$\psi_{i_1, i_2, \ldots, i_{f_1}; i_{f_1+1}, \ldots, i_{f_1+f_2}, \ldots} \tag{4.112}$$

with the following properties.

(1) Indices appearing in the same row of the tableau are first subject to symmetrization.

(2) Subsequent indices appearing in the same column are subject to antisymmetrization.

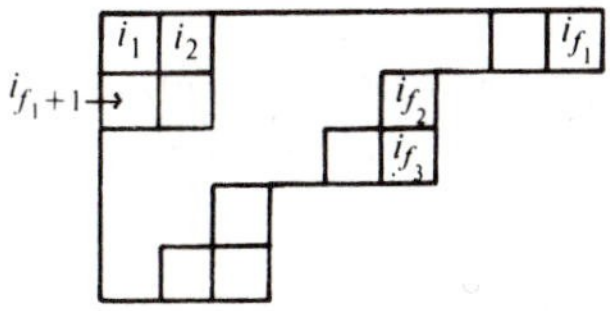

FIG. 4.8.

A tableau where the index numbers do not decrease when going from left to right in a row and always increase from top to bottom is a *standard tableau.* For example, the $n = 3$ mixed-symmetry tensor $\begin{array}{|c|c|}\hline i & j \\ \hline k \\ \cline{1-1}\end{array}$ has the following standard tableaux

$$\begin{array}{|c|c|}\hline 1&1\\\hline 2\\\cline{1-1}\end{array}\quad \begin{array}{|c|c|}\hline 1&1\\\hline 3\\\cline{1-1}\end{array}\quad \begin{array}{|c|c|}\hline 1&2\\\hline 2\\\cline{1-1}\end{array}\quad \begin{array}{|c|c|}\hline 1&2\\\hline 3\\\cline{1-1}\end{array}\quad \begin{array}{|c|c|}\hline 1&3\\\hline 2\\\cline{1-1}\end{array}\quad \begin{array}{|c|c|}\hline 1&3\\\hline 3\\\cline{1-1}\end{array}\quad \begin{array}{|c|c|}\hline 2&2\\\hline 3\\\cline{1-1}\end{array}\quad \begin{array}{|c|c|}\hline 2&3\\\hline 3\\\cline{1-1}\end{array}$$

while tableaux such as $\begin{array}{|c|c|}\hline 1&1\\\hline 1\\\cline{1-1}\end{array}$, $\begin{array}{|c|c|}\hline 2&3\\\hline 1\\\cline{1-1}\end{array}$, and $\begin{array}{|c|c|}\hline 2&1\\\hline 3\\\cline{1-1}\end{array}$ are not standard. The non-standard tableaux give tensors that, by symmetrization or antisymmetrization, either vanish or are not independent of the standard tableaux. Thus for a given pattern of the Young tableaux the number of independent tensors is equal to the number of standard tableaux which can be formed. It is not hard to see that this number for the simplest case of a tensor with k antisymmetric indices is

$$\left.\begin{array}{|c|}\hline \ \\\hline \ \\\hline \ \\\hline \ \\\hline \ \\\hline\end{array}\right\} k, \quad \binom{n}{k} = \frac{n(n-1)\dots(n-k+1)}{1.2\dots k} \tag{4.113}$$

and that for a tensor with k symmetric indices the number is

$$\underbrace{\begin{array}{|c|c|c|}\hline \ & \ & \qquad\qquad \\\hline\end{array}}_{k}, \quad \binom{n+k-1}{k} = \frac{n(n+1)\dots(n+k-1)}{1.2\dots k}. \tag{4.114}$$

One should note that because of antisymmetrization there are not more than n rows in any Young tableau. Also, if there are n rows, we can use $\varepsilon_{i_1 i_2 \dots i_n}$ to contract the indices in the columns with n entries. Pictorially we can simply cross out any column with n rows (see, for example, eqns (4.123) and (4.125)).

Fundamental theorem (See, for example, Hammermesh 1963.) A tensor corresponding to the Young tableau of a given pattern forms the basis of an irreducible representation of SU(n). Moreover if we enumerate all possible Young tableaux under the restriction that there should be no more than $n-1$ rows, the corresponding tensors form a complete set, in the sense that all finite-dimensional irreducible representations of the group are counted only once.

We next give two formulae of the *dimensionality of irreducible representations.* If the Young tableau is characterized by the length of its rows $(f_1, f_2, \dots, f_{n-1})$, define the length differences of adjacent rows as $\lambda_1 = f_1 - f_2$, $\lambda_2 = f_2 - f_3, \dots, \lambda_{n-1} = f_{n-1}$. The dimension of an SU(n) irreducible representation will then be the number of standard tableaux for a given pattern

$$d(\lambda_1, \lambda_2, \dots, \lambda_{n-1}) = (1+\lambda_1)(1+\lambda_2)\dots(1+\lambda_{n-1})$$
$$\times\left(1+\frac{\lambda_1+\lambda_2}{2}\right)\left(1+\frac{\lambda_2+\lambda_3}{2}\right)\dots\left(1+\frac{\lambda_{n-2}+\lambda_{n-1}}{2}\right)$$

$$\times\left(1+\frac{\lambda_1+\lambda_2+\lambda_3}{3}\right)\left(1+\frac{\lambda_2+\lambda_3+\lambda_4}{3}\right)\cdots\left(1+\frac{\lambda_{n-3}+\lambda_{n-2}+\lambda_{n-1}}{3}\right)$$

$$\cdots$$

$$\times\left(1+\frac{\lambda_1+\lambda_2+\ldots\lambda_{n-1}}{n-1}\right). \tag{4.115}$$

One can easily check that the special results of (4.113) and (4.114) for the tableaux $(k, 0, 0, \ldots)$ and $(\underbrace{0, 0, \ldots,}_{k}\ 1, 0, 0, \ldots)$ are recovered.

Example 1. *SU*(2) *group*. The Young tableaux can have only one row: $d(\lambda_1) = (1+\lambda_1)$. Thus $\lambda_1 = 2j$. It follows that a doublet is pictured as □ and a triplet as □□, etc.

Example 2. *SU*(3) *group*. The Young tableaux can have two rows, hence $d(\lambda_1, \lambda_2) = (1+\lambda_1)(1+\lambda_2)(1+(\lambda_1+\lambda_2)/2)$. Thus, $\lambda_1 = p$ and $\lambda_2 = q$ of (4.89).

 (1, 0) **3**, 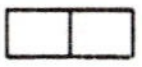 (2, 0) **6**, 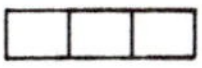 (3, 0) **10**,

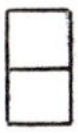 (0, 1) **3***, 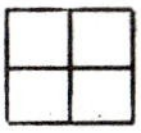 (0, 2) **6***,  (1, 1) **8**. (4.116)

The formula (4.115) is rather cumbersome to use for large values of n; in such cases the second formulation is perhaps more useful. For this we need to introduce two definitions—'hook length' and 'distance to the first box'. For any box in the tableau, draw two perpendicular lines, in the shape of a 'hook', one going to the right and another going downward. The total number of boxes that this hook passes, including the originating box itself, is the *hook length* (h_i) associated with the ith box. For example,

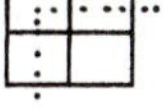 $h_i = 3$, 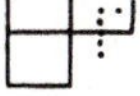 $h_i = 1$. (4.117)

The *distance to the first box* (D_i) is defined to be the number of steps going from the box in the upper left-hand corner of the tableau (the first box) to the ith box with each step towards the right counted as $+1$ unit and each downward step as -1 unit. For example, we have

0	1	2
−1	0	
−2		

(4.118)

The dimension of the SU(n)-irreducible representation associated with the Young tableau is given by

$$d = \prod_i (n + D_i)/h_i \tag{4.119}$$

The products are taken over all boxes in the tableau. For example, for the tableau pattern 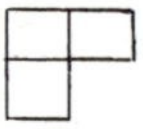, we have hook lengths 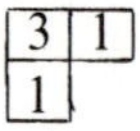 and distances to the first box 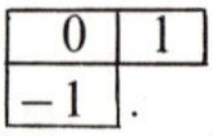. This yields the dimension $d = n(n-1)(n+1)/3$, which gives $d = 8$ for $n = 3$.

Reduction of the product representations

One of the most useful applications of the association of SU(n)-irreducible representations with the Young tableaux is the decomposition of the product representations. To find the irreducible representations in the product of two factors,

(1) In the tableau for the first factor, assign the same symbol, say a, to all boxes in the first row, the same b to all the boxes in the second row, etc.

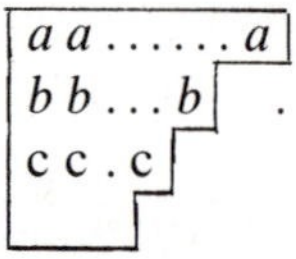 (4.120)

(2) Attach boxes labelled by the symbol a to the tableau of the second factor in all possible ways, subject to the rules that no two as appear in the same column and that the resultant graph is still a Young tableau (i.e. the length of rows does not increase going from top to bottom and there are not more than n rows, etc.). Repeat this process with the bs, ... etc.

(3) After all symbols have been added to the tableau, these added symbols are then read from *right to left* in the first row, then the second row ..., and so forth. This sequence of symbols *aabbac* ... must form a *lattice permutation*. Thus, to the left of any symbol there are no fewer a than b and no fewer b than c, etc.

We consider two examples in the SU(3) group.

Example 1.

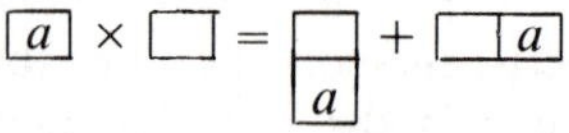 (4.121)

which corresponds to

$$\mathbf{3} \times \mathbf{3} = \mathbf{3}^* + \mathbf{6}. \quad (4.122)$$

Example 2.

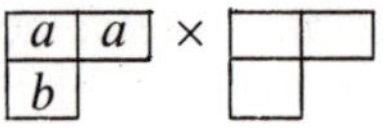 $= \mathbf{8} \times \mathbf{8}.$

First step: 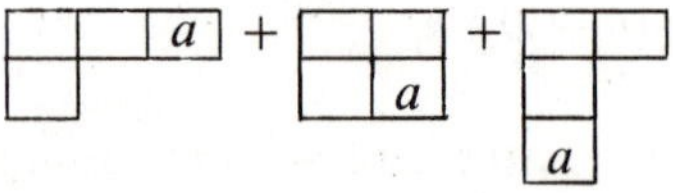

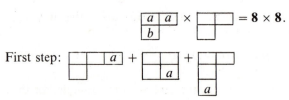

Second step: 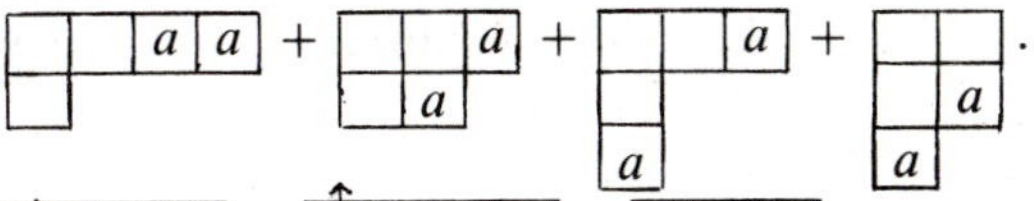

Third step: 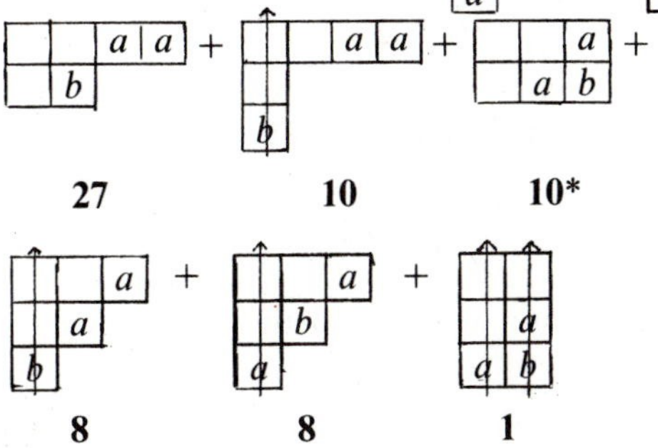

(4.123)

As we have already explained, any column with n boxes in an SU(n) tableau can be 'crossed out'—indicated by a vertical line over the column—thus the last three tableaux yield two octets and one singlet.

Note that tableaux such as 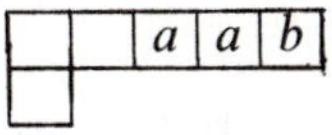 and 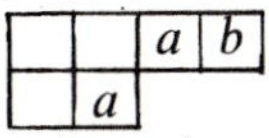 are rejected because the symbols do not form a lattice permutation. Thus

$$\mathbf{8} \times \mathbf{8} = \mathbf{1} + \mathbf{8} + \mathbf{8} + \mathbf{10} + \mathbf{10^*} + \mathbf{27}. \tag{4.124}$$

Young tableaux for conjugate representations. If ψ_i and ψ^i are the bases for the defining representation $\mathbf{n}$ and its complex conjugate $\mathbf{n^*}$. Clearly $\psi_i\psi^i$ is SU(n) invariant. It is not difficult to see from the reduction of product representation that the Young tableau for the conjugate representation is a column of $n-1$ boxes so that there will be an identity representation (a column of n boxes) in the product $\mathbf{n} \times \mathbf{n^*}$.

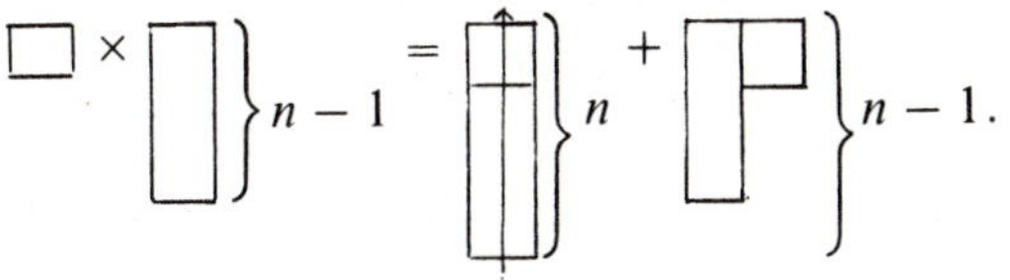

(4.125)

In general if we take the Young tableau for a representation $\psi^{ij\cdots}_{kl\ldots}$ and fill the boxes such that rectangular tableau of n rows is obtained, the additional boxes, when rotated by 180°, form the standard tableau for the complex conjugate representation $(\psi^{ij\cdots}_{kl\ldots})^* = \psi^{kl\cdots}_{ij\ldots}$. For example, in SU(3) we have **6** as pictured by 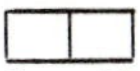 which can be filled in as 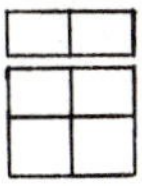. Thus 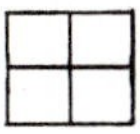 is the Young tableau for **6***.

Group generators in tensor notation

We first concentrate on the (defining) vector representation and later generalize our study to the action of generators on higher-rank tensors.

(1) *Hermitian and real generator matrices.* Any $n \times n$ unitary unimodular matrix U may be written in the form

$$U = \mathrm{e}^{\mathrm{i}H} \tag{4.126}$$

where H is hermitian and traceless. Normally we choose the group parameters to be real. If $\varepsilon_a = \varepsilon_a^*$, $a = 1, 2, \ldots, n^2 - 1$, then

$$H = \varepsilon_a \lambda_a \tag{4.127}$$

where the λ_as are $n \times n$ *hermitian generator matrices*. Those for SU(3) have been displayed in (4.74). From their commutation relation

$$\left[\frac{\lambda_a}{2}, \frac{\lambda_b}{2}\right] = \mathrm{i} f_{abc} \frac{\lambda_c}{2}, \tag{4.128}$$

we can extract the Lie algebra with the identification of $\frac{1}{2}\lambda_a = F_a$ as the generators

$$[F_a, F_b] = \mathrm{i} f_{abc} F_c \tag{4.129}$$

where the f_{abc}s are the structure constants. For the tensor-method approach instead of the form (4.127) we can write the hermitian matrix H as

$$H_i^j = \varepsilon_\beta^\alpha (W_\alpha^\beta)_i^j \tag{4.130}$$

where all indices (α, β, i, j) range from 1 to n. We can choose to have *real generator matrices* which take the form

$$(W_\alpha^\beta)_i^j = \delta_{\alpha i}\, \delta^{\beta j} - \frac{1}{n}\, \delta_\alpha^\beta\, \delta_i^j. \tag{4.131}$$

The hermiticity condition on H is then satisfied by having the hermitian group parameter matrix

$$\varepsilon_\beta^\alpha = \varepsilon_\alpha^{\beta *}.$$

Using (4.131) we can work out the commutator

$$[W_\alpha^\beta, W_\gamma^\delta] = \delta_\gamma^\beta W_\alpha^\delta - \delta_\alpha^\delta W_\gamma^\beta. \tag{4.132}$$

The group generators are defined to satisfy the same commutation relation (the Lie algebra)

$$[F_\alpha^\beta, F_\gamma^\delta] = \delta_\gamma^\beta F_\alpha^\delta - \delta_\alpha^\delta F_\gamma^\beta.$$

The structure constants are simply some combinations of the δ_α^βs. It is not difficult to find the relation between the real generators F_α^β and the hermitian generators F^a,

$$F^a = \tfrac{1}{2}(\lambda_a)_\beta^\alpha F_\alpha^\beta. \tag{4.133}$$

Eqn (4.133) can be inverted by using the identity,

$$\sum_{a=1}^{n^2-1} (\lambda^a)_{\alpha\beta} (\lambda^a)_{\gamma\delta} = 2\left(\delta_{\alpha\delta}\, \delta_{\beta\gamma} - \frac{1}{n}\, \delta_{\alpha\beta}\, \delta_{\gamma\delta}\right) \tag{4.134}$$

which can be derived as follows. Since the $n \times n$ hermitian traceless matrices λ^a, $a = 1, \ldots, (n^2 - 1)$, together with the $n \times n$ identity matrix form a complete set of $n \times n$ hermitian matrices, we can expand an arbitrary hermitian matrix M in terms of them

$$M = m_0 \mathbb{1}_n + \sum_{a=1}^{n2-1} m_a \lambda^a \tag{4.135}$$

where $\mathbb{1}_n$ is the $n \times n$ identity matrix. We normalize the λ^as such that

$$\mathrm{tr}(\lambda^a \lambda^b) = 2\delta^{ab}.$$

The coefficients m_0 and the m_as in eqn (4.135) can be calculated by multiplying matrices of the corresponding bases and taking traces,

$$m_0 = \frac{1}{n} \mathrm{tr}\, M$$

$$m_a = \frac{1}{2} \mathrm{tr}(M\lambda^a).$$

Eqn (4.135) then becomes

$$M = \frac{1}{n}(\mathrm{tr}\, M)\mathbb{1}_n + \sum_{a=1}^{n2-1} \frac{1}{2}(\mathrm{tr}\, M\lambda^a)\lambda^a$$

or, in terms of the components,

$$M_{\alpha\beta} = \frac{1}{n}\left(\sum_{\gamma,\delta} M_{\delta\gamma}\, \delta_{\gamma\delta}\right)\delta_{\alpha\beta} + \frac{1}{2}\sum_a \sum_{\gamma,\delta} (\lambda^a)_{\alpha\beta}(\lambda^a)_{\gamma\delta} M_{\delta\gamma}.$$

Since $M_{\alpha\beta}$ is arbitrary, we get

$$\delta_{\alpha\delta}\, \delta_{\beta\gamma} = \frac{1}{n}\, \delta_{\alpha\beta}\, \delta_{\gamma\delta} + \frac{1}{2}\sum_a (\lambda^a)_{\alpha\beta}(\lambda^a)_{\gamma\delta}$$

which is just the identity (4.134). Using eqn (4.134) we can write the real generator as

$$F_\alpha^\beta = \sum_a F^a (\lambda^a)_\alpha^\beta$$

where we have used the fact that tr $F^a = 0$.

(2) *Real generators in vector representation.* The nondiagonal real generators F_α^β, $\alpha \neq \beta$, are simply raising and lowering operators. For the defining vector representation, W_α^β has a nonzero element only at the αth row and βth column. The infinitesimal SU(n) transformation on the basis $\psi^i \rightarrow \psi'^i = \psi^i + \delta\psi^i$ with $\delta\psi^i = \varepsilon_\beta^\alpha (W_\alpha^\beta)_j^i \psi^j$ shows that

$$(F_\alpha^\beta \psi)^i = \delta_\alpha^i \psi^\beta. \tag{4.136}$$

Thus, F_α^β will take the αth component of ψ and turn it into the βth component, and the result will be zero for all other components. The diagonal generators F_α^αs form a set of mutually commuting operators. Their eigenvalues can be used to characterize the basis functions (states) of irreducible representations. For example, in the defining vector representation, any particular diagonal generator D_l, which is some linear combination of the W^α_αs, may be written

$$(D_l)^i_j = (d_l)_i\,\delta^i_j \tag{4.137}$$

where the $(d_l)_i$s are the eigenvalues of state ψ^i.

(3) *Real generators on higher rank tensors.* The group generator F^β_α acting on any tensor is defined as

$$\delta\psi = \psi' - \psi = \mathrm{i}\varepsilon^\alpha_\beta(F^\beta_\alpha\psi). \tag{4.138}$$

Given (4.99), the general transformation law of ψ with the SU(n) transformation factor given by

$$U_i^k = \delta_i^k + \mathrm{i}\varepsilon^\alpha_\beta(W^\beta_\alpha)^k_i \tag{4.139a}$$

and

$$U_l^j = \delta_l^j - \mathrm{i}\varepsilon^\alpha_\beta(W^\beta_\alpha)^j_l, \tag{4.139b}$$

where the i, j (k, l) indices belong to untransformed (transformed) tensors, we have

$$F^\beta_\alpha\psi^{j_1\ldots j_q}_{i_1\ldots i_p} = \sum_{m=1}^{p} (W^\beta_\alpha)^{k_m}_{i_m}\psi^{j_1\ldots j_q}_{i_1\ldots k_m\ldots i_p}$$
$$- \sum_{n=1}^{q} (W^\beta_\alpha)^{j_n}_{l_n}\psi^{j_1\ldots l_n\ldots j_q}_{i_1\ldots i_p}. \tag{4.140}$$

The presence of the minus sign reflects the fact that the tensors with upper indices correspond to complex conjugate representations, as compared to those with lower indices (see eqns (4.94) and (4.95)). In particular for the diagonal generators $F = D_l$ (4.137) we have

$$D_l\psi^{j_1\ldots j_q}_{i_1\ldots i_p} = \left[\sum_{m=1}^{p} (d_l)_{i_m} - \sum_{n=1}^{q} (d_l)_{j_n}\right]\psi^{j_1\ldots j_q}_{i_1\ldots i_p}. \tag{4.141}$$

Thus the quantum number of the tensor is simply the algebraic sum of the corresponding quantum numbers of the component vectors which make up the tensor.

We will summarize this discussion by working through the simple example of the $j = 1$ representation of SU(2). Instead of using the $m = 1, 0, -1$ states as in (4.51) and as its hermitian generator the 3×3 matrices in (4.52) and (4.55), in the tensor method approach the bases are taken to be $\psi_{ij} \sim \psi^{\mathrm{A}}_i\psi^{\mathrm{B}}_j$. The indices $i, j = 1, 2$ are symmetrized (see (4.68)). The superscripts A and B distinguish the two vectors. The real generators for the SU(2) defining representation are 2×2 matrices:

$$W^1_1 = -W^2_2 = \begin{pmatrix} 1/2 & 0 \\ 0 & -1/2 \end{pmatrix}$$

is the diagonal generator, giving eigenvalues $d(\psi_1) = 1/2$ and $d(\psi_2) = -1/2$ for the two states in ψ_i. We can read off the quantum numbers for the triplet states as

$$d(\psi_{11}) = \tfrac{1}{2} + \tfrac{1}{2} = 1, \qquad d(\psi_{12}) = \tfrac{1}{2} - \tfrac{1}{2} = 0,$$
$$d(\psi_{22}) = -\tfrac{1}{2} - \tfrac{1}{2} = -1. \tag{4.142}$$

$$W^1_2 = \begin{pmatrix} 0 & 0 \\ 1 & 0 \end{pmatrix}$$

is the lowering operator. Thus,

$$W^1_2\psi_{11} = \begin{pmatrix} 0 & 0 \\ 1 & 0 \end{pmatrix}_A \psi^A_1\psi^B_1 + \begin{pmatrix} 0 & 0 \\ 1 & 0 \end{pmatrix}_B \psi^A_1\psi^B_1 = \psi^A_2\psi^B_1 + \psi^A_1\psi^B_2. \tag{4.143}$$

One last comment—the adjoint representation which we have defined in terms of the structure constants of a group (4.19) takes on a particularly simple form in tensor notation. The basis for adjoint representation is simply ψ^i_j with $\psi^i_i = 0$, where its Young tableau is the last one in (4.125) and evidently is self-conjugate. One can show that this is the correct identification by converting ψ^i_j using a method similar to that in eqn (4.133) to an $n^2 - 1$ component vector ϕ^a using

$$\phi^a = \tfrac{1}{2}(\lambda^a)^j_i\psi^i_j. \tag{4.144}$$

The transformation law (4.140) can be used to demonstrate that the matrices for its generators $(F^a)_{bc}$ are indeed the structure constants f_{abc}.

4.4 The quark model

Group theory is relevant in physics because the various symmetry transformations which leave the physical system invariant form a group. The consequence of symmetry can then be deduced through group-theoretical analysis, independent of any detailed dynamical considerations. For example, if a quantum-mechanical system, described by the Hamiltonian $H(\mathbf{r})$, has no preferred direction, all rotation operators $R(\boldsymbol{\theta})$ will leave the Hamiltonian invariant,

$$R(\boldsymbol{\theta})H(\mathbf{r})R^{-1}(\boldsymbol{\theta}) = H(\mathbf{r}). \tag{4.145}$$

Or, in terms of the generators of the rotation, $R(\boldsymbol{\theta}) = \mathrm{e}^{\mathrm{i}\boldsymbol{\theta}\cdot\mathbf{J}}$, this gives

$$[H, J_i] = 0. \tag{4.146}$$

The consequence of this symmetry, i.e. (4.145), is that

$$H(\mathbf{J}_i|n\rangle) = E_n(\mathbf{J}_i|n\rangle) \tag{4.147}$$

if

$$H|n\rangle = E_n|n\rangle. \tag{4.148}$$

Thus, all states connected by a rotation transformation are degenerate. These states form the basis vectors for irreducible representations (j) of the group. From the result on the dimensionality of the irreducible representations in SO(3) we conclude that there is a $(2j + 1)$ degeneracy of energy levels.

In internal symmetries the states are identified with various particles. Such symmetry transformations change particle labels but not the coordinate system, and irreducible representations of the group manifest themselves as degenerate particle multiplets. In this section we shall present a brief discussion of the early successful internal symmetries of the strong interaction. First we have the isotopic spin, or isospin, invariance of SU(2). It is later found to be part of a larger group SU(3). This is the *Eightfold Way* theory of Gell-Mann (1961) and Ne'eman (1961). This in turn led to the proposal by Gell-Mann (1964*b*) and by Zweig (1964*a*) that quarks are constituents of hadrons. Our purpose here is to give an informal historical introduction and to establish, so to speak, the kinematics of the quark model—all in preparation for the study of the dynamics, with quarks being the fundamental matter field.

Isospin invariance—SU(2) symmetry

In the early studies of nuclear reactions it was found that, to a good approximation, the nuclear forces (strong interactions) are independent of the electric charge carried by nucleons. The strong interactions are invariant under a transformation which interchanges proton (p) and neutron (n). More precisely the strong interaction has an SU(2) isospin symmetry in which the p and n states form an isospin doublet. Thus the group structure of isospin symmetry is very similar to that of the usual spin. The isospin generators T_i satisfy the Lie algebra of SU(2)

$$[T_i, T_j] = \mathrm{i}\varepsilon_{ijk}T_k \tag{4.149}$$

where the indices range from 1 to 3. That p and n form a doublet $\begin{pmatrix}\mathrm{p}\\ \mathrm{n}\end{pmatrix}$ means that (see eqn (4.45))

$$T_3|\mathrm{p}\rangle = \tfrac{1}{2}|\mathrm{p}\rangle, \qquad T_3|\mathrm{n}\rangle = -\tfrac{1}{2}|\mathrm{n}\rangle$$

and

$$T_+|\mathrm{n}\rangle = |\mathrm{p}\rangle, \qquad T_-|\mathrm{p}\rangle = |\mathrm{n}\rangle. \tag{4.150}$$

That the strong interaction does not distinguish n from p means that the strong-interaction Hamiltonian H_s has the property

$$[T_i, H_\mathrm{s}] = 0 \qquad i = 1, 2, 3. \tag{4.151}$$

The concept of isospin can be extended to other hadrons. For example, (π^+, π^0, π^-), $(\Sigma^+, \Sigma^0, \Sigma^-)$, and (ρ^+, ρ^0, ρ^-) are $(T = 1)$ isotriplets; $(\mathrm{K}^+, \mathrm{K}^0)$, $(\overline{\mathrm{K}^0}, \mathrm{K}^-)$, and (Ξ^0, Ξ^-) are $(T = 1/2)$ doublets; η, ω, ϕ, and Λ are $(T = 0)$ isosinglets.

Since different members of the isospin multiplet have different electric charges, the electromagnetic interaction clearly does not respect the isospin symmetry. Thus isospin cannot be an exact symmetry. How good a symmetry is it? If the symmetry is an exact one, we have

$$[T_i, H] = 0 \tag{4.152}$$

for the *total* Hamiltonian of the system; all members of an isomultiplet would be strictly degenerate in mass. Thus the mass differences within an isomultiplet are a good measure of the symmetry breaking. Experimentally they are typically at most a few per cent of the masses themselves, e.g.

$$\frac{m_n - m_p}{m_n + m_p} \simeq 0.7 \times 10^{-3} \qquad \frac{m_{\pi^+} - m_{\pi^0}}{m_{\pi^+} + m_{\pi^0}} \simeq 1.7 \times 10^{-2} \text{ etc.} \tag{4.153}$$

We conclude that isospin is a rather good symmetry and write the total Hamiltonian as

$$H = H_0 + H_1$$

where

$$[H_0, T_i] = 0, \qquad [H_1, T_i] \neq 0 \tag{4.154}$$

with

$$H_0 \gg H_1. \tag{4.155}$$

Thus we can treat the symmetry-breaking part (H_1) as a small perturbation. As we have mentioned, the electromagnetic interaction must belong to H_1. It turns out that weak interactions also violate isospin symmetry. One interesting question is whether the strong interaction contains a part that does not respect isospin invariance. We shall return to this question of a possible small isospin violation by the strong interaction in §5.5.

SU(3) symmetry and the quark model

When the Λ and K particles were discovered they were found to be produced copiously but to decay with a long lifetime. It was postulated that these 'new' particles possessed a new additive quantum number, *strangeness* S, which is conserved in the strong interaction (*associated production*) but is violated in the decay of these particles via the weak interaction. For example, the pions and nucleons have zero strangeness but $S(\Lambda^0) = -1$, $S(K^0) = +1$, so that we have the strangeness-conserving strong production $\pi^- + p \to \Lambda^0 + K^0$ which is followed by the strangeness-changing weak decays, $\Lambda^0 \to \pi^- + p$ and $K^0 \to \pi^+ + \pi^-$. The strangeness S, like the electric charge Q, is associated with a U(1) symmetry. In fact it was noted that there is a linear relation, the *Gell-Mann–Nishijima relation*, among S, Q, and the diagonal generator T_3 of the isospin SU(2) (Gell-Mann 1953; Nishijima and Nakano 1953),

$$Q = T_3 + \frac{Y}{2} \tag{4.156}$$

with

$$Y = B + S$$

where B is the baryon number and Y is called the *hypercharge*. Thus isospin and strangeness (or hypercharge) are only approximately conserved, but a certain linear combination, the electric charge, is preserved by all known interactions.

The search continued for 'higher symmetry' that could incorporate isospin T_i and hypercharge Y together in one group by enlarging the multiplet, i.e. to find a larger simple group which contains $SU(2)_T \times U(1)_Y$ as a subgroup. Gell-Mann (1961) and Ne'eman (1961) pointed out that we could group all mesons or baryons with the same spin and parity on the (T_3, Y) plot (Fig. 4.9), and they look very much like representations of the SU(3) group (Fig. 4.6(c), (d)). One sees that 0^-, 1^- mesons and $1/2^+$ baryons fit nicely into the octet representation ($p = q = 1$) while $3/2^+$ baryons fit the decuplet representation ($p = 3$, $q = 0$). The octet particles ψ^i_j being tensors of only two indices can be written in matrices

$$M = \begin{bmatrix} \dfrac{\pi^0}{\sqrt{2}} + \dfrac{\eta^0}{\sqrt{6}} & \pi^+ & K^+ \\ \pi^- & \dfrac{-\pi^0}{\sqrt{2}} + \dfrac{\eta^0}{\sqrt{6}} & K^0 \\ K^- & \bar{K}^0 & \dfrac{-2\eta^0}{\sqrt{6}} \end{bmatrix}$$

$$V = \begin{bmatrix} \dfrac{\rho^0}{\sqrt{2}} + \dfrac{\omega^0}{\sqrt{6}} & \rho^+ & K^{*+} \\ \rho^- & \dfrac{-\rho^0}{\sqrt{2}} + \dfrac{\omega^0}{\sqrt{6}} & K^{*0} \\ K^{*-} & \bar{K}^{*0} & \dfrac{-2\omega^0}{\sqrt{6}} \end{bmatrix} \tag{4.157}$$

$$B = \begin{bmatrix} \dfrac{\Sigma^0}{\sqrt{2}} + \dfrac{\Lambda^0}{\sqrt{6}} & \Sigma^+ & p \\ \Sigma^- & \dfrac{-\Sigma^0}{\sqrt{2}} + \dfrac{\Lambda^0}{\sqrt{6}} & n \\ \Xi^- & \Xi^0 & \dfrac{-2\Lambda^0}{\sqrt{6}} \end{bmatrix}$$

Of course at the time of the *Eightfold-Way* proposal not all the mesons and baryons predicted by this pattern were well established. The discovery of Ω^- (Barnes *et al.* 1964) at the predicted mass value and with the correct decay properties (Gell-Mann 1962*b*) played an important role in convincing a large segment of the physics community as to the correctness of this SU(3) classification scheme. Clearly this SU(3) is not as good a symmetry as the isospin SU(2). A measure of the SU(3) breaking is the mass splitting within the multiplet, e.g. $(m_\Sigma - m_N)/(m_\Sigma + m_N) \approx 0.12$.

One notable feature of the hadron spectrum in the *Eightfold-Way* scheme is that the fundamental (or defining) representation of SU(3) (Fig. 4.6(a), (b)) is not identified with any known particles. The significance of the fundamental representation in any SU(n) group is that all higher-dimensional representations can be built out of the tensor products of the fundamental representation. This property is particularly transparent in the tensor-method approach to SU(n) (see §4.3). On the other hand, more and more strongly interacting 'elementary particles' had been discovered. It is difficult to believe that all these hadrons are truly elementary and devoid of structure.

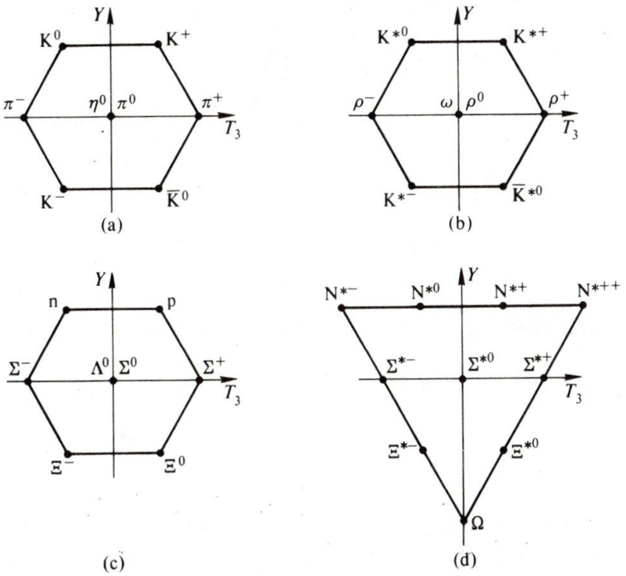

FIG. 4.9. Hadrons in SU(3) representations. Octets for (a) 0^- mesons; (b) 1^- mesons; (c) $\frac{1}{2}^+$ baryons; and decuplet for (d) $3/2^+$ baryons.

Against this background, the *quark model* was proposed, in which all hadrons are built out of spin-1/2 quarks which transform as members of the fundamental representation ($p = 1, q = 0$) of SU(3). (Clearly even if one does not believe in the physical reality of quarks, they are a useful mnemonic device for the less familiar group of SU(3).)

(1) There are three types (*flavours*) of quarks, 'up', 'down', and 'strange', in the fundamental representation, **3**

$$q_i = \begin{pmatrix} q_1 \\ q_2 \\ q_3 \end{pmatrix} = \begin{pmatrix} u \\ d \\ s \end{pmatrix} \tag{4.158}$$

corresponding to a Young tableau of $\square$. The members have quantum numbers

	Q	T	T_3	Y	S	B
u	2/3	1/2	1/2	1/3	0	1/3
d	−1/3	1/2	−1/2	1/3	0	1/3
s	−1/3	0	0	−2/3	−1	1/3

Their antiparticles, called antiquarks, are in the conjugate representation, **3***

$$q^i = \begin{pmatrix} q^1 \\ q^2 \\ q^3 \end{pmatrix} = \begin{pmatrix} \bar{u} \\ \bar{d} \\ \bar{s} \end{pmatrix} \tag{4.159}$$

corresponding to the Young tableau . Their additive quantum numbers are just the negative of those for the quarks

	Q	T	T_3	Y	S	B
$\bar{u}$	$-2/3$	$1/2$	$-1/2$	$-1/3$	0	$-1/3$
$\bar{d}$	$1/3$	$1/2$	$1/2$	$-1/3$	0	$-1/3$
$\bar{s}$	$1/3$	0	0	$2/3$	1	$-1/3$

(2) The mesons ($B = 0$) are $q\bar{q}$ bound states. From $\mathbf{3} \times \mathbf{3^*} = \mathbf{1} + \mathbf{8}$ we have mesons in SU(3) singlets and octets. For the 0^- mesons, we have

$$\pi^+ \sim \bar{d}u, \quad \pi^0 \sim (\bar{u}u - \bar{d}d)/\sqrt{2}, \quad \pi^- \sim \bar{u}d,$$
$$K^+ \sim \bar{s}u, \quad K^0 \sim \bar{s}d, \quad \overline{K^0} \sim \bar{d}s, \quad K^- \sim \bar{u}s,$$
$$\eta^0 \sim (\bar{u}u + \bar{d}d - 2\bar{s}s)/\sqrt{6}. \tag{4.160}$$

Similarly octet 1^- vector mesons have the same quark contents. The 0^- meson η' and the 1^- meson 'ϕ' (more on this later) can be identified with the SU(3) singlet $q^i q_i = (\bar{u}u + \bar{d}d + \bar{s}s)$.

(3) The baryons ($B = 1$) are qqq bound states. From the multiplications

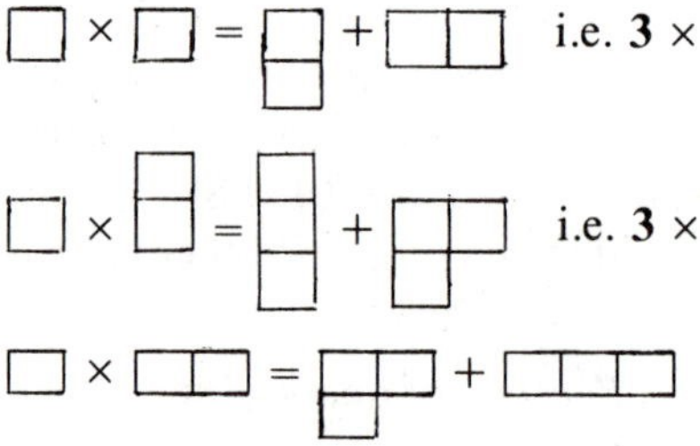

i.e. $\mathbf{3} \times \mathbf{3} = \mathbf{3^*} + \mathbf{6}$

i.e. $\mathbf{3} \times \mathbf{3^*} = \mathbf{1} + \mathbf{8}$

i.e. $\mathbf{3} \times \mathbf{6} = \mathbf{8} + \mathbf{10}$,

we have

$$\mathbf{3} \times \mathbf{3} \times \mathbf{3} = \mathbf{1} + \mathbf{8} + \mathbf{8} + \mathbf{10}. \tag{4.161}$$

The octet parts have the same quantum numbers (T_3, Y) as the octet mesons, even though they have different quark contents, because T_3 and Y (also total isospin T) are generators of the SU(3) group and their eigenvalues for a given representation are uniquely defined. Meson octet states and baryon octet states will have a different baryon number; B is not a generator of SU(3). Specially, for the $1/2^+$ baryons 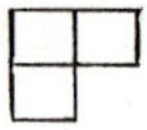

$$p \sim udu, \quad n \sim udd,$$
$$\Sigma^+ \sim suu, \quad \Sigma^0 \sim s(ud + du)/\sqrt{2}, \quad \Sigma^- \sim sdd,$$

$$\Xi^0 \sim \text{ssu}, \quad \Xi^- \sim \text{ssd},$$

$$\Lambda^0 \sim \text{s(ud} - \text{du)}/\sqrt{2} \tag{4.162}$$

and for the $3/2^+$ baryons $\Box\Box\Box$

$$N^{*++} \sim \text{uuu}, \quad N^{*+} \sim \text{uud}, \quad N^{*0} \sim \text{udd}, \quad N^{*-} \sim \text{ddd}$$

$$\Sigma^{*+} \sim \text{suu}, \quad \Sigma^{*0} \sim \text{sud}, \quad \Sigma^{*-} \sim \text{sdd},$$

$$\Xi^{*0} \sim \text{ssu}, \quad \Xi^{*-} \sim \text{ssd},$$

$$\Omega^- \sim \text{sss}. \tag{4.163}$$

The Gell-Mann–Okubo mass formula

Here we study the hadron spectrum in the presence of the SU(3)-breaking $H_1 \ll H_0$ similar to (4.154). The isospin, as mentioned above, seems to be a good symmetry, hence the mass difference in the isospin multiplet can be neglected in discussing the SU(3) breaking. We can proceed in a pure group-theoretical manner. With an assumption about the SU(3) transformation property of H_1 ($\sim Y$, or equivalently $\sim F_8$), the relation among masses of isospin multiplets in a given SU(3) representation can be derived (Gell-Mann 1961; Okubo 1962). Here we will demonstrate this with a simple calculation in the quark model: we assume that the binding energies of quarks are independent of quark flavours (this can be justified later) and that the mass differences in an SU(3) representation are entirely due to the quark mass difference. This is a specific realization of the $H_1 \sim F_8$ assumption. In the approximation of exact SU(2) isospin symmetry, we have $m_u = m_d$. First consider the 0^- meson masses. In terms of the quark masses we have from (4.160)

$$m_\pi^2 = m_0 + 2m_u$$

$$m_K^2 = m_0 + m_u + m_s$$

$$m_\eta^2 = m_0 + \tfrac{2}{3}(m_u + 2m_s) \tag{4.164}$$

where m_0 is the flavour-independent common mass. We have used the quadratic mass for mesons. The principal reason is that this works better (than linear masses). A possible justification is that 0^- meson masses vanish in the SU(3) symmetry limit (see Chapter 5) and perturbation of the energy around such a value automatically leads to a relation among quadratic masses. From (4.164) we obtain

$$4m_K^2 = m_\pi^2 + 3m_\eta^2. \tag{4.165}$$

Experimentally the left-hand side $\simeq 0.98\ \text{GeV}^2$, and the right-hand side $\simeq 0.92\ \text{GeV}^2$. Thus this mass relation is good to a few per cent. Similarly for the $1/2^+$ baryons from (4.162),

$$m_N = m_0 + 3m_u$$

$$m_\Sigma = m_0 + 2m_u + m_s$$

$$m_\Xi = m_0 + m_u + 2m_s$$

$$m_\Lambda = m_0 + 2m_u + m_s.$$

Eliminating the three parameters m_0, m_u, and m_s, we have one relation among the baryon masses

$$\frac{m_\Sigma + 3m_\Lambda}{2} = m_N + m_\Xi \tag{4.166}$$

which is experimentally very well satisfied. We have the left-hand side $\simeq 2.23$ GeV and the right-hand side $\simeq 2.25$ GeV. The quark model allows us to identify the particular mass shifts from the SU(3) average value m_0. This yields an additional relation

$$m_\Lambda = m_\Sigma. \tag{4.167}$$

For the $3/2^+$ baryon decuplet (4.163) we easily derive the equal-spacing rule

$$m_\Omega - m_{\Xi^*} = m_{\Xi^*} - m_{\Sigma^*} = m_{\Sigma^*} - m_{N^*}. \tag{4.168}$$

In fact the mass of Ω^- was first (correctly) predicted from this rule.

ω–ϕ mixings

The Gell-Mann–Okubo mass formula for the 1^- vector meson multiplet does not seem to work. In analogy to the 0^- mesons (4.165), we would have

$$3m_\omega^2 = 4m_{K^*}^2 - m_\rho^2. \tag{4.169}$$

With $m_{K^*} = 890$ MeV and $m_\rho = 770$ MeV, this equation would predict $m_\omega = 926.5$ MeV, while the experimental value is $m_\omega = 783$ MeV. It turns out that there is another 1^- vector meson ϕ with the same quantum numbers as ω (i.e. $T = 0$ and $S = 0$) and it has a mass of 1020 MeV. This leads to the idea that ω is not a pure SU(3) octet state but has a mixture of an SU(3) singlet state (Sakurai 1962). Let V_8 be the $T = Y = 0$ member of the SU(3) octet and V_1 be the SU(3) singlet state, then ω is a linear combination of V_8 and V_1 while ϕ is the other orthogonal combination. More precisely, write the mass matrix in V_8 and V_1 space as

$$M = \begin{pmatrix} m_{88}^2 & m_{81}^2 \\ m_{18}^2 & m_{11}^2 \end{pmatrix}. \tag{4.170}$$

The wavefunctions for V_8 and V_1 are

$$V_8 = (\bar{u}u + \bar{d}d - 2\bar{s}s)/\sqrt{6}$$

$$V_1 = (\bar{u}u + \bar{d}d + \bar{s}s)/\sqrt{3}. \tag{4.171}$$

The eigenvalues of M should be m_ω and m_ϕ

$$R^T M R = \begin{pmatrix} m_\omega^2 & 0 \\ 0 & m_\phi^2 \end{pmatrix} \tag{4.172}$$

where

$$R = \begin{pmatrix} \cos\theta & \sin\theta \\ -\sin\theta & \cos\theta \end{pmatrix}. \tag{4.173}$$

Hence

$$\omega = \cos\theta\, V_8 - \sin\theta\, V_1$$

$$\phi = \sin\theta\, V_8 + \cos\theta\, V_1. \tag{4.174}$$

Since (4.169) predicts that m_{88} is 926.5 MeV, we can calculate the mixing angle. From (4.172),

$$M = R\begin{pmatrix} m_\omega^2 & 0 \\ 0 & m_\phi^2 \end{pmatrix} R^T \tag{4.175}$$

we have

$$\sin\theta = [(m_{88}^2 - m_\omega^2)/(m_\phi^2 - m_\omega^2)]^{1/2}$$
$$= 0.76. \tag{4.176}$$

The fact that $\sin\theta$ is very close to $\sqrt{(2/3)} = 0.81$ has the following significance. If $\sin\theta$ is exactly $\sqrt{(2/3)}$, called '*ideal mixing*', we have

$$\omega = (\bar{u}u + \bar{d}d)/\sqrt{2} \tag{4.177}$$

$$\phi = \bar{s}s. \tag{4.178}$$

Namely, for ideal mixing, ϕ is completely built out of the strange quarks and ω out of the non-strange quarks. Thus with the value of θ in (4.176) it is clear that ϕ is predominantly $\bar{s}s$ and has very few $\bar{u}u$ and $\bar{d}d$ components.

The Zweig rule and the discovery of 'charm'

As ω and ϕ have the same quantum numbers $T = 0$, $Y = 0$, one would expect that they should have very similar strong-interaction properties. In particular their strong decay widths should be comparable. Experimentally this is not so. ω decays predominantly, as it should, into the 3π channel, while $\phi \to 3\pi$ is suppressed relative to $\phi \to \bar{K}K$ even though the phase space for the latter decay is very small (m_ϕ is barely above $2m_K \simeq 998$ MeV). This indicates a strong preference for ϕ to decay into channels involving strange particles rather than into channels without strange particles. To explain this, Zweig (1964*b*) and also others independently (Okubo 1963; Iizuka 1966) suggested that strong processes in which the final states can only be reached through q and $\bar{q}$ annihilations are suppressed. Thus, since ϕ is a predominantly $\bar{s}s$ state, the decay into pions must proceed through the annihilation diagram of Fig. 4.10(a) while the decay into the $\bar{K}K$ channel involves no annihilation of s and $\bar{s}$, as shown in Fig. 4.10(b).

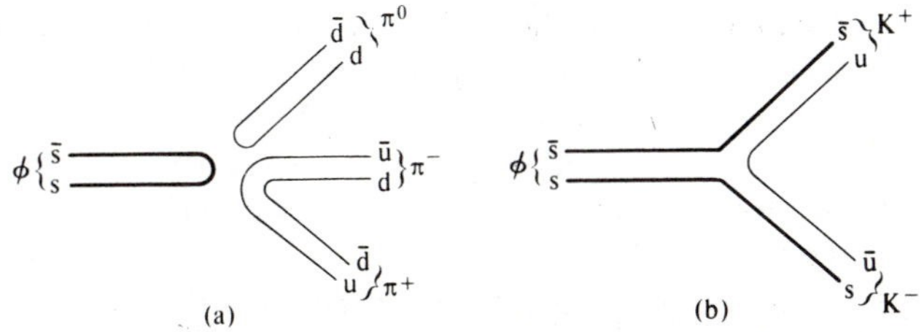

FIG. 4.10. ϕ decays: (a) disallowed; (b) allowed by the Zweig rule.

In 1974 the $\psi/J(3100)$ particle was dicovered (Aubert *et al.* 1974; Augustin *et al.* 1974). It has the rather unusual property of having a width (only about 70 KeV) much narrower than the widths of typical hadrons (e.g. $\Gamma_\rho \sim 150$ MeV, $\Gamma_\omega \sim 10$ MeV). Also width should increase with mass as there will be more decay phase space. The interpretation is that ψ/J is a bound state of a new heavy quark, the *charmed quark* c and its antiparticle $\bar{c}$, i.e. $\psi \sim \bar{c}c$. It is below the threshold for Zweig-rule allowed decays into charmed mesons (bound states involving at least one charmed quark), see Fig. 4.11.

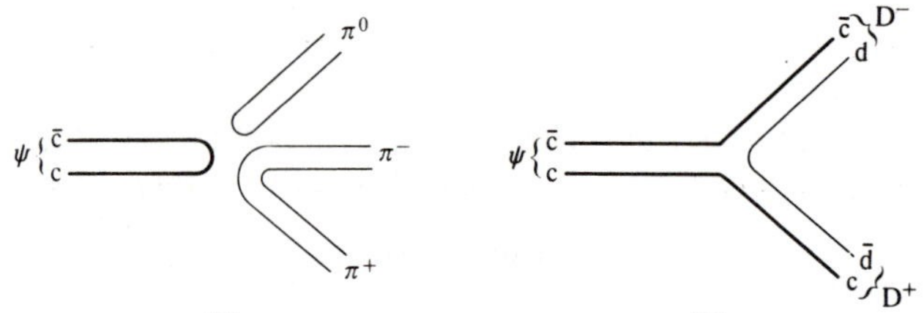

FIG. 4.11. Zweig-rule allowed decays such as (b) $\psi \to D^+D^-$ are forbidden by phase space.

For a detailed discussion on charmed particles see Gaillard, Lee, and Rosner (1975). Here we only note that this new 2/3-charged quark was predicted earlier on the basis of lepton–quark symmetry (Bjorken and Glashow 1964) and, more compellingly, on the basis of the requirement to suppress strangeness-changing neutral-current effects (Glashow, Iliopoulos and Maiani 1970). With this new quantum number the flavour symmetry is enlarged from SU(3) to SU(4). Of course SU(4) is badly broken as m_c is heavy ($\simeq 1.5$ GeV). Consideration of such badly broken symmetry is no longer particularly meaningful. One should go directly to the dynamical consideration of quark models of such hadrons carrying new quantum numbers. For example, we have the additional 0^- meson states

$$\begin{aligned} &D^+ \sim \bar{d}c, \quad D^- \sim \bar{c}d, \quad D^0 \sim \bar{u}c, \quad \overline{D^0} \sim \bar{c}u, \\ &F^+ \sim \bar{s}c, \quad F^- \sim \bar{c}s, \quad \eta_c \sim \bar{c}c. \end{aligned} \tag{4.179}$$

In fact one of the most convincing bits of evidence for the quark model is the detailed verification of level structure and transitions among the various ($\bar{c}c$) 'quarkonium' states.

In 1977 yet another set of narrow-resonance Υs were discovered (Herb *et al.* 1977; Lederman 1978) and they were successfully interpreted as bound states of yet another heavy quark, b (for 'beauty' or 'bottom') carrying charge $-1/3$ with a very large mass $m_b \simeq 5$ GeV.

As we shall see (especially §11.3), from the pattern of fermion family replication in the standard electroweak theory one anticipates at least one more superheavy flavour of quark: this quark, t (for 'truth' or 'top'), should carry charge 2/3. It is to be associated with the b-quark in the same way as pairing of (u, d) and (c, s).

The pioneering paper on heavy quarkonium is Appelquist and Politzer (1975). (For more recent discussions, including ($\bar{b}b$) quarkonium, and reviews see, for example, Appelquist, Barnett, and Lane 1978; Quigg and Rosner 1979; Eichten *et al.* 1980; Shifman 1981.)

Paradoxes of the simple quark model

By simple quark model we mean the model of three, or more, types of quarks as originally invented with no hidden degrees of freedom. This simple model has the following difficulties.

(1) The quarks have fractional electric charges while all the observed hadrons have integer charges. With charge conservation, this implies that at least one of the quarks is absolutely stable. The fractionally charged stable quark has been searched for and so far there is no generally accepted positive evidence for its detection (see however LaRue, Fairbank and Hebard 1977).

(2) Hadrons are seen to be built exclusively out of $\bar{q}q$ and qqq states (and their conjugates). There is no evidence for qq and qqqq bound states. It is difficult to understand the absence of such hadron states with masses comparable to the observed particles.

(3) The most serious problem is that the $J^P = 3/2^+$ decuplet baryon wavefunctions seem to violate the connection between spin and statistics. Take the example of $N^{*++} \sim$ uuu. Since it is a ground state for the system of three u-quarks, the spatial wavefunction has zero total angular momentum and is totally symmetric. But N^{*++} has spin-3/2 and the spins of all u quarks must be lined up in the same direction for the N^{*++} wavefunctions (with the third component of spin $s_3 = +3/2$) so the spin wavefunction is also totally symmetric. Consequently the overall wavefunction is totally symmetric with respect to the interchange of any pairs of constituent quarks. This violates Fermi–Dirac statistics since the u-quark is a spin-1/2 fermion.

Colour degree of freedom

The way out of all these difficulties graduately emerges (Greenberg 1964; Han and Nambu 1965; Nambu 1966). It is to postulate that each quark has a hidden degree of freedom, called *colour*. More specifically each type of quark is assumed to come in three different colours which form a triplet under a colour SU(3) group. Thus for the known quarks we have

$$
\text{flavour}\updownarrow\quad
\begin{aligned}
u_\alpha &= (u_1, u_2, u_3)\\
d_\alpha &= (d_1, d_2, d_3)\\
s_\alpha &= (s_1, s_2, s_3)\\
c_\alpha &= (c_1, c_2, c_3)\\
b_\alpha &= (b_1, b_2, b_3)\\
&\longleftarrow \text{colour} \longrightarrow .
\end{aligned}
\tag{4.180}
$$

The five types (flavours) of quarks correspond to five distinct colour triplets

The colour group operators change the quark from one colour to another but leave the flavour unchanged: $u_1 \leftrightarrow u_2$, $u_2 \leftrightarrow u_3$, $u_3 \leftrightarrow u_1$, or $d_1 \leftrightarrow d_2$, $d_2 \leftrightarrow d_3$, $d_3 \leftrightarrow d_1$, etc. Along with this supposition of an extra degree of freedom, it is further postulated that only colour singlet states are physically observable states.

Since there are colour singlets in the product of $\mathbf{3} \times \mathbf{3}^*$ and $\mathbf{3} \times \mathbf{3} \times \mathbf{3}$, see (4.161), only $q\bar{q}$ and qqq configurations can bind into physically observable hadrons while q, qq, or qqqq states cannot be seen experimentally. The N^{*++} wavefunction is now antisymmetric

$$N^{*++} \sim u_\alpha(x_1)u_\beta(x_2)u_\gamma(x_3)\varepsilon^{\alpha\beta\gamma}. \tag{4.181}$$

where α, β, γ are colour indices 1, 2, 3.

It cannot be overemphasized that this colour SU(3) has nothing to do with the original flavour SU(3) of the Eightfold Way. In fact, unlike all the flavour SU(n) symmetries with $n = 2, 3, 4, \ldots,$ the colour SU(3) symmetry is assumed to be exact.

We now have a very peculiar situation where hadrons are composed of particles which cannot themselves be directly observed. Quarks can exist only inside hadrons and can never be free. This property is usually referred to as *quark confinement*. It should be a part of any viable theory of hadrons as quark bound states.

In recent years physicists have converged on a gauge theory called quantum chromodynamics (QCD) in which the colour quantum number plays a similar role to that of the electric charge in QED. In QCD the coloured quarks will interact with each other through the exchange of the gluons in a manner analogous to the exchange of the photon between charged particles. These interactions are responsible for the (colour-dependent and flavour-independent) binding of quarks into hadrons. Even though QCD has many attractive features (see Chapter 10), quark confinement has not up to now been derived from QCD in a convincing way. But there are many arguments (see for example §10.5), which indicate that it should be a property of QCD.

Chiral symmetry of the strong interaction

In §4.4 we studied the flavour symmetry of the strong interaction and its physical realization in terms of the quark model. The symmetries SU(2), SU(3), etc. are supposed to be manifestations of the quark mass degeneracies $m_u = m_d$ and, to a less good approximation, $m_u = m_d = m_s$. As it turns out, the reason we have such close equalities is not so much that the three quarks happen to have equal masses but that they all are light on the typical strong-interaction energy scale. Thus the symmetry limits should really be $m_u = m_d = 0$, and also to a lesser extent $m_s = 0$, with the corresponding flavour symmetries being $SU(2)_L \times SU(2)_R$ and $SU(3)_L \times SU(3)_R$, the *chiral symmetries*. However, we do not see any particle degeneracy patterns ascribable to such symmetries. The resolution of this paradox lies in the fact that the physical vacuum is not invariant (not a singlet) under these chiral symmetries and, we say, the symmetry is spontaneously broken. The physical manifestation of such symmetry-realization is the presence of a set of near-massless bosons: the three pions, and also the whole octet of 0^- mesons. Here we present the basics of an approach commonly referred to as *current algebra*. The matrix elements of these light pseudoscalar mesons in certain kinematic limits are calculated by a direct application of the commutation and conservation relations of the chiral symmetry currents.

This chapter is organized as follows. In §5.1 we discuss the relation between symmetry and the conservation laws in field theory, and also establish the important result that charge commutation relations are valid even in the presence of symmetry-breaking terms. In §5.2 we emphasize the point that, in field theory, symmetry currents are actually the physical (electromagnetic and weak) currents. Adler's test of the current algebra (of chiral symmetry) in high-energy neutrino scattering is presented. In §5.3 we study spontaneous breakdown of global symmetries and the Goldstone theorem. This introduces us to the subject of partially conserved axial-vector current (PCAC) in §5.4. In the chiral symmetric limit of massless pions we can use PCAC and current algebra to derive a number of low-energy theorems: the Goldberger–Treiman relation, Adler's consistency conditions on the πN scattering amplitude, the Adler–Weisberger sum rules, etc. In §5.5 we study the pattern of (explicit) chiral symmetry breaking as revealed by the pseudoscalar meson masses and the πN σ-term.

The discussion of this chapter will show that the hadronic interactions obey an approximate chiral symmetry, which is realized in the Goldstone mode. Thus any satisfactory theory of the strong interaction must have these flavour-symmetry properties (also confer comments at the end of §5.5). As we shall see, the gauge theory QCD naturally displays such global symmetries.

5.1 Global symmetries in field theory and current commutators

Conservation laws in physics can be attributed to symmetry principles. The invariance of the physical system under certain symmetry transformations implies an appropriate set of conservation laws. In classical physics we have the familiar examples

Translational invariance in time		Energy conservation
$t \to t + a$	$\leftrightarrow$	$\dfrac{dE}{dt} = 0$
Translational invariance in space		Momentum conservation
$r_i \to r_i + b_i$	$\leftrightarrow$	$\dfrac{dp^i}{dt} = 0$
Rotational invariance		Angular momentum conservation
$r_i \to R_{ij}r_j$, with $RR^T = 1$	$\leftrightarrow$	$\dfrac{dJ^i}{dt} = 0.$

In quantum mechanics, observables are associated with operators. Their time evolution in the Heisenberg picture is governed by the commutator with the Hamiltonian

$$\frac{d\mathcal{O}}{dt} = i[H, \mathcal{O}]. \tag{5.1}$$

The conservation law is then equivalent to the statement that the corresponding operator commutes with the Hamiltonian. For example, the angular-momentum conservation $dJ_i/dt = 0$ means that

$$[J_i, H] = 0. \tag{5.2}$$

It follows that the energy levels of the system have a $(2j + 1)$-fold degeneracy, j being the angular momentum eigenvalue. In group theory language the Hamiltonian operator is invariant under the rotation group O(3), which has the generators J_1, J_2, and J_3 satisfying the commutation relation

$$[J_i, J_j] = i\varepsilon_{ijk}J_k \qquad i, j, k = 1, 2, 3. \tag{5.3}$$

The states with definite energy eigenvalues then form representations of the group O(3). The degeneracy of the energy levels is associated with the dimensionalities of the irreducible representations.

Noether's theorem

In field theory, symmetries and conservation laws are related in a similar manner. This connection is made precise by the *Noether theorem* (Noether 1918). For a system described by the Lagrangian

$$L = \int d^3x \mathscr{L}(\phi_i(x), \partial_\mu \phi_i(x)) \tag{5.4}$$

with the equation of motion

$$\partial_\mu \frac{\delta \mathscr{L}}{\delta(\partial_\mu \phi_i)} - \frac{\delta \mathscr{L}}{\delta \phi_i} = 0 \tag{5.5}$$

any continuous symmetry transformation which leaves the action $S = \int L\,\mathrm{d}t$ invariant implies the existence of a conserved current

$$\partial^\mu J_\mu(x) = 0 \tag{5.6}$$

with the charge defined by

$$Q(t) = \int \mathrm{d}^3 x J_0(x) \tag{5.7}$$

which is a constant of motion

$$\frac{\mathrm{d}Q}{\mathrm{d}t} = 0 \tag{5.8}$$

because the surface term at infinity being negligibly small

$$\int \mathrm{d}^3 x\, \partial_0 J_0 = \int \mathrm{d}^3 x \partial^\mu J_\mu = 0. \tag{5.9}$$

Noether's theorem can be illustrated easily in the case of internal symmetry. The Lagrangian density $\mathscr{L}$ is invariant under some symmetry group G, i.e. under the infinitesimal transformation

$$\phi_i(x) \to \phi_i'(x) = \phi_i(x) + \delta\phi_i(x) \tag{5.10}$$

where

$$\delta\phi_i(x) = \mathrm{i}\varepsilon^a t^a_{ij} \phi_j(x)$$

ε^as are (x-independent) small parameters and the t^as are a set of matrices satisfying the Lie algebra of the group G

$$[t^a, t^b] = \mathrm{i}C^{abc} t^c \tag{5.11}$$

where the C^{abc}s are the structure constants of the group G. We have the corresponding change in the Lagrangian density

$$\delta\mathscr{L} = \frac{\delta \mathscr{L}}{\delta \phi_i} \delta\phi_i + \frac{\delta \mathscr{L}}{\delta(\partial_\mu \phi_i)} \delta(\partial_\mu \phi_i). \tag{5.12}$$

Using (5.5) and the fact that

$$\delta(\partial^\mu \phi_i) \equiv \partial_\mu \phi_i' - \partial_\mu \phi_i = \partial_\mu(\delta\phi_i), \tag{5.13}$$

we can write $\delta\mathscr{L}$ as

$$\begin{aligned} \delta\mathscr{L} &= \partial_\mu \frac{\delta \mathscr{L}}{\delta(\partial_\mu \phi_i)} \delta\phi_i + \frac{\delta \mathscr{L}}{\delta(\partial_\mu \phi_i)} \partial_\mu(\delta\phi_i) \\ &= \partial_\mu \left[\frac{\delta \mathscr{L}}{\delta(\partial_\mu \phi_i)} \delta\phi_i \right] \\ &= \varepsilon^a\, \partial_\mu \left[\frac{\delta \mathscr{L}}{\delta(\partial_\mu \phi_i)} \mathrm{i} t^a_{ij} \phi_j \right]. \end{aligned} \tag{5.14}$$

Clearly if the Lagrangian is invariant under the transformation (5.10), i.e. $\delta\mathscr{L} = 0$, eqn (5.14) implies a conserved current

$$\partial^\mu J^a_\mu = 0$$

with

$$J^a_\mu = -\mathrm{i}\frac{\delta\mathscr{L}}{\delta(\partial^\mu\phi_i)}t^a_{ij}\phi_j. \tag{5.15}$$

The conserved charges given by

$$Q^a = \int \mathrm{d}^3x J^a_0(x) \tag{5.16}$$

are the generators of the symmetry group.

These types of symmetries which are characterized by the space–time-independent parameters ε^a in (5.10) are called *global symmetries*. The fields $\phi_i(x)$ are transformed in exactly the same way for all space–time points x.

Example 1. *Abelian* $U(1)$ *symmetry*. The Lagrange density given by

$$\mathscr{L} = \tfrac{1}{2}[(\partial_\mu\phi_1)^2 + (\partial_\mu\phi_2)^2] - \tfrac{1}{2}\mu^2(\phi_1^2 + \phi_2^2) - \tfrac{1}{4}\lambda(\phi_1^2 + \phi_2^2)^2 \tag{5.17}$$

is invariant under the transformation

$$\begin{aligned}\phi_1 \to \phi_1' &= \phi_1 \cos\alpha - \phi_2 \sin\alpha \\ \phi_2 \to \phi_2' &= \phi_1 \sin\alpha + \phi_2 \cos\alpha.\end{aligned} \tag{5.18}$$

It is the O(2) symmetry corresponding to the invariance under rotations in the (ϕ_1, ϕ_2) plane. For infinitesimal transformation, $\alpha \ll 1$,

$$\begin{aligned}\phi_1' &= \phi_1 - \alpha\phi_2 \\ \phi_2' &= \phi_2 + \alpha\phi_1,\end{aligned} \tag{5.19}$$

i.e. $\phi_i' = \phi_i + \mathrm{i}\alpha t_{ij}\phi_j$ with

$$t = \begin{pmatrix} 0 & \mathrm{i} \\ -\mathrm{i} & 0 \end{pmatrix}. \tag{5.20}$$

According to (5.15), the conserved current is

$$J_\mu = -(\partial_\mu\phi_1)\phi_2 + (\partial_\mu\phi_2)\phi_1. \tag{5.21}$$

In terms of the complex fields defined as

$$\begin{aligned}\phi &= \frac{1}{\sqrt{2}}(\phi_1 + \mathrm{i}\phi_2) \\ \phi^* &= \frac{1}{\sqrt{2}}(\phi_1 - \mathrm{i}\phi_2),\end{aligned} \tag{5.22}$$

the Langrangian of (5.17) may be written

$$\mathscr{L} = (\partial_\mu\phi^*)(\partial^\mu\phi) - \mu^2(\phi^*\phi) - \lambda(\phi^*\phi)^2 \tag{5.23}$$

which is invariant under the U(1) transformation

$$\phi \to \phi' = e^{i\alpha}\phi \tag{5.24}$$

giving rise to the conserved current

$$J_\mu = i[(\partial_\mu\phi^*)\phi - (\partial_\mu\phi)\phi^*]. \tag{5.25}$$

We note that ϕ_1 and ϕ_2 (or ϕ and ϕ^*) are degenerate in mass because of the O(2), or U(1), symmetry.

Example 2. *Isospin symmetry SU*(2). We now consider the simple example of non-Abelian symmetry. Let ϕ be an isodoublet

$$\phi = \begin{pmatrix}\phi_1\\ \phi_2\end{pmatrix}. \tag{5.26}$$

The Lagrange density given by

$$\mathscr{L} = (\partial_\mu\phi^\dagger)(\partial^\mu\phi) - \mu^2(\phi^\dagger\phi) - \frac{\lambda}{2}(\phi^\dagger\phi)^2 \tag{5.27}$$

is invariant under the infinitesimal isospin rotation

$$\phi_i \to \phi_i' = \phi_i + i\varepsilon^a \frac{\tau^a_{ij}}{2}\phi_j \tag{5.28}$$

where the τ^as are the standard Pauli matrices. The conserved isospin current is given by

$$J^a_\mu = \frac{-i}{2}(\partial_\mu\phi_i^\dagger\tau^a_{ij}\phi_j - \phi_i^\dagger\tau^a_{ij}\partial_\mu\phi_j). \tag{5.29}$$

The time-components J^a_0 have a simple form

$$\begin{aligned} J^a_0 &= \frac{-i}{2}(\partial_0\phi_i^\dagger\tau^a_{ij}\phi_j - \phi_i^\dagger\tau^a_{ij}\partial_0\phi_j)\\ &= \frac{-i}{2}[\pi_i\tau^a_{ij}\phi_j - \phi_i^\dagger\tau^a_{ij}\pi_j^\dagger] \end{aligned} \tag{5.30}$$

where π_i is the canonical momentum conjugate to ϕ_i. Using the canonical commutators

$$[\pi_i(\mathbf{x}, t), \phi_j(\mathbf{x}', t)] = -i\delta_{ij}\delta^3(\mathbf{x} - \mathbf{x}'), \tag{5.31}$$

we can show (see eqn (5.37) below) that the charges defined by

$$Q^a = \int d^3x J^a_0(x) \qquad a = 1, 2, 3$$

satisfy the commutation relations of SU(2) symmetry

$$[Q_a, Q_b] = i\varepsilon_{abc}Q_c. \tag{5.32}$$

This means that the Q_as are the generators of the SU(2) symmetry. Again, the ϕ_1, ϕ_2 fields have the same mass because of SU(2) symmetry.

Current algebra

In the above we see that the exact symmetries lead to conserved current and their charges generate the group algebra of symmetry transformations. These simple commutation relations are useful for classifying particle states. For instance, the SU(2) transformations in Example 2 give rise to Q^a, $a = 1, 2, 3$ which are generators of isospin SU(2) symmetry (5.32). As we have discussed in §§4.2 and 4.4, they transform a particle into others within a given isospin multiplet. We shall demonstrate below that commutation relations such as (5.32) will hold even in the presence of symmetry-breaking terms. Consider the Lagrangian

$$\mathscr{L} = \mathscr{L}_0 + \mathscr{L}_1 \tag{5.33}$$

where $\mathscr{L}_0$ is invariant under the symmetry group G while $\mathscr{L}_1$ is not. Under the infinitesimal transformation (5.10) we can still define the current J^a_μ as in (5.15) but it will no longer be conserved and the charge defined as

$$Q^a(t) = \int J^a_0(x)\,\mathrm{d}^3x = -\mathrm{i}\int \frac{\delta\mathscr{L}}{\delta(\partial^0\phi_i)} t^a_{ij}\phi_j\,\mathrm{d}^3x \tag{5.34}$$

will not be time-independent. However the factor $\delta\mathscr{L}/\delta(\partial^0\phi_i)$ is still the canonical momentum conjugate to ϕ_i even in the presence of $\mathscr{L}_1$

$$\pi_i(x) = \frac{\delta\mathscr{L}}{\delta(\partial^0\phi_i)} \tag{5.35}$$

and satisfies the canonical commutation relation at equal time

$$[\pi_i(\mathbf{x}, t), \phi_j(\mathbf{y}, t)] = -\mathrm{i}\delta^3(\mathbf{x} - \mathbf{y})\,\delta_{ij}. \tag{5.36}$$

From this we can, without knowing the explicit form of the symmetry-breaking term $\mathscr{L}_1$, calculate the (equal-time) commutator of the charges, by using the identity $[\mathrm{AB}, \mathrm{CD}] = \mathrm{A}[\mathrm{B}, \mathrm{C}]\mathrm{D} - \mathrm{C}[\mathrm{D}, \mathrm{A}]\mathrm{B}$ when $[\mathrm{A}, \mathrm{C}] = [\mathrm{D}, \mathrm{B}] = 0$

$$\begin{aligned}
[Q^a(t), Q^b(t)] &= -\int \mathrm{d}^3x\,\mathrm{d}^3y[\pi_i(\mathbf{x}, t)t^a_{ij}\phi_j(\mathbf{x}, t), \pi_k(\mathbf{y}, t)t^b_{kl}\phi_l(\mathbf{y}, t)] \\
&= -\int \mathrm{d}^3x\,\mathrm{d}^3y(\pi_i(\mathbf{x}, t)t^a_{ij}[\phi_j(\mathbf{x}, t), \pi_k(\mathbf{y}, t)]t^b_{kl}\phi_l(\mathbf{y}, t) \\
&\quad + \pi_k(\mathbf{y}, t)t^b_{kl}[\pi_i(\mathbf{x}, t), \phi_l(\mathbf{y}, t)]t^a_{ij}\phi_j(\mathbf{x}, t)) \\
&= -\int \mathrm{d}^3x(\pi_k(\mathbf{x}, t)\mathrm{i}[t^a, t^b]_{kj}\phi_j(\mathbf{x}, t)).
\end{aligned}$$

Or, using (5.11), we have

$$[Q^a(t), Q^b(t)] = \mathrm{i}C^{abc}Q^c(t). \tag{5.37}$$

Thus even though the $Q^a(t)$s change with time, at any given instant t, the commutation relations of the group algebra will still be satisfied. These relations are usually referred to as *charge algebra*.

Example. *Broken symmetries of the free SU(3) quark model.* Here we are concerned with the flavour SU(3) group (not colour!). The quark fields are in the triplet representation

$$q(x) = \begin{pmatrix} q_1(x) \\ q_2(x) \\ q_3(x) \end{pmatrix} = \begin{pmatrix} u(x) \\ d(x) \\ s(x) \end{pmatrix} \tag{5.38}$$

with the transformation properties

$$q_i \to q_i' = q_i + i\alpha^a(\lambda^a/2)_{ij}q_j, \qquad \alpha^a \ll 1 \tag{5.39}$$

where the λ^as are the eight Gell-Mann matrices

$$\left[\frac{\lambda^a}{2}, \frac{\lambda^b}{2}\right] = if^{abc}\frac{\lambda^c}{2}. \tag{5.40}$$

The f^{abc}s are the SU(3) structure constants. We have the Lagrangian

$$\mathscr{L} = \mathscr{L}_0 + \mathscr{L}_1 \tag{5.41}$$

with

$$\mathscr{L}_0 = i\bar{q}\gamma^\mu\partial_\mu q \tag{5.42}$$

and

$$\mathscr{L}_1 = m_u\bar{u}u + m_d\bar{d}d + m_s\bar{s}s. \tag{5.43}$$

$\mathscr{L}_0$ is SU(3)-invariant while $\mathscr{L}_1$ is not. The currents associated with the SU(3) transformation are given by

$$V_\mu^a(x) = \bar{q}(x)\gamma_\mu(\lambda^a/2)q(x). \tag{5.44}$$

The charges defined by

$$Q^a(t) = \int V_0^a(x)\,d^3x$$

will satisfy the SU(3) algebra

$$[Q^a(t), Q^b(t)] = if^{abc}Q^c(t) \tag{5.45}$$

as a consequence of the canonical commutation relation

$$\{q_{\alpha i}(\mathbf{x}, t), q_{\beta j}^\dagger(\mathbf{y}, t)\} = \delta_{ij}\delta_{\alpha\beta}\delta^3(\mathbf{x} - \mathbf{y}) \tag{5.46}$$

where i, j are the flavour indices and α, β are the Dirac indices. To have exact SU(3) symmetry we actually need $m_u = m_d = m_s$. In the $\mathscr{L}_1 = 0$ limit, $\mathscr{L}_0$ is invariant under transformations of a group larger than SU(3). Besides the transformation of (5.39), $\mathscr{L}_0$ is also unchanged under the axial transformation

$$q_i \to q_i' = q_i + i\beta^a(\lambda^a/2)_{ij}\gamma_5 q_j, \qquad \beta^a \ll 1. \tag{5.47}$$

The corresponding currents are given by

$$A_\mu^a(x) = \bar{q}(x)(\lambda^a/2)\gamma_\mu\gamma_5 q(x) \tag{5.48}$$

which are axial-vector currents. Thus even in the presence of the symmetry-breaking term $\mathscr{L}_1$, we can define the axial charge Q^{5a}

$$Q^{5a}(t) = \int A_0^a(x)\, \mathrm{d}^3x. \tag{5.49}$$

Together with the vector charge $Q^a(x)$, the axial charges generate the following equal-time commutation relations

$$\begin{aligned} [Q^a(t), Q^b(t)] &= \mathrm{i}f^{abc}Q^c(t) \\ [Q^a(t), Q^{5b}(t)] &= \mathrm{i}f^{abc}Q^{5c}(t) \\ [Q^{5a}(t), Q^{5b}(t)] &= \mathrm{i}f^{abc}Q^c(t). \end{aligned} \tag{5.50}$$

These commutation relations correspond to the *chiral* $\mathrm{SU(3)_L \times SU(3)_R}$ *algebra*. To see this, we form left-handed and right-handed charges defined by

$$Q_{\mathrm{L}}^a = \tfrac{1}{2}(Q^a - Q^{5a}) \tag{5.51a}$$

$$Q_{\mathrm{R}}^a = \tfrac{1}{2}(Q^a + Q^{5a}). \tag{5.51b}$$

Eqn (5.50) may then be written as

$$\begin{aligned} [Q_{\mathrm{L}}^a(t), Q_{\mathrm{L}}^b(t)] &= \mathrm{i}f^{abc}Q_{\mathrm{L}}^c(t) \\ [Q_{\mathrm{R}}^a(t), Q_{\mathrm{R}}^b(t)] &= \mathrm{i}f^{abc}Q_{\mathrm{R}}^c(t) \\ [Q_{\mathrm{L}}^a(t), Q_{\mathrm{R}}^a(t)] &= 0. \end{aligned} \tag{5.52}$$

Thus the Q_{L}^as generate the $\mathrm{SU(3)_L}$ algebra while the Q_{R}^as generate the $\mathrm{SU(3)_R}$ algebra.

One can extend the charge algebra (5.37) by considering the equal-time commutators of the charges and their currents. With exactly the same calculation as that leading to (5.37) we can show that

$$[Q^a(t), J_0^b(\mathbf{x}, t)] = \mathrm{i}C^{abc}J_0^c(\mathbf{x}, t). \tag{5.53}$$

Then from Lorentz covariance, we can include the other components of the currents

$$[Q^a(t), J_\mu^b(\mathbf{x}, t)] = \mathrm{i}C^{abc}J_\mu^c(\mathbf{x}, \mathrm{t}). \tag{5.54}$$

Similarly we can go further than (5.53) and have

$$[J_0^a(\mathbf{x}, t), J_0^b(\mathbf{y}, t)] = \mathrm{i}C^{abc}J_0^c(\mathbf{x}, t)\delta^3(\mathbf{x} - \mathbf{y}). \tag{5.55}$$

These relations, and similar extensions of (5.50), are called *current algebra*, see eqn (5.80) below.

If one tries to include spatial components in the current algebra (5.55), one encounters additional terms which vanish upon spatial integration. For example,

$$[J_0^a(\mathbf{x}, t), J_i^b(\mathbf{y}, t)] = \mathrm{i}C^{abc}J_i^c(\mathbf{x}, t)\delta^3(\mathbf{x} - \mathbf{y}) + S_{ij}^{ab}(\mathbf{x})\frac{\partial}{\partial y_j}\delta^3(\mathbf{x} - \mathbf{y}) \tag{5.56}$$

where $S_{ij}^{ab}(\mathbf{x})$ is some operator depending on the explicit form of $J_i^a(\mathbf{x})$. This term vanishes upon integration over space

$$S_{ij}^{ab}(\mathbf{x}) \int \frac{\partial}{\partial y_j} \delta^3(\mathbf{x} - \mathbf{y})\, \mathrm{d}^3 y = 0 \tag{5.57}$$

so it will not modify the charge–current algebra of (5.54). Terms of this type are called *Schwinger terms* (Schwinger 1959). A simple argument will show that they cannot vanish in general. Consider the simplest case of the U(1) symmetry where there is no need of group indices and the structure constants vanish. If we assume the absence of Schwinger terms, the commutator (5.56) becomes

$$[J_0(\mathbf{x}, t), J_i(\mathbf{y}, t)] = 0 \tag{5.58}$$

which implies

$$[J_0(\mathbf{x}, t), \partial_i J_i(\mathbf{y}, t)] = 0. \tag{5.59}$$

From current conservation $\partial^\mu J_\mu = 0$, we obtain

$$[J_0(\mathbf{x}, t), \partial_0 J_0(\mathbf{y}, t)] = 0. \tag{5.60}$$

Taking the vacuum expectation value and inserting a complete set of energy eigenstates, we have

$$\begin{aligned} \langle 0|[J_0(\mathbf{x}, t), \partial_0 J_0(\mathbf{y}, t)]|0\rangle &= \sum_n (\langle 0|J_0(\mathbf{x}, t)|n\rangle\langle n|\partial_0 J_0(\mathbf{y}, t)|0\rangle \\ &\quad - \langle 0|\partial_0 J_0(\mathbf{y}, t)|n\rangle\langle n|J_0(\mathbf{x}, t)|0\rangle) \\ &= \mathrm{i} \sum_n (\mathrm{e}^{\mathrm{i}\mathbf{p}_n \cdot (\mathbf{x}-\mathbf{y})} + \mathrm{e}^{-\mathrm{i}\mathbf{p}_n \cdot (\mathbf{x}-\mathbf{y})}) \\ &\quad \times E_n |\langle 0|J_0(0)|n\rangle|^2. \end{aligned} \tag{5.61}$$

Thus in the limit $\mathbf{x} \to \mathbf{y}$, eqn (5.60) would imply that

$$\sum_n E_n |\langle 0|J_0(0)|n\rangle|^2 = 0. \tag{5.62}$$

From the positivity of energy we must conclude that

$$\langle 0|J_0(0)|n\rangle = 0 \quad \text{for all } |n\rangle. \tag{5.63}$$

Thus, we have $J_0 = 0$ identically and the relation (5.60) is trivial. Therefore, we must have a nonvanishing Schwinger term.

We should also note that the free quark model in fact has two more U(1) symmetries. The first U(1) symmetry corresponds to $\mathscr{L}$ (eqn (5.41)) being invariant under the common phase change for each of the quark fields

$$\mathrm{q}_i(x) \to \mathrm{e}^{\mathrm{i}\beta}\mathrm{q}_i(x) \tag{5.64}$$

with the conserved (baryon-number) current

$$J_\mu^B(x) \sim \bar{\mathrm{q}}_i(x)\gamma_\mu \mathrm{q}_i(x). \tag{5.65}$$

The second U(1) symmetry corresponds to $\mathscr{L}_0$ (eqn (5.42)) being invariant under

$$q_i(x) \to e^{i\beta\gamma_5} q_i(x) \qquad (5.66)$$

with the (partially) conserved ('axial baryon number') current

$$J_\mu^A(x) \sim \bar{q}_i(x)\gamma_\mu\gamma_5 q_i(x). \qquad (5.67)$$

The problems associated with this axial U(1) symmetry will be discussed in §16.3.

5.2 Symmetry currents as physical currents

As we studied in the previous chapter, symmetry groups are of great importance in particle classification. Now we see in field theoretical studies that these symmetry currents (*Noether currents*) and charges satisfy definite commutation relations that are valid even in the presence of symmetry-breaking terms. Another important result of field theoretical studies is that these symmetry currents are just the physical currents appearing in electromagnetic and weak interactions, i.e. the same Noether currents, or some linear combinations thereof, appear in the interaction Lagrangian. Thus current algebra, which represents symmetries of the strong interaction, can be directly tested in electromagnetic or weak-interaction processes involving hadrons. For reviews of applications of current algebra see Adler and Dashen (1968) and de Alfaro *et al.* (1973).

Electromagnetic currents

The most familiar physical current is the electromagnetic current $J_\lambda^{em}(x)$ which is coupled to the photon field $A_\lambda(x)$ in the interaction Lagrangian by

$$\mathscr{L}^{em} = eJ_\lambda^{em}A^\lambda \qquad (5.68)$$

where e is the electromagnetic charge coupling constant. We can decompose the current into leptonic and hadronic parts

$$J_\lambda^{em} = J_{l\lambda}^{em} + J_{h\lambda}^{em}. \qquad (5.69)$$

The leptonic current can be written directly in terms of the charged lepton fields

$$J_{l\lambda}^{em} = -\bar{e}\gamma_\lambda e - \bar{\mu}\gamma_\lambda\mu + \ldots \qquad (5.70)$$

where the fermion field operators are denoted by their particle names. Since the leptons do not have strong interactions and the electromagnetic interaction can be treated perturbatively, we can directly measure $J_{l\lambda}^{em}$ in physical processes. In this respect the hadronic electromagnetic current $J_{h\lambda}^{em}$ is quite different. It cannot be written directly in terms of hadronic fields such as π, K, N, etc. because they are not elementary constituents. While we can express $J_{h\lambda}^{em}$ in terms of the quark fields,

$$J_{h\lambda}^{em} = \tfrac{2}{3}\bar{u}\gamma_\lambda u - \tfrac{1}{3}\bar{d}\gamma_\lambda d - \tfrac{1}{3}\bar{s}\gamma_\lambda s + \ldots, \qquad (5.71)$$

it cannot be used in the same way as the leptonic current. We do not know the hadronic wavefunction in terms of the quark fields and quarks have strong interactions which must be taken into account to all orders. Until this can be done we have to rely on the experimental measurement to learn the structure of the matrix elements of this current. On the other hand, the electromagnetic current is also a symmetry current. In fact it is conserved by all known interactions. The hadronic charge operator

$$Q_{\mathrm{h}}^{\mathrm{em}} = \int J_0^{\mathrm{em}}\, \mathrm{d}^3 x \tag{5.72}$$

obeys the Gell-Mann–Nishijima relation (see §4.4)

$$Q_{\mathrm{h}}^{\mathrm{em}} = T_3 + \frac{Y}{2}.$$

This implies a similar relation for the corresponding currents,

$$J_{\mathrm{h}\lambda}^{\mathrm{em}} = J_\lambda^3 + \tfrac{1}{2} J_\lambda^Y. \tag{5.73}$$

In the quark model this corresponds to

$$\begin{aligned} J_{\mathrm{h}0}^{\mathrm{em}} &= \tfrac{2}{3}\mathrm{u}^\dagger\mathrm{u} - \tfrac{1}{3}\mathrm{d}^\dagger\mathrm{d} - \tfrac{1}{3}\mathrm{s}^\dagger\mathrm{s} + \ldots \\ &= \tfrac{1}{2}(\mathrm{u}^\dagger\mathrm{u} - \mathrm{d}^\dagger\mathrm{d}) + \tfrac{1}{6}(\mathrm{u}^\dagger\mathrm{u} + \mathrm{d}^\dagger\mathrm{d} - 2\mathrm{s}^\dagger\mathrm{s}) + \ldots \\ &= J_0^3 + \tfrac{1}{2} J_0^Y. \end{aligned} \tag{5.74}$$

Weak currents

In weak interactions the currents play a similar role toJ_λ^{em} in electromagnetic interaction. We shall see (cf. Chapter 11) that these two interactions are 'unified' in modern gauge theories and that the weak and electromagnetic currents are members of one multiplet; hence they really have similar theoretical status. In this chapter we shall restrict our discussion to the *charged* weak current J_λ—to the part of J_λ that does not bear any of the new quantum numbers: 'Charm', 'bottom', etc. Similarly to (5.68) it is coupled to the charged intermediate vector boson (IVB) field W_λ in the interaction as

$$\mathscr{L}^W = g J_\lambda W^\lambda + \text{h.c.} \tag{5.75}$$

where g is the coupling constant. From this we have the effective low-energy Lagrangian for a current–current interaction

$$\mathscr{L}_{\mathrm{eff}} = -\frac{G_{\mathrm{F}}}{\sqrt{2}} J_\lambda^\dagger J^\lambda + \text{h.c.} \tag{5.76}$$

where $g^2/M_{\mathrm{W}}^2 = G_{\mathrm{F}}/\sqrt{2}$, as the massive IVB propagator, contributes the M_{W}^2 mass factor in the denominator. $G_{\mathrm{F}} \simeq 10^{-5}$ is the Fermi constant measured in units of inverse proton mass squared. The weak current J_λ can also be separated into leptonic and hadronic parts

$$J^\lambda = J_l^\lambda + J_{\mathrm{h}}^\lambda \tag{5.77}$$

with

$$J_l^\lambda = \bar{\nu}_e\gamma^\lambda(1-\gamma_5)e + \bar{\nu}_\mu\gamma^\lambda(1-\gamma_5)\mu + \ldots \tag{5.78}$$

where $\nu_e(x)$ and $\nu_\mu(x)$ are field operators for the neutrino fields. The hadronic current can be written in the V–A form

$$\begin{aligned} J_h^\lambda = {} & [(V_1^\lambda + iV_2^\lambda) - (A_1^\lambda + iA_2^\lambda)]\cos\theta_c \\ & + [(V_4^\lambda + iV_5^\lambda) - (A_4^\lambda + iA_5^\lambda)]\sin\theta_c \end{aligned} \tag{5.79}$$

where $\theta_c \simeq 0.25$ is the Cabibbo angle (Cabibbo 1963). The subscripts on the vector and axial vector operators are the flavour SU(3) octet indices. The various selection rules and symmetry relations implied by the SU(3) and SU(2) transformation properties of hadronic currents are well tested in semi-leptonic weak processes. For example, the strangeness-conserving vector current $V_1^\lambda + iV_2^\lambda$ and its conjugate $V_1^\lambda - iV_2^\lambda$ are partners of the isovector electromagnetic current V_3^λ in an isospin triplet and the corresponding charges will form the generators of the isospin SU(2) subgroup of SU(3). These isospin currents are of course approximately conserved. This is called the *conserved vector current* (CVC) hypothesis (Feynman and Gell-Mann 1958). Thus from our knowledge of the electromagnetic-current matrix elements (form factors) we can predict the strangeness-conserving weak vector-current form factor by isospin rotations. Similarly the weak form factors of the strangeness-changing vector currents can be fixed by SU(3) rotations since all these vector currents are members of the same SU(3) octet, etc. Furthermore, these vector and axial vector currents are postulated to satisfy the $SU(3)_L \times SU(3)_R$ algebra (Gell-Mann 1964a)

$$\begin{aligned} [V_a^0(\mathbf{x},t), V_b^0(\mathbf{y},t)] &= if_{abc}V_c^0(\mathbf{x},t)\delta^3(\mathbf{x}-\mathbf{y}) \\ [V_a^0(\mathbf{x},t), A_b^0(\mathbf{y},t)] &= if_{abc}A_c^0(\mathbf{x},t)\delta^3(\mathbf{x}-\mathbf{y}) \\ [A_a^0(\mathbf{x},t), A_b^0(\mathbf{y},t)] &= if_{abc}V_c^0(\mathbf{x},t)\delta^3(\mathbf{x}-\mathbf{y}). \end{aligned} \tag{5.80}$$

We note that these relations in which the left-hand sides are quadratic in the currents while the right-hand sides are linear will give rise to non-linear constraints among currents. Thus the normalizations of the currents are fixed by these non-linear commutation relations.

A great triumph of the quark model of hadrons is its successful and simple explanation of all the above symmetry features of the weak hadronic current. Very much like the leptonic current in (5.78), the hadronic weak current in (5.79) can be written directly in terms of quark fields

$$J_h^\lambda = \bar{u}\gamma^\lambda(1-\gamma_5)d\cos\theta_c + \bar{u}\gamma^\lambda(1-\gamma_5)s\sin\theta_c \tag{5.81}$$

and, as we have seen in §5.1, the $SU(3)_L \times SU(3)_R$ algebra (5.80) is also satisfied in a free quark model. The key problem of course remains of finding a fully interacting theory of the quarks in which the strong interaction naturally has this approximate global $SU(3)_L \times SU(3)_R$ chiral symmetry.

Current algebra sum rule

As an illustration of the type of results that can be obtained from the current algebra of (5.80) we will derive the Adler's sum rule for neutrino scattering (Adler 1966).

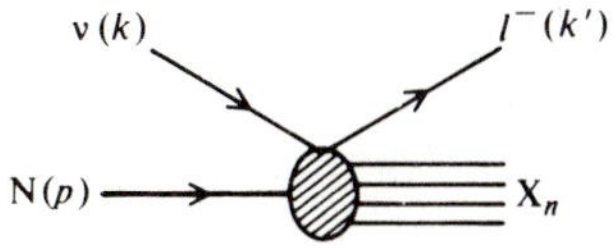

FIG. 5.1.

Consider (Fig. 5.1) the neutrino scattering off a nucleon target producing a charged lepton l^- and some hadron state X_n (n particles with total momentum p_n)

$$\nu(k) + N(p) \to l^-(k') + X(p_n). \tag{5.82}$$

Define

$$\begin{aligned} q &= k - k' \\ \nu &= p \cdot q/M \end{aligned} \tag{5.83}$$

where M is the nucleon mass. In the lab-frame

$$p_\mu = (M, 0, 0, 0), \qquad k_\mu = (E, \mathbf{k}), \qquad k'_\mu = (E', \mathbf{k}'),$$

we have

$$\begin{aligned} q^2 &= -4EE' \sin^2 \frac{\theta}{2} \\ \nu &= E - E' \end{aligned} \tag{5.84}$$

where θ is the angle between $\mathbf{k}$ and $\mathbf{k}'$, and the energy is high enough so that we can make the approximation of taking the charged lepton mass m_l to be zero. From (5.76) and (5.78) the amplitude for this process can be written

$$T_n^{(\nu)} = \frac{G_F}{\sqrt{2}} \bar{u}_l(k', \lambda')\gamma_\lambda(1 - \gamma_5)u_\nu(k, \lambda)\langle n|J_h^\lambda|p, \sigma\rangle \tag{5.85}$$

and the unpolarized differential cross-section as

$$\begin{aligned} d\sigma_n^{(\nu)} = &\frac{1}{|\mathbf{v}|}\frac{1}{2M}\frac{1}{2E}\frac{d^3k'}{(2\pi)^3 2k'_0}\prod_{i=1}^{n}\left[\frac{d^3p_i}{(2\pi)^3 2p_{i0}}\right] \\ &\times \frac{1}{2}\sum_{\sigma\lambda\lambda'}|T_n^{(\nu)}|^2(2\pi)^4\delta(k + p - k' - p_n) \end{aligned} \tag{5.86}$$

where σ, λ, and λ' are spin labels of nucleon, initial and final lepton, respectively, and

$$p_n = \sum_{i=1}^{n} p_i.$$

If we sum over all possible hadronic final states, we obtain the *inclusive cross-section*

$$\frac{d^2\sigma^{(\nu)}}{d|q^2|\,d\nu} = \frac{G_F^2}{32\pi E^2}\, l^{\alpha\beta} W_{\alpha\beta}^{(\nu)} \tag{5.87}$$

with the leptonic tensor

$$\begin{aligned} l_{\alpha\beta} &= \mathrm{tr}[(\not{k}')\gamma_\alpha(1-\gamma_5)(\not{k})\gamma_\beta(1-\gamma_5)] \\ &= 8(k_\alpha k'_\beta + k'_\alpha k_\beta - k\cdot k' g_{\alpha\beta} + i\varepsilon_{\alpha\beta\gamma\delta}k'^\gamma k^\delta) \end{aligned} \tag{5.88}$$

and the hadronic tensor

$$\begin{aligned} W_{\alpha\beta}^{(\nu)}(p,q) &= \frac{1}{4M}\sum_\sigma\sum_n \int \prod_{i=1}^n \left[\frac{d^3p_i}{(2\pi)^3 2p_{i0}}\right] \\ &\quad \times \langle p,\sigma|J_{h\beta}^\dagger(0)|n\rangle\langle n|J_{h\alpha}(0)|p,\sigma\rangle(2\pi)^3\delta^4(p_n-p-q) \\ &= \frac{1}{4M}\sum_\sigma \int \frac{d^4y}{2\pi}\langle p,\sigma|J_{h\beta}^\dagger(y)J_{h\alpha}(0)|p,\sigma\rangle\, e^{iq\cdot y}. \end{aligned} \tag{5.89}$$

Note that $W_{\alpha\beta}^{(\nu)}$ is non-zero only for $q_0 = E - E' > 0$. Since it is a second-rank tensor depending only on momenta p and q, it can be written as

$$\begin{aligned} W_{\alpha\beta}^{(\nu)} = &-W_1^{(\nu)}g_{\alpha\beta} + W_2^{(\nu)}p_\alpha p_\beta/M^2 - iW_3^{(\nu)}\varepsilon_{\alpha\beta\gamma\delta}p^\gamma q^\delta/M^2 \\ &+ W_4^{(\nu)}q_\alpha q_\beta/M^2 + W_5^{(\nu)}(p_\alpha q_\beta + p_\beta q_\alpha)/M^2 \\ &+ iW_6^{(\nu)}(p_\alpha q_\beta - p_\beta q_\alpha)/M^2 \end{aligned} \tag{5.90}$$

where the $W_i^{(\nu)}$s are Lorentz-invariant functions of q^2 and ν of (5.84), called *structure functions*. The cross-section in (5.87) is now

$$\frac{d^2\sigma^{(\nu)}}{d|q^2|\,d\nu} = \frac{G_F^2}{2\pi}\left(\frac{E'}{E}\right)\left[2W_1^{(\nu)}\sin^2\frac{\theta}{2} + W_2^{(\nu)}\cos^2\frac{\theta}{2} - \frac{(E+E')}{M}\sin^2\frac{\theta}{2}\,W_3^{(\nu)}\right]. \tag{5.91}$$

The $W_{4,5,6}^{(\nu)}$s do not appear in the $m_l = 0$ limit.

Similarly for anti-neutrino scattering

$$\bar{\nu} + N \to l^+ + X, \tag{5.92}$$

we have

$$\frac{d^2\sigma^{(\bar{\nu})}}{d|q^2|\,d\nu} = \frac{G_F^2}{2\pi}\left(\frac{E'}{E}\right)\left[2W_1^{(\bar{\nu})}\sin^2\frac{\theta}{2} + W_2^{(\bar{\nu})}\cos^2\frac{\theta}{2} + \frac{(E+E')}{M}\sin^2\frac{\theta}{2}\,W_3^{(\bar{\nu})}\right] \tag{5.93}$$

where the structure functions $W_i^{(\bar{\nu})}$ are defined by

$$\begin{aligned} W_{\alpha\beta}^{(\bar{\nu})}(p,q) &= \frac{1}{4M}\sum_\sigma \int \frac{d^4y}{2\pi}\, e^{iq\cdot y}\langle p,\sigma|J_{h\beta}(y)J_{h\alpha}^\dagger(0)|p,\sigma\rangle \\ &= -W_1^{(\bar{\nu})}g_{\alpha\beta} + W_2^{(\bar{\nu})}p_\alpha p_\beta/M^2 - iW_3^{(\bar{\nu})}\varepsilon_{\alpha\beta\gamma\delta}p^\gamma q^\delta/M^2 \end{aligned}$$

$$+ W_4^{(\bar{v})} q_\alpha q_\beta / M^2 + W_5^{(\bar{v})} (p_\alpha q_\beta + p_\beta q_\alpha)/M^2$$
$$+ \mathrm{i} W_6^{(\bar{v})} (p_\alpha q_\beta - p_\beta q_\alpha)/M^2. \tag{5.94}$$

Again they are non-zero only for $q_0 > 0$. We can alter (5.94) slightly using translational invariance

$$W_{\alpha\beta}^{(\bar{v})}(p, q) = \frac{1}{4M} \sum_\sigma \int \frac{\mathrm{d}^4 y}{2\pi} \mathrm{e}^{\mathrm{i}q\cdot y} \langle p, \sigma | J_{\mathrm{h}\beta}(0) J_{\mathrm{h}\alpha}^\dagger(-y) | p, \sigma \rangle$$
$$= \frac{1}{4M} \sum_\sigma \int \frac{\mathrm{d}^4 y}{2\pi} \mathrm{e}^{-\mathrm{i}q\cdot y} \langle p, \sigma | J_{\mathrm{h}\beta}(0) \mathrm{J}_{\mathrm{h}\alpha}^\dagger(y) | p, \sigma \rangle. \tag{5.95}$$

Now consider the tensor $W_{\alpha\beta}$ defined by

$$W_{\alpha\beta}(p, q) = \frac{1}{4M} \sum_\sigma \int \frac{\mathrm{d}^4 y}{2\pi} \mathrm{e}^{\mathrm{i}q\cdot y} \langle p, \sigma | [J_{\mathrm{h}\beta}^\dagger(y), J_{\mathrm{h}\alpha}(0)] | p, \sigma \rangle. \tag{5.96}$$

When this is compared to (5.89) and (5.95) we have

$$W_{\alpha\beta}(p, q) = W_{\alpha\beta}^{(v)}(p, q) \quad \text{for} \quad q_0 > 0$$
$$= -W_{\beta\alpha}^{(\bar{v})}(p, -q) \quad \text{for} \quad q_0 < 0. \tag{5.97}$$

To derive the sum rule, we take W_{00} and integrate over q_0

$$\int_{-\infty}^{\infty} W_{00}(p, q)\, \mathrm{d}q_0 = \int_0^{\infty} (W_{00}^{(v)}(p, q) - W_{00}^{(\bar{v})}(p, q))\, \mathrm{d}q_0$$
$$= \frac{1}{4M} \sum_\sigma \int \mathrm{d}^3 y\, \mathrm{e}^{-\mathrm{i}\mathbf{q}\cdot\mathbf{y}} \langle p, \sigma | [J_{\mathrm{h}0}^\dagger(\mathbf{y}, 0), J_{\mathrm{h}0}(0)] | p, \sigma \rangle. \tag{5.98}$$

This equal-time commutator can be evaluated using the current algebra of (5.80). The simplest way to calculate this is to use the fact that these commutation relations are also satisfied in the free-quark model (see §5.1) where the current is given by (5.81). Using the canonical anti-commutation relations

$$\{q_i^\dagger(\mathbf{y}, 0), q_j(0)\} = \delta^3(\mathbf{y})\delta_{ij} \tag{5.99}$$

where the indices are those of the Dirac matrix space as well as the flavour space of (5.38), we have

$$[J_{\mathrm{h}0}^\dagger(\mathbf{y},0), J_{\mathrm{h}0}(0)] = [(\cos\theta_\mathrm{c} \mathrm{d}^\dagger(\mathbf{y}, 0) + \sin\theta_\mathrm{c} \mathrm{s}^\dagger(\mathbf{y}, 0))(1 - \gamma_5)\mathrm{u}(\mathbf{y}, 0),$$
$$\mathrm{u}^\dagger(0)(1 - \gamma_5)(\cos\theta_\mathrm{c} \mathrm{d}(0) + \sin\theta_\mathrm{c} \mathrm{s}(0))]$$
$$= 2(\cos\theta_\mathrm{c} \mathrm{d}^\dagger(\mathbf{y}, 0) + \sin\theta_\mathrm{c} \mathrm{s}^\dagger(\mathbf{y}, 0))(1 - \gamma_5)$$
$$\times \{\mathrm{u}(\mathbf{y}, 0), \mathrm{u}^\dagger(0)\}(\cos\theta_\mathrm{c} \mathrm{d}(0) + \sin\theta_\mathrm{c} \mathrm{s}(0))$$
$$- 2\mathrm{u}^\dagger(0)(1 - \gamma_5)(\cos^2\theta_\mathrm{c} \{\mathrm{d}^\dagger(\mathbf{y}, 0), \mathrm{d}(0)\}$$
$$+ \sin^2\theta_\mathrm{c} \{\mathrm{s}^\dagger(\mathbf{y}, 0), \mathrm{s}(0)\})\mathrm{u}(\mathbf{y}, 0)$$

$$\begin{aligned}
&= 2\delta^3(\mathbf{y})(\cos^2\theta_c d^\dagger(0)(1-\gamma_5)d(0) \\
&\quad + \sin^2\theta_c s^\dagger(0)(1-\gamma_5)s(0) \\
&\quad + \sin\theta_c\cos\theta_c(d^\dagger(0)(1-\gamma_5)s(0) + s^\dagger(0)(1-\gamma_5)d(0)) \\
&\quad - u^\dagger(0)(1-\gamma_5)u(0)) \\
&= -\delta^3(\mathbf{y})[4\cos^2\theta_c(V_0^3(0) - A_0^3(0)) \\
&\quad + \sin^2\theta_c(3(V_0^Y(0) - A_0^Y(0)) + 2(V_0^3(0) - A_0^3(0))) \\
&\quad + 4\sin\theta_c\cos\theta_c(V_0^6(0) - A_0^6(0))]
\end{aligned} \tag{5.100}$$

where

$$\begin{aligned}
V_0^Y &= (u^\dagger u + d^\dagger d - 2s^\dagger s)/3 & A_0^Y &= (u^\dagger\gamma_5 u + d^\dagger\gamma_5 d - 2s^\dagger\gamma_5 s)/3 \\
V_0^3 &= (u^\dagger u - d^\dagger d)/2 & A_0^3 &= (u^\dagger\gamma_5 u - d^\dagger\gamma_5 d)/2 \\
V_0^6 &= (d^\dagger s + s^\dagger d)/2 & A_0^6 &= (d^\dagger\gamma_5 s + s^\dagger\gamma_5 d)/2.
\end{aligned} \tag{5.101}$$

Since V_0^6 and A_0^6 are strangeness-changing operators, their matrix elements vanish when taken between nuclear states. Also averaging over nucleon spin, we have

$$\frac{1}{2}\sum_\sigma \langle p, \sigma|A_0^i|p, \sigma\rangle = 0 \quad \text{for all } i. \tag{5.102}$$

Thus (5.98) becomes

$$\int_0^\infty [W_{00}^{(\nu)}(p,q) - W_{00}^{(\bar\nu)}(p,q)]\,dq_0 = \frac{-p_0}{M}[4T_3\cos^2\theta_c + (3Y + 2T_3)\sin^2\theta_c] \tag{5.103}$$

where T_3 and Y are the isospin and hypercharge of the nucleon state

$$\begin{aligned}
\frac{1}{2}\sum_\sigma \langle p, \sigma|V_0^3|p, \sigma\rangle &= 2T_3 p_0 \\
\frac{1}{2}\sum_\sigma \langle p, \sigma|V_0^Y|p, \sigma\rangle &= 2Y p_0.
\end{aligned} \tag{5.104}$$

We now proceed to express the left-hand side of (5.103) in terms of the structure functions

$$\int_0^\infty W_{00}^{(\nu)}\,dq_0 = \int [-W_1^{(\nu)} + W_2^{(\nu)}(p_0/M)^2 + W_4^{(\nu)}(q_0/M)^2 + W_5^{(\nu)}(p_0 q_0/M^2)]\,dq_0. \tag{5.105}$$

A judicial choice of reference frame will simplify this equation. Instead of the nucleon rest frame, we will take the infinite-momentum frame (Fubini and

Furlan 1965) in which the nucleon has infinite momentum orthogonal to **q**

$$|\mathbf{p}| \to \infty \quad \text{with } \mathbf{p}\cdot\mathbf{q} = 0. \tag{5.106}$$

Then in this frame,

$$\begin{aligned} p_0 &= (\mathbf{p}^2 + M^2)^{1/2} \simeq |\mathbf{p}| \to \infty \\ \nu &= p\cdot q/M = p_0 q_0/M \\ q^2 &= q_0^2 - \mathbf{q}^2 = (\nu M/p_0)^2 - \mathbf{q}^2 \to -\mathbf{q}^2 \end{aligned} \tag{5.107}$$

and the largest term on the right-hand side of (5.105) is $W_2^{(\nu)}$

$$\lim_{|\mathbf{p}|\to\infty} \int_0^\infty W_{00}^{(\nu)}\, \mathrm{d}q_0 = \frac{p_0}{M} \int_0^\infty \mathrm{d}\nu\, W_2^{(\nu)}(q^2, \nu) \tag{5.108}$$

where we have assumed that it is valid to intercharge the limit and the integration. Thus the sum rule in (5.103) becomes

$$\int_0^\infty [W_2^{(\bar\nu)}(q^2, \nu) - W_2^{(\nu)}(q^2, \nu)]\, \mathrm{d}\nu = 4T_3 \cos^2\theta_c + (3Y + 2T_3)\sin^2\theta_c$$

$$= \begin{cases} 2\cos^2\theta_c + 4\sin^2\theta_c & \text{for a proton target} \\ -2\cos^2\theta_c + 2\sin^2\theta_c & \text{for a neutron target.} \end{cases} \tag{5.109}$$

This is the *Adler current-algebra sum rule* for neutrino scattering. We remark that even though it is derived in the infinite-momentum frame, the final result expressed in terms of Lorentz invariants should be true in any given frame. It has the notable feature that the q^2-dependence of the left-hand side gets 'integrated away'. This sum rule can be used for any target with appropriate T_3 and Y quantum numbers on the right-hand side. An extension to include other additive flavour quantum rules (beyond strangeness) can be carried out in a straightforward fashion.

.3 Spontaneous breaking of global symmetry, the Goldstone theorem

The $SU(3)_L \times SU(3)_R$ algebra (5.80) generated by the various physical currents suggests that we have a strong-interaction Hamiltonian

$$H = H_0 + \lambda H_1 \tag{5.110}$$

where H_0 is invariant under $SU(3)_L \times SU(3)_R$ transformations and H_1 is not. In the limit of $\lambda = 0$, all generators of the chiral algebra are conserved. We would expect particles to form degenerate multiplets corresponding to irreducible representations of the group $SU(3)_L \times SU(3)_R$. For example, the octet pseudoscalar mesons should be accompanied by an octet of scalar mesons, and the $J^P = (1/2)^+$ baryons should have partners with opposite parities. But in reality there is no evidence for this larger multiplet structure.

This leads to the idea that $SU(3)_L \times SU(3)_R$ symmetry is spontaneously broken and that this symmetry of H_0 is not realized by the particle spectrum.

Non-invariance of the ground state as a symmetry-breaking condition

Let U be an element of the symmetry group which leaves H_0 invariant. Then

$$UH_0U^\dagger = H_0 \tag{5.111}$$

and it connects states that form an irreducible representation (basis) of the group

$$U|A\rangle = |B\rangle. \tag{5.112}$$

From (5.111) and (5.112) it follows immediately that

$$\begin{aligned} E_A &= \langle A|H_0|A\rangle \\ &= \langle B|H_0|B\rangle = E_B. \end{aligned} \tag{5.113}$$

Thus the symmetry of the Hamiltonian H_0 is manifest in the degeneracies of the energy eigenstates corresponding to the irreducible representations of the symmetry group. However implicit in the statement of (5.112) and hence (5.113) is the invariance of the ground state under symmetry transformation. Since $|A\rangle$ and $|B\rangle$ must be related to the ground state $|0\rangle$ through some appropriate creation operators ϕ_A and ϕ_B

$$|A\rangle = \phi_A|0\rangle, \qquad |B\rangle = \phi_B|0\rangle \tag{5.114}$$

and

$$U\phi_A U^\dagger = \phi_B, \tag{5.115}$$

eqn (5.112) follows only if

$$U|0\rangle = |0\rangle. \tag{5.116}$$

When (5.116) is not satisfied, this vitiates (5.113) and the usual symmetry consequence of degenerate energy levels. Such a situation is commonly referred to as a *spontaneous symmetry breakdown*. However, it must be emphasized that, even though the symmetry is not manifest in the degenerate energy levels, there are still symmetry relations coming from the fact the Hamiltonian or the Lagrangian is still invariant under the symmetry transformation.

Ferromagnetism as an example of spontaneous symmetry breakdown

This lack of degeneracy in particle spectra in a symmetry theory may come as a surprise, but there are a number of familiar situations where the ground state is not a symmetric state. A well-known example is ferromagnetism near the Curie temperature T_C. For $T > T_C$, all the dipoles are randomly oriented; the ground state is rotationally invariant. For $T < T_C$, all the dipoles are aligned in some arbitrary direction (spontaneous magnetization) and the rotational symmetry is hidden. Consider the description of this phenomenon by the Ginzburg–Landau theory (1950). For temperatures near the Curie point, the magnetization **M** is expected to be small; a power series expansion of the free energy density can be made, with higher powers of **M** being neglected

$$u(\mathbf{M}) = (\partial_i \mathbf{M})^2 + V(\mathbf{M}) \tag{5.117}$$

where

$$V(\mathbf{M}) = \alpha_1(T)(\mathbf{M}\cdot\mathbf{M}) + \alpha_2(\mathbf{M}\cdot\mathbf{M})^2. \tag{5.118}$$

The energy densities u and V are clearly rotationally symmetric. We have assumed a slowly varying field and kept only the first derivatives. The $(\mathbf{M}\cdot\mathbf{M})^2$ term in (5.118), with a positive coefficient $\alpha_2 > 0$, is included because, at the Curie point, α_1 vanishes

$$\alpha_1 = \alpha(T - T_C) \quad \text{with } \alpha > 0.$$

Since the $(\partial_i \mathbf{M})^2$ term is non-negative, to obtain the ground-state magnetization we simply minimize $V(\mathbf{M})$

$$\delta V/\delta \mathrm{M}_i = 0 \tag{5.119}$$

or

$$\mathbf{M}(\alpha_1 + 2\alpha_2 \mathbf{M}\cdot\mathbf{M}) = 0.$$

For $T > T_C$ (i.e. $\alpha_1 > 0$), the solution is at $\mathbf{M} = 0$. For $T < T_C$ (i.e. $\alpha_1 < 0$) $\mathbf{M} = 0$ is a local maximum and the minimization fixes the magnitude of magnetization (the *order parameter*) to be

$$|\mathbf{M}| = (-\alpha_1/2\alpha_2)^{1/2}. \tag{5.120}$$

But its direction is unspecified by the theory itself. The ground state, having **M** in some definite direction, is one member of this infinitely degenerate set; it is fixed by the boundary condition and is not rotationally symmetric. For temperatures below the Curie point the rotational symmetry of the magnet is spontaneously broken. Thus the symmetry-breaking condition is the non-invariance of the vacuum (ground state)

$$U|0\rangle \neq |0\rangle. \tag{5.121}$$

For $U = \exp(\mathrm{i}\varepsilon^a Q^a)$, where the ε^as are the continuous group parameters, (5.121) can be expressed by the statement that the symmetry charge does not annihilate the vacuum

$$Q^a|0\rangle \neq 0. \tag{5.122}$$

An equivalent statement to (5.121) and (5.122) is that certain field operators have nonvanishing vacuum expectation values

$$\langle 0|\phi_j|0\rangle \neq 0. \tag{5.123}$$

This can be seen easily as a symmetry transformation of the type (5.115) or (5.10) may be represented in terms of generators as

$$[Q^a, \phi_i] = \mathrm{i}t^a_{ij}\phi_j \tag{5.124}$$

where t^a_{ij} is the appropriate representation matrix. Thus (5.122) implies the existence of at least some nonvanishing matrix element $\langle 0|\phi_j|0\rangle$. We should also remark that translational invariance of the vacuum state leads to the conclusion that these matrix elements are space–time-independent constants,

$$\begin{aligned}\langle 0|\phi(x)|0\rangle &= \langle 0|\,\mathrm{e}^{ip\cdot x}\phi(0)\,\mathrm{e}^{-ip\cdot x}|0\rangle \\ &= \langle 0|\phi(0)|0\rangle. \end{aligned} \tag{5.125}$$

The Goldstone theorem

Spontaneous breakdown of a continuous symmetry implies the existence of massless spinless particles. The study of this connection was initiated by Nambu (1960; Nambu and Jona-Lasinio 1961*a*, *b*) and subsequently the proofs, with various degrees of rigour and generality, were provided by Goldstone (1961) and others (Goldstone, Salam, and Weinberg 1962). Such scalar particles are referred to as *Nambu–Goldstone* bosons or simply as *Goldstone bosons.* In the following we shall first give a formal proof (Guralnik *et al.* 1968). This is followed by a number of illustrative examples.

Any continuous symmetry of the Lagrangian, by Noether's theorem, implies the existence of a conserved current

$$\partial_\mu J^\mu(x) = 0. \tag{5.126}$$

Normally we can convert this into the statement of the charge being a constant of motion $\mathrm{d}Q(t)/\mathrm{d}t = 0$ with $Q(t) = \int \mathrm{d}^3x J_0(\mathbf{x}, t)$. However, with spontaneous symmetry breakdown (5.123), $Q(t)$ is not well defined because of the poor convergence property of the field operator in the integrand. Even the weak limit, corresponding to the matrix element $\langle 0|Q^2(t)|0\rangle$, does not exist. The translational invariance property of the vacuum state leads to the result

$$\langle 0|Q^2(t)|0\rangle = \int \mathrm{d}^3x \langle 0|J_0(0)Q(0)|0\rangle$$

which diverges because the integrand is nonvanishing and $\mathbf{x}$-independent. The nonexistence of Q does not really matter since in actual calculation only the commutator of Q need ever appear. For the transformation of some generic field operator $\phi(x)$, we have

$$\begin{aligned}\phi(x) \to \phi'(x) &= \mathrm{e}^{\mathrm{i}\varepsilon Q}\phi(x)\,\mathrm{e}^{-\mathrm{i}\varepsilon Q} \\ &= \phi(x) + \mathrm{i}\varepsilon[Q, \phi(x)] + \dots. \end{aligned}$$

Here we shall only assume that the commutator exists and formulate the proof entirely in terms of its properties. Current conservation (5.126) implies that

$$\begin{aligned} 0 &= \int \mathrm{d}^3x[\partial^\mu J_\mu(\mathbf{x}, t), \phi(0)] \\ &= \partial^0 \int \mathrm{d}^3x[J^0(\mathbf{x}, t), \phi(0)] + \int d\mathbf{S}\cdot[\mathbf{J}(\mathbf{x}, t), \phi(0)]. \end{aligned}$$

For a large enough surface and thus for large space-like separations the second term on the right-hand side vanishes. Hence

$$\frac{\mathrm{d}}{\mathrm{d}t}[Q(t), \phi(0)] = 0. \qquad (5.127)$$

This commutator being some combination of fields, if it has nonvanishing vacuum expectation value

$$\langle 0|[Q(t), \phi(0)]|0\rangle \equiv \eta \neq 0, \qquad (5.128)$$

we say that the symmetry is spontaneously broken. After inserting a complete set of intermediate states and using a translation operator, (5.128) may be written

$$\sum_n (2\pi)^3\delta^3(\mathbf{p}_n)\{\langle 0|J_0(0)|n\rangle\langle n|\phi|(0)|0\rangle\, \mathrm{e}^{-\mathrm{i}E_n t}$$
$$-\langle 0|\phi|(0)|n\rangle\langle n|J_0(0)|0\rangle\, \mathrm{e}^{\mathrm{i}E_n t}\} = \eta. \qquad (5.129)$$

The right-hand side is nonvanishing and time-independent because of (5.128) and (5.127). Since the positive and negative frequency parts do not mutually cancel, (5.129) can be satisfied only if there exists an intermediate state for which

$$E_n = 0 \quad \text{for } \mathbf{p}_n = 0.$$

Thus, it is a massless state (the Goldstone boson). This particle will have the property that

$$\langle n|\phi(0)|0\rangle \neq 0, \qquad \langle 0|J_0(0)|n\rangle \neq 0. \qquad (5.130)$$

Thus it can be connected to vacuum by the current $J_0(0)$ or the operator $\phi(0)$. This theorem is true independently of perturbation theory. We will illustrate it in a few examples.

Discrete symmetry case

Goldstone bosons are not expected to be present in the discrete symmetry case. Our purpose is to illustrate the circumstance in which the symmetry-breaking condition (5.123) can take place. The Lagrange density given by

$$\mathscr{L} = \frac{1}{2}(\partial_\lambda\phi)^2 - \frac{1}{2}\mu^2\phi^2 - \frac{\lambda}{4}\phi^4 \qquad (5.131)$$

has the discrete symmetry

$$\phi \to \phi' = -\phi. \qquad (5.132)$$

Since $\partial_0\phi$ is the momentum field conjugate to ϕ, the Hamiltonian density is given by

$$H = \frac{1}{2}(\partial_0\phi)^2 + \frac{1}{2}(\nabla\phi)^2 + \frac{1}{2}\mu^2\phi^2 + \frac{\lambda}{4}\phi^4.$$

Thus the classical potential (energy density) may be identified as

$$u(\phi) = \frac{1}{2}(\nabla\phi)^2 + V(\phi)$$

with

$$V(\phi) = \frac{1}{2}\mu^2\phi^2 + \frac{1}{4}\lambda\phi^4. \tag{5.133}$$

Since the $(\nabla\phi)^2$ term is non-negative, the minimum of $u(\phi)$ will have the property $\nabla\phi = 0$ with the constant value ϕ given by minimizing $V(\phi)$ of (5.133) directly. Since the coupling λ is positive (so that energy is bound from below), $V(\phi)$ for the two possible cases of $\mu^2 > 0$ and $\mu^2 < 0$ is shown in Fig. 5.2. For $\mu^2 < 0$, the ground-state field is nonvanishing

$$\phi = \pm(-\mu^2/\lambda)^{1/2}. \tag{5.134}$$

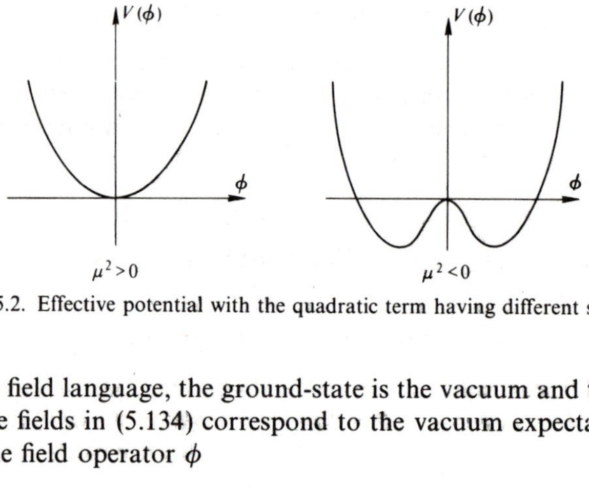

FIG. 5.2. Effective potential with the quadratic term having different signs.

In quantum field language, the ground-state is the vacuum and the classical ground-state fields in (5.134) correspond to the vacuum expectation values (VEV) of the field operator ϕ

$$\langle 0|\phi|0\rangle = v$$

with

$$v = \pm(-\mu^2/\lambda)^{\frac{1}{2}}. \tag{5.135}$$

The two possible values in (5.135) correspond to the two possible vacua. One can choose either one (and only one) to build the theory. Either choice, say $v = +(-\mu^2/\lambda)^{\frac{1}{2}}$, clearly breaks the original reflection symmetry $\phi \to -\phi$ of the theory. (Since the Fock spaces built on the two possible vacua are mutually orthogonal, it is not meaningful to build a theory based on some superposition of the two vacua.) This is the broken symmetry condition.

Since the symmetry in this case is discrete, we do not expect massless Goldstone bosons. To verify this we need to consider small oscillations around the true vacuum. Thus define a new quantum field variable with zero VEV. In terms of this 'shifted field'

$$\phi' = \phi - v, \tag{5.136}$$

the Lagrangian density becomes

$$\mathscr{L} = \frac{1}{2}(\partial_\lambda \phi')^2 - (-\mu^2)\phi'^2 - \lambda v \phi'^3 - \frac{\lambda}{4}\phi'^4. \qquad (5.137)$$

ϕ' describes a particle with mass $(-2\mu^2)^{\frac{1}{2}}$.

Abelian symmetry case

The Lagrange density given by

$$\mathscr{L} = \frac{1}{2}(\partial_\lambda \sigma)^2 + \frac{1}{2}(\partial_\lambda \pi)^2 - V(\sigma^2 + \pi^2) \qquad (5.138)$$

with

$$V(\sigma^2 + \pi^2) = \frac{-\mu^2}{2}(\sigma^2 + \pi^2) + \frac{\lambda}{4}(\sigma^2 + \pi^2)^2 \qquad (5.139)$$

has (continuous) U(1), i.e. O(2), symmetry

$$\begin{pmatrix}\sigma\\ \pi\end{pmatrix} \rightarrow \begin{pmatrix}\sigma'\\ \pi'\end{pmatrix} = \begin{pmatrix}\cos\alpha & \sin\alpha\\ -\sin\alpha & \cos\alpha\end{pmatrix}\begin{pmatrix}\sigma\\ \pi\end{pmatrix}. \qquad (5.140)$$

The extremum of the potential V is determined by

$$\frac{\delta V}{\delta \sigma} = \sigma[-\mu^2 + \lambda(\sigma^2 + \pi^2)] = 0$$

$$\frac{\delta V}{\delta \pi} = \pi[-\mu^2 + \lambda(\sigma^2 + \pi^2)] = 0. \qquad (5.141)$$

For $\mu^2 > 0$, the minimum is at

$$\sigma^2 + \pi^2 = v^2 \qquad (5.142)$$

with

$$v = (\mu^2/\lambda)^{\frac{1}{2}}. \qquad (5.143)$$

A graphic representation of the potential is given in Fig. 5.3. The minima consist of points on a circle with radius v in the (σ, π) plane. They are related

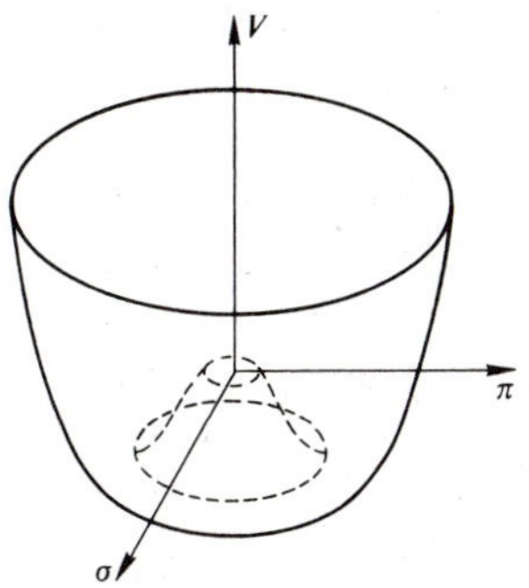

FIG. 5.3. The potential function of eqn (5.139) for $\mu^2 > 0$.

to each other through O(2) rotations. Hence they are all equivalent and there are an infinite number of degenerate vacua. Any point on this circle may be chosen as the true vacuum. We can pick, for example,

$$\langle 0|\sigma|0\rangle = v, \qquad \langle 0|\pi|0\rangle = 0. \tag{5.144}$$

Thus the O(2) symmetry is broken by the vacuum state.

To find the particle spectrum in perturbation theory we consider small oscillations around the true minimum and define a shifted field

$$\sigma' = \sigma - v. \tag{5.145}$$

The Lagrange density may be written

$$\mathscr{L} = \frac{1}{2}[(\partial_\lambda \sigma')^2 + (\partial_\lambda \pi)^2] - \mu^2 \sigma'^2 - \lambda v \sigma'(\sigma'^2 + \pi^2) - \frac{\lambda}{4}(\sigma'^2 + \pi^2)^2. \tag{5.146}$$

There is no quadratic term in the π-field and that in the σ' field has the right sign. Therefore, π is the massless Goldstone boson and σ' is a particle with mass $(2\mu^2)^{\frac{1}{2}}$. In fact it is easy to see this pictorially in Fig. 5.3. Small oscillations around any point of the minimum circle may be decomposed into the radial and polar angle components. The polar-angle oscillations are along an equipotential trajectory and it does not cost any energy and hence corresponds to a massless particle. With our choice of vacuum (5.144) the polar angle oscillation is along the π direction—hence π becomes the Goldstone boson.

We shall also examine these features more formally through the Goldstone theorem and make connection to the proof of eqns (5.128) and (5.130). The conserved current generated by the U(1) symmetry (5.140) is given by

$$J_\lambda(x) = [(\partial_\lambda \pi)\sigma - (\partial_\lambda \sigma)\pi] \tag{5.147}$$

with the associated charge being

$$Q = \int J_0(x)\, \mathrm{d}^3x = \int [(\partial_0 \pi)\sigma - (\partial_0 \sigma)\pi]\, \mathrm{d}^3x. \tag{5.148}$$

Using the canonical commutation relations

$$[\partial_0 \pi(\mathbf{x}, t), \pi(\mathbf{y}, t)] = -\mathrm{i}\delta^3(\mathbf{x} - \mathbf{y}) \tag{5.149a}$$

$$[\partial_0 \sigma(\mathbf{x}, t), \sigma(\mathbf{y}, t)] = -\mathrm{i}\delta^3(\mathbf{x} - \mathbf{y}), \tag{5.149b}$$

we can derive

$$[Q, \pi(0)] = -\mathrm{i}\sigma(0) \tag{5.150a}$$

$$[Q, \sigma(0)] = \mathrm{i}\pi(0). \tag{5.150b}$$

According to the formal proof (eqns (5.128)–(5.130)) of the Goldstone theorem, the symmetry-breaking condition (5.144) implies the existence of a massless particle state, in this case the quanta of the π-field with

$$\langle \pi|\pi(0)|0\rangle \neq 0$$
$$\langle 0|J_0(0)|\pi\rangle \neq 0. \tag{5.151}$$

To see more explicitly that the right-hand side is nonvanishing we note that (5.129) for our case takes on the form

$$\sum_n (2\pi)^3\delta^3(\mathbf{p}_n)\{\langle 0|J_0(0)|n\rangle\langle n|\pi(0)|0\rangle\, e^{-iE_nt} - \langle 0|\pi(0)|n\rangle\langle n|J_0(0)|0\rangle\, e^{iE_nt}\} = -iv. \tag{5.152}$$

The only contribution on the left-hand side is from the massless π state: $|n\rangle = |\pi\rangle$. Thus

$$\int\frac{d^3p}{2p_0}\,\delta^3(\mathbf{p})\{\langle 0|J_0(0)|\pi(p)\rangle\langle\pi(p)|\pi(0)|0\rangle - \langle 0|\pi(0)|\pi(p)\rangle\langle\pi(p)|J_0(0)|0\rangle\} = -iv \tag{5.153}$$

which is satisfied for

$$\langle 0|J_0(0)|\pi(p)\rangle = ivp_0 \tag{5.154}$$

if the normalization condition $\langle 0|\pi(0)|\pi(p)\rangle = 1$ is taken. We note that manifest covariance requires that

$$\langle 0|J_\mu(0)|\pi(p)\rangle = ivp_\mu. \tag{5.155}$$

Thus the matrix element of the current divergence is

$$\langle 0|\partial^\mu J_\mu(0)|\pi(p)\rangle = vm_\pi^2 \tag{5.156}$$

and current conservation implies that either

$$v = \langle 0|\sigma(0)|0\rangle = 0 \tag{5.157}$$

or

$$m_\pi = 0. \tag{5.158}$$

Thus in this example, the nonvanishing VEV, $v = \langle 0|\sigma|0\rangle$, is related to the pion decay constant (see eqn (5.178) below).

Non-Abelian symmetry case: the $SU(2)_L \times SU(2)_R$ σ model

Consider a theory (Schwinger 1958; Polkinghorne 1958; Gell-Mann and Levy 1960) with the following fields: isotriplet of pions $\boldsymbol{\pi} = (\pi_1, \pi_2, \pi_3)$, an isoscalar σ-field and an isodoublet of nucleons N = (p, n). The Lagrangian given by

$$\mathscr{L} = \tfrac{1}{2}[(\partial_\mu\sigma)^2 + (\partial_\mu\boldsymbol{\pi})^2] + \bar{N}i\gamma^\mu\partial_\mu N + g\bar{N}(\sigma + i\boldsymbol{\tau}\cdot\boldsymbol{\pi}\gamma_5)N - V(\sigma^2 + \boldsymbol{\pi}^2) \tag{5.159}$$

with

$$V(\sigma^2 + \boldsymbol{\pi}^2) = \frac{-\mu^2}{2}(\sigma^2 + \boldsymbol{\pi}^2) + \frac{\lambda}{4}(\sigma^2 + \boldsymbol{\pi}^2)^2 \tag{5.160}$$

is invariant under the SU(2) transformations

$$\sigma \to \sigma' = \sigma$$

$$\boldsymbol{\pi} \to \boldsymbol{\pi}' = \boldsymbol{\pi} + \boldsymbol{\alpha} \times \boldsymbol{\pi}$$

$$\mathrm{N} \to \mathrm{N}' = \mathrm{N} - \mathrm{i}\boldsymbol{\alpha} \cdot \frac{\boldsymbol{\tau}}{2} \mathrm{N} \tag{5.161}$$

for $\alpha_i \ll 1$ with the conserved currents given by

$$J^a_\mu = \bar{\mathrm{N}}\gamma_\mu \frac{\tau^a}{2} \mathrm{N} + \varepsilon^{abc}\pi^b \partial_\mu \pi^c \quad \text{for } a = 1, 2, 3. \tag{5.162}$$

The SU(2) generators are

$$Q^a = \int J^a_0(x) \, \mathrm{d}^3 x. \tag{5.163}$$

This Lagrangian (5.159) is also invariant under the axial SU(2) transformations

$$\sigma \to \sigma' = \sigma + \boldsymbol{\beta} \cdot \boldsymbol{\pi}$$

$$\boldsymbol{\pi} \to \boldsymbol{\pi}' = \boldsymbol{\pi} - \boldsymbol{\beta}\sigma$$

$$\mathrm{N} \to \mathrm{N}' = \mathrm{N} + \mathrm{i}\boldsymbol{\beta} \cdot \frac{\boldsymbol{\tau}}{2} \gamma_5 \mathrm{N} \tag{5.164}$$

with the conserved currents given by

$$A^a_\mu = \bar{\mathrm{N}}\gamma_\mu\gamma_5 \frac{\tau^a}{2} \mathrm{N} + (\partial_\mu \sigma)\pi^a - (\partial_\mu \pi^a)\sigma \tag{5.165}$$

and

$$Q^{5a} = \int A^a_0(x) \, \mathrm{d}^3 x. \tag{5.166}$$

These charges generate the $\mathrm{SU(2)_L} \times \mathrm{SU(2)_R}$ algebra

$$[Q^a, Q^b] = \mathrm{i}\varepsilon^{abc} Q^c$$

$$[Q^a, Q^{5b}] = \mathrm{i}\varepsilon^{abc} Q^{5c}$$

$$[Q^{5a}, Q^{5b}] = \mathrm{i}\varepsilon^{abc} Q^c. \tag{5.167}$$

The spontaneous symmetry breakdown will happen for $\mu^2 > 0$ and the minimum of the potential is at

$$\sigma^2 + \boldsymbol{\pi}^2 = v^2 \quad \text{with } v = (\mu^2/\lambda)^{\frac{1}{2}}. \tag{5.168}$$

We can choose $\langle 0|\boldsymbol{\pi}|0\rangle = 0$ and

$$\langle 0|\sigma|0\rangle = v. \tag{5.169}$$

With the shifted field defined as $\sigma' = \sigma - v$ in $V(\sigma^2 + \pi^2)$; we can easily check that $\boldsymbol{\pi}$s are the massless Goldstone bosons. Following a similar procedure to that outlined in the previous example, we can work out commutators such as

$$[Q^{5a}, \pi^b] = -\mathrm{i}\sigma\delta^{ab}. \tag{5.170}$$

And the choice of (5.169) implies that axial charges Q^{5a} do not annihilate the vacuum; in fact

$$\langle 0|A_\mu^a(0)|\pi^a\rangle \neq 0. \tag{5.171}$$

Thus the $SU(2)_L \times SU(2)_R$ symmetry is broken spontaneously into the SU(2) symmetry generated by the charges Q^a of (5.163) because

$$Q^a|0\rangle = 0 \quad \text{for } a = 1, 2, 3. \tag{5.172}$$

We note that in the original Lagrangian (5.159) there is no nucleon mass term because a $m_N\bar{N}N$ term would not be invariant under the axial transformation (5.164). However the chirally symmetric Yukawa coupling of $g\bar{N}(\sigma + \mathrm{i}\boldsymbol{\tau}\cdot\boldsymbol{\pi}\gamma_5)N$ generates a mass term for the nucleon after spontaneous symmetry breakdown

$$g\bar{N}(\sigma + \mathrm{i}\boldsymbol{\tau}\cdot\boldsymbol{\pi}\gamma_5)N \to gv\bar{N}N + g\bar{N}(\sigma' + \mathrm{i}\boldsymbol{\tau}\cdot\boldsymbol{\pi}\gamma_5)N \tag{5.173}$$

and the isodoublet nucleon mass is

$$m_N = gv. \tag{5.174}$$

The meson masses are (isoscalar) $m_\sigma = \sqrt{2}\mu$ and (isotriplet) $m_\pi = 0$. Thus the symmetry of the Lagrangian $SU(2)_L \times SU(2)_R$ is not realized in the particle spectrum which displays only the isospin SU(2) symmetry. If we have $\mu^2 < 0$, then the symmetry is not hidden: σ and **π** will be degenerate in mass and form the **(2, 2)** irreducible representation of the $SU(2)_L \times SU(2)_R$ group.

.4 PCAC and soft pion theorems

The symmetry of the Lagrangian is always reflected in the algebra of currents. But in spontaneous symmetry breakdown the particle spectrum only realizes that portion of the symmetry which is also respected by the ground state. Thus in the $SU(3)_L \times SU(3)_R$ algebra of electromagnetic and weak currents (see §5.2) we want the symmetry to be broken spontaneously in such a way that the vacuum is only SU(3) symmetric. The Goldstone theorem then implies that there must be eight massless pseudoscalar mesons associated with the spontaneously broken axial charges Q^{5a}, $a = 1, \ldots, 8$. Clearly in reality we do not have such massless particles but eight relatively light mesons, **π**, K, and η. We then conclude that the flavour $SU(3)_L \times SU(3)_R$ symmetry must also be broken explicitly and the masses of the 0^- mesons reflect this chiral symmetry breaking (Dashen 1969)

$$\mathscr{H} = \mathscr{H}_0 + \lambda\mathscr{H}' \tag{5.175a}$$

where $\mathscr{H}_0$ is $SU(3)_L \times SU(3)_R$ invariant and $\mathscr{H}'$ is not. Also the pion isotriplet being much lighter than the Ks and η suggests that we can further decompose the symmetry-breaking Hamiltonian into two terms

$$\lambda\mathscr{H}' = \lambda_1\mathscr{H}_1 + \lambda_2\mathscr{H}_2 \tag{5.175b}$$

where $\mathscr{H}_1$ is $SU(2)_L \times SU(2)_R$ invariant and $\lambda_1 \gg \lambda_2$. Thus, $SU(2)_L \times SU(2)_R$ is a much better symmetry than $SU(3)_L \times SU(3)_R$. An example is the free-quark model where chiral symmetry is broken explicitly by the quark mass term (5.43) with $m_s \gg m_u$, m_d; and we identify $\lambda_1\mathscr{H}_1 = m_s\bar{s}s$ and $\lambda_2\mathscr{H}_2 = m_u\bar{u}u + m_d\bar{d}d$. Thus the pions are expected to have masses proportional to the nonstrange quark masses, and the kaons and eta meson to have masses proportional to the strange quark mass (more on this in the next section). In this section we shall derive a number of soft pion theorems which hold in the chiral symmetry limit ($\lambda_2 = 0$) with pions taken to be massless particles. In the next section soft meson theorems sensitive to the structure of the symmetry-breaking term $\lambda_1\mathscr{H}_1 + \lambda_2\mathscr{H}_2$ will be studied.

PCAC

As the discussion in §5.3 indicates, the Goldstone bosons $\pi^a(x)$ have direct couplings to the broken axial charges Q^{5a} and currents A^a_μ as in (5.171):

$$\langle 0|A^a_\mu(0)|\pi^b(p)\rangle = \mathrm{i}f^{ab}p_\mu \qquad a, b = 1, 2, 3 \tag{5.176}$$

where f^{ab} is some nonzero constant. If we assume that the SU(2) isospin symmetry is unbroken, it may be written as

$$f^{ab} = f_\pi\delta^{ab} \tag{5.177}$$

where f_π is the pion decay constant measured in $\pi^+ \to l^+ + \nu_l$. Experimentally we have $f_\pi \simeq 93$ MeV. Taking the divergence of the axial-vector current, we have

$$\langle 0|\partial^\mu A^a_\mu(0)|\pi^b(p)\rangle = \delta^{ab}f_\pi m^2_\pi. \tag{5.178}$$

Thus, if $\lambda_2 = 0$ in (5.175), the $SU(2)_L \times SU(2)_R$ symmetry in the Hamiltonian is exact, and

$$\partial^\mu A^a_\mu = 0 \tag{5.179}$$

which implies that $m^2_\pi = 0$ in (5.178), as required by the Goldstone theorem. However if the symmetry is explicitly broken , $\lambda_2 \neq 0$, we can rewrite (5.178)

$$\langle 0|\partial^\mu A^a_\mu(0)|\pi^b(p)\rangle = f_\pi m^2_\pi\langle 0|\phi^a(0)|\pi^b(p)\rangle \tag{5.180}$$

where ϕ^a is the pion field operator with the normalization

$$\langle 0|\phi^a(0)|\pi^b(p)\rangle = \delta^{ab}.$$

The generalization of (5.180) into an operator relation

$$\partial^\mu A^a_\mu = f_\pi m^2_\pi\phi^a \qquad a = 1, 2,3 \tag{5.181}$$

is known as the *partially conserved axial-vector current* (PCAC) hypothesis (Nambu 1960; Chou 1961; Gell-Mann and Levy 1960). One would think that such a relation which connects the weak currents A^a_μ and the strong interacting pion fields ϕ^a should have a host of experimental implications. In practice, in order to translate this PCAC hypothesis into relations connecting physically measurable quantities, additional assumptions need to be made as we shall see in the following example.

Low energy theorems with one soft pion

(1) *Goldberger–Treiman relation.* Consider the matrix element of the weak axial vector current between nucleon states $\langle \mathrm{p}(k')|(A^1_\mu + \mathrm{i}A^2_\mu)|\mathrm{n}(k)\rangle$ which is measured in the neutron β-decays. From Lorentz covariance we have

$$\langle \mathrm{p}(k')|(A^1_\mu + \mathrm{i}A^2_\mu)|\mathrm{n}(k)\rangle = \bar{u}_\mathrm{p}(k')[\gamma_\mu\gamma_5 g_A(q^2) + q_\mu\gamma_5 h_A(q^2)]u_\mathrm{n}(k) \tag{5.182}$$

where $q = k - k'$ is the momentum transfer between n and p. The form factors g_A, h_A are functions of the invariant q^2. Experimentally we have $g_A(0) \simeq 1.26$. The current divergence has the matrix element

$$\langle \mathrm{p}(k')|\partial^\mu(A^1_\mu + \mathrm{i}A^2_\mu)|\mathrm{n}(k)\rangle = \mathrm{i}\bar{u}_\mathrm{p}(k')\gamma_5 u_n(k)[2m_\mathrm{N} g_A(q^2) + q^2 h_A(q^2)] \tag{5.183}$$

where m_N is the nucleon mass. The PCAC hypothesis of (5.181) with

$$\phi_\pi^+ = (\phi^1 + \mathrm{i}\phi^2)/\sqrt{2} \tag{5.184}$$

yields

$$\begin{aligned}\langle \mathrm{p}(k')|\partial^\mu(A^1_\mu + \mathrm{i}A^2_\mu)|\mathrm{n}(k)\rangle &= \sqrt{2} f_\pi m_\pi^2 \langle \mathrm{p}(k')|\phi_\pi^+|\mathrm{n}(k)\rangle \\ &= \frac{2 f_\pi m_\pi^2}{-q^2 + m_\pi^2} g_{\pi\mathrm{NN}}(q^2)\mathrm{i}\bar{u}_\mathrm{p}(k')\gamma_5 u_\mathrm{n}(k)\end{aligned} \tag{5.185}$$

where $g_{\pi\mathrm{NN}}(q^2)$ is the πNN vertex function. The physical pion–nucleon coupling constant $g_{\pi\mathrm{NN}}$ is defined as

$$g_{\pi\mathrm{NN}} = g_{\pi\mathrm{NN}}(m_\pi^2) \tag{5.186}$$

with an experimental value of $(g_{\pi\mathrm{NN}}/4\pi) \simeq 14.6$. Comparing (5.185) with (5.183) we have

$$\frac{2 f_\pi m_\pi^2}{-q^2 + m_\pi^2} g_{\pi\mathrm{NN}}(q^2) = 2m_\mathrm{N} g_A(q^2) + q^2 h_A(q^2). \tag{5.187}$$

If we set $q^2 = 0$ in this equation, we have

$$f_\pi g_{\pi\mathrm{NN}}(0) = m_\mathrm{N} g_A(0). \tag{5.188}$$

This relates the nucleon axial vector coupling $g_A(0)$ to the πNN vertex at an off-mass-shell point $g_{\pi\mathrm{NN}}(0)$. In order to convert this into a physical relation we need to make an additional assumption that the variation $g_{\pi\mathrm{NN}}(q^2)$ from $q^2 = 0$ to $q^2 = m_\pi^2$ is small, i.e. that $g_{\pi\mathrm{NN}}(q^2)$ is a 'smooth' function

$$g_{\pi\mathrm{NN}}(0) \simeq g_{\pi\mathrm{NN}}(m_\pi^2) = g_{\pi\mathrm{NN}}. \tag{5.189}$$

Only then do we obtain the *Goldberger–Treiman relation* (1958)

$$f_\pi g_{\pi NN} = m_N g_A(0) \tag{5.190}$$

which is satisfied within 10 per cent for the above-quoted measurement values.

We should remark that only the form factor $h_A(q^2)$ has a pion pole term corresponding to the diagram in Fig. 5.4

$$\text{pion pole of } h_A(q^2) = \frac{\sqrt{2}f_\pi}{m_\pi^2 - q^2}\sqrt{2}g_{\pi NN}(q^2). \tag{5.191}$$

At the pion mass shell point $q^2 = m_\pi^2$, eqn (5.187) yields a trivial identity

$$2f_\pi m_\pi^2 g_{\pi NN}(\mathrm{m}_\pi^2) = 2f_\pi m_\pi^2 g_{\pi NN}(\mathrm{m}_\pi^2).$$

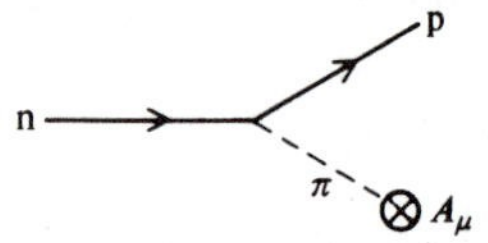

FIG. 5.4.

This example illustrates that the additional smoothness assumption is needed in order to obtain relations among physical quantities from the PCAC hypothesis. In particular we have to extrapolate the pion field from the off-shell point $q^2 = 0$ to the on-shell point $q^2 = m_\pi^2$. Since numerically the pion mass m_π^2 is rather small on the hadronic scale, this extrapolation is believed to cause only a small error. The Goldberger–Treiman relation (5.190) serves as a measure of the typical accuracy of this type of extrapolation. This means that if we extend the PCAC hypothesis to other pseudoscalar mesons, the Ks and η, the extrapolation must be over a much larger kinematical region (from 0 to m_K^2 or m_η^2). Thus the kaon and eta meson PCAC relations are not expected to be as good as those for the pions.

One can also derive the Goldberger–Treiman relation in the chiral $SU(2)_L \times SU(2)_R$ limit ($\lambda_2 = 0$ in (5.175)). In this symmetric limit the currents are conserved

$$\partial^\mu A_\mu = 0 \tag{5.192}$$

which modifies (5.183) to read

$$2M_N g_A(q^2) + q^2 h_A(q^2) = 0 \tag{5.193}$$

and pions are massless Goldstone bosons. Hence the pole term of $h_A(q^2)$ is at $q^2 = 0$. From (5.191) we have

$$\lim_{q^2 \to 0} h_A(q^2) = \frac{-2f_\pi g_{\pi NN}(0)}{q^2} \tag{5.194}$$

which, when combined with (5.193) leads again to (5.190). Thus the deviation from the Goldberger–Treiman relation measures the chiral symmetry-breaking term $\lambda_2\mathcal{H}_2$. Similarly the derivation from the kaon and eta meson

PCAC relations, which we expect to be larger, measured the chiral $SU(3)_L \times SU(3)_R$-breaking term $\lambda_1\mathscr{H}_1 + \lambda_2\mathscr{H}_2$.

(2) *Adler's consistency condition on the* πN *scattering amplitude*. Consider the pion–nucleon scattering amplitude

$$\langle \pi^a(q_2)N(p_2)|\pi^b(q_1)N(p_1)\rangle = i(2\pi)^4\, \delta^4(p_1 + q_1 - p_2 - q_2)\, T^{ab}_{\pi N} \quad (5.195)$$

which has the invariant decomposition

$$T^{ab}_{\pi N} = \bar{u}(p_2)\left[A^{ab} + \gamma\cdot\frac{(q_1+q_2)}{2}B^{ab}\right]u(p_1).$$

The invariant amplitudes A and B are functions of the usual Mandelstam variables s and t or the more symmetric variables

$$\nu \equiv q_1\cdot(p_1+p_2)/2 = q_2\cdot(p_1+p_2)/2$$

$$\nu_B \equiv -q_1\cdot q_2/2.$$

We note that $\nu \to 0$, $\nu_B \to 0$ for either of the pions becoming soft, $q_1 \to 0$ or $q_2 \to 0$. One easily works out the pole-term, i.e. single-nucleon term, contribution to the invariant amplitudes. It can be shown that the combination of invariant amplitudes (which is just the forward scattering for $q_1^2 = q_2^2$)

$$T = A + \nu B \quad (5.196)$$

is nonsingular for either $q_1 \to 0$ and/or $q_2 \to 0$. Furthermore we have the isospin-even and -odd amplitudes

$$A^{ab} = \delta^{ab}A^{(+)} + \tfrac{1}{2}[\tau^a, \tau^b]A^{(-)};$$

similarly for $B^{(\pm)}$ and $T^{(\pm)}$.

To derive the single soft pion theorem we use the standard reduction formula (see, for example, Bjorken and Drell 1965) for the one-pion field operator in (5.195)

$$\begin{aligned} T^{ab}_{\pi N} &= i(-q_2^2 + m_\pi^2)\langle N(p_2)|\phi^a(0)|\pi^b(q_1)N(p_1)\rangle \\ &= \frac{q_2^\mu(-q_2^2+m_\pi^2)}{f_\pi m_\pi^2}\langle N(p_2)|A^a_\mu(0)|\pi^b(q_1)N(p_1)\rangle \end{aligned} \quad (5.197)$$

where we have used the PCAC relation (5.181). Taking the $q_2 \to 0$ limit, the nonsingular amplitude of (5.196) $T(\nu, \nu_B, q_1^2, q_2^2)$ vanishes

$$T^{(+)}(0, 0, m_\pi^2, 0) = 0. \quad (5.198a)$$

This is *Adler's PCAC consistency condition* (Adler 1965*a*). That $T^{(-)}$ is zero in this limit is trivial since it is odd under crossing and we expect it to be proportional to the variable ν. Similarly, we also have the condition,

$$T^{(+)}(0, 0, 0, m_\pi^2) = 0 \quad (5.198b)$$

when taking the $q_1 \to 0$ limit.

Low-energy theorems involving two soft pions

For matrix elements involving more than one current, low-energy theorems may be derived if we combine PCAC and current algebra. In fact the physics and the mathematical procedure to be used are similar to those of the familiar soft-photon theorems which reflects the U(1) gauge invariance of charge conservation (see, for example, Low 1954).

Consider the double divergence of a time-ordered product of two axial vector currents

$$\begin{aligned}\partial_x^\mu \partial_y^\lambda T(A_\mu^a(x)A_\lambda^b(y)) &= \partial_x^\mu \partial_y^\lambda(\theta(x_0 - y_0)A_\mu^a(x)A_\lambda^b(y) \\ &\quad + \theta(y_0 - x_0)A_\lambda^b(y)A_\mu^a(x)) \\ &= \partial_x^\mu(\theta(x_0 - y_0)A_\mu^a(x)\,\partial^\lambda A_\lambda^b(y) \\ &\quad + \theta(y_0 - x_0)\,\partial^\lambda A_\lambda^b(y)A^{a\mu}(x) \\ &\quad - \delta(x_0 - y_0)A_\mu^a(x)A_0^b(y) \\ &\quad + \delta(y_0 - x_0)A_0^b(y)A_\mu^a(x)) \\ &= T(\partial^\mu A_\mu^a(x)\,\partial^\lambda A_\lambda^b(y)) \\ &\quad + \delta(x_0 - y_0)[A_0^a(x), \partial^\lambda A_\lambda^b(y)] \\ &\quad - \partial_x^\mu\,\delta(x_0 - y_0)[A_\mu^a(x), A_0^b(y)]. \qquad (5.199)\end{aligned}$$

Sandwiching this identity between nucleon states and taking the Fourier transform

$$\int d^4x\, d^4y\, e^{iq_1\cdot x}\, e^{-iq_2\cdot y},$$

we obtain

$$\begin{aligned} q_1^\mu q_2^\lambda &\int d^4x\, e^{iq_1\cdot x}\langle N(p_2)|T(A_\mu^a(x)A_\lambda^b(0))|N(p_1)\rangle \\ &= \int d^4x\, e^{iq_1\cdot x}\{\langle N(p_2)|T(\partial^\mu A_\mu^a(x)\,\partial^\lambda A_\lambda^b(0))|N(p_1)\rangle \\ &\quad - iq_1^\mu\langle N(p_2)|\delta(x_0)[A_0^b(0), A_\mu^a(x)]|N(p_1)\rangle \\ &\quad + \langle N(p_2)|\delta(x_0)[A_0^a(x), \partial^\lambda A_\lambda^b(0)]|N(p_1)\rangle\} \qquad (5.200)\end{aligned}$$

where we have used translational invariance and factored out a $(2\pi)^4\,\delta(p_1 + q_1 - p_2 - q_2)$. This relation between the matrix elements of currents and the matrix elements of divergences is an example of the *Ward identities*. It is the starting point for the derivation of low-energy theorems. The question of maintaining the Ward identities in higher-order perturbations will be discussed in §§6.1 and 6.2.

PCAC implies that the first term on the right-hand side of (5.200) is the (nucleon) matrix element of a time-ordered product of two-pion operators, i.e., it is the πN scattering amplitude. The second term can be evaluated from the $\mathrm{SU}(2)_\mathrm{L} \times \mathrm{SU}(2)_\mathrm{R}$ current algebra of (5.80)

$$\delta(x_0)[A_0^b(0), A_\mu^a(x)] = -\mathrm{i}\,\delta(x)\varepsilon^{abc}V_\mu^c(x). \tag{5.201}$$

In principle there is also the contribution from the Schwinger term in this commutator. But as it turns out, the time-ordered product defined in (5.199) is not covariant because of the singularities in the product $T(A_\mu^a(x)A_\nu^b(y))$ as $x_0 \to y_0$ and an extra term should be added to make it covariant. This extra term will cancel the Schwinger term in the commutator of (5.201) (see, for example, Adler and Dashen 1968). The net result is that if one uses the usual time-ordered product one does not have to be concerned about the Schwinger term in the derivation of the Ward identities.

The third term on the right-hand side of (5.200) is an equal-time commutator of a current and a divergence. This commutator, called the *σ-term*, is not governed by the current algebra and it depends on the symmetry-breaking terms (5.175b). In the following application we shall eventually take the limit of $p_1 \to p_2 \equiv p$, $q_1 \to q_2 \equiv q \to 0$. In such a kinematical configuration the σ-term can be shown on general grounds to be symmetric in the indices a and b. To see this, write

$$\lim_{q\to 0} \sigma_\mathrm{N}^{ab}(\mathrm{p}, \mathrm{q}) = \sigma_\mathrm{N}^{ab} = \mathrm{i}\int \mathrm{d}^4x\, \delta(x_0)\langle \mathrm{N}(p)|[A_0^a(\mathbf{x}, x_0), \partial^\lambda A_\lambda^b(0, x_0)]|\mathrm{N}(p)\rangle. \tag{5.202}$$

Using translational invariance and changing variables $\mathbf{x}$ to $-\mathbf{x}$, we can write (5.202) as

$$\begin{aligned}\sigma_\mathrm{N}^{ab} &= \mathrm{i}\int \mathrm{d}^3x \langle \mathrm{N}(p)|[A_0^a(0, x_0), \partial^\lambda A_\lambda^b(\mathbf{x}, x_0)]|\mathrm{N}(p)\rangle \\ &= \mathrm{i}\int \mathrm{d}^3x \langle \mathrm{N}(p)|[A_0^a(\mathbf{x}, x_0), \partial^0 A_0^b(0, x_0)]|\mathrm{N}(p)\rangle \end{aligned}\tag{5.203}$$

where we have used the fact that the spatial divergence vanishes upon integration over all space. We then have

$$\begin{aligned}\sigma_\mathrm{N}^{ab} &= \mathrm{i}\partial_0 \int \mathrm{d}^3x \langle \mathrm{N}(p)|[A_0^a(\mathbf{x}, x_0), A_0^b(0, x_0)]|\mathrm{N}(p)\rangle \\ &\quad - \mathrm{i}\int \mathrm{d}^3x \langle \mathrm{N}(p)|[\partial_0 A_0^a(\mathbf{x}, x_0), A_0^b(0, x_0)]|\mathrm{N}(p)\rangle \end{aligned}\tag{5.204}$$

or

$$\sigma_\mathrm{N}^{ab} - \sigma_\mathrm{N}^{ba} = \mathrm{i}\partial_0 \int \mathrm{d}^3x \langle \mathrm{N}(p)|[A_0^a(\mathbf{x}, x_0), A_0^b(0, x_0)]|\mathrm{N}(p)\rangle.$$

The commutator on the right-hand side will give an isospin charge after the integration over space is performed, and it is time-independent if we neglect isospin-breaking effects. Thus we have

$$\sigma_\mathrm{N}^{ab} = \sigma_\mathrm{N}^{ba}. \tag{5.205}$$

It is also clear from the form of eqn (5.203) that the σ-term, being proportional to $\partial^\lambda A_\lambda$, represents chiral symmetry-breaking effects.

Adler–Weisberger relation. To derive the low-energy theorems for πN amplitudes involving two soft pions we reduce out both of the pions in (5.195) and apply the PCAC formula (5.181)

$$\begin{aligned} T^{ab}_{\pi \mathrm{N}} &= \mathrm{i}\int \mathrm{d}^4x\, \mathrm{e}^{\mathrm{i}q_1\cdot x}(q_1^2 - m_\pi^2)(q_2^2 - m_\pi^2) \\ &\quad \times \langle \mathrm{N}(p_2)|T(\phi^a(x)\phi^b(0))|\mathrm{N}(p_1)\rangle \\ &= \mathrm{i}(q_1^2 - m_\pi^2)(q_2^2 - m_\pi^2)m_\pi^{-4}f_\pi^{-2}\int \mathrm{d}^4x\, \mathrm{e}^{\mathrm{i}q_1\cdot x} \\ &\quad \times \langle \mathrm{N}(p_2)|T(\partial^\mu A^a_\mu(x)\, \partial^\lambda A^b_\lambda(0))|\mathrm{N}(p_1)\rangle. \end{aligned} \tag{5.206}$$

Similarly consider the weak axial-vector current amplitude

$$\begin{aligned} &(2\pi)^4\, \delta^4(p_1 + q_1 - p_2 - q_2)T^{ab}_{\mu\lambda} \\ &\quad = \int \mathrm{d}^4x\, \mathrm{d}^4y\, \mathrm{e}^{\mathrm{i}q_1\cdot x}\, \mathrm{e}^{-\mathrm{i}q_2\cdot y}\langle \mathrm{N}(p_2)|T(A^a_\mu(x)A^b_\lambda(y))|\mathrm{N}(p_1)\rangle. \end{aligned} \tag{5.207}$$

The amplitudes of (5.206) and (5.207) are related by the Ward identity (5.200). With the forward scattering kinematics $p_1 = p_2 \equiv p$ and $q_1 = q_2 \equiv q$, we have

$$q^\mu q^\lambda T^{ab}_{\mu\lambda} = -\mathrm{i}(q^2 - m_\pi^2)^{-2}m_\pi^4 f_\pi^2 T^{ab}_{\pi\mathrm{N}} + \mathrm{i}\nu[\tau^a, \tau^b]/2 - \mathrm{i}\sigma^{ab}_{\mathrm{N}}(p, q) \tag{5.208}$$

where we have used the definitions in (5.206), (5.207), and (5.202). The commutation relation in (5.201) implies that the second term on the right-hand side of (5.200) takes on the form

$$\begin{aligned} -\mathrm{i}q^\mu &\int \mathrm{d}^4x\, \mathrm{e}^{\mathrm{i}q\cdot x}\langle \mathrm{N}(p)|\delta(x_0)[A^b_0(0), A^a_\mu(x)]|\mathrm{N}(p)\rangle \\ &= \varepsilon^{abc}q^\mu \bar{u}(p)\gamma_\mu \tau^c u(p)/2 \\ &= 2p\cdot q\varepsilon^{abc}\tau^c/2 = -\mathrm{i}\nu[\tau^a, \tau^b]/2. \end{aligned} \tag{5.209}$$

The σ-term, as we have demonstrated above, is symmetric in a, b. Since the pion has isospin 1, the isospin symmetric t-channel state must be 0 or 2. Only the isospin-zero state can contribute here since the nucleon has isospin 1/2. Thus we write

$$\sigma^{ab} = \delta^{ab}\sigma_{\mathrm{N}}. \tag{5.210}$$

The left-hand side of (5.208) is quite complicated as it involves contributions from all possible intermediate states that can couple to the nucleon through the axial vector currents. This can be simplified by taking the low-energy limit $q_\mu \to 0$ so that the only surviving terms in $T^{ab}_{\mu\nu}$ are those singular in this limit. It is easy to see that such terms correspond to the one-nucleon pole diagrams in Fig. (5.5).

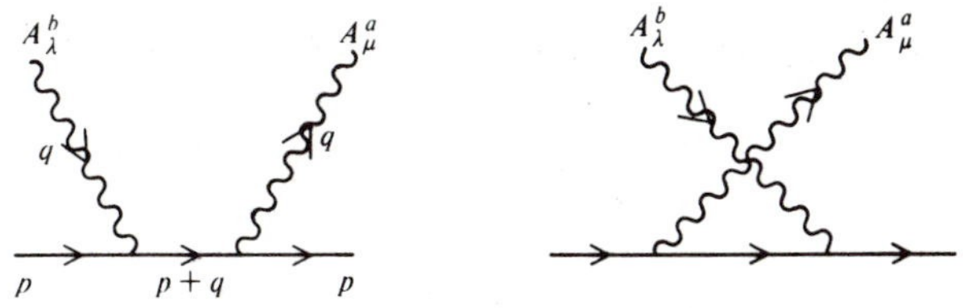

FIG. 5.5.

$$(q^\mu q^\lambda T^{ab}_{\mu\lambda})_{\text{pole}} = 2\mathrm{i}g_A^2\{[\tau^b, \tau^a]\nu - \delta^{ab}q^2\}(\nu^2 - m_N^2 q^2)/(q^4 - 4\nu^2) \tag{5.211}$$

where g_A is defined in (5.182). Since $q^2 \ll \nu = p \cdot q$ for small q, we have

$$(q^\mu q^\lambda T^{ab}_{\mu\lambda})_{\text{pole}} \simeq \mathrm{i}g_A^2\nu[\tau^a, \tau^b]/2. \tag{5.212}$$

Thus in the low-energy limit, the Ward identity (5.208) becomes

$$\mathrm{i}g_A^2\nu[\tau^a, \tau^b]/2 = -\mathrm{i}f_\pi^2 T^{ab}_{\pi\mathrm{N}} + \mathrm{i}\nu[\tau^a, \tau^b]/2 - \mathrm{i}\delta^{ab}\sigma_{\mathrm{N}}. \tag{5.213}$$

The forward amplitude $T_{\pi N}$ is just the combination $A + \nu B$ of (5.196). We have the following soft-pion theorems for the isospin odd and even πN amplitudes $T^{(\pm)}(\nu, \nu_B, q_1^2, q_2^2)$:

$$\lim_{\nu \to 0} \nu^{-1} T^{(-)}(\nu, 0, 0, 0) = (1 - g_A^2)/f_\pi^2 \tag{5.214}$$

and

$$T^{(+)}(0, 0, 0, 0) = -\sigma_{\mathrm{N}}/f_\pi^2. \tag{5.215}$$

To make contact with physical amplitudes ($q_1^2 = q_2^2 = m_\pi^2$) we can extrapolate the result to the physical threshold (scattering length) at $\nu = \nu_0 = m_\pi m_{\mathrm{N}}$, $\nu_B = -m_\pi^2/2$,

$$\nu_0^{-1} T^{(-)}(\nu_0, -m_\pi^2/2, m_\pi^2, m_\pi^2) = f_\pi^{-2}(1 - g_A^2) + O(\lambda_2) \tag{5.216}$$

$$T^{(+)}(\nu_0, -m_\pi^2/2, m_\pi^2, m_\pi^2) = O(\lambda_2) \tag{5.217}$$

where we have used the fact that m_π^2 and the σ-term are chiral symmetry-breaking effects and are of order $O(\lambda_2)$ in (5.175b) (Weinberg 1966). Alternatively we can convert the low-energy theorem to sum rules by using dispersion relations. For example, since $T^{(-)}(\nu, q^2)$ is odd under $\nu \to -\nu$, we can write an unsubtracted dispersion relation for $\nu^{-1}T^{(-)}(\nu, 0)$,

$$\frac{T^{(-)}(\nu, 0)}{\nu} = \frac{2}{\pi} \int_{\nu_0}^{\infty} \frac{\operatorname{Im} T^{(-)}(\nu', 0)\, \mathrm{d}\nu'}{\nu'^2 - \nu^2}. \tag{5.218}$$

Then setting $\nu = 0$ and using (5.214), we get

$$\frac{1-g_A^2}{f_\pi^2} = \frac{2}{\pi}\int_{\nu_0}^{\infty} \frac{\operatorname{Im} T^{(-)}(\nu, 0)\, \mathrm{d}\nu}{\nu^2}. \tag{5.219}$$

We may use the Goldberger–Treiman relation (5.190) to eliminate f_π

$$\frac{1}{g_A^2} = 1 + \frac{2m_N^2}{\pi g_{\pi NN}^2}\int_{\nu_0}^{\infty} \frac{\operatorname{Im} T^{(-)}(\nu, 0)\, \mathrm{d}\nu}{\nu^2}. \tag{5.220}$$

If we make the smoothness assumption

$$\operatorname{Im} T^{(-)}(\nu, 0) \simeq \operatorname{Im} T^{(-)}(\nu, m_\pi^2),$$

we can relate the on-shell amplitude $\operatorname{Im} T^{(-)}(\nu, m_\pi^2)$ to the πN cross-section from the optical theorem

$$\operatorname{Im} T^{(-)}(\nu, m_\pi^2) = \nu\sigma_{\text{tot}}^{(-)}(\nu) = \nu[\sigma_{\text{tot}}^{\pi^- \mathrm{p}}(\nu) - \sigma_{\text{tot}}^{\pi^+ \mathrm{p}}(\nu)]. \tag{5.221}$$

The sum rule in (5.220) becomes

$$\frac{1}{g_A^2} = 1 + \frac{2m_N^2}{\pi g_{\pi NN}^2}\int_{\nu_0}^{\infty} \frac{\mathrm{d}\nu[\sigma_{\text{tot}}^{\pi^- \mathrm{p}}(\nu) - \sigma_{\text{tot}}^{\pi^+ \mathrm{p}}(\nu)]}{\nu}. \tag{5.222}$$

This is the *Adler–Weisberger relation* (Adler 1965*b*; Weisberger 1966). Using experimental values for the πp total cross-sections we get the weak axial nucleon coupling $g_A \simeq 1.24$ which agrees quite well with the experimental value 1.259 ± 0.017.

The isospin-even amplitude is related to the chiral symmetry-breaking σ-term, which will be examined in the next section.

5.5 Pattern of chiral symmetry breaking

Soft-pion theorems such as the Goldberger–Treiman relation and the Adler–Weisberger sum rule are exact chiral $SU(2)_L \times SU(2)_R$ symmetric results. They are not sensitive to the structure of the symmetry-breaking terms in (5.175b). On the other hand, the σ-term represents chiral symmetry-breaking effects. Consider this commutator of axial current with its divergence appearing in (5.199) and (5.203), taken between some general hadronic states of momentum p

$$\begin{aligned}
\sigma_{\alpha\beta}^{ab} &= \mathrm{i}\int \mathrm{d}^3x \langle\beta(p)|[A_0^a(\mathbf{x}, 0), \partial^0 A_0^b(0)]|\alpha(p)\rangle \\
&= -\int \mathrm{d}^3y \langle\beta(p)|[Q^{5a}, [\mathscr{H}(\mathbf{y}, 0), A_0^b(0)]]|\alpha(p)\rangle \\
&= -\int \mathrm{d}^3y \langle\beta(p)|[Q^{5a}, [\mathscr{H}(0), A_0^b(\mathbf{y}, 0)]]|\alpha(p)\rangle \\
&= \langle\beta(p)|[Q^{5a}, [Q^{5b}, \mathscr{H}(0)]]|\alpha(p)\rangle.
\end{aligned} \tag{5.223}$$

Thus the σ-term is simply a double commutator of the Hamiltonian density $\mathscr{H}$ with two axial charges. If the chiral symmetry-breaking term is absent in $\mathscr{H}$, the axial charges are conserved. They commute with the Hamiltonian and the σ commutator vanishes. Thus we can replace $\mathscr{H}$ in (5.223) by the chiral symmetry-breaking term (5.175b).

Measuring the nucleon and vacuum σ-terms

We have already seen (5.215) that the isospin-even πN scattering amplitude in the soft-pion limit is proportional to the σ-commutator between the nucleon states. One must be careful in relating this result for $T^{(+)}(0, 0, 0, 0)$ to the on-shell $q_1^2 = q_2^2 = m_\pi^2$ amplitude since this extrapolation involves a correction term of the same order as the σ-term itself. However a systematic expansion in the chiral $SU(2)_L \times SU(2)_R$ symmetry-breaking parameter λ_2 of (5.175b) is possible (Cheng and Dashen 1971)

$$T^{(+)}(0, 0, m_\pi^2, m_\pi^2) = T^{(+)}(0, 0, 0, 0) + m_\pi^2 \frac{\partial T^{(+)}}{\partial q_1^2} + m_\pi^2 \frac{\partial T^{(+)}}{\partial q_2^2} + O(m_\pi^4). \qquad (5.224)$$

Using Adler's consistency conditions (5.198) such as

$$T^{(+)}(0, 0, m_\pi^2, 0) = T^{(+)}(0, 0, 0, 0) + m_\pi^2 \frac{\partial T^{(+)}}{\partial q_1^2} + O(m_\pi^4) = 0,$$

we have

$$\begin{aligned} T^{(+)}(0, 0, m_\pi^2, m_\pi^2) &= -T^{(+)}(0, 0, 0, 0) + O(m_\pi^4) \\ &= \sigma_N/f_\pi^2 + O(m_\pi^4). \end{aligned} \qquad (5.225)$$

In this expansion in powers of the symmetry-breaking parameter λ_2 (i.e. m_π^2), we have ignored any possible non-analyticity problem (Li and Pagels 1971). It should be noted that even for the on-shell amplitude the kinematic point $\nu = \nu_B = 0$ is still outside the physical region. However the amplitude value at this point can be reliably extrapolated from the physical quantities via ordinary dispersion relations.

So far we have concentrated on the $SU(2)_L \times SU(2)_R$ chiral symmetry. The generalization to $SU(3)_L \times SU(3)_R$ is straightforward. The PCAC relation for octet axial-vector currents reads

$$\partial^\mu A_\mu^a = f_a m_a^2 \phi^a \qquad a = 1, 2, \ldots, 8 \qquad (5.226)$$

where the ϕ^as are the field operators for the octet pseudoscalar mesons. The generalized Goldberger–Treiman relations and soft-meson theorems for meson–baryon scattering amplitudes can be derived in a similar fashion.

We can also obtain more low-energy theorems in the soft-meson limit by considering other matrix elements of the currents. In particular, from (5.226) we have

$$\langle 0|\, \partial^\mu A_\mu^a \,|P_b(k)\rangle = \delta_{ab} m_a^2 f_a. \qquad (5.227)$$

Using the reduction formula and PCAC, we can write this equation as

$$\delta_{ab} m_a^2 f_a = \frac{i(m_b^2 - k^2)}{f_b m_b^2} \int d^4x \, e^{-ik\cdot x} \langle 0|T(\partial^\mu A_\mu^a(0)\, \partial^\nu A_\nu^b(x))|0\rangle$$

$$= \frac{i(m_b^2 - k^2)}{f_b m_b^2} \{ik_\nu \int d^4x \langle 0|T(\partial^\mu A_\mu^a(0)\, A_\nu^b(x))|0\rangle \, e^{-ik\cdot x}$$

$$- \int d^4x \, e^{-ik\cdot x} \langle 0|\delta(x_0)[A_0^b(x), \partial^\mu A_\mu^a(0)]|0\rangle\}. \tag{5.228}$$

The low-energy theorem is then

$$\delta_{ab} m_a^2 f_a^2 = i \int d^4x \langle 0|\delta(x_0)[A_0^b(x), \partial^\mu A_\mu^a(0)]|0\rangle = \sigma_0^{ab} \tag{5.229}$$

where

$$\sigma_0^{ab} = \langle 0|[Q^{5a}, [Q^{5b}, \mathscr{H}(0)]]|0\rangle.$$

Thus the pseudoscalar meson masses are related to the vacuum matrix elements of the σ-term. (This relation (5.229) can also be derived by directly sandwiching eqn (5.199) between the vacuum states.)

The (3, 3*) + (3*, 3) theory of chiral symmetry breaking

The nucleon and vacuum matrix elements of the σ-commutator are related to the experimentally measurable quantities through the relations in (5.225) and (5.229). We now need a theory of chiral symmetry breaking. What is the structure of the $\lambda\mathscr{H}'$ term in (5.175)?

A simple possibility is that the chiral symmetry is broken by the quark mass term only

$$\lambda\mathscr{H}' = m_u \bar{u}u + m_d \bar{d}d + m_s \bar{s}s \tag{5.230}$$

or

$$\lambda_1 \mathscr{H}_1 = m_s \bar{s}s \quad \text{and} \quad \lambda_2 \mathscr{H}_2 = m_u \bar{u}u + m_d \bar{d}d \tag{5.231}$$

since the quark fields transform as

$$q_L = \tfrac{1}{2}(1 - \gamma_5)q \sim (\mathbf{3}, \mathbf{0})$$

$$q_R = \tfrac{1}{2}(1 + \gamma_5)q \sim (\mathbf{0}, \mathbf{3})$$

of the $SU(3)_L \times SU(3)_R$ group. $\lambda\mathscr{H}'$, being of the form $\bar{q}_L q_R + \bar{q}_R q_L$, transforms as a member of the $(\mathbf{3}, \mathbf{3^*}) + (\mathbf{3^*}, \mathbf{3})$ representation. This is the theory of chiral symmetry-breaking proposed by Gell-Mann, Oakes, and Renner (1968), and by Glashow and Weinberg (1968). In group-theoretical language we say

$$\lambda\mathscr{H}' = c_0 u_0 + c_3 u_3 + c_8 u_8 \tag{5.232}$$

where the u_as are a set of scalar densities. In terms of the quark fields (5.38) and Gell-Mann matrices they have the representation

$$u_i = \bar{q}\lambda_i q, \quad i = 0, 1, 2, \ldots, 8 \tag{5.233}$$

with

$$\lambda_0 = (2/3)^{\frac{1}{2}}\mathbf{1}. \tag{5.234}$$

Similarly define the pseudoscalar density as

$$v_i = -\mathrm{i}\bar{\mathrm{q}}\lambda_i\gamma_5\mathrm{q}. \tag{5.235}$$

With the representation of u_i, v_i in (5.233) and (5.235) and Q^a, Q^{5a} in (5.44) and (5.48) one can then work out their commutators using the canonical anti-commutation relations for quark fields,

$$\{\mathrm{q}_\alpha^\dagger(\mathbf{x}, t), \mathrm{q}_\beta(\mathbf{y}, t)\} = \delta_{\alpha\beta}\,\delta^3(\mathbf{x}-\mathbf{y}), \tag{5.236}$$

where α, β label the Dirac and flavour indices. We shall illustrate this procedure for the case $[Q^{5a}(t), u^b(\mathbf{x}, t)]$. Suppressing all space–time dependences we have

$$\begin{aligned}\left[\bar{\mathrm{q}}\frac{\lambda^a}{2}\gamma_0\gamma_5\mathrm{q}, \bar{\mathrm{q}}\lambda^b\mathrm{q}\right] &= [\mathrm{q}_\alpha^\dagger\mathrm{q}_\beta, \mathrm{q}_\gamma^\dagger\mathrm{q}_\delta](\lambda^a\gamma_5)_{\alpha\beta}(\lambda^b\gamma^0)_{\gamma\delta}/2\\ &= (-\mathrm{q}_\alpha^\dagger\mathrm{q}_\gamma^\dagger\{\mathrm{q}_\beta, \mathrm{q}_\delta\} + \mathrm{q}_\alpha^\dagger\{\mathrm{q}_\beta, \mathrm{q}_\gamma^\dagger\}\mathrm{q}_\delta - \mathrm{q}_\gamma^\dagger\{\mathrm{q}_\alpha^\dagger, \mathrm{q}_\delta\}\mathrm{q}_\beta\\ &\quad + \{\mathrm{q}_\alpha^\dagger, \mathrm{q}_\gamma^\dagger\}\mathrm{q}_\beta\mathrm{q}_\delta)(\lambda^a\gamma_5)_{\alpha\beta}(\lambda^b\gamma^0)_{\gamma\delta}/2\\ &= \mathrm{q}^\dagger[\lambda^a\gamma_5, \lambda^b\gamma^0]\mathrm{q}/2 = -\bar{\mathrm{q}}\gamma_5\{\lambda^a, \lambda^b\}\mathrm{q}/2.\end{aligned} \tag{5.237}$$

We can define a totally symmetric symbol d_{abc} by

$$\left\{\frac{\lambda^a}{2}, \frac{\lambda^b}{2}\right\} = \mathrm{i}d^{abc}\frac{\lambda^c}{2}. \tag{5.238}$$

The nonvanishing elements are

$$\begin{aligned}&d_{118} = d_{228} = d_{338} = 1/\sqrt{3}, d_{448} = d_{558} = d_{668} = d_{778} = -\tfrac{1}{2}(3^{\frac{1}{2}}),\\ &d_{344} = d_{355} = -d_{366} = -d_{377} = d_{146} = d_{157} = -d_{247} = d_{256} = 1/2.\end{aligned} \tag{5.239}$$

Furthermore, if we supplement this with (5.234) and

$$d_{0ab} = (\tfrac{2}{3})^{\frac{1}{2}}\,\delta_{ab}, \tag{5.240}$$

we have

$$\delta(x_0)[Q_a^5(x_0), u_j(\mathbf{x}, x_0)] = -\mathrm{i}d_{ajk}v_k(0)\,\delta^4(x). \tag{5.241a}$$

Similarly,

$$\delta(x_0)[Q_a^5(x_0), v_j(\mathbf{x}, x_0)] = \mathrm{i}d_{ajk}u_k(0)\,\delta^4(x) \tag{5.241b}$$

$$\delta(x_0)[Q_a(x_0), u_j(\mathbf{x}, x_0)] = \mathrm{i}f_{ajk}u_k(0)\,\delta^4(x) \tag{5.241c}$$

$$\delta(x_0)[Q_a(x_0), v_j(\mathbf{x}, x_0)] = \mathrm{i}f_{ajk}v_k(0)\,\delta^4(x) \tag{5.241d}$$

with $a = 1, \ldots, 8$ and $j, k = 0, 1, \ldots, 8$. The f_{abc}s are the usual SU(3) structure constants with $f_{ab0} \equiv 0$.

In the quark model language

$$u_0 = (\tfrac{2}{3})^{\frac{1}{2}}(\bar{\mathrm{u}}\mathrm{u} + \bar{\mathrm{d}}\mathrm{d} + \bar{\mathrm{s}}\mathrm{s})$$
$$u_8 = (\tfrac{1}{3})^{\frac{1}{2}}(\bar{\mathrm{u}}\mathrm{u} + \bar{\mathrm{d}}\mathrm{d} - 2\bar{\mathrm{s}}\mathrm{s})$$
$$u_3 = (\bar{\mathrm{u}}\mathrm{u} - \bar{\mathrm{d}}\mathrm{d}). \tag{5.242}$$

The coefficients in (5.232) correspond to quark masses

$$c_0 = \frac{1}{\sqrt{6}}(m_\mathrm{u} + m_\mathrm{d} + m_\mathrm{s})$$
$$c_8 = \frac{1}{\sqrt{3}}\left(\frac{m_\mathrm{u} + \mathrm{m}_\mathrm{d}}{2} - m_\mathrm{s}\right)$$
$$c_3 = \frac{1}{3}(m_\mathrm{u} - m_\mathrm{d}). \tag{5.243}$$

In the following we shall first assume isospin invariance; hence $m_\mathrm{u} = m_\mathrm{d}$ or $c_3 = 0$. The question of isospin violation due to $m_\mathrm{u} \neq m_\mathrm{d}$ will be taken up at the end of this section

$$\lambda\mathscr{H}' = c_0 u_0 + c_8 u_8. \tag{5.244}$$

Thus SU(3) symmetry breaking is due entirely to the $c_8 u_8$ term.

Current quark masses

The double commutator of the σ-term can be calculated using (5.241a) and (5.241b). In actual computation it is simpler if one takes $\lambda\mathscr{H}'$ and the F_a^5s to be 3 × 3 matrices and computes the *anticommutator* of (5.237) directly. One finds (eqn (5.229))

$$f_\pi^2 m_\pi^2 = \frac{(m_\mathrm{u} + m_\mathrm{d})}{2}\langle 0|\bar{\mathrm{u}}\mathrm{u} + \bar{\mathrm{d}}\mathrm{d}|0\rangle$$
$$f_\mathrm{K}^2 m_\mathrm{K}^2 = \frac{(m_\mathrm{u} + m_\mathrm{s})}{2}\langle 0|\bar{\mathrm{u}}\mathrm{u} + \bar{\mathrm{s}}\mathrm{s}|0\rangle$$
$$f_\eta^2 m_\eta^2 = \frac{(m_\mathrm{u} + m_\mathrm{d})}{6}\langle 0|\bar{\mathrm{u}}\mathrm{u} + \bar{\mathrm{d}}\mathrm{d}|0\rangle + \frac{4m_\mathrm{s}}{3}\langle 0|\bar{\mathrm{s}}\mathrm{s}|0\rangle. \tag{5.245}$$

Since the SU(3) symmetry is not spontaneously broken, we will take the vacuum to be SU(3)-symmetric, i.e.

$$\langle 0|\bar{\mathrm{u}}\mathrm{u}|0\rangle = \langle 0|\bar{\mathrm{d}}\mathrm{d}|0\rangle = \langle 0|\bar{\mathrm{s}}\mathrm{s}|0\rangle \equiv \mu^3, \tag{5.246}$$

and, from the definition of decay constants (5.176) and conditions for spontaneous chiral symmetry-breaking ((eqns (5.169) and (5.154)), we have

$$f_\pi = f_\mathrm{K} = f_\eta \equiv f. \tag{5.247}$$

Besides recovering the Gell-Mann–Okubo mass relation $4m_\mathrm{K}^2 = 3m_\eta^2 + m_\pi^2$, we obtain the ratio of quark masses

$$\frac{m_\mathrm{u} + m_\mathrm{d}}{2m_\mathrm{s}} = \frac{m_\pi^2}{2m_\mathrm{K}^2 - m_\pi^2} \simeq \frac{1}{25}. \tag{5.248}$$

The pseudoscalar masses suggest that the strange quark has a much larger mass than the masses of the non-strange quarks u and d. This means that the $SU(2)_L \times SU(2)_R$ symmetry is a much better symmetry than $SU(3)_L \times SU(3)_R$. In terms of the parameters in (5.244), this means that

$$\frac{c_8}{c_0} \simeq -1.25 \tag{5.249}$$

is not far from the $SU(2)_L \times SU(2)_R$ symmetric value of $-2^{\frac{1}{2}}$.

πN σ-term

The πN σ-term for the symmetry-breaking Hamiltonian (5.244) can be similarly computed in the quark model

$$\sigma_N = \tfrac{1}{2}(m_u + m_d)\langle N|\bar{u}u + \bar{d}d|N\rangle. \tag{5.250}$$

At this stage of theoretical development we still do not have a reliable method for calculating such a matrix element. One possible way to estimate this quantity is to invoke the Zweig rule (Cheng 1976)

$$\langle N|\bar{s}s|N\rangle \simeq 0 \tag{5.251}$$

since the nucleon is supposed to contain little strange-quark component (see the discussion following eqn (7.85)). Eqn (5.250) may then be written

$$\begin{aligned}\sigma_N &\simeq \tfrac{1}{2}(m_u + m_d)\langle N|\bar{u}u + \bar{d}d - 2\bar{s}s|N\rangle \\ &= \frac{3(m_u + m_d)}{m_u + m_d - 2m_s}\langle N|c_8 u_8|N\rangle\end{aligned} \tag{5.252}$$

where we have used eqns (5.242) and (5.243). The nucleon mass shift due to the SU(3)-breaking Hamiltonian $c_8 u_8$ is

$$\Delta m_N = \langle N|c_8 u_8|N\rangle \tag{5.253}$$

which can be related to the general baryon octet mass splittings by

$$\Delta m_B = \langle B|c_8 u_8|B\rangle = \alpha\,\mathrm{tr}(\bar{B}u_8 B) + \beta\,\mathrm{tr}(\bar{B}Bu_8). \tag{5.254}$$

On the right-hand side we have written the baryon octet (and u_8) in 3×3 matrices (see (4.157)). That there are two terms on the right-hand side reflects the fact that there are two **8**s in the product **8** × **8** (eqn (4.124)); hence two SU(3) scalars are contained in the product **8** × **8** × **8**. Reading off eqn (5.254) the coefficients α and β are then related to the baryon mass shifts as

$$\begin{aligned}m_N &= m_0 + (\alpha - 2\beta) \\ m_\Sigma &= m_0 + (\alpha + \beta) \\ m_\Xi &= m_0 - (2\alpha - \beta) \\ m_\Lambda &= m_0 - (\alpha + \beta)\end{aligned} \tag{5.255}$$

where we have absorbed an inessential common factor of $\sqrt{3}$ into α and β.

Besides recovering the Gell-Mann–Okubo mass formula eqn (4.166) we can identify

$$\langle \mathrm{N}|c_8 u_8|\mathrm{N}\rangle = (\alpha - 2\beta) = m_\Lambda - m_\Xi. \tag{5.256}$$

Hence, from (5.252),

$$\sigma_\mathrm{N} = \frac{3(m_\mathrm{u} + m_\mathrm{d})}{m_\mathrm{u} + m_\mathrm{d} - 2m_\mathrm{s}}(m_\Lambda - m_\Xi). \tag{5.257}$$

For the ratio (5.248) we obtain $\sigma_\mathrm{N} \approx 30$ MeV. It is clear that there is quite a bit of uncertainty in this estimate, yet the πN σ-term is one of the few places where one can probe the chiral symmetry-breaking pattern with any sort of reliability.

$m_\mathrm{u} \neq m_\mathrm{d}$ and isospin violation by the strong interaction

To close this section we also examine the possibility of isospin symmetry breaking due to $m_\mathrm{u} \neq m_\mathrm{d}$. Since we already know that electromagnetic interaction violates this invariance, we must untangle these two separate mechanisms of isospin-breaking; the $\mathrm{SU}(3)_\mathrm{L} \times \mathrm{SU}(3)_\mathrm{R}$-violating term may be written

$$\lambda\mathscr{H}' = \mathscr{H}_m + \mathscr{H}_\gamma \tag{5.258}$$

with

$$\mathscr{H}_m = m_\mathrm{u}\bar{\mathrm{u}}\mathrm{u} + m_\mathrm{d}\bar{\mathrm{d}}\mathrm{d} + m_\mathrm{s}\bar{\mathrm{s}}\mathrm{s} \tag{5.259}$$

$$\mathscr{H}_\gamma = e^2 \int \mathrm{d}^4x\, T(J^\mu(x)J^\nu(0))D_{\mu\nu}(x) \tag{5.260}$$

where $J^\mu(x)$ is the electromagnetic current operator and $D_{\mu\nu}(x)$ is the photon propagator.

There are two contributions to the pseudoscalar meson masses: one coming from the σ-term due to $\mathscr{H}_m$, the other from the σ-term due to $\mathscr{H}_\gamma$. Eqn (5.229) reads

$$f_a^2 m_a^2\, \delta^{ab} = \sigma_{0m}^{ab} + \sigma_{0\gamma}^{ab} \tag{5.261}$$

where

$$\sigma_{0m}^{ab} = \langle 0|[Q^{5a}, [Q^{5b}, \mathscr{H}_m]]|0\rangle \tag{5.262}$$

$$\sigma_{0\gamma}^{ab} = \langle 0|[Q^{5a}, [Q^{5b}, \mathscr{H}_\gamma]]|0\rangle. \tag{5.263}$$

But

$$[Q^{5a}, \mathscr{H}_\gamma] = 0 \tag{5.264}$$

for any electrically neutral Q^{5a} operator. Thus (Dashen 1969)

$$\sigma_{0\gamma}(\pi^0) = \sigma_{0\gamma}(\mathrm{K}^0) = \sigma_{0\gamma}(\eta^0) = 0. \tag{5.265}$$

Also from SU(3), i.e., U-spin invariance, we have

$$\sigma_{0\gamma}(\pi^+) = \sigma_{0\gamma}(\mathrm{K}^+) \equiv \mu_\gamma^{\sim}. \tag{5.266}$$

After an entirely similar calculation to that which leads to (5.245) and using the SU(3) results (5.246) and (5.247), we have (Weinberg 1977)

$$
\begin{aligned}
f^2 m^2(\pi^+) &= (m_u + m_d)\mu^3 + \mu_\gamma^3 \\
f^2 m^2(\pi^0) &= (m_u + m_d)\mu^3 \\
f^2 m^2(K^+) &= (m_u + m_s)\mu^3 + \mu_\gamma^3 \\
f^2 m^2(K^0) &= (m_d + m_s)\mu^3 \\
f^2 m^2(\eta^0) &= \tfrac{1}{3}(m_u + m_d + 4m_s)\mu^3 .
\end{aligned} \tag{5.267}
$$

We obtain the (improved) Gell-Mann–Okubo relation, as well as

$$\frac{m^2(\pi^0)}{m^2(K^0)} = \frac{m_u + m_d}{m_s + m_d} \tag{5.268}$$

and

$$\frac{m^2(K^+) - m^2(K^0) - m^2(\pi^+)}{m^2(K^0) - m^2(K^+) + m^2(\pi^+) - 2m^2(\pi^0)} = \frac{m_d}{m_u}. \tag{5.269}$$

Besides the quark-mass ratio of (5.248) we also obtain

$$m_d/m_u \simeq 1.8. \tag{5.270}$$

This indicates that u and d quarks are actually not degenerate in mass. This helps us to resolve a number of longstanding difficulties in our picture of isospin invariance being broken only by electromagnetism: the sign puzzle of the proton–neutron and $K^+ - K^0$ mass differences, and the $\eta \to 3\pi$ problem. If one considers this isospin-violating decay $\eta \to 3\pi$ to proceed via a second-order virtual electromagnetic interaction, a straightforward current-algebra calculation then predicts that, in the $SU(2)_L \times SU(2)_R$ chiral symmetric limit, it is strictly forbidden. (For further discussion see §16.3.) It has, of course, been well-known for a long time (Feynman and Speisman 1954) that this picture for isospin breaking produces the 'wrong' sign for the n–p mass difference. Also if electromagnetism is the only source of isospin symmetry-breaking, we will have (through eqns (5.265) and (5.266)) the *Dashen sum rule*

$$m^2(K^+) - m^2(K^0) = m^2(\pi^+) - m^2(\pi^0) \tag{5.271}$$

which again yields a wrong sign for the K^+–K^0 mass difference. But the result in eqn (5.270) shows that there actually is a substantial isospin violation coming from the strong interaction itself.

If m_u and m_d are so different, how do we account for the smallness of the observed isospin violation? This is possible if the u and d mass difference is comparable to m_u and m_d themselves and they are all small on the typical strong-interaction scale. Indeed a modern-day calculation of the n–p mass difference (Gasser and Leutwyler 1975) gives

$$m_u \simeq 4\ \text{MeV}, \qquad m_d \simeq 7\ \text{MeV} \tag{5.272}$$

and we now know (see Chapter 10) that the intrinsic strong-interaction mass–energy scale is about 200 to 400 MeV. Thus SU(2) isotopic spin is a good symmetry because the chiral $SU(2)_L \times SU(2)_R$ is a good symmetry!

To a lesser extent a similar situation holds for the flavour SU(3) and chiral $SU(3)_L \times SU(3)_R$ symmetries. From eqns (5.248) and (5.272) we have

$$m_s \simeq 25(m_u + m_d)/2 \simeq 130 \text{ MeV}. \tag{5.273}$$

We should emphasize the quark masses we have been discussing are *current-algebra quark masses*, or *current quark masses*. They are the parameters of the chiral symmetry breaking. In quantum field theory (e.g. in the free quark model of (5.41)), they are the parameters appearing directly in the Lagrangian. These are different from the *constituent quark masses* $\tilde{m}$ appearing as parameters in (nonrelativistic) bound-state calculations of hadrons: $m_{\text{proton}} \simeq 3\tilde{m}$, $m_\rho \simeq 2\tilde{m}$. Since even in the full strong-interaction theory of quantum chromodynamics (QCD) we have not been able to solve the bound-state problem from first principles, the connection between these two types of quark masses has not been rigorously established. But, as a rule of thumb, the constituent masses $\tilde{m}$ and current masses m_i differ by a common constant

$$\tilde{m}_i = m_i + m_0. \tag{5.274}$$

m_0, being somehow related to the scale parameter of strong interactions, is of order 300 MeV.

Global flavour symmetry of the strong interaction—a summary

The discussion of this chapter leads us to expect that a viable theory of the strong interaction must be approximately chiral $SU(3)_L \times SU(3)_R$ symmetric. The field theory of free quarks (5.41) indeed displays such a global symmetry. It must then be endowed with an interaction that formally (i.e. at the Lagrangian level) preserves this invariance. However the dynamics must be such that this chiral symmetry is spontaneously broken with the vacuum state having quark–antiquark condensate $\langle 0|\bar{q}q|0\rangle \neq 0$, hence not being a chiral singlet. This dynamical breaking should be characterized by a momentum scale comparable to f_π. It should be symmetric with respect to the diagonal subgroup $SU(3)_V$ so that the hadron spectra and interactions exhibit the familiar SU(3) symmetry of *Eightfold-Way* and there is an octet of pseudoscalar Goldstone bosons. This chiral symmetry is also broken explicitly (hence the Goldstone bosons are not strictly massless) with a pattern consistent with the three light quark mass terms in the Lagrangian being the soft symmetry breaking terms. As we shall see in Chapter 10, the gauge theory of the strong interaction (QCD) has all the features compatible with these expectations.

6 Renormalization and symmetry

THE topics of renormalization and symmetry are closely related. The symmetry relations among Green's functions are generally known as the Ward identities (Ward 1950; Takahashi 1957). In a theory with nontrivial symmetry, renormalizability depends critically on the cancellation of divergences from different sectors of the theory as enforced by the Ward identities. This is even more so for gauge theories where we often need to introduce spurious degrees of freedom (e.g., photon longitudinal polarization state, etc.), and one has to be certain that, through the use of Ward identities, these unphysical states are all cancelled in the physical S-matrix elements. On the other hand, one is also concerned with the effects of renormalization on the symmetries themselves. It is this latter aspect of the relation between renormalization and symmetry that we will concentrate on in this chapter.

In §6.1 we first see how the vector-current Ward identity in the $\lambda\phi^4$ theory is maintained up to one-loop diagrams. However similar considerations in §6.2 lead us to the conclusion that the QED Ward identities involving axial-vector currents are spoiled by renormalization, by the triangle fermion loops. There must be 'anomalous terms' in the axial-vector-current divergence equation. This allows us to derive the correct current-algebra low-energy theorem for the $\pi^0 \to 2\gamma$ decay.

The last two sections concern the relationship between spontaneous symmetry breakdown and renormalization. Again we concentrate on the effects of renormalization on symmetry. In §6.3 we present several topics illustrating the renormalization of theories with spontaneous symmetry breaking. Finally in §6.4 we show how the classical potential, which we have used repeatedly to study spontaneous symmetry breaking, can be regarded as the first term in a systematic expansion in powers of Planck's constant and how in certain situations, radiative corrections themselves bring about spontaneous symmetry breakdown.

6.1 The vector-current Ward identity and renormalization

In eqn (5.200) we saw an example of Ward identities. These relations among Green's functions follow from the symmetry properties of the Lagrangian (i.e., current conservation, charge commutator, etc.). They play a crucial role in the derivation of current-algebra low-energy theorems and, as we shall elaborate later on, they are essential in the renormalization programme of any theory with nontrivial symmetries. Thus it is important to check that these Ward identities are not spoiled by higher-order correction terms in perturbation theory. In this section we shall first see, in simple $\lambda\phi^4$ theory, how the vector-current Ward identity is maintained up to one-loop diagrams.

The vector-current Ward identity

The $\lambda\phi^4$ theory of eqn (5.23) has U(1) symmetry with the conserved (vector) current of (5.25). The canonical commutation relation for the complex fields of (5.22)

$$[\partial_0\phi^\dagger(\mathbf{x}, t), \phi(\mathbf{x}', t)] = -\mathrm{i}\delta^3(\mathbf{x} - \mathbf{x}') \tag{6.1}$$

then leads to the commutators

$$\begin{aligned}[J_0(\mathbf{x}, t), \phi(\mathbf{x}', t)] &= \mathrm{i}[\partial_0\phi^\dagger(\mathbf{x}, t), \phi(\mathbf{x}', t)]\phi(\mathbf{x}, t)\\ &= \delta^3(\mathbf{x} - \mathbf{x}')\phi(\mathbf{x}, t)\\ [J_0(\mathbf{x}, t), \phi^\dagger(\mathbf{x}', t)] &= -\delta^3(\mathbf{x} - \mathbf{x}')\phi^\dagger(\mathbf{x}, t).\end{aligned} \tag{6.2}$$

Consider the three-point Green's function given in Fig. 6.1

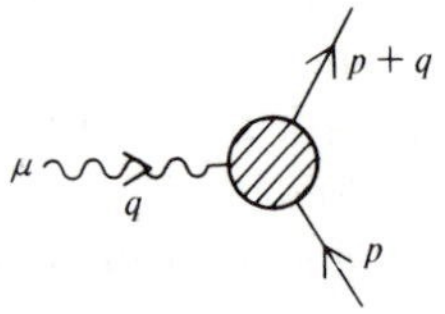

FIG. 6.1. The Green's function of two scalar fields coupled to a vector current.

$$G_\mu(p, q) = \int \mathrm{d}^4x\, \mathrm{d}^4y\, \mathrm{e}^{-\mathrm{i}q\cdot x - \mathrm{i}p\cdot y}\langle 0|T(J_\mu(x)\phi(y)\phi^\dagger(0))|0\rangle. \tag{6.3}$$

We make the standard current-algebra manipulation

$$\begin{aligned}q^\mu G_\mu(p, q) &= -\mathrm{i}\int \mathrm{d}^4x\, \mathrm{d}^4y\, \mathrm{e}^{-\mathrm{i}q\cdot x - \mathrm{i}p\cdot y}\, \partial_x^\mu\langle 0|T(J_\mu(x)\phi(y)\phi^\dagger(0))|0\rangle\\ &= -\mathrm{i}\int \mathrm{d}^4x\, \mathrm{d}^4y\, \mathrm{e}^{-\mathrm{i}q\cdot x - \mathrm{i}p\cdot y}\{\langle 0|T(\partial^\mu J_\mu(x)\phi(y)\phi^\dagger(0))|0\rangle\\ &\quad + \langle 0|T(\delta(x_0 - y_0)[J_0(x), \phi(y)]\phi^\dagger(0)|0\rangle\\ &\quad + \langle 0|\delta(x_0)[J_0(x), \phi^\dagger(0)]\phi(y))|0\rangle\}.\end{aligned} \tag{6.4}$$

The first term on the right-hand side vanishes because of current conservation, $\partial^\mu J_\mu = 0$, and the other terms can be simplified by using (6.2)

$$\begin{aligned}q^\mu G_\mu(p, q) &= -\mathrm{i}\int \mathrm{d}^4x\, \mathrm{e}^{-\mathrm{i}(p+q)\cdot x}\langle 0|T((\phi(x)\phi^\dagger(0))|0\rangle\\ &\quad + \mathrm{i}\int \mathrm{d}^4y\, \mathrm{e}^{-\mathrm{i}p\cdot y}\langle 0|T(\phi^\dagger(0)\phi(y))|0\rangle.\end{aligned} \tag{6.5}$$

The right-hand side is just the propagators for the scalar field

$$\Delta(p) = \int \mathrm{d}^4x\, \mathrm{e}^{-\mathrm{i}p\cdot x}\langle 0|T(\phi(x)\phi^\dagger(0))|0\rangle. \tag{6.6}$$

Eqn (6.5) may be written

$$-iq^{\mu}G_{\mu}(p,q) = \Delta(p+q) - \Delta(p) \tag{6.7}$$

which is an example of vector-current Ward identities.

$Z_J = 1$ and Ward identities

Here we digress briefly to illustrate an application of the Ward identities. They can be used to show that conserved current $J_{\mu}(x)$ is not renormalized as a composite operator. The identity given in (6.7) is for the unrenormalized fields which satisfy the canonical commutation relation (6.1). (For the renormalized fields there will be a factor of Z_{ϕ}^{-1} on the right-hand sides of the commutation relations.) In terms of the renormalized quantities, using the relation (2.167),

$$G_{\mu}^{R}(p,q) = Z_{\phi}^{-1}Z_{J}^{-1}G_{\mu}(p,q)$$
$$\Delta^{R}(p) = Z_{\phi}^{-1}\,\Delta(p).$$

We can express the Ward identity (6.7) as

$$-iZ_{J}q^{\mu}G_{\mu}^{R}(p,q) = \Delta^{R}(p+q) - \Delta^{R}(p).$$

Since the right-hand side of this equation is cutoff-independent, Z_J on the left-hand side must also be cutoff-independent, and we do not need any counter term to renormalize $J_{\mu}(x)$. In other words, the conserved current $J_{\mu}(x)$ is not renormalized as a composite operator, i.e., $Z_J = 1$. Thus the Ward identity (6.7) also holds for the renormalized quantities:

$$-iq^{\mu}G_{\mu}^{R}(p,q) = \Delta^{R}(p+q) - \Delta^{R}(\mathrm{p}). \tag{6.8}$$

From now on we will drop the superscript R in (6.8) with the understanding that these Green's functions refer to the renormalized quantities.

Such a nonrenormalization result holds for all sorts of conserved quantities, in Abelian (as this example shows) and in non-Abelian cases. Actually, for currents associated with non-Abelian symmetries, there is a direct way to understand this result. These currents must obey fundamental commutation relations such as eqn (5.55). The nonlinear nature of commutators fixes their normalizations so that no renormalization is possible.

The vector Ward identity at the one-loop level

It is instructive to see how the one-loop diagrams satisfy the Ward identity of (6.7). In terms of the amputated Green's function $\Gamma_{\mu}(p,q)$ and the 1PI self-energy $\tilde{\Sigma}$ defined by (see eqns (2.48) and (2.24)),

$$\Gamma_{\mu}(p,q) = [i\Delta(p+q)]^{-1}G_{\mu}(p,q)[i\Delta(p)]^{-1}$$
$$[\Delta(p)]^{-1} = p^2 - \mu^2 - \tilde{\Sigma}(p), \tag{6.9}$$

the Ward identity (6.7) takes the form

$$iq^{\mu}\Gamma_{\mu}(p,q) = (p+q)^2 - p^2 - \tilde{\Sigma}(p+q) + \tilde{\Sigma}(p). \tag{6.10}$$

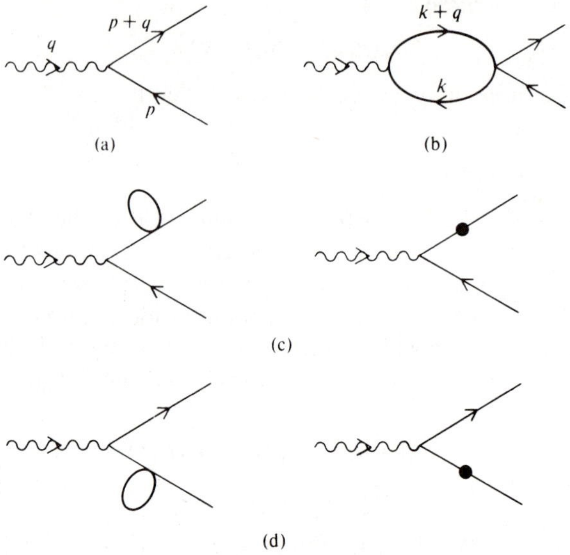

FIG. 6.2. Tree and one-loop diagrams for the vector current and scalar fields vertex function.

For the contributions of $\Gamma_\mu(p, q)$ Fig. 6.2(a) shows the vertex function in the tree approximation

$$iq^\mu\Gamma_\mu^{(a)}(p, q) = iq^\mu(-i)(2p + q)_\mu = (p + q)^2 - p^2 \tag{6.11}$$

which is just the lowest-order Ward identity.

Using dimensional regularization for the one-loop diagram (Fig. 6.2(b)), we have

$$\begin{aligned} iq^\mu\Gamma_\mu^{(b)}(p, q) &= iq^\mu \int \frac{d^n k}{(2\pi)^4}\, i\lambda \frac{i}{k^2 - \mu^2}(-i)(2k + q)_\mu \frac{i}{(k + q)^2 - \mu^2} \\ &= i\lambda \int \frac{d^n k}{(2\pi)^4}\left[\frac{1}{(k + q)^2 - \mu^2} - \frac{1}{k^2 - \mu^2}\right]. \end{aligned} \tag{6.12}$$

For $n < 2$, the first integral on the right-hand side is convergent and we can shift the integration variable k to $k - q$, to get

$$iq^\mu\Gamma_\mu^{(b)}(p, q) = i\lambda \int \frac{d^n k}{(2\pi)^4}\left[\frac{1}{k^2 - \mu^2} - \frac{1}{k^2 - \mu^2}\right] = 0. \tag{6.13}$$

This will still be true when we analytically continue to $n > 2$, in particular to $n = 4$. The contribution of Fig. 6.2(c) is given by

$$iq^\mu\Gamma_\mu^{(c)}(p, q) = iq^\mu(-i)(2p + q)_\mu \frac{i}{(p + q)^2 - \mu^2}[\Sigma(p + q) - \Sigma(0)] \tag{6.14}$$

where

$$-i\Sigma(p + q) = -\frac{i\lambda}{2}\int \frac{d^n k}{(2\pi)^4}\frac{i}{k^2 - \mu^2} \tag{6.15}$$

which is independent of the external momentum. We get

$$\tilde{\Sigma}(p) = \Sigma(p) - \Sigma(0) = 0. \tag{6.16}$$

Hence the right-hand side of (6.14) vanishes and we have

$$\mathrm{i}q^\mu \Gamma^{(c)}_\mu(p, q) = 0. \tag{6.17}$$

Similarly, we also have

$$\mathrm{i}q^\mu \Gamma^{(d)}_\mu(p, q) = 0. \tag{6.18}$$

Thus up to the one-loop order we have, from the sum of Fig. 6.2(a), (b), (c), and (d),

$$\mathrm{i}q^\mu \Gamma_\mu(p, q) = (p + q)^2 - p^2 \tag{6.19}$$

which is the Ward identity of (6.10) because of (6.16).

We note that the above proof of the Ward identity (6.7) (or (6.10) in perturbation theory) involves two important ingredients: (i) the algebraic relation of (6.11); and (ii) the translation of the integration variable in going from (6.12) to (6.13). Generally, in order to maintain the Ward identity, which is the consequence of the symmetry, we must not choose a regularization scheme that will spoil the original symmetry. (It is clear that the dimensional regularization fulfils those requirements.)

.2 Axial-vector-current Ward identity anomaly and $\pi^0 \to 2\gamma$

Following essentially the same steps as in §6.1 we shall see that the validity of the (axial) Ward identity is not automatic when there are fermions in the theory, even after the theory is regularized symmetrically. This is because certain one-loop diagrams introduce anomalous terms which prevent the Ward identities from reproducing themselves recursively at higher orders in the perturbative expansion. Such anomalies were discovered by Adler (1969, 1970) and by Bell and Jackiw (1969) in their current-algebra studies. In the following we shall present an elementary introduction to this subject of *ABJ anomalies*.

The tree-level Ward identities and current divergences from the equation of motion

Consider the three-point functions in electrodynamics

$$T_{\mu\nu\lambda}(k_1, k_2, q) = \mathrm{i} \int \mathrm{d}^4x_1\, \mathrm{d}^4x_2 \langle 0|T(V_\mu(x_1) V_\nu(x_2) A_\lambda(0))|0\rangle\, \mathrm{e}^{\mathrm{i}k_1\cdot x_1 + \mathrm{i}k_2\cdot x_2} \tag{6.20}$$

$$T_{\mu\nu}(k_1, k_2, q) = \mathrm{i} \int \mathrm{d}^4x_1\, \mathrm{d}^4x_2 \langle 0|T(V_\mu(x_1) V_\nu(x_2) P(0))|0\rangle\, \mathrm{e}^{\mathrm{i}k_1\cdot x_1 + \mathrm{i}k_2\cdot x_2} \tag{6.21}$$

where V_μ, A_μ, and P are the vector, axial vector, and pseudoscalar currents, respectively,

$$\begin{aligned} V_\mu(x) &= \bar{\psi}(x)\gamma_\mu\psi(x), \\ A_\mu(x) &= \bar{\psi}(x)\gamma_\mu\gamma_5\psi(x), \\ P(x) &= \bar{\psi}(x)\gamma_5\psi(x), \end{aligned} \tag{6.22}$$

and

$$q = k_1 + k_2.$$

For the Ward identities relating $T_{\mu\nu\lambda}$ and $T_{\mu\nu}$, we need the divergences of V_μ and A_μ which are calculated from the equation of motion

$$\begin{aligned} \partial^\mu V_\mu(x) &= 0 \\ \partial^\mu A_\mu(x) &= 2imP(x) \end{aligned} \tag{6.23}$$

where m is the mass of the fermion field $\psi(x)$. With an elementary application of current-algebra techniques such as

$$\partial_x^\mu(T(J_\mu(x)\mathcal{O}(y))) = T(\partial^\mu J_\mu(x)\mathcal{O}(y)) + [J_0(x), \mathcal{O}(y)]\,\delta(x_0 - y_0) \tag{6.24}$$

for the current $J_\mu(x)$ and the local operator $\mathcal{O}(y)$ and with the knowledge that in our case the equal-time commutators vanish

$$[V_0(x), A_0(y)]\,\delta(x_0 - y_0) = 0, \tag{6.25}$$

we can formally derive the following vector and axial-vector Ward identities

$$k_1^\mu T_{\mu\nu\lambda} = k_2^\nu T_{\mu\nu\lambda} = 0 \tag{6.26}$$

and

$$q^\lambda T_{\mu\nu\lambda} = 2mT_{\mu\nu}. \tag{6.27}$$

Anomalies arising from renormalization

But when we calculate the lowest-order contributions to $T_{\mu\nu\lambda}$ and $T_{\mu\nu}$ (see Figs (6.3) and (6.4) we find that the Ward identities (6.26) and (6.27) are not satisfied

$$T_{\mu\nu\lambda} = i\int\frac{d^4p}{(2\pi)^4}(-1)\left\{\mathrm{tr}\left[\frac{i}{\not{p} - m}\gamma_\lambda\gamma_5\frac{i}{(\not{p} - \not{q}) - m}\gamma_\nu\frac{i}{(\not{p} - \not{k}_1) - m}\gamma_\mu\right] + \begin{pmatrix} k_1 \leftrightarrow k_2 \\ \mu \leftrightarrow \nu \end{pmatrix}\right\} \tag{6.28}$$

$$T_{\mu\nu} = i\int\frac{d^4p}{(2\pi)^4}(-1)\left\{\mathrm{tr}\left[\frac{i}{\not{p} - m}\gamma_5\frac{i}{(\not{p} - \not{q}) - m}\gamma_\nu\frac{i}{(\not{p} - \not{k}_1) - m}\gamma_\mu\right] + \begin{pmatrix} k_1 \leftrightarrow k_2 \\ \mu \leftrightarrow \nu \end{pmatrix}\right\}. \tag{6.29}$$

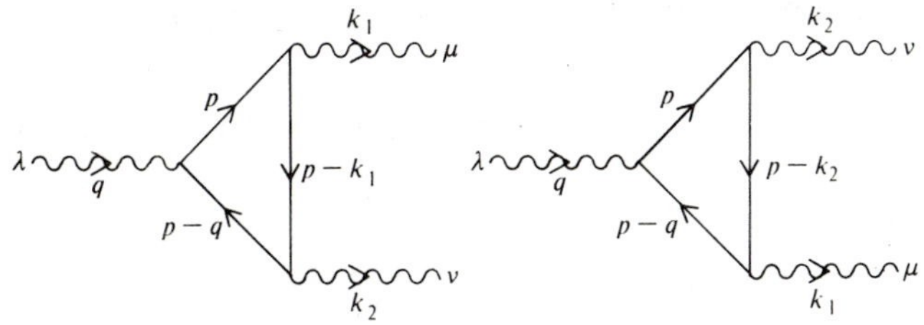

FIG. 6.3. Lowest-order contributions to $T_{\mu\nu\lambda}$ of eqn (6.20).

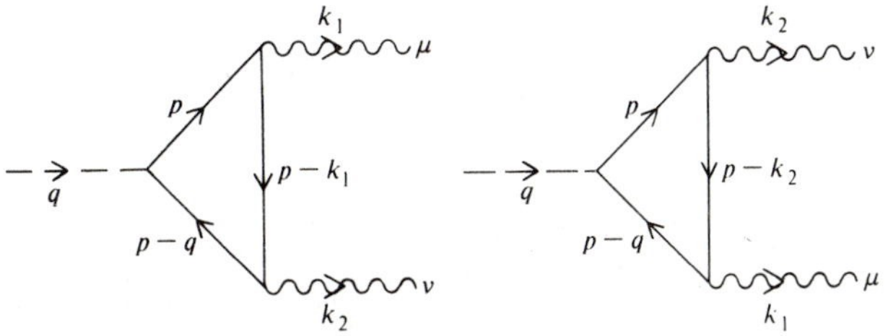

FIG. 6.4. Lowest-order contributions to $T_{\mu\nu}$ of eqn (6.21).

To check the Ward identities, in particular (6.27), we can use the relation

$$\not{q}\gamma_5 = \gamma_5(\not{p} - \not{q} - m) + (\not{p} - m)\gamma_5 + 2m\gamma_5. \tag{6.30}$$

to find that

$$q^\lambda T_{\mu\nu\lambda} = 2mT_{\mu\nu} + \Delta^{(1)}_{\mu\nu} + \Delta^{(2)}_{\mu\nu} \tag{6.31}$$

with

$$\Delta^{(1)}_{\mu\nu} = \int \frac{\mathrm{d}^4 p}{(2\pi)^4} \operatorname{tr}\left\{\frac{\mathrm{i}}{\not{p} - m}\gamma_5\gamma_\nu \frac{\mathrm{i}}{(\not{p} - \not{k}_1) - m}\gamma_\mu - \frac{\mathrm{i}}{(\not{p} - \not{k}_2) - m}\gamma_5\gamma_\nu \frac{\mathrm{i}}{(\not{p} - \not{q}) - m}\gamma_\mu\right\} \tag{6.32a}$$

$$\Delta^{(2)}_{\mu\nu} = \int \frac{\mathrm{d}^4 p}{(2\pi)^4} \operatorname{tr}\left\{\frac{\mathrm{i}}{\not{p} - m}\gamma_5\gamma_\mu \frac{\mathrm{i}}{(\not{p} - \not{k}_2) - m}\gamma_\nu - \frac{\mathrm{i}}{(\not{p} - \not{k}_1) - m}\gamma_5\gamma_\mu \frac{\mathrm{i}}{(\not{p} - \not{q}) - m}\gamma_\nu\right\}. \tag{6.32b}$$

If the integrals $\Delta^{(i)}_{\mu\nu}$ vanish we have the Ward identity in (6.27). Superficially this appears to be the case. The two integrals in $\Delta^{(1)}_{\mu\nu}$ cancel each other if we can shift the integration variable p to $p + k_2$ in the second term. Similarly the other pair of integrals in (6.32b) would cancel. But the integrals are linearly divergent and a translation of integration variable produces extra finite terms with $\Delta^{(1)}_{\mu\nu} \neq 0$ and $\Delta^{(2)}_{\mu\nu} \neq 0$. This ruins the Ward identity.

Shift of integration variable for linearly divergent integrals

It can easily be demonstrated in one dimension that a shift of integration variable may be illegitimate for a divergent integral (see, for example, Jackiw 1972)

$$\Delta(a) = \int_{-\infty}^{\infty} \mathrm{d}x[f(x+a) - f(x)]. \tag{6.33}$$

To see that $\Delta(a)$ may be nonzero, we expand the integrand

$$\begin{aligned}\Delta(a) &= \int_{-\infty}^{\infty} \mathrm{d}x\left[af'(x) + \frac{a^2}{2}f''(x) + \ldots\right]\\ &= a[f(\infty) - f(-\infty)] + \frac{a^2}{2}[f'(\infty) - f'(-\infty)] + \ldots\end{aligned} \tag{6.34}$$

where the primes signify differentiation. When the integral $\int_{-\infty}^{\infty} f(x)\,\mathrm{d}x$ converges (or at most diverges logarithmically) we have $0 = f(\pm\infty) = f'(\pm\infty) = \ldots$, and $\Delta(a)$ vanishes. However, for a linearly divergent integral, $0 \neq f(\pm\infty)$, $0 = f'(\pm\infty) = \ldots$, and $\Delta(a)$ need not vanish

$$\Delta(a) = a[f(\infty) - f(-\infty)]. \tag{6.35}$$

This corresponds to a 'surface term' ('surface' in one dimension is the end-points). The generalization to a linearly divergent integral in n dimensions is straightforward

$$\begin{aligned}\Delta(a) &= \int \mathrm{d}^n r[f(r+a) - f(r)]\\ &= \int \mathrm{d}^n r\left[a^\tau \frac{\partial}{\partial r_\tau} f(r) + a^\tau \frac{\partial}{\partial r_\tau} a^\sigma \frac{\partial}{\partial r_\sigma} f(r) + \ldots\right].\end{aligned} \tag{6.36}$$

After applying Gauss's theorem, all but the first term vanish upon integrating over the surface $r = R \to \infty$

$$\Delta(a) = a^\tau \frac{R_\tau}{R} f(R) S_n(R) \tag{6.37}$$

where $S_n(R)$ is the surface area of the hypersphere with radius R. For the case of four-dimensional Minkowski space, we have

$$\Delta(a) = a^\tau \int \mathrm{d}^4 r\, \partial_\tau f(r) = 2\mathrm{i}\pi^2 a^\tau \lim_{R\to\infty} R^2 R_\tau f(R). \tag{6.38}$$

Ambiguities in $T_{\mu\nu\lambda}$

The one-loop amplitude $T_{\mu\nu\lambda}$ is (superficially) linearly divergent; hence it is not uniquely defined. The expression in (6.28) implies a particular routing of the loop momentum p: the fermion line between the vector and axial vector vertices carries momentum p. We could have chosen to route it differently so that this fermion line carries $p + a$, where a is some (arbitrary) linear combination of k_1 and k_2

$$a = \alpha k_1 + (\alpha - \beta)k_2. \tag{6.39}$$

The fact that integral is linearly divergent implies that $T_{\mu\nu\lambda}$ has an ambiguity in its definition by an amount

$$\begin{aligned}\Delta_{\mu\nu\lambda}(a) &= T_{\mu\nu\lambda}(a) - T_{\mu\nu\lambda}(0)\\ &= (-1)\int \frac{\mathrm{d}^4p}{(2\pi)^4}\\ &\quad\times\left\{\mathrm{tr}\left[\frac{1}{(\not p + \not a) - m}\gamma_\lambda\gamma_5\frac{1}{(\not p+\not a-\not q)-m}\gamma_\nu\frac{1}{(\not p+\not a-\not k_1)-m}\gamma_\mu\right]\right.\\ &\quad\left.-\mathrm{tr}\left[\frac{1}{\not p-m}\gamma_\lambda\gamma_5\frac{1}{(\not p-\not q)-m}\gamma_\nu\frac{1}{(\not p-\not k_1)-m}\gamma_\mu\right]\right\}+\left(\begin{matrix}k_1\leftrightarrow k_2\\ \mu\leftrightarrow\nu\end{matrix}\right)\Bigg)\\ &\equiv \Delta^{(1)}_{\mu\nu\lambda} + \Delta^{(2)}_{\mu\nu\lambda}.\end{aligned} \tag{6.40}$$

Applying the result (6.38), we have

$$\begin{aligned}\Delta^{(1)}_{\mu\nu\lambda} &= (-1)\int\frac{\mathrm{d}^4p}{(2\pi)^4}a^\tau\frac{\partial}{\partial p_\tau}\mathrm{tr}\left[\frac{1}{\not p-m}\gamma_\lambda\gamma_5\frac{1}{(\not p-\not q)-m}\gamma_\nu\frac{1}{(\not p-\not k_1)-m}\gamma_\mu\right]\\ &= \frac{-\mathrm{i}2\pi^2a^\tau}{(2\pi)^4}\lim_{p\to\infty}p^2p_\tau\,\mathrm{tr}(\gamma_\alpha\gamma_\lambda\gamma_5\gamma_\beta\gamma_\nu\gamma_\delta\gamma_\mu)p^\alpha p^\beta p^\delta/p^6\\ &= \frac{\mathrm{i}2\pi^2a_\sigma}{(2\pi)^4}\lim_{p\to\infty}\frac{p^\sigma p^\rho}{p^2}4\mathrm{i}\varepsilon_{\mu\nu\lambda\rho}.\end{aligned} \tag{6.41}$$

After replacing $p^\sigma p^\rho/p^2$ by $g^{\rho\sigma}/4$, we have

$$\Delta^{(1)}_{\mu\nu\lambda} = \varepsilon_{\rho\mu\nu\lambda}a^\rho/8\pi^2. \tag{6.42}$$

Since $\Delta^{(2)}_{\mu\nu\lambda}$ is related to (6.42) by the exchanges $k_1 \leftrightarrow k_2$ and $\mu \leftrightarrow \nu$, we have from eqns (6.40), (6.42), and (6.39)

$$\Delta_{\mu\nu\lambda} = \Delta^{(1)}_{\mu\nu\lambda} + \Delta^{(2)}_{\mu\nu\lambda} = \frac{\beta}{8\pi^2}\varepsilon_{\rho\mu\nu\lambda}(k_1 - k_2)^\rho. \tag{6.43}$$

Thus the definition of $T_{\mu\nu\lambda}$ has an ambiguity signified by the arbitrary parameter β

$$T_{\mu\nu\lambda}(a) = T_{\mu\nu\lambda}(0) - \frac{\beta}{8\pi^2}\varepsilon_{\mu\nu\lambda\rho}(k_1 - k_2)^\rho \equiv T_{\mu\nu\lambda}(\beta). \tag{6.44}$$

Determination of the anomalous term

We now attempt to determine β by imposing the Ward identities. We shall see that no value of β exists such that $T_{\mu\nu\lambda}(a)$ satisfies both the vector and axial-vector Ward identities (eqns (6.26) and (6.27)).

Let us first check the axial Ward identity (6.27). Like those in (6.40) the two surface terms in (6.31) can again be evaluated using (6.38)

$$\Delta^{(1)}_{\mu\nu} = -\frac{k_2^\tau}{(2\pi)^4}\int d^4p\,\frac{\partial}{\partial p_\tau}\left(\frac{\mathrm{tr}[(\not p + m)\gamma_5\gamma_\nu(\not p - \not k_1 + m)\gamma_\mu]}{(p^2-m^2)[(p-k_1)^2-m^2]}\right)$$

$$= -\frac{k_2^\tau}{(2\pi)^4}\,2\mathrm{i}\pi^2 \lim_{p\to\infty}\frac{p_\tau}{p^2}\,\mathrm{tr}(\gamma_\alpha\gamma_5\gamma_\nu\gamma_\beta\gamma_\mu)p^\alpha k_1^\beta$$

$$= \frac{-1}{8\pi^2}\,\varepsilon_{\mu\nu\sigma\rho}k_1^\sigma k_2^\rho \tag{6.45}$$

and

$$\Delta^{(2)}_{\mu\nu} = \Delta^{(1)}_{\mu\nu}. \tag{6.46}$$

Thus from (6.44) and (6.31), we have

$$q^\lambda T_{\lambda\nu\lambda}(\beta) = 2mT_{\mu\nu}(0) - \frac{1-\beta}{4\pi^2}\,\varepsilon_{\mu\nu\sigma\rho}k_1^\sigma k_2^\rho \tag{6.47}$$

For the vector Ward identity (6.26) we have

$$k_1^\mu T_{\mu\nu\lambda}(0) = (-1)\int\frac{d^4p}{(2\pi)^4}\left\{\mathrm{tr}\left[\frac{1}{\not p - m}\gamma_\lambda\gamma_5\frac{1}{(\not p - \not q)-m}\gamma_\nu\frac{1}{(\not p - \not k_1)-m}\not k_1\right]\right.$$

$$\left.+\,\mathrm{tr}\left[\frac{1}{\not p - m}\gamma_\lambda\gamma_5\frac{1}{(\not p - \not q)-m}\not k_1\frac{1}{(\not p - \not k_2)-m}\gamma_\nu\right]\right\}. \tag{6.48}$$

Using

$$\not k_1 = (\not p - m) - [(\not p - \not k_1) - m]$$

$$= [(\not p - \not k_2) - m] - [(\not p - \not q) - m], \tag{6.49}$$

we can rewrite (6.48)

$$k_1^\mu T_{\mu\nu\lambda}(0) = (-1)\int\frac{d^4p}{(2\pi)^4}\,\mathrm{tr}\left[\gamma_\lambda\gamma_5\frac{1}{(\not p - \not q)-m}\gamma_\nu\frac{1}{(\not p - \not k_1)-m}\right.$$

$$\left.-\,\gamma_\lambda\gamma_5\frac{1}{(\not p - \not k_2)-m}\gamma_\nu\frac{1}{\not p - m}\right]. \tag{6.50}$$

Again the right-hand side is a surface term that can be evaluated using (6.38)

$$k_1^\mu T_{\mu\nu\lambda}(0) = \frac{k_1^\tau}{(2\pi)^4}\int d^4p\,\frac{\partial}{\partial p_\tau}\left(\frac{\mathrm{tr}[\gamma_\lambda\gamma_5(\not p - \not k_2 + m)\gamma_\nu(\not p + m)]}{[(p-k_2)^2-m^2](p^2-m^2)}\right)$$

$$= \frac{k_1^\tau}{(2\pi)^4}\,2\mathrm{i}\pi^2\lim_{p\to\infty}\frac{p_\tau}{p^2}\,\mathrm{tr}(\gamma_5\gamma_\lambda\gamma_\alpha\gamma_\nu\gamma_\beta)k_2^\alpha p^\beta$$

$$= \frac{-1}{8\pi^2}\,\varepsilon_{\lambda\sigma\nu\rho}k_1^\rho k_2^\sigma \tag{6.51}$$

or, with (6.44)

$$k_1^\mu T_{\mu\nu\lambda}(\beta) = \frac{(1+\beta)}{8\pi^2}\,\varepsilon_{\nu\lambda\sigma\rho}k_1^\sigma k_2^\rho. \tag{6.52}$$

Thus it is not possible to put (6.47) into the form of (6.27) *and* (6.52) into the form of (6.26) with any choice of β. As it turns out that there is no anomalous term in the Ward identities for $\langle 0|T(\mathrm{VVV})|0\rangle$ and that there are anomalous terms for $\langle 0|T(\mathrm{AAA})|0\rangle$, it is logical to associate the anomaly with the axial-vector current. Thus we fix the momentum routing so that the vector Ward identity (6.26) is maintained: i.e., if $\beta = -1$, the axial Ward identity becomes

$$q^\lambda T_{\mu\nu\lambda} = 2mT_{\mu\nu} - \frac{1}{2\pi^2}\varepsilon_{\mu\nu\sigma\rho}k_1^\sigma k_2^\rho. \tag{6.53}$$

This corresponds to a modification of the axial-vector current divergence eqn (6.23) as

$$\partial^\lambda A_\lambda(x) = 2\mathrm{i}mP(x) + (4\pi)^{-2}\varepsilon^{\mu\nu\rho\sigma}F_{\mu\nu}(x)F_{\rho\sigma}(x) \tag{6.54}$$

where $F_{\mu\nu}(x)$ is the usual electromagnetic field tensor. This extra term, the ABJ anomaly, is thus produced by the renormalization effect and has the following properties:

(1) The anomaly is independent of the fermion masses and should also be present in the massless theory.

(2) Adler and Bardeen (1969) showed that the coefficient in the anomaly term is not affected by higher-order radiative corrections, i.e., triangle diagrams with more than one loop do not contribute to the anomaly term. This can be understood heuristically by noting that the superficial degrees of divergence of the higher-order triangle diagrams are less than one and the momentum-routing ambiguity does not exist for such diagrams.

(3) As our presentation has been in terms of momentum routing and conventional cut-off regularization, the reader may inquire how this anomaly problem rears its head in the dimensional regularization scheme. There the problem shows up as the difficulty of giving a proper definition to the Dirac γ_5 matrix in space–time dimensions other than four.

(4) It was pointed out by Fujikawa (1979) that the ABJ anomalous Ward identity could be formulated rather directly in the path-integral formalism. He showed that the path-integral measure for gauge-invariant fermion theory is not invariant under the γ_5 transformation. The extra Jacobian factor gives rise to the ABJ anomaly.

The ABJ anomaly for non-Abelian cases. In non-Abelian theories, Green's functions with odd number of axial vector couplings up to five-point functions contribute anomalous terms to the divergence of axial-vector current (Bardeen 1969). However the triangle anomaly may be regarded as the basic one since it is the simplest and its absence implies the absence of all other anomalous diagrams. In the following we shall continue to restrict our discussion to the triangle anomaly. Consider

$$T^{abc}_{\mu\nu\lambda}(k_1, k_2; q) = \mathrm{i}\int \mathrm{d}^4x_1\, \mathrm{d}^4x_2 \langle 0|T(V^a_\mu(x_1)V^b_\nu(x_2)A^c_\lambda(0))|0\rangle\, \mathrm{e}^{\mathrm{i}k_1\cdot x_1 + \mathrm{i}k_2\cdot x_2}. \tag{6.55}$$

where

$$V_\mu^a(x) = \bar{\psi}(x)T^a\gamma_\mu\psi(x)$$
$$A_\lambda^c(x) = \bar{\psi}(x)T^c\gamma_\lambda\gamma_5\psi(x) \tag{6.56}$$

and the T^as are the internal symmetry matrices. Also consider

$$T_{\mu\nu}^{abc}(k_1, k_2; q) = \int d^4x_1 d^4x_2 \langle 0|T(V_\mu^a(x_1)V_\nu^b(x_2)P^c(0))|0\rangle\, e^{ik_1\cdot x_1 + ik_2\cdot x_2} \tag{6.57}$$

with

$$P^c(x) = \bar{\psi}(x)T^c\gamma_5\psi(x). \tag{6.58}$$

The anomaly in the axial-vector Ward identity is

$$q^\lambda T_{\mu\nu\lambda}^{abc} = 2mT_{\mu\nu}^{abc} - \frac{1}{2\pi^2}\varepsilon_{\mu\nu\rho\sigma}k_1^\rho k_2^\sigma D^{abc} + \text{commutator terms} \tag{6.59}$$

where

$$D^{abc} = \tfrac{1}{2}\,\mathrm{tr}(\{T^a, T^b\}T^c). \tag{6.60}$$

In the non-Abelian situation the Ward identity usually also involves equal-time commutators (see eqn (6.65) below for an example).

$\pi^0 \to 2\gamma$

An important application of the ABJ anomaly in current algebra is in the derivation of the soft-pion theorem for the $\pi^0 \to 2\gamma$ decay. This amplitude is defined as

$$\langle\gamma(k_1\varepsilon_1)\gamma(k_2\varepsilon_2)|\,\pi^0(q)\rangle = i(2\pi)^4\delta^4(q - k_1 - k_2)\varepsilon_1^\mu(k_1)\varepsilon_2^\nu(k_2) \times \Gamma_{\mu\nu}(k_1, k_2, q) \tag{6.61}$$

with

$$\Gamma_{\mu\nu}(k_1, k_2, q) = e^2\int d^4z\, d^4y\, e^{ik_1\cdot z + ik_2\cdot y}\langle 0|T(J_\mu^{\mathrm{em}}(z)\, J_\nu^{\mathrm{em}}(y))|\pi^0(q)\rangle \tag{6.62}$$

which has the Lorentz covariant structure

$$\Gamma^{\mu\nu}(k_1, k_2, q) = i\varepsilon_{\mu\nu\sigma\rho}k_1^\sigma k_2^\rho\Gamma(q^2). \tag{6.63}$$

To derive the low-energy theorem, consider the amplitude

$$\Gamma_{\mu\nu\lambda}(k_1, k_2, q) = \int d^4x\, d^4y\, e^{ik_2\cdot y - iq\cdot x}\langle 0|T(A_\lambda^3(x)J_\nu^{\mathrm{em}}(y)J_\mu^{\mathrm{em}}(0))|0\rangle \tag{6.64}$$

which satisfies the Ward identity

$$q^\lambda \Gamma_{\mu\nu\lambda}(k_1, k_2, q) = -\mathrm{i} \int \mathrm{d}^4x\, \mathrm{d}^4y\, \mathrm{e}^{\mathrm{i}k_2\cdot y - \mathrm{i}q\cdot x}$$
$$\times \{\langle 0|T(\partial^\lambda A^3_\lambda(x) J^{\mathrm{em}}_\nu(y) J^{\mathrm{em}}_\mu(0))|0\rangle$$
$$+ \langle 0|T(\delta(x_0 - y_0)[A^3_0(x), J^{\mathrm{em}}_\nu(y)] J^{\mathrm{em}}_\mu(0))|0\rangle$$
$$+ \langle 0|T(\delta(x_0)[A^3_0(x), J^{\mathrm{em}}_\mu(0)] J^{\mathrm{em}}_\nu(y))|0\rangle\}. \quad (6.65)$$

From the current algebra of eqn (5.80) the commutator terms vanish. Naively one would identify the first term on the right-hand side as the $\pi^0 \to 2\gamma$ amplitude via PCAC (5.181)

$$\Gamma^{\mu\nu}(k_1, k_2, q) = \frac{-\mathrm{i}e^2(-q^2 + m_\pi^2)}{f_\pi m_\pi^2} \int \mathrm{d}^4x\, \mathrm{d}^4y\, \mathrm{e}^{\mathrm{i}k_2\cdot y - \mathrm{i}q\cdot x}$$
$$\times \langle 0|T(\partial^\lambda A^3_\lambda(x) J^{\mathrm{em}}_\nu(y) J^{\mathrm{em}}_\mu(0))|0\rangle. \quad (6.66)$$

Then there should be the soft-pion result of $\Gamma(q^2 = 0) = 0$ (Sutherland 1967; Veltman 1967) since the left-hand side of (6.65) vanishes when $q^\lambda \to 0$ as $\Gamma_{\mu\nu\lambda}$ does not have intermediate states degenerate with the vacuum and coupling to the vacuum through the axial-vector current $A^3_\lambda(x)$. However one must include the anomaly term in the Ward identity (i.e., PCAC is modified in this case)

$$q^\lambda \Gamma_{\mu\nu\lambda}(k_1, k_2, q) = \frac{f_\pi m_\pi^2}{e^2(m_\pi^2 - q^2)} \Gamma_{\mu\nu}(k_1, k_2, q) - \frac{\mathrm{i}D}{2\pi^2} \varepsilon_{\mu\nu\sigma\rho} k_1^\sigma k_2^\rho \quad (6.67)$$

where D is the anomaly coefficient (6.60) (see eqn (6.72)). We then obtain the low-energy theorem

$$\lim_{q\to 0} \Gamma_{\mu\nu}(k_1, k_2, q) = \frac{\mathrm{i}e^2 D}{2\pi^2 f_\pi} \varepsilon_{\mu\nu\sigma\rho} k_1^\sigma k_2^\rho \quad (6.68)$$

or

$$\Gamma(0) = \frac{e^2 D}{2\pi^2 f_\pi}. \quad (6.69)$$

Thus in the soft-pion limit the contribution to the $\pi^0 \to 2\gamma$ amplitude comes entirely from the anomaly (Adler 1969). To calculate D, let us first assume the simple quark model (without the colour degrees of freedom) where the currents are given by

$$J^{\mathrm{em}}_\mu(x) = \bar{\mathrm{q}}(x)\gamma_\mu Q \mathrm{q}(x)$$
$$A^3_\mu(x) = \bar{\mathrm{q}}(x)\gamma_\mu\gamma_5 \frac{\lambda^3}{2} \mathrm{q}(x) \quad (6.70)$$

with

$$Q = \frac{1}{3}\begin{pmatrix} 2 & & \\ & -1 & \\ & & -1 \end{pmatrix} \qquad \lambda_3 = \begin{pmatrix} 1 & & \\ & -1 & \\ & & 0 \end{pmatrix}. \quad (6.71)$$

The anomaly coefficient (6.60) takes on the value

$$D = \frac{1}{2}\,\mathrm{tr}\left(\{Q, Q\}\frac{\lambda_3}{2}\right) = \frac{1}{6} \tag{6.72}$$

yielding $\Gamma(0) = 0.0123 m_\pi^{-1}$ which is about a factor of 3 smaller than the experimental value of $\Gamma(m_\pi^2) \simeq 0.0375 m_\pi^{-1}$. This lends support to the idea that quarks carry colour degrees of freedom. The anomaly coefficient D is proportional to the trace of the fermion loops and there will be an additional factor of 3 coming from summing over the three colours

$$\Gamma(0) = 0.037 m_\pi^{-1}. \tag{6.73}$$

We note that in the above calculation the strangeness flavour does not play a role as λ_3 is nonzero only for the first two components. Physically this corresponds to the statement that the pion is composed of nonstrange quarks. Clearly (6.73) is not modified when other flavours (c, b, ...) are included.

6.3 Renormalization in theories with spontaneous symmetry breaking

In this section we discuss two topics related to the renormalization of theories with spontaneous symmetry breaking. First we study the one-loop renormalization of the simplest $\lambda\phi^4$ theory (5.131) with spontaneous breaking of its discrete symmetry. We show how 'tadpole diagrams' contribute to a shift in the vacuum expectation value (VEV) of the scalar field and that the counterterms are the same as those of the symmetric theory. We next study a case of spontaneously broken continuous symmetry and show how the Goldstone particles remain massless even in the presence of higher-order radiative corrections.

One-loop renormalization and the VEV shift

We return to the theory considered in eqn (5.137). The original Lagrangian (5.131)

$$\mathscr{L} = \frac{1}{2}(\partial_\lambda\phi)^2 + \frac{1}{2}\mu^2\phi^2 - \frac{\lambda}{4}\phi^4 \tag{6.74}$$

has the discrete symmetry $\phi \to \phi' = -\phi$. When $\mu^2 > 0$, this symmetry is broken by the vacuum with ϕ developing VEV

$$\langle 0|\phi|0\rangle = v$$
$$v = (\mu^2/\lambda)^{\frac{1}{2}}. \tag{6.75}$$

Perturbing around this vacuum, we define a shifted field

$$\phi' = \phi - v.$$

In terms of this field we have the potential (5.137)

$$V(\phi') = \mu^2\phi'^2 + \lambda v\phi'^3 + \frac{\lambda}{4}\phi'^4 \tag{6.76}$$

which corresponds to a scalar with mass $2\mu^2$ with the self-interacting vertices shown in Fig. 6.5.

FIG. 6.5. Interaction vertices for the Lagrangian in eqn (6.76).

We now study the renormalization effects at the one-loop level.

(1) *One-point function* (the tadpole diagram in Fig. 6.6). We have

FIG. 6.6.

$$\begin{aligned} T &= (-6\mathrm{i}\lambda v)\frac{1}{2}\int\frac{\mathrm{d}^4k}{(2\pi)^4}\frac{\mathrm{i}}{k^2-2\mu^2} \\ &= -3\mathrm{i}\lambda v I_2 \end{aligned} \tag{6.77}$$

where

$$I_2 = \int\frac{\mathrm{d}^4k}{(2\pi)^4}\frac{\mathrm{i}}{k^2-2\mu^2}. \tag{6.78}$$

To cancel this divergence we need a counterterm $-D\phi'$ in the Lagrangian with

$$D = -3\lambda v I_2. \tag{6.79}$$

(2) *Two-point function* (Fig. 6.7). Diagrams (a) and (b) give rise to self-energy terms

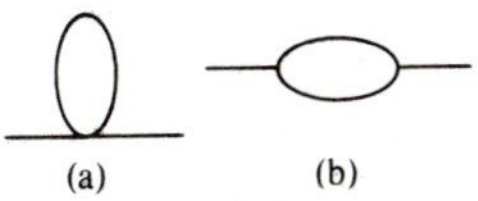

FIG. 6.7.

$$\Sigma_a(0) = (-6\mathrm{i}\lambda)\frac{1}{2}\int\frac{\mathrm{d}^4k}{(2\pi)^4}\frac{\mathrm{i}}{k^2-2\mu^2} = -3\mathrm{i}\lambda I_2 \tag{6.80}$$

$$\Sigma_b(0) = (-6\mathrm{i}\lambda v)^2\frac{1}{2}\int\frac{\mathrm{d}^4k}{(2\pi)^4}\left(\frac{\mathrm{i}}{k^2-2\mu^2}\right)^2 = 18\mathrm{i}\lambda^2v^2I_4 \tag{6.81}$$

where

$$I_4 = \mathrm{i}\int\frac{\mathrm{d}^4k}{(2\pi)^4}\left(\frac{\mathrm{i}}{k^2-2\mu^2}\right)^2 \tag{6.82}$$

We have

$$\Sigma(0) = \Sigma_a(0) + \Sigma_b(0) = -3i\lambda I_2 + 18i\lambda^2 v^2 I_4. \tag{6.83}$$

This requires a counterterm $-A\phi'^2$ with

$$A = -\frac{3}{2}\lambda I_2 + 9\lambda^2 v^2 I_4. \tag{6.84}$$

(3) *Three-point function* (Fig. 6.8).

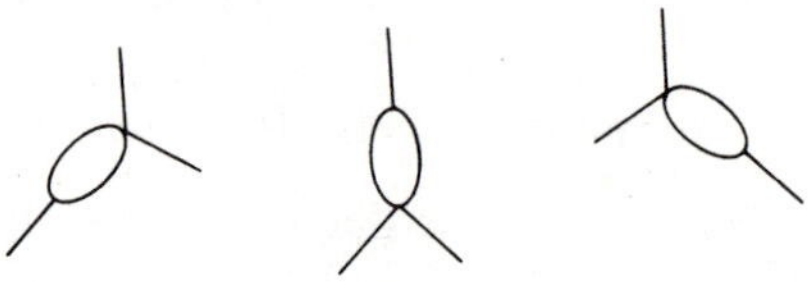

FIG. 6.8.

$$\Gamma_3(0) = 3(-6i\lambda v)(-6i\lambda)\frac{1}{2}\int\frac{d^4k}{(2\pi)^4}\left(\frac{i}{k^2-2\mu^2}\right)^2$$
$$= 54i\lambda^2 v I_4. \tag{6.85}$$

For this we need counterterm $-B\phi'^3$ with

$$B = -\frac{i}{3!}(54i\lambda^2 v I_4) = 9\lambda^2 v I_4. \tag{6.86}$$

(4) *Four-point function* (Fig. 6.9).

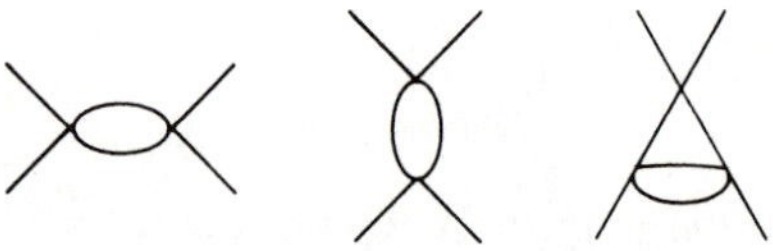

FIG. 6.9.

$$\Gamma_4(0) = 3(-6i\lambda)^2\frac{1}{2}\int\frac{d^4k}{(2\pi)^4}\left(\frac{i}{k^2-2\mu^2}\right)^2$$
$$= 54i\lambda^2 I_4 \tag{6.87}$$

which requires the counterterm $-\frac{1}{4}C\phi'^4$ with

$$C = -\frac{i}{3!}(54i\lambda^2 I_4) = 9\lambda^2 I_4. \tag{6.88}$$

Therefore, the one-loop counterterms are

$$\delta V(\phi') = A\phi'^2 + B\phi'^3 + \frac{C}{4}\phi'^4 + D\phi' \tag{6.89}$$

and the effective potential to the one-loop level is given by

$$V_1(\phi') = V(\phi') + \delta V(\phi') = (\mu^2 + A)\phi'^2 + (\lambda v + B)\phi'^3 + \tfrac{1}{4}(\lambda + C)\phi'^4 + D\phi'. \tag{6.90}$$

Let δv be the shift in VEV due to the one-loop contribution

$$\left.\frac{\delta V_1(\phi')}{\delta\phi'}\right|_{\phi'=\delta v} = 0$$

or

$$2(\mu^2 + A)\,\delta v + 3(\lambda v + B)(\delta v)^2 + (\lambda + C)(\delta v)^3 + D = 0. \tag{6.91}$$

Since δv is small, we can neglect higher-order terms so that

$$\delta v = -\frac{D}{2(\mu^2 + A)} \simeq \frac{-D}{2\mu^2}. \tag{6.92}$$

Thus to eliminate the linear term in ϕ', we define a shifted field

$$\phi'' = \phi' - \delta v \tag{6.93}$$

$$\langle 0|\phi''|0\rangle = 0 \quad \text{to one loop.} \tag{6.94}$$

In terms of this new field, the potential can be written as

$$V_1(\phi'' + \delta v) = a\phi''^2 + b\phi''^3 + \frac{c}{4}\phi''^4 \tag{6.95}$$

where

$$c = \lambda + C = \lambda + 9\lambda^2 I_4$$

$$b = \lambda v + B + \lambda\,\delta v = \lambda v + 9\lambda^2 v I_4 + \frac{3\lambda^2 v}{2\mu^2} I_2$$

$$a = \mu^2 + A + 3\,\delta v \lambda v = \mu^2 + 3\lambda I_2 + 9\lambda^2 v^2 I_4. \tag{6.96}$$

From (6.96) we can check that

$$b^2 - ac = 0. \tag{6.97}$$

Throughout the above computation we have consistently dropped higher powers of δv. Eqn (6.97) means that we have

$$V_1(\phi') = V_1(\phi'' + \delta v) = a\phi''^2 + (ac)^{\frac{1}{2}}\phi''^3 + \frac{c}{4}\phi''^4$$

$$= \frac{1}{4}\frac{a^2}{c} - \frac{a}{2}\left(\phi'' + \sqrt{\frac{a}{c}}\right)^2 + \frac{c}{4}\left(\phi'' + \sqrt{\frac{a}{c}}\right)^4. \tag{6.98}$$

This means the effective potential still has reflection symmetry in terms of

$$\phi = \phi'' + \sqrt{\frac{a}{c}} \tag{6.99}$$

in spite of the addition of counterterms (6.89). In other words the counterterms still have the original symmetry as they reflect the ultraviolet properties of the theory which are insensitive to soft symmetry breaking. Therefore, even for broken symmetry theory, we only need the counterterms of the symmetric theory. Further discussion on this point will be given below in the study of U(1) symmetric theory.

Goldstone bosons remain massless in higher orders

We now take up theories with continuous symmetries where spontaneous symmetry breaking is accompanied by the presence of Goldstone bosons. We show how the Goldstone particles remain massless even in the presence of higher-order radiative corrections. Again we will illustrate the point with the simplest example of U(1) symmetry at the one-loop level.

The U(1) symmetric Lagrangian (eqn (5.138))

$$\mathscr{L} = \frac{1}{2}(\partial_\lambda \sigma)^2 + \frac{1}{2}(\partial_\lambda \pi)^2 + \frac{\mu^2}{2}(\sigma^2 + \pi^2) + \frac{\lambda}{4}(\sigma^2 + \pi^2)^2 \tag{6.100}$$

when expressed in terms of the shifted field

$$\sigma' = \sigma - v \tag{6.101}$$

with the VEV

$$\langle 0|\sigma|0\rangle = v = (\mu^2/\lambda)^{\frac{1}{2}} \tag{6.102}$$

gives rise to the Lagrangian (eqn (5.146))

$$\mathscr{L} = \frac{1}{2}[(\partial_\lambda \sigma')^2 + (\partial_\lambda \pi)^2] - \mu^2 \sigma'^2 - \lambda v \sigma'(\sigma'^2 + \pi^2) - \frac{\lambda}{4}(\sigma'^2 + \pi^2)^2 \tag{6.103}$$

which has a massless field π and massive $(2\mu^2)$ field σ' interacting through the vertices shown in Fig. 6.10. To see how the π-field remains massless at the one-loop level we need to check that the mass renormalization counterterm δm vanishes. This must be so since it cannot be absorbed in any redefinition of the physical π-mass. We recall that the self-energy diagrams have the expansion

$$\Sigma(p^2) = \Sigma(0) + \Sigma'(0)p^2 + \tilde{\Sigma}(p^2) \tag{6.104}$$

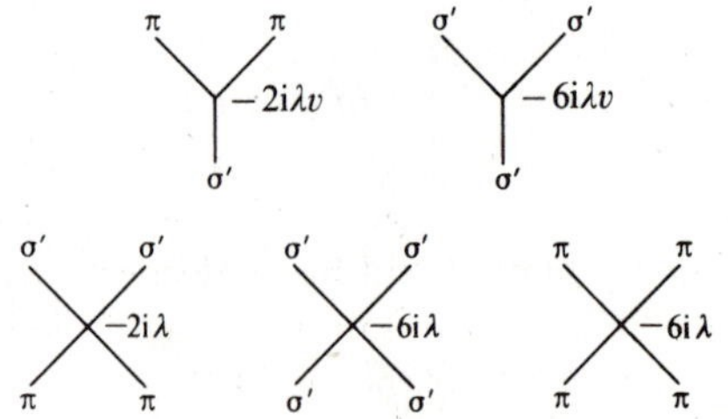

FIG. 6.10. Interaction vertices for the Lagrangian of eqn (6.103).

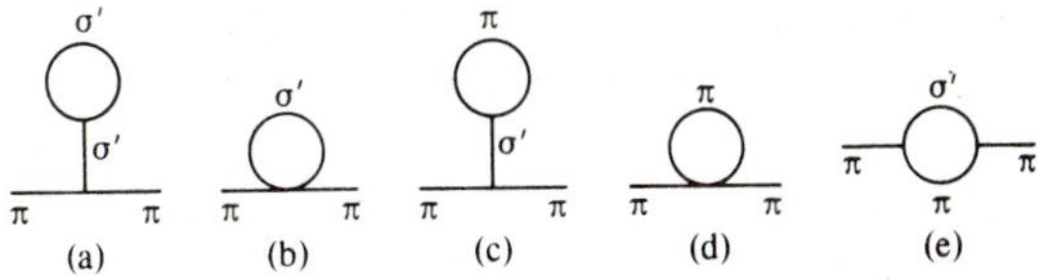

FIG. 6.11. Self-energy diagrams for the π-particle.

where δm^2 is identified with $\Sigma(0)$. The one-loop diagrams shown in Fig. 6.11 can be calculated using the Feynman rules of Fig. (6.10)

$$\Sigma_a(0) = (-2i\lambda v)\frac{i}{-2\mu^2}(-6i\lambda v)\frac{1}{2}\int\frac{d^4k}{(2\pi)^4}\frac{i}{k^2-2\mu^2} = -3\lambda I(2\mu^2)$$

$$\Sigma_b(0) = (-2i\lambda)\frac{1}{2}\int\frac{d^4k}{(2\pi)^4}\frac{i}{k^2-2\mu^2} = \lambda I(2\mu^2)$$

$$\Sigma_c(0) = (-2i\lambda v)\frac{i}{-2\mu^2}(-2i\lambda v)\frac{1}{2}\int\frac{d^4k}{(2\pi)^4}\frac{i}{k^2} = -\lambda I(0)$$

$$\Sigma_d(0) = (-6i\lambda)\frac{1}{2}\int\frac{d^4k}{(2\pi)^4}\frac{i}{k^2} = 3\lambda I(0)$$

$$\Sigma_e(0) = (-2i\lambda v)^2\int\frac{d^4k}{(2\pi)^4}\frac{i}{k^2}\frac{i}{k^2-2\mu^2} = 2\lambda[I(2\mu^2)-I(0)] \tag{6.105}$$

where the subscripts on the Σs indicate the diagram number in Fig. 6.11 and

$$I(m^2) = \int\frac{d^4k}{(2\pi)^4}\frac{1}{k^2-m^2}. \tag{6.106}$$

Thus $I(2\mu^2) = iI_2$ of (6.78). Again we have the usual symmetry factors 1/2. Clearly the contributions to $\Sigma(0)$ coming from all five diagrams sum up to zero

$$\Sigma(0) = \Sigma_a(0) + \Sigma_b(0) + \Sigma_c(0) + \Sigma_d(0) + \Sigma_e(0) = 0. \tag{6.107}$$

This is the promised result: the π-particle remains massless as required by the Goldstone theorem

Soft symmetry breaking and renormalizability

We note that the Feynman rule of Fig. 6.10 shows that there are five vertices which in turn depend only on the two parameters λ and v. This means that there are three relations among these five couplings (five Green's functions). These Ward identities are consequences of the original U(1) symmetry. The counterterms for these vertices must satisfy the same relations in order that they can be absorbed into the redefinitions of σ and v (Lee 1972*a*). This can be checked explicitly in the one-loop approximation, as was done in the earlier example of discrete symmetry. Again this illustrates that the counterterms in the spontaneously broken theory have the same structure as in the corresponding symmetric theory. In fact this is the key point which explains why spontaneous symmetry breaking ultimately does not spoil the renormalizability of the theory.

We will amplify briefly the important point that the renormalizability of a spontaneously broken theory depends only on the renormalizability of the symmetric theory. This is in fact a slightly stronger version of a theorem (Symanzik 1970*a*) which states that soft symmetry-breaking terms do not destroy the renormalizability of a symmetric theory. By *soft symmetry breaking* we mean asymmetric terms of dimension less than four, i.e., they correspond to vertices with a negative index of divergence (eqn (2.134))

$$\delta_i = \mathrm{d}_i - 4 < 0. \tag{6.108}$$

From the result of eqn (2.144) for the index of divergence of the counterterm

$$\delta_{\mathrm{ct}} \leq \sum_i n_i \delta_i \tag{6.109}$$

and from the fact that symmetry-breaking counterterms can arise only from diagrams that involve at least one symmetry-breaking interaction, one deduces immediately that the index for asymmetric counterterms must be negative

$$\delta_{\mathrm{ct}}^{\mathrm{SB}} < 0. \tag{6.110}$$

We can illustrate this theorem with two simple examples in the U(1) theory of eqn (6.100).

(1) $\mathscr{L}_{\mathrm{SB}} = c\sigma$, *which has dimension one or* $\delta_{\mathrm{SB}} = -3$. Thus the only counterterm which satisfies $\delta_{\mathrm{ct}}^{\mathrm{SB}} \leq -3$ is $\mathscr{L}_{\mathrm{ct}}^{\mathrm{SB}} = -A\sigma$ (as the π-term can be excluded by the reflection symmetry $\pi \to -\pi$). This does not destroy the renormalizability of the theory.

(2) $\mathscr{L}_{\mathrm{SB}} = c(\sigma^2 - \pi^2)$, *which has dimension two or* $\delta_{\mathrm{SB}} = -2$. Since the only interactions with dimension less than two or $\delta_{\mathrm{ct}} \leq -2$ are $(\sigma^2 + \pi^2)$ and $(\sigma^2 - \pi^2)$ (as terms linear in the fields can be excluded by reflection symmetry), the renormalizability is again maintained.

Spontaneous symmetry breaking is clearly of the soft variety as shifting the fields only changes the terms with dimension less than four. The remarkable point is that these breaking terms not only do not induce asymmetric terms having dimension equal to or greater than four but the lower-dimensional terms maintain the same algebraic relations as the original theory. Thus, the process of renormalization does not introduce additional symmetry breaking, in the sense that symmetric counterterms suffice to remove infinities from the theory whether or not the symmetry is realized in the 'conventional' or Goldstone modes.

.4 The effective potential and radiatively induced spontaneous symmetry breakdown

In previous discussions of spontaneous symmetry breaking (SSB) in field theory, we used the classical potential part of the Lagrangian to decide at each level of perturbation theory which is the true vacuum. We have to shift the field at each order (see calculations in the previous section). We need a more systematic method for treating SSB which enables us to survey all possible vacua at once, and to compute higher-order correction before deciding which vacuum the theory finally picks. The formalism is the effective potential (Schwinger 1951*a*,*b*; Goldstone *et al.* 1962; Jona-Lasinio 1964) and the appropriate approximation scheme is the loop expansion (Nambu 1968). Here we follow the presentation of Coleman and Weinberg (1973).

The effective potential formalism

To illustrate this approach in path-integral formalism, we first consider the simple case of one scalar field. The generating functional for the Green's function is given by eqn (1.74)

$$W[J] = \int [\mathrm{d}\phi] \exp\left\{\mathrm{i}\int \mathrm{d}^4x[\mathscr{L}(\phi(x)) + J(x)\phi(x)]\right\}. \tag{6.111}$$

We can also think of this as the vacuum-to-vacuum transition amplitude in the presence of the external source $J(x)$

$$W[J] = \langle 0|0\rangle_J. \tag{6.112}$$

When we expand $\ln W[J]$ in a functional Taylor series in $J(x)$, the coefficients will be the connected Green's functions (1.76)

$$\ln W[J] = \sum_n \frac{1}{n!}\int \mathrm{d}^4x_1 \ldots \mathrm{d}^4x_n G^{(n)}(x_1 \ldots x_n)J(x_1)\ldots J(x_n). \tag{6.113}$$

We define the classical field ϕ_c as the vacuum expectation value (VEV) of the operator ϕ in the presence of the source $J(x)$

$$\phi_c(x) = \frac{\delta \ln W}{\delta J(x)} = \left[\frac{\langle 0|\phi(x)|0\rangle}{\langle 0|0\rangle}\right]_J. \qquad ((6.114)$$

The *effective action* of the classical field $\Gamma(\phi_c)$ is defined by the functional Legendre transform

$$\Gamma(\phi_c) = \ln W[J] - \int \mathrm{d}^4x J(x)\phi_c(x). \tag{6.115}$$

From this definition, it follows that

$$\frac{\delta\Gamma(\phi_c)}{\delta\phi_c} = -J(x). \tag{6.116}$$

We can also expand $\Gamma(\phi_c)$ in powers of ϕ_c

$$\Gamma(\phi_c) = \sum_n \frac{1}{n!}\int d^4x_1 \ldots d^4x_n \Gamma^{(n)}(x_1 \ldots x_n)\phi_c(x_1)\ldots\phi_c(x_n). \quad (6.117)$$

It is possible to show that $\Gamma^{(n)}(x_1 \ldots x_n)$ is the sum of all 1PI Feynman diagrams with n external lines. Alternatively we can expand the effective action $\Gamma(\phi_c)$ in powers of momentum. In position space this expansion takes on the form

$$\Gamma(\phi_c) = \int d^4x[-V(\phi_c) + \tfrac{1}{2}(\partial_\mu\phi_c)^2 Z(\phi_c) + \ldots]. \quad (6.118)$$

The term without derivatives $V(\phi_c)$ is called the *effective potential.* To express V in terms of 1PI Greens functions, we first write $\Gamma^{(n)}$ in momentum space

$$\Gamma^{(n)}(x_1 \ldots x_n) = \int \frac{d^4k_1}{(2\pi)^4}\cdots\frac{d^4k_n}{(2\pi)^4}(2\pi)^4\,\delta^4(k_1 + \ldots k_n)$$
$$\times\, e^{i(k_1\cdot x_1 + \ldots k_n\cdot x_n)}\Gamma^{(n)}(k_1 \ldots k_n). \quad (6.119)$$

Putting this into (6.117) and expanding in powers of k_i, we get

$$\Gamma(\phi_c) = \sum_n \frac{1}{n!}\int d^4x_1 \ldots d^4x_n \int\frac{d^4k_1}{(2\pi)^4}\cdots\frac{d^4k_n}{(2\pi)^4}$$
$$\times \int d^4x\, e^{i(k_1 + \ldots k_n)\cdot x}\, e^{i(k_1\cdot x_1 + \ldots k_n\cdot x_n)}$$
$$\times\,[\Gamma^{(n)}(0, \ldots 0)\phi_c(x_1)\ldots\phi_c(x_n) + \ldots]$$
$$= \int d^4x \sum_n \frac{1}{n!}\{\Gamma^{(n)}(0, \ldots 0)[\phi_c(x)]^n + \ldots\}. \quad (6.120)$$

Comparing (6.118) and (6.120) we see that the nth derivative of $V(\phi_c)$ is just the sum of all 1PI diagrams with n external lines carrying zero momenta

$$V(\phi_c) = -\sum_n \frac{1}{n!}\Gamma^{(n)}(0, \ldots 0)[\phi_c(x)]^n. \quad (6.121)$$

The usual renormalization conditions of the perturbation theory can be expressed in terms of functions occurring in eqn (6.118). For example in $\lambda\phi^4$ theory we can define the mass squared as the value of the inverse propagator at zero momentum

$$\Gamma^{(2)}(0) = -\mu^2. \quad (6.122)$$

Then we have

$$\mu^2 = \left.\frac{d^2V}{d\phi_c^2}\right|_{\phi_c=0}. \quad (6.123)$$

Similarly, if we define the four-point function at zero external momenta to be the coupling constant

$$\Gamma^{(4)}(0) = -\lambda, \quad (6.124)$$

then

$$\lambda = \frac{d^4 V}{d\phi_c^4}\bigg|_{\phi_c = 0}. \tag{1.125}$$

Similarly, the standard condition for wavefunction renormalization becomes

$$Z(\phi_c)|_{\phi_c = 0} = 1. \tag{6.126}$$

Consider now the SSB of a theory with some internal symmetry. SSB occurs if the quantum field ϕ develops a non-zero VEV even when the source $J(x)$ vanishes. In this language with eqns (6.114) and (6.116), this means that it occurs if

$$\frac{\delta\Gamma(\phi_c)}{\delta\phi_c} = 0 \tag{6.127}$$

for some nonzero value of ϕ_c. Furthermore, since we are typically only interested in cases where VEV is translationally invariant, we can simplify this to

$$\frac{\delta V(\phi_c)}{\delta\phi_c} = 0 \qquad \text{for} \quad \phi_c \neq 0. \tag{6.128}$$

The value of ϕ_c, for which the minimum of $V(\phi_c)$ occurs will be denoted by $\langle\phi\rangle$, which is the expectation value of ϕ in the new vacuum.

Loop expansion

To calculate $V(\phi_c)$ we need an approximation scheme which preserves the main advantage of this effective potential formalism, i.e., the capability to survey all vacua at once before deciding which is the true ground state. Clearly ordinary perturbation theory with its expansion in coupling constants is not appropriate as we need to, at each order, identify the true vacuum state and shift the field.

Instead, we will here organize perturbation theory in the form of *loop expansion*. This is an expansion according to the increasing number of independent loops of connected Feynman diagrams. Thus the lowest-order graphs will be the Born diagrams or tree graphs. The next order consists of the one-loop diagrams which have one integration over the internal momenta, etc. For the effective potential (6.121) each loop level still involves an infinite summation corresponding to all possible external lines. The usual classical potential we have been working with is simply the first term (the tree graphs) of $V(\phi_c)$ in this loop expansion. In fact the loop expansion can be identified as an expansion in powers of the Planck's constant $\hbar$. This can be seen as follows. Let I be the number of internal lines and V the number of vertices in a given Feynman diagram. The number of independent loops L will be the number of independent internal momenta after the momentum conservation at each vertex is taken into account. Since one combination of these momentum conservations corresponds to the overall conservation of external momenta, the number of independent loops in a given Feynman diagram is given by

$$L = I - (V - 1). \tag{6.129}$$

To relate L to the powers of $\hbar$, we have to keep track of the factor $\hbar$ in the standard quantization procedure. First there is one power of $\hbar$ in the canonical commutation relation

$$[\phi(\mathbf{x}, t), \pi(\mathbf{y}, t)] = \mathrm{i}\hbar\, \delta^3(\mathbf{x} - \mathbf{y}). \tag{6.130}$$

This will give rise to a factor of $\hbar$ in the free propagator in momentum space

$$\langle 0|T(\phi(x)\phi(0))|0\rangle = \int \frac{\mathrm{d}^4 k}{(2\pi)^4}\, \mathrm{e}^{\mathrm{i}k\cdot x} \frac{\mathrm{i}\hbar}{k^2 - m^2 + \mathrm{i}\varepsilon}. \tag{6.131}$$

The other place where $\hbar$ appears is in the evolution operator $\exp[-\mathrm{i}Ht/\hbar]$ which gives rise to the operator $\exp[\mathrm{i}/\hbar \int \mathscr{L}_{\mathrm{int}}(\phi)\, \mathrm{d}^4 x]$ in the interaction picture. This means that there will be a factor of $1/\hbar$ for each vertex. Thus, for a given Feynman diagram, we have P powers of $\hbar$ with

$$P = I - V = L - 1. \tag{6.132}$$

Thus the number of loops and the power of $\hbar$ are directly correlated. The statement that loop expansion corresponds to an expansion in Planck's constant is really a statement that it is an expansion in some parameter a that multiplies the *total* Lagrange density

$$\mathscr{L}(\phi, \partial_\mu \phi, a) \equiv a^{-1} \mathscr{L}(\phi, \partial_\mu \phi). \tag{6.133}$$

The above counting of the $\hbar$ powers (P) reflects the fact that while every vertex carries a factor a^{-1}, the propagator carries a factor a because it is the *inverse* of the differential operator occurring in the quadratic terms in $\mathscr{L}$. Because $\hbar$, or a, is a parameter that multiplies the total Lagrangian, it is unaffected by shifts of fields and by the redefinition or division of $\mathscr{L}$ into free and interacting parts associated with such shifts. In short it allows us to compute $V(\phi_c)$ before the shift; thus it is an appropriate perturbation scheme for our purpose.

We should remark that this loop expansion is certainly not a worse approximation scheme than the ordinary coupling-constant expansion perturbation theory since the loop expansion includes the latter as a subset at a given loop level. In fact if we fix the number of the external lines (which we do not in the calculation of $V(\phi_c)$) these two expansions are identical for the simple case of one coupling constant. For example, in $\lambda\phi^4$ theory we have (eqn (2.58)) for Green's functions with E external lines

$$4V = E + 2I \tag{6.134}$$

which can be converted into a relation between the powers of the coupling constant (V) and the number of loops by using (6.129) to eliminate I

$$V = \tfrac{1}{2}E - L + 1. \tag{6.135}$$

The effective potential of the $\lambda\phi^4$ theories

We now illustrate the calculation of the effective potential in the simple case of $\lambda\phi^4$ theories in the one-loop approximation. The Lagrangian density is given by

$$\mathscr{L} = \mathscr{L}_0 + \mathscr{L}_I \tag{6.136}$$

with

$$\mathscr{L}_0 = \tfrac{1}{2}(\partial_\mu\phi)^2 - \tfrac{1}{2}\mu^2\phi^2$$

$$\mathscr{L}_I = -\frac{\lambda}{4!}\phi^4.$$

We shall study the three cases corresponding to $\mu^2 > 0$, $\mu^2 < 0$, and $\mu^2 = 0$.

(1) $\mu^2 > 0$ *case* (*no SSB*). To calculate the effective potential in eqn (6.121), we must sum all one-loop diagrams with an even number of external lines having zero momenta (see Fig. 6.12).

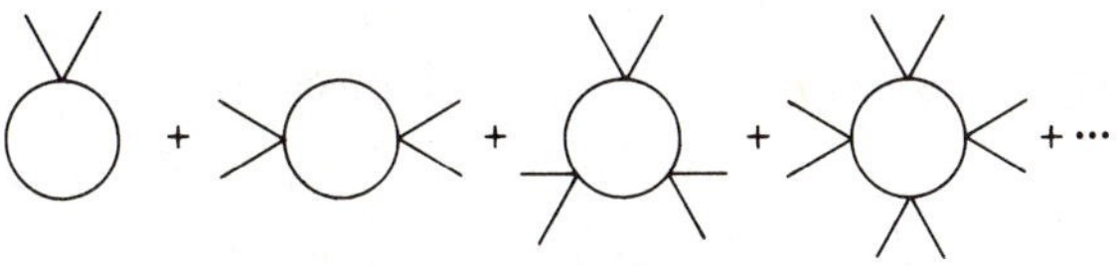

FIG. 6.12.

The 1PI Green's function is given by

$$\Gamma^{(2n)}(0, \ldots 0) = \mathrm{i}S_n \int \frac{\mathrm{d}^4k}{(2\pi)^4}\left[(-\mathrm{i}\lambda)\frac{\mathrm{i}}{k^2 - \mu^2 + \mathrm{i}\varepsilon}\right]^n \tag{6.137}$$

where S_n is the symmetry factor

$$S_n = \frac{(2n)!}{2^n 2n} \tag{6.138}$$

corresponding to the fact that there are $(2n)!$ ways to distribute $2n$ particles to the external lines of the diagram and that interchanges of any two external lines at a given vertex or reflections and rotations of n vertices on the ring do not lead to new contributions. The no-loop and one-loop effective potential is then given by

$$V(\phi_c) = \frac{1}{2}\mu^2\phi_c^2 + \frac{\lambda}{4!}\phi_c^4 + \mathrm{i}\int \frac{\mathrm{d}^4k}{(2\pi)^4}\sum_{n=1}^{\infty}\frac{1}{2n}\left[\frac{(\lambda/2)\phi_c^2}{k^2 - \mu^2 + \mathrm{i}\varepsilon}\right]^n$$

$$= \frac{1}{2}\mu^2\phi_c^2 + \frac{\lambda}{4!}\phi_c^4 + \frac{\mathrm{i}}{2}\int \frac{\mathrm{d}^4k}{(2\pi)^4}\ln\left[1 + \frac{\lambda\phi_c^2/2}{k^2 - \mu^2 + \mathrm{i}\varepsilon}\right]. \tag{6.139}$$

The integral is divergent. If it is cut off at some large momentum, we obtain

$$V(\phi_c) = \frac{1}{2}\mu^2\phi_c^2 + \frac{\lambda}{4!}\phi_c^4 + \frac{\Lambda^2}{32\pi^2}\left(\mu^2 + \frac{\lambda}{2}\phi_c^2\right) + \frac{1}{64\pi^2}\left(\mu^2 + \frac{\lambda}{2}\phi_c^2\right)^2\left[\ln\left(\frac{\mu^2 + \lambda\phi_c^2/2 + i\varepsilon}{\Lambda^2}\right) - \frac{1}{2}\right]. \tag{6.140}$$

The appearance of the combination $\mu^2 + (\lambda/2)\phi_c^2$ results from the summation

$$\frac{1}{k^2 - \mu^2} + \frac{1}{k^2 - \mu^2}\frac{1}{2}\lambda\phi_c^2\frac{1}{k^2 - \mu^2} + \ldots = \frac{1}{k^2 - (\mu^2 + \lambda\phi_c^2/2)}, \tag{6.141}$$

i.e., the $\lambda\phi_c^2/2$ term acts effectively as a mass insertion (for further remarks see eqn (6.147) below).

To remove the cut-off dependence we introduce counterterms which have the same structure as the original potential

$$V_{ct}(\phi_c) = \frac{A}{2}\phi_c^2 + \frac{B}{4!}\phi_c^4. \tag{6.142}$$

so that the renormalized effective potential, given by

$$V_r(\phi_c) = V(\phi_c) + V_{ct}(\phi_c), \tag{6.143}$$

is finite and cut-off-independent. The coefficients A and B in (6.142) can be determined by the renormalization conditions (6.123)–(6.126). In this way we have

$$V_r = \frac{1}{2}\mu^2\phi_c^2 + \frac{\lambda}{4!}\phi_c^4 + \frac{1}{64\pi^2}\left[(\mu^2 + \frac{\lambda}{2}\phi_c^2)^2\ln\left(\frac{\mu^2 + \lambda\phi_c^2/2}{\mu^2}\right) - \frac{\lambda\mu^2}{2}\phi_c^2 - \frac{3}{8}\lambda^2\phi_c^4\right]. \tag{6.144}$$

We see that the large ϕ_c behaviour of $V(\phi_c)$ is modified by radiative corrections.

(2) $\mu^2 < 0$ *case (with SSB)*. For illustrative purposes we choose this time to separate the Lagrangian differently

$$\mathscr{L} = \tfrac{1}{2}(\partial_\mu\phi)^2 - U(\phi) \tag{6.145}$$

where

$$U(\phi) = \frac{1}{2}\mu^2\phi^2 + \frac{\lambda}{4!}\phi^4 \tag{6.146}$$

and take $U(\phi)$ as a perturbation. There are two vertices: μ^2 and $\frac{1}{2}\lambda\phi^2$, shown in Fig. 6.13(a). Their combination is the second derivative $U''(\phi)$ which is just

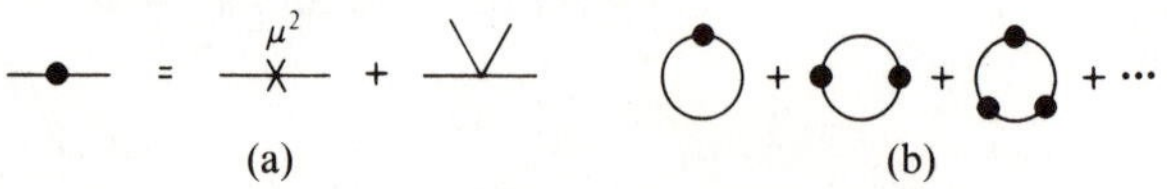

FIG. 6.13.

the scalar mass squared for $\phi = \langle\phi\rangle$. Thus we use the notation

$$m_s^2(\phi) = \mu^2 + \tfrac{1}{2}\lambda\phi^2. \tag{6.147}$$

The effective potential calculated by summing up the diagrams in Fig. 6.13(b), with a massless propagator and $m_s^2(\phi)$ vertices, has the form

$$\begin{aligned} V(\phi_c) &= \frac{1}{2}\mu^2\phi_c^2 + \frac{\lambda}{4!}\phi_c^4 + \mathrm{i}\int\frac{\mathrm{d}^4k}{(2\pi)^4}\sum_{n=1}^{\infty}\frac{1}{2n}\left[\frac{m_s^2(\phi_c)}{k^2+\mathrm{i}\varepsilon}\right]^n \\ &= \frac{1}{2}\mu^2\phi_c^2 + \frac{\lambda}{4!}\phi_c^4 + \frac{\mathrm{i}}{2}\int\frac{\mathrm{d}^4k}{(2\pi)^4}\ln\left[1+\frac{m_s^2(\phi_c)}{k^2+\mathrm{i}\varepsilon}\right] \\ &= \frac{1}{2}\mu^2\phi_c^2 + \frac{\lambda}{4!}\phi_c^4 + \frac{m_s^4(\phi_c)}{64\pi^2}\left[\ln\frac{m_s^2(\phi_c)}{\mu^2}+\ldots\right]. \end{aligned} \tag{6.148}$$

We see that, apart from an irrelevant constant, this is the same result as case (1) with $\mu^2 > 0$. Thus, if we choose the same renormalization condition as in case (1), we arrive at the renormalized potential (6.144). In this case the quantity μ^2 has a different physical meaning and is not the mass of the particle in the zeroth order. Nevertheless it is still a finite quantity which serves to parametrize the theory.

The calculations in cases (1) and (2) illustrate the advantage of this perturbative approach: the same result can be obtained whether there is SSB or not; we need not shift the field beforehand. Also, this approach is insensitive to how we divide up the Lagrangian, whether as in (6.136) or (6.145). More importantly, the calculations illustrate the feature that even in the presence of SSB the counterterms are still the same as those of the symmetric theory. In other words the ultraviolet divergences respect the symmetry of the Lagrangian even if the vacuum does not. This reflects the fact that SSB is generated by the nonzero VEV $\langle\phi\rangle$ which has the dimensions of mass and the ultraviolet divergences are insensitive to finite mass scales.

(3) $\mu^2 = 0$ *case* (*SSB driven by radiative corrections*?). In this case $V(\phi)$ is flat at $\phi = 0$ and the usual procedure of using the classical potential is inadequate to determine whether SSB is induced or not, and we have to go to higher orders to see the pattern of symmetry realization. Coleman and Weinberg (1973) were the first ones to point out this interesting phenomenon of radiatively induced SSB. We cannot take the $\mu^2 \to 0$ limit of eqn (6.144) because of the infrared singularity. To get around this difficulty, we will choose a renormalization condition for the coupling constant different from that of (6.125). Instead, at $\phi_c = 0$, we have

$$\lambda = \left.\frac{\mathrm{d}^4V}{\mathrm{d}\phi_c^4}\right|_{\phi_c = M} \tag{6.149}$$

where M is an arbitrary mass parameter. The effective potential now takes, for $\mu \ll M$, the form

$$V_r(\phi_c) = \frac{1}{2}\mu^2\phi_c^2 + \frac{\lambda}{4!}\phi_c^4 + \frac{1}{64\pi^2}\left\{\left(\mu^2 + \frac{\lambda}{2}\phi_c^2\right)^2 \ln\left[\frac{\mu^2 + \lambda\phi_c^2/2}{\mu^2}\right] - \frac{1}{2}\lambda\mu^2\phi_c^2 - \frac{25}{24}\lambda^2\phi_c^4 + \frac{1}{4}\lambda^2\phi_c^4 \ln\left(\frac{2\mu^2}{\lambda M^2}\right)\right\}. \tag{6.150}$$

This has the $\mu^2 = 0$ limit

$$V_r(\phi_c) = \frac{1}{4!}\lambda\phi_c{}^4 + \frac{\lambda^2\phi_c{}^4}{256\pi^2}\left[\ln\frac{\phi_c{}^2}{M^2} - \frac{25}{6}\right]. \tag{6.151}$$

Since the second term in eqn (6.151) is negative for small ϕ_c, it has the effect of turning the zeroth-order minimum at the origin into a local maximum and producing a new minimum at some point away from the origin. In other words, the one-loop correction has generated SSB. However this conclusion is unwarranted as the new minimum is located at

$$\lambda \ln\frac{\langle\phi\rangle^2}{M^2} = \frac{-32}{3}\pi^2 + O(\lambda). \tag{6.152}$$

The fact that $\lambda \ln(\langle\phi^2\rangle/M^2)$ is bigger than one (i.e., the loop contribution is larger than that of the tree) means that the new minimum lies outside the validity of the one-loop approximation. In this simple theory with one coupling constant such a result is inevitable. Since we want the one-loop contribution to compete with the three contribution, $\lambda \ln\langle\phi^2\rangle$ must be large, yet λ is the only parameter in the theory. This implies that to avoid this difficulty we should examine theories with more than one coupling.

Massless scalar QED (dimensional transmutation)

The Lagrange density is given by

$$\mathscr{L} = -\frac{1}{4}F_{\mu\nu}F^{\mu\nu} + \frac{1}{2}|(\partial_\mu - \mathrm{i}eA_\mu)\phi|^2 - \frac{\lambda}{4!}(\phi^*\phi)^2 \tag{6.153}$$

where A_μ is the photon field with $F_{\mu\nu} = \partial_\mu A_\nu - \partial_\nu A_\mu$. ϕ is a complex scalar field

$$\phi = \phi_1 + \mathrm{i}\phi_2, \tag{6.154}$$

where $\phi_{1,2}$ is real. Now we have two coupling constants λ and e and it will be possible to obtain a small expansion parameter.

The calculation will be considerably simplified in the *Landau gauge*, where the photon propagator is

$$\mathrm{i}\Delta_{\mu\nu}(k) = -\mathrm{i}\frac{g_{\mu\nu} - k_\mu k_\nu/k^2}{k^2 + \mathrm{i}\varepsilon}. \tag{6.155}$$

Because we always work with zero external momenta and because $k^{\mu}\Delta_{\mu\nu}(k) = 0$, all graphs of the type shown in Fig. 6.14 make no contribution. Also note that the effective potential can depend only on $\phi_c^2 = \phi_c^*\phi_c = \phi_{1c}^2 + \phi_{2c}^2$.

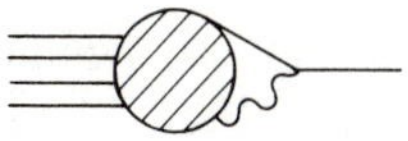

FIG. 6.14.

We need only compute diagrams having ϕ_1 external lines; there are three basic classes of graphs with different particles running in the loop (Fig. 6.15).

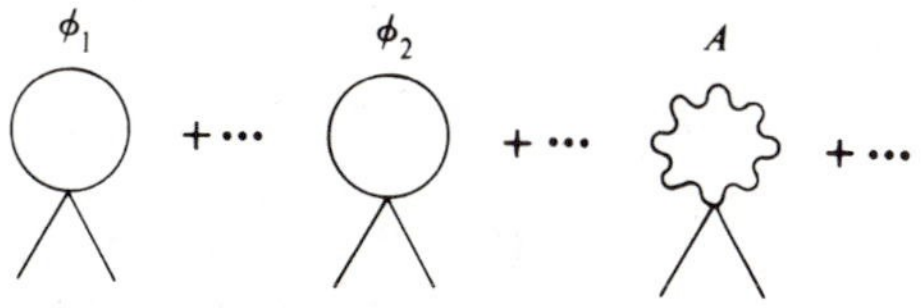

FIG. 6.15.

In this way we obtain

$$V(\phi_c) = \frac{1}{4!}\lambda\phi_c^4 + \frac{\phi_c^4}{64\pi^2}\left\{\left(\frac{\lambda}{2}\right)^2\left[1 + \left(\frac{1}{3}\right)^2\right] + 3e^4\right\}\left[\ln\frac{\phi_c^2}{M^2} - \frac{25}{6}\right]. \quad (6.156)$$

The calculation is entirely analogous to the previous case; we only need to note that the $\phi_1^2\phi_2^2$ coupling is 1/3 of the ϕ_1^4 coupling because of the different Wick contractions and that the extra factor 3 in the photon loop comes from the trace of Landau-gauge propagator. If we assume that λ is of order e^4, we can neglect the λ^2 term in (6.156). Since M is arbitrary, we take it to be the actual location of the new minimum, $M = \langle\phi\rangle$. The effective potential becomes

$$V(\phi_c) = \frac{1}{4!}\lambda\phi_c{}^4 + \frac{3e^4}{64\pi^2}\phi_c{}^4\left(\ln\frac{\phi_c{}^2}{\langle\phi\rangle^2} - \frac{25}{6}\right). \quad (6.157)$$

We also have the relation

$$0 = V'(\langle\phi\rangle) = \left(\frac{\lambda}{6} - \frac{11e^4}{16\pi^2}\right)\langle\phi\rangle^3$$

or

$$\lambda = \frac{33}{8\pi^2}e^4. \quad (6.158)$$

Surprisingly, the two independent coupling constants are related. This can be understood from the fact that we start out with two dimensionless parameters e and λ; we end up also with two parameters e and $\langle\phi\rangle$. In other words we have traded a dimensionless parameter λ for a dimensional parameter $\langle\phi\rangle$. This has been called *dimensional transmutation* (Coleman and Weinberg 1973) which is a general feature of the theory without any mass scale. Changes in M always involve a new definition of the coupling constant. With (6.158), the potential in (6.157) can be written simply as

$$V(\phi_c) = \frac{3e^4}{64\pi^2}\,\phi_c^4 \ln\left(\frac{\phi_c^2}{\langle\phi\rangle^2} - \frac{1}{2}\right) \tag{6.159}$$

which has a minimum at $\phi_c = \langle\phi\rangle$ and symmetry is spontaneously broken. This time the loop expansion is valid as the one-loop contribution can be smaller than that of the tree. Incidently since the mass of the 'photon' after SSB, $m_v = e\langle\phi\rangle$, we may use the notation

$$m_v(\phi) = e\phi \tag{6.160}$$

and write (6.157) as

$$V(\phi_c) = \frac{3m_v^4(\phi_c)}{64\pi^2} \ln\left(\frac{\phi_c^2}{\langle\phi\rangle^2} - \frac{1}{2}\right). \tag{6.161}$$

This is to be contrasted with eqn (6.148) for the scalar loop.

We should also remark that, if there are fermions in the theory, there will be, in addition to the scalar and vector loops, fermion loops. We can calculate their contribution in an analogous manner. If we follow the procedure of case (2) above and work with massless propagators, we will have the general vertex $m_f(\phi) = m + h\phi$, where m is the bare fermion mass and h is the Yukawa coupling. Since the trace of an odd number of Dirac γ-matrices is zero, we must have an even number of fermion propagators. Then we can group terms as

$$\ldots m_f(\phi)\frac{1}{\not{k}}\,m_f(\phi)\frac{1}{\not{k}}\ldots = \ldots \frac{1}{k^2}\,m_f^2(\phi)\ldots. \tag{6.162}$$

Then the calculation becomes exactly the same as the scalar case except the important difference of an overall minus sign for the loop integral. And we have the fermion loop contribution

$$\delta V(\phi_c) = \frac{-4}{64\pi^2}\,m_f^4(\phi_c)[\ln \phi_c^2/M^2 + \ldots]. \tag{6.163}$$

The factor of 4 arises from the trace of Dirac matrices. Thus in a theory having vector, scalar, and fermion loops, we can combine (6.161), (6.148), and (6.163) to obtain the one-loop contribution

$$\delta V(\phi_c) = \frac{1}{64\pi^2}\,[3m_v^4(\phi_c) + m_s^4(\phi_c) - 4m_f^4(\phi_c)] \ln \phi_c^2/M^2 + \ldots. \tag{6.164}$$

7 The parton model and scaling

Lepton–nucleon scatterings at high energy and large momentum transfer exhibit the remarkable phenomenon known as *Bjorken scaling*. This correlation pattern of the energy and angular distribution of the scattered leptons in these deep inelastic processes can be described simply by Feynman's parton model (Feynman 1972). At short distances hadrons may be viewed as composed of (almost free) point-like spin 1/2 constituents, the *partons*. It is natural to identify them as quarks. After a description of the quark–parton model and some of its applications in the first two sections we will then present the formal field theoretical apparatus required to describe the short-distance behaviour. This is Wilson's operator product expansion (Wilson 1969) with coefficients satisfying the renormalization group equations. Thus the stage is set for a field theory of strong interactions with the quarks being identified as the fundamental matter fields.

7.1 The parton model of deep inelastic lepton–hadron scattering

Kinematics and Bjorken scaling

The leptons used in deep inelastic processes are either charged leptons (electron or muon) or neutrinos which scatter off the target nucleons via the electromagnetic or weak interactions, respectively.

Electron–nucleon case. The momenta for the reaction

$$\mathrm{e}(k) + \mathrm{N}(p) \to \mathrm{e}(k') + \mathrm{X}(p_n) \tag{7.1}$$

are shown in Fig. 7.1, where X is some hadronic final state with total four-momentum p_n.

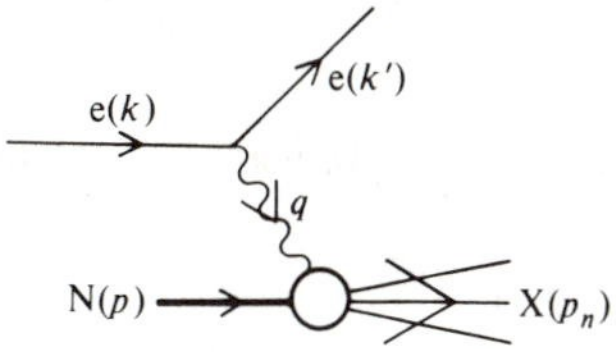

FIG. 7.1.

We define the kinematic variables by

$$q = k - k', \quad \nu = p \cdot q / M, \quad W^2 = p_n^2 = (p + q)^2 \tag{7.2}$$

In the lab-frame we have

$$p_\mu = (M, 0, 0, 0), \quad k_\mu = (E, \mathbf{k}), \quad k'_\mu = (E', \mathbf{k}'). \tag{7.3}$$

Then

$$\nu = E - E' \tag{7.4}$$

is the energy loss of the lepton and, when the lepton mass m_l is neglected

$$q^2 = (k - k')^2 = -4EE' \sin^2 \frac{\theta}{2} \leq 0, \quad Q^2 = -q^2 \tag{7.5}$$

where θ is the scattering angle of the lepton. The amplitude is given by

$$T_n = e^2 \bar{u}(k', \lambda')\gamma^\mu u(k, \lambda) \frac{1}{q^2} \langle n | J_\mu^{\text{em}}(0) | p, \sigma \rangle \tag{7.6}$$

where J_μ^{em} is the hadronic electromagnetic current. The unpolarized differential cross-section is given by

$$\begin{aligned} \mathrm{d}\sigma_n = {} & \frac{1}{|\mathbf{v}|} \frac{1}{2M} \frac{1}{2E} \frac{\mathrm{d}^3 k'}{(2\pi)^3 2k'_0} \prod_{i=1}^{n} \left[\frac{\mathrm{d}^3 p_i}{(2\pi)^3 2p_{i0}} \right] \\ & \times \frac{1}{4} \sum_{\sigma\lambda\lambda'} |T_n|^2 (2\pi)^4 \, \delta^4(p + k - k' - p_n). \end{aligned} \tag{7.7}$$

where $p_n = \Sigma_{i=1}^n p_i$. If we sum over all possible hadronic final states (i.e., they are not observed) we obtain the *inclusive cross-section*

$$\frac{\mathrm{d}^2\sigma}{\mathrm{d}\Omega \, \mathrm{d}E'} = \frac{\alpha^2}{q^4} \left(\frac{E'}{E} \right) l^{\mu\nu} W_{\mu\nu} \tag{7.8}$$

where $\alpha = e^2/4\pi$ is the fine structure constant. The leptonic tensor corresponds to

$$l_{\mu\nu} = \frac{1}{2} \operatorname{tr}(\not{k}' \gamma_\mu \not{k} \gamma_\nu) = 2(k_\mu k'_\nu + k'_\mu k_\nu + \frac{q^2}{2} g_{\mu\nu}) \tag{7.9}$$

and the hadronic tensor is given by

$$\begin{aligned} W_{\mu\nu}(p, q) &= \frac{1}{4M} \sum_\sigma \sum_n \int \prod_{i=1}^{n} \left[\frac{\mathrm{d}^3 p_i}{(2\pi)^3 2p_{i0}} \right] \\ & \quad \times \langle p, \sigma | J_\mu^{\text{em}}(0) | n \rangle \langle n | J_\nu^{\text{em}}(0) | p, \sigma \rangle (2\pi)^3 \, \delta^4(p_n - p - q) \\ &= \frac{1}{4M} \sum_\sigma \int \frac{\mathrm{d}^4 x}{2\pi} \, \mathrm{e}^{\mathrm{i}q \cdot x} \langle p, \sigma | J_\mu^{\text{em}}(x) J_\nu^{\text{em}}(0) | p, \sigma \rangle. \end{aligned} \tag{7.10}$$

Sometimes it is more convenient to rewrite this in the form of a commutator; we observe that

$$\begin{aligned} & \int \frac{\mathrm{d}^4 x}{2\pi} \, \mathrm{e}^{\mathrm{i}q \cdot x} \langle p, \sigma | J_\nu(0) J_\mu(x) | p, \sigma \rangle \\ &= \sum_n \int \frac{\mathrm{d}^4 x}{2\pi} \, \mathrm{e}^{\mathrm{i}(p_n - p + q) \cdot x} \langle p, \sigma | J_\nu(0) | n \rangle \langle n | J_\mu(0) | p, \sigma \rangle \\ &= \sum_n (2\pi)^3 \, \delta^4(p_n - p + q) \langle p, \sigma | J_\nu(0) | n \rangle \langle n | J_\mu(0) | p, \sigma \rangle. \end{aligned} \tag{7.11}$$

In the lab frame $q_0 = \nu > 0$ there is no intermediate state $|n\rangle$ with energy $E_n = M - \nu \leq M$ which can contribute; thus the above term vanishes. We can therefore write

$$W_{\mu\nu}(p, q) = \frac{1}{4M} \sum_{\sigma} \int \frac{d^4x}{2\pi} e^{iq\cdot x} \langle p, \sigma | [J_{\mu}^{\text{em}}(x), J_{\nu}^{\text{em}}(0)] | p, \sigma \rangle. \qquad (7.12)$$

From current conservation $\partial^{\mu} J_{\mu}^{\text{em}} = 0$ we have

$$q^{\mu} \langle p, \sigma | J_{\mu}^{\text{em}} | n \rangle = 0$$

or

$$q^{\mu} W_{\mu\nu} = q^{\nu} W_{\mu\nu} = 0. \qquad (7.13)$$

From (7.13) and the fact that $W_{\mu\nu}$ is a second-rank Lorentz tensor depending on the momenta p_{μ} and q_{μ}, one can deduce its covariant decomposition

$$W_{\mu\nu}(p, q) = \left[-W_1 \left(g_{\mu\nu} - \frac{q_{\mu} q_{\nu}}{q^2} \right) + \frac{W_2}{M^2} \left(p_{\mu} - \frac{p \cdot q}{q^2} q_{\mu} \right) \left(p_{\nu} - \frac{p \cdot q}{q^2} q_{\nu} \right) \right] \qquad (7.14)$$

where $W_{1,2}$ are Lorentz-invariant *structure functions* of the target nucleon depending on the invariant variables q^2 and ν of (7.2) and (7.5). We can then write (7.8)

$$\frac{d^2\sigma}{d\Omega\, dE'} = \frac{\alpha^2}{4E^2 \sin^4 \frac{\theta}{2}} \left(2W_1 \sin^2 \frac{\theta}{2} + W_2 \cos^2 \frac{\theta}{2} \right). \qquad (7.15)$$

A measurement of the inclusive cross-section yields information about the structure functions $W_{1,2}(q^2, \nu)$ which are the strong-interaction quantities characterizing the response (and hence the structure) of the target nucleon to electromagnetic probes.

To get some feeling about the structure functions, we first consider the special case where the final hadronic state $X(p_n)$ is also a nucleon. The matrix element of the electromagnetic current between either the proton or the neutron states can be written as

$$\langle N(p') | J_{\mu}^{\text{em}}(0) | N(p) \rangle = \bar{u}(p')[\gamma_{\mu} F_1(q^2) + i\sigma_{\mu\nu} q^{\nu} F_2(q^2)/2M] u(p) \qquad (7.16)$$

with $q = p - p'$. $F_{1,2}(q^2)$ are Lorentz-invariant *form factors*. For the case of the proton, $F_1^p(0) = 1$ and $F_2^p(0) = 1.79$ (nucleon magnetons) measure the total charge and anomalous magnetic moment, respectively, of the proton; $F_1^n(0) = 0$ and $F_2^n(0) = -1.91$ measure the total charge and anomalous magnetic moment, respectively, of the neutron. To check that $F_1^p(0) = 1$ does give the total charge of the proton as $+1$, we note that the charge operator

$$Q|p\rangle = |p\rangle \qquad (7.17)$$

implies

$$\langle p'|Q|p\rangle = \langle p'|p\rangle = 2E(2\pi)^3\,\delta^3(\mathbf{p}'-\mathbf{p}). \tag{7.18}$$

On the other hand from (7.16) we have

$$\begin{aligned}\langle p'|Q|p\rangle &= \int \mathrm{d}^3x \langle p'|J_0^{\mathrm{em}}(x)|p\rangle \\ &= \int \mathrm{d}^3x\, \mathrm{e}^{\mathrm{i}(p'-p)\cdot x}\langle p'|J_0^{\mathrm{em}}(0)|p\rangle \\ &= (2\pi)^3\,\delta^3(\mathbf{p}'-\mathbf{p})\bar{u}(p')\gamma_0 u(p)F_1(0) \\ &= 2E(2\pi)^3\,\delta^3(\mathbf{p}'-\mathbf{p})F_1(0).\end{aligned} \tag{7.19}$$

Thus (7.17) implies that $F_1(0) = 1$, as promised. It is straightforward to take the elastic limit $p_n^2 = M^2$ in (7.10) and obtain

$$W_1^{\mathrm{el}}(q^2, \nu) = \delta(q^2 + 2M\nu)\frac{q^2}{2M}G_{\mathrm{M}}^2(q^2)$$

$$W_2^{\mathrm{el}}(q^2, \nu) = \delta(q^2 + 2M\nu)\frac{2M}{(1 - q^2/4M^2)}\left[G_{\mathrm{E}}^2(q^2) - \frac{q^2}{4M^2}G_{\mathrm{M}}^2(q^2)\right] \tag{7.20}$$

where

$$G_{\mathrm{E}}(q^2) = F_1(q^2) + \frac{q^2}{4M^2}F_2(q^2)$$

$$G_{\mathrm{M}}(q^2) = F_1(q^2) + F_2(q^2) \tag{7.21}$$

are the electric and magnetic form factors, respectively. The elastic electron–nucleon cross-section is then

$$\frac{\mathrm{d}\sigma^{\mathrm{el}}}{\mathrm{d}\Omega} = \frac{\alpha}{4E^2} \times \frac{\cos^2\frac{\theta}{2}(1 - q^2/4M^2)^{-1}[G_{\mathrm{E}}^2 - (q^2/4M^2)G_{\mathrm{M}}^2] - \sin^2\frac{\theta}{2}(q^2/2M^2)G_{\mathrm{M}}^2}{\left[1 + (2E/M)\sin^2\frac{\theta}{2}\right]\sin^4\frac{\theta}{2}} \tag{7.22}$$

Thus measurements of the elastic eN differential cross-section yield information about the electric and magnetic form factors. Experimentally G_{E} and G_{M} for the proton are given by (for a review see Taylor 1975)

$$G_{\mathrm{E}}(q^2) \approx \frac{G_{\mathrm{M}}(q^2)}{\kappa_{\mathrm{p}}} \approx \frac{1}{(1 - q^2/0.7\ \mathrm{GeV}^2)^2} \tag{7.23}$$

where $\kappa_{\mathrm{p}} = 2.79$ is the magnetic moment of the proton. If the proton were a point-like (structureless) particle, we would have $G_{\mathrm{M}}(q^2) = G_{\mathrm{E}}(q^2) = 1$. Thus the nontrivial dependence of q^2 in (7.23) indicates the structure of the proton. Also for large q^2, the elastic cross-section falls off rapidly as $G_{\mathrm{E}} \approx G_{\mathrm{M}} \sim q^{-4}$.

If the inelastic cross-sections for final states other than the nucleon all behaved much like the elastic cross-section, we would expect them to fall off rapidly for large q^2. The surprise is that experimentally these cross-sections for large final-state invariant mass $W \gg M$ seem to have a much more moderate dependence on q^2 (for a review see Panofsky 1968). This leads to the idea that there must be some point-like constituents inside the nucleon much as the large-angle scattering of the α-particle in Rutherford's experiments suggested that the charge of the target atom was concentrated in the 'point-like' nucleus. These structureless particles inside the nucleon are called *partons*. A proper description of the parton model will be given after we have made a more precise statement of the deep inelastic scattering behaviour in terms of *Bjorken scaling* (Bjorken 1969).

Define the dimensionless scaling variables

$$x = \frac{-q^2}{2M\nu} = \frac{Q^2}{2M\nu}. \tag{7.24}$$

The range of x

$$0 \leq x \leq 1 \tag{7.25}$$

is given by the fact that the invariant mass of the unobserved final hadronic state is larger than the nucleon mass

$$W^2 = (p+q)^2 = q^2 + 2M\nu + M^2 \geq M^2. \tag{7.26}$$

Note that the elastic scattering corresponds to $x = 1$. Also define the variable

$$y = \frac{\nu}{E} = 1 - \frac{E'}{E} \tag{7.27}$$

which is the fraction of the initial energy transfered to the hadrons. From the fact that $0 \leq E' \leq E$ we obtain the range of y

$$0 \leq y \leq 1. \tag{7.28}$$

It is convenient to express the cross-section in terms of the x and y variables. Using the relation

$$\mathrm{d}x\,\mathrm{d}y = \frac{E'\,\mathrm{d}\Omega\,\mathrm{d}E'}{E\;2\pi y M} \tag{7.29}$$

and the definitions

$$\begin{aligned} MW_1(q^2, \nu) &= G_1(x, q^2/M^2) \\ \nu W_2(q^2, \nu) &= G_2(x, q^2/M^2), \end{aligned} \tag{7.30}$$

we can write (7.15) in the form

$$\frac{\mathrm{d}^2\sigma}{\mathrm{d}x\,\mathrm{d}y} = \frac{8\pi\alpha^2}{MEx^2y^2}\left[xy^2G_1 + \left(1 - y - \frac{M}{2E}xy\right)G_2\right]. \tag{7.31}$$

Bjorken scaling is the statement that in the large Q^2 limit with x fixed, the G_is are functions of x only. Thus,

$$\lim_{\substack{|q^2|\to\infty \\ x \text{ fixed}}} G_i\left(x, \frac{q^2}{M^2}\right) = F_i(x). \tag{7.32}$$

The dimensionless structure functions become independent of any mass scale. The $F_i(x)$s are called the *scaling functions*. Experimentally Bjorken scaling seems to be obtained for a rather modest value of $Q^2 \geq 2(\text{GeV})^2$ in ep scattering.

Neutrino–nucleon scattering. Next we come to the case of the charged weak-current process

$$\nu_l(k) + \text{N}(p) \to l^-(k') + \text{X}(p_n). \tag{7.33}$$

Since the basic idea is exactly the same as for the electromagnetic lN scattering considered above and since the reaction has also been presented in our discussion of the Adler-current-algebra sum rule in §5.2, we merely summarize the results. We will assume the current–current interaction for the weak effective Lagrangian

$$\mathscr{L}_{\text{eff}} = -\frac{G_{\text{F}}}{\sqrt{2}} J_\lambda^\dagger J^\lambda + \text{h.c.} \tag{7.34}$$

where G_{F} is the Fermi constant. The (charged) weak current J^λ can be separated into the leptonic and hadronic parts

$$J^\lambda = J_l^\lambda + J_{\text{h}}^\lambda \tag{7.35}$$

The leptonic weak current has the explicit form

$$J_l^\lambda = \bar{\nu}_e \gamma^\lambda (1 - \gamma_5)\, \text{e} + \bar{\nu}_\mu \gamma^\lambda (1 - \gamma_5) \mu + \dots . \tag{7.36}$$

The cross-section for neutrino and antineutrino scatterings can be written as

$$\frac{\text{d}^2\sigma^{(\nu)}}{\text{d}\Omega\,\text{d}E'} = \frac{G_{\text{F}}^2}{2\pi^2} E'^2 \left[2\sin^2\frac{\theta}{2}\, W_1^{(\nu)} + \cos^2\frac{\theta}{2}\, W_2^{(\nu)} - \frac{(E+E')}{M} \sin^2\frac{\theta}{2}\, W_3^{(\nu)} \right] \tag{7.37}$$

$$\frac{\text{d}^2\sigma^{(\bar{\nu})}}{\text{d}\Omega\,\text{d}E'} = \frac{G_{\text{F}}^2}{2\pi^2} E'^2 \left[2\sin^2\frac{\theta}{2}\, W_1^{(\bar{\nu})} + \cos^2\frac{\theta}{2}\, W_2^{(\bar{\nu})} + \frac{(E+E')}{M} \sin^2\frac{\theta}{2}\, W_3^{(\bar{\nu})} \right] \tag{7.38}$$

where the structure functions $W_i^{(\nu)}$ are defined as

$$\begin{aligned} W_{\alpha\beta}^{(\nu)} &= \frac{1}{4M} \sum_\sigma \int \frac{\text{d}^4x}{2\pi}\, \text{e}^{\text{i}q\cdot x} \langle p, \sigma | [J_{\text{h}\beta}(x), J_{\text{h}\alpha}^\dagger(0)] | p, \sigma \rangle \\ &= -W_1^{(\nu)} g_{\alpha\beta} + W_2^{(\nu)} p_\alpha p_\beta / M^2 - \text{i} W_3^{(\nu)} \varepsilon_{\alpha\beta\gamma\delta} p^\gamma q^\delta / M^2 \\ &\quad + W_4^{(\nu)} q_\alpha q_\beta / M^2 + W_5^{(\nu)} (p_\alpha q_\beta + p_\beta q_\alpha)/M^2 \\ &\quad + \text{i} W_6^{(\nu)} (p_\alpha q_\beta - p_\beta q_\alpha)/M^2. \end{aligned} \tag{7.39}$$

The $W_i^{(\bar{\nu})}$'s can be obtained from (7.39) using $J_{h\alpha} \leftrightarrow J^{\dagger}_{h\alpha}$ (also see §5.2). We also define dimensionless structure functions

$$\begin{aligned} MW_1^{(\nu)}(q^2, \nu) &= G_1^{(\nu)}(x, q^2/M^2) \\ \nu W_2^{(\nu)}(q^2, \nu) &= G_2^{(\nu)}(x, q^2/M^2) \\ \nu W_3^{(\nu)}(q^2, \nu) &= G_3^{(\nu)}(x, q^2/M^2). \end{aligned} \tag{7.40}$$

These structure functions will also satisfy the Bjorken scaling

$$\lim_{\substack{|q^2| \to \infty \\ x \text{ fixed}}} G_i^{(\nu)}(x, q^2/M^2) = F_i^{(\nu)}(x). \tag{7.41}$$

It is often useful to have structure functions with definite helicities. They can be obtained as follows. In the lab frame, choose the z-axis such that

$$p_\mu = (M, 0, 0, 0) \quad \text{and} \quad q_\mu = (q_0, 0, 0, q_3). \tag{7.42}$$

The longitudinal polarization vector is then of the form

$$\varepsilon_\mu^{(S)} = \frac{1}{\sqrt{-q^2}}(q_3, 0, 0, q_0) \tag{7.43}$$

and the corresponding structure function is

$$\begin{aligned} W_S(q^2, \nu) &= \varepsilon_\mu^{(S)*} W^{\mu\nu} \varepsilon_\nu^S \\ &= -W_1 - \frac{q_3^2}{q^2} W_2 = (1 - \nu^2/q^2) W_2 - W_1 \end{aligned} \tag{7.44}$$

where we have suppressed the neutrino superscript (ν). The right- and left-handed transverse polarization vectors are

$$\varepsilon_\mu^{(R)} = \frac{1}{\sqrt{2}}(0, 1, \mathrm{i}, 0) \tag{7.45}$$

$$\varepsilon_\mu^{(L)} = \frac{1}{\sqrt{2}}(0, 1, -\mathrm{i}, 0) \tag{7.46}$$

and their structure functions are

$$\begin{aligned} W_R &= W_1 + \frac{1}{2M}(\nu^2 - q^2)^{\frac{1}{2}} W_3 \\ W_L &= W_1 - \frac{1}{2M}(\nu^2 - q^2)^{\frac{1}{2}} W_3. \end{aligned} \tag{7.47}$$

Note that these structure functions with definite helicities, W_L, W_R, and W_S, have to be positive. In the scaling limit, the following helicity functions are functions of x only

$$\begin{aligned} 2MW_S &\to F_S = \frac{1}{x} F_2 - 2F_1 \\ MW_L &\to F_L = F_1 - \tfrac{1}{2} F_3 \\ MW_R &\to F_R = F_1 + \tfrac{1}{2} F_3. \end{aligned} \tag{7.48}$$

The differential cross-section can be written

$$\frac{d^2\sigma^{(\nu)}}{dx\,dy} = G_F^2 \frac{MEx}{\pi}[(1-y)F_S^{(\nu)} + F_L^{(\nu)} + (1-y)^2 F_R^{(\nu)}] \tag{7.49}$$

$$\frac{d^2\sigma^{(\bar{\nu})}}{dx\,dy} = G_F^2 \frac{MEx}{\pi}[(1-y)F_S^{(\bar{\nu})} + (1-y)^2 F_L^{(\bar{\nu})} + F_R^{(\bar{\nu})}]. \tag{7.50}$$

Note that these equations imply that the total cross-section will grow linearly with energy. (This is also the typical behaviour of the neutrino scattering off the point-like lepton.) That this is indeed the behaviour observed experimentally also suggests that there are point-like constituents in the nucleon.

The parton model

We shall now calculate the lepton–nucleon structure functions in the *parton model* (Feynman 1972; Bjorken and Paschos 1969) which is the subnucleon version of the familiar impulse approximation of high-energy scattering of composite particles with weakly bound constituents. The inclusive scattering is viewed as due to incoherent elastic scattering from point-like constituents of the nucleon, the partons, depicted in Fig. (7.2). The final-state partons then recombine (fragment) somehow into hadronic states. Thus we are making the physical assumptions that (1) during the time of current-parton interaction one can ignore interactions among partons themselves and (2) the final-state interactions (necessary for partons to fragment into hadrons) take place on such a relatively long time-scale that they can be ignored in the calculation of the inclusive cross-sections. For a more detailed presentation the reader is referred to Close (1979).

Specifically, each of the spin-1/2 partons is hypothesized to carry a fraction of the original nucleon momentum ξp with $0 \leq \xi \leq 1$, i.e., we neglect any parton momentum transverse to the nucleon momentum. Then the contribution to the hadronic tensor (7.10) from such a spin-1/2 parton can be immediately worked out as

$$K_{\mu\nu}(\xi) = \frac{1}{4\xi M}\sum_{\text{spin}} \frac{d^3p'}{(2\pi)^3 2p'_0}$$
$$\times \langle \xi p, \sigma | J_\mu^{\text{em}}(0) | p', \sigma' \rangle \langle p', \sigma' | J_\nu^{\text{em}}(0) | \xi p, \sigma \rangle (2\pi)^3\, \delta^4(p' - \xi p - q)$$
$$= \frac{1}{4\xi M}\sum_{\text{spin}} \bar{u}(\xi p)\gamma_\mu u(p')\bar{u}(p')\gamma_\nu u(\xi p)\, \delta(p'_0 - \xi p_0 - q_0)/2p'_0. \tag{7.51}$$

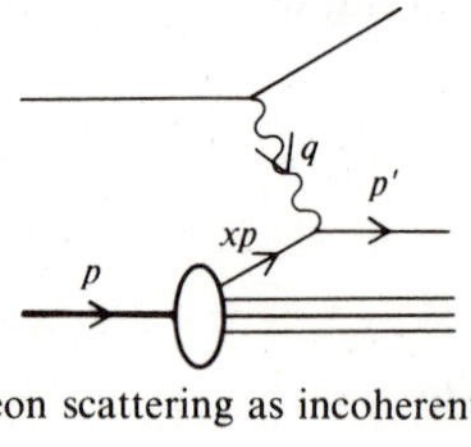

FIG. 7.2. Inelastic lepton–nucleon scattering as incoherent elastic scattering from partons.

The factor of ξ in the denominator appears because of the change in the relative flux from p to ξp. The delta function may be written as

$$\begin{aligned}\delta(p'_0 - \xi p_0 - q_0)/2p'_0 &= \theta(p'_0)\,\delta[p'^2_0 - (\xi p_0 + q_0)^2] \\ &= \theta(p'_0)\,\delta[p'^2 - (\xi p + q)^2] \\ &= \theta(\xi p_0 + q_0)\,\delta(2M\nu\xi + q^2) \\ &= \theta(\xi p_0 + q_0)\,\frac{\delta(\xi - x)}{2M\nu}.\end{aligned} \tag{7.52}$$

For the spin sum, we have

$$\begin{aligned}&\frac{1}{2}\sum_{\text{spin}} \bar{u}(\xi p)\gamma_\mu u(\xi p + q)\bar{u}(\xi p + q)\gamma_\nu u(\xi p) \\ &= \frac{\xi}{2}\,\mathrm{tr}[\not{p}\gamma_\mu(\xi\not{p} + \not{q})\gamma_\nu] \\ &= 2\xi[p_\mu(\xi p + q)_\nu + (\xi p + q)_\mu p_\nu - p\cdot(\xi p + q)g_{\mu\nu}] \\ &= 4M^2\xi^2(p_\mu p_\nu/M^2) - 2M\nu\xi g_{\mu\nu} + \ldots\end{aligned} \tag{7.53}$$

where we have neglected the parton mass. The parton tensor (7.51) is then

$$K_{\mu\nu}(\xi) = \delta(\xi - x)\left(\frac{\xi}{\nu}\frac{p_\mu p_\nu}{M^2} - \frac{1}{2M}g_{\mu\nu} + \ldots\right). \tag{7.54}$$

Let $f(\xi)\,\mathrm{d}\xi$ be the number of partons with momenta between ξ and $\xi + \mathrm{d}\xi$ (weighted by the squared charge). Then we can calculate the hadronic tensor in terms of an integral over $K_{\mu\nu}(\xi)$

$$\begin{aligned}W_{\mu\nu} &= \int_0^1 f(\xi)K_{\mu\nu}(\xi)\,\mathrm{d}\xi \\ &= \frac{xf(x)}{\nu}\frac{p_\mu p_\nu}{M^2} - \frac{f(x)}{2M}g_{\mu\nu} + \ldots.\end{aligned} \tag{7.55}$$

In this way the delta function that enforces the mass-shell condition of the final parton leads to the structure-function dependence on $x = -q^2/2M\nu$ alone,

$$MW_1 \to F_1(x) = \tfrac{1}{2}f(x) \tag{7.56}$$

$$\nu W_2 \to F_2(x) = xf(x). \tag{7.57}$$

Thus the scaling functions of (7.32) are measures of the momentum distribution of the the parton in the target nucleon.

We also note, from eqns (7.56) and (7.57), that

$$2xF_1(x) = F_2(x) \tag{7.58}$$

which is known as the *Callan–Gross relation* (1969). This is a direct consequence of the assumed spin-1/2 nature of partons. For example if we had used spin-0 partons, we would have had

$$K_{\mu\nu} \propto \langle xp|J_\mu|xp+q\rangle\langle xp+q|J_\nu|xp\rangle$$
$$\propto (2xp+q)_\mu(2xp+q)_\nu.$$

With the absence of a $g_{\mu\nu}$ term this would lead to

$$F_1(x) = 0. \tag{7.59}$$

There is a simple interpretation of (7.58) and (7.59) in terms of the helicity structure functions. If we define photon polarization vectors as in (7.43) and (7.46), we have eqn (7.48)

$$F_S = \frac{1}{x}F_2 - 2F_1 \tag{7.60}$$

$$F_T = F_L + F_R = 2F_1. \tag{7.61}$$

Thus (7.58) and (7.59) can be translated into

$$F_S = 0 \qquad \text{for a spin-1/2 parton} \tag{7.62}$$

$$F_T = 0 \qquad \text{for a spin-0 parton.} \tag{7.63}$$

To see that these results follow from angular momentum conservation, we go to the *Breit frame* of reference in which the momentum of the parton just reverses its direction without changing its magnitude upon collision with the virtual photon

$$q_\mu = (0, 0, 0, -2xp),$$
$$xp_\mu = (xp, 0, 0, xp),$$
$$p'_\mu = (xp, 0, 0, -xp). \tag{7.64}$$

If the parton has spin-0, only the virtual photon with zero helicity (ε^S) can contribute while the helicity ± 1 states ($\varepsilon^L, \varepsilon^R$) do not conserve angular momentum along the direction of motion. On the other hand, for the spin-1/2 parton (with negligible mass) the spin component also gets reversed upon collision and this will require the virtual photon to be in the ± 1 helicity state; hence $F_S = 0$. Experimentally (7.58) or (7.62) is reasonably satisfied in the scaling region and we can conclude that nucleons do indeed have charged spin-1/2 point-like constituents.

7.2 Sum rules and applications of the quark–parton model

It is natural to identify these charged spin-1/2 partons with the quarks which were first invented to account for the spectroscopic properties of hadrons. Eventually we shall develop a field theory of the strong interaction, quantum chromodynamics (QCD), which is a non-Abelian generalization of the familiar theory of quantum electrodynamics (QED). In QCD the quarks, like the electrons in QED, are the basic matter fields with interactions mediated by the electrically neutral vector fields, the *gluons*, much as the photon mediates the electromagnetic interactions among electrons. With this picture we can also have a qualitative understanding of the form of the parton distribution function $f(x)$. Experimentally it has the shape shown in Fig. (7.3). To understand this we follow the presentation by Close (1979) and start with a primitive model of three free quarks of the nucleon (see Fig. 7.4) for which the parton distribution function is essentially a delta function at $x = 1/3$, i.e., $f(x) \sim \delta(x - 1/3)$. We turn on the interaction of the quarks with gluons; this distribution will be smeared by the gluon exchange between quarks (Fig. 7.5). Then, just as the case of QED where the virtual photon (emitted with a bremsstrahlung momentum spectrum of dk/k) can create e^+e^- pairs, the gluons (emitted with a probability $\sim dx/x$) can produce $q\bar{q}$ pairs. These processes of internal conversion and bremsstrahlung will produce a '$q\bar{q}$ *sea*' at small x to give the final distribution (Fig. 7.6).

We would like to see whether the high-energy lepton–nucleon scattering data are consistent with the quark quantum number fixed by the spectroscopic phenomenology. Working with the quark model with only light

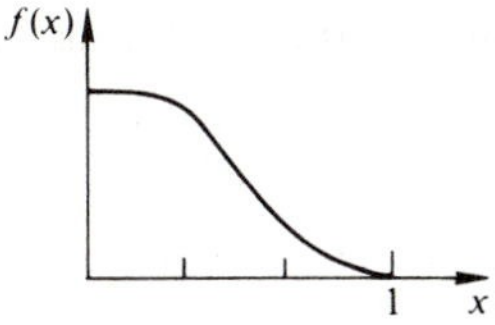

FIG. 7.3.

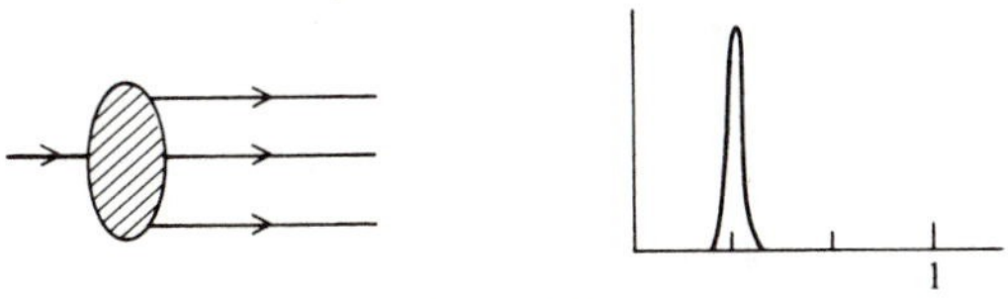

FIG. 7.4.

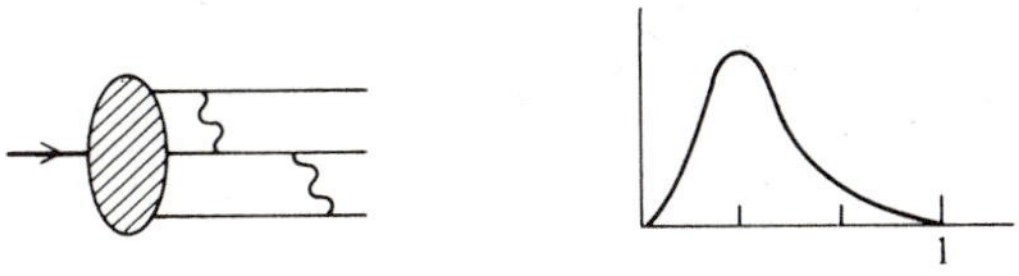

FIG. 7.5.

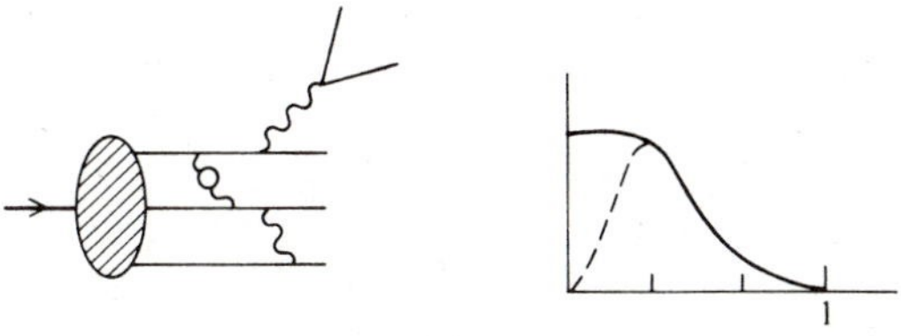

FIG. 7.6.

quarks u, d, and s, the hadronic electromagnetic current has the explicit form

$$J_\mu^{\text{em}} = \tfrac{2}{3}\bar{u}\gamma_\mu u - \tfrac{1}{3}\bar{d}\gamma_\mu d - \tfrac{1}{3}\bar{s}\gamma_\mu s \tag{7.65}$$

and, below the charm threshold, the hadronic (charged) weak current has the form

$$J_h^\mu = \bar{u}\gamma^\mu(1-\gamma_5)(d\cos\theta_c + s\sin\theta_c) \tag{7.66}$$

We shall work in the approximation of vanishing Cabibbo angle ($\theta_c \approx 0$); hence

$$J_h^\mu \approx \bar{u}\gamma^\mu(1-\gamma_5)\,d. \tag{7.67}$$

Here we have used the quark labels as particle field operators. In the following we shall use these names to denote the quark–parton distribution function instead. With the squared electric charges factored out explicitly, we have, through eqns (7.56) and (7.61),

$$F_T^{\text{ep}}(x) = f_p(x) = \tfrac{4}{9}(u_p + \bar{u}_p) + \tfrac{1}{9}(d_p + \bar{d}_p) + \tfrac{1}{9}(s_p + \bar{s}_p) \tag{7.68}$$

$$F_T^{\text{en}}(x) = f_n(x) = \tfrac{4}{9}(u_n + \bar{u}_n) + \tfrac{1}{9}(d_n + \bar{d}_n) + \tfrac{1}{9}(s_n + \bar{s}_n) \tag{7.69}$$

where $q_N(x)$ denotes the probability of finding a parton with longitudinal momentum fraction x carrying the quantum number of quark q in the target nucleon N. They are constrained by the quantum number of nucleon. For example,

Isospin:

$$\frac{1}{2}\int_0^1 \{[u_p(x) - \bar{u}_p(x)] - [d_p(x) - \bar{d}_p(x)]\}\,dx = \frac{1}{2} \tag{7.70}$$

Strangeness:

$$\int_0^1 [s_p(x) - \bar{s}_p(x)]\,dx = 0 \tag{7.71}$$

Charge:

$$\int_0^1 \frac{2}{3}[u_p(x) - \bar{u}_p(x)]\,dx - \int_0^1 \frac{1}{3}[d_p(x) - \bar{d}_p(x)]\,dx - \int_0^1 \frac{1}{3}[s_p(x) - \bar{s}_p(x)]\,dx = 1.$$

Using isospin symmetry (i.e., the invariance under the interchanges $p \leftrightarrow n$ and $u \leftrightarrow d$), we have

$$u_p(x) = d_n(x) \equiv u(x)$$

$$d_p(x) = u_n(x) \equiv d(x)$$

$$s_p(x) = s_n(x) \equiv s(x). \tag{7.72}$$

Thus (7.68) and (7.69) may be written

$$F_T^{ep}(x) = f_p(x) = \tfrac{4}{9}(u + \bar{u}) + \tfrac{1}{9}(d + \bar{d}) + \tfrac{1}{9}(s + \bar{s}) \tag{7.73}$$

$$F_T^{en}(x) = f_n(x) = \tfrac{4}{9}(d + \bar{d}) + \tfrac{1}{9}(u + \bar{u}) + \tfrac{1}{9}(s + \bar{s}) \tag{7.74}$$

The ratio of the proton and neutron structure function is then

$$\frac{F_T^{ep}(x)}{F_T^{en}(x)} = \frac{4(u + \bar{u}) + (d + \bar{d}) + (s + \bar{s})}{(u + \bar{u}) + 4(d + \bar{d}) + (s + \bar{s})}. \tag{7.75}$$

Since all $q(x)$s are positive, we must have the bounds (Nachtmann 1972)

$$\frac{1}{4} \leq \frac{F_T^{en}(x)}{F_T^{ep}(x)} \leq 4 \tag{7.76}$$

which is in fact consistent with experimental data.

Furthermore, our discussion of the quark–parton model at the beginning of this section also suggests (e.g., Kuti and Weisskopf 1971; Landshoff and Polkinghorne 1971) that the quark distribution function can be usefully decomposed into *valence quarks* and *sea quarks*

$$q(x) = q_v(x) + q_s(x). \tag{7.77}$$

The presence of the valence quarks is already indicated by the original quark model. Thus protons and neutrons have valence quarks of (uud) and (udd), respectively. The sea quarks correspond to those quark pairs produced by the gluons: they are symmetric with respect to the flavour SU(3) group and, as indicated by our discussion above, they should be concentrated in the small-x region. For the proton target we have

$$\begin{aligned} &u_v(x) = 2d_v(x) \\ &s_v(x) = \bar{u}_v(x) = \bar{d}_v(x) = \bar{s}_v(x) = 0 \\ &u_s(x) = \bar{u}_s(x) = d_s(x) = \bar{d}_s(x) = s_s(x) = \bar{s}_s(x) \equiv G(x). \end{aligned} \tag{7.78}$$

Thus eqns (7.73) and (7.74) may be written as

$$\begin{aligned} F_T^{ep}(x) &= \tfrac{1}{2}u_v(x) + \tfrac{4}{3}G(x) \\ F_T^{en}(x) &= \tfrac{1}{3}u_v(x) + \tfrac{4}{3}G(x). \end{aligned} \tag{7.79}$$

Their difference directly measures the valence quark distribution

$$F_T^{ep}(x) - F_T^{en}(x) = \tfrac{1}{6}u_v(x) \tag{7.80}$$

which should have a peak approximately around $x = 1/3$, as suggested by Fig. 7.5. Also, since we expect $G(x)$ to be important only in the $x \to 0$ region, we should see that ratio behave as

$$F_T^{ep}(x)/F_T^{en}(x) \to \begin{cases} 1 & \text{as} \quad x \to 0 \\ 3/2 & \text{as} \quad x \to 1. \end{cases} \tag{7.81}$$

These expectations of the quark–parton model are again corroborated by experimental observations.

For the neutrino–nucleon structure functions, corresponding to eqns (7.73) and (7.74), we will merely list the basic results

$$
\begin{aligned}
F_{\mathrm{L}}^{\nu\mathrm{p}}(x) &= 2\mathrm{d}(x) & F_{\mathrm{L}}^{\nu\mathrm{n}}(x) &= 2\mathrm{u}(x)\\
F_{\mathrm{R}}^{\nu\mathrm{p}}(x) &= 2\bar{\mathrm{u}}(x) & F_{\mathrm{R}}^{\nu\mathrm{n}}(x) &= 2\bar{\mathrm{d}}(x)\\
F_{\mathrm{S}}^{\nu\mathrm{p}}(x) &= 0 & F_{\mathrm{S}}^{\nu\mathrm{n}}(x) &= 0\\
F_{\mathrm{L}}^{\bar{\nu}\mathrm{p}}(x) &= 2\mathrm{u}(x) & F_{\mathrm{L}}^{\bar{\nu}\mathrm{n}}(x) &= 2\mathrm{d}(x)\\
F_{\mathrm{R}}^{\bar{\nu}\mathrm{p}}(x) &= 2\bar{\mathrm{d}}(x) & F_{\mathrm{R}}^{\bar{\nu}\mathrm{n}}(x) &= 2\bar{\mathrm{u}}(x)\\
F_{\mathrm{S}}^{\bar{\nu}\mathrm{p}}(x) &= 0 & F_{\mathrm{S}}^{\bar{\nu}\mathrm{n}}(x) &= 0.
\end{aligned}
\tag{7.82}
$$

The functions on the right-hand sides are distribution functions for proton targets and the factors of 2 reflect the presence of both vector and axial-vector parts in the weak currents. We can then isolate the strange-quark distribution as follows. Using $F_2 = x(F_{\mathrm{L}} + F_{\mathrm{R}} + F_{\mathrm{S}})$ and (7.82), we have

$$F_2^{\nu\mathrm{p}}(x) + F_2^{\nu\mathrm{n}}(x) = 2x(\mathrm{u} + \bar{\mathrm{u}} + \mathrm{d} + \bar{\mathrm{d}})\,. \tag{7.83}$$

From $F_2 = xF_{\mathrm{T}}$ and eqns (7.73) and (7.74), we have

$$F_2^{\mathrm{ep}}(x) + F_2^{\mathrm{en}}(x) = x[5(\mathrm{u} + \bar{\mathrm{u}} + \mathrm{d} + \bar{\mathrm{d}})/9 + 2(\mathrm{s} + \bar{\mathrm{s}})/9]\,. \tag{7.84}$$

They imply that

$$F_2^{\mathrm{ep}}(x) + F_2^{\mathrm{en}}(x) - \frac{5}{18}[F_2^{\nu\mathrm{p}}(x) + F_2^{\nu\mathrm{n}}(x)] = \frac{2x}{9}[\mathrm{s}(x) + \bar{\mathrm{s}}(x)]\,. \tag{7.85}$$

The experimental data are consistent with a vanishing right-hand side, except for the small x (<0.2) region. In other words the strange quark and antiquark content of the nucleon is very small. Furthermore, if we assume that the sea-quark distributions are SU(3) symmetric, the $\bar{\mathrm{u}}(x)$ and $\bar{\mathrm{d}}(x)$ contents should also be small.

We next consider a number of sum rules; their validity strongly supports the quark–parton picture we have presented.

The Adler sum rule

This sum rule has already been derived in our discussion of current algebra. Eqn (5.109) takes on the following ($\theta_{\mathrm{c}} = 0$) form in the scaling limit $-q^2 \to \infty$, $\nu \to \infty$, with x fixed

$$\int_0^1 \frac{\mathrm{d}x}{x}[F_2^{\bar{\nu}\mathrm{p}}(x) - F_2^{\nu\mathrm{p}}(x)] = 2. \tag{7.86}$$

We can obtain the same result directly from the quark–parton model. Since

$$F_2 = x(F_{\mathrm{L}} + F_{\mathrm{R}} + F_{\mathrm{S}}), \tag{7.87}$$

the combination of structure functions appearing on the left-hand side of

(7.86) can be expressed in terms of the quark distribution function through (7.82)

$$F_2^{\bar{\nu}p}(x) - F_2^{\nu p}(x) = 2x\{[u(x) - \bar{u}(x)] - [d(x) - \bar{d}(x)]\} \qquad (7.88)$$
$$= 4xT_3(x)$$

where T_3 is the third component of the isospin density. Using the fact that proton has isospin-1/2 (eqn (7.70)) we immediately recover (7.86)

$$\int_0^1 \frac{dx}{x}[F_2^{\bar{\nu}p}(x) - F_2^{\nu p}(x)] = 4\int_0^1 T_3(x)\,dx = 2. \qquad (7.89)$$

Gross–Llewellyn Smith sum rule

The sum of the scaling functions $F_3 = F_R - F_L$

$$F_3^{\nu p}(x) + F_3^{\nu n}(x) = -2[u(x) + d(x) - \bar{u}(x) - \bar{d}(x)] \qquad (7.90)$$

can be written as a combination of baryon number and strangeness densities

$$F_3^{\nu p}(x) + F_3^{\nu n}(x) = -6[B(x) + \tfrac{1}{3}S(x)] \qquad (7.91)$$

with

$$B(x) = \tfrac{1}{3}[u(x) + d(x) + s(x) - \bar{u}(x) - \bar{d}(x) - \bar{s}(x)] \qquad (7.92)$$

$$S(x) = -[s(x) - \bar{s}(x)]. \qquad (7.93)$$

Since the proton has baryon number 1 and zero strangeness, we obtain the Gross–Llewellyn Smith (1969) sum rule

$$\int_0^1 dx[F_3^{\nu p}(x) + F_3^{\nu n}(x)] = -6. \qquad (7.94)$$

The momentum sum rule

If the quarks were to carry all the momentum of the target nucleon, we would have the sum rule

$$\int_0^1 [u(x) + d(x) + s(x) + \bar{u}(x) + \bar{d}(x) + \bar{s}(x)]x\,dx = 1. \qquad (7.95)$$

Since the $x \approx 0$ region is not important to this integral, we can drop all the sea-quark contributions

$$\int_0^1 [u(x) + d(x)]\,x\,dx = 1. \qquad (7.96)$$

The quark density functions on the left-hand side can be expressed directly in terms of the measurable structure functions (7.84) and we obtain

$$\int_0^1 [F_2^{ep}(x) + F_2^{en}(x)]\, dx = \tfrac{5}{9}. \tag{7.97}$$

Similarly,

$$\int_0^1 [F_2^{\nu p}(x) + F_2^{\nu n}(x)]\, dx = 2. \tag{7.98}$$

Experimentally, however, we find the integrals in (7.97) to be approximately 0.28. This indicates that almost 50 per cent of the nucleon momentum is carried by some constituents which do not interact with the electromagnetic or the weak currents (Llewellyn Smith 1974). This again is in accord with the expectations of the QCD–parton model where one identifies these neutral constituents with the gluons.

Other applications of the quark–parton model

For the remaining part of this section we shall briefly touch upon two other topics of the quark–parton model—its applications in the description of high energy e^+e^- annihilations and the Drell–Yan process of lepton pair production in hadron–hadron collisions. We will follow the presentation of (Close 1979) and (Aitchson and Hey 1982).

e^+e^- annihilation

(1) $e^+e^- \to \mu^+\mu^-$. We shall use e^+e^- annihilation through the one-photon intermediate state into a $\mu^+\mu^-$ pair (Fig. 7.7(a)) as the 'reference reaction' in

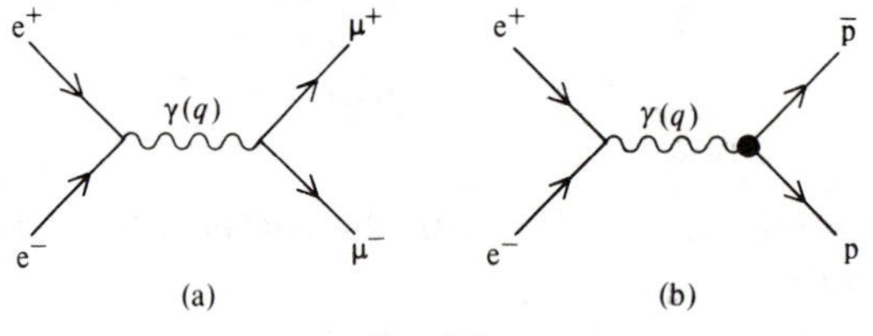

FIG. 7.7.

describing annihilations into other final states. The total cross section for $e^+e^- \to \mu^+\mu^-$ may be calculated in QED as

$$\sigma(e^+e^- \to \mu^+\mu^-) = \frac{4\pi\alpha^2}{3q^4}\left(1 - \frac{4m_\mu^2}{q^2}\right)^{1/2}(2m_\mu^2 + q^2) \tag{7.99}$$

where q is the intermediate photon momentum

$$q^2 = (p_+ + p_-)^2 \equiv s \geq 0. \tag{7.100}$$

For high energies $s \gg m_\mu^2$, we obtain

$$\sigma(e^+e^- \to \mu^+\mu^-) = \frac{4\pi\alpha^2}{3s}. \tag{7.101}$$

The fact that the cross-section falls as s^{-1} is typical for e^+e^- annihilation into point-like particles.

(2) $e^+e^- \to p\bar{p}$. The amplitude for this process (Fig. 7.7(b)) is related to that of ep elastic scattering by crossing symmetry. We calculate the cross-section to be

$$\sigma(e^+e^- \to p\bar{p}) = \frac{4\pi\alpha^2}{3q^4}\left(1 - \frac{4M_p^2}{q^2}\right)^{1/2}[2M_p^2 G_E^2(q^2) + q^2 G_M^2(q^2)] \tag{7.102}$$

where $G_E(q^2)$ and $G_M(q^2)$ are the electric and magnetic form factors, respectively (see (7.21) with $q^2 \geq 0$. For large q^2, $G_E(q^2) \propto G_M(q^2) \sim q^{-4}$. Thus for high energies $s \gg M_p^2$, the cross-section $\sigma(e^+e^- \to p\bar{p}) \sim s^{-5}$ falls off rapidly as is typical for annihilations into any given final state of hadrons with structure.

(3) $e^+e^- \to$ hadrons. We now consider the inclusive process of e^+e^- annihilation into all possible hadronic final states. In the quark–parton model we expect this to take place via $e^+e^- \to q_i\bar{q}_i$ and the quarks then fragment into free hadrons (Fig. 7.8(a)). The subscript i ranges over all possible (flavour and colour) labels of the produced quarks. Thus

$$\sigma(e^+e^- \to \text{hadrons}) = \sum_i \sigma(e^+e^- \to q_i\bar{q}_i). \tag{7.103}$$

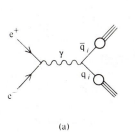

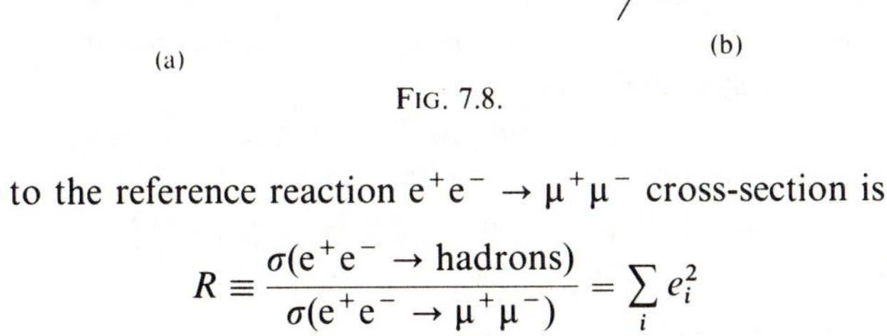

FIG. 7.8.

The ratio to the reference reaction $e^+e^- \to \mu^+\mu^-$ cross-section is then

$$R \equiv \frac{\sigma(e^+e^- \to \text{hadrons})}{\sigma(e^+e^- \to \mu^+\mu^-)} = \sum_i e_i^2 \tag{7.104}$$

so the ratio R measures the sum of squared quark charges (Cabibbo, Parisi, and Testa 1970). Thus for energies below the charm threshold we sum over three colours of the u, d, and s quarks

$$R = 3(\tfrac{4}{9} + \tfrac{1}{9} + \tfrac{1}{9}) = 2 \qquad \text{for } \sqrt{s} < 2m_c \tag{7.105}$$

and above charm and bottom thresholds

$$R = 2 + 3(\tfrac{4}{9} + \tfrac{1}{9}) = \tfrac{11}{3} \qquad \text{for } \sqrt{s} > 2m_b. \tag{7.106}$$

The data seems to support this scaling behaviour with three colours. (See Fig. 7.9 taken from a recent review by Felst (1981).)

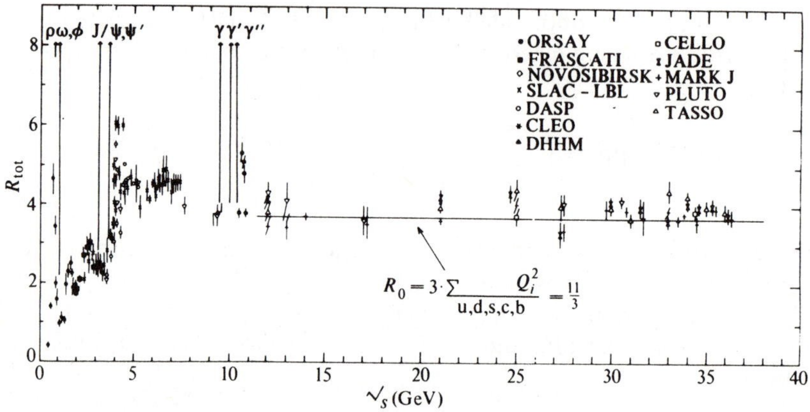

FIG. 7.9. Compilation of R-values from different e^+e^- experiments. Only statistical errors are shown. The quark–parton model prediction (eqn (7.106)) is also indicated.

One should also mention that e^+e^- annihilations have many other detailed features that check well with the quark–parton model description. For example, one finds Bjorken scaling in the differential cross-section for the inclusive reaction $e^+e^- \to hX$ (i.e., one of the final hadrons is detected) which is related by crossing to the inelastic $ep \to eX$ scattering. Also, there is strong experimental evidence that the e^+e^- annihilation final-state hadrons form *jets*, i.e., they tend to flow in preferred cones of small width. For 4 GeV $< \sqrt{s} <$ 7.5 GeV one finds two-jet events (Hanson *et al.* 1975; Fig. 7.8(b)) with the jet axes having the angular distribution $\sim(1 + \cos^2\theta)$, where θ is the polar angle of the jet axis with respect to the e^+e^- beam. Such a distribution is characteristic of an e^+e^- final state of spin-1/2 point-like particles. This is clearly in agreement with the expectation of the quark–parton model with its implicit assumption of transverse momentum cut-off. Finally at even higher energies ($>$7.5 GeV) corresponding to the gluon bremsstrahlung as expected in the QCD–parton model, one begins to find evidence for three-jet events (Brandelik *et al.* 1979; Barber *et al.* 1979; Berger *et al.* 1979).

The Drell–Yan process

As we shall see in the next section, it is possible to provide a more formal basis for parton-model descriptions of the deep inelastic lN scattering and high-energy e^+e^- annihilation in terms of the light-cone and short-distance operator product expansions. However this cannot be done for other high-energy and large-momentum transfer processes. The parton formulation has the advantage of suggesting parton descriptions for which direct formal operator argument may not be possible. The most important example is the Drell–Yan process (1971) (Fig. 7.10(a))

$$\mathrm{pp} \to \mu^+\mu^- \mathrm{X} \tag{7.107}$$

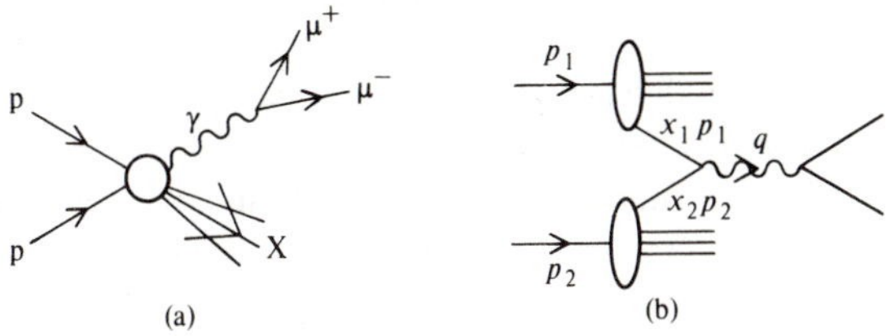

FIG. 7.10.

where a $\mu^+\mu^-$ pair is produced in hadron–hadron (usually proton–proton) collisions along with the unobserved hadron state X. The parton model leads us to expect that in the limit of large $s \equiv (p_1 + p_2)^2 \to \infty$ and large virtual photon mass $q^2 \to \infty$, with the ratio q^2/s fixed, the reaction can be assumed to proceed via the annihilation of a parton and antiparton, each coming from one of the initial hadrons, into a massive virtual photon which then decays into the observed $\mu^+\mu^-$ pair (Fig. 7.10(b)). In the centre-of-mass system, neglecting all masses, we have

$$p_1^\mu = (p, 0, 0, p), \quad p_2^\mu = (p, 0, 0, -p)$$

and

$$s = 4p^2. \tag{7.108}$$

Also neglecting the parton masses and transverse momenta, the parton momenta have the form

$$k_1^\mu = x_1 p_1^\mu, \quad k_2^\mu = x_2 p_2^\mu \tag{7.109}$$

leading to the photon momentum

$$q^\mu = ((x_1 + x_2)p, 0, 0, (x_1 - x_2)p). \tag{7.110}$$

Thus we have

$$q^2 = 4x_1x_2p^2 = x_1x_2s. \tag{7.111}$$

The probabilities of a quark and antiquark pair of the ith type with momentum fractions x_1 and x_2 in the initial protons are given by

$$\mathrm{q}_i(x_1)\,\mathrm{d}x_1\bar{\mathrm{q}}_i(x_2)\,\mathrm{d}x_2 + \bar{\mathrm{q}}_i(x_1)\,\mathrm{d}x_1\mathrm{q}_i(x_2)\,\mathrm{d}x_2. \tag{7.112}$$

This is to be multiplied by the cross-section for the basic parton process of $\mathrm{q}_i\bar{\mathrm{q}}_i \to \mu^+\mu^-$ of

$$\frac{\mathrm{d}\sigma}{\mathrm{d}q^2}(\mathrm{q}_i\bar{\mathrm{q}}_i \to \mu^+\mu^-) = \frac{4\pi\alpha^2}{3q^2} e_i^2\,\delta(q^2 - x_1x_2s) \tag{7.113}$$

to obtain the cross-section for the Drell–Yan process

$$\frac{d\sigma}{dq^2}(\mathrm{pp}\to\mu^+\mu^-X)=\frac{4\pi\alpha^2}{3q^4}\sum_i e_i^2\int_0^1[\mathrm{q}_i(x_1)\bar{\mathrm{q}}_i(x_2)$$

$$+\bar{\mathrm{q}}_i(x_1)\mathrm{q}_i(x_2)]\frac{dx_1\,dx_2}{x_1x_2}\delta\left(\frac{1}{x_1x_2}-\frac{s}{q^2}\right). \quad (7.114)$$

Thus the parton model will have $q^4(d\sigma/dq^2)$ to scale as a function of q^2/s. The experimental support for this prediction is quite convincing. Note that the same quark distribution functions are measured in the deep inelastic lepton proton scatterings, so one can also make a prediction of the absolute magnitude.

7.3 Free-field light-cone singularities and Bjorken scaling

In this section we study Bjorken scaling in the framework of field theory, thus giving some of the parton model results a more formal foundation. For a general introduction see, for example, Gross (1976) and Ellis (1977).

The deep inelastic limit and the light cone

First we will demonstrate that the deep inelastic lN processes of §7.1, with $-q^2$, $\nu\to\infty$, and $-q^2/2M\nu$ fixed, probe the light-cone behaviour of the current commutator. We recall (7.12) that the hadronic tensor in the differential cross-section can be expressed as a current commutator

$$W_{\mu\nu}(p,q)=\frac{1}{4M}\sum_\sigma\int\frac{d^4x}{2\pi}\,e^{iq\cdot x}\langle p,\sigma|[J_\mu(x),J_\nu(0)]|p,\sigma\rangle. \quad (7.115)$$

The scalar product in the exponential may be written

$$q\cdot x=\frac{(q_0+q_3)}{\sqrt{2}}\frac{(x_0-x_3)}{\sqrt{2}}+\frac{(q_0-q_3)}{\sqrt{2}}\frac{(x_0+x_3)}{\sqrt{2}}-\mathbf{q}_T\cdot\mathbf{x}_T, \quad (7.116)$$

where $\mathbf{q}_T=(q_1,q_2)$ and $\mathbf{x}_T=(x_1,x_2)$. In the rest frame of the target nucleon, the momenta are given by

$$p_\mu=(M,0,0,0,),\quad q_\mu=(\nu,0,0,(\nu^2-q^2)^{\frac{1}{2}}). \quad (7.117)$$

In the deep inelastic limit ($-q^2$, $\nu\to\infty$ with $-q^2/2M\nu$ fixed) we observe that

$$q_0+q_3\sim 2\nu\quad\text{and}\quad q_0-q_3\sim q^2/2\nu. \quad (7.118)$$

We expect that the dominant contribution to the integral (7.115) comes from regions with less rapid oscillations, i.e., $q\cdot x=O(1)$; hence

$$x_0-x_3\sim O(1/\nu)\quad\text{and}\quad x_0+x_3\sim O(1/xM) \quad (7.119)$$

or

$$x_0^2 - x_3^2 \sim O\left(\frac{1}{-q^2}\right). \tag{7.120}$$

Thus $x^2 = x_0^2 - x_3^2 - x_T^2 \leq x_0^2 - x_3^2 \sim O(1/-q^2)$ which vanishes as $-q^2 \to \infty$. In other words, in the scaling limit we are probing the structure of the current product near the light cone. This reduces the study of Bjorken scaling in field theory to the study of the light-cone behaviour of the current product.

Free-field light-cone singularities

As it turns out, Bjorken scaling corresponds to the statement that the current commutator has the light-cone behaviour of a free-field theory (for a review see Frishman 1974). To pave the way, we shall first study the free-field light-cone behaviour of some simpler products.

(1) *Products of fields.* In free field theories, the products of fields such as commutators and propagators are singular on the light cone ($x^2 \approx 0$) and the leading singularities are independent of the masses. Consider, for example, the propagator of the scalar field given by

$$\langle 0|T(\phi(x)\phi(0))|0\rangle = \mathrm{i}\Delta_{\mathrm{F}}(x) = \mathrm{i}\int \frac{\mathrm{d}^4k}{(2\pi)^4}\frac{\mathrm{e}^{-\mathrm{i}k\cdot x}}{k^2 - m^2 + \mathrm{i}\varepsilon}. \tag{7.121}$$

The Fourier transform in (7.121) can be calculated to give (see, for example, Bogoliubov and Shirkov 1959)

$$\Delta_{\mathrm{F}}(x) = \frac{-1}{4\pi}\delta(x^2) + \frac{m}{8\pi\sqrt{x^2}}\theta(x^2)[\mathrm{J}_1(m\sqrt{x^2}) - \mathrm{i}\mathrm{N}_1(m\sqrt{x^2})]$$
$$- \frac{\mathrm{i}m}{4\pi^2\sqrt{-x^2}}\theta(-x^2)\mathrm{K}_1(m\sqrt{-x^2}) \tag{7.122}$$

where J_n, N_n, and K_n are Bessel functions. For $x^2 \approx 0$, we have

$$\Delta_{\mathrm{F}}(x) = \frac{-1}{4\pi}\delta(x^2) + \frac{\mathrm{i}}{4\pi^2}\frac{1}{x^2} - \frac{\mathrm{i}m^2}{8\pi^2}\ln\frac{m\sqrt{x^2}}{2} - \frac{m^2}{16\pi^2}\theta(x^2)$$
$$= \frac{\mathrm{i}}{4\pi^2}\frac{1}{(x^2 - \mathrm{i}\varepsilon)} + O(m^2x^2). \tag{7.123}$$

The leading singularity is independent of the masses as the $x^2 \approx 0$ region corresponds to the large k^2 in the momentum space. Thus we can also calculate this mass-independent singularity directly from a simpler object, the massless propagator

$$\Delta_{\mathrm{F}}(x) = \int \frac{\mathrm{d}^4k}{(2\pi)^4}\frac{\mathrm{e}^{-\mathrm{i}k\cdot x}}{k^2 + \mathrm{i}\varepsilon}$$
$$= \int \frac{\mathrm{d}^3k}{(2\pi)^4}\mathrm{e}^{\mathrm{i}\mathbf{k}\cdot\mathbf{x}}\int_{-\infty}^{\infty}\frac{\mathrm{d}k_0\,\mathrm{e}^{-\mathrm{i}k_0x_0}}{k_0^2 - \mathbf{k}^2 + \mathrm{i}\varepsilon}. \tag{7.124}$$

The k_0-integration can be performed by the standard contour method

$$\int_{-\infty}^{\infty} \frac{dk_0\, e^{-ik_0x_0}}{k_0^2 - \mathbf{k}^2 + i\varepsilon} = \frac{-i\pi}{|\mathbf{k}|}[\theta(x_0)\, e^{-i|\mathbf{k}|x_0} + \theta(-x_0)\, e^{i|\mathbf{k}|x_0}]. \tag{7.125}$$

We then have

$$\Delta_F(x) = -\frac{1}{8\pi^2 r}\int_0^{\infty} d\kappa(e^{i\kappa r} - e^{-i\kappa r})[\theta(x_0)\, e^{-i\kappa x_0} + \theta(-x_0)\, e^{i\kappa x_0}] \tag{7.126}$$

where $\kappa = |\mathbf{k}|$ and $r = |\mathbf{x}|$. Using the identity

$$\int_0^{\infty} e^{\pm i\alpha\tau}\, d\tau = \int_0^{\infty} e^{\pm i(\alpha \pm i\varepsilon)\tau}\, d\tau = \frac{\mp 1}{i(\alpha \pm i\varepsilon)}, \tag{7.127}$$

we obtain

$$\begin{aligned}\Delta_F(x) &= \frac{-1}{8\pi^2 r}\left[\theta(x_0)\left(\frac{1}{r - x_0 + i\varepsilon} + \frac{1}{r + x_0 - i\varepsilon}\right)\right.\\ &\quad \left. + \theta(-x_0)\left(\frac{1}{r + x_0 + i\varepsilon} + \frac{1}{r - x_0 - i\varepsilon}\right)\right]\\ &= \frac{-i}{4\pi^2}\left[\frac{\theta(x_0)}{r^2 - x_0^2 + i\varepsilon x_0} + \frac{\theta(-x_0)}{r^2 - x_0^2 - i\varepsilon x_0}\right]\\ &= \frac{i}{4\pi^2}\frac{1}{x^2 - i\varepsilon}\end{aligned} \tag{7.128}$$

which agrees with (7.123). One can do a similar calculation of the leading singularity for the commutator of two scalar fields

$$[\phi(x), \phi(0)] = i\Delta(x) = \frac{1}{(2\pi)^3}\int d^4k\, e^{-ik\cdot x}\varepsilon(k_0)\,\delta(k^2 - m^2) \tag{7.129}$$

with the result

$$\Delta(x) \simeq \frac{-1}{2\pi}\varepsilon(x_0)\,\delta(x^2) \qquad \text{for } x^2 \approx 0. \tag{7.130}$$

Thus, we have the singular-function identity

$$i\int d^4k\, e^{-ik\cdot x}\varepsilon(k_0)\,\delta(k^2) = (2\pi)^2\varepsilon(x_0)\,\delta(x^2). \tag{7.131}$$

The result (7.130) can be viewed in another way: the light-cone singularity of the commutator $\Delta(x)$ and that of the propagator function $\Delta_F(x)$ are directly related

$$\Delta(x) = 2\varepsilon(x_0)\,\mathrm{Im}(i\Delta_F(x)). \tag{7.132}$$

This reflects the singular-function identity

$$\frac{1}{-x^2 + \mathrm{i}\varepsilon} - \frac{1}{-x^2 - \mathrm{i}\varepsilon} = -2\pi\mathrm{i}\varepsilon(x_0)\,\delta(x^2) \tag{7.133}$$

which is a special case of the general identity

$$\left(\frac{1}{-x^2 + \mathrm{i}\varepsilon}\right)^n - \left(\frac{1}{-x^2 - \mathrm{i}\varepsilon}\right)^n = \frac{-2\pi\mathrm{i}}{(n-1)!}\,\varepsilon(x_0)\,\delta^{(n-1)}(x^2). \tag{7.134}$$

In the following calculations we shall obtain the commutator singularities through those of the propagators by making the replacement

$$(-x^2 - \mathrm{i}\varepsilon)^{-n} \to 2\pi\mathrm{i}\varepsilon(x_0)\,\delta^{(n-1)}(x^2)/(n-1)! \tag{7.135}$$

For the fermions, the results are summarized as

$$\begin{aligned}\{\psi_\alpha(x), \bar{\psi}_\beta(y)\} &= \mathrm{i}S_{\alpha\beta}(x-y),\\ S_{\alpha\beta}(x) &= (\mathrm{i}\gamma\cdot\partial + m)_{\alpha\beta}\,\Delta(x),\\ \langle 0|T(\psi_\alpha(x)\bar{\psi}_\beta(y))|0\rangle &= \mathrm{i}S^{\mathrm{F}}_{\alpha\beta}(x-y),\\ S^{\mathrm{F}}_{\alpha\beta}(x) &= (\mathrm{i}\gamma\cdot\partial + m)_{\alpha\beta}\,\Delta_{\mathrm{F}}(x).\end{aligned} \tag{7.136}$$

For $x^2 \approx 0$, we have

$$S_{\alpha\beta}(x) \approx (\mathrm{i}\gamma\cdot\partial)_{\alpha\beta}\left[\frac{1}{2\pi}\,\varepsilon(x_0)\,\delta(x^2)\right] \tag{7.137}$$

$$S^{\mathrm{F}}_{\alpha\beta}(x) \approx (\mathrm{i}\gamma\cdot\partial)_{\alpha\beta}\left[\frac{\mathrm{i}}{4\pi^2}\frac{1}{(x^2 - \mathrm{i}\varepsilon)}\right]. \tag{7.138}$$

(2) *Product of scalar currents.* We can extend this analysis to the case of composite operators. Consider for example the scalar current

$$J(x) = {:}\phi^2(x){:}. \tag{7.139}$$

Note that the effect of the normal ordering is to remove the singularities which occur in the product $\phi(x+\zeta)\phi(x-\zeta)$ as $\zeta^\mu \to 0$. The singularities in the product $T(J(x)J(0))$ can be worked out by using Wick's theorem

$$\begin{aligned}T(J(x)J(0)) &= T({:}\phi^2(x){:}{:}\phi^2(0){:})\\ &= 2[\langle 0|T(\phi(x)\phi(0))|0\rangle]^2\\ &\quad + 4\langle 0|T(\phi(x)\phi(0))|0\rangle{:}\phi(x)\phi(0){:}\\ &\quad + {:}\phi^2(x)\phi^2(0){:}\\ &= -2[\Delta_{\mathrm{F}}(x, m)]^2 + 4\mathrm{i}\Delta_{\mathrm{F}}(x, m^2){:}\phi(x)\phi(0){:}\\ &\quad + {:}\phi^2(x)\phi^2(0){:}.\end{aligned} \tag{7.140}$$

Hence for $x^2 \approx 0$, we get

$$\begin{aligned}T(J(x)J(0)) \approx{}& \frac{1}{8\pi^4(x^2 - \mathrm{i}\varepsilon)^2} - \frac{{:}\phi(x)\phi(0){:}}{\pi^2(x^2 - \mathrm{i}\varepsilon)}\\ &+ {:}\phi^2(x)\phi^2(0){:}.\end{aligned} \tag{7.141}$$

If we take (7.141) between two arbitrary states $|A\rangle$ and $|B\rangle$,

$$\langle A|T(J(x)J(0))|B\rangle \approx \frac{\langle A\,|\,B\rangle}{8\pi^4(x^2-\mathrm{i}\varepsilon)^2} - \frac{\langle A|:\phi(x)\phi(0):|B\rangle}{\pi^2(x^2-\mathrm{i}\varepsilon)} + \langle A|:\phi^2(x)\phi^2(0):|B\rangle \tag{7.142}$$

which corresponds to the diagrams in Fig. 7.11.

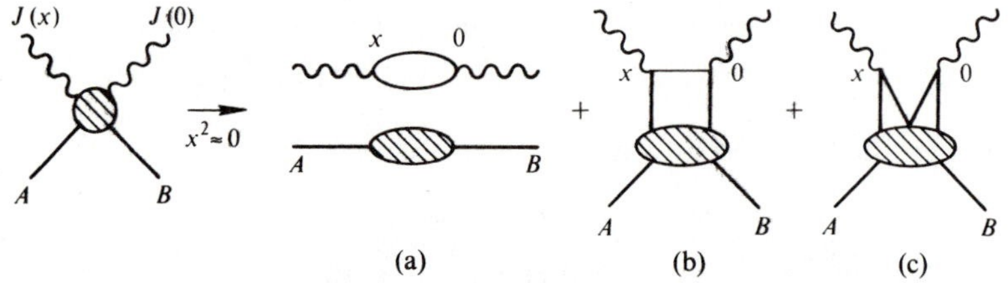

FIG. 7.11. Diagrams that are singular on the light cone. Free massless propagators are represented by light straight lines: (a) has two; (b) has one; (c) has none. Note that only (a) will contribute if $|A\rangle = |B\rangle = |0\rangle$.

To calculate the singularities of the commutator $[J(x), J(0)]$ we only need use the identity given in (7.135).

Free-field singularities and scaling

Now we are ready to demonstrate **Bjorken** scaling in free-field theory. Consider the electromagnetic current given by

$$J_\mu(x) = :\bar{\psi}(x)\gamma_\mu Q\psi(x): \tag{7.143}$$

where Q is the charge operator. Following the same procedure used in the above case of scalar current densities, instead of the commutator $[J_\mu(x), J_\nu(0)]$, we will first calculate the time-ordered product by using Wick's theorem

$$\begin{aligned} T(J_\mu(x)J_\nu(0)) &= T(:\bar{\psi}(x)\gamma_\mu Q\psi(x)::\bar{\psi}(0)\gamma_\nu Q\psi(0):) \\ &= \mathrm{tr}[\mathrm{i}S_\mathrm{F}(-x)\gamma_\mu \mathrm{i}S_\mathrm{F}(x)\gamma_\nu Q^2] \\ &\quad + :\bar{\psi}(x)\gamma_\mu Q\mathrm{i}S_\mathrm{F}(x)\gamma_\nu Q\psi(0): \\ &\quad + :\bar{\psi}(0)\gamma_\nu Q\mathrm{i}S_\mathrm{F}(-x)\gamma_\mu Q\psi(x): \\ &\quad + :\bar{\psi}(x)\gamma_\mu Q\psi(x)\bar{\psi}(0)\gamma_\nu Q\psi(0): \end{aligned} \tag{7.144}$$

where $S_\mathrm{F}(x)$ was defined in (7.136). Using (7.138) and the identity

$$\gamma_\mu\gamma_\nu\gamma_\lambda = (S_{\mu\nu\lambda\rho} + \mathrm{i}\varepsilon_{\mu\nu\lambda\rho}\gamma_5)\gamma^\rho \tag{7.145}$$

where $S_{\mu\nu\lambda\rho} = g_{\mu\nu}g_{\lambda\rho} + g_{\mu\rho}g_{\nu\lambda} - g_{\mu\lambda}g_{\nu\rho}$, we can write (7.144) in the limit $x^2 \approx 0$ as

$$
\begin{aligned}
T(J_\mu(x)J_\nu(0)) \approx (\mathrm{tr}\, Q^2)\,&\frac{(x^2 g_{\mu\nu} - 2x_\mu x_\nu)}{\pi^4(x^2 - \mathrm{i}\varepsilon)^4} \\
&+ \frac{\mathrm{i}x^\alpha}{2\pi^2(x^2 - \mathrm{i}\varepsilon)^2}\{S_{\mu\alpha\nu\beta}[V^\beta(x,0) - V^\beta(0,x)] \\
&+ \mathrm{i}\varepsilon_{\mu\alpha\nu\beta}[A^\beta(x,0) - A^\beta(0,x)]\} \\
&+ :\bar{\psi}(x)\gamma_\mu Q\psi(x)\bar{\psi}(0)\gamma_\nu Q\psi(0): \qquad (7.146)
\end{aligned}
$$

where

$$V^\beta(x, y) = :\bar{\psi}(x)\gamma^\beta Q^2\psi(y): \qquad (7.147)$$

$$A^\beta(x, y) = :\bar{\psi}(x)\gamma^\beta\gamma_5 Q^2\psi(y):. \qquad (7.148)$$

If we write

$$\frac{x^2 g_{\mu\nu} - 2x_\mu x_\nu}{(x^2 - \mathrm{i}\varepsilon)^4} = \frac{2}{3}\frac{g_{\mu\nu}}{(x^2 - \mathrm{i}\varepsilon)^3} - \frac{1}{12}\partial_\mu\,\partial_\nu\frac{1}{(x^2 - \mathrm{i}\varepsilon)^2} \qquad (7.149)$$

and

$$\frac{x^\alpha}{(x^2 - \mathrm{i}\varepsilon)^2} = \frac{-1}{2}\partial^\alpha\left(\frac{1}{x^2 - \mathrm{i}\varepsilon}\right) \qquad (7.150)$$

and use the substitution (7.135), we obtain the leading light-cone singularities of the current commutator

$$
\begin{aligned}
[J_\mu(x), J_\nu(0)] \approx \frac{\mathrm{i}\,\mathrm{tr}\,Q^2}{\pi^3}&\{\tfrac{2}{3}g_{\mu\nu}\,\delta''(x^2)\varepsilon(x_0) + \tfrac{1}{6}\,\partial_\mu\,\partial_\nu[\delta'(x^2)\varepsilon(x_0)]\} \\
&+ \{S_{\mu\alpha\nu\beta}[V^\beta(x,0) - V^\beta(0,x)] \\
&+ \mathrm{i}\varepsilon_{\mu\alpha\nu\beta}[A^\beta(x,0) - A^\beta(0,x)]\}\,\partial^\alpha\frac{[\delta(x^2)\varepsilon(x_0)]}{2\pi} \\
&+ :\bar{\psi}(x)\gamma_\mu Q\psi(x)\bar{\psi}(0)\gamma_\nu Q\psi(0):. \qquad (7.151)
\end{aligned}
$$

We can then translate this explicit form of the current commutator into statements on the cross-sections for e^+e^- annihilation and for inelastic lN scatterings.

(1) e^+e^- *annihilation*. Following the procedure of §7.1 it is straightforward to show that the total hadronic cross-section for e^+e^- annihilation can be written as a current commutator

$$\sigma(e^+e^- \to \text{hadrons}) = \frac{8\pi^2\alpha^2}{3(q^2)^2}\int \mathrm{d}^4x\, \mathrm{e}^{\mathrm{i}q\cdot x}\langle 0|[J_\mu(x), J^\mu(0)]|0\rangle. \qquad (7.152)$$

The most singular light-cone term comes from the first one on the right-hand side of (7.151). (Thus this actually probes the short distance behaviour of the current commutator.)

$$\sigma(e^+e^- \to \text{hadrons}) \approx \frac{8\pi^2\alpha^2 i\,\mathrm{tr}\,Q^2}{3\pi^3(q^2)^2}\int d^4x\, e^{iq\cdot x}\{\tfrac{8}{3}\,\delta''(x^2)\varepsilon(x_0) + \tfrac{1}{6}\,\partial^2[\delta'(x^2)\varepsilon(x_0)]\}. \tag{7.153}$$

Using the identity in (7.131), we get

$$\sigma(e^+e^- \to \text{hadrons}) \approx \frac{8\pi^2\alpha^2\,\mathrm{tr}\,Q^2}{3\pi(q^2)^2}\left(\frac{8}{3}\frac{q^2}{4} - \frac{q^2}{6}\right)\theta(q^2)\varepsilon(q_0) = \frac{4\pi\alpha^2}{3q^2}\,\mathrm{tr}\,Q^2 \tag{7.154}$$

or

$$R = \frac{\sigma(e^+e^- \to \text{hadrons})}{\sigma(e^+e^- \to \mu^+\mu^-)} = \mathrm{tr}\,Q^2. \tag{7.155}$$

This justifies the results of the parton model (7.104) if the leading short distance singularity is that of the free-field theory. We next consider the genuine light-cone process of deep inelastic lepton–hadron scattering.

(2) *Lepton–hadron scattering*. For deep inelastic lN scattering (7.115), the first term on the right-hand side of (7.151) will not contribute since it is a c-number; thus the nontrivial leading singular term will be the second one

$$\left[J_\mu\left(\frac{x}{2}\right), J_\nu\left(-\frac{x}{2}\right)\right] \approx \left\{S_{\mu\alpha\nu\beta}\left[:\bar{\psi}\left(\frac{x}{2}\right)\gamma^\beta Q^2\psi\left(-\frac{x}{2}\right): - :\bar{\psi}\left(-\frac{x}{2}\right)\gamma^\beta Q^2\psi\left(\frac{x}{2}\right):\right] + i\varepsilon_{\mu\alpha\nu\beta}\left[:\bar{\psi}\left(\frac{x}{2}\right)\gamma^\beta\gamma_5 Q^2\psi\left(-\frac{x}{2}\right): - :\bar{\psi}\left(-\frac{x}{2}\right)\gamma^\beta\gamma_5 Q^2\psi\left(\frac{x}{2}\right):\right]\right\}\partial^\alpha\left[\frac{\partial(x^2)\varepsilon(x_0)}{2\pi}\right]. \tag{7.156}$$

We can expand the bilocal operator

$$\bar{\psi}\left(\frac{x}{2}\right)\psi\left(-\frac{x}{2}\right) = \bar{\psi}(0)\left[1 + \overleftarrow{\partial}_{\mu_1}\frac{x^{\mu_1}}{2} + \frac{1}{2!}\overleftarrow{\partial}_{\mu_1}\overleftarrow{\partial}_{\mu_2}\frac{x^{\mu_1}}{2}\frac{x^{\mu_2}}{2} + \ldots\right] \times\left[1 - \frac{x^{\nu_1}}{2}\overrightarrow{\partial}_{\nu_1} + \frac{1}{2!}\frac{x^{\nu_1}}{2}\frac{x^{\nu_2}}{2}\overrightarrow{\partial}_{\nu_1}\overrightarrow{\partial}_{\nu_2} - \ldots\right]\psi(0) = \sum_n \frac{1}{n!}\frac{x^{\mu_1}}{2}\frac{x^{\mu_2}}{2}\cdots\frac{x^{\mu_n}}{2}\bar{\psi}(0)\overleftrightarrow{\partial}_{\mu_1}\overleftrightarrow{\partial}_{\mu_2}\ldots\overleftrightarrow{\partial}_{\mu_n}\psi(0) \tag{7.157}$$

to get

$$\left[J_\mu\left(\frac{x}{2}\right), J_\nu\left(-\frac{x}{2}\right)\right] = \sum_{\text{odd } n} \frac{1}{n!} \frac{x^{\mu_1}}{2} \frac{x^{\mu_2}}{2} \cdots \frac{x^{\mu_n}}{2} \mathcal{O}^{(n+1)}_{\beta\mu_1\mu_2\ldots\mu_n}(0)$$

$$\times S_{\mu\alpha\nu\beta}\, \partial^\alpha \left[\frac{\delta(x^2)\varepsilon(x_0)}{2\pi}\right]$$

$$+ \sum_{\text{even } n} \frac{1}{n!} \frac{x^{\mu_1}}{2} \frac{x^{\mu_2}}{2} \cdots \frac{x^{\mu_n}}{2} \mathcal{O}'^{(n+1)}_{\beta\mu_1\mu_2\ldots\mu_n}(0)$$

$$\times \mathrm{i}\varepsilon_{\mu\alpha\nu\beta}\, \partial^\alpha \left[\frac{\delta(x^2)\varepsilon(x_0)}{2\pi}\right] \tag{7.158}$$

where

$$\mathcal{O}^{(n+1)}_{\beta\mu_1\mu_2\ldots\mu_n}(0) = \bar{\psi}(0)\, \overleftrightarrow{\partial}_{\mu_1} \overleftrightarrow{\partial}_{\mu_2} \cdots \overleftrightarrow{\partial}_{\mu_n} \gamma^\beta Q^2 \psi(0) \tag{7.159a}$$

$$\mathcal{O}'^{(n+1)}_{\beta\mu_1\mu_2\ldots\mu_n}(0) = \bar{\psi}(0)\, \overleftrightarrow{\partial}_{\mu_1} \overleftrightarrow{\partial}_{\mu_2} \cdots \overleftrightarrow{\partial}_{\mu_n} \gamma^\beta \gamma_5 Q^2 \psi(0). \tag{7.159b}$$

To calculate the structure functions, we write

$$\frac{1}{2}\sum_\sigma \langle p\sigma | \mathcal{O}^{(n+1)}_{\beta\mu_1\mu_2\ldots\mu_n}(0) | p\sigma \rangle = A^{(n+1)} p^\beta p_{\mu_1} p_{\mu_2} \cdots p_{\mu_n} + \text{trace terms} \tag{7.160}$$

where $A^{(n+1)}$ some constant and where the trace terms, which contain one or more factors of $g_{\mu_i\mu_j}$, will produce powers of x^2 when contracted with $x^{\mu_1}x^{\mu_2}\ldots x^{\mu_n}$ in (7.158) and are less important near the light cone $x^2 \approx 0$. Also the $\mathcal{O}'^{(n+1)}$ term will not contribute to the spin-averaged structure functions due to the antisymmetry property of $\varepsilon_{\mu\alpha\nu\beta}$. We then have for (7.115)

$$W_{\mu\nu}(p, q) \approx \frac{1}{2M} \int \frac{\mathrm{d}^4 x}{2\pi}\, \mathrm{e}^{\mathrm{i}x\cdot q} \sum_{\text{odd } n}^{\infty} \left(\frac{x\cdot p}{2}\right)^n \frac{p^\beta}{n!} A^{(n+1)}$$

$$\times S_{\mu\alpha\nu\beta}\, \partial^\alpha \left[\frac{\delta(x^2)\varepsilon(x_0)}{2\pi}\right]. \tag{7.161}$$

Define

$$\sum_{\text{odd } n}^{\infty} \left(\frac{x\cdot p}{2}\right)^n \frac{A^{(n+1)}}{n!} = \int \mathrm{d}\xi\, \mathrm{e}^{\mathrm{i}x\cdot \xi p} f(\xi). \tag{7.162}$$

then

$$W_{\mu\nu}(p, q) \approx \frac{\mathrm{i}}{2M} \int \frac{\mathrm{d}^4 x}{2\pi}\, \mathrm{e}^{\mathrm{i}x\cdot q} \int \mathrm{d}\xi\, \mathrm{e}^{\mathrm{i}x\cdot \xi p} f(\xi)$$

$$\times S_{\mu\alpha\nu\beta} (q + \xi p)^\alpha p^\beta \frac{\delta(x^2)\delta(x_0)}{2\pi}. \tag{7.163}$$

Using the identity (7.131)

$$i\int \frac{d^4x}{(2\pi)^2}\, e^{ix\cdot(q+\xi p)}\, \delta(x^2)\varepsilon(x_0) = \delta((q+\xi p)^2)\varepsilon(q_0+\xi p_0), \qquad (7.164)$$

we have

$$W_{\mu\nu} \approx \frac{1}{M}\int d\xi f(\xi)\, \delta(q^2 + 2M\nu\xi)$$

$$\times\, (g_{\mu\alpha}g_{\nu\beta} + g_{\mu\beta}g_{\nu\alpha} - g_{\mu\nu}g_{\alpha\beta})(q+\xi p)^\alpha p^\beta$$

$$= \frac{1}{2M^2\nu}\int d\xi f(\xi)\, \delta(\xi + q^2/2M\nu)(-M\nu g_{\mu\nu} + 2\xi p_\mu p_\nu + \ldots)$$

$$= f(x)\left[-\frac{g_{\mu\nu}}{2M} + \frac{x}{\nu}\frac{p_\mu p_\nu}{M^2} + \ldots\right] \qquad (7.165)$$

for $x = -q^2/2M\nu$. Thus we recover the parton-model results of eqns (7.56) and (7.57)

$$MW_1 \to F_1(x) = \tfrac{1}{2}f(x)$$

$$\nu W_2 \to F_2(x) = xf(x). \qquad (7.166)$$

This implies that the assumption of canonical free-field light-cone structure is equivalent to that of the parton model.

PART II

8 Gauge symmetries

THE symmetries we have discussed up to this point are *global symmetries*. The parameters ε^a of the symmetry transformation in eqn (5.10) are independent of space–time; thus fields at different space–time points are all supposed to transform by the same amount. We now consider theories where the symmetry transformations are space–time dependent, i.e., $\varepsilon^a = \varepsilon^a(x)$. They are called *local symmetries* or *gauge symmetries* (Weyl 1929). We shall see that such symmetries may be used to generate dynamics, the *gauge interactions*. The prototype gauge theory is quantum electrodynamics. It is now believed that all fundamental interactions are described by some form of gauge theory. In the first section, after an introductory discussion of QED with its Abelian U(1) local symmetry, we study the fundamentally richer systems of non-Abelian gauge theories, the *Yang–Mills theories* (1954). After an elementary geometric look at gauge invariance, we present in the last section the subject of spontaneous symmetry breakdown in a gauge theory.

8.1 Local symmetries in field theory

Abelian gauge theory

As we have already stated, QED is an Abelian gauge theory. It is instructive to show that the theory can actually be 'derived' by requiring the Dirac free electron theory to be gauge invariant and renormalizable.

Consider the Lagrangian for a free-electron field $\psi(x)$

$$\mathscr{L}_0 = \bar{\psi}(x)(\mathrm{i}\gamma^\mu \partial_\mu - m)\psi(x). \tag{8.1}$$

Clearly it has a global U(1) symmetry corresponding to the invariance of the theory under a phase change

$$\begin{aligned}\psi(x) &\to \psi'(x) = \mathrm{e}^{-\mathrm{i}\alpha}\psi(x)\\ \bar{\psi}(x) &\to \bar{\psi}'(x) = \mathrm{e}^{\mathrm{i}\alpha}\bar{\psi}(x).\end{aligned} \tag{8.2}$$

We are going to turn this symmetry into a local symmetry, i.e., 'to gauge the symmetry' by replacing α with $\alpha(x)$. Thus we are going to construct a theory which will be invariant under a space–time dependent phase change,

$$\begin{aligned}\psi(x) &\to \psi'(x) = \mathrm{e}^{-\mathrm{i}\alpha(x)}\psi(x).\\ \bar{\psi}(x) &\to \bar{\psi}'(x) = \mathrm{e}^{\mathrm{i}\alpha(x)}\bar{\psi}(x).\end{aligned} \tag{8.3}$$

The derivative term will now have a rather complicated transformation

$$\begin{aligned}\bar{\psi}(x)\,\partial_\mu\psi(x) \to \bar{\psi}'(x)\,\partial_\mu\psi'(x) &= \bar{\psi}(x)\,\mathrm{e}^{\mathrm{i}\alpha(x)}\,\partial_\mu(\mathrm{e}^{-\mathrm{i}\alpha(x)}\psi(x))\\ &= \bar{\psi}(x)\,\partial_\mu\psi(x) - \mathrm{i}\bar{\psi}(x)\,\partial_\mu\alpha(x)\psi(x).\end{aligned} \tag{8.4}$$

The second term spoils the invariance. We need to form a *gauge-covariant derivative* D_μ, to replace ∂_μ, and $D_\mu\psi(x)$ will have the simple transformation

$$D_\mu\psi(x) \to [D_\mu\psi(x)]' = e^{-i\alpha(x)}D_\mu\psi(x) \tag{8.5}$$

so that the combination $\bar{\psi}(x)D_\mu\psi(x)$ is gauge invariant. In other words, the action of the covariant derivative on the field will not change the transformation property of the field. This can be realized if we enlarge the theory with a new vector field $A_\mu(x)$, the *gauge field*, and form the covariant derivative as

$$D_\mu\psi = (\partial_\mu + ieA_\mu)\psi \tag{8.6}$$

where e is a free parameter which we eventually will identify with the electric charge. Then the transformation law for the covariant derivative (8.5) will be satisfied if the gauge field $A_\mu(x)$ has the transformation property

$$A_\mu(x) \to A'_\mu(x) = A_\mu(x) + \frac{1}{e}\partial_\mu\alpha(x) \tag{8.7}$$

From (8.1) we now have

$$\mathscr{L}'_0 = \bar{\psi}i\gamma^\mu(\partial_\mu + ieA_\mu)\psi - m\bar{\psi}\psi. \tag{8.8}$$

To make the gauge field a true dynamical variable we need to add a term to the Lagrangian involving its derivatives. The simplest gauge-invariant term of dimension-four or less (with a conventional normalization) is

$$\mathscr{L}_A = -\tfrac{1}{4}F_{\mu\nu}F^{\mu\nu} \tag{8.9}$$

where

$$F_{\mu\nu} = \partial_\mu A_\nu - \partial_\nu A_\mu. \tag{8.10}$$

By direct substitution of (8.7) we see that $F_{\mu\nu}$ is in fact gauge invariant by itself. It is useful to see this in another way—the antisymmetric tensor $F_{\mu\nu}$ is related to the covariant derivative as

$$(D_\mu D_\nu - D_\nu D_\mu)\psi = ieF_{\mu\nu}\psi. \tag{8.11}$$

From (8.5) is it easy to see that

$$[(D_\mu D_\nu - D_\nu D_\mu)\psi]' = e^{-i\alpha}[(D_\mu D_\nu - D_\nu D_\mu)\psi] \tag{8.12}$$

or

$$F'_{\mu\nu}\psi' = (F_{\mu\nu}\psi)e^{-i\alpha} \tag{8.13}$$

or

$$F'_{\mu\nu} = F_{\mu\nu}. \tag{8.14}$$

Combining (8.8) and (8.9) we arrive at the QED Lagrangian

$$\mathscr{L} = \bar{\psi}i\gamma^\mu(\partial_\mu + ieA_\mu)\psi - m\bar{\psi}\psi - \tfrac{1}{4}F_{\mu\nu}F^{\mu\nu}. \tag{8.15}$$

The following features of (8.15) should be noted

(1) The photon is massless because a $A_\mu A^\mu$ term is not gauge invariant.

(2) The minimal coupling of photon to the electron is contained in the covariant derivative $D_\mu\psi$ which can be constructed from the transformation property of the electron field. In other words, the coupling of the photon to any matter field is determined by its transformation property under the symmetry group. This is usually referred to as *universality*. Other (higher-dimensional) gauge-invariant couplings such as $\bar{\psi}\sigma_{\mu\nu}\psi F^{\mu\nu}$ are ruled out by the requirement of renormalizability.

(3) The Lagrangian of (8.15) does not have a gauge-field self-coupling, because the photon does not carry a charge (or U(1) quantum number). Thus, without a matter field, the theory is a free-field theory.

We shall see that the first two features will still hold for non-Abelian gauge theories but the last will not. The presence of gauge-field self-coupling will make such non-Abelian theories highly nonlinear and will give rise to a number of fundamentally distinctive properties.

Non-Abelian gauge symmetry—Yang–Mills fields

In 1954 Yang and Mills extended the gauge principle to non-Abelian symmetry. (For subsequent development of the Yang–Mills theories see Utiyama 1956; Gell-Mann and Glashow 1961.) We shall illustrate the construction for the simplest case of isospin SU(2).

Let the fermion field be an isospin doublet,

$$\psi = \begin{pmatrix}\psi_1\\ \psi_2\end{pmatrix}. \tag{8.16}$$

Under an SU(2) transformation, we have

$$\psi(x) \to \psi'(x) = \exp\left\{\frac{-\mathrm{i}\boldsymbol{\tau}\cdot\boldsymbol{\theta}}{2}\right\}\psi(x) \tag{8.17}$$

where $\boldsymbol{\tau} = (\tau_1, \tau_2, \tau_3)$ are the usual Pauli matrices, satisfying

$$\left[\frac{\tau_i}{2}, \frac{\tau_j}{2}\right] = \mathrm{i}\varepsilon_{ijk}\frac{\tau_k}{2} \qquad i, j, k = 1, 2, 3 \tag{8.18}$$

and $\boldsymbol{\theta} = (\theta_1, \theta_2, \theta_3)$ are the SU(2) transformation parameters. The free Lagrangian

$$\mathscr{L}_0 = \bar{\psi}(x)(\mathrm{i}\gamma^\mu\,\partial_\mu - m)\psi(x) \tag{8.19}$$

is invariant under the global SU(2) symmetry with $\{\theta_i\}$ being space–time independent. However under the local symmetry transformation

$$\psi(x) \to \psi'(x) = U(\theta)\psi(x) \tag{8.20}$$

with

$$U(\theta) = \exp\left\{\frac{-\mathrm{i}\boldsymbol{\tau}\cdot\boldsymbol{\theta}(x)}{2}\right\}, \tag{8.21}$$

the free Lagrangian $\mathscr{L}_0$ is no longer invariant because the derivative term transforms as

$$\bar{\psi}(x)\,\partial_\mu\psi(x) \to \bar{\psi}'(x)\,\partial_\mu\psi'(x) = \bar{\psi}(x)\,\partial_\mu\psi(x) + \bar{\psi}(x)U^{-1}(\theta)[\partial_\mu U(\theta)]\psi(x). \qquad (8.22)$$

To construct a gauge-invariant Lagrangian we follow a procedure similar to that of the Abelian case. First we introduce the vector gauge fields A_μ^i, $i = 1, 2, 3$ (one for each group generator) to form the gauge-covariant derivative through the minimal coupling

$$\mathrm{D}_\mu\psi = \left(\partial_\mu - \mathrm{i}g\frac{\boldsymbol{\tau}\cdot\mathbf{A}_\mu}{2}\right)\psi \qquad (8.23)$$

where g is the coupling constant in analogy to e in (8.6). We demand that $\mathrm{D}_\mu\psi$ have the same transformation property as ψ itself, i.e.

$$\mathrm{D}_\mu\psi \to (\mathrm{D}_\mu\psi)' = U(\theta)\mathrm{D}_\mu\psi. \qquad (8.24)$$

This implies that

$$\left(\partial_\mu - \mathrm{i}g\frac{\boldsymbol{\tau}\cdot\mathbf{A}'_\mu}{2}\right)(U(\theta)\psi) = U(\theta)\left(\partial_\mu - \mathrm{i}g\frac{\boldsymbol{\tau}\cdot\mathbf{A}_\mu}{2}\right)\psi. \qquad (8.25)$$

or

$$\left[\partial_\mu U(\theta) - \mathrm{i}g\frac{\boldsymbol{\tau}\cdot\mathbf{A}'_\mu}{2}\,U(\theta)\right]\psi = -\mathrm{i}gU(\theta)\frac{\boldsymbol{\tau}\cdot\mathbf{A}_\mu}{2}\,\psi$$

or

$$\frac{\boldsymbol{\tau}\cdot\mathbf{A}'_\mu}{2} = U(\theta)\frac{\boldsymbol{\tau}\cdot\mathbf{A}_\mu}{2}\,U^{-1}(\theta) - \frac{\mathrm{i}}{g}[\partial_\mu U(\theta)]U^{-1}(\theta) \qquad (8.26)$$

which defines the transformation law for the gauge fields. For an infinitesimal change $\boldsymbol{\theta}(x) \ll 1$,

$$U(\theta) \cong 1 - \mathrm{i}\frac{\boldsymbol{\tau}\cdot\boldsymbol{\theta}(x)}{2} \qquad (8.27)$$

and (8.26) becomes

$$\frac{\boldsymbol{\tau}\cdot\mathbf{A}'_\mu}{2} = \frac{\boldsymbol{\tau}\cdot\mathbf{A}_\mu}{2} - \mathrm{i}\theta^j A_\mu^k\left[\frac{\tau_j}{2}, \frac{\tau_k}{2}\right] - \frac{1}{g}\left(\frac{\boldsymbol{\tau}}{2}\cdot\partial_\mu\boldsymbol{\theta}\right)$$

$$= \frac{\boldsymbol{\tau}\cdot\mathbf{A}_\mu}{2} + \frac{1}{2}\varepsilon^{ijk}\tau^i\theta^j A_\mu^k - \frac{1}{g}\left(\frac{\boldsymbol{\tau}}{2}\cdot\partial_\mu\boldsymbol{\theta}\right)$$

or

$$A_\mu^{i\prime} = A_\mu^i + \varepsilon^{ijk}\theta^j A_\mu^k - \frac{1}{g}\partial_\mu\theta^i. \qquad (8.28)$$

The second term is clearly the transformation for a triplet (the adjoint) representation under SU(2). Thus the A_μ^is carry charges, in contrast to the Abelian gauge field.

To obtain the antisymmetric second-rank tensor of the gauge fields we can follow (8.11) and study the combination

$$(\mathrm{D}_\mu \mathrm{D}_\nu - \mathrm{D}_\nu \mathrm{D}_\mu)\psi \equiv \mathrm{i}g\left(\frac{\tau^i}{2} F^i_{\mu\nu}\right)\psi \tag{8.29}$$

with

$$\frac{\boldsymbol{\tau}\cdot\mathbf{F}_{\mu\nu}}{2} = \partial_\mu \frac{\boldsymbol{\tau}\cdot\mathbf{A}_\nu}{2} - \partial_\nu \frac{\boldsymbol{\tau}\cdot\mathbf{A}_\mu}{2} - \mathrm{i}g\left[\frac{\boldsymbol{\tau}\cdot\mathbf{A}_\mu}{2}, \frac{\boldsymbol{\tau}\cdot\mathbf{A}_\nu}{2}\right] \tag{8.30}$$

or

$$F^i_{\mu\nu} = \partial_\mu A^i_\nu - \partial_\nu A^i_\mu + g\varepsilon^{ijk} A^j_\mu A^k_\nu. \tag{8.31}$$

From the fact that $\mathrm{D}_\mu\psi$ has the same gauge transformation property as ψ, we see that

$$[(\mathrm{D}_\mu \mathrm{D}_\nu - \mathrm{D}_\nu \mathrm{D}_\mu)\psi]' = U(\theta)(\mathrm{D}_\mu \mathrm{D}_\nu - \mathrm{D}_\nu \mathrm{D}_\mu)\psi. \tag{8.32}$$

Substituting the definition (8.29) of $F^i_{\mu\nu}$ on both sides of (8.32), we have

$$\boldsymbol{\tau}\cdot\mathbf{F}'_{\mu\nu} U(\theta)\psi = U(\theta)\boldsymbol{\tau}\cdot\mathbf{F}_{\mu\nu}\psi$$

or

$$\boldsymbol{\tau}\cdot\mathbf{F}'_{\mu\nu} = U(\theta)(\boldsymbol{\tau}\cdot\mathbf{F}_{\mu\nu})U^{-1}(\theta). \tag{8.33}$$

For the infinitesimal transformation $\theta_i \ll 1$, this translates into

$$F^{i\prime}_{\mu\nu} = F^i_{\mu\nu} + \varepsilon^{ijk}\theta^j F^k_{\mu\nu}. \tag{8.34}$$

Unlike the Abelian case, $F^i_{\mu\nu}$ transform nontrivially, like a triplet under SU(2). However the product

$$\mathrm{tr}\{(\boldsymbol{\tau}\cdot\mathbf{F}_{\mu\nu})(\boldsymbol{\tau}\cdot\mathbf{F}^{\mu\nu})\} \propto F^i_{\mu\nu}F^{i\mu\nu}$$

is gauge invariant.

We can summarize the above discussion by displaying the complete gauge-invariant Lagrangian which describes the interaction between gauge fields A^i_μ and the SU(2) doublet fields

$$\mathscr{L} = -\tfrac{1}{4}F^i_{\mu\nu}F^{i\mu\nu} + \bar{\psi}\mathrm{i}\gamma^\mu \mathrm{D}_\mu\psi - m\bar{\psi}\psi \tag{8.35}$$

where

$$F^i_{\mu\nu} = \partial_\mu A^i_\nu - \partial_\nu A^i_\mu + g\varepsilon^{ijk}A^j_\mu A^k_\nu \tag{8.36}$$

$$\mathrm{D}_\mu\psi = \left(\partial_\mu - \mathrm{i}g\frac{\boldsymbol{\tau}\cdot\mathbf{A}_\mu}{2}\right)\psi. \tag{8.37}$$

The SU(2) gauge transformations of fields are

$$\psi(x) \to \psi'(x) = \exp\left\{-\mathrm{i}\frac{\boldsymbol{\tau}\cdot\boldsymbol{\theta}(x)}{2}\right\}\psi(x) = U(\theta)\psi(x) \tag{8.38}$$

$$\frac{\boldsymbol{\tau}\cdot\mathbf{A}_\mu}{2} \to \frac{\boldsymbol{\tau}\cdot\mathbf{A}'_\mu}{2} = U(\theta)\left(\frac{\boldsymbol{\tau}\cdot\mathbf{A}_\mu}{2}\right)U^{-1}(\theta) - \frac{\mathrm{i}}{g}[\partial_\mu U(\theta)]U^{-1}(\theta) \tag{8.39}$$

with infinitesimal forms

$$\psi \to \psi' = \psi - \mathrm{i}\frac{\boldsymbol{\tau}\cdot\boldsymbol{\theta}}{2}\psi \tag{8.40}$$

$$A^i_\mu \to A^{i\prime}_\mu = A^i_\mu + \varepsilon^{ijk}\theta^j A^k_\mu - \frac{1}{g}\partial_\mu\theta^i. \tag{8.41}$$

Generalization to higher groups and arbitrary representation for ψ is straightforward. Let G be some simple Lie group with generators satisfying the algebra

$$[F^a, F^b] = \mathrm{i}C^{abc}F^c \tag{8.42}$$

where the C^{abc}s are the totally antisymmetric structure constants. ψ is supposed to belong to some representation with representation matrices T^a. Thus

$$[T^a, T^b] = \mathrm{i}C^{abc}T^c. \tag{8.43}$$

The covariant derivative is then

$$\mathrm{D}_\mu\psi = (\partial_\mu - \mathrm{i}gT^aA^a_\mu)\psi \tag{8.44}$$

and the second-rank tensor for gauge fields is

$$F^a_{\mu\nu} = \partial_\mu A^a_\nu - \partial_\nu A^a_\mu + gC^{abc}A^b_\mu A^c_\nu \tag{8.45}$$

$$(\mathbf{T}\cdot\mathbf{F})_{\mu\nu} = \partial_\mu(\mathbf{T}\cdot\mathbf{A}_\nu) - \partial_\nu(\mathbf{T}\cdot\mathbf{A}_\mu) - \mathrm{i}g[\mathbf{T}\cdot\mathbf{A}_\mu, \mathbf{T}\cdot\mathbf{A}_\nu] \tag{8.46}$$

$$\mathscr{L} = -\tfrac{1}{4}F^a_{\mu\nu}F^{a\mu\nu} + \bar{\psi}(\mathrm{i}\gamma^\mu\mathrm{D}_\mu - m)\psi. \tag{8.47}$$

The Lagrangian is then invariant under the transformation of the group G

$$\psi(x) \to \psi'(x) = U(\mathbf{T}\cdot\boldsymbol{\theta}(x))\psi(x) \equiv U(\theta_x)\psi(x) \tag{8.48}$$

$$\mathbf{T}\cdot\mathbf{A}_\mu(x) \to \mathbf{T}\cdot\mathbf{A}'_\mu(x) = U(\theta_x)\mathbf{T}\cdot\mathbf{A}_\mu U^{-1}(\theta_x) - \frac{\mathrm{i}}{g}[\partial_\mu U(\theta_x)]U^{-1}(\theta_x) \tag{8.49}$$

with the infinitesimal variations taking on the forms

$$\psi(x) \to \psi'(x) = \psi(x) - \mathrm{i}T^a\theta^a(x)\psi(x) \tag{8.50}$$

$$A^a_\mu(x) \to A^{a\prime}_\mu(x) = A^a_\mu(x) + C^{abc}\theta^b(x)A^c_\mu(x) - \frac{1}{g}\partial_\mu\theta^a(x). \tag{8.51}$$

The pure Yang–Mills term, $-\frac{1}{4}F^a_{\mu\nu}F^{a\mu\nu}$, contains factors that are trilinear and quadrilinear in A^a_μ,

$$-gC^{abc}\,\partial_\mu A^a_\nu A^{b\mu}A^{c\nu} - \frac{g^2}{4}C^{abc}C^{ade}A^b_\mu A^c_\nu A^{d\mu}A^{e\nu}, \tag{8.52}$$

which correspond to self-couplings of non-Abelian gauge fields. They are brought about by the nonlinear terms in $F^a_{\mu\nu}$ (8.45), because the gauge fields A^a_μ themselves transform nontrivially, like the generators, as members of the adjoint representation. While this is fundamentally different from (rather, richer than) in the Abelian case, the properties of universality and masslessness of the gauge fields remain essentially the same. We note that the number of massless gauge fields is equal to the number of generators of gauge symmetry. Concerning universality we should make two comments about the coupling strength.

(1) In the case of Abelian gauge theories there is no restriction on the coupling strength between the gauge field A_μ and other fields. Thus the electron carries charge e and another particle can in principle carry any charge λe with an arbitrary λ (for instance $\lambda = \pi$). In a non-Abelian gauge theory such as the SU(2) case considered above with a doublet ψ, the situation is more restrictive. If one tries, for example, to couple the gauge field to an extra doublet ϕ with a coupling λg, the commutation relation (8.18) insuring gauge invariance gets rescaled and implies $\lambda^2 = \lambda$ or $\lambda = 1$. Basically in non-Abelian theories the normalization of the generators are fixed by the non-linear relation of commutator, hence g cannot be scaled arbitrarily.

(2) Can there be different gauge couplings associated with different gauge fields? If the group is *simple*, as just stated, there can be only one coupling constant. However if the group is a product of simple groups such as SU(2) × SU(3) where each set of generators closes under commutation and commutes with other sets, there will be an independent coupling for each factor group.

8.2 Gauge invariance and geometry

Einstein's successful formulation of general relativity in 1916 unveiled a profound connection between gravitation and geometry. This discovery inspired Weyl (1919, 1921) to incorporate electromagnetism into geometry through the concept of a space–time dependent (local) *scale* transformation. Namely, at a neighbouring point, a distance $\mathrm{d}x^\mu$ away, the scale is changed (from one) to $(1 + S_\mu\,\mathrm{d}x^\mu)$, and thus a space–time dependent function is changed according to

$$f(x) \to (f + (\partial_\mu f)\,\mathrm{d}x^\mu)(1 + S_\mu\,\mathrm{d}x^\mu) \simeq f + [(\partial_\mu + S_\mu)f]\,\mathrm{d}x^\mu. \quad (8.53)$$

Weyl tried to derive electromagnetism by requiring invariance under this local scale transformation and by identifying the scale factor with the vector potential: $S_\mu \leftrightarrow A_\mu$. His initial attempt was not successful. By 1925 modern quantum mechanics has emerged. Here a key concept was to identify the momentum with the operator $(-\mathrm{i}\partial_\mu)$, and the canonical momentum in the presence of electromagnetic field with $(-\mathrm{i}\partial_\mu + eA_\mu)$. It was then realized that the correct identification of Weyl's scale factor should be $S_\mu \leftrightarrow \mathrm{i}A_\mu$, and that what would be required would be invariance of the theory under space–time dependent *phase* transformation, see eqn (8.3). However when Weyl (1929) finally worked out this approach he retained his original terminology of 'gauge invariance', the invariance under a change of the scale, a change of the gauge. For a concise history of the gauge field concept, the reader is referred to the lectures by Yang (1975).

The framework for a proper geometrical discussion of gauge fields is the modern theory of fibre bundles. However such a differential geometry study would be beyond the scope of this presentation. It suffices for us to note the existence of a deep geometric foundation to the gauge field concept. Here we merely present an elementary geometric look at gauge invariance. We will show how gauge fields A_μ describe parallel transport in charge space with the curvature tensor being the field intensity $F_{\mu\nu}$.

We briefly review some of the basic geometric concepts for a curved space. First, there is the notion of *parallel transport*. In any space, to compare two vectors (or any tensors) at two different space points $V_\mu(x)$ and $V_\mu(x')$, we must first move (parallel transport) V_μ from x to x', i.e. the two vectors must be in the same coordinate system before we can take their difference. Thus there are always two steps in such a comparison

$$\mathrm{D}V_\mu = \delta V_\mu + \mathrm{d}V_\mu \tag{8.54}$$

where δV_μ is the (apparent) change due to moving these two vectors to the same coordinate origin and $\mathrm{d}V_\mu$ is their difference measured in the same coordinate system. The operational definition of parallel transport is such as to keep the vector, throughout the transport, at a fixed angle to the tangent of the trajectory. Clearly parallel transport is a trivial operation in flat (Euclidean) space as it does not introduce any change of the vector $\delta V_\mu = 0$, and covariant differentiation is simply ordinary differentiation

$$\mathrm{D}V^\mu = \mathrm{d}V^\mu = (\partial_\lambda V^\mu)\,\mathrm{d}x^\lambda \quad \text{in flat space.} \tag{8.55}$$

However, in a curvilinear system there will be an apparent change in such a translation, as the coordinate axes differ from point to point in such a system (i.e. the metric is position-dependent). For x and x' infinitesimally separated by a distance $\mathrm{d}x^\lambda$, we expect δV_μ to be linear in $\mathrm{d}x^\lambda$ and V^μ

$$\delta V^\mu = -\Gamma^\mu_{\nu\lambda} V^\nu\,\mathrm{d}x^\lambda$$

and

$$\delta V_\mu = \Gamma^\nu_{\mu\lambda} V_\nu\,\mathrm{d}x^\lambda \tag{8.56}$$

since $\delta(V^\mu V_\mu) = 0$. The coefficient $\Gamma^\mu_{\nu\lambda}$ is the *Affine connection* or the *Christoffel symbol* and may be shown to be (some combination of) the derivative of the metric (and hence vanishes in a space with constant metric). The comparison of the original vector at x' after parallel transport from x to x' results in

$$\begin{aligned}\mathrm{D}V^\mu &= V^\mu(x') - [V^\mu(x) + \delta V^\mu]\\ &= (\partial_\lambda V^\mu + \Gamma^\mu_{\nu\lambda} V^\nu)\,\mathrm{d}x^\lambda\end{aligned} \tag{8.57}$$

where the combination in the parenthesis is the covariant derivative. This contrast between flat and curved spaces is illustrated in Fig. 8.1.

Another important concept in non-Euclidean geometry is the *curvature tensor* which can be best introduced through the notion of parallel transport of a vector around a closed path (see Fig. 8.2 for an illustrative comparison). Consider the apparent variation of the vector after being moved around a small parallelogram $PP_1P_2P_3$ composed of two vectors a^μ and b^μ (and their parallel displacements) (Fig. 8.3). Let δV_μ and $\delta V'_\mu$ be the apparent changes along the paths PP_1P_2 and PP_3P_2, respectively. Thus the total apparent change for a round trip is

$$\Delta V_\mu = \delta V_\mu - \delta V'_\mu \tag{8.58}$$

with

$$\delta V_\mu = (\Gamma^\nu_{\mu\alpha} V_\nu)_P a^\alpha + (\Gamma^\nu_{\mu\beta} V_\nu)_{P_1}(b^\beta + \delta b^\beta) \tag{8.59a}$$

$$\delta V'_\mu = (\Gamma^\nu_{\mu\beta} V_\nu)_P b^\beta + (\Gamma^\nu_{\mu\alpha} V_\nu)_{P_3}(a^\alpha + \delta a^\alpha) \tag{8.59b}$$

We can expand the quantities evaluated at points P_1 and P_3 so that all tensors are measured at a common point P

$$(\Gamma^\nu_{\mu\beta} V_\nu)_{P_1} = (\Gamma^\nu_{\mu\beta} + \partial_\alpha \Gamma^\nu_{\mu\beta} a^\alpha)(V_\nu + \Gamma^\sigma_{\nu\alpha} V_\sigma a^\alpha); \tag{8.60}$$

similarly for $(\Gamma^\nu_{\mu\alpha} V_\nu)_{P_3}$. Substituting these results and

$$a^\alpha + \delta a^\alpha = a^\alpha - \Gamma^\alpha_{\tau\eta} a^\tau b^\eta \tag{8.61}$$

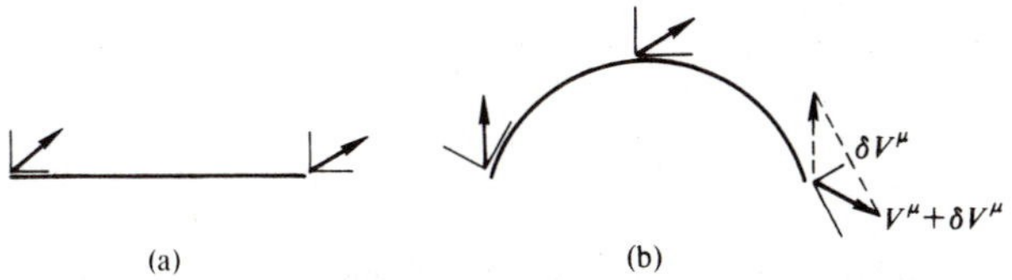

FIG. 8.1. Apparent changes induced by parallel transport (a) in flat space (no apparent change); and (b) in curved space.

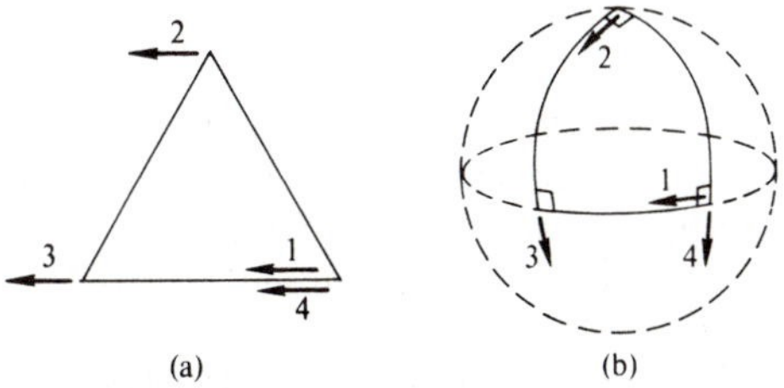

FIG. 8.2. Apparent changes induced by parallel transport (a) around a closed path 1–2–3–4 in flat space; and (b) on a spherical surface.

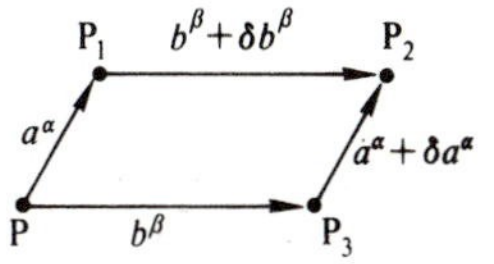

FIG. 8.3.

and a similar expression for $(b^\beta + \delta b^\beta)$ into eqns (8.59) and (8.58), we have

$$\Delta V_\mu = R^{\nu}_{\mu\alpha\beta} V_\nu \sigma^{\alpha\beta}. \tag{8.62}$$

The apparent change around a closed path is proportional to the vector itself, to the area (tensor) bounded by the path $\sigma^{\alpha\beta} = a^\alpha b^\beta$, and to the curvature tensor

$$R^{\nu}_{\mu\alpha\beta} = \partial_\alpha \Gamma^{\nu}_{\mu\beta} - \partial_\beta \Gamma^{\nu}_{\mu\alpha} + \Gamma^{\lambda}_{\mu\beta}\Gamma^{\nu}_{\lambda\alpha} - \Gamma^{\lambda}_{\mu\alpha}\Gamma^{\nu}_{\lambda\beta}. \tag{8.63}$$

For the rather simple example of a spherical triangle with 90° at each vertex (Fig. 8.2), the apparent change is clearly $\pi/2$. This agrees with eqn (8.62) as the curvature tensor reduces to a curvature $R = 1/r^2$ where r is the radius of the sphere. When multiplied by the area of the triangle $\pi r^2/2$ one gets an apparent change of $\pi/2$.

A direct comparison of (8.57) with the gauge covariant derivative $D_\mu\psi$ (8.44) indicates that $\mathbf{T}\cdot\mathbf{A}_\mu$ has a geometrical interpretation as the 'connection' (i.e. Christoffel symbol) in the internal charge space. As in (8.56), under a parallel transport the field $\psi(x)$ undergoes, because of the local change of axes, an apparent change

$$\psi(x) \to \psi(x + dx) = \psi(x) + \delta\psi(x)$$

with

$$\delta\psi(x) = ig\mathbf{T}\cdot\mathbf{A}_\mu\psi \, dx^\mu \tag{8.64}$$

where $\mathbf{T}$ is the set of representation matrices of the symmetry generators and ψ is the basis vector.

For parallel transport of a finite interval from x to x' we can exponentiate (8.64) and obtain

$$P(x', x) = \exp\left\{ig \int_x^{x'} \mathbf{T}\cdot\mathbf{A}_\mu(y)\, dy^\mu\right\} \tag{8.65}$$

where the line integral is taken along a path joining x and x'. Thus for every path we can associate a group element. Let us check that such an interpretation is compatible with the transformation properties of gauge fields. For simplicity consider infinitesimal parallel transport (8.64)

$$\psi(x + dx) = (1 + ig\mathbf{T}\cdot\mathbf{A}_\mu\, dx^\mu)\psi(x). \tag{8.66}$$

Now for a different choice of frame at each point, we make a gauge transformation, a rotation of axes, at each point by $\mathbf{T}\cdot\boldsymbol{\theta}(x)$ as in (8.48)

$$\psi(x) \to \psi'(x) = U(\theta_x)\psi(x) \tag{8.67}$$

and

$$\psi(x') \to \psi'(x') = U(\theta_{x'})\psi(x'). \tag{8.68}$$

In order to keep the product $\bar{\psi}(x + dx)P(x + dx, x)\psi(x)$ invariant, parallel transport 'connection' must transform according to

$$1 + \mathrm{i}g\mathbf{T}\cdot\mathbf{A}'_\mu\,\mathrm{d}x^\mu = U(\theta_{x+\mathrm{d}x})(1 + \mathrm{i}g\mathbf{T}\cdot\mathbf{A}_\mu\,\mathrm{d}x^\mu)U^{-1}(\theta_x)$$
$$= (U(\theta_x) + \partial_\nu U(\theta_x)\,\mathrm{d}x^\nu)(1 + \mathrm{i}g\mathbf{T}\cdot\mathbf{A}_\mu\,\mathrm{d}x^\mu)U^{-1}(\theta_x) \qquad (8.69)$$

or

$$\mathbf{T}\cdot\mathbf{A}'_\mu(x) = U(\theta_x)\mathbf{T}\cdot\mathbf{A}_\mu U^{-1}(\theta_x) - \frac{\mathrm{i}}{g}[\partial_\mu U(\theta_x)]U^{-1}(\theta_x) \qquad (8.70)$$

which is the required transformation property of gauge fields as given in (8.49). This result clearly holds also for finite separations. Thus the parallel transport operator (8.65) has the gauge transformation

$$P(x', x) \to P'(x', x) = U(\theta_{x'})P(x', x)U^{-1}(\theta_x). \qquad (8.71)$$

A direct comparison of $\mathbf{T}\cdot\mathbf{F}_{\mu\nu}$ (8.46) with $R^\rho_{\mu\nu\lambda}$ (8.63) suggests that $\mathbf{T}\cdot\mathbf{F}_{\mu\nu}$ be interpreted as the curvature of internal charge space. This can be checked explicitly by considering parallel transport around a closed path C. For simplicity we choose C to be a parallelogram with one corner at x_μ and two sides $\mathrm{d}x_\mu$ and δx_μ.

$$P_\square = P(x, x + \mathrm{d}x)P(x + \mathrm{d}x, x + \mathrm{d}x + \delta x)$$
$$\times P(x + \mathrm{d}x + \delta x, x + \delta x)P(x + \delta x, x) \qquad (8.72)$$

where the Ps are the parallel transport matrices of (8.65). Using the matrix identity

$$\mathrm{e}^{\lambda A}\,\mathrm{e}^{\lambda B} = \mathrm{e}^{\lambda(A+B)+\lambda^2/2[A,B]} + O(\lambda^3), \qquad (8.73)$$

we find

$$P(x, x + \mathrm{d}x)P(x + \mathrm{d}x, x + \mathrm{d}x + \delta x)$$
$$= \exp[\mathrm{i}gA_\mu(x)\,\mathrm{d}x^\mu]\exp[\mathrm{i}gA_\nu(x + \mathrm{d}x)\,\delta x^\nu]$$
$$= \exp\{\mathrm{i}g(A_\mu\,\mathrm{d}x^\mu + A_\nu\,\delta x^\nu + \partial_\mu A_\nu\,\mathrm{d}x^\mu\,\delta x^\nu) - \frac{g^2}{2}[A_\mu, A_\nu]\,\mathrm{d}x^\mu\,\delta x^\nu\} \qquad (8.74)$$

and

$$P(x + \mathrm{d}x + \delta x, x + \delta x)P(x + \delta x, x)$$
$$= \exp[-\mathrm{i}gA_\mu(x + \delta x)\,\mathrm{d}x^\mu]\exp[-\mathrm{i}gA_\nu(x)\,\delta x^\nu]$$
$$= \exp\{-\mathrm{i}g(A_\mu\,\mathrm{d}x^\mu + A_\nu\,\delta x^\nu + \partial_\nu A_\mu\,\mathrm{d}x^\mu\,\delta x^\nu) - \frac{g^2}{2}[A_\mu, A_\nu]\,\mathrm{d}x^\mu\,\delta x^\nu\}. \qquad (8.75)$$

Thus (8.72) may be written as

$$P_\square = \exp\{\mathrm{i}g(\partial_\mu A_\nu - \partial_\nu A_\mu - \mathrm{i}g[A_\mu, A_\nu])\,\mathrm{d}x^\mu\,\delta x^\nu\}. \qquad (8.76)$$

In eqns (8.74)–(8.76) we have simplified the notation by writing $\mathbf{T}\cdot\mathbf{A}_\mu$ as A_μ.

The above calculation justifies the identification of

$$\mathbf{T}\cdot\mathbf{F}_{\mu\nu} = \partial_\mu \mathbf{T}\cdot\mathbf{A}_\nu - \partial_\nu \mathbf{T}\cdot\mathbf{A}_\mu - ig[\mathbf{T}\cdot\mathbf{A}_\mu, \mathbf{T}\cdot\mathbf{A}_\nu] \tag{8.77}$$

as the curvature tensor of the internal charge space.

In summary, the essential point is that, in any (physical or internal) space where the coordinates are position-dependent, the significance of comparing two vectors (or any two tensors) at different points is lost. The standard way of dealing with this problem is to introduce the notion of parallel transport or affine connection. In the case of physical curved space–time the Christoffel symbol is introduced and in the case of internal charge space the gauge fields are introduced. They 'compensate' for the change of local frames at each space time point.

8.3 Spontaneous breaking of gauge symmetry, the Higgs phenomenon

We saw in §8.1 that the imposition of local symmetry implies the existence of massless vector particles. If we want to avoid this feature of the gauge theory and obtain massive vector bosons, the gauge symmetry must be broken somehow. If we introduce explicit breaking terms in the form of arbitrary gauge boson masses we alter the high-energy behaviour of the theory in such a way that the renormalizability of the theory is lost (see the discussion in §2.4). We may contemplate the possibility of spontaneous breaking of the symmetry, as discussed in §5.3. Thus, we have the situation of a hidden symmetry: the Lagrangian is still fully invariant under the symmetry transformations but the dynamics are such that the vacuum, the ground state, is not a singlet of the symmetry group. The choice of one from all the possible degenerate ground states as the physical vacuum breaks the symmetry. This spoils the usual symmetry consequence of energy-level degeneracies. But, according to the Goldstone theorem of §5.3, this would imply the existence of a set of massless scalar bosons. Thus either way it would seem that we run into undesirable massless particles.

As it turns out the Goldstone theorem is evaded in gauge theories as the proof of the theorem requires the validity of all the usual field theory axioms: manifest Lorentz covariance, positivity of the norm, etc. There is no gauge-fixing condition we can impose for which a gauge theory obeys *all* the axioms of the usual field theories. In covariant gauges we have states of negative norm (longitudinal photons); in the radiation or axial gauges, we do not have manifest Lorentz covariance. If we regard the massless gauge bosons and massless Goldstone bosons as diseases of the theory, each turns out to be the cure of the other. They both disappear from the physical spectrum of the theory by combining to form massive vector particles, without ruining the good high-energy behaviour of the symmetric theory. This remarkable phenomenon was first suggested by Anderson (1958, 1963) who pointed out that several cases in nonrelativistic condensed-matter physics may be interpreted as due to massive photons. Particularly in superconductivity we have the phenomenon of magnetic flux exclusion (the Meissner effect) and this corresponds to a finite-range electromagnetic field, hence a 'massive photon'. The extra longitudinal component is in fact coupled to the collective density fluctuation of the electron system—the plasma oscillation. The proper generalizations to relativistic field theory were carried out by Englert and Brout (1964), and by Guralnik *et al.* (1964), and more completely, by Higgs (1964*a*, 1966). In the literature it is commonly referred to as the *Higgs phenomenon*. 't Hooft (1971*b*) first showed that gauge theories are renormalizable even in the presence of spontaneous symmetry breakdown.

Abelian case

Consider the simple case of Abelian U(1) gauge theory.

$$\mathscr{L} = (\mathrm{D}_\mu\phi)^\dagger(\mathrm{D}^\mu\phi) + \mu^2\phi^\dagger\phi - \lambda(\phi^\dagger\phi)^2 - \tfrac{1}{4}F_{\mu\nu}F^{\mu\nu} \tag{8.78}$$

where

$$\begin{aligned} \mathrm{D}_\mu\phi &= (\partial_\mu - \mathrm{i}gA_\mu)\phi \\ F_{\mu\nu} &= \partial_\mu A_\nu - \partial_\nu A_\mu. \end{aligned} \tag{8.79}$$

The Lagrangian is invariant under the local gauge transformation

$$\begin{aligned} \phi(x) &\to \phi'(x) = \mathrm{e}^{-\mathrm{i}\alpha(x)}\phi(x) \\ A_\mu(x) &\to A'_\mu(x) = A_\mu(x) - \frac{1}{g}\partial_\mu\alpha(x). \end{aligned} \tag{8.80}$$

When $\mu^2 > 0$, the minimum of the potential

$$V(\phi) = -\mu^2\phi^\dagger\phi + \lambda(\phi^\dagger\phi)^2 \tag{8.81}$$

is at

$$|\phi| = v/\sqrt{2} \tag{8.82}$$

with

$$v = (\mu^2/\lambda)^{\frac{1}{2}}. \tag{8.83}$$

This means that the field operator ϕ develops a vacuum expectation value

$$|\langle 0|\phi|0\rangle| = v/\sqrt{2}. \tag{8.84}$$

If we write ϕ in terms of the real fields ϕ_1 and ϕ_2

$$\phi = \frac{1}{\sqrt{2}}(\phi_1 + \mathrm{i}\phi_2), \tag{8.85}$$

we can choose

$$\langle 0|\phi_1|0\rangle = v \quad \text{and} \quad \langle 0|\phi_2|0\rangle = 0. \tag{8.86}$$

Thus, the Lagrangian (and the potential) have U(1) symmetry and the minimization can only fix the modulus of ϕ. To pick one as in (8.86) out of this infinite number of possible minimum values as the physical vacuum breaks the symmetry. (This example is essentially the one given in §5.3 with a change of notation: $\sigma \to \phi_1$ and $\pi \to \phi_2$.) Thus with the shifted fields

$$\phi_1' = \phi_1 - v \quad \text{and} \quad \phi_2' = \phi_2 \tag{8.87}$$

we would conclude that ϕ_2' corresponds to the massless Goldstone boson. The added feature of the case at hand is of course that we have a local gauge symmetry. The ordinary derivative is replaced by the covariant derivative. That term will yield

$$\begin{aligned}|D_\mu\phi|^2 &= |(\partial_\mu - igA_\mu)\phi|^2\\ &= \tfrac{1}{2}(\partial_\mu\phi_1' + gA_\mu\phi_2')^2 + \tfrac{1}{2}(\partial_\mu\phi_2' - gA_\mu\phi_1')^2\\ &\quad - gvA^\mu(\partial_\mu\phi_2' + gA_\mu\phi_1') + \frac{g^2v^2}{2}A^\mu A_\mu.\end{aligned} \tag{8.88}$$

The last term can be interpreted as a mass term for A_μ. Thus the gauge boson acquires a mass $M = gv$.

Unitary gauge (Abelian case). The presence of the term

$$gvA^\mu\,\partial_\mu\phi_2' \tag{8.89}$$

in eqn (8.88) will bring about a mixing between A_μ and ϕ_2' to make this interpretation less clear. To remove this mixing term, we will parametrize the complex field in polar variables and shift only the modulus field

$$\begin{aligned}\phi(x) &= \frac{1}{\sqrt{2}}[v + \eta(x)]\exp(i\xi(x)/v)\\ &= \frac{1}{\sqrt{2}}[v + \eta(x) + i\xi(x) + \ldots].\end{aligned} \tag{8.90}$$

Thus, for small oscillations, $\eta(x)$ and $\xi(x)$ are really $\phi_1'(x)$ and $\phi_2'(x)$, respectively. The free Lagrangian also keeps the same form

$$\mathscr{L}_0 = \frac{1}{2}[(\partial_\mu\eta)^2 - (\partial_\mu\xi)^2] - \frac{\mu^2}{2}(\eta^2 + \xi^2). \tag{8.91}$$

The canonical quantization conditions are not changed; $\eta(x)$ and $\xi(x)$ have the same particle interpretation as ϕ_1 and ϕ_2.

We can now remove the unwanted term (8.89) by transforming $\phi_2'(x)$ or $\xi(x)$ away or, more accurately speaking, by fixing the gauge (the *unitary gauge*). To do this, we define new fields

$$\begin{aligned}\phi^u(x) &= \exp(-i\xi/v)\phi(x) = \frac{1}{\sqrt{2}}(v + \eta(x))\\ B_\mu(x) &= A_\mu(x) - \frac{1}{gv}\partial_\mu\xi(x).\end{aligned} \tag{8.92}$$

From the property of gauge transformation (8.80), we have

$$\begin{aligned}D_\mu\phi &= \exp(-i\xi/v)(\partial_\mu\phi^u - igB_\mu\phi^u)\\ &= \exp(-i\xi/v)(\partial_\mu\eta - igB_\mu(v + \eta))/\sqrt{2}\end{aligned}$$

and

$$|D_\mu\phi|^2 = \tfrac{1}{2}|\partial_\mu\eta - igB_\mu(v+\eta)|^2 \tag{8.93}$$

and also

$$F_{\mu\nu} = \partial_\mu B_\nu - \partial_\nu B_\mu.$$

The Lagrangian of (8.78) may be written

$$\begin{aligned}\mathscr{L} &= \frac{1}{2}|\partial_\mu\eta - igB_\mu(v+\eta)|^2 + \frac{\mu^2}{2}(v+\eta)^2 - \frac{\lambda}{4}(v+\eta)^4 \\ &\quad -\frac{1}{4}(\partial_\mu B_\nu - \partial_\nu B_\mu)^2 \\ &= \mathscr{L}_0 + \mathscr{L}_1\end{aligned}$$

with

$$\begin{aligned}\mathscr{L}_0 &= \frac{1}{2}(\partial_\mu\eta)^2 - \frac{1}{2}\mu^2\eta^2 - \frac{1}{4}(\partial_\mu B_\nu - \partial_\nu B_\mu)^2 + \frac{1}{2}(gv)^2 B_\mu B^\mu \\ \mathscr{L}_1 &= \frac{1}{2}g^2 B_\mu B^\mu \eta(2v+\eta) - \lambda v^2\eta^3 - \frac{1}{4}\lambda\eta^4.\end{aligned} \tag{8.94}$$

It is clear that $\mathscr{L}_0$ is the free Lagrangian density for a massive vector boson with mass $M = gv$ and a scalar meson with mass $m = \sqrt{2}\mu$. The field $\xi(x)$ has disappeared from the Lagrangian. This may be less surprising when we count the degrees of freedom. Before spontaneous symmetry breaking, we had two scalar fields ϕ_1 and ϕ_2 and one massless gauge boson A_μ (with only two polarization states). After the symmetry breaking, we have only one scalar field η and one massive gauge boson B_μ (with three polarization states). Thus the massless gauge field A_μ combines with the scalar field ξ to become a massive vector field B_μ in (8.92). This is the Higgs mechanism for the Abelian case. The $\xi(x)$ field is called a *would-be-Goldstone boson.*

Non-Abelian case

It is straightforward to generalize the Higgs mechanism to theories with non-Abelian gauge symmetry. Consider the case of an SU(2) gauge theory with a complex doublet of scalar fields $\phi = \begin{pmatrix}\phi_1 \\ \phi_2\end{pmatrix}$

$$\mathscr{L} = (D_\mu\phi)^\dagger(D_\mu\phi) - V(\phi) - \tfrac{1}{4}F^a_{\mu\nu}F^{a\mu\nu} \tag{8.95}$$

where

$$\begin{aligned}D_\mu\phi &= \left(\partial_\mu - ig\frac{\boldsymbol{\tau}}{2}\cdot\mathbf{A}_\mu\right)\phi \\ F^a_{\mu\nu} &= \partial_\mu A^a_\nu - \partial_\nu A^a_\mu + g\varepsilon^{abc}A^b_\mu A^c_\nu \\ V(\phi) &= -\mu^2(\phi^\dagger\phi) + \lambda(\phi^\dagger\phi)^2.\end{aligned} \tag{8.96}$$

For $\mu^2 > 0$ the classical potential is a minimum at

$$\langle\phi^\dagger\phi\rangle_0 = v^2/2 \quad \text{with} \quad v = (\mu^2/\lambda)^{\frac{1}{2}}. \tag{8.97}$$

We can choose the physical vacuum corresponding to the expectation value having the form

$$\langle\phi\rangle_0 = \frac{1}{\sqrt{2}}\begin{pmatrix}0\\ v\end{pmatrix}. \tag{8.98}$$

If we define the new field

$$\phi' = \phi - \langle\phi\rangle_0, \tag{8.99}$$

then $\langle\phi'\rangle_0 = 0$. The covariant derivative term will generate a mass for the vector boson field since

$$(\mathrm{D}_\mu\phi)^\dagger(\mathrm{D}_\mu\phi) = \left(\left(\partial_\mu - \mathrm{i}g\frac{\boldsymbol{\tau}}{2}\cdot\mathbf{A}_\mu\right)(\phi' + \langle\phi\rangle_0)\right)^\dagger \times \left(\left(\partial^\mu - \mathrm{i}g\frac{\boldsymbol{\tau}}{2}\cdot\mathbf{A}^\mu\right)(\phi' + \langle\phi\rangle_0)\right) \tag{8.100}$$

contains the factor

$$\frac{1}{4}g^2\langle\phi\rangle_0^\dagger\boldsymbol{\tau}\cdot\mathbf{A}_\mu\boldsymbol{\tau}\cdot\mathbf{A}_\mu\langle\phi\rangle_0 = \frac{1}{2}\left(\frac{gv}{2}\right)^2\mathbf{A}_\mu\mathbf{A}^\mu \tag{8.101}$$

corresponding to $\mathbf{A}_\mu$ having a mass

$$M_A = \frac{gv}{2}. \tag{8.102}$$

In the scalar sector, we have

$$\begin{aligned}\phi^\dagger\phi &= \phi'^\dagger\phi' + \langle\phi^\dagger\rangle_0\phi' + \phi'^\dagger\langle\phi\rangle_0 + \langle\phi^\dagger\rangle_0\langle\phi\rangle_0\\ (\phi^\dagger\phi)^2 &= v^2\phi'^\dagger\phi' + (\langle\phi^\dagger\rangle_0\phi' + \phi'^\dagger\langle\phi\rangle_0)^2 + \ldots.\end{aligned} \tag{8.103}$$

Writing $\phi' = \begin{pmatrix}\phi_1'\\ \phi_2'\end{pmatrix}$, the term quadratic in ϕ' is

$$\frac{\lambda v^2}{2}(\phi_2' + \phi_2'^\dagger)^2 = \frac{\mu^2}{2}(\phi_2' + \phi_2'^\dagger)^2. \tag{8.104}$$

This means that only the combination $(\phi_2' + \phi_2'^\dagger)/\sqrt{2}$ is massive (physical Higgs particle). The other three states ϕ_1', $\phi_1'^\dagger$, and $(\phi_2' - \phi_2'^\dagger)/\sqrt{2}$ are the would-be-Goldstone bosons, which will combine with the original three massless gauge bosons to become three massive vector bosons.

Unitary gauge (non-Abelian case). To see this explicitly we go to the unitary gauge. We parametrize the scalar doublet

$$\phi(x) = \exp\left\{\mathrm{i}\frac{\boldsymbol{\tau}}{v}\cdot\boldsymbol{\xi}(x)\right\}\begin{pmatrix}0\\ \dfrac{v + \eta(x)}{\sqrt{2}}\end{pmatrix} \tag{8.105}$$

where $\langle \xi^a \rangle_0 = \langle \eta \rangle_0 = 0$. We can then define the new fields by

$$\phi^u(x) = U(x)\phi(x) = \frac{1}{\sqrt{2}}\begin{pmatrix} 0 \\ v + \eta \end{pmatrix} \tag{8.106}$$

$$\frac{\boldsymbol{\tau}}{2}\cdot \mathbf{B}_\mu = U(x)\frac{\boldsymbol{\tau}}{2}\cdot \mathbf{A}_\mu U^{-1}(x) - \frac{\mathrm{i}}{g}[\partial_\mu U(x)]U^{-1}(x) \tag{8.107}$$

where

$$U(x) = \exp\left\{-\mathrm{i}\frac{\boldsymbol{\tau}}{v}\cdot \boldsymbol{\xi}(x)\right\}. \tag{8.108}$$

From the properties of the gauge transformation, we obtain

$$\mathrm{D}_\mu \phi = U^{-1}(x)\mathrm{D}_\mu \phi^u \tag{8.109}$$

$$F^a_{\mu\nu}F^{a\mu\nu} = G^a_{\mu\nu}G^{a\mu\nu} \tag{8.110}$$

where

$$\mathrm{D}_\mu \phi^u = \left(\partial_\mu - \mathrm{i}g\frac{\boldsymbol{\tau}}{2}\cdot \mathbf{B}_\mu\right)\phi^u \tag{8.111}$$

$$G^a_{\mu\nu} = (\partial_\mu B^a_\nu - \partial_\nu B^a_\mu + g\varepsilon^{abc}B^b_\mu B^c_\nu). \tag{8.112}$$

Then the Lagrangian density in the unitary gauge is simply

$$\mathscr{L} = (\mathrm{D}_\mu \phi^u)^\dagger(\mathrm{D}^\mu \phi^u) + \frac{\mu^2}{2}(v+\eta)^2 - \frac{\lambda}{4}(v+\eta)^4 - \frac{1}{4}G^a_{\mu\nu}G^{a\mu\nu}. \tag{8.113}$$

The first term contains a term quadratic in $\mathbf{B}_\mu$

$$\frac{g^2}{8}(0, v)(\boldsymbol{\tau}\cdot \mathbf{B}_\mu\, \boldsymbol{\tau}\cdot \mathbf{B}^\mu)\begin{pmatrix} 0 \\ v \end{pmatrix} = \frac{1}{2}\left(\frac{gv}{2}\right)^2 \mathbf{B}_\mu \mathbf{B}^\mu. \tag{8.114}$$

We have vector particles with mass $M_B = gv/2$. Thus the original SU(2) gauge symmetry is completely broken; all three gauge fields acquire mass.

Pattern of symmetry breaking

It is important to keep in mind that the pattern of symmetry breaking is not arbitrary but depends on the structure of the theory in particular the (group) representation content of the scalar field (Kibble 1967; Li 1974). For example, if we have a triplet of real scalar fields $\boldsymbol{\phi}$ instead of the complex doublet, the gauge symmetry SU(2) will be broken down to a residual U(1) gauge symmetry with one massless vector boson remaining. To see that this is the case, start with the scalar potential

$$V(\boldsymbol{\phi}) = -\mu^2\boldsymbol{\phi}^2 + \lambda(\boldsymbol{\phi}^2)^2. \tag{8.115}$$

Again, minimization of $V(\boldsymbol{\phi})$ only determines the magnitude

$$|\langle \boldsymbol{\phi} \rangle_0| = v/\sqrt{2} \quad \text{with} \quad v = (\mu^2/\lambda)^{\frac{1}{2}}. \tag{8.116}$$

We are free to choose the vacuum state so that

$$\langle \boldsymbol{\phi} \rangle_0 = \frac{1}{\sqrt{2}} \begin{pmatrix} 0 \\ 0 \\ v \end{pmatrix}. \tag{8.117}$$

The ground state $\boldsymbol{\phi}$ points in the 3-direction, the symmetry is spontaneously broken. The covariant derivative term

$$[\mathrm{D}_\mu \boldsymbol{\phi}]^2 = [\partial_\mu \boldsymbol{\phi}' - ig(\boldsymbol{\phi}' + \langle \boldsymbol{\phi} \rangle_0) \times \mathbf{A}_\mu]^2 \tag{8.118}$$

will not contain any term quadratic in A_μ^3. Thus the A_μ^3 field continues to describe a massless vector boson. We note that this pattern of symmetry breaking is related to the fact that $\langle \boldsymbol{\phi} \rangle_0$ of (8.117) is still invariant under an O(2), i.e. a U(1), rotation in (1, 2) space.

The number of massive gauge bosons

Since the number of massless gauge bosons corresponds to the number of generators of the (unbroken) gauge symmetry group, the number of gauge bosons that become massive (or the number of would-be-Goldstone bosons) is equal to the difference in the number of generators of the original symmetry and of the final symmetry. We shall present a proof of this statement; this also will provide us with a chance to introduce some general formalism.

Consider the general Lagrangian density

$$\mathscr{L} = \tfrac{1}{2}[(\partial_\mu \phi_i + igT_{ij}^a A_\mu^a \phi_j)][(\partial^\mu \phi_i - igT_{ik}^a A^{a\mu} \phi_k)] \\ - V(\phi_i) - \tfrac{1}{4} F_{\mu\nu}^a F^{a\mu\nu} \tag{8.119}$$

where ϕ_i is a set of real fields, transforming according to some (possibly reducible) representation of the gauge symmetry group G with n generators

$$\phi_i(x) \to \phi_i'(x) = \phi_i(x) + \mathrm{i}\varepsilon^a(\mathrm{x})\mathrm{T}_{ij}^a \phi_j(\mathrm{x}), \quad a = 1, 2, \ldots, n. \tag{8.120}$$

Given that the potential in (8.119) is invariant under an arbitrary group transformation (if the potential is invariant under a larger group, there will be scalars which become massive only through radiative corrections. They are often referred to as the *pseudo-Goldstone bosons* (Weinberg 1972*b*)),

$$0 = \delta V = \frac{\partial V}{\partial \phi_i} \delta\phi_i = \varepsilon^a \frac{\partial V}{\partial \phi_i} T_{ij}^a \phi_j$$

or

$$\frac{\partial V}{\partial \phi_i} T_{ij}^a \phi_j = 0, \qquad a = 1, \ldots, n. \tag{8.121}$$

Differentiation gives

$$\frac{\partial^2 V}{\partial \phi_i \, \partial \phi_k} T_{ij}^a \phi_j + \frac{\partial V}{\partial \phi_i} T_{ik}^a = 0. \tag{8.122}$$

If V is a minimum at $\phi_i = v_i$, then the second term in (8.122) vanishes, and

$$\left.\frac{\partial^2 V}{\partial\phi_i\,\partial\phi_k}\right|_{\phi_i=v_i} T^a_{ij}v_j = 0. \tag{8.123}$$

We should remark that, in the global symmetry case, (8.123) corresponds to the statement that a number of scalars are massless. That this second derivative matrix corresponds to a mass matrix can be seen by an expansion of $V(\phi_i)$ around the minimum

$$V(\phi_i) = V(v_i) + \frac{1}{2}\left.\frac{\partial^2 V}{\partial\phi_i\,\partial\phi_k}\right|_{\phi_i=v_i}(\phi_i - v_i)(\phi_k - v_k) + \ldots. \tag{8.124}$$

Thus the mass matrix is

$$(M^2)_{ik} = \left.\frac{\partial^2 V}{\partial\phi_i\,\partial\phi_k}\right|_{\phi_i=v_i}. \tag{8.125}$$

Now suppose G has a subgroup G' with n' generators that leaves the vacuum invariant

$$T^b_{ij}v_j = 0 \quad \text{for} \quad b = 1, 2, \ldots n' \tag{8.126a}$$

and

$$T^c_{ij}v_j \neq 0 \quad \text{for} \quad c = n' + 1, \ldots, n. \tag{8.126b}$$

If we choose T^a to be linearly independent, eqns (8.126), (8.125), and (8.123) clearly imply that M^2 has $n - n'$ zero eigenvalues and hence $(n - n')$ Goldstone bosons. In the gauge symmetry case, these $(n - n')$ massless states correspond to the $(n - n')$ would-be-Goldstone bosons. In preparation for gauging them away, we re-parametrize ϕ_i by

$$\phi_i = \exp\{\mathrm{i}T^c_{ij}\xi^c(x)/v\}(v_j + \eta_j(x)) \tag{8.127}$$

where $c = 1, 2, \ldots, (n - n')$, i.e. we sum over the broken generators. v in the exponent is the magnitude of v_i. The $\eta_i(x)$s are the remaining scalar fields which are orthogonal to ξ^cs. After a gauge transformation. $A_\mu \to A'_\mu$ with gauge function $\theta(x) = -\mathrm{i}T^c\xi^c(x)/v$. The Lagrangian has a quadratic term in A'^c_μ

$$-\frac{g^2}{2}(T^c v, T^d v)A'^c_\mu A'^{d\mu}. \tag{8.128}$$

After diagonalization this leads to $(n - n')$ massive vector bosons. To summarize, the number of would-be-Goldstone bosons is equal to the difference in the number of generators of the original and the final gauge symmetries.

In this section we have chosen to fix the gauge so that the particle content of the theory is obvious (the unitary gauge). In the next chapter, at the end of §9.2, we shall also discuss another class of gauge choices (the renormalizable gauge or the R gauge) where the would-be-Goldstone bosons are not eliminated explicitly but the gauge-field propagators manifestly have good high-energy behaviour and the renormalizability of the theory is more transparent.

9 Quantum gauge theories

WE now proceed to quantize the gauge theories, explain their perturbative solutions, and discuss the generalized Ward identities of such theories.

9.1 Path-integral quantization of gauge theories

Gauge theories, being gauge invariant, represent systems with constrained dynamical variables, i.e., there are variables that do not represent true dynamical degrees of freedom. The quantization procedure of such theories is more involved than that for the scalar field theory discussed in Chapter 1. For gauge theories the path-integral formalism provides the most direct quantization procedure.

Difficulties of gauge theory quantization

We are already familiar with the problem of quantizing the electromagnetic field $A_\mu(x)$. In the canonical formalism one identifies the canonical variables $A_\mu(x)$ and their conjugate momenta $\pi_\mu(x) = \delta\mathscr{L}(x)/\delta(\partial_0 A^\mu(x))$ as operators and postulates their commutation relations. One immediately discovers that $\pi_0(x)$ and $\nabla\cdot\boldsymbol{\pi}(x)$ vanish, which implies that $A_0(x)$ and $\nabla\cdot\mathbf{A}(x)$ commute with all canonical operators. They are really c-numbers. The four-vector field $A_\mu(x)$ actually represents only two independent dynamical degrees of freedom. The canonical commutation relations for these transverse fields $\mathbf{A}_\perp(x)$ and $\boldsymbol{\pi}_\perp(x)$ have to be formulated so that they are compatible with the above-mentioned constraints. For example, we can take the constraint in the form of $\nabla\cdot\mathbf{A}(x) = 0$ (*radiation gauge*) or $A_3(x) = 0$ (*axial gauge*). In such formulations one sacrifices manifest Lorentz covariance. Alternatively, one maintains explicit Lorentz covariance and introduces spurious degrees of freedom into the theory. This brings about a Hilbert space with indefinite metric (the *Gupta–Bleuler formulation*). A physically sensible theory is recovered only after we restrict the admissible states to those satisfying (the *Lorentz gauge*) $\partial^\mu A_\mu|\Psi\rangle = 0$. The key point in all these formulations is that one must remove the redundant degrees of freedom (resulting from gauge invariance) of the theory by some acceptable gauge-fixing conditions. In the language of path-integral quantization formalism, one must restrict the functional integration to reflect these gauge-fixing conditions. A consistent implementation of such constraints for non-Abelian theories is a highly nontrivial matter, and the problem was finally solved through the work of Feynman (1963), DeWitt (1967), Faddeev and Popov (1967), and many others (Mandelstam 1968; Popov and Faddeev 1967; Veltman 1970; 't Hooft 1971*a*, *b*).

We shall restate the difficulty of quantization directly in terms of the path-integral formulation of gauge theories. To be specific, we consider the case of SU(2) Yang–Mills fields

$$\mathscr{L} = -\tfrac{1}{4}F^a_{\mu\nu}F^{a\mu\nu}, \quad a = 1, 2, 3 \tag{9.1}$$

with

$$F^a_{\mu\nu} = \partial_\mu A^a_\nu - \partial_\nu A^a_\mu + g\varepsilon^{abc}A^b_\mu A^c_\nu. \tag{9.2}$$

If we write the generating functional as

$$W[J] = \int [\mathrm{d}\mathbf{A}_\mu] \exp\left\{\mathrm{i} \int \mathrm{d}^4x[\mathscr{L}(x) + \mathbf{J}_\mu(x)\cdot \mathbf{A}^\mu(x)]\right\}. \tag{9.3}$$

the free-field part is then

$$W_0[J] = \int [\mathrm{d}\mathbf{A}_\mu] \exp\left\{\mathrm{i} \int \mathrm{d}^4x[\mathscr{L}_0(x) + \mathbf{J}_\mu(x)\cdot \mathbf{A}^\mu(x)]\right\} \tag{9.4}$$

with

$$\begin{aligned}\int \mathrm{d}^4x \mathscr{L}_0(x) &= -\frac{1}{4}\int \mathrm{d}^4x(\partial_\mu A^a_\nu - \partial_\nu A^a_\mu)(\partial^\mu A^{a\nu} - \partial^\nu A^{a\mu}) \\ &= \frac{1}{2}\int \mathrm{d}^4x A^a_\mu(x)(g^{\mu\nu}\,\partial^2 - \partial^\mu\,\partial^\nu)A^a_\nu(x).\end{aligned} \tag{9.5}$$

Now we have a situation very similar to the scalar field theory and we would like to proceed and perform the Gaussian integration as in eqns (1.79) and (1.81)

$$\int [\mathrm{d}\phi] \exp[-\tfrac{1}{2}\langle\phi K\phi\rangle + \langle J\phi\rangle] \sim \frac{1}{\sqrt{\det K}} \exp\langle JK^{-1}J\rangle. \tag{9.6}$$

However this is not possible, because the operator

$$K_{\mu\nu} \equiv g_{\mu\nu}\,\partial^2 - \partial_\mu\,\partial_\nu \tag{9.7}$$

in (9.5) does not have an inverse, as we shall demonstrate below.

Assuming $G^{\nu\lambda}(x-y)$ is the inverse of $K_{\mu\nu}$,

$$(g_{\mu\nu}\,\partial^2 - \partial_\mu\,\partial_\nu)G^{\nu\lambda}(x-y) = g^\lambda_\mu\,\delta^4(x-y). \tag{9.8}$$

Using the Fourier transform

$$G^{\nu\lambda}(x) = \int \frac{\mathrm{d}^4k}{(2\pi)^4}\,\mathrm{e}^{-ik\cdot x}G^{\nu\lambda}(k), \tag{9.9}$$

we have

$$(-k^2 g_{\mu\nu} + k_\mu k_\nu)G^{\nu\lambda}(k) = g^\lambda_\mu. \tag{9.10}$$

With the invariant decomposition

$$G^{\nu\lambda}(k) = a(k^2)g^{\nu\lambda} + b(k^2)k^\nu k^\lambda, \tag{9.11}$$

it is clear that the left-hand side of eqn (9.10) $= -a(k^2)(k^2 g_\mu^\lambda - k_\mu k^\lambda)$ cannot be equal to the right-hand side. Thus $K_{\mu\nu}$ does not have an inverse. Another way to see this is that the operator $K_{\mu\nu}$ given in (9.7) satisfies the relation $K_{\mu\nu}(x)K^\nu_\lambda(x) \propto K_{\mu\lambda}(x)$. This means that it is a projection operator (which projects out the transverse degree of freedom of the gauge field) and clearly does not have an inverse. Equivalently, for the case det K vanishes and eqn (9.6) is not applicable. This singular nature of the path integral, i.e. this extra infinity, is related to the gauge invariance of the theory. In eqns (9.3) and (9.4) we have summed over all the field configurations, including 'orbits' that are related by gauge transformations. This overcounting is at the root of the divergent functional integral. We need to seek a prescription to divide out this infinite (functional) volume of the orbit. To quantize a gauge theory, it is necessary to fix the gauge.

Isolating the path-integral volume factor

(1) *A two-dimensional case as illustrative example.* Before launching into the actual calculation that will isolate this volume factor from the functional integration (hence in infinite-dimensional space) we shall use a two-dimensional integral to illustrate our strategy

$$W = \int \mathrm{d}x\, \mathrm{d}y\, \mathrm{e}^{\mathrm{i}S(x,y)}$$

$$= \int \mathrm{d}\mathbf{r}\, \mathrm{e}^{\mathrm{i}S(\mathbf{r})} \tag{9.12}$$

where $\mathbf{r} = (r, \theta)$ is the label in the polar coordinate system. $S(\mathbf{r})$ is supposed to be invariant under a rotation in two-dimensional space

$$S(\mathbf{r}) = S(\mathbf{r}_\phi) \tag{9.13}$$

for

$$\mathbf{r} = (r, \theta) \to \mathbf{r}_\phi = (r, \theta + \phi). \tag{9.14}$$

Thus $S(\mathbf{r})$ is a constant over the (circular) orbit. In this simple case if we only wish to sum over the contribution from the inequivalent $S(\mathbf{r})$s we can simply divide out the 'volume factor' corresponding to the polar angle integration, $\int \mathrm{d}\theta = 2\pi$. To do this we adopt the following procedure which can be generalized to more complicated situations. First we insert

$$1 = \int \mathrm{d}\phi\, \delta(\theta - \phi) \tag{9.15}$$

into the original expression for W

$$W = \int \mathrm{d}\phi \int \mathrm{d}\mathbf{r}\, \mathrm{e}^{\mathrm{i}S(\mathbf{r})}\, \delta(\theta - \phi) = \int \mathrm{d}\phi\, W_\phi \tag{9.16}$$

where $W_\phi = \int \mathrm{d}\mathbf{r}\, \delta(\theta - \phi)\, \mathrm{e}^{\mathrm{i}S(\mathbf{r})}$ is evaluated for a given angle ϕ. Thus, we first calculate W along a fixed angle $\theta = \phi$, then integrate over the contributions

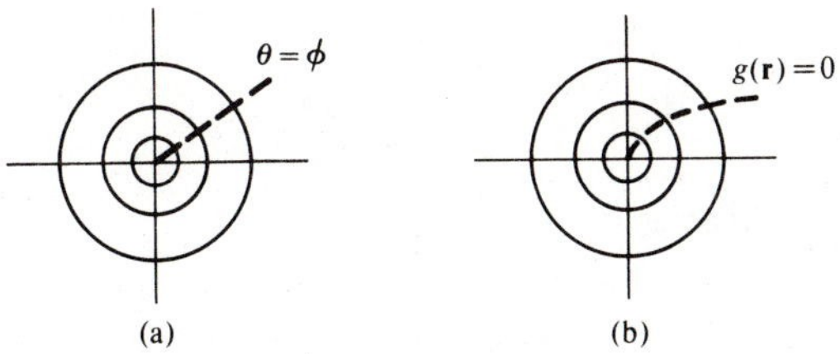

FIG. 9.1.

for all values of ϕ (see Fig. 9.1(a)). Using the invariance property of S in (9.13) we have

$$W_\phi = W_{\phi'}. \tag{9.17}$$

Thus the volume of the orbit can be factored out

$$\begin{aligned} W &= \int \mathrm{d}\phi W_\phi = W_\phi \int \mathrm{d}\phi \\ &= 2\pi W_\phi. \end{aligned} \tag{9.18}$$

Generally a constraint that is more complicated than $\theta = \phi$ may be chosen, and we represent this by

$$g(\mathbf{r}) = 0 \tag{9.19}$$

which intersects each of the orbits once as shown in Fig. 9.1(b), i.e. the equation $g(\mathbf{r}_\phi) = 0$ must have a unique solution ϕ for a given value of r. For this general constraint (9.19), instead of the simple eqn (9.15), we need (to define) a function $\Delta_g(\mathbf{r})$ such that

$$[\Delta_g(\mathbf{r})]^{-1} = \int \mathrm{d}\phi\, \delta[g(\mathbf{r}_\phi)]. \tag{9.20}$$

Hence

$$\Delta_g(\mathbf{r}) = \left.\frac{\partial g(\mathbf{r})}{\partial \theta}\right|_{g=0}, \tag{9.21}$$

and $\Delta_g(\mathbf{r})$ is itself invariant under the two-dimensional rotation (9.14) since

$$\begin{aligned} [\Delta_g(\mathbf{r}_{\phi'})]^{-1} &= \int \mathrm{d}\phi\, \delta[g(\mathbf{r}_{\phi+\phi'})] \\ &= \int \mathrm{d}\phi''\, \delta[g(\mathbf{r}_{\phi''})] \\ &= [\Delta_g(\mathbf{r})]^{-1}. \end{aligned} \tag{9.22}$$

Repeating steps (9.16)–(9.18) the volume factor in W can then be isolated

$$W = \int \mathrm{d}\phi W_\phi \tag{9.23}$$

with

$$W_\phi = \int \mathrm{d}\mathbf{r}\, \mathrm{e}^{\mathrm{i}S(\mathbf{r})}\, \Delta_g(\mathbf{r})\, \delta[g(\mathbf{r}_\phi)]. \tag{9.24}$$

W_ϕ is rotationally invariant

$$\begin{aligned} W_{\phi'} &= \int \mathrm{d}\mathbf{r}\, \mathrm{e}^{\mathrm{i}S(\mathbf{r})}\, \Delta_g(\mathbf{r})\, \delta[g(\mathbf{r}_{\phi'})] \\ &= \int \mathrm{d}\mathbf{r}'\, \mathrm{e}^{\mathrm{i}S(\mathbf{r}')}\, \Delta_g(\mathbf{r}')\, \delta[g(\mathbf{r}'_\phi)] \\ &= W_\phi. \end{aligned} \tag{9.25}$$

where we have introduced the variable $\mathbf{r}' = (r, \phi')$ and have used the fact that $S(\mathbf{r})$, $\Delta_g(\mathbf{r})$, and the integration measure $\mathrm{d}\mathbf{r}$, are invariant under rotations. Thus to remove the 'volume factor', we can insert a constraining δ-function and multiply it by a function Δ_g defined by (9.20).

(2) *The PI volume factor in gauge theories.* We now return to the task of isolating the actual volume factor in the functional integration of the generating functional in gauge theory. The procedure will be exactly the same as for the simple case we have just discussed. The action is invariant under the gauge transformation

$$\mathbf{A}_\mu \to \mathbf{A}_\mu^\theta$$

where

$$\mathbf{A}_\mu^\theta \cdot \boldsymbol{\tau}/2 = U(\theta)\left[\mathbf{A}_\mu \cdot \boldsymbol{\tau}/2 + \frac{1}{\mathrm{i}g} U^{-1}(\theta)\, \partial_\mu U(\theta)\right] U^{-1}(\theta)$$

with

$$U(\theta) = \exp[-\mathrm{i}\boldsymbol{\theta}(x) \cdot \boldsymbol{\tau}/2]. \tag{9.26}$$

$\boldsymbol{\theta}$s are the space–time dependent parameters of the group. The $\boldsymbol{\tau}$s are the Pauli matrices. Thus the action is constant on the orbit of the gauge group formed out of all the $\mathbf{A}_\mu^\theta$s for some fixed $\mathbf{A}_\mu$ with $U(\theta)$ ranging over all elements of the group SU(2). A proper quantization procedure must restrict the path integration to a 'hypersurface' which intersects each orbit only once. Thus, if we write the equation for the hypersurface as

$$f_a(\mathbf{A}_\mu) = 0 \qquad a = 1, 2, 3, \tag{9.27}$$

then the equation

$$f_a(\mathbf{A}_\mu^\theta) = 0 \tag{9.28}$$

must have an unique solution $\boldsymbol{\theta}$ for a given $\mathbf{A}_\mu$. Eqn (9.27) is clearly a gauge-fixing condition.

We also need to define the integration over the group space. Let θ and θ' be elements of an SU(2) group. In terms of the representation matrices $U(\theta)$ the multiplication of group elements takes on the form

$$U(\theta)U(\theta') = U(\theta\theta'). \tag{9.29}$$

In the neighbourhood of identity, we can write

$$U(\theta) = 1 + \mathrm{i}\boldsymbol{\theta}\cdot\boldsymbol{\tau}/2 + \mathrm{O}(\boldsymbol{\theta}^2). \tag{9.30}$$

The integration measure over group space can be chosen as

$$[\mathrm{d}\theta] = \prod_{a=1}^{3} \mathrm{d}\theta_a \tag{9.31}$$

which is invariant in the sense that

$$\mathrm{d}(\theta\theta') = \mathrm{d}\theta'.$$

We can now isolate the desired volume factor by defining a function $\Delta_f[\mathbf{A}_\mu]$

$$\Delta_f^{-1}[\mathbf{A}_\mu] = \int [\mathrm{d}\theta(x)]\, \delta[f_a(\mathbf{A}_\mu^\theta)]. \tag{9.32}$$

Thus

$$\Delta_f[\mathbf{A}_\mu] = \det M_f \tag{9.33}$$

where

$$(M_f)_{ab} = \frac{\delta f_a}{\delta\theta_b}. \tag{9.34}$$

Thus, M_f is just the response of $f_a[\mathbf{A}_\mu]$ to the infinitesimal gauge transformation. More precisely, from (9.30), the infinitesimal gauge transformation is of the form

$$A_\mu^{\theta a} = A_\mu^a + \varepsilon^{abc}\theta^b A_\mu^c - \frac{1}{g}\,\partial_\mu\theta^a \tag{9.35}$$

and the response of $f_a[\mathbf{A}_\mu]$ is

$$f_a[\mathbf{A}_\mu^\theta(x)] = f_a[\mathbf{A}_\mu(x)] + \int \mathrm{d}^4y[M_f(x, y)]_{ab}\theta_b(y) + O(\theta^2). \tag{9.36}$$

Because of the requirement that eqn (9.28) have a unique solution, $(\det M_f)$ does not vanish.

$\Delta_f[\mathbf{A}_\mu]$ has the important property that it is gauge invariant. To see this, we write (9.32) as

$$\Delta_f^{-1}[\mathbf{A}_\mu] = \int [\mathrm{d}\theta'(x)]\, \delta[f_a(\mathbf{A}_\mu^{\theta'})]; \tag{9.37}$$

then

$$\begin{aligned}
\Delta_f^{-1}[\mathbf{A}_\mu^\theta] &= \int [\mathrm{d}\theta'(x)]\, \delta[f_a(\mathbf{A}_\mu^{\theta\theta'}(x))] \\
&= \int [\mathrm{d}(\theta(x)\theta'(x))]\, \delta[f_a(\mathbf{A}_\mu^{\theta\theta'}(x))] \\
&= \int [\mathrm{d}\theta''(x)]\, \delta[f_a(\mathbf{A}_\mu^{\theta''}(x))] \\
&= \Delta_f^{-1}[\mathbf{A}_\mu].
\end{aligned} \tag{9.38}$$

We now substitute (9.32) into the path-integral representation of the vacuum-to-vacuum amplitude

$$\int [\mathrm{d}\mathbf{A}_\mu] \exp\left\{\mathrm{i}\int \mathrm{d}^4x \mathscr{L}(x)\right\} = \int [\mathrm{d}\boldsymbol{\theta}(x)][\mathrm{d}\mathbf{A}_\mu(x)]\, \Delta_f[\mathbf{A}_\mu] \times \delta[f_a(\mathbf{A}_\mu^\theta)] \exp\left\{\mathrm{i}\int \mathrm{d}^4x \mathscr{L}(x)\right\}$$
$$= \int [\mathrm{d}\boldsymbol{\theta}(x)][d\mathbf{A}_\mu(x)]\, \Delta_f[\mathbf{A}_\mu] \times \delta[f_a(\mathbf{A}_\mu)] \exp\left\{\mathrm{i}\int \mathrm{d}^4x \mathscr{L}(x)\right\}. \tag{9.39}$$

To arrive at the last line we used the fact that both $\Delta_f[\mathbf{A}_\mu]$ and $\exp\{\mathrm{i}\int \mathrm{d}^4x\, \mathscr{L}(x)\}$ are invariant under a gauge transformation $\mathbf{A}_\mu^\theta \to \mathbf{A}_\mu$. Now, the integrand is independent of $\theta(x)$ and the integration over $\Pi_x\, \mathrm{d}\theta(x)$ is the infinite orbit volume we have been seeking to identify. This suggests that the prescription for the generating functional of the gauge field $\mathbf{A}_\mu$ (after applying eqn (9.33) and eqn (9.39)), should be

$$W_f[\mathbf{J}] = \int [\mathrm{d}\mathbf{A}_\mu](\det M_f)\, \delta[f_a(\mathbf{A}_\mu)] \exp\left\{\mathrm{i}\int \mathrm{d}^4x[\mathscr{L}(x) + \mathbf{J}_\mu \cdot \mathbf{A}^\mu]\right\}. \tag{9.40}$$

This is the *Faddeev–Popov ansatz* (1967). In other words, we can get rid of the unwanted redundancy in the quantization procedure by restricting the functional measure with $\det|\delta f/\delta\theta|\, \delta[f(A_\mu)]$.

Consistency check of the FP ansatz in axial gauge

Before proceeding further with the formalism we shall make an elementary check of the FP ansatz with a specific example. Consider the following choice of the gauge-fixing condition (9.27), the *axial gauge* (Arnowitt and Fickler 1962)

$$f_a = A_3^a = 0. \tag{9.41}$$

Under the gauge transformation (9.35) we have for (9.36)

$$f_a(\mathbf{A}_\mu^\theta) = A_3^a + \varepsilon^{abc}\theta^b A_3^c - \frac{1}{g}\partial_3\theta^a$$
$$= -\frac{1}{g}\partial_3\theta^a \tag{9.42}$$

because of (9.41). Thus we have the response matrix $M_f = (-1/g)\,\partial_3\,\delta_{ab}$, which is independent of the gauge field. For this choice of the (axial) gauge we can therefore ignore the (det M_f) factor in $W_f[\mathbf{J}]$

$$W_f[\mathbf{J}] = \int [\mathrm{d}\mathbf{A}_\mu]\,\delta(\mathbf{A}_3)\exp\{\mathrm{i}S[\mathbf{J}]\} \tag{9.43}$$

$$S[\mathbf{J}] = \int \mathrm{d}^4x[-\tfrac{1}{4}(F^a_{\mu\nu})^2 + J^a_\mu A^{a\mu}]$$

It is more convenient to work with an alternative form of the generating functional

$$W'_f[\mathbf{J}] = \int [\mathrm{d}\mathbf{F}_{\mu\nu}][\mathrm{d}\mathbf{A}_\mu]\,\delta(\mathbf{A}_3)\exp\{\mathrm{i}S'[\mathbf{J}]\} \tag{9.44}$$

with

$$S'[\mathbf{J}] = \int \mathrm{d}^4x[-\tfrac{1}{2}(F^a_{\mu\nu})^2 + \tfrac{1}{4}F^{\mu\nu a}(\partial_\mu A^a_\nu - \partial_\nu A^a_\mu + g\varepsilon^{abc}A^b_\mu A^c_\nu) + J^a_\mu A^{\mu a}]. \tag{9.45}$$

If we integrate over $F^a_{\mu\nu}$, $W'_f[\mathbf{J}]$ reduces to $W_f[\mathbf{J}]$ of (9.43).

Let us check the compatibility of the FP formulation eqns ((9.43) and (9.44)) with the canonical quantization, to see whether it does restrict the functional integration to the same dynamical variables as deduced with the canonical procedure.

We first identify the independent canonical variables in the axial gauge $A^a_3 = 0$. The Lagrangian in (9.45) becomes

$$\begin{aligned}\mathscr{L}' = &-\tfrac{1}{2}(F^a_{\mu\nu})^2 + \tfrac{1}{2}F^{ija}(\partial_i A^a_j - \partial_j A^a_i + g\varepsilon^{abc}A^b_i A^c_j)\\ &+ F^{0ia}(\partial_0 A^a_i - \partial_i A^a_0 + g\varepsilon^{abc}A^b_0 A^c_i)\\ &+ F^{i3a}(-\partial_3 A^a_i) + F^{03a}(-\partial_3 A^a_0)\end{aligned} \tag{9.46}$$

where $i, j = 1,2$.

The Euler–Lagrange equations

$$\partial^\lambda \frac{\delta\mathscr{L}'}{\delta(\partial^\lambda F^a_{\mu\nu})} = \frac{\delta\mathscr{L}'}{\delta F^a_{\mu\nu}} \tag{9.47}$$

and

$$\partial^\lambda \frac{\delta\mathscr{L}'}{\delta(\partial^\lambda A^a_\mu)} = \frac{\delta\mathscr{L}'}{\delta A^a_\mu} \tag{9.48}$$

give rise to the following constraint equations (having no time derivatives)

$$\begin{aligned}F^a_{ij} &= \partial_i A^a_j - \partial_j A^a_i + g\varepsilon^{abc}A^b_i A^c_j\\ F^a_{i3} &= -\partial_3 A^a_i\\ F^a_{03} &= -\partial_3 A^a_0\\ \partial^i F^a_{0i} - \partial^3 F^a_{03} &= -g\varepsilon^{abc}F^b_{0i}A^{ic}\end{aligned} \tag{9.49}$$

and to the following dynamical equations

$$F^a_{0i} = \partial_0 A^a_i - \partial_i A^a_0 + g\varepsilon^{abc} A^b_0 A^c_i \tag{9.50a}$$

$$\partial^\mu F^a_{\mu i} = -g\varepsilon^{abc}(F^b_{ij} A^{jc} + F^b_{0i} A^{0c}). \tag{9.50b}$$

Thus A^a_0, F^a_{ij}, F^a_{i3}, and F^a_{03} are constraint variables; they can be eliminated from S' (resulting in S'') by the constraint equations (9.49) in terms of the remaining variables A^a_i and F^a_{0i}. This identification of independent canonical variables leads us to construct the generating functional

$$W''_f[\mathbf{J}] = \int [\mathrm{d}\mathbf{F}_{01}][\mathrm{d}\mathbf{F}_{02}][\mathrm{d}\mathbf{A}_1][\mathrm{d}\mathbf{A}_2] \exp\{\mathrm{i}S''[\mathbf{J}]\}. \tag{9.51}$$

Our consistency check of the FP ansatz now consists in showing the equivalence of this functional integral to that in (9.44) and thus (9.43). We need to show that, if a dynamical variable appears at most quadratically (with a constant coefficient) in the action, then integrating over the variable is the same as eliminating it from the action by the Euler–Lagrange equation. This is indeed the case and we can illustrate this theorem as follows. Consider the functional (Gaussian) integral

$$\begin{aligned}\int [\mathrm{d}\phi] \exp\{\mathrm{i}S[\phi]\} &= \int [\mathrm{d}\phi] \exp\left\{\mathrm{i}\int \mathrm{d}^4x[\tfrac{1}{2}a\phi^2(x) + f(x)\phi(x)]\right\}\\ &= \exp\left\{-\frac{\mathrm{i}}{2a}\int \mathrm{d}^4x[f(x)]^2\right\}.\end{aligned} \tag{9.52}$$

On the other hand, the Euler–Lagrange equation from S in (9.52) yields

$$a\phi(x) + f(x) = 0. \tag{9.53}$$

Thus eliminating $\phi(x)$ in S, we have

$$S = -\frac{1}{2a}\int \mathrm{d}^4x[f(x)]^2 \tag{9.54}$$

which is the same as eqn (9.52). This also completes our demonstration that the FP ansatz indeed provides the correct restriction (i.e. the same as canonical procedure) on the integration measure. As we illustrated in the introduction to PI formalism in Chapter 1, the Hamiltonian PI formalism where

$$W[J] \sim \int [\mathrm{d}\phi\, \mathrm{d}\pi] \exp\left\{\mathrm{i}\int \mathrm{d}^4x[\pi\partial_0\phi - \mathscr{H}(\pi, \phi) + J\phi]\right\} \tag{9.55}$$

is equivalent to the Lagrangian PI where

$$W[J] \sim \int [\mathrm{d}\phi] \exp\left\{\mathrm{i}\int \mathrm{d}^4x[\mathscr{L}(\phi, \partial_\mu\phi) + J\phi]\right\}. \tag{9.56}$$

It is not difficult to check that (9.51) is the Hamiltonian formulation with F^a_{0i} being the transverse canonical momenta.

Because we can drop the FP determinant with this choice of axial gauge, the quantization is particularly simple. But in this gauge we lose manifest Lorentz invariance and Feynman rules are complicated.

Abelian gauge theory

We should remark that all the formalism developed in this section also encompasses the simpler case of Abelian gauge theory. Under a U(1) gauge transformation, eqn (9.35) reads

$$A_\mu^\theta(x) = A_\mu(x) - \frac{1}{g}\,\partial_\mu\theta(x). \tag{9.57}$$

It is then clear that for any choice of linear gauge-fixing condition of (9.27) the response matrix M_f in (9.34) or (9.36), like the special case of non-Abelian theory in the axial gauge just considered, will be independent of $A_\mu(x)$. The FP factor ($\det M_f$) plays no physical role and can be dropped from the generating functional,

$$W_f[J] = \int [\mathrm{d}A_\mu]\,\delta[f(A_\mu)]\exp\left\{\mathrm{i}\int \mathrm{d}^4x[\mathscr{L}(x) + J_\mu(x)A^\mu(x)]\right\}. \tag{9.58}$$

9.2 Feynman rules in covariant gauges

For practical calculations it is more convenient to use the covariant gauges where unlike the axial gauge unphysical 'ghost fields' are needed. We start with the generating functional (9.40) in the form

$$W[\mathbf{J}] = \int [\mathrm{d}\mathbf{A}_\mu]\exp\left\{\mathrm{i}S_{\text{eff}} + \mathrm{i}\int \mathrm{d}^4x\mathbf{J}^\mu\cdot\mathbf{A}_\mu\right\}. \tag{9.59}$$

Thus, the FP modification of the integration measure $\det M_f\,\delta[f_a(A_\mu)]$ can be exponentiated and expressed as additional terms in the action, leading to a new S_{eff}. In this language, the problem of gauge field quantization is solved because these new factors lead to a new K operator for the prototype Gaussian integrand in (9.6), which will have a nonvanishing determinant and possess an inverse.

Faddeev–Popov ghosts

It is straightforward to write $\det M_f$ in an exponential form,

$$\det M_f = \exp\{\mathrm{tr}(\ln M_f)\}. \tag{9.60}$$

If we further write

$$M_f \equiv 1 + L, \tag{9.61}$$

then

$$\exp\{\mathrm{tr}(\ln M_f)\} = \exp\left\{\mathrm{tr}\, L + \frac{1}{2}\mathrm{tr}\, L^2 + \ldots \frac{1}{n}\mathrm{tr}\, L^n + \ldots\right\}$$
$$= \exp\left\{\int d^4x L_{aa}(x, x) + \frac{1}{2}\int d^4x\, d^4y L_{ab}(x, y)L_{ba}(y, x) + \ldots\right\} \tag{9.62}$$

which is represented diagramatically in Fig. 9.2.

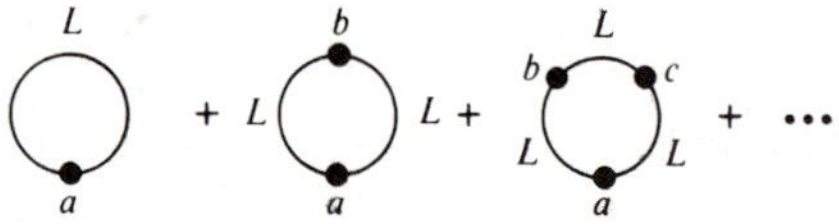

FIG. 9.2. Diagrammatical representation of the Faddeev–Popov determinant.

This series may be viewed as arising from loops generated by a fictitious isotriplet of the complex scalar fields $\mathbf{c}(x)$. Their presence and interactions can be described by the generating functional

$$\det M_f \sim \int [d\mathbf{c}][d\mathbf{c}^\dagger] \exp\left\{i \int d^4x\, d^4y \sum_{a,b} c_a^\dagger(x)[M_f(x, y)]_{ab} c_b(y)\right\}. \tag{9.63}$$

Because the Gaussian integral is proportional to det M_f as in the case for the Grassmann number, rather than $(\det M_f)^{-1}$, we see that the scalar fields $\mathbf{c}(x)$ must obey Fermi statistics (recall the discussion in §1.3, especially eqns (1.139)–(1.141)). They are referred to as *Faddeev–Popov ghost fields*.

Gauge-fixing terms

We next attempt to convert the delta function $\delta[f_a(\mathbf{A}_\mu)]$ into an exponential factor. This can be accomplished by first generalizing the gauge-fixing condition $f_a(\mathbf{A}_\mu) = 0$ to

$$f_a[\mathbf{A}_\mu] = B_a(x) \tag{9.64}$$

where $B_a(x)$ is an arbitrary function of space and time, independent of the gauge field. The definition (9.32) of Δ_f is correspondingly generalized

$$\int [d\theta(x)]\, \Delta_f[\mathbf{A}_\mu]\, \delta[f_a(\mathbf{A}_\mu^\theta) - B_a(x)] = 1. \tag{9.65}$$

Clearly this definition yields the same Δ_f as in (9.32). And we can extract the infinite-orbit volume factor as before and prescribe a generating functional as

$$W[\mathbf{J}] = \int [d\mathbf{A}_\mu][d\mathbf{B}](\det M_f)\, \delta[f_a(\mathbf{A}_\mu) - B_a]$$
$$\times \exp\left\{i \int d^4x \left[\mathscr{L}(x) - \mathbf{J}^\mu \cdot \mathbf{A}_\mu - \frac{1}{2\xi}\mathbf{B}^2(x)\right]\right\} \tag{9.66}$$

where we have inserted a constant

$$\text{constant} \sim \int [\mathrm{d}\mathbf{B}] \exp\left\{-\frac{\mathrm{i}}{2\xi}\int \mathrm{d}^4x\mathbf{B}^2(x)\right\}$$

where ξ is some arbitrary constant coefficient, the *gauge parameter*. The generating functional of (9.66) differs from that of (9.40) by an immaterial normalization factor. We can then use the delta functional to perform the integration over $[\mathrm{d}\mathbf{B}(x)]$. Also, substituting in (9.63), we have

$$W[\mathbf{J}] = \int [\mathrm{d}\mathbf{A}_\mu][\mathrm{d}\mathbf{c}][\mathrm{d}\mathbf{c}^\dagger] \exp\{\mathrm{i}S_{\mathrm{eff}}[\mathbf{J}]\}$$

with

$$S_{\mathrm{eff}}[\mathbf{J}] = S[\mathbf{J}] + S_{\mathrm{gf}} + S_{\mathrm{FPG}} \tag{9.67}$$

where the additional terms are the gauge-fixing term

$$S_{\mathrm{gf}} = -\frac{1}{2\xi}\int \mathrm{d}^4x\{f_a[\mathbf{A}_\mu(x)]\}^2 \tag{9.68}$$

and the FP ghost term

$$S_{\mathrm{FPG}} = \int \mathrm{d}^4x\,\mathrm{d}^4y \sum_{a,b} c_a^\dagger(x)[M_f(x,y)]_{ab}c_b(y). \tag{9.69}$$

Covariant gauges in symmetric gauge theories

Here we shall make a specific choice for the condition in (9.27), i.e. (9.64) (the *covariant*, or *Lorentz*, *gauges*)

$$f_a(\mathbf{A}_\mu) = \partial^\mu A_\mu^a = 0 \qquad a = 1, 2, 3. \tag{9.70}$$

Under the infinitesimal gauge transformation

$$U(\theta(x)) = 1 + \mathrm{i}\boldsymbol{\theta}(x)\cdot\boldsymbol{\tau}/2 + O(\theta^2) \tag{9.71}$$

$$A_\mu^{a\theta}(x) = A_\mu^a(x) + \varepsilon^{abc}\theta^b(x)A_\mu(x) - \frac{1}{g}\,\partial_\mu\theta^a(x), \tag{9.72}$$

we have

$$\begin{aligned} f^a(\mathbf{A}_\mu^\theta) &= f^a(\mathbf{A}_\mu) + \partial^\mu\left[\varepsilon^{abc}\theta^b(x)A_\mu^c(x) - \frac{1}{g}\,\partial_\mu\theta^a(x)\right] \\ &= f^a(\mathbf{A}_\mu) + \int \mathrm{d}^4y[M_f(x,y)]_{ab}\theta^b(y) \end{aligned} \tag{9.73}$$

with

$$[M_f(x,y)]_{ab} = -\frac{1}{g}\,\partial^\mu[\delta^{ab}\,\partial_\mu - g\varepsilon^{abc}A_\mu^c]\,\delta^4(x-y). \tag{9.74}$$

From (9.70) and (9.74) we can calculate the extra terms (9.68) and (9.69) in the effective action S_{eff},

$$S_{\text{gf}} = -\frac{1}{2\xi}\int \mathrm{d}^4x(\partial^\mu \mathbf{A}_\mu)^2 \tag{9.75}$$

$$S_{\text{FPG}} = \frac{1}{g}\int \mathrm{d}^4x \sum_{a,b} c_a^\dagger(x)\,\partial^\mu[\delta_{ab}\,\partial_\mu - g\varepsilon_{abc}A_\mu^c]c_b(x). \tag{9.76}$$

Introducing the source functions $\eta_a^\dagger$, η_a for the ghost fields c_a and $c_a^\dagger$, we can write the generating functional

$$\begin{aligned} W_f[\mathbf{J}, \boldsymbol{\eta}, \boldsymbol{\eta}^\dagger] = \int [\mathrm{d}\mathbf{A}_\mu\,\mathrm{d}\mathbf{c}\,\mathrm{d}\mathbf{c}^\dagger]\exp\Bigg\{\mathrm{i}\int \mathrm{d}^4x\Bigg[&\mathscr{L}(x) - \frac{1}{2\xi}(\partial^\mu A_\mu^a)^2 \\ &+ c_a^\dagger\,\partial^\mu[\delta_{ab}\,\partial_\mu - g\varepsilon_{abc}A_\mu^c]c_b \\ &+ J_\mu^a A^{\mu a} + \eta^{a\dagger}c^a + \eta^a c^{a\dagger}\Bigg]\Bigg\} \end{aligned} \tag{9.77}$$

where we have redefined $\mathbf{c}$ and $\mathbf{c}^\dagger$ to absorb the $1/g$ factor in M_f.

(A) **Perturbation expansion in covariant gauges.** To do the perturbation expansion first for a pure Yang–Mills theory, we decompose $S_{\text{eff}} = S_0 + S_{\text{I}}$ where the free action is quadratic in the fields,

$$\begin{aligned} S_0 = \int \mathrm{d}^4x\Bigg[&-\frac{1}{4}(\partial_\mu A_\nu^a - \partial_\nu A_\mu^a)^2 - \frac{1}{2\xi}(\partial^\mu A_\mu^a)^2 \\ &+ c_a^\dagger\,\partial^2 c_a + J_\mu^a A^{\mu a} + \eta^{a\dagger}c^a + \eta^a c^{a\dagger}\Bigg], \end{aligned} \tag{9.78}$$

and the remainder is the interaction term

$$\begin{aligned} S_{\text{I}}[\mathbf{A}_\mu, \mathbf{c}, \mathbf{c}^\dagger] = \int \mathrm{d}^4x[&-\tfrac{1}{2}(\partial_\mu A_\nu^a - \partial_\nu A_\mu^a)g\varepsilon^{abc}A^{b\mu}A^{c\nu} \\ &+ \tfrac{1}{4}g^2\varepsilon^{abc}\varepsilon^{ade}A_\mu^b A_\nu^c A^{d\mu}A^{e\nu} \\ &- \mathrm{i}gc^{a\dagger}\,\partial^\mu\varepsilon^{abc}A_\mu^c c^b]. \end{aligned} \tag{9.79}$$

The generating functional can then be written

$$W[\mathbf{J}, \boldsymbol{\eta}, \boldsymbol{\eta}^\dagger] = \exp\left\{\mathrm{i}S_{\text{I}}\left[\frac{\delta}{\mathrm{i}\delta\mathbf{J}_\mu}, \frac{\delta}{\mathrm{i}\delta\boldsymbol{\eta}}, \frac{\delta}{\mathrm{i}\delta\boldsymbol{\eta}^\dagger}\right]\right\}W_A^0[\mathbf{J}]W_c^0[\boldsymbol{\eta}, \boldsymbol{\eta}^\dagger] \tag{9.80}$$

with

$$\begin{aligned} W_A^0[\mathbf{J}] = \int [\mathrm{d}\mathbf{A}_\mu]\exp\Bigg\{\mathrm{i}\int \mathrm{d}^4x\Bigg[&-\frac{1}{4}(\partial_\mu A_\nu^a - \partial_\nu A_\mu^a)^2 \\ &-\frac{1}{2\xi}(\partial^\mu A_\mu^a)^2 + J_\mu^a A^{a\mu}\Bigg]\Bigg\} \\ W_c^0[\boldsymbol{\eta}, \boldsymbol{\eta}^\dagger] = \int [\mathrm{d}\mathbf{c}^\dagger][\mathrm{d}\mathbf{c}]\exp\Bigg\{-\mathrm{i}\int \mathrm{d}^4x[&c^{a\dagger}\,\partial^2 c^a \\ &- \eta^{a\dagger}c^a - \eta^a c^{a\dagger}]\Bigg\}. \end{aligned} \tag{9.81}$$

(B) Propagators. To calculate the propagator for the A_μ field, we rewrite W_A^0 as

$$W_A^0[\mathbf{J}] = \int [\mathrm{d}\mathbf{A}_\mu] \exp\left\{\mathrm{i}\int \mathrm{d}^4x\left[\frac{1}{2}A_\mu^a(g^{\mu\nu}\,\partial^2 - \frac{\xi-1}{\xi}\partial^\mu\,\partial^\nu)\,\delta_{ab}A_\nu^b + J_\mu^a A^{a\mu}\right]\right\}$$

$$= \int [\mathrm{d}\mathbf{A}_\mu] \exp\left\{\mathrm{i}\int \mathrm{d}^4x\left[\frac{1}{2}A_\mu^a K_{ab}^{\mu\nu} A_\nu^b + J_\mu^a A^{\mu a}\right]\right\} \tag{9.82}$$

with

$$K_{ab}^{\mu\nu} = \left[g^{\mu\nu}\,\partial^2 - \left(1 - \frac{1}{\xi}\right)\partial^\mu\,\partial^\nu\right]\delta_{ab}$$

which possesses an inverse and we can use (9.6) to integrate over $[\mathrm{d}\mathbf{A}_\mu]$,

$$W_A^0[J] = \exp\left\{-\frac{\mathrm{i}}{2}\int \mathrm{d}^4x\,\mathrm{d}^4y J_\mu^a(x) G_{ab}^{\mu\nu}(x-y) J_\nu^b(y)\right\} \tag{9.83}$$

where

$$G_{ab}^{\mu\nu}(x-y) = \delta_{ab}\int \frac{d^4k}{(2\pi)^4}\,\mathrm{e}^{-\mathrm{i}k(x-y)}\left[-\left(g^{\mu\nu} - \frac{k^\mu k^\nu}{k^2}\right) - \xi\frac{k^\mu k^\nu}{k^2}\right]\frac{1}{k^2+\mathrm{i}\varepsilon}.$$

It is easy to check that

$$\int \mathrm{d}^4y K_{ab}^{\mu\nu}(x-y) G_{\nu\lambda}^{bc}(y-z) = g_\lambda^\mu\,\delta_a^c\,\delta^4(x-z). \tag{9.84}$$

Similarly, we find

$$W_c^0[\boldsymbol{\eta},\boldsymbol{\eta}^\dagger] = \exp\left\{-\mathrm{i}\int \mathrm{d}^4x\,\mathrm{d}^4y\,\eta^{a\dagger}(x) G^{ab}(x-y)\eta^a(y)\right\} \tag{9.85}$$

where

$$G^{ab}(x-y) = -\int \frac{\mathrm{d}^4k}{(2\pi)^4}\frac{\mathrm{e}^{-\mathrm{i}k\cdot(x-y)}}{k^2+\mathrm{i}\varepsilon}\delta_{ab}.$$

Thus we have the Feynman rules.

(i) Vector boson propagator

$$\mathrm{i}\Delta_{\mu\nu}^{ab}(k) = -\delta_{ab}\mathrm{i}\left[g^{\mu\nu} - (1-\xi)\frac{k_\mu k_\nu}{k^2}\right]\frac{1}{k^2+\mathrm{i}\varepsilon};$$

(ii) FP ghost propagator

$$\mathrm{i}\Delta^{ab}(k) = -\mathrm{i}\delta_{ab}\frac{1}{k^2+\mathrm{i}\varepsilon}.$$

A ghost field line, like that for a fermion, has directions. Thus a ghost is distinct from its antiparticle.

(C) **Gauge field couplings.** For non-Abelian theories there are self-couplings among the gauge fields, with polarization vectors $\varepsilon^{\mu}(k)$

$$\varepsilon^{\mu}(k_1)\varepsilon^{\nu}(k_2)\varepsilon^{\lambda}(k_3)\Gamma^{abc}_{\mu\nu\lambda}(k_1, k_2, k_3) \tag{9.86}$$

and

$$\varepsilon^{\mu}(k_1)\varepsilon^{\nu}(k_2)\varepsilon^{\lambda}(k_3)\varepsilon^{\rho}(k_4)\Gamma^{abcd}_{\mu\nu\lambda\rho}(k_1, k_2, k_3, k_4). \tag{9.87}$$

The Feynman rules for the vertices (Γs) follow from S_{I} in eqn (9.79). One can work them out by a straightforward application of the procedure outlined in Chapter 1 (for the $\lambda\phi^4$ theory). But they can just as easily be deduced from their symmetry properties under the interchange of gauge fields. Such a derivation also helps us to remember their structure. In momentum space the first term of (9.79) has the form

$$\frac{1}{3!}\tilde{A}^{a\mu}(k_1)\tilde{A}^{b\nu}(k_2)\tilde{A}^{c\lambda}(k_3)\Gamma^{abc}_{\mu\nu\lambda}(k_1, k_2, k_3) \tag{9.88}$$

where the $\tilde{A}$s are the Fourier transform of the gauge fields, and $\Gamma^{abc}_{\mu\nu\lambda}$ is the Feynman rule vertex of (9.86) which must be totally symmetric under the interchange of As. The SU(2) structure is already fixed,

$$\Gamma^{abc}_{\mu\nu\lambda}(k_1, k_2, k_3) = \varepsilon^{abc}\Gamma_{\mu\nu\lambda}(k_1, k_2, k_3). \tag{9.89}$$

The Lorentz structure can then be deduced. It is clear from (9.79) that $\Gamma_{\mu\nu\lambda}(k_1, k_2, k_3)$ is made up of terms like $k_{2\mu}g_{\nu\lambda}$. The precise combination can be worked out from the condition that $\Gamma_{\mu\nu\lambda}(k_1, k_2, k_3)$ must be antisymmetric with respect to index interchanges: $\mu \leftrightarrow \nu$, $1 \leftrightarrow 2$, etc. since ε^{abc} is totally antisymmetric. In this way we find

(iii)

$$\mathrm{i}\Gamma^{abc}_{\mu\nu\lambda} = \mathrm{i}g\varepsilon^{abc}[(k_1 - k_2)_{\lambda}g_{\mu\nu} + (k_2 - k_3)_{\mu}g_{\nu\lambda} + (k_3 - k_1)_{\nu}g_{\mu\lambda}]$$

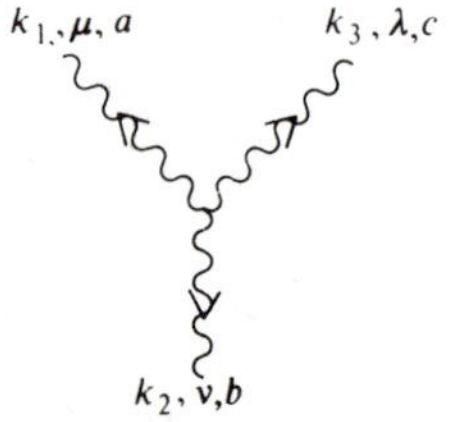

with

$$k_1 + k_2 + k_3 = 0.$$

Similarly for the quartic gauge-field self-coupling in (9.79) we have the vertex

(iv)

$$\mathrm{i}\Gamma^{abcd}_{\mu\nu\lambda\rho} = \mathrm{i}g^2[\varepsilon^{abe}\varepsilon^{cde}(g_{\mu\lambda}g_{\nu\rho} - g_{\nu\lambda}g_{\mu\rho}) + \varepsilon^{ace}\varepsilon^{bde}(g_{\mu\nu}g_{\lambda\rho} - g_{\lambda\nu}g_{\mu\rho}) + \varepsilon^{ade}\varepsilon^{cbe}(g_{\mu\lambda}g_{\rho\nu} - g_{\rho\lambda}g_{\mu\nu})]$$

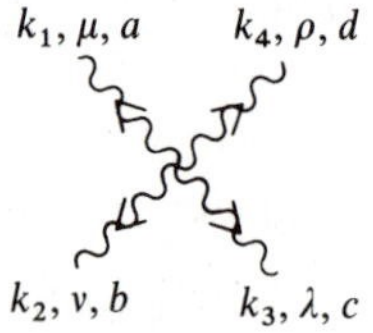

with

$$k_1 + k_2 + k_2 + k_4 = 0.$$

For the covariant gauge vertex which couples the ghost fields to a gauge field with polarization vector $\varepsilon^{\mu}(k_1 + k_2)$ we have

(v)

$$\mathrm{i}\Gamma^{abc}_{\mu} = g\varepsilon^{abc}k_{1\mu}$$

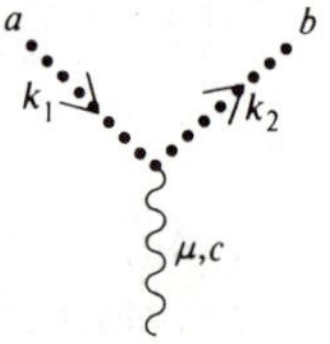

Note the asymmetric appearance of this vertex. One should preserve a consistent convention of entering the momentum of either the left or the right ghost line at every vertex. The ghost only enters in closed loops. Topologically for every diagram with a gauge-field closed loop there is one with a ghost loop in the same place. Most importantly, like the case of a fermion loop, we must insert an extra minus sign for every ghost loop.

We note that only the propagator depends on the gauge parameter ξ and we can make suitable choices of ξ for specific purposes. Within this class of covariant gauges, the choice of $\xi = 1$ is called the *'t Hooft–Feynman gauge*; $\xi = 0$ is the *Landau gauge*.

(D) **Fermions.** It is straightforward to add fermions to the pure Yang–Mills theory considered above: we merely insert in the Lagrangian all the possible gauge-invariant terms that have dimension less than four

$$\mathscr{L}_f = \bar{\psi}(\mathrm{i}\gamma^\mu \mathrm{D}_\mu - m)\psi \tag{9.90}$$

where

$$\mathrm{D}_\mu \psi = \partial_\mu \psi - \mathrm{i}gT^a A^a_\mu \psi .$$

T^a is the representation matrix. For example, if ψ is an SU(2) doublet, $T^a = \tau^a/2$. We then have the additional Feynman rules involving fermions (with group indices $n, m, \ldots$).

(vi) Fermion propagator

$$\mathrm{i}\Delta_{nm}(k) = \delta_{nm} \frac{\mathrm{i}}{\gamma \cdot k - m + \mathrm{i}\varepsilon}$$

(vii) Fermion gauge boson vertex

$$\mathrm{i}\Gamma^{a\mu}_{nm} = \mathrm{i}g(T^a)_{nm}\gamma^\mu$$

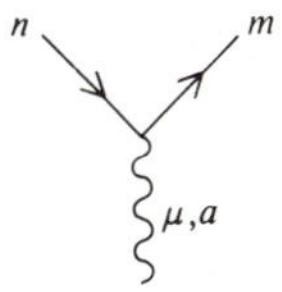

R_ξ gauges in spontaneously broken gauge theories

Finally we come to the covariant-gauge Feynman rules for gauge theories with spontaneous symmetry breakdown. First, let us consider the case of Abelian symmetry. Recalling the discussion of §8.3, it is desirable to eliminate the mixing terms of eqn (8.89)

$$gvA^\mu \, \partial_\mu \phi_2 \tag{9.91}$$

where ϕ_2 is the would-be-Goldstone boson field. There, we choose a gauge, the unitary gauge, so that ϕ_2 is absent from the theory. Thus, ϕ_2 can be identified with the phase of the complex scalar field and this fictitious degree of freedom can be eliminated by a gauge transformation. The advantage of the unitary gauge is that the particle content of the theory is manifest; all we have in the theory are the physical states of the real Higgs particle and the massive gauge boson, which has the propagator normally expected for a massive vector field (with mass $M = gv$)

$$i\Delta_{\mu\nu}(k) = \frac{-i(g_{\mu\nu} - k_\mu k_\nu/M^2)}{k^2 - M^2 + i\varepsilon}. \tag{9.92}$$

We have already mentioned in Chapter 2 that a theory with such a spin-1 propagator seems to be unrenormalizable from the power-counting arguments—unless there are hidden cancellations among Green's functions. However, from the observation that the original Lagrangian *before* spontaneous symmetry breaking is renormalizable by power counting, 't Hooft (1971*a*,*b*) proved that the theory remains renormalizable even *after* the symmetry breakdown. The key is to choose another set of gauges, *the renormalizable gauges*, in which the theory has good high-energy behaviour. The point is that in the unitary gauge, although the particle content is simple, renormalizability is not transparent, as the finite S-matrix only results from cancellations among divergent Green's functions. But the theory should be equivalent to that in the renormalizable gauge, where we obtain propagators with mild high-energy behaviour at the expense of introducing fictitious particles (the would-be-Goldstone bosons). Thus in the renormalizable gauge, unitarity is not manifest and we have to check that the spurious degrees of freedom do cancel in the physical amplitude. Such theories have been described as being 'cryptorenormalizable'.

A general class of renormalizable gauges may be represented by choosing the gauge-fixing condition (9.64) as

$$f(A_\mu, \phi) = (\partial^\mu A_\mu + \xi M\phi_2) = 0 \tag{9.93}$$

where ξ is an arbitrary parameter. Then the gauge-fixing Lagrange density (9.68) is generalized to

$$\mathscr{L}_{\text{gf}} = -\frac{1}{2\xi}(\partial^\mu A_\mu + \xi M\phi_2)^2 \tag{9.94}$$

which is added to the original Lagrangian. In this way the mixing term (9.91) is eliminated without transforming away the ϕ_2 field. Not displaying the FP ghost part, the free Lagrange density is given below where the ϕ_1' and ϕ_2' are the shifted scalar fields eqn (8.87),

$$\mathscr{L}_0 = \frac{1}{2}[(\partial^\mu\phi_1')^2 - 2\mu^2\phi_1'^2] + \frac{1}{2}[(\partial^\mu\phi_2')^2 - \xi M^2\phi_2'^2]$$
$$-\frac{1}{4}(\partial_\mu A_\nu - \partial_\nu A_\mu)^2 + \frac{1}{2}M^2 A_\mu A^\mu - \frac{1}{2\xi}(\partial^\mu A_\mu)^2. \tag{9.95}$$

This yields

(viii) Higgs scalar propagator

$$i\Delta_1(k) = \frac{i}{k^2 - 2\mu^2 + i\varepsilon}; \qquad \phi_1' \text{ - - -}\!\!\rightarrow\!\text{- - -}$$

(ix) Would-be-Goldstone boson propagator

$$i\Delta_2(k) = \frac{i}{k^2 - \xi M^2 + i\varepsilon};$$

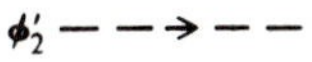

(x) Gauge boson propagator in the R_ξ gauge

$$i\Delta_{\mu\nu}(k) = \frac{-i}{k^2 - M^2 + i\varepsilon}\left[g_{\mu\nu} - (1-\xi)\frac{k_\mu k_\nu}{k^2 - \xi M^2}\right]$$

$$= -i\left[\frac{g_{\mu\nu} - k_\mu k_\nu/M^2}{k^2 - M^2 + i\varepsilon} + \frac{k_\mu k_\nu/M^2}{k^2 - \xi M^2 + i\varepsilon}\right].$$

The interaction Lagrangian is given by

$$\mathscr{L}_1 = gA_\mu(\partial^\mu \phi'_1 \phi'_2 - \partial^\mu \phi'_2 \phi'_1) + \frac{1}{2}g^2 A_\mu A^\mu(\phi'^2_1 + \phi'^2_2)$$

$$- \frac{\lambda}{4}(\phi'^2_1 + \phi'^2_2)^2 + g^2 v A^\mu A_\mu \phi'_1 - \lambda v \phi'_1(\phi'^2_1 + \phi'^2_2) \quad (9.96)$$

which yields the vertices

(xi) $\phi'_1\phi'_2A$-vertex

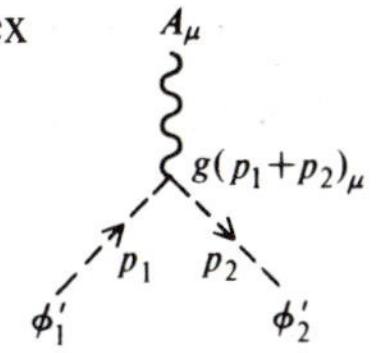

$AA\phi'_1$-vertex

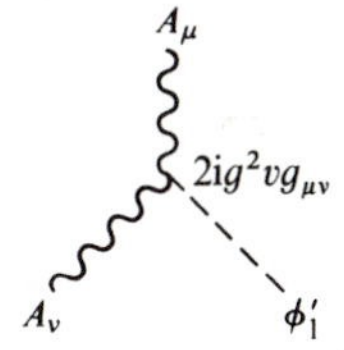

$AA\phi\phi$-vertex

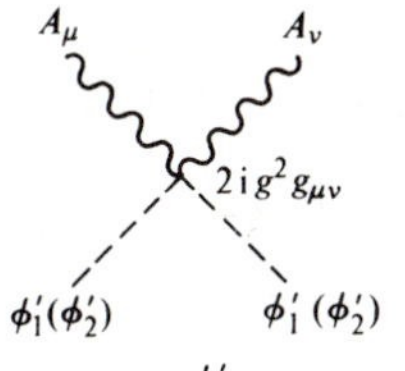

$\phi\phi\phi$-vertices

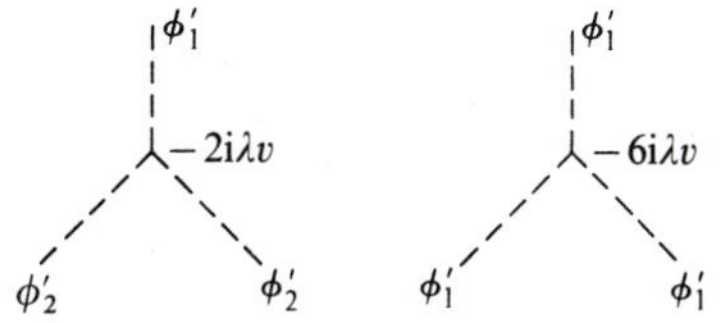

$\phi\phi\phi\phi$-vertices.

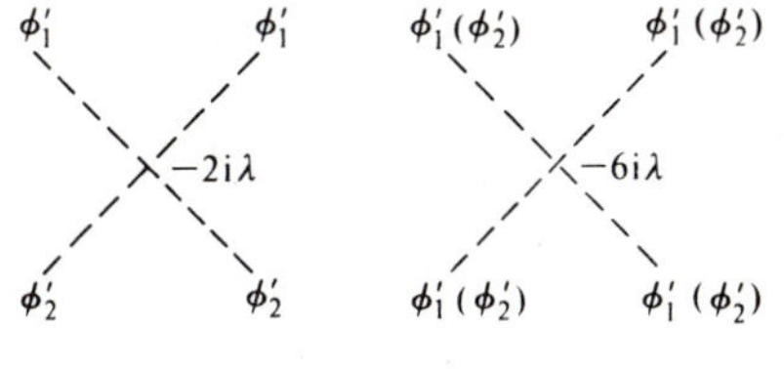

This class of gauges, called the R_ξ-*gauge* (Fujikawa, Lee, and Sanda 1972; Yao 1973), is characterized by a gauge parameter ξ. For any finite values of ξ the vector boson propagator has the asymptotic behaviour

$$\Delta_{\mu\nu}(k) \to O(k^{-2}) \quad \text{as} \quad k \to \infty.$$

Hence, from the fact that the coupling constants in the interaction Lagrangian have dimension 0 or 1 in units of mass, the theory is renormalizable by power counting. For example the particular choice $\xi = 1$ leads to the propagator form

$$\mathrm{i}\Delta_{\mu\nu}(k) = \frac{-\mathrm{i}g_{\mu\nu}}{k^2 - M^2 + \mathrm{i}\varepsilon}.$$

On the other hand in the limit $\xi \to \infty$, the propagator for the ϕ'_2-field (ix) vanishes and the would-be-Goldstone boson decouples. The vector meson propagator (x) reduces to the standard form for a massive spin-1 particle (9.92). Thus we recover the unitary gauge where there are no unphysical fields.

For any finite values of ξ we have the unphysical singularities in the ϕ'_2-propagator (ix) at $k^2 = \xi M^2$ and in the gauge boson propagator (x) also at $k^2 = \xi M^2$. In order to preserve unitarity, these unphysical poles must cancel in the S-matrix element involving only physical particles: A_μ and ϕ'_1. This is indeed the case. We can illustrate this with the following example (Fujikawa *et al.* 1972). For the process

$$\phi'_1(k_1) + \phi'_1(k_2) \to A(k_3) + A(k_4),$$

among the tree-level diagrams, we have those in Fig. 9.3 (as well as those with the final A-lines crossed). Fig. 9.3(a) due to A-exchange is given by

$$\mathrm{i}T^{(A)} = \varepsilon^\mu(k_3)(2\mathrm{i}g^2 v)^2\varepsilon^\nu(k_4)(-\mathrm{i})\left[\frac{g_{\mu\nu} - q_\mu q_\nu/M^2}{q^2 - M^2} + \frac{q_\mu q_\nu/M^2}{q^2 - \xi M^2}\right].$$

At the unphysical pole $q^2 = \xi M^2$, it may be written as

$$\mathrm{i}T^{(A)}_{\text{pole}} = \frac{-4\mathrm{i}g^2}{q^2 - \xi M^2}[k_{1\mu}\varepsilon^\mu(k_3)][k_{2\nu}\varepsilon^\nu(k_4)]. \tag{9.97}$$

But the diagram in Fig. 9.3(b) due to ϕ'_2-exchange contributes as

$$\mathrm{i}T^{(\phi_2)} = \varepsilon^\mu(k_3)g(k_1 + q)_\mu\varepsilon^\nu(k_4)g(k_2 - q)_\nu\frac{\mathrm{i}}{q^2 - \xi M^2}$$

$$= \frac{4\mathrm{i}g^2}{q^2 - \xi M^2}[k_{1\mu}\varepsilon^\mu(k_3)][k_{2\nu}\varepsilon^\nu(k_4)]$$

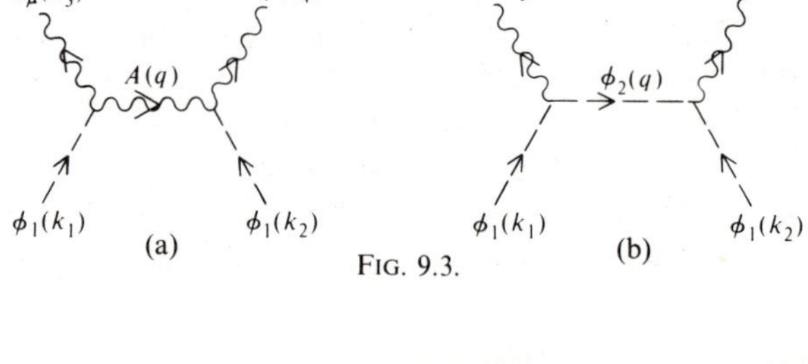

FIG. 9.3.

which will cancel the unphysical pole due to A-exchange in (9.97). This type of cancellation of unphysical poles is very general and can be proven to all orders in perturbation theory by using the generalized Ward identities which are a consequence of the gauge invariance of the theory. In practice a good check on the gauge invariance of a calculation of some physical S-matrix element is the disappearance of the arbitrary ξ-parameter in the final result. An example of such an R_ξ gauge calculation will be given in §13.3.

The generalization to non-Abelian symmetries is straightforward. For example, the corresponding R_ξ gauge-fixing term in the Lagrangian for the SU(2) example with complex doublet ϕ will be

$$\mathscr{L}_{\text{gf}} = -\frac{1}{2\xi}\left[\partial^\mu \mathbf{A}_\mu - ig\xi\left(\langle\phi\rangle_0^\dagger \frac{\boldsymbol{\tau}}{2}\phi' - \phi'^\dagger \frac{\boldsymbol{\tau}}{2}\langle\phi\rangle_0\right)\right]^2. \tag{9.98}$$

The R_ξ-gauge Feynman rule for the SU(2) × U(1) standard theory of the electroweak interaction is given in Appendix B.

9.3 The Slavnov–Taylor identities

Having outlined the quantization procedure and Feynman rules for gauge theories in §§9.1 and 9.2, we can proceed to make perturbative calculations. The regularization and renormalization procedure reviewed in Chapter 2 can be applied. From the Feynman rules of §9.2 we see that all couplings are dimensionless and that the high-energy behaviour of the propagators is such that the theory should be renormalizable by power counting. The divergent higher-order diagrams can be regularized by the dimension (d) continuation scheme. We can thus identify the appropriate counterterms to be inserted in the Lagrangian. They are of the same form as those in the original Lagrangian but are multiplied by coefficients which diverge in the limit $\varepsilon = (4 - d) \to 0$. After the addition of these counterterms, the resultant Lagrangian will generate, to all orders, Green's functions that are finite when $\varepsilon \to 0$. In practice, it is a very complicated programme. The resurgence of field theoretical studies of particle interactions in recent years was to a large extent brought about by 't Hooft's proof (1971*a*,*b*) that non-Abelian gauge theory is renormalizable, and that renormalizability is not spoiled even if the gauge symmetry is spontaneously broken. A detailed discussion of gauge theory renormalization (Lee and Zinn-Justin 1972, 1973) is beyond the scope of this book. Here we only study the generalized Ward identities of the Yang–Mills theory, sometimes referred to in the literature as the Slavnov (1972) and Taylor (1971) identities, which play an important role in the renormalization programme.

The Ward identities are relations among different Green's functions. They reflect the theory's (nontrivial) symmetry (here the gauge invariance of the original action). These relations are important to the renormalization programme as they restrict the number of independent ultraviolet divergences to ensure that gauge-noninvariant counterterms are absent. Recall that in the simple Abelian gauge theory of QED we have, because of the Ward identities, the equality $Z_1 = Z_2$ which ensures that if two particles have the same bare charges then they will also have the same renormalized charges. In non-Abelian theory we have many more renormalization constants and many more such equalities are required to ensure that the renormalized Lagrangian is still gauge invariant. Furthermore, as we shall presently illustrate, Ward identities ensure that all the unphysical singularities are cancelled in the physical amplitudes. The formal derivation of the generalized Ward identities for non-Abelian gauge theory will be given at the end of the section.

The Ward identities and unitarity

Consider a simple SU(2) gauge theory with fermions (f) in a doublet representation. The requirement that the S-matrix must be unitary,

$$SS^\dagger = S^\dagger S = 1 \quad \text{or} \quad \sum_c S_{ac} S^*_{bc} = \delta_{ab}$$

implies that the scattering amplitude T_{ab}, which is related to S_{ab} by

$$S_{ab} = \delta_{ab} + \mathrm{i}(2\pi)^4\, \delta^4(p_a - p_b) T_{ab}, \tag{9.99}$$

will satisfy the relation

$$\operatorname{Im} T_{ab} = \frac{1}{2} \sum_c T_{ac} T^*_{bc} (2\pi)^4\, \delta^4(p_a - p_c). \tag{9.100}$$

In other words, the requirement that the S-matrix must be unitary implies that the imaginary part of the scattering amplitude T_{ab} is directly related to a sum over products of matrix elements connecting the initial and final states to all physical states with the same energy–momentum as the initial and final states. For our calculation we shall consider the fermion and anti-fermion scattering amplitude $T(\mathrm{f\bar{f}} \to \mathrm{f\bar{f}})$ with the intermediate states being the two gauge boson states (see, for example, Feynman 1977; Aitchison and Hey 1982). This is represented schematically in Fig. 9.4.

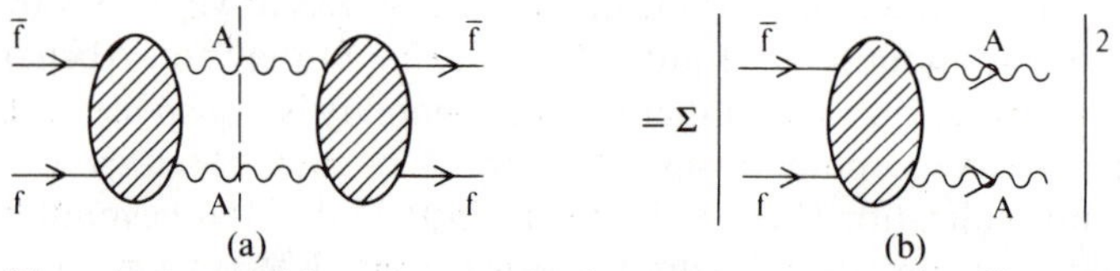

FIG. 9.4. The unitarity condition relates (a) the absorptive part of the $\mathrm{f\bar{f}} \to \mathrm{f\bar{f}}$ amplitude to (b) the sum of the squared amplitude for $\mathrm{f\bar{f}} \to AA$ in the physical region of two gauge bosons.

The imaginary part of the scattering amplitude on the left-hand side of eqn (9.100) can be calculated by replacing the propagators in the intermediate states by their imaginary parts and multiplying them by the on-shell scattering amplitudes $T(\mathrm{f\bar{f}} \to AA)$ and $T^*(AA \to \mathrm{f\bar{f}})$ (Cutkosky 1960).

For the vector boson propagator (i) of §9.2 we take the 't Hooft–Feynman gauge with the gauge parameter $\xi = 1$,

$$\Delta^{ab}_{\mu\nu} = \delta^{ab}(-g_{\mu\nu})/(k^2 + \mathrm{i}\varepsilon). \tag{9.101}$$

It has the imaginary part

$$\pi\, \delta^{ab} g_{\mu\nu}\, \delta(k^2)\theta(\omega) \tag{9.102}$$

where $\omega = |\mathbf{k}|$. Similarly the imaginary part of the ghost propagator (ii) of §9.2 is

$$\pi\, \delta^{ab}\, \delta(k^2)\theta(\omega). \tag{9.103}$$

The step functions in (9.102) and (9.103) have the effect of constraining the intermediate gauge particle states and ghost states to the same physical region. The unitarity condition for the fourth-order amplitude then reads

$$\int \mathrm{d}\rho_2 [\tfrac{1}{2} T^{ab}_{\mu\nu} T^{ab*}_{\mu'\nu'} g^{\mu\mu'} g^{\nu\nu'} - S^{ab} S^{ab*}]$$
$$= \tfrac{1}{2} \int \mathrm{d}\rho_2\, T^{ab}_{\mu\nu} T^{ab*}_{\mu'\nu'} P^{\mu\mu'}(k_1) P^{\nu\nu'}(k_2) \tag{9.104}$$

where $T^{ab}_{\mu\nu}$ and S^{ab} are the $\mathrm{f}\bar{\mathrm{f}} \to A^a_\mu A^b_\nu$ and $\mathrm{f}\bar{\mathrm{f}} \to c^{a\dagger} c^b$ amplitudes where A^a_μ and c^a are the gauge and ghost fields, respectively. The $\mathrm{d}\rho_2$ integration is over the two (massless)-particle phase space. The $P_{\mu\nu}$s are polarization sums of gauge particles

$$P^{\mu\mu'}(k_1) = \sum_{\sigma=1,2} \varepsilon^\mu_1(k_1, \sigma)\varepsilon^{\mu'}_1(k_1, \sigma)$$
$$P^{\nu\nu'}(k_2) = \sum_{\sigma=1,2} \varepsilon^\nu_2(k_2, \sigma)\varepsilon^{\nu'}_2(k_2, \sigma) \tag{9.105}$$

where $\varepsilon^\mu_1(k_1, \sigma)$ and $\varepsilon^\nu_2(k_2, \sigma)$ are polarization four-vectors of the two gauge particles with momenta k_1 and k_2 respectively.

We note that in this case the left-hand side of (9.104) receives a contribution coming from the ghost fields while the right-hand side does not because ghosts are not physical states. This is the feature that makes the demonstration of the unitarity relation nontrivial. As we shall see, what ultimately allows the unitarity relation to hold is that the polarization sum $P^{\mu\nu}$ in (9.105) is not just $g^{\mu\nu}$ and the effect of the ghost fields is just to make up the difference.

We shall carry out the lowest nontrivial order calculation as in eqn (9.104). The imaginary part of the amplitude $\mathrm{f}\bar{\mathrm{f}} \to \mathrm{f}\bar{\mathrm{f}}$ of eqn (9.100) (the cut-diagrams of Fig. 9.5) has been written via eqns (9.102) and (9.103) as squares of the $\mathrm{f}\bar{\mathrm{f}} \to AA$ amplitude (Fig. 9.6) and of the $\mathrm{f}\bar{\mathrm{f}} \to c^\dagger c$ amplitude (Fig. 9.7). The factor of 1/2 on the left-hand side in (9.104) arises because there are nine diagrams when one squares the amplitude in Fig. 9.6, eight of them are just twice those of Fig. 9.5(a)–(d) and the ninth one corresponds to Fig. 9.5(e) with the closed gauge boson loop having a symmetry factor of 1/2. The FP ghost field c behaves like a fermion with $c \neq c^\dagger$; hence there is a minus sign and no symmetry factor in front of the SS^* term.

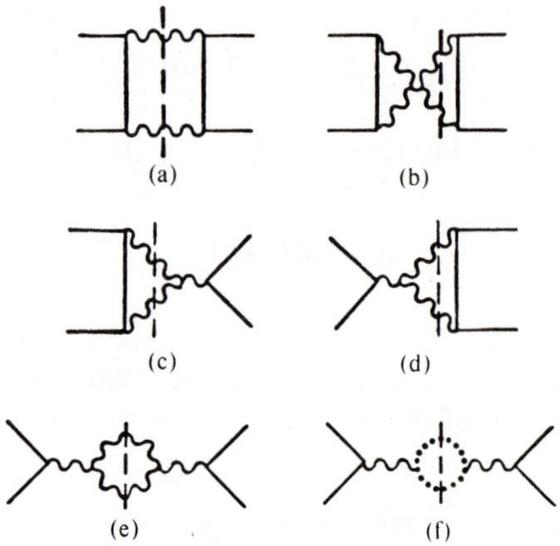

FIG. 9.5. Fourth-order cut-diagrams for $f\bar{f} \to f\bar{f}$ where the intermediate-state particles are gauge particles and FP ghosts.

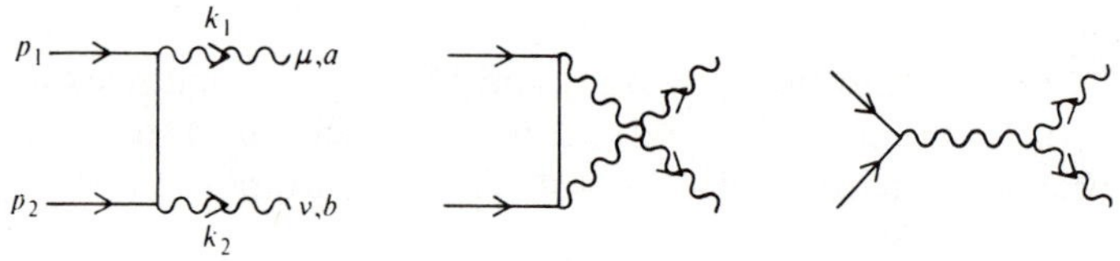

FIG. 9.6. Diagrams for $T^{ab}_{\mu\nu}$ where the final state is two gauge bosons.

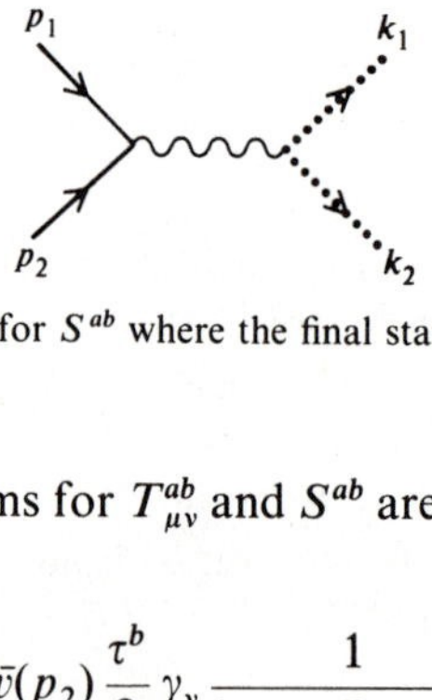

FIG. 9.7. Diagram for S^{ab} where the final state is two FP ghosts.

The lowest-order diagrams for $T^{ab}_{\mu\nu}$ and S^{ab} are shown in Figs. 9.6 and 9.7, respectively

$$\begin{aligned} T^{ab}_{\mu\nu} = & - \mathrm{i}g^2 \bar{v}(p_2) \frac{\tau^b}{2} \gamma_\nu \frac{1}{(\not{p}_1 - \not{k}_1) - m} \frac{\tau^a}{2} \gamma_\mu u(p_1) \\ & - \mathrm{i}g^2 \bar{v}(p_2) \frac{\tau^a}{2} \gamma_\mu \frac{1}{(\not{k}_1 - \not{p}_2) - m} \frac{\tau^b}{2} \gamma_\nu u(p_1) \\ & - g^2 \varepsilon^{abc} [(k_1 - k_2)_\lambda g_{\mu\nu} + (k_1 + 2k_2)_\mu g_{\nu\lambda} \\ & - (2k_1 + k_2)_\nu g_{\mu\lambda}] \frac{1}{(k_1 + k_2)^2} \bar{v}(p_2) \frac{\tau^c}{2} \gamma^\lambda u(p_1) \end{aligned} \tag{9.106}$$

$$S^{ab} = - \mathrm{i}g^2 \varepsilon^{abc} \frac{1}{(k_1 + k_2)^2} \bar{v}(p_2) \frac{\tau^c}{2} \not{k}_1 u(p_1). \tag{9.107}$$

Gauge-particle polarization. We now digress to a discussion on gauge-particle polarization. First, concentrate on one of the particles. The gauge particle being massless has only two physical polarization states, $\varepsilon^{\mu}(k, \sigma)$, $\sigma = 1, 2$. Thus the three four-vectors k_{μ}, $\varepsilon_{\mu}(k, 1)$, $\varepsilon_{\mu}(k, 2)$ do not completely span the four-dimensional space. We can furnish another vector η_{μ} such that

$$\eta \cdot \varepsilon(k, \sigma) = 0, \qquad \sigma = 1,2, \tag{9.108}$$

where $\varepsilon_{\mu}(\mathrm{k}, \sigma)$ satisfies the orthogonality condition

$$\varepsilon(k, 1) \cdot \varepsilon(k, 2) = 0, \tag{9.109}$$

and the transverse condition

$$k \cdot \varepsilon(k, \sigma) = 0. \tag{9.110}$$

Since $k^2 = 0$ and η_{μ} cannot be proportional to k_{μ}, we must have $k \cdot \eta \neq 0$. By the usual procedure of establishing completeness relations, these orthogonality conditions and the normalizations, $\varepsilon^2(k, \sigma) = -1$, yield the polarization sum

$$P_{\mu\nu} = -g_{\mu\nu} + Q_{\mu\nu}$$

with

$$Q_{\mu\nu} = [(k \cdot \eta)(k_{\mu}\eta_{\nu} + k_{\nu}\eta_{\mu}) - \eta^2 k_{\mu}k_{\nu}]/(k \cdot \eta)^2. \tag{9.111}$$

Clearly the extra term $Q_{\mu\nu}$ subtracts out the nontransverse polarization states. The task of checking the unitarity condition of (9.104) involves verifying that the FP ghost term precisely compensates for the extra projection terms in the polarization sum. Our calculation will be simplified if we adopt the convenient choice $\eta^2 = 0$. Then the extra terms in the two polarization sums of (9.105) take on the forms

$$Q^{\mu\mu'}(k_1, \eta_1) = (k_1^{\mu}\eta_1^{\mu'} + k_1^{\mu'}\eta_1^{\mu})/(k_1 \cdot \eta_1) \tag{9.112a}$$

and

$$Q^{\nu\nu'}(k_2, \eta_2) = (k_2^{\nu}\eta_2^{\nu'} + k_2^{\nu'}\eta_2^{\nu})/(k_2 \cdot \eta_2). \tag{9.112b}$$

Ward identities from lowest-order diagrams. To evaluate the right-hand side of (9.104) we need to study the contractions $k_1^{\mu}T_{\mu\nu}^{ab}$, etc. The first two terms of (9.106) do not vanish

$$\begin{aligned}
&-ig^2\bar{v}(p_2)\left(\frac{\tau^b}{2}\frac{\tau^a}{2}\gamma_{\nu}\frac{(\not{p}_1 - \not{k}_1) + m}{(p_1 - k_1)^2 - m^2}\not{k}_1\right. \\
&\qquad\left. + \frac{\tau^a}{2}\frac{\tau^b}{2}\not{k}_1\frac{(\not{k}_1 - \not{p}_2) + m}{(k_1 - p_2)^2 - m^2}\gamma_{\nu}\right)u(p_1) \\
&= -ig^2\bar{v}(p_2)\left[\frac{\tau^a}{2}, \frac{\tau^b}{2}\right]\gamma_{\nu}u(p_1) \\
&= g^2\varepsilon^{abc}\bar{v}(p_2)\frac{\tau^c}{2}\gamma_{\nu}u(p_1)
\end{aligned} \tag{9.113}$$

where we have used

$$(\not{p}_1 + m)\not{k}_1 u(p_1) = [2p_1k_1 - \not{k}_1(\not{p}_1 - m)]u(p_1) = (2p_1k_1)u(p_1),$$

etc. We make the parenthetical remark that, for the case of Abelian gauge theory (QED) in the covariant gauge, we would obtain a null result in this contraction because $\varepsilon^{abc} = 0$, yielding the familiar Ward identities: $k_1^\mu T_{\mu\nu} = T_{\mu\nu}k_2^\nu = 0$. It follows immediately that for Abelian theory the $Q^{\mu\mu'}$ and $Q^{\nu\nu'}$ terms in the polarization sums will not contribute, and unitarity can be maintained without the presence of the FP ghosts in closed loops. The last term in $T^{ab}_{\mu\nu}$ of (9.106), which is present only in non-Abelian theories, when contracted with k_1^μ yields

$$-g^2\varepsilon^{abc}[2k_1\cdot k_2 g_{\nu\lambda} + (k_1 - k_2)_\lambda k_{1\nu}$$

$$-(2k_1 + k_2)_\nu k_{1\lambda}]\frac{1}{(k_1 + k_2)^2}\bar{v}(p_2)\frac{\tau^c}{2}\gamma^\lambda u(p_1)$$

$$= -g^2\varepsilon^{abc}\bar{v}(p_2)\frac{\tau^c}{2}\gamma_\nu u(p_1) \tag{9.114}$$

$$-g^2\varepsilon^{abc}\frac{k_{1_\nu}}{(k_1 + k_2)^2}\bar{v}(p_2)\frac{\tau^c}{2}(\not{k}_1 + \not{k}_2)u(p_1) \tag{9.115}$$

$$-g^2\varepsilon^{abc}\frac{k_{2\nu}}{(k_1 + k_2)^2}\bar{v}(p_2)\frac{\tau^c}{2}\not{k}_1 u(p_1). \tag{9.116}$$

Now (9.114) cancels (9.113), and the second term (9.115) vanishes because $p_1 + p_2 = k_1 + k_2$ and because of Dirac's equation. The third term (9.116) is proportional to the ghost amplitude S^{ab} of (9.107). Thus we have

$$k_1^\mu T^{ab}_{\mu\nu} = -\mathrm{i}S^{ab}k_{2\nu}. \tag{9.117}$$

Similarly,

$$T^{ab}_{\mu\nu}k_2^\nu = -\mathrm{i}S^{ab}k_{1\mu}. \tag{9.118}$$

From these relations we can also deduce $k_1^\mu T^{ab}_{\mu\nu}k_2^\nu = 0$. These are examples of Ward identities for non-Abelian theories. (Their formal derivation will be presented later on.)

It is then simple to check that the unitarity condition (9.104) is indeed satisfied as the right-hand side reads

$$\frac{1}{2}\int \mathrm{d}\rho_2\{T_{\mu\nu}T^*_{\mu'\nu'}[-g^{\mu\mu'} + (k_1^\mu\eta_1^{\mu'} + k_1^{\mu'}\eta_1^\mu)(k_1\eta_1)^{-1}]$$

$$\times [-g^{\nu\nu'} + (k_2^\nu\eta_2^{\nu'} + k_2^{\nu'}\eta_2^\nu)(k_2\eta_2)^{-1}]$$

$$= \frac{1}{2}\int \mathrm{d}\rho_2\{TT^*gg + [(k_1T\eta_2)(\eta_1T^*k_2)$$

$$+ (\eta_1Tk_2)(k_1T^*\eta_2)](k_1\eta_1)^{-1}(k_2\eta_2)^{-1}$$

$$- [(k_1T)\cdot(\eta_1T^*) + (\eta_1T)\cdot(k_1T^*)](k_1\eta_1)^{-1}$$

$$-[(Tk_2)\cdot(T^*\eta_2) + (T\eta_2)\cdot(T^*k_2)](k_2\eta_2)^{-1}\}$$

$$= \frac{1}{2} \int \mathrm{d}\rho_2 \{TT^* gg + 2SS^* - 2SS^* - 2SS^*\}$$

$$= \frac{1}{2} \int \mathrm{d}\rho_2 \{TT^* gg - 2SS^*\} \tag{9.119}$$

where we have used eqns (9.117) and (9.118).

To summarize, the unitary condition (9.104) relates the left-hand side, where we have used the covariant gauge Feynman rules of §9.2 with their spurious states of longitudinal polarization and FP ghosts, to the right-hand side where only the physical transverse polarization states appear because of the (axial) gauge conditions of (9.108). The spurious states of covariant gauge on the left-hand side do cancel among themselves and in the axial gauge on the right-hand side there are only physical states. In short, the FP ghost fields are needed in order to maintain the unitarity condition.

The BRS transformation and the Ward identities

In non-Abelian gauge theories with their FP ghost terms, the most efficient way to derive the Ward identities is through the use of the BRS (Becchi, Rouet, and Stora 1974) generalized gauge transformations. Again consider simple SU(2) theory with a set of fermions in doublet representation

$$\mathscr{L} = -\tfrac{1}{4} F^a_{\mu\nu} F^{a\mu\nu} + \bar{\psi} i\gamma^\mu \mathrm{D}_\mu \psi - m\bar{\psi}\psi \tag{9.120}$$

where

$$\mathrm{D}_\mu \psi = (\partial_\mu - \mathrm{i} g A^a_\mu T^a)\psi \tag{9.121}$$

$$F^a_{\mu\nu} = \partial_\mu A^a_\nu - \partial_\nu A^a_\mu + g\varepsilon^{abc} A^b_\mu A^c_\nu \tag{9.122}$$

with

$$a = 1, 2, 3 \quad \text{and} \quad T^a = \tau^a/2. \tag{9.123}$$

The Lagrangian is invariant under the local gauge transformation

$$\delta\psi = -\mathrm{i} T^a \theta^a \psi \tag{9.124}$$

$$\delta A^a_\mu = \varepsilon^{abc} \theta^b A^c_\mu - \frac{1}{g} \partial_\mu \theta^a. \tag{9.125}$$

When we include the gauge-fixing term and the Faddeev–Popov ghost term according to (9.75) and (9.76), the effective Lagrangian density in the covariant gauge (9.70) becomes

$$\mathscr{L}_{\text{eff}} = \mathscr{L} + \mathscr{L}_{\text{gf}} + \mathscr{L}_{\text{FPG}} \tag{9.126}$$

with

$$\mathscr{L}_{\text{gf}} = -\frac{1}{2\xi} (\partial^\mu \mathrm{A}^a_\mu)^2 \tag{9.127}$$

$$\mathscr{L}_{\text{FPG}} = \mathrm{i} c^\dagger_a \, \partial^\mu [\delta_{ab} \, \partial_\mu - g\varepsilon_{abc} A^c_\mu] c_b. \tag{9.128}$$

Instead of the complex ghost fields c and $c^\dagger$ it turns out to be more convenient to work with real Grassmann fields ρ and σ defined by

$$c_a = (\rho_a + \mathrm{i}\sigma_a)/\sqrt{2}$$
$$c_a^\dagger = (\rho_a - \mathrm{i}\sigma_a)/\sqrt{2}. \tag{9.129}$$

Using the anticommutivity properties of the Grassmann fields $\rho^2 = \sigma^2 = 0$, $\rho\sigma = -\sigma\rho$, etc. we have

$$\mathscr{L}_{\mathrm{FPG}} = -\mathrm{i}\partial^\mu \rho_a (\mathrm{D}_\mu \sigma_a) \tag{9.130}$$

with

$$\mathrm{D}_\mu \sigma^a = \partial_\mu \sigma^a - g\varepsilon^{abc}\sigma^b A_\mu^c. \tag{9.131}$$

S_{eff} is not invariant under the general gauge transformation (eqns (9.124) and (9.125)) with an arbitrary θ^a, but it is invariant under the BRS transformation

$$\delta A_\mu^a = \omega \mathrm{D}_\mu \sigma^a \tag{9.132a}$$
$$\delta\psi = \mathrm{i}g\omega(T^a\sigma^a)\psi \tag{9.132b}$$
$$\delta\rho^a = -\mathrm{i}\omega\, \partial^\mu A_\mu^a/\xi \tag{9.132c}$$
$$\delta\sigma^a = -g\omega\varepsilon^{abc}\sigma^b\sigma^c/2 \tag{9.132d}$$

where ω is a space–time-independent anticommuting Grassmann variable and ξ is the usual (covariant) gauge parameter.

S_{eff} is invariant under BRS transformations. As (9.132a) may be written as

$$\delta A_\mu^a = \mathrm{i}\partial_\mu \omega\sigma^a - \mathrm{i}g\varepsilon^{abc}\omega\sigma^b A_\mu^c, \tag{9.133}$$

the BRS transformation is in fact a gauge transformation with a specific choice of the gauge function

$$\theta^a = -g\omega\sigma^a \tag{9.134}$$

Thus the original action $S = \int \mathrm{d}^4x\mathscr{L}$ is unchanged under this transformation, $\delta S = 0$. We need to show that, in eqn (9.67), $\delta(S_{\mathrm{gf}} + S_{\mathrm{FPG}}) = 0$ also

$$\delta\left[\frac{1}{2\xi}(\partial^\mu A_\mu^a)^2 + \mathrm{i}\partial^\mu\rho^a(\mathrm{D}_\mu\sigma^a)\right] = \frac{1}{\xi}(\partial^\lambda A_\lambda^a)\,\partial_\mu(\delta A^{a\mu}) + \mathrm{i}\partial^\mu(\delta\rho^a)(\mathrm{D}_\mu\sigma^a)$$
$$+ \mathrm{i}\partial^\mu\rho^a\,\delta(\mathrm{D}_\mu\sigma^a). \tag{9.135}$$

Concentrate first on the change of the covariant derivative σ^a

$$\begin{aligned}\delta(\mathrm{D}_\mu\sigma^a) &= \delta(\partial_\mu\sigma^a - g\varepsilon^{abc}\sigma^b A_\mu^c)\\ &= -g\omega\varepsilon^{abc}\,\partial_\mu(\sigma^b\sigma^c)/2\\ &\quad + g\varepsilon^{abc}(\omega \mathrm{D}_\mu\sigma^b)\sigma^c + g\varepsilon^{abc}A_\mu^b(-g\omega\varepsilon^{cde}\sigma^d\sigma^e/2).\end{aligned} \tag{9.136}$$

Terms linear in g and in g^2 separately cancel

$$-g\omega\varepsilon^{abc}(\partial_\mu\sigma^b)\sigma^c + g\omega\varepsilon^{abc}(\partial_\mu\sigma^b)\sigma^c = 0 \tag{9.137}$$

and

$$g^2\omega\sigma^c\sigma^d A^e_\mu(\varepsilon^{abc}\varepsilon^{bde} - \varepsilon^{cbd}\varepsilon^{bea} - \varepsilon^{dbe}\varepsilon^{bac}) = 0. \tag{9.138}$$

The last combination vanishes because of the Jacobi identity. Thus

$$\delta(\mathrm{D}_\mu\sigma^a) = 0. \tag{9.139}$$

Using (9.132a) and (9.132c) in (9.135) we then have,

$$\delta(S_{\mathrm{gf}} + S_{\mathrm{FPG}}) = \int\left[\frac{1}{\xi}(\partial^\lambda A^a_\lambda)\,\partial_\mu(\omega\mathrm{D}^\mu\sigma^a) + \partial^\mu\left(\frac{\omega}{\xi}\,\partial^\lambda A^a_\lambda\right)(\mathrm{D}_\mu\sigma^a)\right]\mathrm{d}^4x$$

$$= \int\partial^\mu\left[\frac{1}{\xi}(\partial^\lambda A^a_\lambda)(\omega\mathrm{D}_\mu\sigma^a)\right]\mathrm{d}^4x = 0. \tag{9.140}$$

This completes the proof that S_{eff} is invariant under the BRS generalized gauge transformation (9.132).

Derivation of the generalized Ward identities. The generalized Ward identities reflect the symmetry corresponding to the invariance of the effective action under the BRS transformation. To obtain these relations among Green's functions we study the generating functional of the Green's functions by introducing the sources $\mathbf{J}_\mu$, $\boldsymbol{\alpha}$, $\boldsymbol{\beta}$, $\bar{\chi}$ and χ for the fields $\mathbf{A}_\mu$, $\boldsymbol{\rho}$, $\boldsymbol{\sigma}$, $\bar{\psi}$, and ψ respectively. It turns out that to obtain identities that are linear in derivatives with respect to the sources, it is convenient to also introduce the source terms $\boldsymbol{\kappa}_\mu$, $\mathbf{v}$, λ, and $\bar{\lambda}$ for the composite operators $\mathrm{D}_\mu\boldsymbol{\sigma}$, $\frac{1}{2}\boldsymbol{\sigma}\times\boldsymbol{\sigma}$, $\mathbf{T}\cdot\boldsymbol{\sigma}\psi$ and $\bar{\psi}\mathbf{T}\cdot\boldsymbol{\sigma}$ which appear in the BRS transformation (9.132). Thus the generating functional is of the form

$$W[\mathbf{J}, \boldsymbol{\alpha}, \boldsymbol{\beta}, \chi, \bar{\chi}, \boldsymbol{\kappa}, \mathbf{v}, \lambda, \bar{\lambda}]$$

$$= \int[\mathrm{d}\mathbf{A}^\mu][\mathrm{d}\boldsymbol{\rho}][\mathrm{d}\boldsymbol{\sigma}][\mathrm{d}\psi][\mathrm{d}\bar{\psi}]\exp\left\{\mathrm{i}\int\mathrm{d}^4x(\mathscr{L}_{\mathrm{eff}} + \Sigma)\right\} \tag{9.141}$$

where the source term Σ is given by

$$\Sigma = \mathbf{J}_\mu\cdot\mathbf{A}^\mu + \boldsymbol{\alpha}\cdot\boldsymbol{\rho} + \boldsymbol{\beta}\cdot\boldsymbol{\sigma} + \bar{\chi}\psi + \bar{\psi}\chi + \boldsymbol{\kappa}_\mu\cdot\mathrm{D}^\mu\boldsymbol{\sigma}$$

$$+ \tfrac{1}{2}\mathbf{v}\cdot(\boldsymbol{\sigma}\times\boldsymbol{\sigma}) + \bar{\lambda}\mathbf{T}\cdot\boldsymbol{\sigma}\psi + \bar{\psi}\mathbf{T}\cdot\boldsymbol{\sigma}\lambda. \tag{9.142}$$

Since S_{eff} is invariant under the BRS transformation, so is the generating functional of the Green's functions $\delta W = 0$, which implies

$$\int\mathrm{d}^4x\int[\mathrm{d}\mathbf{A}_\mu][\mathrm{d}\boldsymbol{\rho}][\mathrm{d}\boldsymbol{\sigma}][\mathrm{d}\psi][\mathrm{d}\bar{\psi}](\delta\Sigma)$$

$$\times\exp\left\{\mathrm{i}\int\mathrm{d}^4x'[\mathscr{L}_{\mathrm{eff}}(x') + \Sigma(x')]\right\} = 0 \tag{9.143}$$

where $\delta\Sigma$ is the infinitesimal change of the source term due to the BRS transformation

$$\delta\Sigma = \mathbf{J}_\mu\cdot\delta\mathbf{A}^\mu + \boldsymbol{\alpha}\cdot\delta\boldsymbol{\rho} + \boldsymbol{\beta}\cdot\delta\boldsymbol{\sigma} + \bar{\chi}\,\delta\psi + \delta\bar{\psi}\chi$$

$$+ \boldsymbol{\kappa}_\mu\cdot\delta(\mathrm{D}^\mu\boldsymbol{\sigma}) + \frac{\mathbf{v}}{2}\cdot\delta(\boldsymbol{\sigma}\times\boldsymbol{\sigma}) + \bar{\lambda}\,\delta(\mathbf{T}\cdot\boldsymbol{\sigma}\psi) + \delta(\bar{\psi}\mathbf{T}\cdot\boldsymbol{\sigma})\lambda. \tag{9.144}$$

We shall first demonstrate that not only $\delta(\mathrm{D}_\mu\boldsymbol{\sigma}) = 0$ as in (9.139); the changes of the composite operators all vanish

$$\delta(\sigma^a\sigma^b - \sigma^b\sigma^a)$$
$$= (-g\omega/2)(\varepsilon^{acd}\sigma^c\sigma^d\sigma^b + \varepsilon^{bcd}\sigma^a\sigma^c\sigma^d - \varepsilon^{bcd}\sigma^c\sigma^d\sigma^a - \varepsilon^{acd}\sigma^b\sigma^c\sigma^d) = 0 \quad (9.145)$$

and

$$\delta(\mathbf{T}\cdot\boldsymbol{\sigma})\psi = \mathbf{T}\cdot\delta\boldsymbol{\sigma}\psi + \mathbf{T}\cdot\boldsymbol{\sigma}\,\delta\psi$$
$$= \mathbf{T}\cdot\left(-\frac{g}{2}\omega\boldsymbol{\sigma}\times\boldsymbol{\sigma}\right)\psi - \mathrm{i}g\omega(\mathbf{T}\cdot\boldsymbol{\sigma})(\mathbf{T}\cdot\boldsymbol{\sigma})\psi = 0 \quad (9.146)$$

because

$$\mathbf{T}\cdot\boldsymbol{\sigma}\mathbf{T}\cdot\boldsymbol{\sigma} = T^aT^b\sigma^a\sigma^b = \frac{1}{2}(T^aT^b - T^bT^a)\sigma^a\sigma^b$$
$$= \frac{\mathrm{i}}{2}\varepsilon^{abc}T^c\sigma^a\sigma^b = \frac{\mathrm{i}}{2}\mathbf{T}\cdot(\boldsymbol{\sigma}\times\boldsymbol{\sigma}). \quad (9.147)$$

Thus eqn (9.143) may be written

$$\omega\int \mathrm{d}^4x \int[\mathrm{d}\mathbf{A}_\mu]\ldots[\mathrm{d}\bar{\psi}](\mathbf{J}_\mu\cdot\mathrm{D}^\mu\boldsymbol{\sigma} + \boldsymbol{\alpha}\cdot\partial^\mu\mathbf{A}_\mu/\xi - g\boldsymbol{\beta}\cdot\boldsymbol{\sigma}\times\boldsymbol{\sigma}/2$$
$$+ \mathrm{i}g\bar{\chi}\mathbf{T}\cdot\boldsymbol{\sigma}\psi - \mathrm{i}g\bar{\psi}\mathbf{T}\cdot\boldsymbol{\sigma}\chi)\exp\left(\mathrm{i}\int\mathrm{d}^4x'[\mathscr{L}_{\mathrm{eff}} + \Sigma]\right) = 0 \quad (9.148)$$

or

$$\omega\int\mathrm{d}^4x\left(J^a_\mu\frac{\delta}{\delta\kappa^a_\mu} + \frac{\alpha^a}{\xi}\partial^\mu\frac{\delta}{\delta J^a_\mu} - \frac{g}{2}\beta^a\frac{\delta}{\delta v^a}\right.$$
$$\left. + \mathrm{i}g\bar{\chi}\frac{\delta}{\delta\lambda} - \mathrm{i}g\frac{\delta}{\delta\bar{\lambda}}\chi\right)W[\mathbf{J},\ldots,\bar{\lambda}] = 0. \quad (9.149)$$

This equation is the generalized Ward identity which relates different types of Green's functions. To obtain relations for any particular set of Green's functions we simply differentiate (9.149) with respect to the external source functions J_μ, α, β, ... and set them equal to zero afterwards. This is a rather tedious procedure. A simpler way to get the content of (9.149) for the Green's functions at hand is to use the fact that $\delta W = 0$ also implies that Green's functions are invariant under the BRS transformation (see, for example, Llewellyn Smith 1980). This can then be used to give relations that are equivalent to (9.149). For instance,

$$\delta\langle 0|T(A^a_\mu(x)A^b_\nu(0))|0\rangle = 0 \quad (9.150)$$

implies that

$$\langle 0|T(\delta A^a_\mu(x)A^b_\nu(0))|0\rangle + \langle 0|T(A^a_\mu(x)\,\delta A^b_\nu(0))|0\rangle = 0 \quad (9.151)$$

or

$$\omega\langle 0|T(\mathrm{D}_\mu\sigma^a(x)A^b_\nu(0))|0\rangle + \omega\langle 0|T(A^a_\mu(x)\mathrm{D}_\nu\sigma^b(0))|0\rangle = 0. \quad (9.152)$$

Clearly, if we differentiate (9.149) with respect to $J^a_\mu(x)$ and $J^b_\nu(0)$ and set all source terms equal to zero, we also obtain (9.152) as the result. In practice this method used in (9.150) and (9.151) is quite direct. However all the information about gauge invariance is contained in eqn (9.149), which is more compact and is very useful for formal manipulation.

Formal derivation of the Ward identity (eqn (9.117)). Finally we shall demonstrate that the Ward identities (9.117) and (9.118) used in proving unitarity in the illustrative example are contained in eqn (9.149). Consider the four-point function $\langle 0|T(\rho A_\nu \bar{\psi}\psi)|0\rangle$. Its invariance under the BRS transformation yields

$$\langle 0|T(\delta\rho A_\nu \bar{\psi}\psi)|0\rangle + \langle 0|T(\rho\, \delta A_\nu \bar{\psi}\psi)|0\rangle + \langle 0|T(\rho A_\nu\, \delta\bar{\psi}\psi)|0\rangle + \langle 0|T(\rho A_\nu \bar{\psi}\, \delta\psi)|0\rangle = 0. \quad (9.153)$$

But when the ψs are on-shell the composite operators corresponding to $\delta\bar{\psi}$ and $\delta\psi$ will not contribute because they do not have a one-particle pole. To see this more explicitly,

$$\langle 0|T(\rho(x_1)A_\nu(x_2)\bar{\psi}(x_3)\,\delta\psi(x_4))|0\rangle = \mathrm{i}g\omega\langle 0|T(\rho(x_1)A_\nu(x_2)\bar{\psi}(x_3)\mathbf{T}\cdot\boldsymbol{\sigma}(x_4)\psi(x_4))|0\rangle \quad (9.154)$$

which has the momentum space representation shown in Fig. 9.8. Clearly it does not have a one-particle pole in the variable k_4 and will vanish when we put ψ on the mass shell by multiplying the inverse propagator $(\gamma\cdot k_4 - m)$

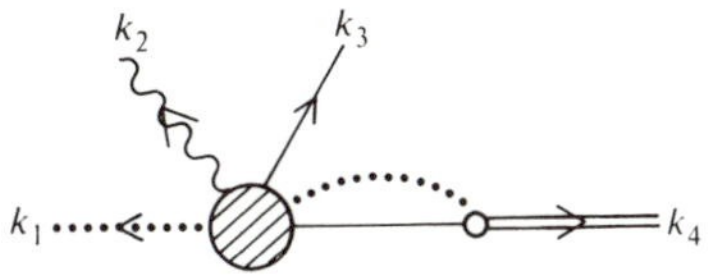

FIG. 9.8.

with $\not{k}_4 \to m$. Therefore when all particles are on-shell, only those terms that are linear in the field will survive. Eqn (9.153) reduces to the form

$$\frac{\omega}{\xi}\langle 0|T(\partial^\mu A^a_\mu A^b_\nu \bar{\psi}\psi)|0\rangle + \omega\langle 0|T(\rho^a\, \partial_\nu \sigma^b \bar{\psi}\psi)|0\rangle = 0.$$

For the choice of 't Hooft–Feynman gauge (9.101) with the gauge parameter $\xi = 1$, this corresponds to

$$k_1^\mu T^{ab}_{\mu\nu} = -\mathrm{i}S^{ab}k_{2\nu},$$

just the Ward identity of (9.117).

A final comment. We have illustrated the importance of Ward identities in checking the proper cancellation of the unphysical singularities (the longitudinal component of the gauge fields and FP ghosts) in the physical amplitudes. This issue of unitarity is particularly relevant for the spontaneously broken gauge theories. In a class of gauge choices for such theories, one encounters further unphysical particles, the would-be-Goldstone-bosons. One needs to check their decoupling by using Ward identities. Thus we must be sure that Ward identities are satisfied to all orders in perturbation theory. Since they are reflections of the theory's symmetries, it is important that we adopt regularization procedures that respect these symmetries. One of the virtues of the dimensional regularization scheme is that it clearly preserves invariance under the BRS transformation—hence its consequence, the generalized Ward identities. However, as we studied in §6.2 the validity of certain axial-vector-current Ward identities is not automatic even after the theory is regularized symmetrically. Thus for theories with fermions one must check that the theory is free of the ABJ anomaly which would spoil the renormalizability of the theory.

10 Quantum chromodynamics

Historically the first successful application of the Yang–Mills theory was the unification of the weak and electromagnetic interactions (see Chapter 11). We choose to present first, however, the gauge theory of the strong interaction, quantum chromodynamics (QCD), since the basic structure of this theory is a somewhat simpler introduction to the subject as it does not involve spontaneous breaking of the gauge symmetry.

QCD represents a remarkable synthesis of the various ideas we have developed about hadronic physics: quarks, partons, colour, current algebra, etc. The simple quark model was initially developed in early-1960s to account for the regularities observed in the hadron spectrum, with hadrons interpreted as bound states of localized but essentially noninteracting quarks (§4.4). This view of quarks as the fundamental constituents became more plausible as relations abstracted from the quantum field theory of quarks, i.e. the algebra of quark currents and their divergences were successfully applied in the late 1960s (Chapter 5). It was also gradually realized that the above picture needed to be augmented with quarks having a hidden three-valued quantum number called colour. Then came a series of important experimental measurements, starting with the ones performed by the SLAC–MIT group at the end of the decade, on deep inelastic lepton–nucleon scatterings. The cross-sections were revealed to satisfy Bjorken scaling which could be successfully interpreted by Feynman's parton model (Chapter 7). The significance of scaling and the parton model picture is that although the hadron constituents (quarks) are not produced as free particles in the final states of deep inelastic scatterings, they behave as if they were weakly bound inside the target nucleon. As we shall see, all these features can be elegantly combined in the theory of QCD.

The property of QCD that led directly to its discovery in 1973 as a candidate theory of the strong interaction is asymptotic freedom, i.e. coupling strength decreases at short distances. In this chapter we shall concentrate mainly on the short-distance properties where perturbative QCD is applicable. Only in the last section will we touch upon the long-distance feature of quark confinement as analysed by the non-perturbative method of lattice gauge theory. It is a remarkable fact that here we have a theory of the strong interaction in which we are reasonably confident as to the correctness of the Lagrangian, but do not know how to deduce many of its dynamical implications for low energy–momentum scales: confinement, spontaneous breaking of chiral symmetries, and the hadron mass spectrum. It should be pointed out that all indications are that the assumed properties are indeed consistent with QCD.

10.1 The discovery of asymptotic freedom

Gross and Wilczek (1973*a*,*b*) and Politzer (1973, 1974) discovered that, for non-Abelian gauge theories, the origin of the coupling constant is a stable fixed point in the deep Euclidean limit. ('t Hooft (1972) also noticed that in Yang–Mills theories the slope of the renormalization group β-function at the origin is negative.) Theories having this property are referred to as being *asymptotically free*. This is remarkable as we shall show that *no* renormalizable field theory can be asymptotically free without non-Abelian gauge fields (Zee 1973*a*; Coleman and Gross 1973). Thus in Yang–Mills theories, contrary to the case in all other field theories, the coupling constant decreases at short distances. In the familiar Abelian theory of QED, one has an intuitive understanding of the decrease of the effective coupling constant at long distance as being due to dielectric screening by the cloud of virtual electron–positron pairs. Thus, for non-Abelian gauge theories, we have to understand an anti-screening effect. As we shall discuss, the cloud of virtual gauge particles, which are bosons carrying (colour) charge and spin, makes the Yang–Mills vacuum behave like a paramagnetic substance and, through relativistic invariance, this implies that the vacuum anti-screens charges.

Theories without Yang–Mills fields are not asymptotically free

Let us recapitulate some of the relevant points made in Chapter 3 where the renormalization group was illustrated with the simple $\lambda\phi^4$ theory

$$\mathscr{L} = \frac{1}{2}[(\partial_\mu\phi)^2 - m^2\phi^2] - \frac{\lambda}{4!}\phi^4. \tag{10.1}$$

When all energy–momenta are scaled up σp_i, $\sigma \to \infty$ (the deep Euclidean region), apart from trivial dimension factors, the Green's function depends on $t = \ln \sigma$ only through the effective coupling constant $\bar{\lambda}(\lambda, t)$, which is in turn governed by the renormalization group β-function (see eqns (3.104) and (3.105))

$$\frac{\mathrm{d}\bar{\lambda}}{\mathrm{d}t} = \beta(\bar{\lambda}). \tag{10.2}$$

We are interested in the slope of the β-function at the origin $\lambda = 0$ because, for small couplings, β can be calculated perturbatively and because the sign of the slope determines whether $\lambda = 0$ is an ultraviolet or an infrared fixed point. If the theory is asymptotically free with $\lambda = 0$ being a ultraviolet fixed point, we recover the canonical (i.e. free field theory) light cone singularities and the parton model with its free quarks at short distances as given in Chapter 7.

The β-function can be calculated as follows. In a massless theory, the only scale parameter μ appears in the subtraction point which is needed to define all the renormalized quantities. From eqns (3.32) or (3.71), the β-function in this case is given by

$$\beta(\lambda) = \mu \frac{d\lambda}{d\mu} \tag{10.3}$$

where λ is the renormalized coupling constant related to the bare coupling as given by eqn (2.50)

$$\lambda = Z_\lambda^{-1} Z_\phi^2 \lambda_0. \tag{10.4}$$

Z_ϕ is the scalar wavefunction renormalization constant,

$$\phi = Z_\phi^{-1/2} \phi_0 \tag{10.5}$$

or, equivalently, is defined in terms of the unrenormalized scalar propagator at the subtraction point chosen to be, for example, some Euclidean point $p^2 = -\mu^2$,

$$\mathrm{i}\Delta(p^2)\Big|_{p^2=-\mu^2} = \frac{-\mathrm{i}Z_\phi}{\mu^2}. \tag{10.6}$$

Similarly, the vertex renormalization constant Z_λ can be defined through the unrenormalized four-point vertex function as in eqn (2.38)

$$\Gamma^{(4)}(p_1, p_2, p_3, p_4)\Big|_{p_i^2=-\mu^2} = -\mathrm{i}\lambda_0/Z_\lambda. \tag{10.7}$$

The one-loop contributions to the two-point and four-point Green's functions and hence to Z_ϕ and Z_λ are shown in Fig. 10.1. Thus the renormalization constants are functions of the bare coupling λ_0 and the ratio of the cut-off Λ to the subtraction parameter μ.

(a) (b)

FIG. 10.1. Lowest-order scalar meson (a) self-energy and (b) vertex radiative correction graphs.

To obtain the β-function as in (10.3) one merely has to calculate the divergent part of the Zs and differentiate with respect to the cut-off (see eqn (3.44))

$$\beta = -\lambda \frac{\partial}{\partial \ln \Lambda} [2 \ln Z_\phi(\lambda_0, \Lambda/\mu) - \ln Z_\lambda(\lambda_0, \Lambda/\mu)]. \tag{10.8}$$

The one-loop contribution to Z_ϕ in Fig. 10.1(a) vanishes ($Z_\phi = 1$); the only nontrivial diagrams are those in Fig. 10.1(b) (see eqn (3.47)). We obtain

$$\beta(\lambda) = \frac{+3\lambda^2}{16\pi^2} + O(\lambda^3). \tag{10.9}$$

As the $\lambda < 0$ region is not allowed (since the Hamiltonian is unbounded from below), the positive slope for $\lambda > 0$ means that the simple $\lambda\phi^4$ theory of (10.1) does not exhibit free-field asymptotic behaviour at large Euclidean momenta. This situation actually holds for the entire class of scalar field theories with internal symmetries, i.e. eqn (10.1) is generalized to

$$\phi \to \phi_i$$

$$\lambda\phi^4 \to \lambda_{ijkl}\phi_i\phi_j\phi_k\phi_l$$

where λ_{ijkl} is symmetric in its indices. The scalar fields $\phi_i(x)$ belong to some (possibly reducible) representation of the symmetry group and in every term all the internal symmetry indices are contracted. Now we have a whole set of quartic couplings, satisfying equations generalized from (10.9).

$$\beta_{ijkl} = \frac{\mathrm{d}\lambda_{ijkl}}{\mathrm{d}t} = \frac{+1}{16\pi^2}[\lambda_{ijmn}\lambda_{mnkl} + \lambda_{ikmn}\lambda_{mnjl} + \lambda_{ilmn}\lambda_{mnjk}]. \quad (10.10)$$

The theory is still not asymptotically free because one can easily find that there are β-functions having positive slopes. For example,

$$\beta_{1111} = \frac{3}{16\pi^2}\lambda_{11mn}\lambda_{mn11} > 0. \quad (10.11)$$

We now consider theories with scalar bosons and fermions interacting through the renormalizable Yukawa coupling

$$\mathscr{L} = \bar{\psi}(\mathrm{i}\gamma^\mu\,\partial_\mu - m_\psi)\psi + \tfrac{1}{2}[(\partial_\mu\phi)^2 - m_\phi^2\phi^2] - \lambda\phi^4 + \rho\bar{\psi}\psi\phi. \quad (10.12)$$

We have two coupled renormalization group equations corresponding to the graphs in Figs 10.2 and 10.3

$$\beta_\lambda = \frac{\mathrm{d}\lambda}{\mathrm{d}t} = A_\lambda\lambda^2 + B_\lambda\lambda\rho^2 + C_\lambda\rho^4 \quad (10.14)$$

$$\beta_\rho = \frac{\mathrm{d}\rho}{\mathrm{d}t} = A_\rho\rho^3 + B_\rho\lambda^2\rho. \quad (10.15)$$

It should be noted that the lowest-order terms are not necessarily all single-loop diagrams. In particular, Fig. 10.3(b) is a two-loop term. However, in

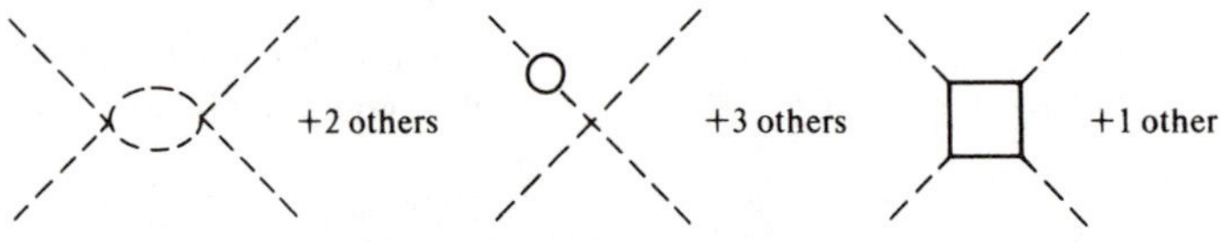

FIG. 10.2. Lowest-order contribution to the λ-coupling renormalization constant.

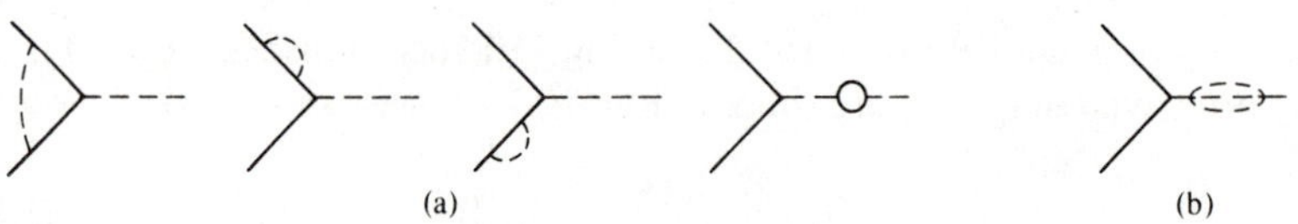

FIG. 10.3. Lowest-order contribution to the Yukawa coupling renormalization constant.

order to have a negative β_λ in eqn (10.14) with A_λ positive, we must have ρ^2 at least of the order of λ, hence the $\lambda^2\rho$ term of Fig. 10.3(b) may be dropped when compared to the ρ^3 term of Fig. 10.3(a) which yields the result

$$\beta_\rho = \frac{+1}{16\pi^2}(2 + \tfrac{1}{2} + \tfrac{1}{2} + 2)\rho^3 > 0. \tag{10.16}$$

Hence the theory (10.12) is also not asymptotically free. Again this statement can be generalized to the entire class of renormalizable theories with spin-0 and spin-1/2 fields having all possible internal symmetries. Thus we have the interaction Lagrange density

$$\mathscr{L}_1 = -\lambda_{ijkm}\phi_i\phi_j\phi_k\phi_m + \bar{\psi}_a(A^k_{ab} + \mathrm{i}B^k_{ab}\gamma_5)\psi_b\phi_k \tag{10.17}$$

The combination $\rho^k_{ab} = A^k_{ab} + \mathrm{i}B^k_{ab}$ satisfies the renormalization group equation generalized from (10.16) as

$$\beta_\rho = \frac{\mathrm{d}\rho^i}{\mathrm{d}t} = \frac{1}{16\pi^2}[2\rho^j\rho^{i\dagger}\rho^j + \tfrac{1}{2}\rho^j\rho^{j\dagger}\rho^i + \tfrac{1}{2}\rho^i\rho^j\rho^{j\dagger}$$
$$+ (\mathrm{tr}\,\rho^i\rho^{j\dagger})\rho^j + (\mathrm{tr}\,\rho^{i\dagger}\rho^j)\rho^j], \tag{10.18}$$

repeated indices being summed over. We have used the property that, when a (massless) fermion propagator is moved over, the gamma matrix commutation is such that one has $\rho \to \rho^\dagger$. From this, we get

$$\frac{\mathrm{d}}{\mathrm{d}t}(\mathrm{tr}\,\rho^{\dagger i}\rho^i) = \frac{1}{8\pi^2}[2(\mathrm{tr}\,\rho^{i\dagger}\rho^j\rho^{i\dagger}\rho^j)$$
$$+\frac{1}{2}(\mathrm{tr}\,\rho^{i\dagger}\rho^j\rho^{j\dagger}\rho^i) + \frac{1}{2}(\mathrm{tr}\,\rho^{i\dagger}\rho^i\rho^j\rho^{j\dagger})$$
$$+ \mathrm{Re}(\mathrm{tr}\,\rho^{i\dagger}\rho^j)(\mathrm{tr}\,\rho^{i\dagger}\rho^j) + (\mathrm{tr}\,\rho^{i\dagger}\rho^j)(\mathrm{tr}\,\rho^i\rho^{j\dagger})]. \tag{10.19}$$

The second and third terms on the right-hand side are positive definite since they are traces of the square of hermitian matrices $(\rho^i\rho^{i\dagger})$. The fourth term is less than the last term because

$$\mathrm{Re}[(\mathrm{tr}\,\rho^{i\dagger}\rho^j)(\mathrm{tr}\,\rho^{i\dagger}\rho^j)] \leq |\mathrm{tr}(\rho^{i\dagger}\rho^j)|^2 = (\mathrm{tr}\,\rho^{i\dagger}\rho^j)(\mathrm{tr}\,\rho^i\rho^{j\dagger}). \tag{10.20}$$

Hence,

$$8\pi^2\frac{\mathrm{d}}{\mathrm{d}t}(\mathrm{tr}\,\rho^{i\dagger}\rho^i) \geq 2[(\mathrm{tr}\,\rho^{i\dagger}\rho^j\rho^{i\dagger}\rho^j) + (\mathrm{tr}\,\rho^{i\dagger}\rho^j)(\mathrm{tr}\,\rho^i\rho^{j\dagger})]$$
$$= (\rho^i_{ab}\rho^i_{cd} + \rho^i_{ad}\rho^i_{cb})(\rho^{j\dagger}_{ba}\rho^{j\dagger}_{dc} + \rho^{j\dagger}_{da}\rho^{j\dagger}_{bc}) \geq 0. \tag{10.21}$$

to reach the last line we have written out the trace terms explicitly and relabelled some of their indices. Thus *all* renormalizable theories with only spin-0 and spin-1/2 fields are not asymptotically free. Finally we have the familiar result that the QED β-function has a positive slope at the origin of coupling space, since

$$e = Z_1^{-1}Z_2Z_3^{1/2}e_0 = Z_3^{1/2}e_0 \tag{10.22}$$

where we have used the Ward identity $Z_1 = Z_2$ and Z_3 is the photon wavefunction renormalization constant that can be calculated from the vacuum polarization diagram Fig. 10.4(a)

$$Z_3 = 1 - \frac{e^2}{6\pi^2}\ln(\Lambda/\mu). \tag{10.23}$$

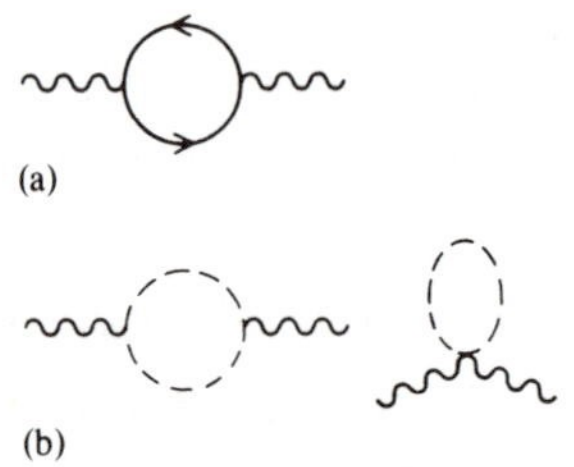

FIG. 10.4.

Thus, just as in (10.8), we have

$$\beta_e = -e\frac{\partial[\ln Z_3^{1/2}(e_0, \Lambda/\mu)]}{\partial \ln \Lambda}$$

$$= \frac{+e^3}{12\pi^2} + O(e^5). \tag{10.24}$$

Similarly, for scalar QED with Z_3 receiving a contribution from the charged scalar meson loop diagram of Fig. 10.4(b), we obtain

$$\beta_e = \frac{+e^3}{48\pi^2} + O(e^5). \tag{10.25}$$

Non-Abelian gauge theories are asymptotically free

The general Yang–Mills theory has the Lagrange density given in §8.1

$$\mathscr{L} = -\tfrac{1}{2}\,\mathrm{tr}(F_{\mu\nu}F^{\mu\nu}) \tag{10.26}$$

$$F_{\mu\nu} = \partial_\mu A_\nu - \partial_\nu A_\mu - \mathrm{i}g[A_\mu, A_\nu]$$

where

$$A_\mu = T^a A_\mu^a$$

is a matrix of hermitian vector fields with

$$[T^a, T^b] = \mathrm{i}C^{abc}T^c$$

$$\mathrm{tr}(T^aT^b) = \tfrac{1}{2}\,\delta^{ab}. \tag{10.27}$$

To quantize the theory we must fix the gauge. In §9.2 the covariant-gauge Feynman rules are given, with its gauge-fixing terms and the Faddeev–Popov ghosts. We have in particular the gauge boson propagator in the form

$$\mathrm{i}\Delta_{\mu\nu}^{ab}(k) = \mathrm{i}[-g_{\mu\nu} + (1-\xi)k_\mu k_\nu/k^2]\frac{\delta^{ab}}{k^2 + \mathrm{i}\varepsilon} \tag{10.28}$$

where ξ is the gauge parameter. The β-function can then be calculated much in the same manner as in all the other renormalizable theories considered above. If we choose to define the gauge coupling through the vector three-point function according to rule (iii) of §9.2

$$\mathrm{i}\Gamma^{abc}_{\mu\nu\lambda}(k_1, k_2, k_3) = \mathrm{i}gC^{abc}[g_{\mu\nu}(k_1 - k_2)_\lambda + g_{\nu\lambda}(k_2 - k_3)_\mu + g_{\lambda\mu}(k_3 - k_1)_\nu]. \tag{10.29}$$

Then the renormalized coupling constant is related to the bare coupling as

$$g = Z_A^{3/2} Z_g^{-1} g_0 \tag{10.30}$$

where Z_A is the vector wavefuntion renormalization constant

$$A_\mu = Z_A^{-1/2} A_{0\mu} \tag{10.31}$$

or, equivalently, is defined in terms of the unrenormalized (transverse) vector propagator at the (Euclidean) subtraction point $k^2 = -\mu^2$

$$[\mathrm{i}\Delta^{ab}_{\mu\nu}(k)]_0^{\mathrm{tr}}|_{k^2=-\mu^2} = \mathrm{i}Z_A\left(g_{\mu\nu} + \frac{k_\mu k_\nu}{\mu^2}\right)\delta^{ab}/\mu^2. \tag{10.32}$$

Similarly, the vertex renormalization constant Z_g can be defined through the unrenormalized three-point vertex (10.29)

$$[\Gamma^{abc}_{\mu\nu\lambda}(k_1, k_2, k_3)]_0 = Z_g^{-1} g_0 C^{abc}[g_{\mu\nu}(k_1 - k_2)_\lambda + g_{\nu\lambda}(k_2 - k_3)_\mu + g_{\lambda\mu}(k_3 - k_1)_\nu] \quad \text{at} \quad k_i^2 = -\mu^2. \tag{10.33}$$

One may find it helpful to compare eqns (10.30), (10.31), (10.32), and (10.33) to their $\lambda\phi^4$ counterparts in eqns (2.40), (2.23), (2.21), and (2.38). The one-loop contributions to Z_A and Z_g are shown in Figs. 10.5 and 10.6, respectively. After a tedious calculation one finds

$$Z_A = 1 + \frac{g_0^2}{16\pi^2}\left(\frac{13}{3} - \xi\right)t_2(\mathrm{V}) \ln \Lambda/\mu \tag{10.34}$$

FIG. 10.5. Vector self-energy graphs (the dotted loop is that of the FP ghosts).

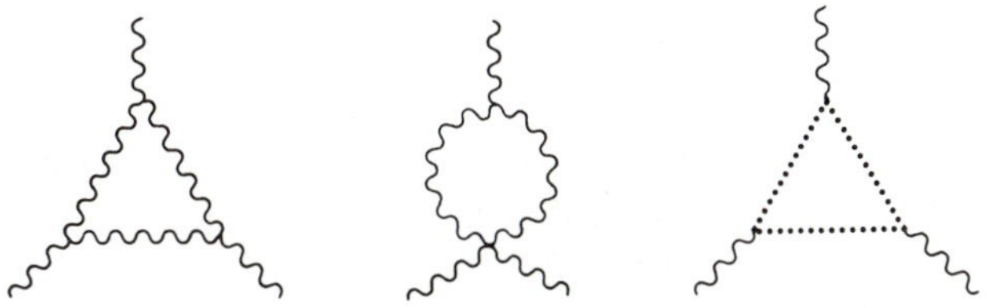

FIG. 10.6. Trilinear gauge-boson vertex correction.

and

$$Z_g = 1 + \frac{g_0^2}{16\pi^2}\left(\frac{17}{6} - \frac{3\xi}{2}\right)t_2(\text{V}) \ln \Lambda/\mu \tag{10.35}$$

where

$$t_2(\text{V})\, \delta^{ab} = C^{acd}C^{bcd}. \tag{10.36}$$

As we have seen in §4.1 the structure constants C^{abc} themselves form the adjoint representation matrix

$$C^{abc} = [T^a(\text{V})]_{bc} \tag{10.37}$$

and (10.36) may be written as

$$t_2(\text{V})\, \delta^{ab} = \text{tr}\{T^a(\text{V})T^b(\text{V})\}. \tag{10.38}$$

We have labelled these quantities with V, for vector, since the vector gauge fields A^a_μ belong to the adjoint representation of the group. Hence (10.38) shows that $t_2(\text{V})$ can be interpreted as the sum of the squared symmetry charges of the vector gauge particles. Also from the simple property of the SU(n) adjoint representation discussed in §4.1, we obtain by using eqns (4.21) and (4.136)

$$t_2(\text{V}) = \text{n} \quad \text{for} \quad \text{SU}(n). \tag{10.39}$$

From (10.34) and (10.35) we immediately obtain the famous result

$$\beta_g = -g\frac{\partial}{\partial \ln \Lambda}\left[\frac{3}{2}\ln Z_A - \ln Z_g\right]$$

$$= -\frac{g^3}{16\pi^2}\frac{11}{3}t_2(\text{V}) < 0 \tag{10.40}$$

which, at this one-loop level, turns out to be independent of the gauge parameter ξ. If the gauge fields are coupled to fermions and scalar mesons with representation matrices $T^a(\text{F})$ and $T^a(\text{S})$, respectively, then we can make use of results (10.24) and (10.25) directly

$$\beta_g = \frac{g^3}{16\pi^2}\left[-\frac{11}{3}t_2(\text{V}) + \frac{4}{3}t_2(\text{F}) + \frac{1}{3}t_2(\text{S})\right] \tag{10.41}$$

where

$$t_2(\text{F})\, \delta^{ab} = \text{tr}\{T^a(\text{F})T^b(\text{F})\}$$

and

$$t_2(\text{S})\, \delta^{ab} = \text{tr}\{T^a(\text{S})T^b(\text{S})\} \tag{10.42}$$

are the sums of the squared symmetry charges for the fermions and scalars, respectively. For fermions and scalars in the fundamental representation of SU(n), we have $T^a(\text{S}) = T^a(\text{F}) = \lambda^a/2$ and

$$t_2(\text{F}) = t_2(\text{S}) = 1/2. \tag{10.43}$$

If one uses the two component fermion fields or real scalars the coefficient of $t_2(F)$ and $t_2(S)$ in (10.41) should have an additional factor of 1/2.

The Yang–Mills vacuum as a paramagnetic medium

One of the most remarkable features of quantum field theory is that Yang–Mills theories are the only asymptotically free theories in four dimensions. As it turns out there is a simple physical explanation of this phenomenon (Nielsen 1981; Hughes 1981). This explanation ultimately has to do with the fact that non-Abelian gauge fields have spin and obey Bose–Einstein statistics and, unlike the Abelian photon, they carry the gauge symmetry charges themselves.

As we have mentioned at the beginning of this section, asymptotic freedom means that the vacuum anti-shields charges, i.e. it acts like a dielectric medium with dielectric constant

$$\varepsilon < 1. \tag{10.44}$$

Also the quantum field theory vacuum differs from the ordinary polarizable medium on a very important point: it is relativistically invariant. This means that the (relative) magnetic permeability μ is related to the dielectric constant by

$$\mu\varepsilon = 1 \tag{10.45}$$

so that the velocity of light is 1 in the vacuum. This allows us to translate the electric responses into its magnetic responses, which have two elements.

(1) *Landau diamagnetism* ($\mu < 1$). The charged particles in the medium, in response to the external magnetic field, produce a current which itself induces a magnetic field opposing the original field.

(2) *Pauli paramagnetism* ($\mu > 1$). If the particles have magnetic moments they tend to align with the external field.

It turns out to be easier to visualize the magnetic response of the Yang–Mills vacuum; the anti-screening of (10.44) means

$$\mu > 1.$$

The Yang–Mills vacuum acts like a paramagnetic medium. We note that such a correspondence does not exist for ordinary polarizable material which can, for example, have both the properties of ($\varepsilon > 1$) dielectric screening and of ($\mu > 1$) paramagnetism.

It should be emphasized that the electromagnetic terminology is used here only as an analogue to ordinary U(1) gauge theory. Thus by charge we really mean the gauge symmetry charges. For example, in the SU(3) gauge theory of QCD, they are the colour charges; by electric and magnetic responses we mean the colour electric and magnetic responses, and so on. When we say that the Yang–Mills fields of QCD (gluons) carry charge, magnetic moment, electric quadrupole moment, etc. we mean they carry colour charge, colour magnetic moment, etc. (In actual fact gluons are electrically neutral.) What are then the charge, magnetic moment, etc. of the Yang–Mills particle? Consider the simple SU(2) theory. It has a trilinear gauge field coupling given by the Feynman rule (iii) of §9.2.

$$\mathrm{i}\Gamma^{abc}_{\mu\nu\lambda}(k_1, k_2, k_3) = \mathrm{i}\varepsilon^{abc}[g_{\mu\nu}(k_1 - k_2)_\lambda + g_{\nu\lambda}(k_2 - k_3)_\mu + g_{\lambda\mu}(k_3 - k_1)_\nu] \tag{10.46}$$

which can be thought of as the vertex of a charged spin-1 particle and a photon. For the purpose of this interpretation we assume that the SU(2) gauge fields A^a_μ have the correspondence

$$A^3_\mu \to A_\mu \quad \text{corresponds to the photon field;}$$

$$\frac{1}{\sqrt{2}}(A^1_\mu \pm A^2_\mu) \to V^\pm_\mu \text{ corresponds to the charged spin-1 fields with mass M.} \tag{10.47}$$

Then the ε^{abc} factor gives

$$\varepsilon^{-+3} = \tfrac{1}{2}\mathrm{i}(\varepsilon^{123} - \varepsilon^{213}) = \mathrm{i}. \tag{10.48}$$

We pick the momentum configuration such that (recall that all k_is are supposed to point into the vertex)

$$k^\mu_1 \simeq \left(M, -\frac{\mathbf{k}}{2}\right), \quad k^\mu_2 \simeq \left(-M, -\frac{\mathbf{k}}{2}\right), \quad k_3 \simeq (0, \mathbf{k})$$

for $M \gg |\mathbf{k}|$; the polarization vectors for the charged particles are

$$\varepsilon^\mu_1 = \left(-\frac{\mathbf{k}\cdot\mathbf{e}}{2M}, \mathbf{e}\right), \quad \varepsilon^\mu_2 = \left(\frac{\mathbf{k}\cdot\mathbf{e}'}{2M}, \mathbf{e}'\right)$$

so that $\varepsilon_1 \cdot k_1 \simeq \varepsilon_2 \cdot k_2 \simeq 0$. After contracting ε^μ_1, ε^ν_2 and A^λ into (10.46) we obtain

$$\mathrm{i}\Gamma^{-+3}_{\mu\nu\lambda}\varepsilon^\mu_1\varepsilon^\nu_2 A^\lambda = 2M\left[g\mathbf{e}\cdot\mathbf{e}'A_0 + \frac{g}{M}(\mathbf{e}\times\mathbf{e}')\cdot(\mathbf{k}\times\mathbf{A})\right] + O(k^2). \tag{10.49}$$

Thus we can identify g with the electric charge and g/M with the magnetic moment corresponding to a gyromagnetic ratio of

$$\gamma_V = 2. \tag{10.50}$$

We can calculate the vacuum energy density in the presence of an external magnetic field

$$u_0 = \frac{1}{2\mu} B^2_{\mathrm{ext}}. \tag{10.51}$$

From this the magnetic permeability μ can be extracted. Nielsen (1981) and Hughes (1981) have shown that, for $\mu \equiv 1 + \chi$ where χ is the magnetic susceptibility,

$$\chi \sim (-1)^{2s} q^2 \sum_{s_3} \left(-\frac{1}{3} + \gamma^2 s_3^2\right) \tag{10.52}$$

where q, γ, and s_3 are the charge, gyromagnetic ratio, and the projection of spin in the direction of the external magnetic field, respectively. The two terms correspond to diamagnetic and paramagnetic responses, respectively. The factor $(-1)^{2s}$ in front means that there is an extra minus sign for a fermion system. When normal-ordering the creation and annihilation operators in the Hamiltonian to isolate the vacuum energy term, the anticommutation relations of the fermion fields give rise to this extra minus sign. (This is the same reason that a fermion loop in Feynman graphs is accompanied by a minus sign.) As a simple check one sees that for fermion ($\gamma_F = 2$)

$$\chi_F \sim -q_F^2 2(-\tfrac{1}{3} + 1) = -\tfrac{4}{3} q_F^2. \tag{10.53}$$

That the susceptibility is negative means that the system is diamagnetic ($\mu_F < 1$) hence has the property of dielectric screening $\varepsilon_F > 1$ as in QED. Also note the well-known ratio of 3 for the relative paramagnetic and diamagnetic contributions. Keeping in mind that the massless vector gauge particles have only two helicity states $s_3 = \pm 1$, we obtain, for the vector, fermion, and scalar particles

$$\chi \sim \tfrac{22}{3} q_V^2 - \tfrac{4}{3} q_F^2 - \tfrac{1}{3} q_S^2. \tag{10.54}$$

To convert these to the gauge charges, the squared charge factors are identified with the trace terms of (10.38) and (10.42)

$$\begin{aligned} q_V^2 &\to \tfrac{1}{2} t_2(V) \\ q_F^2 &\to t_2(F) \\ q_S^2 &\to t_2(S). \end{aligned} \tag{10.55}$$

We then obtain a result identical to (10.41). The factor 1/2 in (10.55) reflects the fact that in gauge theories the vector particles have been represented by hermitian fields and each complex charged field actually has two real components (see, for example, eqn (10.47)).

Gauge theories with scalar mesons

We now consider the possibility of giving *all* Yang–Mills vector bosons masses through the Higgs mechanism (as in §8.3) without destroying asymptotic freedom (Gross and Wilczek 1973*b*; Cheng, Eichten, and Li 1974). This appears to be very difficult. The problem has to do with the quartic couplings of the Higgs scalars which tend to be ultraviolet unstable. The modification of eqn (10.10) involves adding contributions from the diagrams in Fig. 10.7

$$-\frac{1}{16\pi^2}\left[12 s_2(S) g^2 \lambda_{ijkl} - 3 A_{ijkl} g^4\right] \tag{10.56}$$

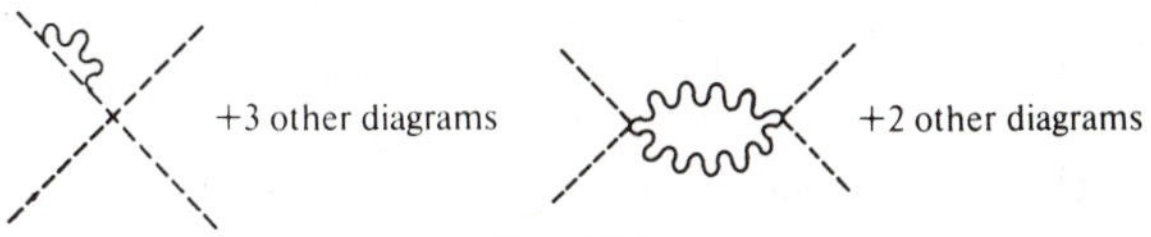

FIG. 10.7.

where

$$s_2(S)\,\delta_{ij} \equiv (T^a(S)T^a(S))_{ij}$$
$$A_{ijkl} \equiv \{T^a(S), T^b(S)\}_{ij}\{T^a(S), T^b(S)\}_{kl}$$
$$+ \text{ two other terms by permutation.} \tag{10.57}$$

Thus basically we have renormalization group equations of the form

$$\frac{dg^2}{dt} = -b_0 g^4 \qquad b_0 > 0$$

$$\frac{d\lambda}{dt} = A\lambda^2 + B'\lambda g^2 + Cg^4. \tag{10.58}$$

Introducing the variable $v = \lambda/g^2$, eqn (10.58) becomes

$$\frac{1}{g^2}\frac{dv}{dt} = Av^2 + Bv + C \equiv \beta_v \tag{10.59}$$

with $B = B' - b_0$. Asymptotic freedom requires that $g \to 0$ and $\lambda \to 0$, i.e. v approaches a fixed point in the ultraviolet limit. Since the right-hand side of (10.59) is a second-order polynomial, the condition for $\beta_v = 0$ to have real roots is simply

$$\Delta = B^2 - 4AC \geq 0. \tag{10.60}$$

Let us call these two roots v_1 and v_2 with $v_2 > v_1$. Since the slopes at these two points

$$\left.\frac{d\beta_v}{dv}\right|_{v_1} = A(v_1 - v_2) < 0$$

and

$$\left.\frac{d\beta_v}{dv}\right|_{v_2} = A(v_2 - v_1) > 0, \tag{10.61}$$

the smaller v_1 is a stable fixed point. But λ and v_1 are required to remain positive (so that the classical potential is bound from below); this requires

$$B < 0 \tag{10.62}$$

because both A and C are positive and $v_1 = (-B - \sqrt{\Delta})/4A < -B/2A$. In all the cases examined, these asymptotic conditions (eqns (10.60) and (10.62)) always imply that only a small number of scalar mesons are allowed in the theory, too small a number to do the job of breaking down the gauge symmetry completely and giving all gauge bosons nonzero masses. This situation is not changed even in the presence of Yukawa couplings: their contributions to (10.58) are generally small. (An important exception is the supersymmetric theory.) Thus gauge symmetry is not broken spontaneously in an asymptotically free theory and this suggests that one should work with such theories omitting elementary scalars altogether.

Product groups

Up to this point we have restricted our considerations to simpe Lie groups, i.e. theories with only one gauge coupling constant. The more general case involves direct products of simple groups $G_1 \times G_2 \times \ldots G_n$, each with its own coupling constants g_i. To lowest-order in g_i^3 the β-functions are independent of each other and, therefore, the results can be deduced directly from those for simple groups. In particular, if one of these factors, G_i, is an Abelian U(1) group, the associated gauge couplings will not be driven to zero and the theory is not asymptotically free.

10.2 The QCD Lagrangian and the symmetries of the strong interaction

The success of the quark–parton model in describing Bjorken scaling as observed in deep inelastic lepton–hadron scattering clearly suggests that the field theory of the strong interaction should be asymptotically free so that the quark can interact weakly at short distances. We have shown in the last section that only Yang–Mills theories can exhibit free-field asymptotic behaviour at large Euclidean momenta.

Which symmetry of the quark model should be gauged? We have already seen in §4.4 how, by postulating that quarks have a hidden three-valued quantum number called colour, one can overcome the paradoxes of the simple quark model. This idea of exact colour symmetry is strengthened by the agreement with experimental measurements of the anomaly calculation of the $\pi^\circ \to 2\gamma$ rate (§6.2) and of the parton-model calculation of $\sigma(e^+e^- \to \text{hadrons})$ (§7.2). Furthermore, since we also need to assume that only colour singlets are observable, it suggests that the forces between the coloured quarks must be colour-dependent. In fact a colour-independent strong interaction would imply the phenomenologically unacceptable result that every hadron should have degenerate partners having different colours. All this leads to the idea that it is the colour symmetry of the quark model that should be gauged. Thus, the strong interaction should be described by an SU(3) colour Yang–Mills theory with each flavour of quarks transforming as the fundamental triplet representation. This, together with our requirement that the strong interaction theory be renormalizable, fixes (almost) completely the form of the Lagrangian. The theory is called *quantum chromodynamics* (Gross and Wilczek 1973*a*; Weinberg 1973*b*; Fritzsch, Gell-Mann, and Leutweyler 1973) with a Lagrangian usually written as

$$\mathscr{L}_{\text{QCD}} = -\tfrac{1}{2}\,\text{tr}\, G_{\mu\nu}G^{\mu\nu} + \sum_k^{n_f} \bar{q}_k(i\gamma^\mu D_\mu - m_k)q_k \tag{10.63}$$

where

$$G_{\mu\nu} = \partial_\mu A_\nu - \partial_\nu A_\mu - ig[A_\mu, A_\nu]$$
$$D_\mu q_k = (\partial_\mu - igA_\mu)q_k$$
$$A_\mu = \sum_{a=1}^{8} A_\mu^a \lambda^a/2 \tag{10.64}$$

where the λ^as are the Gell-Mann matrices that satisfy the SU(3) commutation relations

$$\left[\frac{\lambda_a}{2}, \frac{\lambda_b}{2}\right] = \mathrm{i}f^{abc}\frac{\lambda^c}{2} \tag{10.65}$$

and the normalization condition

$$\mathrm{tr}(\lambda^a \lambda^b) = 2\,\delta^{ab}. \tag{10.66}$$

The strong interaction gauge fields A^a_μ are called *gluons* and the q_ks are the quark fields with the subscript k being the flavour index $k = 1, 2, \ldots, n_f$ (n_f is the number of quark flavours)

$$q_k\colon \mathrm{u, d, s, c, b}, \ldots \tag{10.67}$$

In (10.63) we have left out a possible SU(3)-invariant and dimension-4 renormalizable term

$$\mathrm{tr}\, G_{\mu\nu}\tilde{G}^{\mu\nu} \quad \text{with} \quad \tilde{G}^{\mu\nu} = \tfrac{1}{2}\varepsilon^{\mu\nu\lambda\rho}G_{\lambda\rho}.$$

Such a term can be written as the divergence of a current $\mathrm{tr}\,\tilde{G}G \sim \partial^\mu K_\mu$ hence it contributes only as a surface term in the action. Making the usual assumption of fields vanishing at infinity $A^a_\mu \to 0$, one is normally justified in discarding this term. As it turns out, for a class of gauge fields with nontrivial topological properties, this justification may not hold. Experimentally we know that, if the $\mathrm{tr}\,\tilde{G}G$ term exists in the QCD Lagrangian, it must be multiplied by an extremely small coefficient. For the time being we shall decree its absence and shall take up this whole area of instanton problems in Chapter 16.

The QCD Lagrangian (10.63) clearly possesses all the well-known strong-interaction symmetries. It conserves charge conjugation and parity. Because the gluons are flavour-independent it conserves strangeness, etc. In fact eqn (10.63) has all the flavour symmetries of a free quark model, particularly the SU(3) × SU(3) chiral symmetry, broken explicitly by the quark mass term, as discussed in Chapter 5 (see comments at the end of §5.5). If QCD dynamics is such that chiral symmetry is realized in the Goldstone mode then all the successes of PCAC and current algebra can be accounted for.

Gauge invariance, renormalizability, and QCD symmetries

It is important to realize that these symmetry properties are not put in eqn (10.63) by hand; they are the consequences of gauge invariance and renormalizability. In the following we shall show that (10.63) *is* equivalent to the most general renormalizable SU(3) Yang–Mills theory of quarks and gluons

$$\mathscr{L}^{(0)} = \bar{q}(A + B\gamma_5)\mathrm{i}\gamma^\mu \mathrm{D}_\mu q + \bar{q}(C + iD\gamma_5)q - \frac{Z}{2}\,\mathrm{tr}\,GG \tag{10.68}$$

where A, B, C, and D are all hermitian matrices in the flavour space and Z is a constant. We can rewrite (10.68) in terms of the left-handed and right-handed quark fields

$$\mathrm{q_L} = \tfrac{1}{2}(1 - \gamma_5)\mathrm{q}, \qquad \mathrm{q_R} = \tfrac{1}{2}(1 + \gamma_5)\mathrm{q}$$

with the result

$$\mathscr{L}^{(0')} = \bar{\mathrm{q}}_\mathrm{L}(A + B)\mathrm{i}\gamma^\mu \mathrm{D}_\mu \mathrm{q}_\mathrm{L} + \bar{\mathrm{q}}_\mathrm{R}(A - B)\mathrm{i}\gamma^\mu \mathrm{D}_\mu \mathrm{q}_\mathrm{R}$$

$$+ \bar{\mathrm{q}}_\mathrm{L}(C + \mathrm{i}D)\mathrm{q}_\mathrm{R} + \bar{\mathrm{q}}_\mathrm{R}(C - \mathrm{i}D)\mathrm{q}_\mathrm{L} - \frac{Z}{2}\,\mathrm{tr}\, GG. \qquad (10.69)$$

This can be transformed into eqn (10.63) by the following two steps.

(1) We will first rescale the gluon field and gauge coupling using

$$A^a_\mu \to Z^{1/2} A^a_\mu, \qquad g \to Z^{-1/2} g$$

so that $Z \to 1$ in the gluon kinetic energy term without affecting the covariant derivatives of the quark fields. Then introduce new $\mathrm{q_L}$ and $\mathrm{q_R}$ quark fields so that the two independent matrices $A + B$ and $A - B$ both become unit matrices. Now the Lagrangian takes on the form

$$\mathscr{L}^{(1)} = \bar{\mathrm{q}}_\mathrm{L}\mathrm{i}\gamma^\mu \mathrm{D}_\mu \mathrm{q}_\mathrm{L} + \bar{\mathrm{q}}_\mathrm{R}\mathrm{i}\gamma^\mu \mathrm{D}_\mu \mathrm{q}_\mathrm{R} + \bar{\mathrm{q}}_\mathrm{L} M \mathrm{q}_\mathrm{R} + \bar{\mathrm{q}}_\mathrm{R} M^\dagger \mathrm{q}_\mathrm{L} - \tfrac{1}{2}\,\mathrm{tr}\, GG \qquad (10.70)$$

where

$$M = C + \mathrm{i}D \qquad (10.71)$$

or, equivalently

$$\mathscr{L}^{(1')} = \bar{\mathrm{q}}\mathrm{i}\gamma^\mu \mathrm{D}_\mu \mathrm{q} + \bar{\mathrm{q}}(C + \mathrm{i}D\gamma_5)\mathrm{q} - \tfrac{1}{2}\,\mathrm{tr}\, GG. \qquad (10.72)$$

We have not bothered to introduce new labels for the new fields and matrices.

(2) We now make use of an important result of the linear algebra. (A proof of this theorem will be presented in §11.3.) It states that a general matrix such as M in (10.71), which is neither diagonal nor symmetric, can always be diagonalized with positive eigenvalues by a bi-unitary transformation. Thus,

$$SMT^\dagger = M_\mathrm{d} \qquad (10.73)$$

where S and T are unitary matrices and M_d is diagonal with positive elements

$$\mathrm{tr}\, M_\mathrm{d} = \sum_k m_k. \qquad (10.74)$$

We see immediately that this allows us to transform $\mathscr{L}^{(1)}$ of (10.70) into the canonical form of (10.63). Thus, we can redefine the quark fields

$$\mathrm{q_L} \to S\mathrm{q_L} \quad \text{and} \quad \mathrm{q_R} \to T\mathrm{q_R} \qquad (10.75)$$

so that the mass term in (10.72) is diagonalized and free of γ_5 without at the same time introducing γ_5s into the $\mathrm{D}_\mu\mathrm{q}$ terms. The physics that makes this possible is that strong interactions are mediated by flavour-neutral vector gluons. Had it not been for the flavour-independence of the colour gluon fields, we would not have the matrices A, B, C, and D in (10.68) to commute with the generators of the gauge group; had it not been for the spin-1 nature of the gluon, the symmetries of the strong interactions would be controlled by other terms besides the quark mass matrices. Thus this is another argument (besides the difficulties discussed in the last section on the Higgs mechanism in an asymptotically free theory) against the presence of elementary scalar fields in strong interaction theory.

Two more general comments on QCD symmetries follow.

(A) **Chiral symmetries of QCD.** We have already stated that, in the limit $m_u = m_d = m_s = 0$, $\mathscr{L}_{QCD}$, like that of the free-quark model, is invariant under the chiral unitary transformations of eqns (5.39) and (5.47). In other words, in the absence of the quark mass matrix, the theory is invariant under the unitary transformations of (10.75), and we have a $U(3)_L \times U(3)_R$ symmetry. The diagonal subgroups SU(3) and U(1) are realized in the normal mode; i.e. the vacuum is also invariant under the $U(3)_{L+R}$ transformations. The hadrons form degenerate SU(3) multiplets and baryon number is conserved. The remaining symmetries—the axial SU(3) and U(1) symmetries, corresponding to the $U(3)_{L-R}$ transformations—are not manifest in particle degeneracies. Since we are not using any elementary scalar fields in the theory, we must assume that the dynamics is such that the QCD vacuum breaks these axial symmetries. (Whether this actually takes place is a difficult dynamical problem that is still not completely settled yet; but all indications are that this indeed takes place according to our expectation.) The Goldstone theorem then informs us that there should be approximately massless pseudoscalar mesons in the hadron spectrum. Eight of them can indeed be identified readily 3 πs, 4 Ks, and 1 η. However, since we need to break an axial U(3) symmetry, we are still one pseudo-scalar short. This is the famous axial U(1) problem. Namely, in the massless limit QCD (in fact any quark model) is invariant under the phase rotation

$$q_k \to e^{i\gamma_5\theta} q_k \tag{10.76}$$

where we have the same θ for all k, i.e. u_L, d_L, s_L are multiplied by a common phase $e^{-i\theta}$ and u_R, d_R, s_R by $e^{i\theta}$. This approximate symmetry is not observed in the strong interaction: it is not realized either in the normal or the Goldstone mode. The resolution of this U(1) problem will be discussed in Chapter 16 in connection with the instanton solutions of QCD mentioned at the beginning of this section.

(B) **Stability of QCD symmetries against weak radiative corrections.** We now discuss briefly the problem of symmetry violation terms as induced by weak radiative corrections (Weinberg 1973*b*). Although the subject of gauge theories of electroweak interactions has not been introduced, we can still discuss this problem since we only need a few general properties of such theories.

(i) The generators of the electroweak gauge group commute with all those of QCD, i.e. gluons are flavour-neutral and weak intermediate vector bosons (W-bosons) and currents are all colour singlets.

(ii) The weak interactions have an energy scale set by the W-boson masses $M_W = O(10^2 \text{ GeV})$.

(iii) The weak gauge coupling constants are of order e and are related to the familiar weak Fermi constant by $G_F = O(\alpha/M_W^2)$, where $\alpha = e^2/4\pi$ is the fine structure constant.

Knowing that weak couplings are $O(e)$, *a priori* one would fear that weak radiative corrections would induce unacceptably large $O(\alpha)$ violations of parity and strangeness conservations. However property (i) implies that these radiative corrections themselves are all invariant under $SU(3)_{\text{colour}}$. In particular the order α additions to the Lagrangian must be dimension-4 operators; hence (as we have demonstrated above), with suitable re-definitions of the gluon and quark fields, the Lagrangian can be restored to the canonical form of (10.63) with all its symmetries. Terms involving operators of dimension $D > 4$ will, by dimensional analysis, be multiplied by coefficients $(M_W)^{-(D-4)}$. For example, a term of the form $\bar{q}\gamma_\mu(1-\gamma_5)q\bar{q}\gamma^\mu(1-\gamma_5)q$ has $D = 6$; hence it must have a coefficient $O(\alpha/M_W^2)$ and, by property (iii), of order G_F.

Thus QCD has the attractive feature that in zeroth order it automatically possesses a set of global symmetries which match perfectly with the known strong-interaction symmetries and which are stable against weak radiative corrections.

10.3 Renormalization group analysis of scaling and scaling violation

For the QCD Lagrangian (10.63), the renormalization group β-function (eqn (10.41)) with $t_2(V) = 3$, $t_2(F) = 1/2$, and $t_2(S) = 0$ takes on the value

$$\beta_g = \frac{-1}{16\pi^2}(11 - \tfrac{2}{3}n_f)g^3 \equiv -bg^3 \tag{10.77}$$

where n_f is the number of quark flavours. As one changes the momentum scale $p_i \to \lambda p_i$, the effective gauge coupling $\bar{g}(g, t)$ obeys the equation

$$\frac{d\bar{g}}{dt} = -b\bar{g}^3 \tag{10.78}$$

with

$$t = \ln \lambda. \tag{10.79}$$

One can integrate eqn (10.78) to obtain

$$\bar{g}^2(t) = \frac{g^2}{1 + 2bg^2 t} \tag{10.80}$$

where $g = \bar{g}(g, 0)$. Thus for $n_f < 17$, i.e. $b > 0$, the denominator of (10.80) cannot vanish. For large momenta λp_i with $\lambda \to \infty$, we have $\bar{g}(t) \to 0$ and asymptotic freedom. But we should note that the effective coupling decreases to zero very slowly, as a logarithm, $\bar{g} \sim (2b \ln \lambda)^{-1}$. For convenience, we can choose the scaling parameter λ as the ratio of the momentum of interest Q to the subtraction scale μ, i.e. $\lambda^2 = Q^2/\mu^2$ or $t = \frac{1}{2} \ln Q^2/\mu^2$. Then we can rewrite eqn (10.80)

$$\alpha_s(Q^2) = \frac{\alpha_s(\mu^2)}{1 + 4\pi b \alpha_s(\mu^2) \ln Q^2/\mu^2} \tag{10.81}$$

where $\alpha_s(Q^2) = \bar{g}^2(t)/4\pi$ and $\alpha_s(\mu^2) = g^2/4\pi$. We can further simplify this equation by defining the parameter Λ through the equation

$$\ln \Lambda^2 = \ln \mu^2 - \frac{1}{\alpha_s(\mu^2) 4\pi b}$$

to get

$$\alpha_s(Q^2) = \frac{4\pi}{(11 - \frac{2}{3}n_f) \ln Q^2/\Lambda^2}. \tag{10.82}$$

In this form, the strong gauge coupling constant $\alpha_s(Q^2)$ is expressed in terms of one single parameter Λ. From this we see that for small momenta, $\alpha_s(Q^2)$ increases and in fact it diverges at $Q^2 = \Lambda^2$. Even though eqn (10.82) is a perturbative formula and breaks down for large couplings, the value of Λ is still a useful measure for the energy scale where the strong-interaction coupling constant becomes large. Hence Λ is the fundamental momentum scale of the theory and is called the *QCD scale parameter*.

Since QCD is asymptotically free, at first sight one would think that this allows us to use the renormalization group and perturbation theory to calculate a large number of high-energy processes. Actually this is not the case: the renormalization group analysis is a theory of scale transformations and this involves uniform multiplication of all components of the four-momenta; the ultraviolet asymptotic limit is the deep Euclidean region where all particles are far away from their mass shell. Fortunately, there are physical situations where some of the 'external particles' are infinitely off their mass shell. In the lowest-order electroweak coupling approximation, the semileptonic inclusive processes can be factorized into a known leptonic part and a hadronic quantity that corresponds to a forward scattering amplitude of a photon (or W-boson) with variable mass $-q^2$. In particular the cross-sections of e^+e^- annihilation and lepton–hadron scatterings measure the absorptive part of the electroweak current product matrix elements (see eqns (7.10) and (7.152)) between some state $|A\rangle$

$$\langle A|T(J_\mu(x)J_\nu(0))|A\rangle. \tag{10.83}$$

The high-energy and high-$(-q^2)$ limit does correspond to the deep Euclidean region. (For general descriptions of applications of asymptotic freedom see Politzer 1974 and Gross 1976.)

e^+e^- annihilation

According to eqn (7.152), for the case of e^+e^- annihilation, $|A\rangle$ of eqn

(10.83) is the vacuum state. Thus the photon mass q^2 is the only scale and renormalization group analysis can be applied directly. Consider the inverse photon propagator in QED

$$\Gamma^{\mu\nu}(q) = (-g^{\mu\nu} + q^\mu q^\nu/q^2)\Pi(q^2) \tag{10.84}$$

where the vacuum polarization $\Pi(q^2)$ has the naive dimension 2, so that the relevant renormalization group equation (3.58) becomes

$$\left[\frac{\partial}{\partial t} - \beta_g \frac{\partial}{\partial g} + 2\gamma_A - 2\right]\Pi(q^2) = 0 \tag{10.85}$$

where γ_A is the anomalous dimension of the photon field,

$$\gamma_A = -\frac{1}{2}\frac{\partial}{\partial \ln \Lambda}[\ln Z_3(e, g, \Lambda/\mu)]. \tag{10.86}$$

Z_3 is the usual photon wavefunction (i.e. vacuum polarization) renormalization constant (with its one- and two-loop graphs shown in Fig. 10.8) which yields

$$\gamma_A = C\left(3\sum_k^{n_f} e_k^2\right)\left[1 + \frac{3s_2(V)}{16\pi^2}g^2 + \ldots\right]. \tag{10.87}$$

FIG. 10.8.

For $T^a(V)$, the representation matrices of the vector gauge fields, we have

$$\begin{aligned} s_2(V)\,\delta_{ij} &= (T^a(V)T^a(V))_{ij} \\ &= (n^2-1)/2n \quad \text{for} \quad SU(n). \end{aligned} \tag{10.88}$$

For SU(3) gluons $s_2(V) = 4/3$. e_k is the electric charge of quark flavour k. We have not written out the precise form of the proportional constant C since it will be cancelled in the result that we shall quote. Solving (10.85) as in eqn (3.68), we have for $Q^2 = -q^2$,

$$\begin{aligned} \Pi(Q^2) &\simeq Q^2 \exp\left[-2\int_0^t \gamma_A(g(t'))\,dt'\right] \\ &\simeq Q^2\left[1 - 2C\left(3\sum e_k^2\right)\left(t + \frac{3s_2(V)}{16\pi^2 b}\ln t + \ldots\right)\right]. \end{aligned} \tag{10.89}$$

The $\sigma(e^+e^- \to \text{hadrons})$ cross-section can be obtained by taking the absorptive part

$$R(Q^2) \equiv \frac{\sigma(e^+e^- \to \text{hadrons})}{\sigma(e^+e^- \to \mu^+\mu^-)} = \left(3\sum_k e_k^2\right)\left[1 + \frac{\alpha_s(Q^2)}{\pi} + \ldots\right]. \tag{10.90}$$

Therefore the simple parton scaling result of eqn (7.104) is recovered together with a QCD correction term (Appelquist and Georgi 1973; Zee 1973*b*) with $\alpha_s(Q^2)$ given by (10.81). Thus the ratio $R(Q^2)$ approaches $R(Q^2 = \infty)$ from above. This subasymptotic correction term, at least for the region above the charm threshold, is still probably less than the experimental uncertainties, so while the e^+e^- annihilation total cross-section is not an ideal place to measure $\alpha_s(Q^2)$, the overall experimental data is consistent with the QCD prediction of eqn (10.90).

Inelastic *l*N scatterings

According to eqn (7.10), $|A\rangle$ of (10.83) is the nucleon state for this case of *l*N scatterings. Hence we will be studying a physical quantity with two mass scales: the variable photon mass $Q^2 = -q^2 \to \infty$, but the nucleon must be on-shell $p^2 = M^2$. One must devise methods to factorize the matrix into a product of momentum-independent quantities (which will be identified with the structure functions and the parton distribution function) and q^2-dependent functions which scale according to the renormalization group. (For more explicit discussions of this factorization see §10.4.)

(A) **Operator product expansions.** The technique effecting such a factorization is the operator-product expansion (Wilson 1969) in which the singularities of the operator products are expressed as a sum of nonsingular operators with the coefficients being singular c-number functions. The physical basis for this expansion is that a product of local operators at distances small compared to the characteristic length of the system should look like a local operator.

(A1) *Short-distance expansion.*

$$A(x)B(y) \underset{(x-y)_\mu \to 0}{\approx} \sum_i C_i(x-y)\mathcal{O}_i(\tfrac{1}{2}(x+y)) \tag{10.91}$$

where A, B, and $\mathcal{O}_i$ are local operators. The $\mathcal{O}_i$s that can appear must have quantum numbers which match those of AB on the left-hand side. The $C_i(x)$s are singular c-number functions called the *Wilson coefficients*. It has been proven for renormalizable theories that such expansions are valid as $x \to y$ to any finite order of perturbation theory. The short-distance behaviour of the Wilson coefficients is expected to be that obtained, up to a logarithmic multiplicative factor, by naive dimensional counting

$$C_i(x) \underset{x \ll 1/m}{\rightarrow} (x)^{d_i - d_A - d_B}(\ln xm)^p[1 + O(xm)] \tag{10.92}$$

where d_A, d_B, and d_i are the dimensions (in units of mass) of A, B, and $\mathcal{O}_i$, respectively. The higher the dimension of $\mathcal{O}_i$ the less singular are the coefficients $C_i(x)$; hence the dominant operators at a short distance are those with the smallest dimensions.

The usefulness of this expansion derives from its universality—the Wilson coefficients are independent of the processes under consideration. Process dependence is exhibited in the matrix element of the local operator $\mathcal{O}_i$ which is nonsingular at short distances. Another advantage is that in a given theory the expansion usually involves a rather small number of operators. Hence the ensuing calculation is relatively simple.

(A2) *Light-cone expansion.* We already encountered this type of expansion in §7.3. Eqns (7.141) and (7.146) are examples of the generic light-cone expansion

$$A\left(\frac{x}{2}\right)B\left(-\frac{x}{2}\right) \approx \sum_i C_i(x)\mathcal{O}_i\left(\frac{x}{2}, -\frac{x}{2}\right) \quad \text{for} \quad x^2 \approx 0 \tag{10.93}$$

with singular c-number functions and regular bilocal operators $\mathcal{O}_i(x, y)$. Then one can expand the bilocal operators in a Taylor series (as in eqn (7.157)) to write

$$\mathcal{O}_i\left(\frac{x}{2}, -\frac{x}{2}\right) = \sum_j x^{\mu_1} \ldots x^{\mu_j} \mathcal{O}^{(j,i)}_{\mu_1 \ldots \mu_j}(0) \tag{10.94}$$

so that the product of two local operators can also be expanded in terms of local operators on the light cone

$$A\left(\frac{x}{2}\right)B\left(-\frac{x}{2}\right) \underset{x^2 \to 0}{\approx} \sum_{j,i} C_i^{(j)}(x^2) x^{\mu_1} \ldots x^{\mu_j} \mathcal{O}^{(j,i)}_{\mu_1 \ldots \mu_j}(0). \tag{10.95}$$

If we take the bases $\mathcal{O}^{(j,i)}_{\mu_1 \ldots \mu_j}$ to be symmetric traceless tensors with j indices, they correspond to operators of spin j. The light-cone, $x^2 \to 0$, behaviour of the Wilson coefficients (just as in (10.92)) can be obtained by naive dimensional counting

$$C_i^{(j)}(x) \underset{x^2 \to 0}{\to} (\sqrt{x^2})^{d_{j,i} - j - d_A - d_B} (\ln x^2 m^2)^p \tag{10.96}$$

where $d_{j,i}$ is the dimension of $\mathcal{O}^{(j,i)}_{\mu_1 \ldots \mu_j}(0)$. Hence unlike the case in (10.92) the leading term corresponds to the lowest value of $(d_{j,i} - j)$, i.e. the dimension of $\mathcal{O}^{(j,i)}_{\mu_1 \ldots \mu_j}$ minus the spin of $\mathcal{O}^{(j,i)}_{\mu_1 \ldots \mu_j}$. Such a combination is called the *twist* of an operator

$$\tau = d - j \tag{10.97}$$

which denotes twist = dimension − spin. The operators with lowest twist dominate in the light-cone expansion.

The scalar field ϕ, the fermion field ψ, and the gauge field $F_{\mu\nu}$ all have twist-one. Taking the derivative of these fields cannot reduce the twist and at best leaves it unchanged because taking the derivative will increase the dimension by one unit while changing the spin by 1 or 0. Thus the minimum twist of an operator which involves m fields is m. The most important light-cone operators have twist-two, examples of which are listed below

scalar:

$$\mathcal{O}^{(j,s)}_{\mu_1 \ldots \mu_j} = \phi^* \overleftrightarrow{\partial}_{\mu_1} \overleftrightarrow{\partial}_{\mu_2} \ldots \overleftrightarrow{\partial}_{\mu_j} \phi \tag{10.98a}$$

fermions:

$$\mathcal{O}^{(j,f)}_{\mu_1 \ldots \mu_j} = \bar{\psi} \gamma_{\mu_1} \overleftrightarrow{\partial}_{\mu_2} \ldots \overleftrightarrow{\partial}_{\mu_j} \psi + \text{permutations} \tag{10.98b}$$

vector:

$$\mathcal{O}^{(j,g)}_{\mu_1\ldots\mu_j} = F_{\mu_1\alpha}\overleftrightarrow{D}_{\mu_2}\ldots\overleftrightarrow{D}_{\mu_{j-1}}F^{\alpha}_{\ \mu_j} + \text{permutations}. \tag{10.98c}$$

The derivatives in (10.98a,b) will be replaced by covariant derivatives $\overleftrightarrow{D}_\mu$ in gauge theory. We have seen in §7.3 that dominance of the canonical twist-two operator in free-field theory leads to Bjorken scaling.

Now we can begin to see how the deep inelastic lN scattering cross-sections can be factorized into two parts—one being momentum-independent and the other scaling in a way controlled by the renormalization group, where the cross-section is related to the absorptive part of the forward current–nucleon scattering amplitude for which one then makes a light-cone expansion; the maxtrix elements of the local operators will then give rise to the momentum-independent part and the c-number Wilson coefficients satisfy the renormalization group equation. We shall first show that these Wilson coefficients are related to the integrated moments of the lN structure functions.

(B) **Moments of structure functions and Wilson coefficients.** In order that the principal festures of our manipulations not be obscured by complicated Lorentz structures we shall first illustrate our procedure with scalar currents $J(x)$. Consider the forward scattering amplitude (Fig. 10.9)

$$T(q^2, \nu) = \int \mathrm{d}^4x\, \mathrm{e}^{-\mathrm{i}q\cdot x}\langle p|T\left(J\left(\frac{x}{2}\right)J\left(-\frac{x}{2}\right)\right)|p\rangle \tag{10.99}$$

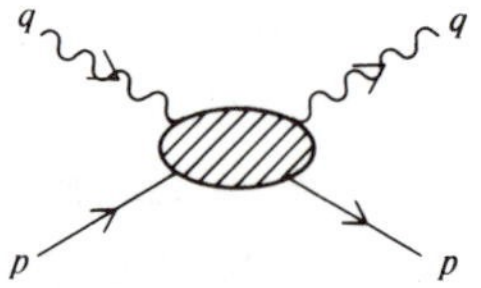

FIG. 10.9.

where $\nu = p\cdot q/M$. (Details of the kinematics may be found in §7.1.) Writing the operator-product expansion

$$T\left(J\left(\frac{x}{2}\right)J\left(-\frac{x}{2}\right)\right) \underset{x^2\to 0}{\approx} \sum_{i,j} C^j_i(x^2)x^{\mu_1}\ldots x^{\mu_j}\mathcal{O}^{(j,i)}_{\mu_1\ldots\mu_j}(0) \tag{10.100}$$

(with the index i ranging over all twist-2 operators), then the amplitude in (10.99) becomes

$$\begin{aligned} T(q^2, \nu) &\approx \sum_{i,j} \int \mathrm{d}^4x\, \mathrm{e}^{-\mathrm{i}q\cdot x}x^{\mu_1}\ldots x^{\mu_j}C^j_i(x^2)\langle p|\mathcal{O}^{(j,i)}_{\mu_1\ldots\mu_j}(0)|p\rangle \\ &= \sum_{i,j} (\mathrm{i})^j \frac{\partial}{\partial q_{\mu_1}}\frac{\partial}{\partial q_{\mu_2}}\cdots\frac{\partial}{\partial q_{\mu_j}} \\ &\quad\times\left[\int \mathrm{d}^4x\, \mathrm{e}^{-\mathrm{i}q\cdot x}C^j_i(x^2)\right]\langle p|\mathcal{O}^{(j,i)}_{\mu_1\ldots\mu_j}(0)|p\rangle. \end{aligned} \tag{10.101}$$

The matrix element of the local operator $\mathcal{O}^{(j,i)}_{\mu_1\ldots\mu_j}$ which is symmetric and traceless can be parametrized as

$$\langle p|\mathcal{O}^{(j,i)}_{\mu_1\ldots\mu_j}(0)|p\rangle = O^{(j)}_i[p_{\mu_1}p_{\mu_2}\ldots p_{\mu_j} - \text{trace terms}] \tag{10.102}$$

where $O^{(j)}_i$ is a constant and the trace term will contain at least one $g_{\mu_m\mu_n}$ factor. Replacing

$$\frac{\partial}{\partial q_{\mu_1}}\frac{\partial}{\partial q_{\mu_2}}\cdots\frac{\partial}{\partial q_{\mu_j}} = 2^j q^{\mu_1}q^{\mu_2}\ldots q^{\mu_j}\left(\frac{\partial}{\partial q^2}\right)^j + \text{trace terms}, \tag{10.103}$$

we obtain for large $-q^2$, with $-q^2/2M\nu$ fixed,

$$\begin{aligned} T(q^2,\nu) &\underset{-q^2\to\infty}{\approx} \sum_{i,j} (2\mathrm{i})^j(p\cdot q)^j\left(\frac{\partial}{\partial q^2}\right)^j\left[\int \mathrm{d}^4x\, \mathrm{e}^{-\mathrm{i}q\cdot x}C^j_i(x^2)\right]O^{(j)}_i \\ &= \sum_j\sum_i\left(\frac{2p\cdot q}{-q^2}\right)^j \tilde{C}^j_i(q^2)O^{(j)}_i \end{aligned} \tag{10.104}$$

where

$$\tilde{C}^j_i(q^2) = (-\mathrm{i}q^2)^j\left(\frac{\partial}{\partial q^2}\right)^j\int \mathrm{d}^4x\, \mathrm{e}^{-\mathrm{i}q\cdot x}C^{(j)}_i(x^2) \tag{10.105}$$

which is essentially the Fourier transform of $x^{\mu_1}\ldots x^{\mu_j}C^j_i(x^2)$. Note that the trace terms in (10.102) and (10.103) will have lower powers of $(2p\cdot q)$ and can be safely neglected in the scaling limit. Thus for the amplitude $T(q^2,\nu)$, decomposed in terms of spin projections

$$T(q^2,\nu) = \sum_j T_j(q^2,\nu), \tag{10.106}$$

we have from (10.104) that

$$T_j(q^2,\nu) \underset{-q^2\to\infty}{\approx} x^{-j}\sum_i \tilde{C}^{(j)}_i O^{(j)}_i \tag{10.107}$$

for $x = -q^2/2M\nu$. This implies that to isolate an operator of a given spin j we need just expand $T(q^2,\nu)$ in powers of x^{-1} for large $-q^2$. In deep inelastic scattering, since one actually measures the absorptive part of the forward amplitude $T(q^2,\nu)$,

$$W(q^2,\nu) = \frac{1}{\pi}\,\mathrm{Im}\, T(q^2,\nu). \tag{10.108}$$

The amplitude can be reconstructed from the measured quantities by using the dispersion relation

$$T(q^2,\nu) = \int\frac{\nu^s\,\mathrm{d}\nu'}{\nu'^s(\nu'-\nu)}\, W(q^2,\nu') + P_{s-1}(q^2,\nu) \tag{10.109}$$

where we have assumed s number of subtractions with $P_{s-1}(q^2,\nu)$ being a polynomial in ν of order $s-1$ for fixed q^2. If we further assume that, for large $-q^2$, we have $P_{s-1}(q^2, \nu) \to P_{s-1}(x)$, a polynomial in x^{-1} of order $s-1$, then

$$T(q^2, \nu) = \int_{-1}^{+1} \frac{(x')^{s-1} W(q^2, x')\,dx'}{(x')^{s-1}(x'-x)} + P_{s-1}(x)$$

$$\underset{x\to\infty}{\approx} P_{s-1}(x) + \sum_{J=s}^{\infty} x^{-J} \int_{-1}^{+1} dx'(x')^{J-1} W(q^2, x'). \quad (10.110)$$

The first s terms in this expansion are undetermined because of the unknown subtraction constants. However for $J \geq s$ (comparing eqn (10.110) to eqns (10.106) and (10.107)), we have

$$\int_{-1}^{+1} dx x^{J-1} W(q^2, x) \underset{-q^2\to\infty}{\approx} \sum_i \tilde{C}_i^{(J)}(q^2) O_i^{(J)} \quad \text{for} \quad J \geq s. \quad (10.111)$$

Thus the moments of the structure functions measure the (Fourier transforms of) Wilson coefficient functions $\tilde{C}_i^{(J)}(Q^2)$.

For the more realistic case of the electromagnetic current $J \to J_\mu^{\text{em}}$, we can make a similar analysis. With the usual assumption about the high-energy behaviour of the forward Compton amplitude, the relation (10.111) will hold for all $J \geq 2$, when the t-channel of the Compton scattering (i.e. the current $\times$ current channel) has the quantum number of the vacuum, and for $J \geq 1$ in the non-vacuum channels. When decomposed in terms of the two invariant eN inelastic structure functions $F_{1,2}(x, Q^2)$ as in eqn (7.30) (we have changed notation from that used in §7.1: $G_i(x, q^2) \to F_i(x, Q^2)$), the result corresponding to eqn (10.111) reads

$$\int_{-1}^{1} dx x^{J-2} F_2(x, Q^2) \approx \frac{1}{4} \sum_i \tilde{C}_i^{(J)}(Q^2) O_i^{(J)} \quad (10.112a)$$

$$\int_{-1}^{1} dx x^{J-1} F_1(x, Q^2) \approx \frac{1}{2} \sum_i \tilde{C}_i^{(J)}(Q^2) O_i^{(J)}. \quad (10.112b)$$

We have succeeded in isolating from the cross-section, which has two mass scales $p^2 = M^2$ and q^2, a factor which depends only on q^2, to which we can apply the scale transformation $q_\mu \to \lambda q_\mu$ and renormalization group analysis. We note that exact Bjorken scaling (eqn (7.32)) $F_i(x, q^2) \to F_i(x)$ corresponds to free-field behaviour

$$\tilde{C}_i^{(J)}(Q^2) \to \text{constant as } Q^2 \to \infty. \quad (10.113)$$

In general we expect this to be modified by the interaction. The simplest possible deviation from this scaling behaviour would be such that the $\ln(Q^2/m^2)$ powers in every order of the perturbation are summed up into some Q^2 powers as

$$\tilde{C}_i^{(J)}(Q^2) \sim \left(\frac{1}{Q^2}\right)^{\gamma_J/2}. \tag{10.114}$$

This can be interpreted as an anomalous dimension γ_J acquired by the operator $\mathcal{O}^{(J,i)}_{\mu_1\mu_2\ldots\mu_J}$ due to the interaction. As conserved quantities are finite, their renormalization constants do not depend on cut-off and their anomalous dimension vanishes. The electromagnetic currents and the energy momentum tensors are such quantities. If they can appear on the right-hand side of the operator-product expansion, then (the Fourier transforms of) their Wilson coefficients will scale as in free-field theory without anomalous dimensions. For the general moments, eqn (10.114) however implies that

$$\int_0^1 F_2(x, Q^2)x^{J-2}\,\mathrm{d}x \underset{Q^2\to\infty}{\approx} \left(\frac{1}{Q^2}\right)^{\gamma_J/2}. \tag{10.115}$$

Since the structure functions F_i are positive definite, the anomalous dimension γ_J must monotonically increase with J and, since the energy–momentum tensor $\theta_{\mu\nu}$ does appear in vacuum channel of the current operator-product expansion, we have $\gamma_2 = 0$ and

$$\gamma_j \geq 0 \quad \text{for} \quad J \geq 2. \tag{10.116}$$

This means that the moments of the structure functions should in the vacuum channel decrease with increasing Q^2. Also experimentally observed approximate scaling implies that the anomalous dimensions must be very small.

(C) **Renormalization group equations for the Wilson coefficients.** In Chapter 3 we introduced the renormalization group equation for a general Green's function. We can obtain similar equations for the Wilson coefficients by comparing the renormalization group equations satisfied by Green's functions containing the operator product itself and containing the local operators appearing in the expansion of the operator product (Christ, Hasslacher, and Mueller 1972).

Schematically the operator-product expansion is of the form

$$A(x)B(0) \underset{x\to 0}{\approx} \sum_i C_i(x, g, \mu)\mathcal{O}_i(0) \tag{10.117}$$

where g is the coupling constant and μ is the reference (subtraction) point for the renormalization. Or, in terms of n-point Green's functions with insertions of AB and $\mathcal{O}_i$,

$$\Gamma^{(n)}_{AB} \underset{x\to o}{\approx} \sum_i C_i(x, g, \mu)\Gamma^{(n)}_{\mathcal{O}_i} \tag{10.118}$$

where

$$\Gamma^{(n)}_{AB} = \langle 0|T(A(x)B(0)\prod_{k=1}^{n}\phi_k(y_k))|0\rangle \tag{10.118a}$$

$$\Gamma^{(n)}_{\mathcal{O}_i} = \langle 0|T(\mathcal{O}_i(0)\prod_{k=1}^{n}\phi_k(y_k))|0\rangle. \tag{10.118b}$$

The Green's functions separately satisfy the renormalization group equations (see eqn (3.58))

$$\left[\mathrm{D} + \sum_k \gamma_k(g) - \gamma_A(g) - \gamma_B(g)\right]\Gamma^{(n)}_{AB}(g, \mu) = 0$$

and

$$\left[\mathrm{D} + \sum_k \gamma_k(g) - \gamma_i(g)\right]\Gamma^{(n)}_{\mathcal{O}_i}(g, \mu) = 0$$

where

$$\mathrm{D} = \mu\frac{\partial}{\partial\mu} + \beta(g)\frac{\partial}{\partial g} \tag{10.119}$$

and γ_A, γ_B, and γ_i are the anomalous dimensions of the operators of A, B, and $\mathcal{O}_i$, respectively; we have assumed that they do not mix under renormalization. Using (10.118), clearly we have

$$[\mathrm{D} + \gamma_A(g) + \gamma_B(g) - \gamma_i(g)]C_i(x, g, \mu) = 0. \tag{10.120}$$

Thus, the Wilson coefficient C_i behaves as if it were a Green's function of the operators A, B, and $\mathcal{O}_i$. The solution as given by eqn (3.68) takes on the form

$$C_i(\mathrm{e}^{-t}x_0, g, \mu) = \mathrm{e}^{t(d_A + d_B - d_i)}\exp\left\{\int_0^t \mathrm{d}t'[\gamma_A(\bar{g}(t')) + \gamma_B(\bar{g}(t')) - \gamma_i(\bar{g}(t'))]\right\}C_i(x_0, g(t), \mu) \tag{10.121}$$

where d_A, d_B, and d_i are the naive dimensions of A, B, and $\mathcal{O}_i$. Similar equations and solutions of course hold for the Fourier transforms of the Wilson coefficients $\tilde{C}_i(q^2, g, \mu)$. For the case of deep inelastic scattering we have the light-cone expansion which can be turned into sums of infinite towers of local operators of increasing spin n with Wilson coefficients $\tilde{C}^n_i(q^2, g, \mu)$, which are related to the moments of the structure functions by (10.112)

$$M_n(Q^2) = \int_0^1 \mathrm{d}x x^{n-2}F_2(x, Q^2) \approx \frac{1}{8}\sum_i \tilde{C}^n_i(Q^2, g, \mu)\mathcal{O}^n_i$$

$$= \frac{1}{8}\sum_i \tilde{C}^n_i(Q_0^2, \bar{g}(t), \mu)\exp\left[-\int_0^t \gamma^i_n(\bar{g}(t'))\,\mathrm{d}t'\right]\mathcal{O}^n_i \tag{10.122}$$

where γ^i_n is the anomalous dimension of the operator $\mathcal{O}^{(n,i)}_{\mu_1\ldots\mu_n}$.

(D) **Deep inelastic scattering in QCD.** We will now apply this analysis of deep inelastic scattering in QCD (Gross and Wilczek 1974; Georgi and

Politzer 1974), where the electromagnetic current is given by

$$J_\mu(x) = \sum_{k=1}^{n_f} :\bar{q}_k(x)\gamma_\mu e_k q_k(x): \tag{10.123}$$

where $q_k(x)$ is the quark field operator with flavour index k, e_k is the charge carried by q_k, and the sum over colour is implicit. Consider the forward Compton scattering amplitude (averaging over nucleon spin is understood)

$$T^{(p,q)}_{\mu\nu} = \frac{\mathrm{i}}{2M}\int \frac{\mathrm{d}^4 y}{2\pi}\, \mathrm{e}^{\mathrm{i}p\cdot y}\langle p|T(J_\mu(y)J_\nu(0))|p\rangle \tag{10.124}$$

$$= \left(-g_{\mu\nu} + \frac{q_\mu q_\nu}{q^2}\right)T_1 + \frac{1}{M^2}\left(p_\mu - \frac{p\cdot q}{q^2}q_\mu\right)\left(p_\nu - \frac{p\cdot q}{q^2}q_\nu\right)T_2. \tag{10.125}$$

The absorptive parts of the invariant amplitudes T_i are just the structure functions W_i measured in deep inelastic scattering

$$W_{1,2}(\nu, q_2) = \frac{1}{\pi}\,\mathrm{Im}\, T_{1,2}(\nu, q^2). \tag{10.126}$$

(D1) *Operator-product expansion and moments of structure functions.* The operator-product expansion on the light cone is of the form

$$\mathrm{i}T(J_\mu(y)J_\nu(0)) = \sum_{n,i}\{-g_{\mu\nu}y_{\mu_1}y_{\mu_2}\ldots y_{\mu_n}\mathrm{i}^n C^{(n)}_{1,i}(y^2, g, \mu)$$
$$+ g_{\mu\mu_1}g_{\nu\mu_2}y_{\mu_3}\ldots y_{\mu_n}\mathrm{i}^{n-2}C^{(n)}_{2,i}(y^2, g, \mu)\}\mathcal{O}^{(n)\mu_1\mu_2\ldots\mu_n}_i(0). \tag{10.127}$$

In QCD there are three sets of gauge-invariant twist-2 operators (see eqns (10.128a, b, c) below) which dominate expansion near the light cone $y^2 \approx 0$

$$\mathcal{O}^{(n)\mu_1\ldots\mu_n}_{\mathrm{NS}}(x) = \frac{1}{2}\frac{\mathrm{i}^{n-1}}{n!}\left\{\bar{q}(x)\frac{\lambda^a}{2}\gamma^{\mu_1}\mathrm{D}^{\mu_2}\ldots\mathrm{D}^{\mu_n}q(x)\right.$$
$$\left. + \text{ permutations of vector indices}\right\} \tag{10.128a}$$

where the λ^as are the standard $n_f \times n_f$ hermitian traceless matrices in the flavour group $\mathrm{SU}(n_f)$; thus for a theory with three flavours u, d, and s they are just the familiar SU(3) Gell-Mann matrices. This set of operators will contribute to operator-product expansions for the flavour non-singlet combinations of structure functions such as $F_2^{\mathrm{ep}} - F_2^{\mathrm{en}}$ or $F_3^{\nu A}$, for neutrino scattering off an isoscalar target A. Here we will devote most of our effort to the study of the more involved case of the flavour–singlet combinations. They can receive contributions from two sets of operators

$$\mathcal{O}^{(n)\mu_1\ldots\mu_n}_q(x)$$
$$= \frac{1}{2}\frac{\mathrm{i}^{n-1}}{n!}\{\bar{q}(x)\gamma^{\mu_1}\mathrm{D}^{\mu_2}\ldots\mathrm{D}^{\mu_n}q(x) + \text{permutations}\} \tag{10.128b}$$

$$\mathcal{O}^{(n)\mu_1\ldots\mu_n}_{\mathrm{G}}(x)$$
$$= \frac{\mathrm{i}^{n-2}}{n!}\,\mathrm{tr}\{G^{\mu_1\nu}\mathrm{D}^{\mu_2}\ldots\mathrm{D}^{\mu_{n-1}}G^{\mu_n}_\nu + \text{permutations}\}. \tag{10.128c}$$

Because $\mathcal{O}_{\rm q}^{(n)}$ and $\mathcal{O}_{\rm G}^{(n)}$ have identical quantum numbers, they can mix under renormalization (see §2.4).

Substituting (10.127) into (10.124), we obtain

$$T_{\mu\nu}(p, q) = \frac{\mathrm{i}}{2M}\int\frac{\mathrm{d}^4 y}{2\pi}\,\mathrm{e}^{\mathrm{i}q\cdot y}\sum_{n,i}\{-g_{\mu\nu}y_{\mu_1}y_{\mu_2}\dots y_{\mu_n}\mathrm{i}^n C_{1,i}^{(n)}(y^2, g, \mu)$$
$$+ g_{\mu\mu_1}g_{\nu\mu_2}\mathrm{i}^{n-2}(y_{\mu_3}\dots y_{\mu_n})C_{2,i}^{(n)}(y^2, g, \mu)\}$$
$$\times \langle p|\mathcal{O}_i^{(n)\mu_1\dots\mu_n}(0)|p\rangle + \dots. \tag{10.129}$$

Writing the spin-averaged matrix element as

$$\langle p|\mathcal{O}_i^{(n)\mu_1\dots\mu_n}(0)|p\rangle = A_i^{(n)}(p^{\mu_1}p^{\mu_2}\dots p^{\mu_n} + \text{trace terms}) \tag{10.130}$$

where $A_i^{(n)}$ is a constant, eqn (10.129) becomes

$$T_{\mu\nu}(p, q) \sim \frac{1}{2M}\sum_{n,i}\left\{-g_{\mu\nu}(2p\cdot q)^n\left(\frac{\partial}{\partial q^2}\right)^n\int \mathrm{d}^4 y\,\mathrm{e}^{\mathrm{i}q\cdot y}C_{1,i}^{(n)}(y^2, g, \mu)\right.$$
$$\left.+ p_\mu p_\nu(2p\cdot q)^{n-2}\left(\frac{\partial}{\partial q^2}\right)^n\int \mathrm{d}^4 y\,\mathrm{e}^{\mathrm{i}q\cdot y}C_{2,i}^{(n)}(y^2, g, \mu)\right\}A_i^{(n)} + \dots$$
$$= \frac{1}{2M}\sum_{n,i}\left\{-g_{\mu\nu}\left(\frac{2p\cdot q}{-q^2}\right)^n\tilde{C}_{1,i}^{(n)}(q^2, g, \mu)\right.$$
$$\left.+ p_\mu p_\nu\left(\frac{2p\cdot q}{-q^2}\right)^{n-1}\frac{1}{2p\cdot q}\tilde{C}_{2,i}^{(n)}(q^2, g, \mu)\right\}A_i^{(n)} + \dots \tag{10.131}$$

where

$$\tilde{C}_{1,i}^{(n)}(Q^2, g, \mu) = (Q^2)^n\left(\frac{\partial}{\partial q^2}\right)^n\int \mathrm{d}^4 y\,\mathrm{e}^{\mathrm{i}q\cdot y}C_{1,i}^{(n)}(y^2, g, \mu)$$
$$\tilde{C}_{2,i}^{(n)}(Q^2, g, \mu) = (Q^2)^{n-1}\left(\frac{\partial}{\partial q^2}\right)^{n-2}\int \mathrm{d}^4 y\,\mathrm{e}^{\mathrm{i}q\cdot y}C_{2,i}^{(n)}(y^2, g, \mu). \tag{10.132}$$

From (10.131) we can immediately read out the invariant amplitudes

$$T_1(x, Q^2) = \frac{1}{2M}\sum_{n,i}x^{-n}\tilde{C}_{1,i}^{(n)}(Q^2, g, \mu)A_i^{(n)} \tag{10.133a}$$

$$T_2(x, Q^2) = \frac{1}{2M}\sum_{n,i}x^{-n+1}\tilde{C}_{2,i}^{(n)}(Q^2, g, \mu)A_i^{(n)}. \tag{10.133b}$$

By the same route we took going from (10.106) to (10.112), we can obtain relation between the moments of the structure functions and the (Fourier transforms of) Wilson coefficients

$$\int_0^1 \mathrm{d}x x^{n-2}F_2(x, Q^2) \approx \frac{1}{8}\sum_i \tilde{C}_{1,i}^{(n)}(Q^2, g, \mu)A_i^{(n)} \tag{10.134a}$$

$$\int_0^1 \mathrm{d}x x^{n-1}F_1(x, Q^2) \approx \frac{1}{4}\sum_i \tilde{C}_{2,i}^{(n)}(Q^2, g, \mu)A_i^{(n)}. \tag{10.134b}$$

(D2) *Renormalization group analysis and anomalous dimension matrix for the singlet case.* We shall concentrate on the flavour singlet case. Because of mixing among twist-2 operators, the renormalization group equation for the Wilson coefficients takes on the matrix form

$$(\mathrm{D}\,\delta_{jk} - \gamma_{kj}^{(n)})\tilde{C}_{\alpha,\,j}^{(n)}(q^2, g, \mu) = 0, \quad \alpha = 1, 2 \tag{10.135}$$

where $\gamma^{(n)}$ is the 2×2 anomalous dimension matrix for the flavour singlet operators $\mathcal{O}_{\mathrm{q}}^{(n)}$ and $\mathcal{O}_{\mathrm{G}}^{(n)}$. The solution to eqn (10.135) is

$$\tilde{C}_{\alpha,i}^{(n)}(Q^2/\mu^2, g) \sim \sum_i \tilde{C}_{\alpha,j}^{(n)}(1, \bar{g}(t)) \exp\left\{-\int_0^t \mathrm{d}t' \gamma_{ji}^{(n)}(\bar{g}(t'))\right\} \tag{10.136}$$

where

$$t = \tfrac{1}{2}\ln(Q^2/\mu^2). \tag{10.137}$$

In asymptotically free QCD one can calculate $\gamma^{(n)}$ perturbatively according to the one-loop diagrams of Fig. 10.10

$$\gamma_{ij}^{(n)} = d_{ij}^{(n)} g^2 + O(g^3) \tag{10.138}$$

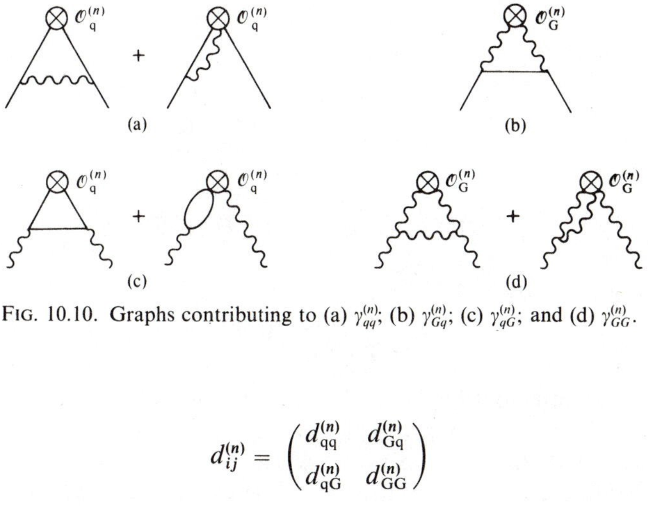

FIG. 10.10. Graphs contributing to (a) $\gamma_{qq}^{(n)}$; (b) $\gamma_{Gq}^{(n)}$; (c) $\gamma_{qG}^{(n)}$; and (d) $\gamma_{GG}^{(n)}$.

with

$$d_{ij}^{(n)} = \begin{pmatrix} d_{\mathrm{qq}}^{(n)} & d_{\mathrm{Gq}}^{(n)} \\ d_{\mathrm{qG}}^{(n)} & d_{\mathrm{GG}}^{(n)} \end{pmatrix}$$

where

$$d_{\mathrm{qq}}^{(n)} = \frac{1}{16\pi^2}\left[\frac{8}{3}\left(1 - \frac{2}{n(n+1)} + 4\sum_{j=2}^{n}\frac{1}{j}\right)\right]$$

$$d_{\mathrm{qG}}^{(n)} = \frac{1}{16\pi^2}\left[-\frac{16}{3}\frac{n^2+n+2}{n(n^2-1)}\right]$$

$$d_{\mathrm{Gq}}^{(n)} = \frac{1}{16\pi^2}\left[-4n_{\mathrm{f}}\frac{n^2+n+2}{n(n+1)(n+2)}\right]$$

$$d_{\mathrm{GG}}^{(n)} = \frac{1}{16\pi^2}\left\{6\left[\frac{1}{3} - \frac{4}{n(n-1)} - \frac{4}{(n+1)(n+2)} + 4\sum_{j=2}^{n}\frac{1}{j}\right] + \frac{4}{3}n_{\mathrm{f}}\right\} \tag{10.139}$$

n_f is the number of quark flavours. This matrix can be diagonalized by taking appropriate linear combinations of $\mathcal{O}_G^{(n)}$ and $\mathcal{O}_q^{(n)}$ and its eigenvalues correspond to the anomalous dimensions of the linear combinations that are multiplicatively renormalizable. Using the lowest-order expression (10.80) of the effective coupling constant in (10.138), we have

$$-\int_0^t \mathrm{d}t'\gamma_{ji}^{(n)}(\bar{g}(t')) = -\int_0^t \frac{g^2 d_{ji}^{(n)}\,\mathrm{d}t'}{1+2bg^2t'}$$

$$= \frac{-d_{ji}^{(n)}}{2b}\ln(1+2bg^2t) \to \frac{-d_{ji}^{(n)}}{2b}\ln t. \qquad (10.140)$$

The Wilson coefficient functions have the large-q^2 behaviour

$$\tilde{C}_{\alpha,i}^{(n)}(Q^2/\mu^2, g) \sim \sum_j \tilde{C}_{\alpha,j}^{(n)}(1,0)\left[\ln\left(\frac{Q^2}{\mu^2}\right)\right]^{-d_{ji}^{(n)}/2b} \qquad (10.141)$$

where the $\tilde{C}_{\alpha,j}^{(n)}(-1,0)$s are the Wilson coefficients in free-field theory. For large Q^2, (10.134) now reads

$$M_2^{(n)}(Q^2) = \int_0^1 \mathrm{d}x x^{n-2}F_2(x, Q^2/\mu^2)$$

$$\approx \frac{1}{8}\sum_i \tilde{C}_{2,j}^{(n)}(1,0)A_i^{(n)}[\ln(Q^2/\mu^2)]^{-d_{ji}^{(n)}/2b} \qquad (10.142a)$$

$$M_1^{(n)}(Q^2) = \int_0^1 \mathrm{d}x x^{n-1}F_1(x, Q^2/\mu^2)$$

$$\approx \frac{1}{4}\sum_i \tilde{C}_{1,j}^{(n)}(1,0)A_i^{(n)}[\ln(Q^2/\mu^2)]^{-d_{ji}^{(n)}/2b}. \qquad (10.142b)$$

These are the principal QCD results on deep inelastic scattering. We have obtained them by factoring the inclusive cross-section into a momentum-independent part (the local operator) and a part that scales according to the renormalization group (the Wilson coefficients). In the asymptotically free QCD the leading singular ($Q^2 \to \infty$) behaviour of the Wilson coefficients can be calculated in terms of the renormalization group β-function and the anomalous dimensions $\gamma_{ij}^{(n)}$, while the matrix elements $A_i^{(n)}$ of the local operator cannot be obtained without solving the (long-distance) bound-state problem in QCD. While we cannot calculate the scaling functions themselves, nevertheless we do have enough information on the pattern of scaling and scaling violation. QCD predicts that moments of the structure functions have a very weak dependence on q^2 as in eqn (10.142). So we have approximate Bjorken scaling with logarithmic violations.

(D3) *Momentum sum rule revisited.* As an illustration we shall work out the flavour-singlet second moment ($n = 2$) of the structure function, say

$F_2(x, Q^2)$. In this case one of the spin-2 and twist-2 local operators will be the energy–momentum tensor $\theta_{\mu\nu}$. Its conservation implies no renormalization $Z_\theta = 1$; hence we should find a vanishing anomalous dimension $\gamma_\theta = 0$. (For a related discussion on the renormalization of the conserved vector current operator $Z_J = 1$, see §6.1.) For $n = 2$, eqn (10.128) indicates that there are two twist-2 operators

$$[\mathcal{O}_q^{(2)}]_{\mu\nu} = \frac{i}{4} \sum_{k=1}^{n_f} \bar{q}_k(\gamma_\mu \overleftrightarrow{\partial}_\nu + \gamma_\nu \overleftrightarrow{\partial}_\mu) q_k \tag{10.143a}$$

$$[\mathcal{O}_G^{(2)}]^{\mu\nu} = \mathrm{tr}(G^{\mu\lambda} G_\lambda^\nu). \tag{10.143b}$$

We are taking only the lowest-order terms; thus $G_{\mu\nu} = \partial_\mu A_\nu - \partial_\nu A_\mu$. The matrix in (10.139) takes on the value

$$d_{ij}^{(2)} = \frac{1}{36\pi^2} \begin{pmatrix} 16 & -3n_f \\ -16 & 3n_f \end{pmatrix}. \tag{10.144}$$

It clearly has a zero eigenvalue $\tilde{d}_\theta = 0$ corresponding to the left eigenvector $(1, 1) = (1, 0) + (0, 1)$, i.e. the combination

$$\theta_{\mu\nu} = [\mathcal{O}_q^{(2)} + \mathcal{O}_G^{(2)}]_{\mu\nu} \tag{10.145}$$

which is just the energy-momentum tensor. The other eigenvalue is $\tilde{d}_{\theta'} = (16 + 3n_f)/36\pi^2$. We can in general express $d_{ij}^{(n)}$ in terms of its eigenvalues $\tilde{d}_i$ and their projection operators P_i with respect to the eigenvectors

$$d_{ij}^{(n)} = \sum_k \tilde{d}_k (P_k)_{ij}, \quad P_k P_l = \delta_{kl} P_k, \quad \sum_k P_k = 1. \tag{10.146}$$

It is easy to see that for our $n = 2$ case, we have the projection operators (for left vectors)

$$P_\theta = \frac{1}{16 + 3n_f} \begin{pmatrix} 3n_f & 3n_f \\ 16 & 16 \end{pmatrix}, \quad P_{\theta'} = \frac{1}{16 + 3n_f} \begin{pmatrix} 16 & -3n_f \\ -16 & 3n_f \end{pmatrix}. \tag{10.147}$$

The second moment of the flavour singlet structure function can be written as

$$\int_0^1 dx F_2(x, Q^2) = \frac{1}{8} \sum_{i,j} \{ \tilde{C}_{2,i}^{(2)}(1, 0)(P_\theta)_{ij} A_j^{(2)} + \tilde{C}_{2,i}^{(2)}(1, 0)(P_{\theta'})_{ij} A_j^{(2)} [\ln(Q^2/\mu^2)]^{-\tilde{d}_{\theta'}/2b} \}. \tag{10.148}$$

Since $\tilde{d}_{\theta'} > 0$, we have for $Q^2 \to \infty$

$$\int_0^1 dx F_2(x, Q^2) = \frac{1}{8} \sum_{i,j} \{ \tilde{C}_{2,i}^{(2)}(1, 0)(P_\theta)_{ij} A_j^{(2)} \}. \tag{10.149}$$

To calculate the right-hand side we note that, to lowest-order in $\bar{g}(t)$, only the quarks contribute to the (current) operator-product expansion; hence the free-field values of the Wilson coefficients are

$$\tilde{C}^{(2)}_{2,\mathrm{G}}(1,0) = 0, \quad \tilde{C}^{(2)}_{2,q}(1,0) = \langle e_{\mathrm{q}}^2 \rangle \tag{10.150}$$

where

$$\langle e_{\mathrm{q}}^2 \rangle = \frac{1}{n_{\mathrm{f}}} \sum_k e_k^2 \tag{10.151}$$

is the average quark charge squared. Thus the right-hand side involves only $\Sigma_j\,(P_\theta)_{\mathrm{q}j} A_j^{(2)}$. To evaluate this we first note

$$(1,0)\frac{1}{16+3n_{\mathrm{f}}}\begin{pmatrix} 3n_{\mathrm{f}} & 3n_{\mathrm{f}} \\ 16 & 16 \end{pmatrix}\begin{pmatrix} \mathcal{O}_{\mathrm{q}}^{(2)} \\ \mathcal{O}_{\mathrm{G}}^{(2)} \end{pmatrix}$$

$$= \frac{3n_{\mathrm{f}}}{16+3n_{\mathrm{f}}}(\mathcal{O}_{\mathrm{q}}^{(2)} + \mathcal{O}_{\mathrm{G}}^{(2)}) \equiv r\theta \tag{10.152}$$

where θ is the energy–momentum tensor operator (10.145) and where

$$r = \frac{3n_{\mathrm{f}}}{16+3n_{\mathrm{f}}} \tag{10.153}$$

can be interpreted as the fraction of momentum carried by quarks. For three flavours of quarks, $r = 9/25$ which is in accord with the experimental results when interpreted in the quark–parton model (momentum sum rule) of §7.2. Unlike in the general case, the matrix element of $\theta_{\mu\nu}$ is known

$$\langle p|\theta_{\mu\nu}|p\rangle = p_\mu p_\nu - g_{\mu\nu}p^2. \tag{10.154}$$

When eqn (10.154) is compared with (10.128) and (10.130), we have $(P_\theta)_{\mathrm{q}j} A_j^{(2)} = 8r$; in this way eqn (10.149) becomes

$$\left[\int_0^1 \mathrm{d}x F_2(x, q^2)\right]_{\text{singlet}} = r\langle e_{\mathrm{q}}^2\rangle. \tag{10.155}$$

This means that the area under the structure-function curve scales without deviation.

(D4) *Pattern of scaling violation.* For general n, the diagonalized $d_i^{(n)}$'s have the property of increasing slowly with n. For $n > 2$, where the large x (~ 1) region is important, the moment $M^{(n)}(Q^2)$ decreases as $[\ln(Q^2)]^{-|d_n|/2b}$; for $n < 2$, where the small-x (~ 0) region is important, the moment $M_2^{(n)}(Q^2)$ increases as $[\ln(Q^2)]^{+|d_n|/2b}$. This implies the following pattern of scaling violation; as $Q^2 \to \infty$, the large-x part of $F_2(x, Q^2)$ decreases while the small-x part increases, while the area under the curve remains unchanged throughout (see Fig. 10.11). In the infinite Q^2 limit, the structure function approaches a sharp spike at $x = 0$.

In concluding this section we will briefly discuss a type of experimental check on QCD predictions that has been performed. To this end we will consider the simpler case of a *non-singlet* combination of structure functions

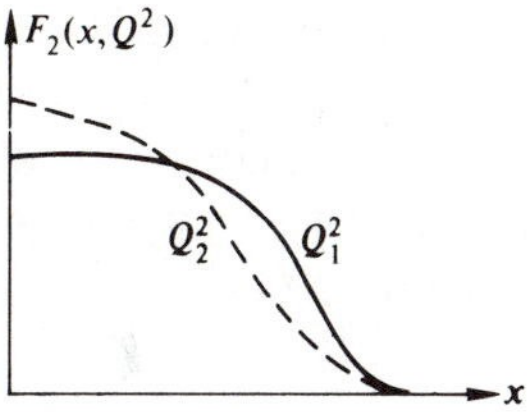

FIG. 10.11. The pattern of scaling violation of $F_2(x, Q^2)$ for $Q_2^2 > Q_1^2$.

where there is no complication due to operator mixing. Then the result corresponding to (10.142) reads

$$[M_2^{(n)}(Q^2)]_{\text{non-singlet}} = B^{(n)}[\ln(Q^2/\mu^2)]^{-d^{(n)}/2b} \tag{10.156}$$

where $B^{(n)}$ is an unpredicted constant and $d^{(n)}$ is the g^2 coefficient of the anomalous dimension for the operator in (10.128),

$$d^{(n)} = \frac{1}{6\pi^2}\left[1 - \frac{2}{n(n+1)} + 4\sum_{j=2}^{n}\frac{1}{j}\right]. \tag{10.157}$$

which is the same as $d_{\text{qq}}^{(n)}$ as one would expect. Now we take logarithms of the moments

$$\ln M^{(n)}(Q^2) = \frac{-d^{(n)}}{2b}\ln\ln(Q^2/\mu^2) + \ldots$$

$$\ln M^{(n')}(Q^2) = \frac{-d^{(n')}}{2b}\ln\ln(Q^2/\mu^2) + \ldots. \tag{10.158}$$

The dots on the right-hand sides of (10.158) represent non-leading terms (which could still be significant at the present energy level).

If we plot these two logarithms against each other, we should get a straight line with slope $d^{(n)}/d^{(n')}$. Such experimental plots have been found to agree with the QCD prediction within limits of experimental error, see, for example, Bosetti *et al.* (1978) and de Groot *et al.* (1979).

0.4 The parton model and perturbative QCD

In this section we present a brief introduction to the study of perturbative QCD. We shall concentrate again on the prototype processes of deep inelastic lepton–hadron scattering. First we present the parton-model probabilistic interpretation of the QCD result obtained in the previous section. We then show in what ways the same result can be recovered by summing the leading logarithms in perturbation theory.

Let us first recapitulate the QCD result for deep inelastic lN scatterings obtained by using the operator-product expansion and the renormalization group equation. For the nth moment of the flavour non-singlet combinations of the structure functions $F_2(x, Q^2)$, the result (10.156) may be written

$$M^{(n)}(t) = M^{(n)}(0)\left[\frac{\alpha_s(t)}{\alpha_s(0)}\right]^{d^{(n)}/2b} \tag{10.159}$$

where $t = \frac{1}{2}\ln Q^2/Q_0^2$, $\alpha_s(t)$ is the effective strong gauge-coupling constant squared (10.82) and b and $d^{(n)}$ are the leading coefficients of the renormalization group β-function (10.77) and anomalous dimension (10.157), respectively. Thus QCD predicts 'almost scaling' with a logarithmical violation. This pattern is illustrated in Fig. 10.11.

The parton picture of scaling violations

How are we to understand this QCD pattern of scaling violation in the parton model or, for that matter, patterns of scaling violation in general? As we have discussed in Chapter 7, strict Bjorken scaling can be obtained if we are allowed to make an impulse approximation and the reaction may be viewed as an incoherent sum of scatterings off the (free) constituent of the target. Thus this picture is applicable to any weakly bound system, and can be viewed as a process in which the virtual photon probes the structure of the target. From the uncertainty principle a virtual photon with mass $(Q^2)^{1/2}$ will resolve structure on the length scale of $(Q^2)^{-1/2}$. When we increase the virtual photon mass, then the structure at shorter distances will be revealed. Typically this picture of matter as 'a box within a box, etc.' (i.e. discrete levels) will lead to scaling violations at particular intervals of Q^2 (when the next level is reached). Thus we would generally observe scaling and rescaling as layers of matter are unravelled. At these scaling violation junctures, the structure function will change with Q^2 and get redistributed towards the region corresponding to smaller values of the scaling variable x, as each constituent will carry a smaller fraction of the target momentum.

In such a parton picture, the QCD result can be interpreted in terms of hadrons having a continuous set of constituent layers. At any particular Q^2 the shape of the nucleon structure function can be understood schematically as in Figs. 7.4–7.6. As we increase Q^2 and penetrate deeper into the dressed quark we will find more virtual quarks and gluons and the valence quarks will have to share the original nucleon momentum more and more with the gluons and sea quarks. Ultimately the structure function approaches a delta function at $x = 0$

$$F_2(x, Q^2) \underset{Q^2 \to \infty}{\approx} \delta(x). \qquad (10.160)$$

The Altarelli–Parisi equation

Thus, according to the above picture, the QCD scaling violation comes from the fact that the effective strong coupling $\alpha_s(t)$ does not vanish fast enough as $Q^2 \to \infty$. Even though the theory is asymptotically free there are still some residual interactions at short distances. All this can be cast in well-defined quantitative expressions as first suggested by Altarelli and Parisi (1977). For a simple presentation see (Close 1979).

At $t = \frac{1}{2}\ln Q^2/Q_0^2$, let the probability of a quark carrying the fraction x of nucleon momentum by $q(x)$. As we increase the virtual photon mass t to $t + \delta t$ an additional probability $\delta q(x)$ may be revealed corresponding to the

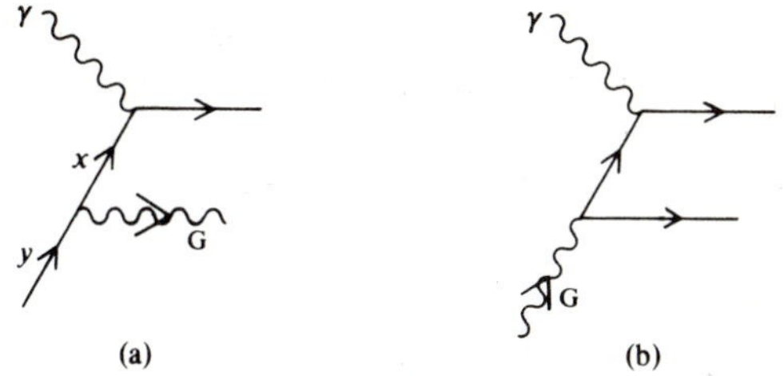

FIG. 10.12.

possibility of another quark with momentum fraction $y > x$ radiating away a gluon and reducing its momentum from y to x (Fig. 10.12(a)). Thus we can think of this quark as being contained in the one originally observed at a lower value of t. This suggests that the quark distribution is actually t-dependent and that we can define a quantity $P_{qq}(z)$ corresponding to the variation of the probability distribution (per unit t) of finding a quark in a quark with $z = x/y$ fraction of its momentum (y). This variation is clearly an order-g^2 effect; consequently we have the following evolution equation

$$\frac{\mathrm{d}q(x,t)}{\mathrm{d}t} = \bar{g}^2(t)\int_x^1 \mathrm{d}y \int_0^1 \mathrm{d}z\, \delta(x-yz) P_{qq}(z) q(y,t)$$

$$= \bar{g}^2(t)\int_x^1 \frac{\mathrm{d}y}{y} P_{qq}(x/y) q(y,t). \tag{10.161}$$

This convolution integral can be disentangled by a Mellin transform as the moment of the product is the product of the moments of the functions. Define the moments of the distribution (i.e. structure) function

$$M^{(n)}(t) = \int_0^1 \mathrm{d}x x^n q(x,t). \tag{10.162}$$

Eqn (10.161) then reads

$$\frac{\mathrm{d}M^{(n)}(t)}{\mathrm{d}t} = \bar{g}^2(t)\int_0^1 \mathrm{d}x x^n \int_x^1 \frac{\mathrm{d}y}{y} P_{qq}(x/y) q(y,t)$$

$$= \bar{g}^2(t)\int_0^1 \mathrm{d}y y^n q(y,t) \int_0^1 \mathrm{d}z z^n P_{qq}(z)$$

$$= \bar{g}^2(t) M^{(n)}(t) D^{(n)} \tag{10.163}$$

where

$$D^{(n)} = \int_0^1 \mathrm{d}z z^n P_{qq}(z). \tag{10.164}$$

Thus the parton picture of scaling violations leads to this differential equation of the moment of the structure function; its evolution is seen to be controlled by the t-dependent quark–gluon coupling $\alpha_s(t)$. The solution to (10.163) can be easily obtained by integration

$$\frac{dM^{(n)}(t)}{M^{(n)}(t)} = \frac{D^{(n)}}{2b}\frac{dt}{t}$$

or

$$\frac{M^n(t)}{M^n(t_0)} = \left[\frac{\alpha_s(t)}{\alpha_s(t_0)}\right]^{-D^{(n)}/2b}. \tag{10.165}$$

This is precisely the QCD result for non-singlet moments (10.159) if we make the identification

$$D^{(n)} = -d^{(n)}. \tag{10.166}$$

Thus the moments of the Altarelli–Parisi function P_{qq} are simply the non-singlet anomalous dimension coefficients $d_{qq}^{(n)}$.

When we extend the above analysis to the flavour-singlet combination of structure functions, we have the additional feature that the (singlet) gluon distribution function $G(x, t)$ also contributes (see Fig. 10.12(b)). Corresponding to (10.161), we now have

$$\frac{dq(x,t)}{dt} = \bar{g}^2(t)\int_x^1 \frac{dy}{y}\,[P_{qq}(x/y)q(y,t) + P_{Gq}(x/y)G(y,t)] \tag{10.167}$$

$$\frac{dG(x,t)}{dt} = \bar{g}^2(t)\int_x^1 \frac{dy}{y}\,[P_{qG}(x/y)q(y,t) + P_{GG}(x/y)G(y,t)]. \tag{10.168}$$

(Appropriate summation over the quark flavour indices to obtain the singlet combination is assumed.) Eqn (10.167) shows that the singlet quark distribution varies with t, not only because of gluon bremsstrahlung, but also because gluons can convert into quark and antiquark pairs in a flavour-independent way, etc. The coupled evolution equations then correspond to operator-mixing in the more formal approach of using the operator product expansion and the renormalization group equation. The moments P_{qq}, P_{qG}, P_{Gq}, and P_{GG} can be identified with (negative) anomalous dimension coefficients $d_{qq}^{(n)}$, $d_{qG}^{(n)}$, $d_{Gq}^{(n)}$, and $d_{GG}^{(n)}$.

Perturbation theory and the parton model

Now we see that QCD results such as (10.159) have a simple interpretation in the parton model with quarks and gluons. This suggests that they can be obtained directly in perturbation theory without invoking the formal apparatus of the operator-product expansion and the renormalization group equation. Before performing this perturbative calculation we shall give an overview of the problem of obtaining the parton-model result in perturbation theory.

As we mentioned in the last section, although QCD is asymptotically free with a decreasing effective coupling constant $\bar{g}^2(q^2) \sim (\ln q^2)^{-1}$, it is still difficult to apply perturbation theory directly to many high-energy processes. There are at least two reasons for this situation.

(1) Perturbation theory deals with quarks and gluons, which are not physical asymptotic states. Thus, by and large, straightforward applications of asymptotic freedom are restricted to calculating gross features that are independent of the detailed final-state properties. Quantities such as high-energy e^+e^- hadronic total cross-sections and deep inelastic lN inclusive cross-sections are determined by the initial response of the system at short distance–time and are insensitive to the complicated nonperturbative process which turns quarks and gluons into hadrons with unit probability.

(2) Physical quantities usually depend on, in addition to the effective coupling $g(q^2)$, some mass parameter m. Thus if one attempts to expand physical quantities in powers of $g(q^2)$ in the large-q^2 limit, one generally also encounters terms of the form $[g^2(q^2) \ln q^2/m^2]^n$ which spoil the expansion. Physically, sensitivity to m indicates that the large-distance properties of the theory are involved. Thus the use of perturbation theory in QCD requires that either of the following conditions be satisfied.

(A) $\ln q^2/m^2$ terms do not occur. This is the case for $\sigma_{\text{tot}}(e^+e^- \to \text{hadrons})$, and $\sigma(e^+e^- \to \text{jets})$ where groups of particles within some narrow cones are summed (see §7.2). They are finite in the $m \to 0$ limit and therefore are free of $\ln q^2/m^2$ factors.

(B) $\ln q^2/m^2$ terms can be summed up and the sensitivity to m can be somehow factored out. This 'factorization of mass singularities' has already been briefly mentioned in the previous section and it occurs in the case of deep inelastic lepton scatterings. To see the usefulness of the factorization property more explicitly, consider a dimensionless observable $\Psi(g(\mu^2), Q^2/\mu^2, m^2/\mu^2)$. Since the value of Ψ cannot depend on the scale μ chosen to define the coupling constant g, we have

$$\mu \frac{d\Psi}{d\mu} = \left(\mu \frac{\partial}{\partial \mu} + \mu \frac{\partial g}{\partial \mu} \frac{\partial}{\partial g} + \mu \frac{\partial m}{\partial \mu} \frac{\partial}{\partial m}\right)\Psi = 0. \tag{10.169}$$

This is the renormalization group equation for the observable. If we can establish that Ψ factorizes as $Q^2 \to \infty$

$$\psi \underset{Q^2 \to \infty}{\to} C(g(\mu^2), Q^2/\mu^2) D(g(\mu^2), m^2/\mu^2) + O(m^2/Q^2), \tag{10.170}$$

then (10.169) can be used to determine the Q^2-dependence

$$\mu \frac{d\Psi}{d\mu} = \left(\mu \frac{dC}{d\mu}\right) D + \left(\mu \frac{dD}{d\mu}\right) C = 0 \tag{10.171}$$

or

$$\frac{\mu}{C}\frac{dC}{d\mu} = -\frac{\mu}{D}\frac{dD}{d\mu} \equiv \gamma(g) \tag{10.172}$$

where the separation constant can depend only on g and C satisfies the renormalization group equation in the standard form

$$\left[\mu\frac{\partial}{\partial\mu} + \beta(g)\frac{\partial}{\partial g} - \gamma(g)\right]C(g(\mu^2), Q^2/\mu^2) = 0. \tag{10.173}$$

From this and $\gamma(g) = g^2 d$ we obtain the familiar solution (e.g. see eqn (10.159))

$$C(g(\mu), Q^2/\mu^2) = C(0, 1)\left[\frac{g^2(Q^2)}{g^2(\mu^2)}\right]^{d/2b}$$
$$= C(0, 1)[1 + bg^2(\mu^2)\ln(Q^2/\mu^2)]^{-d/2b} \tag{10.174}$$

where we have used the explicit form of $g^2(Q^2)$ as given in eqn (10.80). We note that the function $C(g(\mu), Q^2/\mu^2)$ depends on $g(\mu)$ and Q^2/μ^2 only through the combination $g^2(\mu)\ln(Q^2/\mu^2)$. That each factor of $g^2(\mu^2)$ is accompanied by $\ln(Q^2/\mu^2)$ clearly indicates a correspondence to a summation in perturbation theory of the leading logarithmical terms, which appear as $g^{2n}(\mu^2)(\ln(Q^2/\mu^2))^n$.

We saw in the previous section that factorization of mass singularities is accomplished using the operator-product expansion; the c-number singular functions then satisfy the renormalization group equations. The present discussion then suggests the following correspondence between the formal field-theoretical apparatus and the simpler perturbative procedures

Formal apparatus		*Perturbative procedures*
operator product expansion	$\leftrightarrow$	factorization of mass singular diagrams
renormalization group equation	$\leftrightarrow$	summation of leading logarithms

It should be pointed out that, while this translation table may be helpful in clarifying the meaning of our calculational procedures, in most cases no such simple separation of steps is possible. In fact factorization is usually not achieved until summation of logarithms have been performed.

Before further discussion of perturbation theory and the parton-model results we will first carry out an explicit calculation illustrating the points made above.

Perturbative calculation of deep inelastic scattering

We are interested in the forward Compton scattering amplitude of a virtual photon (with momentum q_μ) by a quark (momentum p_μ). Corresponding to the operator-product expansion, here we need to identify and factorize diagrams containing powers of $\ln p^2$. Thus we are looking for diagrams that are divergent as the mass $p^2 = m^2 \to 0$. There are two types of divergences, which are often called *infrared divergences* and *mass singularities*.

The first type of divergence is brought about in the phase-space region when the momentum of the (real or virtual) massless particles vanishes. The Bloch–Nordsieck theorem (1937) assures us that such infrared divergences cancel in the inclusive cross-section. But all cross-sections are in fact inclusive because of the finiteness of energy resolution in all experiments. Essentially this situation still holds for the deep inelastic scattering.

We will concentrate on the divergence resulting from mass singularities. These mass singularities occur in theories with coupled massless particles, and are due simply to the kinematical fact that two massless particles (with momenta p and p', say) which are moving parallel to each other have combined invariant mass equal to zero

$$k^2 = (p + p')^2 = 2EE'(1 - \cos\theta) \to 0 \quad \text{as} \quad \theta \to 0 \tag{10.175}$$

even though neither p_μ nor p'_μ are soft. This is sometimes called *collinear divergence*. There is also a theorem, by Kinoshita (1962) and by Lee and Nauenberg (1964), which can be roughly stated as follows. For inclusively enough cross-sections the mass singularities also cancel. The physical basis of this theorem is very similar to that of Bloch and Nordsieck; i.e. in physical measurements the angular resolution is not perfect and we sum over all states within the finite angular resolution. Thus in our calculation we will not be concerned with mass singularities coming from undetected final-state particles moving parallel to each other, since they cancel. In deep inelastic scattering, however, there are still mass singularities left over. They arise from regions of phase space where the internal momentum of a massless particle is parallel to that of an external massless particle to which it is coupled. Consider the one-loop diagrams in Fig. 10.13. Let $k = p + p'$ be the

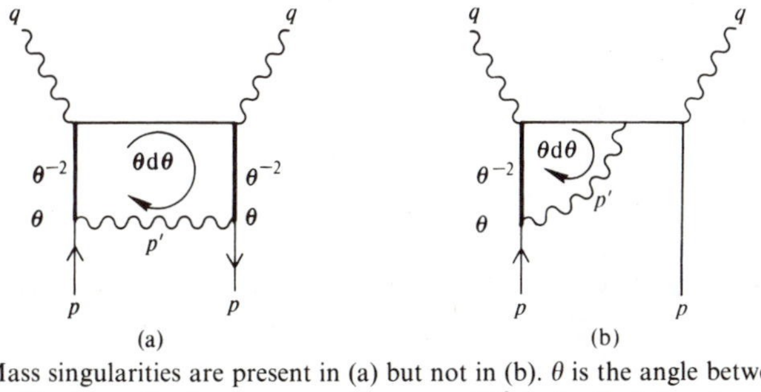

FIG. 10.13. Mass singularities are present in (a) but not in (b). θ is the angle between $\mathbf{p}$ and $\mathbf{p}'$. The heavy-lined propagator diverges as θ^{-2} in the limit $\theta \to 0$.

loop momentum. Hence we have a factor of θ^{-2} from each of the $1/k^2$ propagators, and $\theta\,d\theta$ from the $d(\cos\theta)$ factor in d^4k, and finally each massless particle vertex contributes a power of θ. Thus Fig. 10.13(a) has mass singularity $d\theta/\theta \sim \ln p^2$ and Fig. 10.13(b) is finite. We shall therefore concentrate on Fig. 10.13(a) and its generalizations to higher orders.

One-loop diagrams. It is convenient to use the axial gauge, eqns (9.41) and (9.108),

$$\eta_\mu A^\mu = 0 \tag{10.176}$$

where η_μ is some four-vector. The gluon propagator in this gauge is given by

$$D_{\mu\nu}(k) = \frac{d_{\mu\nu}(k)}{k^2 + \mathrm{i}\varepsilon}$$

with

$$d_{\mu\nu}(k) = g_{\mu\nu} - (k_\mu \eta_\nu + k_\nu \eta_\mu)(k \cdot \eta)^{-1} + \eta^2 k_\mu k_\nu (k \cdot \eta)^{-2}. \quad (10.177)$$

There is no Faddeev–Popov ghost in this gauge, and the calculation is greatly simplified. To take the absorptive part of the amplitude we replace the propagator $(k^2 + i\varepsilon)^{-1}$ by $\pi\,\delta(k^2)$. Thus the diagram in Fig. 10.14(a)

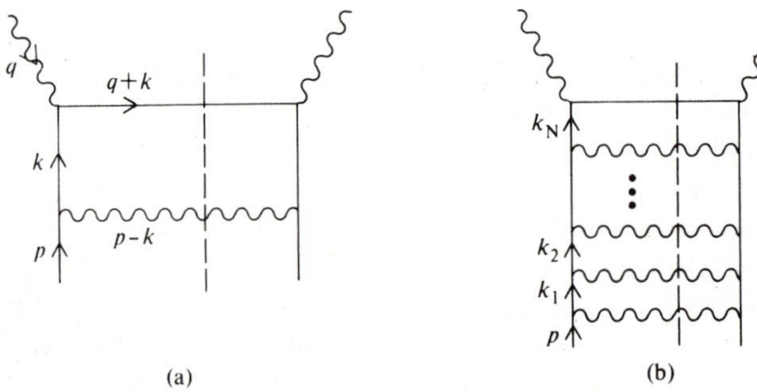

FIG. 10.14. The absorptive part of the forward Compton scattering amplitudes.

contributes to the hadronic tensor of the inelastic lepton scattering cross-section (eqn (7.10)) as

$$W_{\mu\nu}(p, q) = \frac{g^2 s_2(\mathrm{V})}{2M} \int \frac{\mathrm{d}^4 k}{(2\pi)^3} \frac{\delta((p-k)^2)\,\delta((q+k)^2)}{(k^2)^2} d^{\rho\sigma}(p-k) T_{\rho\mu\nu\sigma}$$

where

$$T_{\rho\mu\nu\sigma} = \tfrac{1}{2}\,\mathrm{tr}[\not{p}\gamma_\rho \not{k}\gamma_\mu(\not{k} + \not{q})\gamma_\nu \not{k}\gamma_\sigma] \quad (10.178)$$

and $s_2(\mathrm{V})$ is the usual colour factor of eqn (10.88) and equal 4/3 for $\mathrm{SU}(3)_{\mathrm{colour}}$. It is convenient to use the Sudakov variables (1956)

$$k_\mu = \xi p_\mu + \beta q'_\mu + k_{\perp\mu} \quad (10.179)$$

with

$$q' = q + xp \quad \text{and} \quad k_\perp \cdot p = k_\perp \cdot q' = 0. \quad (10.180)$$

We assume that $q'^2 \approx p^2 \approx 0$ and that $p \cdot q' \approx p \cdot q$, together with $-q^2$, are large.

$$\mathrm{d}^4 k = 2\pi(p \cdot q)\,\mathrm{d}\beta\,\mathrm{d}\xi\,\mathrm{d}k_\perp^2 \quad (10.181)$$

As we shall see, the final $\ln(-q^2/p^2)$ term comes from regions of phase space corresponding to $\beta \approx 0$ and $\xi \approx x$. Thus the variable x introduced in (10.180) has the usual parton-model interpretation as the fraction of the longitudinal target momentum. (This also *a posteriori* justifies taking $q'^2 \approx 0$.) In terms of these variables we have

$$\int \mathrm{d}^4k\, \delta((p-k)^2)\, \delta((q+k)^2) = 2\pi p\cdot q \int \mathrm{d}\beta\, \mathrm{d}\xi\, \mathrm{d}k_\perp^2$$
$$\times \frac{1}{2p\cdot q(1-\xi)}\delta\left(\beta - \frac{k_\perp^2}{2p\cdot q(1-\xi)}\right)\frac{1}{2p\cdot q}\delta\left(\xi - x + \frac{1-x}{1-\xi}\frac{k_\perp^2}{2p\cdot q}\right). \tag{10.182}$$

Squaring (10.179) we obtain

$$k^2 = k_\perp^2 + 2\beta\xi(p\cdot q) = k_\perp^2 + \frac{\xi}{1-\xi}k_\perp^2 = k_\perp^2/(1-\xi). \tag{10.183}$$

Therefore (10.182) can be rewritten

$$\int \mathrm{d}^4k\, \delta((p-k)^2)\, \delta((q+k)^2) = \frac{\pi}{2p\cdot q}\int \mathrm{d}\beta\, \mathrm{d}\xi\, \mathrm{d}k^2\, \delta\left(\beta - \frac{k^2}{2p\cdot q}\right)$$
$$\times\, \delta\left(\xi - x + (1-x)\frac{k^2}{2p\cdot q}\right). \tag{10.184}$$

Contracting the gluon polarization tensor $d^{\rho\sigma}(p-k)$ with the trace term $T_{\rho\mu\nu\sigma}$ (taking $\eta = q'$) and doing the Dirac algebra yields a leading term proportional to the tree graph (eqn (7.53))

$$d^{\rho\sigma}(p-k)T_{\rho\mu\nu\sigma} = (4k^2)\frac{1+\xi^2}{1-\xi}\frac{1}{2}\mathrm{tr}[x\not{p}\gamma_\mu(\not{q}+x\not{p})\gamma_\nu]. \tag{10.185}$$

This results from the relative contributions of the first and second terms of (10.177) ($(1-\xi)k^2$ and $2\xi k^2/(1-\xi)$, respectively) and from dropping the contribution of the third term which is of the order of $(k^2)^2$ and thus does not contribute to the final $\ln(Q^2/p^2)$. Collecting terms (10.184) and (10.185),

$$W_{\mu\nu}(p,q) = \frac{g^2}{6\pi^2}\int\frac{\mathrm{d}k^2}{k^2}\left(\frac{1+x^2}{1-x}\right)\left[\frac{x}{\nu}\frac{p_\mu p_\nu}{p^2} + \ldots\right]. \tag{10.186}$$

Comparing (10.186) with the simple parton-model result of eqn (7.54), we see that the one-loop diagram makes the following contribution to the parton distribution function $f(x)$ (which is one originally),

$$\mathrm{d}f(x) = \tfrac{1}{2}g^2 P(x)\frac{\mathrm{d}k^2}{k^2} \tag{10.187}$$

with

$$P(x) = \left(\frac{1+x^2}{1-x}\right)\frac{1}{3\pi^2}. \tag{10.188}$$

We choose to write it in a form that will be more useful for interpretation and for higher-order calculations

$$\mathrm{d}f = \left[\frac{1}{2}g^2\frac{\mathrm{d}k^2}{k^2}\right]\left\{\int\frac{\mathrm{d}\xi}{\xi}\delta\left(1-\frac{x}{\xi}\right)P(x)\right\}. \tag{10.189}$$

Thus the diagram in Fig. 10.14(a) with its gluon bremsstrahlung introduces a momentum dependence in the quark distribution function having an *Altarelli–Parisi function* of $P(x)$. If we have been careful in keeping track of p^2 terms, we will find the limits of integration for the $\mathrm{d}k^2$ integral to be

$$\int_{p^2}^{Q^2} \frac{\mathrm{d}k^2}{k^2} = \ln(Q^2/m^2) \quad \text{for} \quad p^2 \approx m^2. \tag{10.190}$$

Thus the one-loop diagram (Fig. 10.14(a)) has a mass-singular form

$$f(x, Q^2) \sim [g^2 \ln(Q^2/m^2)]\left(\frac{1+x^2}{1-x}\right). \tag{10.191}$$

Higher-order diagrams. Features similar to those in one-loop diagrams also appear in higher-order contributions. The general feature (Gribov and Lipatov 1972) is that the dominant contributions in the axial gauge come from the ladder diagrams (Fig. 10.14(b)) from the region of phase space corresponding to

$$p^2 \ll k_1^2 \ll k_2^2 \ldots \ll k_N^2 \ll Q^2. \tag{10.192}$$

It is understood that in Fig. 10.14(b) the vertex and self-energy insertions are included. One of the effects of these insertions is that at each vertex we should use the momentum-dependent effective couplings $g(k_i^2)$.

For each rung of the ladder we have a box diagram similar to the one we have just evaluated with the result written in the form of (10.189). Now for all the transverse momentum integrals $\mathrm{d}k_i^2$, corresponding to the square brackets of (10.189), we have

$$I_{\mathrm{T}}^{(N)}(Q^2/m^2) = \int_{m^2}^{Q^2} \frac{\mathrm{d}k_1^2}{2k_1^2} g^2(k_1^2) \int_{m^2}^{k_1^2} \frac{\mathrm{d}k_2^2}{2k_2^2} g^2(k_2^2) \ldots$$

$$\times \int_{m^2}^{k_{N-1}^2} \frac{\mathrm{d}k_N^2}{2k_N^2} g^2(k_N^2)$$

$$\approx \frac{1}{N!}\left[\int_{m^2}^{Q^2} \frac{\mathrm{d}k^2}{2k^2} g^2(k^2)\right]^N \equiv \frac{1}{N!}\rho^N. \tag{10.193}$$

Using the asymptotic form of $g^2(k^2)$, we have

$$\rho = \int_{m2}^{Q2} \frac{\mathrm{d}\ln(k^2)}{2b \ln k^2/\mu^2} = \frac{1}{2b}[\ln\ln(Q^2/\mu^2) - \ln\ln(m^2/\mu^2)]. \tag{10.194}$$

The integral over the longitudinal components of momentum, corresponding to the curly brackets of (10.189), is of the form

$$I_{\mathrm{L}}^{(N)} = \int_{\xi_2}^{1} \frac{\mathrm{d}\xi_1}{\xi_1} \int_{\xi_3}^{1} \frac{\mathrm{d}\xi_2}{\xi_2} \cdots \int_{0}^{1} \frac{\mathrm{d}\xi_N}{\xi_N}\, \delta\!\left(1 - \frac{x}{\xi_N}\right)$$

$$\times\, P(\xi_1)P(\xi_2/\xi_1)\ldots P(\xi_N/\xi_{N-1}). \tag{10.195}$$

The fraction of momentum p carried by quark i is ξ_i and that by quark $i+1$ is ξ_{i+1}. The argument of the Altarelli–Parisi function is the ratio of these two fractions ξ_{i+1}/ξ_i. This equation is in a multiple-convolution form and we can take moments to simplify the result

$$\int_0^1 \mathrm{d}x x^{n-1} I_{\mathrm{L}}^{(N)}(x) = \left[\int_0^1 \mathrm{d}z z^n P(z)\right]^N \equiv [d^{(n)}]^N. \tag{10.196}$$

Then the moment of the structure function coming from the N-rung ladder diagram is

$$M_N^{(n)}(q^2) = \frac{1}{N!}\,[\rho d^{(n)}]^N \tag{10.197}$$

Summing over N we get

$$\begin{aligned} M^{(n)}(Q^2) &= \sum_N M_N^{(n)}(Q^2) = \exp[\rho d^{(n)}] \\ &= \exp\left\{\frac{d^{(n)}}{2b}\,[\ln \ln(Q^2/\mu^2) - \ln \ln(m^2/\mu^2)]\right\} \\ &= [\ln(Q^2/\mu^2)]^{d(n)/2b}[\ln(m^2/\mu^2)]^{-d(n)/2b}. \end{aligned} \tag{10.198}$$

This shows the factorization property that $M^{(n)}(Q^2)$ decomposes into a function of Q^2/μ^2 multiplied by a function of m^2/μ^2. Note that this factorization takes place only after diagram summation; individual terms do not factorize.

Finally let us check to see whether $d^{(n)}$ agrees with the result obtained previously. From (10.196) and (10.188),

$$d^{(n)} = \int_0^1 \mathrm{d}z z^n P(z) = \frac{1}{3\pi^2} \int_0^1 \mathrm{d}z\, \frac{z^n(1+z^2)}{1-z} \tag{10.199}$$

which actually diverges as $z \to 1$. More carefully analysis shows that when we did the calculations for Fig. 10.15(a) we should also have included the vertex correction diagram (Fig. 10.15(b)) for the Bloch–Nordsieck infrared divergence cancellation to take place. This corresponds to the replacement

$$\int_0^1 \frac{f(z)}{1-z}\,\mathrm{d}z \to \int_0^1 \frac{f(z)-f(1)}{1-z}\,\mathrm{d}z. \tag{10.200}$$

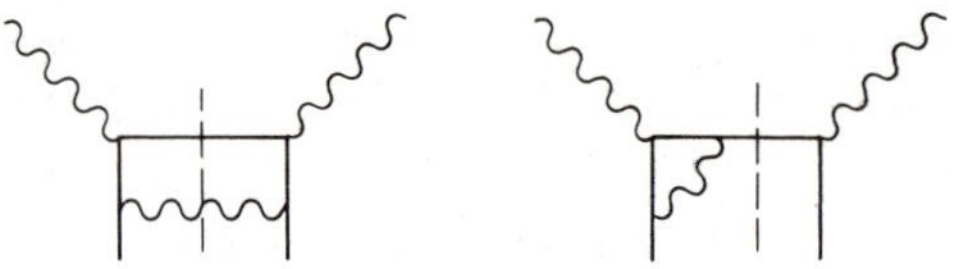

FIG. 10.15.

Thus $d^{(n)}$ can be calculated

$$\begin{aligned} d^{(n)} &= \frac{1}{3\pi^2}\int_0^1 \mathrm{d}z \frac{z^n(1+z^2)-2}{1-z} \\ &= \frac{1}{3\pi^2}\int_0^1 \frac{\mathrm{d}x}{x}[(1-x)^n(2-2x+x^2)-2] \\ &= \frac{1}{6\pi^2}\left[1-\frac{2}{n(n+1)}+4\sum_{j=2}^{n}\frac{1}{j}\right]. \end{aligned} \tag{10.201}$$

Thus by summing leading logarithms we obtain exactly the same results as obtained by using the more formal procedures of renormalization group equations.

Once we understand the ideas and calculations for the familiar deep inelastic case, it is conceptually fairly straightforward to generalize the scheme to other hard scattering processes. The real advantage is that in most such hard scattering reactions a direct application of the operator-product expansion is not feasible. Thus we can use the perturbative QCD approach to calculate the Drell–Yan process of inclusive production of lepton pairs with large invariant mass in hadron collision (see §7.3). Here factorization is achieved and one finds the same parton distribution functions there with momentum dependences as those in the deep inelastic case. Another outstanding example of the perturbative QCD calculation of physical quantities that are not 'infrared sensitive' is the prediction of the final-state angular distribution (jets) (Sternman and Weinberg 1977) and energy flows in high-energy e^+e^- annihilations into hadrons (Brown and Ellis 1981).

10.5 Lattice gauge theory and colour confinement

Quantum chromodynamics has the remarkable property of being asymptotically free. The vanishing for short distances of the effective coupling gives the correct description of Bjorken scaling (and its violation by logarithms). The same statement also suggests that the effective coupling increases for long distances and this points towards a possible resolution of the central paradox in the phenomenological quark picture, i.e. quarks must behave like free particles for short distances while they must also be completely confined on long time and length scales. Of course strong coupling itself is not enough to explain quark confinement. To do so one must show that in QCD the particle spectrum is realized in terms of bound states of quarks and gluons, etc. In the following we shall first give a brief qualitative discussion of how quark confinement might come about in the asymptotically free gauge theory. This should motivate the lattice method of defining such a theory and we will then demonstrate that the confinement property can be obtained in a strong coupling approximation of the lattice gauge theory.

Qualitative picture of confinement

Qualitative ideas about the nature of confinement that have some correct physical consequences tend to picture quarks as being bound by 'strings' (Nambu 1974) or tubes of colour flux. It has been suggested that the QCD vacuum is a condensate of gluons and as well as light quark–antiquark pairs. (For a review see, for example, Mandelstam 1979.) This is somewhat analogous to the ground state of a superconductor. There the condensate of paired electrons gives rise to the Meissner effect of magnetic flux exclusion from the condensate unless the energy balance favours a local breakdown to the normal phase. And one can imagine placing a pair of magnetic monopole and antimonopole into this superconducting medium; the magnetic flux will be confined to a string-like configuration joining the pair of monopoles. Analogous to this situation the energetically favoured configuration of, say, a quark and an antiquark, has a strongly localized normal region connecting the pair in which the colour (electric) flux lines are restricted (see Fig. 10.16). By translational invariance the energy density of

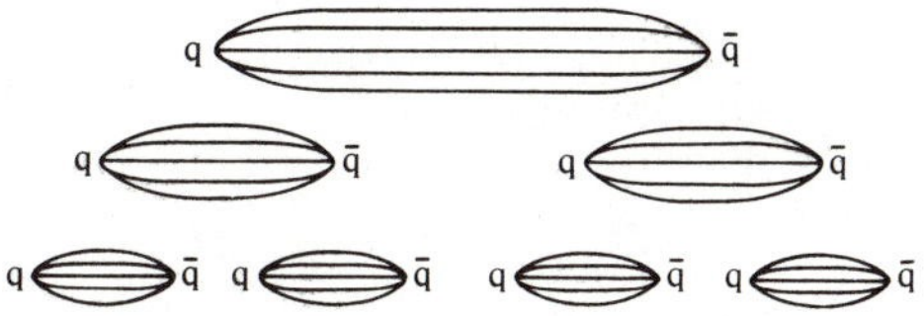

FIG. 10.16. Lines of force between a quark and an antiquark. When the quarks move away from each other, the breaking of the string is accompanied by further quark–antiquark pair production.

the gluon field along the flux tube is a constant. Hence the total field energy is linearly proportional to distance. This means the quarks are confined by a linearly rising long-distance potential

$$E(r) \to Kr \tag{10.202}$$

where r is the separation between the sources and K is a constant referred to as the *string tension*. Thus after the production of a quark-antiquark pair (say, in e^+e^- annihilation) it becomes energetically favourable with increasing separation for the string to break and produce another quark–antiquark pair, and so on. This proceeds until the original string is broken down into several strings of length typical of the hadronic size of ~ 1 fm ($\sim 5\ \text{GeV}^{-1}$) corresponding to the original quark pair being converted into a whole set of hadrons having a typical energy-mass of 1 GeV. Thus we can crudely estimate the string tension to be

$$K \approx \tfrac{1}{5}\,\mathrm{GeV}^2. \tag{10.203}$$

The string model for hadrons has another consequence in agreement with observation: hadrons lie on *Regge trajectories* (Regge 1959) with slopes $\approx 1\ \mathrm{GeV}^{-2}$. A decade of intense study of hadron dynamics through their pattern of particle exchange in which unitarity and the analyticity property of the scattering amplitudes play a crucial role has led to many insights into particle interactions. One of the important discoveries in this S-matrix approach (Chew 1962) was that hadrons of a given internal symmetry quantum number but different spins obey a simple spin(J) – mass(M_J) relation (Chew and Frautschi 1961); we say they lie on Regge trajectory

$$J = \alpha_0 + \alpha' M_J^2,$$

with

$$\alpha' \simeq 1\ \mathrm{GeV}^{-2}. \tag{10.204}$$

Now imagine two massless (and, for simplicity, spinless) quarks, connected by a string of length d, rotating with the speed of light (Gasiorowicz and Rosner 1981). Thus each point, at a distance r from the centre, has the local velocity $v/c = 2r/d$. The total mass is then

$$M = 2\int_0^{d/2} \frac{K\,\mathrm{d}r}{(1 - v^2/c^2)^{1/2}} = \frac{\pi K\,\mathrm{d}}{2} \tag{10.205}$$

and the total angular momentum is

$$J = 2\int_0^{d/2} \frac{Krv\,\mathrm{d}r}{(1 - v^2/c^2)^{1/2}} = \frac{\pi K\,\mathrm{d}^2}{8}. \tag{10.206}$$

Thus the string tension K of (10.202) can be expressed directly in terms of the Regge slope α' of (10.204)

$$K = \frac{1}{2\pi\alpha'}. \tag{10.207}$$

The experimental value α' in (10.204) leads to a $K \approx 0.2\ \mathrm{GeV}^2$ in qualitative agreement with the rough estimate of (10.203).

Field theories on the lattice

In order to study these long-distance properties of QCD we need to regularize the theory that is independent of the usual Feynman diagram expansion, which is appropriate for weak couplings. For this purpose Wilson (1974, 1975) introduced the lattice gauge theory in which the space–time continuum is discretized. This provides a natural cut-off scheme as wavelengths shorter than twice the lattice spacing, a, have no meaning and this restricts the domain of momenta to a region bounded by π/a. Also this formulation of the field theory allows for a close analogy with a statistical mechanics system. So we can call upon all our experience and intuition of statistical mechanics to solve problems in quantum field theory. With a cut-off on high momenta the kinetic energy is bounded; we can then treat it as a perturbation in the strong coupling limit. This corresponds to the method of high-temperature expansion in statistical mechanics. With a finite lattice there are a finite number of variables. It is then possible to study various physically interesting quantities (e.g. energy spectrum, correlation functions, etc.) in the path-integral formalism by computer simulation based on the Monte Carlo method.

As with any cut-off prescription, considerable freedom remains in the lattice formulation. In the limit of vanishing lattice spacing, the physics of a renormalizable field theory should be independent of the details of the regulator; it should, so to speak, lose the memory of the lattice spacing. This means that in this limit the coherence length of the theory should be infinite when compared to the lattice spacing. In the language of statistical mechanics the divergence of the correlation length corresponds to a second-order, or continuous, *phase transition*. If a model has only a first-order transition, the coherence length never becomes infinite and the desired continuum theory does not exist. Furthermore, analytic results in the strong and weak coupling regimes can easily be established. We need to ascertain whether these properties are connected continuously in the theory, i.e. whether there are phase transitions at intermediate couplings. Thus it is important to study the phase structure of lattice field theory. At present most of our knowledge of the phases of QCD is obtained in various numerical studies of the theory.

There are two popular methods for introducing the lattice in field theory. In the Euclidean lattice formulation both space and time are discretized (Wilson 1974). Here after a Wick rotation (to Euclidean space) the quantization is performed via the path-integral formalism. This method has the advantage of keeping some vestige of the original Lorentz symmetry and this allows for a particularly elegant formulation. Also, as we shall see in the case of lattice gauge theory, there is no need to introduce the gauge-fixing terms. The other method is the Hamiltonian lattice formulation in which only the spatial dimensions are discretized in a Minkowski space–time (Kogut and Susskind 1975*a*). The theory can be canonically quantized with the usual Hamiltonian formalism. This has the advantage that some physical quantities (especially the mass spectrum) are more directly accessible to computation. Our discussion will follow the Euclidean formulation with a space–time lattice spacing a. The lattice site will be labelled by a four-vector n. The four-dimensional integration will then be replaced by a sum

$$\int d^4x \to a^4 \sum_n \cdot \qquad (10.208)$$

Scalar fields. We first study the simplest case of scalar field $\phi(x)$. The continuum field theory in Euclidean space has the action

$$S(\phi) = \int \mathrm{d}^4x[\tfrac{1}{2}(\partial_\mu\phi)^2 + V(\phi)] \tag{10.209}$$

where

$$V(\phi) = \frac{1}{2}m^2\phi^2 + \frac{\lambda}{4}\phi^4. \tag{10.210}$$

The scalar field ϕ exists at every lattice site n,

$$\phi(x) \to \phi_n. \tag{10.211}$$

The derivative is replaced by

$$\partial_\mu\phi(x) \to \frac{1}{a}(\phi_{n+\hat{\mu}} - \phi_n) \tag{10.212}$$

where $\hat{\mu}$ is a four-vector of length a in the direction of μ. The lattice action is then

$$S(\phi) = \sum_n \left\{\frac{a^2}{2}\sum_{\mu=1}^{4}(\phi_{n+\hat{\mu}} - \phi_n)^2 + a^4\left(\frac{m^2}{2}\phi_n^2 + \frac{\lambda}{4}\phi_n^4\right)\right\}. \tag{10.213}$$

It is instructive to go over to momentum space to see the spectrum of the ($\lambda = 0$) free-field theory. For this we take the Fourier transform

$$\phi_n = \int \frac{\mathrm{d}^4k}{(2\pi)^4}\,\mathrm{e}^{\mathrm{i}k\cdot n}\phi(k). \tag{10.214}$$

Since it is meaningless to consider wavelengths less than twice the lattice spacing, the above integral is taken over only one '*Brillouin zone*' of the reciprocal lattice. Thus,

$$-\frac{\pi}{a} \le k_\mu \le \frac{\pi}{a} \quad \text{for each } \mu \tag{10.215}$$

where $k_\mu \equiv k \cdot \mu$. After substituting eqn (10.214) into eqn (10.213), we have factors such as those coming from the kinetic energy term

$$\begin{aligned} a^4\sum_n \int \frac{\mathrm{d}^4k}{(2\pi)^4}\frac{\mathrm{d}^4k'}{(2\pi)^4}\,\mathrm{e}^{\mathrm{i}(k+k')\cdot n}(\mathrm{e}^{\mathrm{i}ak_\mu} - 1)(\mathrm{e}^{\mathrm{i}ak'_\mu} - 1) \\ = \int \frac{\mathrm{d}^4k}{(2\pi)^4}(\mathrm{e}^{\mathrm{i}ak_\mu} - 1)(\mathrm{e}^{-\mathrm{i}ak_\mu} - 1) \\ = 4\int \frac{\mathrm{d}^4k}{(2\pi)^4}\sin^2(ak_\mu/2). \end{aligned}$$

The free field action can then be written

$$S_0(\phi) = \frac{1}{2}\int \frac{\mathrm{d}^4k}{(2\pi)^4}\left[\sum_\mu \frac{4}{a^2}\sin^2\left(\frac{ak_\mu}{2}\right) + m^2\right]\phi(-k)\phi(k). \tag{10.216}$$

Thus each mode contributes to the action in the momentum space a quantity

$$S(k) \equiv m^2 + \sum_\mu \frac{4}{a^2} \sin^2\left(\frac{ak_\mu}{2}\right)$$

rather than the standard $m^2 + k^2$. Nevertheless, these two expressions have the same continuum limit as they coincide at the minimum value of $k = 0$ (see Fig. 10.17(a)).

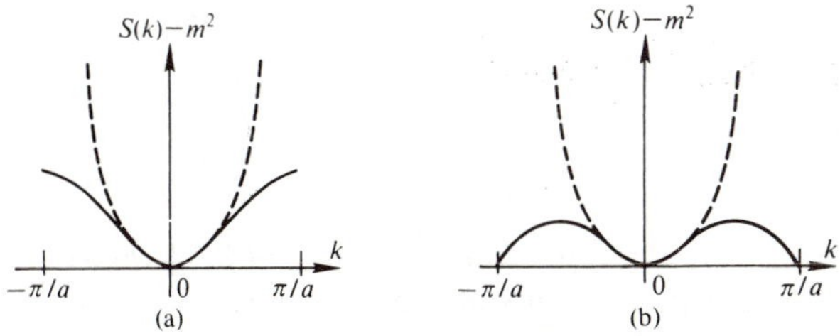

FIG. 10.17. The dispersion relation $S(k)$ for free (a) scalar and (b) fermion systems. The dashed curves k^2 are for familiar continuum theories. The solid curves correspond to the latticized systems.

The lattice action (10.213) can be quantized by using the Feynman path-integral formalism in which the expectation value of a product of fields is given by

$$\langle 0|\phi_{n_1}\phi_{n_2}\dots\phi_{n_l}|0\rangle = \frac{1}{Z}\int \prod_n [\mathrm{d}\phi_n](\phi_{n_1}\phi_{n_2}\dots\phi_{n_l})\,\mathrm{e}^{-S(\phi)} \quad (10.217)$$

where

$$Z = \int \prod_n [\mathrm{d}\phi_n]\,\mathrm{e}^{-S(\phi)}. \quad (10.218)$$

The meaning of the integrals should be clear as we may recall that the usual functional integrals are actually defined on a discretized space–time lattice and an appropriate continuum limit is taken at the end. If we rescale the fields as

$$\phi'_n = \sqrt{\lambda}\phi_n, \quad (10.219)$$

the lattice action becomes

$$S(\phi) = \frac{1}{\lambda} S'(\phi') \quad (10.220)$$

where

$$S'(\phi') = \sum_n \left\{\frac{a^2}{2}\sum_\mu (\phi'_{n+\hat{\mu}} - \phi'_n)^2 + a^4\left(\frac{m^2}{2}\phi'^2_n + \frac{1}{4}\phi'^4_n\right)\right\}, \quad (10.221)$$

i.e. the coupling constant λ has become an overall factor in the action. In this way eqns (10.217) and (10.218) may be written as

$$\langle 0|\phi'_{n_1}\phi'_{n_2}\dots\phi'_{n_l}|0\rangle = \frac{1}{Z'}\int\prod_n [\mathrm{d}\phi'_n](\phi'_{n_1}\phi'_{n_2}\dots\phi'_{n_l}) \times \exp\left\{-\frac{1}{\lambda}S'(\phi')\right\} \tag{10.222}$$

$$Z' = \int\prod_n [\mathrm{d}\phi'_n]\exp\left\{-\frac{1}{\lambda}S'(\phi')\right\}. \tag{10.223}$$

We note that (10.223) has the same structure as the partition function in statistical mechanics if we make the identification

$$\frac{1}{\lambda}\to\beta\equiv\frac{1}{kT}. \tag{10.224}$$

Thus the strong coupling expansion (in powers of λ^{-1}) will correspond to the high-temperature expansion in statistical systems.

Fermion fields. We now consider fermion fields on a lattice. The same procedure will yield a Euclidean lattice action for the free-fermion system

$$S_0(\psi) = \sum_n\left\{\frac{a^3}{2}\sum_{\mu=1}^{4}\bar{\psi}_n\gamma_\mu(\psi_{n+\hat{\mu}}-\psi_{n-\hat{\mu}}) + ma^4\bar{\psi}_n\psi_n\right\}. \tag{10.225}$$

The γ matrices are Euclidean, that is

$$\{\gamma_\mu,\gamma_\nu\} = 2\,\delta_{\mu\nu}. \tag{10.226}$$

Eqn (10.225) has the momentum space form

$$S_0(\psi) = \int\frac{\mathrm{d}^4k}{(2\pi)^4}\bar{\psi}(-k)\left\{\mathrm{i}\sum_\mu\gamma_\mu\frac{\sin ak_\mu}{a} + m\right\}\psi(k) \tag{10.227}$$

which yields a dispersion relation

$$S(k) = \frac{\sin^2 ak_\mu}{a^2} + m^2 \tag{10.228}$$

shown in Fig. 10.17(b). There are now two equal minima in a Brillouin zone. One is located at $k = 0$ and gives the correct continuum limit. The other mode at $k = \pm\pi/a$ carries an infinite momentum as $a \to 0$ and yet can be excited for finite a. The fermion degeneracy, i.e. the doubling of the number of fermionic states, must be suppressed with some appropriate modification of the latticized theory. This is permissible as long as the continuum limit is not affected. Clearly many degeneracy regularization procedures are possible; we will present the one due to Wilson, who proposed the simple addition of a non-local factor of

$$\Delta\mathscr{L} = \frac{1}{2a}\bar{\psi}_n(\psi_{n+\hat{\mu}}+\psi_{n-\hat{\mu}}-2\psi_n) \tag{10.229}$$

to the lattice Lagrange density. Thus the free fermion action takes on the form in Euclidean space

$$S_0(\psi) = \sum_n \left\{ \frac{a^3}{2} \sum_\mu \bar{\psi}_n[(1 + \gamma_\mu)\psi_{n+\hat{\mu}} + (1 - \gamma_\mu)\psi_{n-\hat{\mu}} - 2\psi_n] + ma^4\bar{\psi}_n\psi_n \right\} \tag{10.230}$$

which has the momentum-space representation

$$S_0(\psi) = \int \frac{d^4k}{(2\pi)^4} \bar{\psi}(-k) \left\{ i \sum_\mu \gamma_\mu \frac{\sin ak_\mu}{a} + m - \sum_\mu \frac{\cos ak_\mu - 1}{a} \right\} \psi(k). \tag{10.231}$$

This increases the unwanted minimum without affecting the small-k behaviour and the $k = 0$ minimum will be only one to survive on the continuum limit.

Local gauge invariance and the QCD action

As we have illustrated above, considerable freedom exists in lattice formulation. One is free to add to the Lagrangian terms which will not contribute in the continuum limit. Using this freedom, Wilson has presented a particularly elegant lattice formulation for gauge theories. His prescription keeps local gauge invariance as an exact symmetry in a mathematically well-defined system.

Recall our discussion in §8.2 of the geometric interpretation of gauge invariance. When a material particle undergoes a parallel transport along a world line C from x to x', it can be represented by a 'non-integrable phase factor' of the wavefunction (i.e. for every path we can associate a group element as in eqn (8.65))

$$U(x', x) = \exp\left\{ ig \int_C \mathbf{T} \cdot \mathbf{A}_\mu(y)\, dy^\mu \right\}. \tag{10.232}$$

Thus, for a gauge transformation, specified by the gauge function

$$\Phi(\theta_x) = \exp\{i\mathbf{T} \cdot \boldsymbol{\theta}(x)\} \tag{10.233}$$

we have

$$\psi(x) \to \Phi(\theta_x)\psi(x) \tag{10.234}$$

$$\bar{\psi}(x) \to \bar{\psi}(x)\Phi^\dagger(\theta_x)$$

and

$$U(x', x) \to \Phi(\theta_{x'})U(x', x)\Phi^\dagger(\theta_x). \tag{10.235}$$

(Please note the notation changes for the gauge transformation and parallel transport matrices from the notation used in §8.2.)

The lattice version of gauge transformation in (10.234) and (10.235) may be written as

$$\psi_n \to \Phi_n \psi_n$$
$$\bar{\psi}_n \to \bar{\psi}_n \Phi_n^\dagger \tag{10.236}$$

and

$$U(n+\hat{\mu}, n) \to \Phi_{n+\hat{\mu}} U(n+\hat{\mu}, n) \Phi_n^\dagger \tag{10.237}$$

where, for the SU(3) gauge symmetry,

$$\Phi_n = \exp\left\{ \mathrm{i} \frac{\lambda^i}{2} \theta_n^i \right\} \tag{10.238}$$

with λ^i, $i = 1, 2, \ldots, 8$, being the usual Gell–Mann matrices and

$$U(n+\hat{\mu}, n) = \exp\left\{ \mathrm{i}ag \frac{\lambda^i}{2} A_{n,\mu}^i \right\}. \tag{10.239}$$

This is the lattice version of the parallel transport matrix between adjacent sites $n \to n + \hat{\mu}$ and is usually called a *link variable*. From (10.236) and (10.237) we have $\bar{\psi}_n U(n, n+\hat{\mu}) \psi_{n+\hat{\mu}}$ as a gauge-invariant combination. This suggests the modification of (10.230) to obtain the quark part of the SU(3) gauge-invariant QCD action

$$S_{\mathrm{QCD}} = S(q) + S(A) \tag{10.240}$$

$$S(q) = \sum_n \left\{ \frac{a^3}{2} \sum_\mu \bar{\psi}_n [(1+\gamma_\mu) U(n, n+\hat{\mu}) \psi_{n+\hat{\mu}} \cdot \right.$$
$$\left. + (1-\gamma_\mu) U(n, n-\hat{\mu}) \psi_{n-\hat{\mu}} + 2\psi_n] - ma^4 \bar{\psi}_n \psi_n \right\}. \tag{10.241}$$

What about the lattice action for the gluon field $S(A)$? Clearly this term must be composed of link variables only. The simplest gauge-invariant combination will be four-link variables. (As they are matrices in SU(3) space two- and three-link combinations are trivial because of unitarity and determinantal constraints.) This suggests that

$$S(A) = \frac{-1}{2g^2} \sum_p \mathrm{tr}\, U_p \tag{10.242}$$

with

$$U_p = U(n, n+\hat{\mu}) U(n+\hat{\mu}, n+\hat{\mu}+\hat{\nu}) U(n+\hat{\mu}+\hat{\nu}, n+\hat{\nu}) U(n+\hat{\nu}, n) \tag{10.243}$$

which is a product of four-link variables around an elementary square (called a *plaquette*; see Fig. 10.18). The sum in (10.242) is over all plaquettes of the lattice. From our discussion of a parallel transport around a square given in §8.2, it follows that (10.242) may be written as

$$S(A) = \frac{-1}{2g^2} \sum_p \mathrm{tr}\{\exp(\mathrm{i}a^4 g^2 F_{n,\mu\nu})\} \tag{10.244}$$

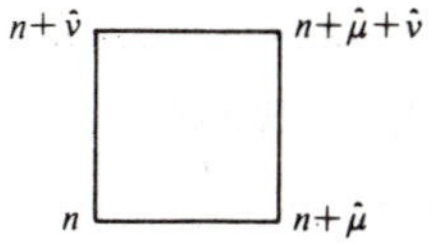

FIG. 10.18. A lattice plaquette.

with

$$F_{n,\mu\nu} = \partial_\mu A_{n,\nu} - \partial_\nu A_{n,\mu} - ig[A_{n,\mu}, A_{n,\nu}]$$

$$\partial_\mu A_{n,\nu} \equiv \frac{1}{a}(A_{n+\hat{\mu},\nu} - A_{n,\nu}) \tag{10.245}$$

where $A_{n,\mu}$ is the gluon field $A^i_\mu \lambda^i/2$ at the site n. The continuum limit can be recovered immediately as

$$S(A) = \frac{-1}{2g^2}\sum_p \left\{1 - \frac{a^4}{2}g^2 F^i_{\mu\nu}F^{i\mu\nu} + \ldots\right\}$$

$$\to \frac{1}{4}\int d^4x F^i_{\mu\nu}F^{i\mu\nu} \tag{10.246}$$

where we have used the fact that the λs are traceless (so the linear $F_{\mu\nu}$ term disappears) and $\text{tr}(\lambda^i\lambda^j) = 2\,\delta^{ij}$.

Confinement criteria, the Wilson loop

To see whether QCD has confinement one can study the energy of a system composed of a quark at $x = (t, \mathbf{0})$ and an antiquark at $x = (t, \mathbf{R})$. If there is no confinement, we expect

$$E(R) \to 2m \quad \text{for} \quad R \to \infty \tag{10.247}$$

where m is the quark mass. Confinement implies that the interquark potential energy grows without bound,

$$E(R) \to \infty \quad \text{for} \quad R \to \infty. \tag{10.248}$$

We can represent the $q\bar{q}$ state at time t as (see, for example, Bander 1981)

$$|q(t, \mathbf{0})\bar{q}(t, \mathbf{R})\rangle = \sum_C f(C)\Gamma[(t, \mathbf{R}), (t, \mathbf{0}); C]|0\rangle \tag{10.249}$$

where $\Gamma[x', x; C]$ is the gauge-invariant $q\bar{q}$ operator

$$\Gamma[x', x; C] = \bar{q}(x')U(x', x; C)q(x) \tag{10.250}$$

with

$$U(x', x; C) = \exp\left\{ig\int_x^{x'} \frac{1}{2}\lambda^i A^i(y)\,dy^\mu\right\} \tag{10.251}$$

and C is a path joining x and x'. Now consider the overlap between the $q\bar{q}$

state at $t = 0$ and the $q\bar{q}$ state at $t = T$,

$$\Omega(T, R) = \langle 0|\Gamma^{\dagger}[(0, \mathbf{0}), (0, \mathbf{R}); \mathrm{C}]\Gamma[(T, \mathbf{0}), (T, \mathbf{R}); \mathrm{C}]|0\rangle . \qquad (10.252)$$

Inserting a complete set of energy eigenstates, we obtain in the Euclidean space

$$\Omega(T, R) = \sum |\langle 0|\Gamma^{\dagger}[(0, \mathbf{0}), (0, \mathbf{R}); \mathrm{C}]|n\rangle|^2 \, \mathrm{e}^{-E_n T}. \qquad (10.253)$$

For large T, the smallest E_n will dominate. This smallest energy eigenvalue corresponds to the potential energy of the $q\bar{q}$ system separated by a distance R

$$\lim_{T\to\infty} \Omega(T, R) \sim \mathrm{e}^{-E(R)T}. \qquad (10.254)$$

In terms of the quark fields the overlap function $\Omega(\mathrm{T, R})$ may be written as

$$\Omega(T, R) = \langle 0|\bar{q}(0, \mathbf{R})U[(0, \mathbf{R}), (0, \mathbf{0}); \mathrm{C}]q(0, \mathbf{0})$$

$$\bar{q}(T, \mathbf{0})U[(T, \mathbf{0}), (T, \mathbf{R}); \mathrm{C}]q(T, \mathbf{R})|0\rangle . \qquad (10.255)$$

If we treat the quark fields as external sources (as in the case of heavy quark states), then the quark propagator (in a background of gluon fields) can be expressed as

$$\langle 0|q^{\beta}(t', \mathbf{x})\bar{q}^{\alpha}(t, \mathbf{x})|0\rangle = \exp\left[\mathrm{i}\int_t^{t'} A_0(\tau, \mathbf{x})\,\mathrm{d}\tau\right]\langle 0|q^{\beta}(t', \mathbf{x})\bar{q}^{\alpha}(t, \mathbf{x})|0\rangle_{\text{free}}$$

$$\sim U[(t', \mathbf{x}), (t, \mathbf{x}); \mathrm{C}]\,\delta^{\alpha\beta}\,\mathrm{e}^{-m|t'-t|}. \qquad (10.256)$$

Combining (10.256) and (10.255), we have

$$\Omega(T, R) \sim \mathrm{e}^{-2mT}W(\mathrm{C}) \qquad (10.257)$$

where

$$W(\mathrm{C}) = \langle 0|\mathrm{tr}\, U[x, x; \mathrm{C}]|0\rangle . \qquad (10.258)$$

Here C is the rectangular path of Fig. 10.19. The correlation function $W(\mathrm{C})$

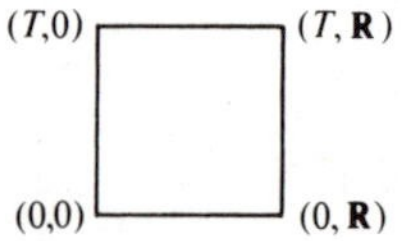

FIG. 10.19. A Wilson loop.

is called the *Wilson loop* and its behaviour determines the confinement property, as a comparison of (10.257) with (10.254) shows that

$$\lim_{T\to\infty} W(\mathrm{C}) \sim \mathrm{e}^{-T[E(R)-2m]}. \qquad (10.259)$$

As we shall illustrate presently, in the strong coupling limit, the Wilson loop of lattice gauge theory obeys an *area law* so that for a large contour

$$W(\mathrm{C}) \sim \exp\{-KA(\mathrm{C})\} \qquad (10.260)$$

where K is a constant and $A(\mathrm{C})$ is the area of the surface enclosed by path C.

Thus for the rectangular path of Fig. 10.19

$$A(\mathrm{C}) = TR. \tag{10.261}$$

Substituting eqns (10.260) and (10.261) into eqn (10.259) we have the linearly rising potential (10.202) for the q$\bar{\text{q}}$ system. Also, we can identify the constant K in (10.260) as the string tension. Note that we treat the quarks as a mere external colour source. Thus using the Wilson loop it is possible to study the question of colour confinement in a pure gluon theory without quark fields.

The area law in the strong coupling limit

In eqns (10.222) and (10.223) we displayed the path-intergral formulation of the correlation function in a scalar field theory. The Wilson loop may be similarly expressed in a pure gluon QCD; it will be a functional integral over gluon fields. Since the (infinitesimal) link variables are directly related to the gauge fields, we can choose to work with the link variables as the basic dynamic degrees of freedom in a lattice gauge theory. As we shall see, this allows for a simpler formulation of eqn (10.258)

$$W(\mathrm{C}) = \frac{1}{Z}\int \prod_{n,\mu} \mathrm{d}U(n, n+\hat{\mu})\, \mathrm{tr}\, U(x, x; \mathrm{C}) \exp\left\{-\frac{1}{2g^2}\sum_p \mathrm{tr}\, U_p\right\}. \tag{10.262}$$

where

$$Z = \int \prod_{n,\mu} \mathrm{d}U(n, n+\hat{\mu}) \exp\left\{-\frac{1}{2g^2}\sum_p \mathrm{tr}\, U_p\right\}. \tag{10.263}$$

Note that no gauge-fixing term has been added to the action because the link variable (i.e. lattice gauge field) has only finite range. The volume in the path-integral space generated by all possible gauge transformations is finite. Hence no convergence factor corresponding to the gauge-fixing term will be necessary before passing on to the continuum limit (see §9.1). The link variables are group elements of SU(3). We may set up a generalized Euler angle representation for these SU(3) unitary matrices and determine explicitly the form of the group integral in terms of the eight Euler angles. In any case we have the orthogonality properties

$$\int \mathrm{d}U(n, n+\hat{\mu})[U(n, n+\hat{\mu})]_{ij} = 0 \tag{10.264}$$

$$\int \mathrm{d}U(n, n+\hat{\mu})[U(n, n+\hat{\mu})]_{ij}[U^{\dagger}(n, n+\hat{\mu})]_{kl} = \tfrac{1}{3}\,\delta_{il}\,\delta_{jk} \tag{10.265}$$

$$\int \mathrm{d}U(n, n+\hat{\mu})[U(n, n+\hat{\mu})]_{ij}[U(n, n+\hat{\mu})]_{kl} = 0. \tag{10.266}$$

The identity (10.265) implies that two links going in opposite directions will cancel each other after integration. Thus, if we have two adjacent plaquettes oriented in the same direction, they will merge into a rectangular path after integrating over the common link as illustrated in Fig. 10.20. In the strong

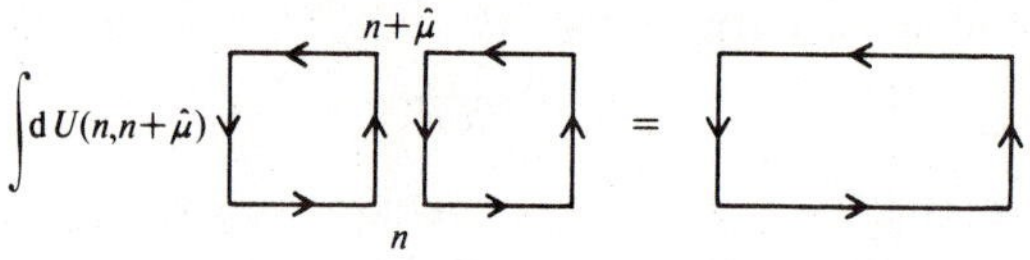

FIG. 10.20.

coupling limit we have $1/g^2$ as a small parameter and the exponential in (10.262) can be expanded

$$W(\mathrm{C}) = \frac{1}{Z}\int \mathrm{d}U(n, n+\hat{\mu})\,\mathrm{tr}\,U(x, x; \mathrm{C})$$

$$\times\left[1 - \frac{1}{2g^2}\sum_p \mathrm{tr}\,U_p + \frac{1}{2!}\left(\frac{1}{2g^2}\right)^2 \sum_p\sum_{p'} \mathrm{tr}\,U_p\,\mathrm{tr}\,U_{p'}, + \ldots\right]. \tag{10.267}$$

For simplicity we will take a planar rectangular path C as in Fig. 10.19. It is clear that in this strong coupling limit the orgothonality properties of the link variables given in eqns (10.264)–(10.266) imply that the lowest-order nonvanishing contribution to $W(\mathrm{C})$ is the $(1/g^2)^{N_\mathrm{p}}$ term where N_p is the minimal number of plaquettes required to cover the area enclosed by the path C,

$$W(\mathrm{C}) \sim \left(\frac{1}{g^2}\right)^{N_\mathrm{p}}. \tag{10.268}$$

This corresponds to the area law since the area enclosed by the path C is given by

$$A(\mathrm{C}) = a^2 N_\mathrm{p}. \tag{10.269}$$

Hence,

$$W(\mathrm{C}) \sim (g^2)^{-A(\mathrm{C})/a^2} = \exp\{-(TR\ln g^2)/a^2\} \tag{10.270}$$

or, comparing this with (10.260) we have the linearly arising potential

$$E(R) = KR \tag{10.271}$$

with

$$K = \frac{\ln g^2}{a^2}. \tag{10.272}$$

The weak coupling expansion of the Wilson loop can also be considered by first passing to the continuum limit and replacing the action by the Gaussian approximation. We then obtain a *perimeter law*. It turns out that this is the familiar Coulomb's law, $E(R) \sim R^{-1}$, in disguise. With the appropriate renormalization effects taken into account the property of asymptotic freedom is recovered.

Does this mean that we have proven that QCD possesses both properties of asymptotic freedom and colour confinement? No, not yet, as we note that the U(1) lattice gauge theory also has both the perimeter (i.e. Coulomb's) law and the area law (confinement)—the above strong-coupling result is derived without using the non-Abelian nature of the theory. The catch is that the weak and strong coupling regimes may be separated by one or more discontinuous phase transitions. It has been proven that in four dimensions the Abelian gauge theory has a non-trivial phase structure, but similar analytic proof that QCD does not undergo a phase transition at some finite couplings has not been obtained. However this problem can be investigated using numerical analysis. In particular two approaches have been very successful. The first involves the use of Pade techniques to extrapolate from the strong coupling expansion to a regime where weak coupling predictions become valid (Kogut, Pearson, and Shigemitsu 1979). The second approach is Monte Carlo simulation in which the path integral (10.263) is considered as a partition function for a statistical system (Creutz 1979). Various correlations can be calculated by first generating configurations typical of the system in thermal equalibrium. The results from both methods indicate that no phase transitions occur in the intermediate coupling region. The strong coupling behaviour (10.272) $g^2(a) \sim e^{Ka^2}$ does go into the weak coupling $g^2(a) \sim 1/(\ln a^{-1})$ as $a \to 0$. It is found that the transition is smooth and rapid (see also Kogut *et al.* 1981), just as a similar calculation for U(1) theory shows convincing evidence for a phase transition. Furthermore, these methods are able to produce satisfactory numerical relations between the long-distance parameter string tension and the gauge coupling at short distance (or the conventional QCD scale parameter, $100\ \text{MeV} \le \Lambda \le 300\ \text{MeV}$). Thus these numerical results are encouraging in indicating that asymptotic freedom and colour confinement do coexist in a single phase of QCD.

11 Standard electroweak theory I: basic structure

IN this chapter and the following one we shall present the standard gauge theory of weak and electromagnetic interactions. It combines in one framework quantum electrodynamics and the low-energy V–A theory of weak interactions for charged currents. The unified theory is renormalizable. It also predicts a new set of neutral currents; its successful experimental confrontation in recent years strengthens our confidence as to the correctness of the theory. Chapter 11 will serve as an introduction and will emphasize the theoretical structure of the model. In Chapter 12 the phenomenological implications of the model will be presented. Possible extension and modification of the standard electroweak theory will be discussed in Chapter 13.

11.1 Weak interactions before gauge theory

In this section we shall provide a brief review of weak-interaction theories before the advent of gauge models: the four-fermion theory as well as the intermediate vector boson theory. We shall discuss in particular the difficulties encountered if they were taken to be self-contained theories.

Four-fermion interactions

Soon after Pauli's neutrino postulate, Fermi (1934*a*,*b*) proposed his theory for the β-decay $n \to p\, e\, \bar{\nu}$:

$$\mathscr{L}_F(x) = -\frac{G_F}{\sqrt{2}}[\bar{p}(x)\gamma_\lambda n(x)][\bar{e}(x)\gamma^\lambda \nu(x)] + \text{h.c.} \tag{11.1}$$

where the fermion field operators are denoted by their particle names, and

$$G_F \simeq 10^{-5}/m_p^2 \tag{11.2}$$

is the Fermi coupling constant with m_p being the proton mass.

In the ensuing years other processes such as the π–μ and μ–e decays have been discovered and found, like β-decays, to have a comparatively long lifetime. The concept of a distinctive class of interactions, the weak interactions, began to emerge. The surprising discovery of parity non-conservation (Lee and Yang 1956; Wu *et al.* 1957) stimulated a great deal of research and the eventual formulation of the V–A theory (Feynman and Gell-Mann 1958; Sudarshan and Marshak 1958; Sakurai 1958). It is suggested that an effective Lagrangian, very much like the one in eqn (11.1), describes the weak interactions

$$\mathscr{L}_{\text{eff}}(x) = -\frac{G_F}{\sqrt{2}} J^\dagger_\lambda(x) J^\lambda(x) + \text{h.c.} \tag{11.3}$$

with the weak current $J_\lambda(x)$ being of the vector-minus-axial-vector form.

If we separate in the current the leptonic and hadronic parts,

$$J_\lambda(x) = J_{l\lambda}(x) + J_{h\lambda}(x) \tag{11.4}$$

the leptonic current $J^\lambda_l(x)$ can be written directly in terms of lepton fields

$$J^\lambda_l(x) = \bar{\nu}_e \gamma^\lambda (1 - \gamma_5)\, e + \bar{\nu}_\mu \gamma^\lambda (1 - \gamma_5) \mu \tag{11.5}$$

and the hadronic current $J^\lambda_h(x)$ can be decomposed into parts having definite flavour SU(3) transformation properties as in eqn (5.79) and can be neatly summarized by writing the current directly in terms of quark fields

$$J^\lambda_h(x) = \bar{u} \gamma^\lambda (1 - \gamma_5)\, d_\theta \tag{11.6}$$

with

$$d_\theta = \cos\theta_c\, d + \sin\theta_c\, s \tag{11.7}$$

where θ_c is the Cabibbo angle $\approx 13°$. When compared eqn (11.6) to (11.5), lepton–quark symmetry then suggests the generalization of (11.6) to

$$J^\lambda_h(x) = \bar{u} \gamma^\lambda (1 - \gamma_5)\, d_\theta + \bar{c} \gamma^\lambda (1 - \gamma_5) s_\theta \tag{11.8}$$

with

$$s_\theta = \cos\theta_c\, s - \sin\theta_c\, d \tag{11.9}$$

where the c-field is the postulated new heavy quark, the charmed quark (Bjorken and Glashow 1964). More importantly, it has been shown that any sensible weak-interaction theory must have this extra hadronic current in order to suppress to an acceptable level the induced strangeness-changing neutral-current effects (Glashow, Iliopoulos and Maiani 1970). This suppression mechanism, although invented before the general acceptance of gauge theories, can best be explained in this new context and we shall do so in §11.3 and in more detail in §12.2.

We should take note of some of the common properties of the weak currents as given by eqns (11.5) and (11.8). They are charged currents, all with one unit of charge; in the lowest order there are no neutral-current processes such as the reaction $\nu_\mu N \to \nu_\mu N$. They are bilinear in the fundamental fermion fields involving the helicity projection operator $(1 - \gamma_5)$. We can rephrase this by saying that only the left-handed (LH) fermions are present in the weak currents. Thus,

$$\bar{\psi} \gamma^\lambda (1 - \gamma_5) \psi = 2 \bar{\psi}_L \gamma^\lambda \psi_L \tag{11.10}$$

with

$$\psi_L = \tfrac{1}{2}(1 - \gamma_5)\psi, \quad \psi_R = \tfrac{1}{2}(1 + \gamma_5)\psi \tag{11.11}$$

and

$$\psi = \psi_L + \psi_R. \tag{11.12}$$

The rule that 'parity violation is maximal in weak interactions' has a simple interpretation in terms of the lepton and quark fields: in charged-current weak interactions the fundamental dynamical degrees of freedom for matter are the two-component left-handed fermion fields.

If we restrict our applications to the leading order in G_F, the V–A Lagrangian of (11.3) is very successful in describing a vast amount of low-energy weak-interaction experimental data. The exceptions are few. It is not clear whether nonleptonic weak decays are correctly described by a $J_{h\lambda}(x)J_h^{\lambda}(x)$ coupling as required by (11.3) because of our inability to do reliable strong-interaction calculations. The CP violation discovered in the neutral K-meson system (Christenson, Cronin, Fitch, and Turlay 1964) cannot be incorporated in any simple fashion. The most successful phenomenological theory of CP nonconservation postulates the existence of a new super-weak interaction (Wolfenstein 1964). In short, the lowest-order V–A theory correctly describes the domain of weak-interaction phenomena in which one believes it should be applicable.

Nevertheless the Lagrangian in (11.3) cannot be taken as a self-consistent quantum field theory of weak interactions: it is not renormalizable and even in the lowest order of G_F it violates unitarity at high energies.

(1) *Lack of renormalizability.* The interaction in (11.3) is not renormalizable. It is a dimension-six operator, or, more transparently, the coupling constant G_F has dimension $(\text{mass})^{-2}$. Thus the higher-order contributions are increasingly divergent and they cannot be organized in such a way as to be absorbed in a few 'bare' quantities (see Chapter 2, especially §2.4). In a nonrenormalizable theory, even though the coupling may be small, there is no guarantee that the higher-order terms are not large. Then the whole corpus of selection rules (e.g. the absence of $\Delta S > 1$, $\Delta S = -\Delta Q$ charged-current processes) which is the basis of the successful V–A theory will be made meaningless if the lowest-order Born terms do not dominate.

(2) *Violation of unitarity.* Even if we restrict ourselves to the Born approximation there are certain processes which will violate unitarity. Consider the reaction of $\nu_\mu e \to \mu\nu_e$ as described by the effective Lagrangian of (11.3): the amplitude has only the $J = 1$ partial wave and the high-energy cross-section $\sigma \sim G_F^2 s$, with $s = 2m_e E$ (E being the ν_μ energy in the laboratory frame). However unitarity requires that $\sigma(J = 1)$ be bounded by s^{-1}. Thus for energies above $\sqrt{s} \sim G_F^{-1/2} \simeq 300$ GeV the theoretical cross-section from (11.3) would violate unitarity. (For details see §11.2.)

It turns out that these two problems are closely related. If the lower-order diagrams have bad high-energy behaviour to violate unitarity, the higher-order contributions, which are integrals over lower-order diagrams, cannot be renormalized.

Intermediate vector boson theory (IVB)

Like the electromagnetic current, the weak current in $\mathscr{L}_{\text{eff}}$ (eqn (11.3)) transforms as a four-vector under the Lorentz transformation and one may imitate the successful QED by introducing a new massive field W_μ and write the basic interaction as

$$\mathscr{L}_{\rm I} = g(J_\mu W^\mu + \text{h.c.}). \tag{11.13}$$

Then the four-fermion Lagrangian (11.3) can be viewed as the effective low-energy theory generated by $\mathscr{L}_{\rm I}$ in second order (see Fig. 11.1) with the

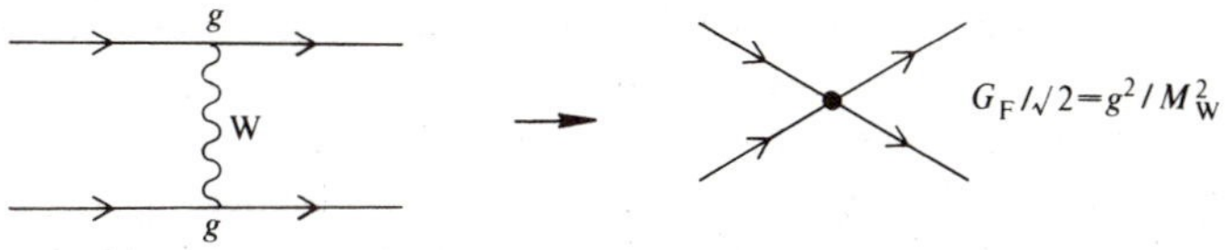

FIG. 11.1. The four-fermion interaction as the low-energy approximation to the g^2 diagram in IVB theory.

identification $g^2 M_W^{-2} = 2^{-1/2} G_F$, as the massive IVB propagator contributes the M_W^{-2} factor. Since QED is a gauge theory, one may interpret this interaction Lagrangian as resulting from a Yang–Mills construction—but with gauge bosons being massive.

Let us now examine the problems of unitarity and renormalizability in IVB theory. The problem with unitarity remains: but it is shifted to some other processes. For example the reaction $\nu\bar{\nu} \to W^+W^-$ with longitudinally polarized Ws can be shown (see §11.2 below) in the IVB theory to have a high-energy amplitude $\sim G_F E^2$ (E being the neutrino energy) in the pure $J = 1$ partial wave, which unitarity requires to be bounded by a constant.

Even though the coupling g is now dimensionless, this theory is still not renormalizable. The free massive vector boson Lagrangian is

$$\mathscr{L}_W = -\tfrac{1}{4}(\partial_\mu W_\nu^\dagger - \partial_\nu W_\mu^\dagger)(\partial^\mu W^\nu - \partial^\nu W^\mu) + M_W^2 W_\mu^\dagger W^\mu \tag{11.14}$$

which gives the propagator in momentum space

$$\mathrm{i}\Delta_{\mu\nu}(k) = -\mathrm{i}\,\frac{g_{\mu\nu} - k_\mu k_\nu/M_W^2}{k^2 - M_W^2 + \mathrm{i}\varepsilon}. \tag{11.15}$$

This propagator behaves like a constant as $k \to \infty$ and the interaction is not renormalizable by power counting. The problem of course lies in the IVB mass, which gives the term $k_\mu k_\nu/M_W^2$ in the propagator. However just this massiveness of W is required in order to yield the desired low-energy four-fermion theory. The key problem is then how to introduce gauge-boson mass terms in the Yang–Mills theory without spoiling renormalizability.

1.2 Construction of the standard SU(2) × U(1) theory

As discussed in the last section, weak interactions must involve massive intermediate vector bosons and yet their mass terms spoil renormalizability. This serious contradiction was finally resolved with the emergence of the spontaneously broken gauge theory. Also, the correct renormalizable theory displays the unity of weak interactions with electromagnetism. The required gauge symmetry is SU(2) × U(1). This model is now the standard theory of the electroweak interaction.

Noting the vectorial nature of both interactions, Schwinger (1957) was the first to advance the idea of weak and electromagnetic unification. Glashow (1958) suggested that the desired renormalizable theory of weak interactions would involve this unification, and later he (Glashow 1961) proposed such a model which has the SU(2) × U(1) gauge symmetry. Renormalizability is not preserved in his theory as the IVB masses were inserted by hand. A similar attempt was also made by Salam and Ward (1964). Finally the renormalizable theory with IVB masses generated by the Higgs mechanism was proposed by Weinberg (1967) and also discussed independently by Salam a year later (1968). Thus the standard theory is often referred to as the Weinberg–Salam model or the Glashow–Weinberg–Salam model. However, the importance of this approach was not recognized by the general community of high-energy theoretical physicists until 't Hooft (1971*a*,*b*) transformed the subject by proving the renormalizability of gauge theories, with and without spontaneous symmetry breaking.

In the following discussion we shall present the Weinberg–Salam theory. For simplicity we shall restrict our initial consideration to the lightest fermion 'family' or 'generation' (ν_e, e, u, d). As we shall show, a theory in the one-family approximation is completely self-contained and self-consistent. The heavier families (ν_μ, μ, c, s) and (ν_τ, τ, t, b) will be given structures identical to the light one. Their incorporation into the standard theory is discussed in §11.3.

Choice of the group SU(2) × U(1)

The algebraic approach. To motivate the choice of the gauge group we need only to consider an IVB theory with an electron and its neutrino. The weak-interaction Lagrangian is given by eqns (11.13) and (11.5)

$$\mathscr{L}_{\mathrm{W}} = g(J_\lambda \mathrm{W}^\lambda + \mathrm{h.c.}) \tag{11.16}$$

where

$$J_\lambda = \bar{\nu}_e \gamma_\lambda (1 - \gamma_5)\, \mathrm{e} \tag{11.17}$$

is the V–A charged current. On the other hand the electromagnetic interaction of these leptons is given by

$$\mathscr{L}_{\mathrm{em}} = \mathrm{e}\, J_\lambda^{\mathrm{em}} A^\lambda \tag{11.18}$$

where

$$J_\lambda^{\mathrm{em}} = -\bar{\mathrm{e}}\, \gamma_\lambda\, \mathrm{e} \tag{11.19}$$

is the electromagnetic current. In a unified gauge theory of weak and electromagnetic interactions, we need at least three vector gauge bosons ($\mathrm{W}^\pm$ and the photon) to couple to the currents J, $J^\dagger$, and J^{em}. The simplest group with three generators is SU(2). However, as we shall demonstrate immediately, the three currents listed in eqns (11.17) and (11.19) do not close (to form an algebra) under commutation. Define the weak and electric charges as

$$T_+(t) = \tfrac{1}{2}\int d^3x J_0(x) = \tfrac{1}{2}\int d^3x \nu_e^\dagger(1-\gamma_5)\,e$$

$$T_-(t) = T_+^\dagger(t)$$

$$Q(t) = \int d^3x J_0^{em}(x) = -\int d^3x\, e^\dagger\, e. \tag{11.20}$$

Using the canonical commutation relations for fermions

$$\{\psi_i^\dagger(\mathbf{x},t), \psi_j(\mathbf{x}',t)\} = \delta_{ij}\,\delta^3(\mathbf{x}-\mathbf{x}'),$$

we can show that

$$[T_+(t), T_-(t)] = 2T_3(t) \tag{11.21}$$

with

$$T_3(t) = \tfrac{1}{4}\int d^3x[\nu_e^\dagger(1-\gamma_5)\nu_e - e^\dagger(1-\gamma_5)\,e]. \tag{11.22}$$

$T_3 \neq Q$ and thus $T_\pm$, Q do not form a closed algebra. The reason behind this is not difficult to see. In order for Q to be a generator of SU(2) the charges of a complete multiplet must add up to zero, corresponding to the requirement that the generators for SU(2) must be traceless. In the case at hand we are attempting to form a doublet out of ν_e and e which clearly do not satisfy this condition. Also, $T_\pm(t)$ are of the V–A form while Q is purely vector.

There are two alternatives at this point.

(i) We can introduce another gauge boson coupled to T_3 as given in eqn (11.21). These four generators can now form the group SU(2) × U(1). This is the choice we shall adopt eventually.

(ii) We can add new fermions to the multiplet and thus modify the currents in order that the new set of $T_\pm$ and Q will be closed to form SU(2) under commutation. In our case we may attempt to form a triplet with e, ν_e, and a new charged heavy lepton E^+. Such a theory has in fact been constructed by Georgi and Glashow (1972*a*). (They also introduced a neutral heavy lepton N in order to obtain the V–A form for the weak current at low energies.) In this model,

$$\tfrac{1}{2}(1-\gamma_5)\begin{pmatrix} E^+ \\ \nu_e \cos\alpha + N \sin\alpha \\ e \end{pmatrix},$$

$$\tfrac{1}{2}(1+\gamma_5)\begin{pmatrix} E^+ \\ N \\ e \end{pmatrix},$$

and

$$\tfrac{1}{2}(1+\gamma_5)(\mathrm{N}\cos\alpha - \nu_e \sin\alpha)$$

are the two triplets and one singlet in the model. This yields the weak charge

$$T_+(t) = \tfrac{1}{2}\int \mathrm{d}^3x[\mathrm{E}^+(1-\gamma_5)(\nu_e\cos\alpha + \mathrm{N}\sin\alpha) + (\nu_e^\dagger\cos\alpha + \mathrm{N}^\dagger\sin\alpha)(1-\gamma_5)\,\mathrm{e} + \mathrm{E}^\dagger(1+\gamma_5)\mathrm{N} + \mathrm{N}^\dagger(1+\gamma_5)\,\mathrm{e}]. \tag{11.23}$$

It is then straightforward to see that

$$[T_+(t), T_-(t)] = 2Q(t)$$

with

$$Q = \int \mathrm{d}^3x[\mathrm{E}^\dagger\mathrm{E} - \mathrm{e}^\dagger\,\mathrm{e}]. \tag{11.24}$$

Clearly the only neutral current in this case is the electromagnetic current and the model was ruled out by the discovery of weak neutral-current effects in 1973. Furthermore, it is difficult to incorporate fractionally charged quarks into such a model.

The unitarity argument. Equivalently we can argue from unitarity that it is necessary to introduce either a new charged lepton or a new neutral gauge-boson. Consider the reaction $\nu\bar{\nu} \to \mathrm{W}^+\mathrm{W}^-$ with both Ws being longitudinally polarized. The lowest-order amplitude is given by Fig. 11.2

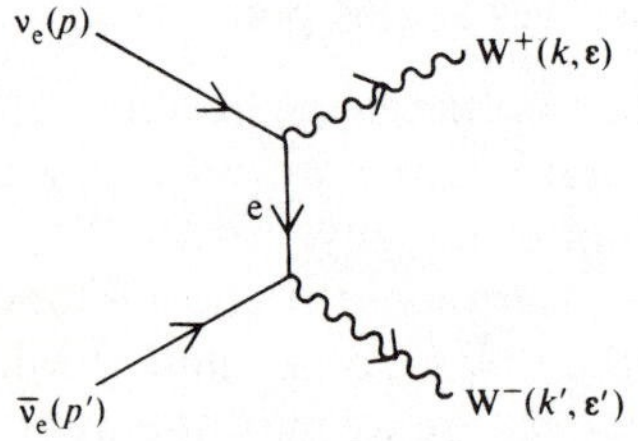

FIG. 11.2. $\nu\bar{\nu} \to \mathrm{W}^+\mathrm{W}^-$ with a t-channel exchange of leptons.

$$T_t(\nu\bar{\nu} \to \mathrm{W}^+\mathrm{W}^-) = -\mathrm{i}\bar{v}(p')(-\mathrm{i}g\not{\varepsilon}')(1-\gamma_5) \times \frac{\mathrm{i}}{\not{p} - \not{k} - m_e}(-\mathrm{i}g\not{\varepsilon})(1-\gamma_5)u(p)$$

$$= -2g^2\bar{v}(p')\frac{\not{\varepsilon}'(\not{p}-\not{k})\not{\varepsilon}(1-\gamma_5)}{(p-k)^2 - m_e^2}u(p). \tag{11.25}$$

The polarization vectors $\varepsilon_\mu^{(i)}(\mathrm{k})$ with $\varepsilon^{(i)}\cdot\varepsilon^{(j)} = -\delta_{ij}$ and $k\cdot\varepsilon^{(i)} = 0$ may be chosen in the rest frame of the W boson as $\varepsilon_0^{(i)} = 0$ and $\varepsilon_j^{(i)} = \delta_{ij}$. To obtain $\varepsilon^{(i)}$ for a moving W boson: $k_\mu = (E, 0, 0, k)$ with $k = (E^2 - M_W^2)^{\frac{1}{2}}$ we can make the appropriate Lorentz boost along the z-axis. The transverse polarizations will not change under such a Lorentz transformation while the longitudinal polarization vector becomes $\varepsilon_\mu^{(3)} = M_W^{-1}(k, 0, 0, E)$. In the high-energy limit with $k = E - M_W^2/2E + \ldots$ the $\varepsilon_\mu^{(3)}$ vector can be approximated as

$$\varepsilon_\mu^{(3)} = k_\mu/M_W + O(M_W/E). \tag{11.26}$$

Thus, substituting (11.26) into (11.25), we have

$$\begin{aligned} T_t &\simeq -2g^2/(k^2 - 2p\cdot k)\bar{v}(p')(\not{k}'/M_W) \\ &\quad \times (\not{p} - \not{k})(\not{k}/M_W)(1 - \gamma_5)u(p) \\ &\simeq \frac{2g^2}{M_W^2}\bar{v}(p')\not{k}'(1-\gamma_5)u(p) \end{aligned} \tag{11.27}$$

where we have used the result $(\not{p} - \not{k})\not{k}u(p) = (2p\cdot k - k^2)u(p)$. To show more explicitly that this amplitude is a pure $J = 1$ partial wave, we can choose the following momentum configurations

$$\begin{aligned} p_\mu &= (E, 0, 0, E), & p'_\mu &= (E, 0, 0, -E) \\ k_\mu &= (E, k\mathbf{e}), & k'_\mu &= (E, -k\mathbf{e}) \end{aligned}$$

with

$$\mathbf{e} = (\sin\theta, 0, \cos\theta). \tag{11.28}$$

Since ν and $\bar{\nu}$ have opposite helicities, we have

$$u(p) = \sqrt{E}\begin{pmatrix} 1 \\ \dfrac{\boldsymbol{\sigma}\cdot\mathbf{p}}{E} \end{pmatrix}\chi_{-1/2} = \sqrt{E}\begin{pmatrix} 1 \\ \sigma_z \end{pmatrix}\chi_{-1/2} \tag{11.29}$$

$$\bar{v}(p') = \sqrt{E}\,\chi_{1/2}^\dagger\left(\frac{\boldsymbol{\sigma}\cdot\mathbf{p}'}{E}, -1\right) = \sqrt{E}\,\chi_{1/2}^\dagger(-\sigma_z, -1) \tag{11.30}$$

where

$$\chi_{1/2} = \begin{pmatrix} 1 \\ 0 \end{pmatrix}, \qquad \chi_{-1/2} = \begin{pmatrix} 0 \\ 1 \end{pmatrix}$$

and thus the combination in eqn (11.27) becomes

$$\begin{aligned} \bar{v}(p')\not{k}'(1-\gamma_5)u(p) &= E\chi_{1/2}^\dagger(-1, -1) \\ &\quad \times \begin{pmatrix} E & k\boldsymbol{\sigma}\cdot\mathbf{e} \\ -k\boldsymbol{\sigma}\cdot\mathbf{e} & -E \end{pmatrix}\begin{pmatrix} 1 & -1 \\ -1 & 1 \end{pmatrix}\begin{pmatrix} 1 \\ -1 \end{pmatrix}\chi_{-1/2} \\ &= -4E\chi_{1/2}^\dagger(E - k\boldsymbol{\sigma}\cdot\mathbf{e})\chi_{-1/2} = 4Ek\sin\theta. \end{aligned} \tag{11.31}$$

We have $T_t \simeq G_F E^2 \sin\theta$ as $E \to \infty$. The partial-wave expansion for the helicity amplitude is (Jacob and Wick 1959; see also Frazer 1966) in this case

$$T_{\lambda_3\lambda_4,\lambda_1\lambda_2}(E,\theta) = \sum_{J=M}^{\infty}(2J+1)T^J_{\lambda_3\lambda_4,\lambda_1\lambda_2}(E)d^J_{\mu\lambda}(\theta)$$

where $\lambda_1 = -\lambda_2 = 1/2$ and $\lambda_3 = \lambda_4 = 0$ are the helicities of the initial and final particles with $\lambda = \lambda_1 - \lambda_2 = 1$, $\mu = \lambda_3 - \lambda_4 = 0$ and $M = \max(\lambda, \mu) = 1$. $d^J_{\mu\lambda}(\theta)$ is the usual rotation matrix with $d^1_{10}(\theta) = \sin\theta$. It is clear that T_t corresponds to a pure $J = 1$ partial wave and violates the unitarity bound of $T^{J=1}(E) <$ constant. To cancel this bad high-energy behaviour, we need other diagrams for this reaction. There are two possibilities: *s*-channel or *u*-channel exchange diagrams. (The *t*-channel diagram will not help as it gives the similar contribution as before with the same sign.)

The heavy-lepton alternative; the *u*-channel exchange diagram in Fig. 11.3(a) yields the amplitude

$$T_u(\nu\bar{\nu} \to W^+W^-) = -2g'^2\bar{v}(p')\frac{\not\varepsilon(\not p - \not k')\not\varepsilon'(1-\gamma_5)}{(p-k')^2 - m_E^2}u(p)$$

$$\simeq \frac{-2g'^2}{M_W^2}\bar{v}(p')\not k'(1-\gamma_5)u(p). \tag{11.32}$$

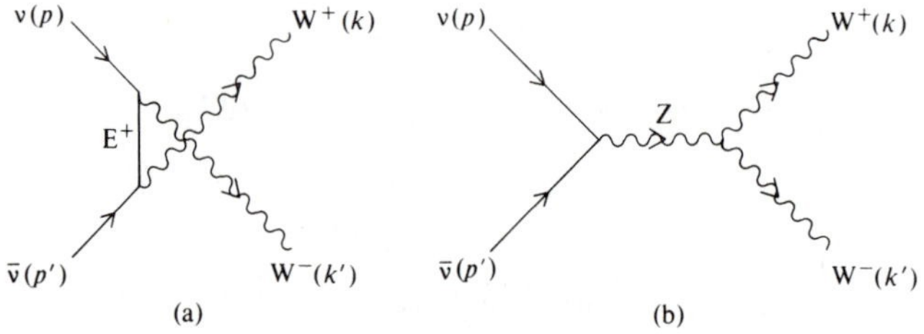

FIG. 11.3. $\nu\bar{\nu} \to W^+W^-$ with (a) u-channel and (b) s-channel exchanges.

If $g^2 = g'^2$, this will cancel the bad high-energy behaviour given in eqn (11.27).

The neutral vector boson alternative; the *s*-channel exchange diagram in Fig. 11.3(b) yields the amplitude

$$T_s(\nu\bar{\nu} \to W^+W^-) = -i\bar{v}(p')(-if\gamma_\beta)(1-\gamma_5)u(p)L_{\alpha\mu\nu}\varepsilon'^\mu(k')\varepsilon^\nu(k)$$

$$\times\, i[-g^{\alpha\beta} + (k+k')^\alpha(k+k')^\beta/M_z^2]/[(k+k')^2 - M_z^2]. \tag{11.33}$$

Choosing the ZWW coupling to have the Yang–Mills structure

$$L_{\alpha\mu\nu} = -if'[(k'-k)_\alpha g_{\mu\nu} - (2k'+k)_\nu g_{\alpha\mu} + (2k+k')_\mu g_{\alpha\nu}], \tag{11.34}$$

we get

$$\varepsilon'^\mu\varepsilon^\nu L_{\alpha\mu\nu} = -if'[(k'-k)_\alpha(\varepsilon'\cdot\varepsilon) - 2(k'\cdot\varepsilon)\varepsilon'_\alpha + 2(k\cdot\varepsilon')\varepsilon_\alpha]$$

$$\simeq -\frac{if'}{M_W^2}[(k-k')_\alpha(k\cdot k')]. \tag{11.35}$$

Hence

$$T_s \simeq \frac{-ff'}{M_W^2}\bar{v}(p')\not k'(1-\gamma_5)u(p). \tag{11.36}$$

If we choose $ff' = 2g^2$, this will also cancel the amplitude in (11.27).

In fact if one demands that all the amplitudes which violate unitarity be cancelled out, one ends up with a renormalizable Lagrangian which is the same as the one derived formally from the algebraic approach. (For details of such a construction see Llewellyn Smith 1973; Bell 1973; Cornwall, Levin, and Tiktopoulous 1974, 1975.)

With the choice of the group SU(2) × U(1), it is straightforward to write down the gauge-invariant Lagrangian

$$\mathscr{L}_1 = -\tfrac{1}{4}F^i_{\mu\nu}F^{i\mu\nu} - \tfrac{1}{4}G_{\mu\nu}G^{\mu\nu} \tag{11.37}$$

where

$$F^i_{\mu\nu} = \partial_\mu A^i_\nu - \partial_\nu A^i_\mu + g\varepsilon^{ijk}A^j_\mu A^k_\nu, \quad i = 1, 2, 3 \tag{11.38}$$

and

$$G_{\mu\nu} = \partial_\mu B_\nu - \partial_\nu B_\mu \tag{11.39}$$

are the SU(2) and U(1) gauge-field tensors, respectively. Clearly before spontaneous symmetry breakdown this corresponds to four massless gauge bosons.

Fermions (in the one-family approximation)

In this section we shall study the standard model with its fermionic sector composed of e, ν_e leptons and u, d quarks only. As it turns out such a simplified theory is completely self-contained. Heavier fermions such as μ, s, c, . . ., etc. will be incorporated in §11.3.

As we have already mentioned in §11.1 the basic fermion dynamic degrees of freedom are the two-component fields with definite helicities. More pertinently, since gauge interactions conserve helicity, we can and should make independent choices for left-handed and right-handed fermions. Thus the first family is composed of the following 15 two-component fermions

$$\psi = \mathrm{\nu_{eL}, e_L, e_R, u_L, u_R, d_L, d_R} \tag{11.40}$$

with

$$\mathrm{e_L} = \tfrac{1}{2}(1 - \gamma_5)\,\mathrm{e}$$
$$\mathrm{e_R} = \tfrac{1}{2}(1 + \gamma_5)\,\mathrm{e}, \text{ etc.} \tag{11.41}$$

The colour indices $\alpha = 1, 2, 3$ on the quark fields have been suppressed.

SU(2) × U(1) quantum number assignment. From eqns (11.5), (11.7), (11.20), and (11.22) we have learned that the SU(2) group is generated by the weak charges

$$T_+ = \int (\nu^\dagger_{\mathrm{eL}}\mathrm{e_L} + \mathrm{u^\dagger_L d_L})\,\mathrm{d}^3x$$

$$T_- = (T_+)^\dagger$$

$$T_3 = \tfrac{1}{2}\int (\nu^\dagger_{\mathrm{eL}}\nu_{\mathrm{eL}} - \mathrm{e^\dagger_L e_L} + \mathrm{u^\dagger_L u_L} - \mathrm{d^\dagger_L d_L})\,\mathrm{d}^3x. \tag{11.42}$$

From these expressions for the SU(2) generators, it is clear that

$$l_L \equiv \begin{pmatrix} \nu_{eL} \\ e_L \end{pmatrix} \quad \text{and} \quad q_L \equiv \begin{pmatrix} u_L \\ d_L \end{pmatrix} \tag{11.43}$$

are SU(2) doublets and e_R, u_R, and d_R are singlets. The U(1) group should be chosen in such a way that the electric charge Q

$$\begin{aligned} Q &= \int (-e^\dagger e + \tfrac{2}{3}u^\dagger u - \tfrac{1}{3}d^\dagger d)\, d^3x \\ &= \int (-e_L^\dagger e_L - e_R^\dagger e_R + \tfrac{2}{3}u_L^\dagger u_L + \tfrac{2}{3}u_R^\dagger u_R - \tfrac{1}{3}d_L^\dagger d_L \\ &\quad -\tfrac{1}{3}\, d_R^\dagger d_R)\, d^3x \end{aligned} \tag{11.44}$$

can be a linear combination of the U(1) generator and T_3 of SU(2) in eqn (11.42). We observe that the combination

$$\begin{aligned} Q - T_3 = \int [&-\tfrac{1}{2}(\nu_{eL}^\dagger \nu_{eL} + e_L^\dagger e_L) + \tfrac{1}{6}(u_L^\dagger u_L + d_L^\dagger d_L) \\ &- e_R^\dagger e_R + \tfrac{2}{3}u_R^\dagger u_R - \tfrac{1}{3}d_R^\dagger d_R]\, d^3x \end{aligned} \tag{11.45}$$

has the property of giving the same quantum number to all members of an SU(2) doublet in (11.43). Clearly it commutes with all the SU(2) generators, i.e.

$$[Q - T_3, T_i] = 0, \quad i = 1, 2, 3. \tag{11.46}$$

We then choose

$$Y = 2(Q - T_3) \tag{11.47}$$

as the generator of the U(1) group and refer to Y as the *weak hypercharge*. Unlike the T_is, Y does not obey any nonlinear commutation relations. Its scale and hence the proportional constant between it and $(Q - T_3)$ is strictly a convention. To obtain the correct electric charges for particles we must use eqns (11.45) and (11.47) and make the assignments

$$\begin{aligned} &Y(l_L) = -1, \quad Y(e_R) = -2, \\ &Y(q_L) = 1/3, \quad Y(u_R) = 4/3, \quad Y(d_R) = -2/3. \end{aligned} \tag{11.48}$$

These hypercharge values may be remembered as twice the average charges of each multiplet, as the average T_3 value is always zero. We should note that the group structure allows for any hypercharge assignment. Thus 'charge quantization' (i.e. particle electric charges are integral multiples of some basic unit) is not automatic in the $SU(2)_L \times U(1)_Y$ theory. This can be obtained only if the gauge group is semi-simple. We shall see in Chapter 14 that this feature will be obtained when we combine this electroweak theory with SU(3) quantum chromodynamics into a 'grand unified gauge theory'.

Anomaly cancellation. As we have discussed in §§6.2 and 9.3, the ABJ anomaly spoils the renormalizability of a gauge theory (Gross and Jackiw 1972; Georgi and Glashow 1972*b*; Bouchiat, Iliopoulos, and Meyer 1972). The fermionic gauge couplings must not introduce anomalous Ward identities. Thus, for the fermion representation R with representation matrix $T^a(R)$, the trace $\mathrm{tr}(\{T^a(R), T^b(R)\}T^c(R))$ in eqn (6.60) must vanish.

In the present case, the fermions are either doublets or singlets under SU(2). The matrix T^a will be either the Pauli matrix τ^a or the U(1) hypercharge Y. Since the group SU(2) is anomaly-free

$$\mathrm{tr}(\{\tau^i, \tau^j\}\tau^k) = 2\,\delta^{ij}\,\mathrm{tr}(\tau^k) = 0, \tag{11.49}$$

we will consider cases where at least one of the Ts is the hypercharge Y. Because every member of a given SU(2) multiplet has the same hypercharge value, for the case of two Ts being a Y we have,

$$\mathrm{tr}(\tau^i YY) \propto \mathrm{tr}\,\tau^i = 0,$$

and for the case of one T being a Y we have,

$$\mathrm{tr}(\{\tau^i, \tau^j\}Y) = 2\,\delta^{ij}\,\mathrm{tr}\,Y. \tag{11.49a}$$

Thus this anomaly contribution is proportional to the trace of Y (the sum of all fermionic hypercharge values)

$$\mathrm{tr}\,Y = \sum_i Y_i = \sum_{\text{lepton}} Y + \sum_{\text{quark}} Y.$$

But this vanishes by explicit calculation for the fermion assignments in each generation

$$\sum_{\text{lepton}} Y = -1 \times 2 - 2 = -4$$

$$\sum_{\text{quark}} Y = 3(\tfrac{1}{3} \times 2 + \tfrac{4}{3} - \tfrac{2}{3}) = 4. \tag{11.49b}$$

The factor 3 in front of the parenthesis is due to the colour degree of freedom. For the case when all Ts are the hypercharge, we have from eqn (11.47),

$$\mathrm{tr}(YYY) = 8\,\mathrm{tr}(Q^3 - 3Q^2T_3 + 3QT_3^2 - T_3^3)$$

$$\propto \mathrm{tr}(Q^2T_3 - QT_3^2) \tag{11.50a}$$

because $\mathrm{tr}\,T_3^3 = 0$, and because we can ignore the Q^3 term as the electromagnetic current is a vector (V) and the VVV type of triangular fermion loops does not have anomaly. Explicit calculation of the right-hand side of eqn (11.50a) yields

$$\sum_{\text{lepton}} (Q^2T_3 - QT_3^2) = -\tfrac{1}{2} + \tfrac{1}{4} = -\tfrac{1}{4}$$

$$\sum_{\text{quark}} (Q^2T_3 - QT_3^2) = (\tfrac{2}{3} - \tfrac{1}{6} - \tfrac{1}{2} + \tfrac{1}{4}) = \tfrac{1}{4}. \tag{11.50b}$$

Thus the anomalies again cancel as in eqn (11.49) and we can conclude that with the fermion assignments of eqns (11.43) and (11.48) the SU(2) × U(1) theory is free of ABJ anomaly.

In fact there is a convenient way to remember the cancellation displayed by eqns (11.49) and (11.50). Using eqn (11.47) we can express eqns (11.49a) and (11.50a) as

$$\mathrm{tr}\, Y \propto \mathrm{tr}\, Q$$

$$\mathrm{tr}(Q^2 T_3 - Q T_3^2) \propto \mathrm{tr}(T_3 Q Y) \propto \mathrm{tr}(T_3^2 Y) \propto \mathrm{tr}\, Q \tag{11.51}$$

because $\mathrm{tr}(T_3 Q^2) = \mathrm{tr}(T_3 Y^2) = \mathrm{tr}(T_3^3) = 0$. Thus the nontrivial contribution to the ABJ anomaly in SU(2) × U(1) theory is proportional to

$$\mathrm{tr}\, Q = \sum_i Q = 0 \tag{11.52}$$

Lepton and quark charges cancel when the three colours are taken into account. [Remark: Given (11.52) one can easily show that the entire $\mathrm{SU(3)_c} \times \mathrm{SU(2)_L} \times \mathrm{U(1)_Y}$ standard model is free of anomaly. Namely, the additional triangle diagram contributions involving gluons and electroweak gauge bosons also mutually cancel.]

With the fermions given in (11.40) the gauge-invariant Lagrangian takes on the form

$$\mathscr{L} = \mathscr{L}_1 + \mathscr{L}_2$$

where

$$\mathscr{L}_2 = \bar{\psi} \mathrm{i} \gamma^\mu \mathrm{D}_\mu \psi \tag{11.53}$$

with the covariant derivative

$$\mathrm{D}_\mu \psi = \left(\partial_\mu - \mathrm{i} g \mathbf{T} \cdot \mathbf{A}_\mu - \mathrm{i} g' \frac{Y}{2} B_\mu \right) \psi . \tag{11.54}$$

For example,

$$\mathrm{D}_\mu l_\mathrm{L} = \left(\partial_\mu - \mathrm{i} \frac{g}{2} \boldsymbol{\tau} \cdot \mathbf{A}_\mu + \mathrm{i} \frac{g'}{2} B_\mu \right) l_\mathrm{L} \tag{11.55}$$

and

$$\mathrm{D}_\mu e_\mathrm{R} = (\partial_\mu + \mathrm{i} g' B_\mu) e_\mathrm{R} . \tag{11.56}$$

We should note in particular that there are no gauge-invariant terms in $\mathscr{L}$ that are bilinear in the fermion fields. Hence there are no SU(2) × U(1) symmetric fermion mass terms.

Symmetry breaking via the Higgs mechanism

We need to introduce a set of scalar fields Φ and this set develops a $U(1)_{\mathrm{em}}$ symmetric vacuum expectation value $\langle \Phi \rangle_0$ so that we have the following pattern of symmetry breaking

$$\mathrm{SU(2)_L} \times \mathrm{U(1)_Y} \overset{\langle \Phi \rangle_0}{\rightarrow} U(1)_{\mathrm{em}}$$

Three of the original four SU(2) × U(1) gauge bosons will become massive and one, corresponding to the photons, remains massless.

As we have seen in §8.3, a complex scalar doublet breaks the SU(2) symmetry completely and one member of this doublet must be neutral in order to have possibility of a $U(1)_{\mathrm{em}}$-invariant $\langle\Phi\rangle_0$. We need a complex doublet with the charge assignment

$$\Phi = \begin{pmatrix}\phi^+ \\ \phi^0\end{pmatrix}, \qquad Y(\Phi) = 1. \tag{11.57}$$

The complete gauge-invariant Lagrangian including the scalar fields is then

$$\mathscr{L} = \mathscr{L}_1 + \mathscr{L}_2 + \mathscr{L}_3 + \mathscr{L}_4 \tag{11.58}$$

with

$$\mathscr{L}_3 = (\mathrm{D}_\mu\Phi)^\dagger(\mathrm{D}^\mu\Phi) - V(\Phi) \tag{11.59}$$

where

$$\mathrm{D}_\mu\Phi = \left(\partial_\mu - \frac{\mathrm{i}}{2}g\boldsymbol{\tau}\cdot\mathbf{A}_\mu - \frac{\mathrm{i}}{2}g' B_\mu\right)\Phi \tag{11.60}$$

and

$$V(\Phi) = -\mu^2\Phi^\dagger\Phi + \lambda(\Phi^\dagger\Phi)^2. \tag{11.61}$$

The most general SU(2) × U(1) Yukawa coupling between scalars and fermions is given by

$$\mathscr{L}_4 = f^{(e)}\bar{l}_{\mathrm{L}}\Phi e_{\mathrm{R}} + f^{(u)}\bar{q}_{\mathrm{L}}\tilde{\Phi}u_{\mathrm{R}} + f^{(d)}\bar{q}_{\mathrm{L}}\Phi d_{\mathrm{R}} + \mathrm{h.c.} \tag{11.62}$$

with the isodoublet

$$\tilde{\Phi} = \mathrm{i}\tau_2\Phi^*.$$

having hypercharge $Y(\tilde{\Phi}) = -1$.

As we have already discussed in §§5.3 and 8.3, for positive values of μ^2 and λ in eqn (11.61), we have spontaneous symmetry breakdown as the scalar develops VEV

$$\langle\Phi\rangle_0 \equiv \langle 0|\Phi|0\rangle = \begin{pmatrix} 0 \\ \dfrac{v}{\sqrt{2}} \end{pmatrix} \text{with} \quad v = (\mu^2/\lambda)^{\frac{1}{2}}. \tag{11.63}$$

Particle spectra and interactions in the unitary gauge

Using the polar variables for the scalar fields

$$\Phi(x) = U^{-1}(\boldsymbol{\zeta})\begin{pmatrix} 0 \\ \dfrac{v+\eta(x)}{\sqrt{2}} \end{pmatrix}$$

with

$$U(\boldsymbol{\zeta}) = \exp[\mathrm{i}\boldsymbol{\zeta}(x)\cdot\boldsymbol{\tau}/v]. \tag{11.64}$$

Thus the original two complex fields $\phi^+(x)$ and $\phi^0(x)$ in (11.57) are parametrized in terms of four real fields $\zeta_i(x)$ and $\eta(x)$. These shifted fields have zero VEV

$$\langle 0|\zeta_i|0\rangle = \langle 0|\eta|0\rangle = 0. \tag{11.65}$$

In order to display the particle spectra, we then make a gauge transformation, i.e., go to the unitary gauge, by defining new fields. [Remark: We have simplified the expression in (11.64). In principle ζ_i should be multiplied by all the broken symmetry generators, as in (8.127). For the case at hand, however, the difference is immaterial.]

$$\Phi' = U(\zeta)\Phi = \begin{pmatrix} 0 \\ \dfrac{v+\eta(x)}{\sqrt{2}} \end{pmatrix}$$

or

$$\Phi'(x) = \frac{v+\eta(x)}{\sqrt{2}}\chi \quad \text{with} \quad \chi = \begin{pmatrix} 0 \\ 1 \end{pmatrix} \tag{11.66}$$

and

$$l'_{\mathrm{L}} = U(\zeta)l_{\mathrm{L}}, \quad e'_{\mathrm{R}} = e_{\mathrm{R}}$$

$$q'_{\mathrm{L}} = U(\zeta)q_{\mathrm{L}}, \quad u'_{\mathrm{R}} = u_{\mathrm{R}}, \quad d'_{\mathrm{R}} = d_{\mathrm{R}}$$

$$\frac{\boldsymbol{\tau}\cdot\mathbf{A}'_\mu}{2} = U(\zeta)\left(\frac{\boldsymbol{\tau}\cdot\mathbf{A}_\mu}{2}\right)U^{-1}(\zeta) - \mathrm{i}/g[\partial_\mu U(\zeta)]U^{-1}(\zeta)$$

$$B'_\mu = B_\mu. \tag{11.67}$$

We next express each $\mathscr{L}_i$ of the Lagrangian (11.58) in terms of the new fields. Consider first those containing Φs and use eqn (11.66)

$$\mathscr{L}_3 = (\mathrm{D}_\mu\Phi')^\dagger(\mathrm{D}^\mu\Phi') - V(\Phi') \tag{11.68}$$

with

$$\mathrm{D}_\mu\Phi' = \left(\partial_\mu - \mathrm{i}\frac{g}{2}\boldsymbol{\tau}\cdot\mathbf{A}'_\mu - \mathrm{i}\frac{g'}{2}B'_\mu\right)\left[\frac{v+\eta(x)}{\sqrt{2}}\right]\chi \tag{11.69}$$

$$V(\Phi') = \mu^2\eta^2 + \lambda v\eta^3 + \frac{\lambda}{4}\eta^4 \tag{11.70}$$

and

$$\mathscr{L}_4 = \frac{\eta(x)}{\sqrt{2}}[f^{(\mathrm{e})}\bar{e}'_{\mathrm{L}}e'_{\mathrm{R}} + f^{(\mathrm{u})}\bar{u}'_{\mathrm{L}}u'_{\mathrm{R}} + f^{(\mathrm{d})}\bar{d}'_{\mathrm{L}}d'_{\mathrm{R}}]$$

$$+ \frac{v}{\sqrt{2}}[f^{(\mathrm{e})}\bar{e}'_{\mathrm{L}}e'_{\mathrm{R}} + f^{(\mathrm{u})}\bar{u}'_{\mathrm{L}}u'_{\mathrm{R}} + f^{(\mathrm{d})}\bar{d}'_{\mathrm{L}}d'_{\mathrm{R}}] + \text{h.c.} \tag{11.71}$$

Mass spectrum. From the above equations one can easily read off the mass terms which are bilinear in the fields:

(i) *Scalar mass:* (the physical Higgs particle)

$$m_\eta = \sqrt{2}\,\mu. \tag{11.72}$$

(ii) *Fermion masses:*

$$m_e = f^{(e)}v/\sqrt{2},$$
$$m_u = f^{(u)}v/\sqrt{2}, \qquad m_d = f^{(d)}v/\sqrt{2}. \tag{11.73}$$

(iii) *Vector meson masses:* The three 'would-be-Goldstone bosons' $\zeta(x)$ are transformed away in eqn (11.66). (We say they are 'eaten' by the gauge bosons to form three massive IVBs.) The vector meson masses are contained in the $(D_\mu\Phi')^2$ term in $\mathscr{L}_3$.

$$\begin{aligned}\mathscr{L}_{\text{VMM}} &= \frac{v^2}{2}\chi^\dagger\left(\frac{g}{2}\boldsymbol{\tau}\cdot\mathbf{A}'_\mu + \frac{g'}{2}B'_\mu\right)\left(\frac{g}{2}\boldsymbol{\tau}\cdot\mathbf{A}'^\mu + \frac{g'}{2}B'^\mu\right)\chi \\ &= \frac{v^2}{8}\{g^2[(A'^1_\mu)^2 + (A'^2_\mu)^2] + (gA'^3_\mu - g'B'_\mu)^2\} \\ &\equiv M_W^2 W^+_\mu W^{-\mu} + \tfrac{1}{2}M_Z^2 Z_\mu Z^\mu.\end{aligned} \tag{11.74}$$

We need to make the following identifications of mass and charge eigenstates. For the charged vector mesons,

$$M_W^2 W^+_\mu W^{-\mu} = \frac{g^2v^2}{8}[(A'^1_\mu)^2 + (A'^2_\mu)^2]. \tag{11.75}$$

Thus,

$$W^\pm_\mu = (A'^1_\mu \mp iA'^2_\mu)/\sqrt{2} \tag{11.76}$$

and

$$M_W^2 = g^2v^2/4. \tag{11.77}$$

For the neutral vector mesons, because $U(1)_{em}$ is unbroken $Q\langle\Phi\rangle_0 = 0$, the associated gauge boson (the photon) will remain massless, we have

$$\begin{aligned}\frac{1}{2}M_Z^2 Z_\mu Z^\mu &= \frac{v^2}{8}(gA'^3_\mu - g'B'_\mu)^2 \\ &= \frac{v^2}{8}(A'^3_\mu, B'_\mu)\begin{pmatrix} g^2 & -gg' \\ -gg' & g'^2\end{pmatrix}\begin{pmatrix} A'^{3\mu} \\ B'^\mu\end{pmatrix} \\ &= \frac{1}{2}(Z_\mu, A_\mu)\begin{pmatrix} M_Z^2 & 0 \\ 0 & 0\end{pmatrix}\begin{pmatrix} Z^\mu \\ A^\mu\end{pmatrix}.\end{aligned} \tag{11.78}$$

We have diagonalized the mass matrix by an orthogonal transformation

$$\begin{aligned} Z_\mu &= \cos\theta_W A'^3_\mu - \sin\theta_W B'_\mu \\ A_\mu &= \sin\theta_W A'^3_\mu + \cos\theta_W B'_\mu \end{aligned} \tag{11.79}$$

with

$$\tan\theta_W = g'/g \tag{11.80}$$

and

$$M_Z^2 = v^2(g^2 + g'^2)/4. \tag{11.81}$$

The angle of rotation θ_W is generally referred to as the *Weinberg angle*. Using eqns (11.27), (11.80), and (11.81) we also have the ratio between vector meson masses

$$\rho = M_W^2/(M_Z^2\cos^2\theta_W) = 1 \tag{11.82}$$

Doublet Higgs and $\rho = 1$. The derivation of (11.82) is predicated on our use of the Higgs scalar in the doublet representation. To see this we retrace some of the steps in its derivation. Gauge invariance requires that the scalar potential in eqn (11.61) is a function of $|\Phi|^2$, which may be viewed as the length of a four-vector made up of the four real components of Φ. Thus $V(\Phi) = V(|\Phi|^2)$ has a larger $O(4) \approx SU(2)_L \times SU(2)'$ symmetry and an $O(3) \approx SU(2)$ symmetry after SSB (11.63). We have already encountered such a symmetry-breaking pattern back in §5.3 with the SU(2) × SU(2) σ-model. The multiplet $(\sigma, \boldsymbol{\pi})$ transforms as a $(\frac{1}{2}, \frac{1}{2})$ under SU(2) × SU(2) and as a four-vector under O(4) symmetry. The true vacuum singles out a direction (chosen to be σ in that case) and the theory breaks down to an O(3) symmetry with $\boldsymbol{\pi}$ being still a degenerate triplet. In the case here $\mathbf{A}_\mu$ remains to be a degenerate triplet under the remaining O(3) symmetry. Thus we have the crucial equality for the three terms that are bilinear in $\mathbf{A}_\mu$ fields (11.74)

$$\left(\frac{g}{2}\boldsymbol{\tau}\langle\Phi\rangle_0\mathbf{A}_\mu\right)^2 = \frac{1}{2}M_W^2[(A_\mu^1)^2 + (A_\mu^2)^2 + (A_\mu^3)^2]. \tag{11.83}$$

Once given that A_μ^3 also has the mass term $\frac{1}{2}M_W^2$, the form for the neutral IVB mass matrix in (11.78) is then completely fixed by the trace ($=M_Z^2/2$) and the determinant ($=0$) conditions

$$\tfrac{1}{2}(A_\mu^3, B_\mu)\begin{pmatrix} M_W^2 & M_W(M_Z^2 - M_W^2)^{1/2} \\ M_W(M_Z^2 - M_W^2)^{1/2} & M_Z^2 - M_W^2 \end{pmatrix}\begin{pmatrix} A_\mu^3 \\ B_\mu \end{pmatrix}. \tag{11.84}$$

The mass relation of eqn (11.82) follows immediately.

Charged current. In order to identify the currents with those of V–A theory, let us next examine the fermion gauge interactions as contained in $\mathscr{L}_2$ of eqns (11.53) and (11.54)

$$\begin{aligned}
&\bar{l}'_L\left(\frac{g}{2}\boldsymbol{\tau}\cdot\not{\mathbf{A}}' - \frac{g'}{2}\not{B}'\right)l'_L + \bar{q}'_L\left(\frac{g}{2}\boldsymbol{\tau}\cdot\not{\mathbf{A}}' + \frac{g'}{6}\not{B}'\right)q'_L \\
&\quad - \bar{e}'_R g'\not{B}'e'_R + \bar{u}'_R\frac{2g'}{3}\not{B}'u'_R - \bar{d}'_R\frac{g'}{3}\not{B}'d'_R \\
&= g(\tfrac{1}{2}\bar{l}'_L\boldsymbol{\tau}\gamma_\mu l'_L + \tfrac{1}{2}\bar{q}'_L\boldsymbol{\tau}\gamma_\mu q'_L)\mathbf{A}'^\mu \\
&\quad + \tfrac{1}{2}g(-\bar{l}'_L\gamma_\mu l'_L + \tfrac{1}{3}\bar{q}'_L\gamma_\mu q'_L - 2\bar{e}'_R\gamma_\mu e'_R + \tfrac{4}{3}\bar{u}'_R\gamma_\mu u_R - \tfrac{2}{3}\bar{d}'_R\gamma_\mu d'_R)B'^\mu \\
&\equiv (gJ_\mu^1 A'^{1\mu} + gJ_\mu^2 A'^{2\mu}) + (gJ_\mu^3 A'^{3\mu} + \tfrac{1}{2}g'J_\mu^Y B'^\mu)
\end{aligned} \tag{11.85}$$

corresponding to charged- and neutral-current interactions.

For the charged current interaction in (11.85), we have

$$\mathscr{L}_{\mathrm{CC}} = (gJ^1_\mu A'^{1\mu} + gJ^2_\mu A'^{2\mu})$$

$$= \frac{g}{\sqrt{2}}(J^+_\mu W^{+\mu} + J^-_\mu W^{-\mu}) \tag{11.86}$$

with $W^\pm_\mu$ given by eqn (11.76) and

$$J^+_\mu = J^1_\mu + \mathrm{i}J^2_\mu = \bar{\nu}'_{\mathrm{L}}\gamma_\mu e'_{\mathrm{L}} + \bar{u}'_{\mathrm{L}}\gamma_\mu d'_{\mathrm{L}}$$

$$= \tfrac{1}{2}\bar{\nu}'\gamma_\mu(1-\gamma_5)e' + \tfrac{1}{2}\bar{u}'\gamma_\mu(1-\gamma_5)d'. \tag{11.87}$$

For the low-energy four-fermion interaction shown in Fig. 11.1 we can generate the following effective Lagrangian

$$\mathscr{L}^{\mathrm{CC}}_{\mathrm{eff}} = \frac{-g^2}{2M^2_{\mathrm{W}}} J^+_\mu J^{-\mu}. \tag{11.88}$$

This is just the V–A theory. When eqn (11.88) is compared to eqns (11.3), (11.5), and (11.87), we can make the identification

$$\frac{G_{\mathrm{F}}}{\sqrt{2}} = \frac{g^2}{8M^2_{\mathrm{W}}}. \tag{11.89}$$

Using eqn (11.77), this implies that the vacuum expectation value in (11.63) has the size

$$v = 2^{-1/4}G_{\mathrm{F}}^{-1/2} \simeq 250\ \mathrm{GeV}. \tag{11.90}$$

Neutral currents. For the neutral-current interaction in (11.85), we have

$$\mathscr{L}_{\mathrm{NC}} = gJ^3_\mu A'^{3\mu} + \tfrac{1}{2}g'J^Y_\mu B'^\mu$$

$$= gJ^3_\mu(\cos\theta_{\mathrm{W}} Z^\mu + \sin\theta_{\mathrm{W}} A^\mu)$$

$$+ \tan\theta_{\mathrm{W}}\, g(J^{\mathrm{em}}_\mu - J^3_\mu)(\cos\theta_{\mathrm{W}} A^\mu - \sin\theta_{\mathrm{W}} Z^\mu) \tag{11.91}$$

where we have used eqns (11.79), (11.80), and (11.47). Thus

$$\mathscr{L}_{\mathrm{NC}} = eJ^{\mathrm{em}}_\mu A^\mu + (g/\cos\theta_{\mathrm{W}})J^0_\mu Z^\mu \tag{11.92}$$

with

$$e = g\sin\theta_{\mathrm{W}} \tag{11.93}$$

$$J^0_\mu = J^3_\mu - \sin^2\theta_{\mathrm{W}} J^{\mathrm{em}}_\mu. \tag{11.94}$$

The neutral currents can be written out explicitly in terms of the fermion fields

$$J^0_\mu = \sum_f [g^f_{\mathrm{L}}\bar{f}_{\mathrm{L}}\gamma_\mu f_{\mathrm{L}} + g^f_{\mathrm{R}}\bar{f}_{\mathrm{R}}\gamma_\mu f_{\mathrm{R}}]$$

$$= \frac{1}{2}\sum [g^f_{\mathrm{L}}\bar{f}\gamma_\mu(1-\gamma_5)f + g^f_{\mathrm{R}}\bar{f}\gamma_\mu(1+\gamma_5)f] \tag{11.95a}$$

where $f = \nu_e$, e, u, and d. The weak neutral-current couplings are

$$g^f_{\mathrm{L,R}} = T_3(f_{\mathrm{L,R}}) - Q(f)\sin^2\theta_{\mathrm{W}}. \tag{11.95b}$$

From eqns (11.42) and (11.44) we then have

$$\begin{aligned} g_L^\nu &= \tfrac{1}{2}, & g_R^\nu &= 0 \\ g_L^e &= -\tfrac{1}{2} + \sin^2\theta_W, & g_R^e &= \sin^2\theta_W \\ g_L^u &= \tfrac{1}{2} - \tfrac{2}{3}\sin^2\theta_W, & g_R^u &= -\tfrac{2}{3}\sin^2\theta_W \\ g_L^d &= -\tfrac{1}{2} + \tfrac{1}{3}\sin^2\theta_W, & g_R^d &= \tfrac{1}{3}\sin^2\theta_W. \end{aligned} \qquad (11.95c)$$

Starting from the coupling (11.92) we can then generate low-energy four-fermion interactions corresponding to the products of neutral currents

$$\begin{aligned} \mathscr{L}_{\text{eff}}^{\text{NC}} &= \frac{-g^2}{2\cos^2\theta_W M_Z^2} J_\mu^0 J^{0\mu} \\ &= \frac{-g^2}{2M_W^2} J_\mu^0 J^{0\mu} \end{aligned} \qquad (11.96)$$

where eqn (11.82) has been used. The factor of 2 in the denominator comes from the symmetry factor in Feynman rule for two identical currents in (11.96). Thus the SU(2) × U(1) theory predicts a set of new weak interactions which are of comparable strength to the more familiar Fermi interactions of charged currents. As an illustration, consider the ν_ee elastic scattering; both charged and neutral currents contribute. The low-energy amplitude (given by Fig. 11.4) is

$$\begin{aligned} T(\nu_e e \to \nu_e e) = \frac{g^2}{8M_W^2} \{ &[\bar{\nu}_e\gamma_\mu(1-\gamma_5)e][\bar{e}\gamma^\mu(1-\gamma_5)\nu_e] \\ &-\tfrac{1}{2}[\bar{\nu}_e\gamma_\mu(1-\gamma_5)\nu_e][\bar{e}\gamma^\mu(1-\gamma_5)e - 4\sin^2\theta_W\bar{e}\gamma^\mu e]\}. \end{aligned} \qquad (11.97)$$

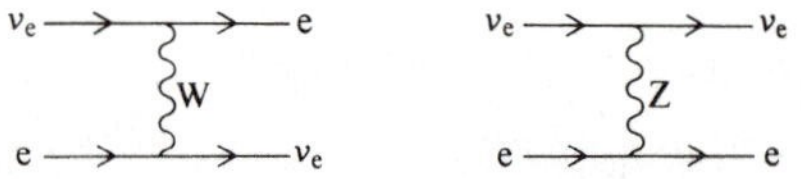

FIG. 11.4. Charged- and neutral-current contributions to elastic ν_ee scattering.

After a simple Fierz rearrangement (see Appendix A) these two contributions can be combined,

$$T(\nu_e e \to \nu_e e) = \frac{G_F}{\sqrt{2}}[\bar{\nu}_e\gamma_\mu(1-\gamma_5)\nu_e][\bar{e}\gamma^\mu(a - b\gamma_5)e]$$

where

$$a = 2\sin^2\theta_W + \tfrac{1}{2} \quad \text{and} \quad b = \tfrac{1}{2}. \qquad (11.98)$$

Of course the old V–A theory would have predicted $a = b = 1$.

As we shall discuss in the next chapter, such a class of neutral-current processes has been discovered in high-energy experiments. All data are consistent with a value for the Weinberg's angle in the neighbourhood of (for a review see Kim, Langacker, Levine, and Williams 1981)

$$\sin^2\theta_W \simeq 0.22 \qquad (11.99)$$

Discussions. The next major step in testing the standard theory will clearly be the discovery of charged and neutral gauge bosons having the prescribed mass values and couplings. From eqns (11.89) and (11.93) we have

$$M_W = 2^{-5/4} e G_F^{-1/2}/\sin\theta_W$$
$$\simeq 37 \text{ GeV}/\sin\theta_W. \qquad (11.100)$$

With the value of θ_W given in eqn (11.99) we anticipate $M_W \simeq 80$ GeV and, with eqn (11.82), $M_Z \simeq 90$ GeV. Another crucial test will be the discovery of the Higgs particle η. However the numerical value of m_η as given in eqn (11.72) is not fixed by any previously measured quantities. This makes a search for the η difficult, particularly in view of its very weak couplings to fermions, being of the order of $(m_f/M_W)e$, as given by eqn (11.71).

In fact the standard theory, even with only one family of fermions, has seven arbitrary parameters e, $\sin\theta_W$, M_W, m_η, m_e, m_u, m_d; or in terms of the original symmetric Lagrangian the seven arbitrary parameters are the two gauge couplings g and g', the two scalar self-couplings μ^2 and λ, and the three Yukawa couplings: $f^{(e)}$, $f^{(u)}$, and $f^{(d)}$. We note that the electroweak unification is in a sense not complete: we need to insert two gauge coupling constants g and g' to account for these two classes of interactions. The quest for further unification will be discussed in Chapter 14.

Finally we emphasize that the guiding principle in constructing gauge models is gauge symmetry and renormalizability. We must include in the Lagrangian *all* gauge-invariant terms of dimension-four or less. Terms may be excluded without destroying renormalizability only with the imposition of the appropriate global symmetries. In the case of the standard electroweak theory no *ad hoc* global symmetry has been imposed. The conservation laws the theory possesses, such as those of baryon number and lepton number, are consequences of gauge invariance and renormalizability once the representation content of fermion and scalar particles are given. Thus both leptons and quarks in each fermion family must be present in order that renormalizability is not spoiled by the ABJ anomalies. On the other hand cross-couplings between leptons and quarks are absent in eqn (11.62) because they are forbidden by gauge symmetry. For instance, if there were a set of scalars transforming as a doublet under the SU(2) symmetry and as a triplet under $SU(3)_{\text{colour}}$: h^i_α $(i = 1, 2;\ \alpha = 1, 2, 3)$, then we would have the gauge-invariant Yukawa couplings $\bar{l}_{iL} h^i_\alpha q^\alpha_R$ and $\bar{q}_{i\alpha} h^i_\beta q^c_\gamma \varepsilon^{\alpha\beta\gamma}$ corresponding to baryon- and lepton-number nonconserving terms. Thus the structure of the Weinberg–Salam theory automatically leads to B and L conservations. [Here we ignore the unobservably small instanton effects ('t Hooft 1976).] Lepton number violation by Majorana neutrino mass terms will be studied in §13.2. Baryon- and lepton-number conservation laws are both violated in grand unified gauge theories (see Chapter 14).

11.3 Fermion family replication

In §11.2 we have shown how to construct the basic version of the standard electroweak gauge theory with a restricted fermion content: the 15 two-component fields of eqn (11.40). The left-hand (LH) fields transform as doublets and the right-hand (RH) fields are singlets. Now we would like to remove this restriction. Eqns (11.5) and (11.9) would lead us to expect further LH fermion doublets

$$\begin{pmatrix} \nu_e \\ e \end{pmatrix}_L \quad \begin{pmatrix} \nu_\mu \\ \mu \end{pmatrix}_L \qquad \begin{pmatrix} u \\ d_\theta \end{pmatrix}_L \quad \begin{pmatrix} c \\ s_\theta \end{pmatrix}_L \tag{11.101}$$

with

$$d_\theta = \cos\theta_c d + \sin\theta_c s$$

$$s_\theta = \cos\theta_c s - \sin\theta_c d, \tag{11.102}$$

again with all RH fields being singlets. In other words there is a 'duplication' of the fermionic structure in the theory. With the discovery of τ, ν_τ leptons (Perl *et al.* 1975, 1977) and the b quark (Herb *et al.* 1977; Lederman 1978) we see that this replication continues: the experimental data do not contradict the expectation that $(\nu_\tau, \tau)_L$ is an SU(2) doublet and that there exists an even heavier t quark which will complete the doublet with b. The task of this section is to study the systematic incorporation of these fermions into the standard model. We would like to understand the presence of the Cabibbo angle in quark doublets and the absence of a corresponding mixing angle in the lepton sector. Other important issues related to the multifamilies of fermions are the suppression of flavour-changing neutral-current effects and the CP-violation phases in fermion gauge couplings (also see §12.2).

Gauge vs. mass eigenstates

The presence of the Cabibbo angle in (11.102) already indicates that we must distinguish between two types of fermionic states: gauge interaction eigenstates (having definite gauge transformation properties, e.g. the d_θ and s_θ fields) and mass eigenstates (e.g. the d and s fields); they are related by some linear transformations. This is because the fermions are massless before the spontaneous symmetry breaking and the fermion mass eigenstates are determined by the Yukawa coupling after the spontaneous symmetry breaking (SSB). The replication of fermionic structure alluded to in the introductory remarks means that there are several groups of identical gauge eigenstates. Thus the fermionic states actually carry a 'family' index or 'generation' index: $A = e, \mu, \tau, \ldots$. In §11.2 we simplified our construction by taking only one family of fermions. In this approximation there is no distinction between gauge eigenstates and mass eigenstates. Since only the gauge transformation properties were used, the fermion fields in §11.2 should be regarded as those of the gauge eigenstates. The proper multi-family generalization would then involve replacing each of these gauge eigenstates in §11.2 by a vector in the family index space. We need to make the following substitution:

$$\begin{aligned} \mathrm{e} &\to \mathrm{e}'_A = (\mathrm{e}', \mu', \tau') \\ \nu_\mathrm{e} &\to \nu'_A = (\nu'_\mathrm{e}, \nu'_\mu, \nu'_\tau) \\ \mathrm{u} &\to \mathrm{p}'_A = (\mathrm{u}', \mathrm{c}', \mathrm{t}') \\ \mathrm{d} &\to \mathrm{n}'_A = (\mathrm{d}', \mathrm{s}', \mathrm{b}'). \end{aligned} \tag{11.103}$$

The SU(2) doublets are then

$$l_{A\mathrm{L}} = \begin{pmatrix} \nu'_A \\ \mathrm{e}'_A \end{pmatrix}_\mathrm{L} \quad \text{and} \quad \mathrm{q}_{A\mathrm{L}} = \begin{pmatrix} \mathrm{p}'_A \\ \mathrm{n}'_A \end{pmatrix}_\mathrm{L}. \tag{11.104}$$

The prime on the field indicates it is a gauge eigenstate. (This represents a change of notation from that in §11.2 where the primed states are fields in the unitary gauge.) The fermionic couplings in the Lagrangian also involve a contraction of family indices. The generalizations of eqns (11.53) and (11.62) are

$$\begin{aligned} \mathscr{L}_2 = {} & \bar{l}_{A\mathrm{L}}\mathrm{i}\left(\not\partial - \frac{\mathrm{i}g}{2}\tau\cdot\not{A} + \frac{\mathrm{i}g'}{2}\not{B}\right)l_{A\mathrm{L}} + \bar{\mathrm{e}}'_{A\mathrm{R}}\mathrm{i}(\not\partial + \mathrm{i}g'\not{B})\,\mathrm{e}'_{A\mathrm{R}} \\ & + \bar{\mathrm{q}}_{A\mathrm{L}}\mathrm{i}\left(\not\partial - \frac{\mathrm{i}g}{2}\tau\cdot\not{A} - \frac{\mathrm{i}g'}{6}\not{B}\right)\mathrm{q}_{A\mathrm{L}} \\ & + \bar{\mathrm{p}}'_{A\mathrm{R}}\mathrm{i}\left(\not\partial - \frac{2\mathrm{i}g'}{3}\not{B}\right)\mathrm{p}'_{A\mathrm{R}} + \bar{\mathrm{n}}'_{A\mathrm{R}}\mathrm{i}\left(\not\partial + \frac{\mathrm{i}g'}{3}\not{B}\right)\mathrm{n}'_{A\mathrm{R}} \end{aligned} \tag{11.105}$$

and

$$\begin{aligned} \mathscr{L}_4 = {} & f^{(\mathrm{e})}_{AB}\bar{l}_{A\mathrm{L}}\Phi\,\mathrm{e}'_{B\mathrm{R}} + f^{(p)}_{AB}\bar{\mathrm{q}}_{A\mathrm{L}}\tilde{\Phi}\mathrm{p}'_{B\mathrm{R}} \\ & + f^{(\mathrm{n})}_{AB}\bar{\mathrm{q}}_{A\mathrm{L}}\Phi\mathrm{n}'_{B\mathrm{R}} + \mathrm{h.c.} \end{aligned} \tag{11.106}$$

Thus the Yukawa coupling constants in eqn (11.62) are replaced by coupling matrices in the family index space. After SSB, given in (11.63), we have

$$\begin{aligned} \mathscr{L}_4 = {} & \frac{\eta(x)}{\sqrt{2}}\left[f^{(\mathrm{e})}_{AB}\,\bar{\mathrm{e}}'_{A\mathrm{L}}\,\mathrm{e}'_{B\mathrm{R}} + f^{(\mathrm{p})}_{AB}\bar{\mathrm{p}}'_{A\mathrm{L}}\,\mathrm{p}'_{B\mathrm{R}} + f^{(\mathrm{n})}_{AB}\bar{\mathrm{n}}'_{A\mathrm{L}}\,\mathrm{n}'_{B\mathrm{R}}\right] \\ & + \frac{v}{\sqrt{2}}\left[f^{(\mathrm{e})}_{AB}\bar{\mathrm{e}}'_{A\mathrm{L}}\,\mathrm{e}'_{B\mathrm{R}} + f^{(\mathrm{p})}_{AB}\bar{\mathrm{p}}'_{A\mathrm{L}}\,\mathrm{p}'_{B\mathrm{R}} + f^{(\mathrm{n})}_{AB}\bar{\mathrm{n}}'_{A\mathrm{L}}\,\mathrm{n}'_{B\mathrm{R}}\right] + \mathrm{h.c.} \end{aligned} \tag{11.107}$$

which is a generalization of (11.71). Thus in the gauge-eigenstate basis the fermionic mass matrices are

$$M^{(i)}_{AB} = \frac{-v}{\sqrt{2}}f^{(i)}_{AB} \qquad i = \mathrm{e}, \mathrm{p}, \mathrm{n}. \tag{11.108}$$

Biunitary transformations. The important point is that there is no reason for these mass matrices to be diagonal; in fact generally they are neither symmetric nor hermitian. We shall now demonstrate that this type of matrices can be diagonalized by biunitary transformations, i.e. given M_{AB} there exist unitary matrices S and T such that

$$S^{\dagger}MT = M_{\mathrm{d}} \tag{11.109}$$

where M_{d} is diagonal with positive eigenvalues. Basically the points are that any matrix M can always be written as the product of a hermitian matrix (H) and a unitary matrix (V)

$$M = HV \tag{11.110}$$

and the hermitian H can then be diagonalized by some unitary matrix.

The proof proceeds as follows. $MM^{\dagger}$ is hermitian and positive; it can be diagonalized by an unitary matrix S

$$S^{\dagger}(MM^{\dagger})S = M_{\mathrm{d}}^2 \tag{11.111}$$

with

$$M_{\mathrm{d}}^2 = \begin{pmatrix} m_1^2 & & \\ & m_2^2 & \\ & & m_3^2 \end{pmatrix}.$$

The matrix S is unique up to a diagonal phase matrix; i.e. if eqn (11.111) holds, then

$$(SF)^{\dagger}(MM^{\dagger})(SF) = M_{\mathrm{d}}^2$$

with

$$F = \begin{pmatrix} \mathrm{e}^{\mathrm{i}\phi_1} & & \\ & \mathrm{e}^{\mathrm{i}\phi_2} & \\ & & \mathrm{e}^{\mathrm{i}\phi_3} \end{pmatrix}. \tag{11.112}$$

These phase degrees of freedom will be studied in more detail when we take up the question of CP-violating gauge couplings. Here they can be used to ensure that all eigenvalues of M_{d} in (11.109) are positive

$$M_{\mathrm{d}} = \begin{pmatrix} m_1 & & \\ & m_2 & \\ & & m_3 \end{pmatrix} \qquad m_i > 0. \tag{11.113}$$

Define a hermitian matrix H by

$$H = SM_{\mathrm{d}}S^{\dagger}. \tag{11.114}$$

Then we can show that V defined by

$$V \equiv H^{-1}M \quad \text{and} \quad V^{\dagger} = M^{\dagger}H^{-1} \tag{11.115}$$

is a unitary matrix because of eqns (11.111) and (11.114)

$$\begin{aligned} VV^{\dagger} &= H^{-1}MM^{\dagger}H^{-1} \\ &= H^{-1}SM_{\mathrm{d}}^2S^{\dagger}H^{-1} \\ &= H^{-1}(SM_{\mathrm{d}}S^{\dagger})(SM_{\mathrm{d}}S^{\dagger})H^{-1} \\ &= H^{-1}HHH^{-1} = 1. \end{aligned} \tag{11.116}$$

From the definitions of H and V (eqns (11.114) and (11.115)) we have

$$S^{\dagger}HS = S^{\dagger}MV^{\dagger}S = M_{\mathrm{d}} \tag{11.117}$$

or

$$S^{\dagger}MT = M_{\mathrm{d}} \tag{11.118}$$

where $T \equiv V^{\dagger}S$ is also unitary; this is the promised result. Thus the relation between gauge eigenstates and mass eigenstates follows

$$\begin{aligned}\bar{\psi}'_{\mathrm{L}}M\psi'_{\mathrm{R}} &= (\bar{\psi}'_{\mathrm{L}}S)(S^{\dagger}MT)(T^{\dagger}\psi'_{\mathrm{R}})\\ &= \bar{\psi}_{\mathrm{L}}M_{\mathrm{d}}\psi_{\mathrm{R}}\end{aligned} \tag{11.119}$$

with

$$\begin{aligned}\psi'_{\mathrm{L}} &= S\psi_{\mathrm{L}}\\ \psi'_{\mathrm{R}} &= T\psi_{\mathrm{R}}.\end{aligned} \tag{11.120}$$

Mixing matrix in the quark charged current couplings

We now apply this result to the charged weak current for quarks as derived from eqn (11.105) or from eqn (11.87) by the substitution of eqn (11.103)

$$\begin{aligned}J^{+}_{\mathrm{q}\mu} &= \bar{\mathrm{q}}_{A\mathrm{L}}\gamma_{\mu}\tau^{+}\mathrm{q}_{A\mathrm{L}}\\ &= \bar{\mathrm{p}}'_{A\mathrm{L}}\gamma_{\mu}\mathrm{n}'_{A\mathrm{L}}\\ &= \bar{\mathrm{p}}_{A\mathrm{L}}\gamma_{\mu}[S^{\dagger}_{(\mathrm{p})}S_{(\mathrm{n})}]_{AB}\mathrm{n}_{B\mathrm{L}}\end{aligned} \tag{11.121}$$

where

$$\begin{aligned}\mathrm{p}'_{\mathrm{L}} &= S_{(\mathrm{p})}\mathrm{p}_{\mathrm{L}}\\ \mathrm{n}'_{\mathrm{L}} &= S_{(\mathrm{n})}\mathrm{n}_{\mathrm{L}}.\end{aligned} \tag{11.122}$$

Thus, in terms of the mass eigenstates, the quark doublets of the three families are

$$\mathrm{q}_{A\mathrm{L}}:\ \begin{pmatrix}\mathrm{u}\\ \mathrm{d}''\end{pmatrix}_{\mathrm{L}}\ \begin{pmatrix}\mathrm{c}\\ \mathrm{s}''\end{pmatrix}_{\mathrm{L}}\ \begin{pmatrix}\mathrm{t}\\ \mathrm{b}''\end{pmatrix}_{\mathrm{L}} \tag{11.123}$$

where

$$\begin{pmatrix}\mathrm{d}''\\ \mathrm{s}''\\ \mathrm{b}''\end{pmatrix} = U\begin{pmatrix}\mathrm{d}\\ \mathrm{s}\\ \mathrm{b}\end{pmatrix} \tag{11.124}$$

with $U = S^{\dagger}_{(\mathrm{p})}S_{(\mathrm{n})}$. Clearly U is also a unitary matrix.

CP violation phases. If the mass matrices were real, U would be an orthogonal matrix, and all fermion gauge couplings would be real and they could not induce CP violations. On the other hand a general 3×3 unitary matrix, can be parameterized by the three real rotational angles plus six complex phases. Not all complex phases have physical meaning however, as some of them can be removed by a redefinition of quark fields (except for the charged current all other terms in the Lagrangian are diagonal in quark flavours). At first sight one would think that six quark flavours can absorb six phases. (Keep in mind that q_L and q_R must be rotated by the same phase to keep the masses real.) This is actually not correct as the mixing matrix U is invariant when we change all quarks by the same phase. Thus only five phases can be removed by redefinitions. To see this explicitly, let us start with the first doublet in (11.123)

$$q_{1L} = \begin{pmatrix} u \\ U_{11}d + U_{12}s + U_{13}b \end{pmatrix}. \tag{11.125}$$

If U_{11} has phase δ, i.e.

$$U_{11} = R_{11}e^{i\delta} \qquad R_{11} \text{ real}, \tag{11.126}$$

then this δ can be pulled out by a redefinition of the u-quark field

$$u \to u' = ue^{-i\delta} \tag{11.127}$$

and

$$q_{1L} = e^{i\delta}\begin{pmatrix} u' \\ R_{11}d + U'_{12}s + U'_{13}b \end{pmatrix}. \tag{11.128}$$

Similarly, we can factor out the complex phases of U_{21} and U_{31} by a redefinition of the c and t quark fields in eqn (11.123). These overall doublet phases are immaterial because there are no gauge couplings in eqn (11.105) between doublets with different family indices. Finally we can absorb two more phases of U_{12} and U_{13} by a redefinition of the s and b fields. The doublets of (11.123) take on the form

$$\begin{pmatrix} u' \\ R_{11}d + R_{12}s' + R_{13}b' \end{pmatrix}_L \begin{pmatrix} c' \\ R_{21}d + R_{22}e^{i\delta_1}s' + R_{23}e^{i\delta_2}b' \end{pmatrix}_L$$
$$\begin{pmatrix} t' \\ R_{31}d + R_{32}e^{i\delta_3}s' + R_{33}e^{i\delta_4}b' \end{pmatrix}_L \tag{11.129}$$

We have now reduced the number of parameters to 13, (9 R_{ij}s and 4 δ_is). The normalization of each state gives three conditions and the orthogonality conditions among the different states yield six conditions. Therefore we are left with four independent parameters. Since we need three parameters for the real 3×3 orthogonal matrix, we end up with one independent phase.

We shall summarize the above counting procedure by applying it to the general case of n-doublets: A complex $n \times n$ matrix has $2n^2$ real parameters which are reduced down to n^2 when the unitary condition is imposed. $(2n - 1)$ phases can be removed by redefinitions of quark states. Keeping in mind that there are $n(n - 1)/2$ parameters (angles) in an $n \times n$ orthogonal matrix, we arrive at the number of independent physical phases

$$n^2 - (2n - 1) - n(n - 1)/2 = (n - 1)(n - 2)/2. \tag{11.130}$$

Thus for a two-family theory there is one mixing angle—the Cabbibo angle—and no phase. Only for a theory with $2 \times 3 = 6$ quark flavours can one obtain a nontrivial CP-violating phase in the quark gauge couplings. This observation was first made by Kobayashi and Maskawa (1973) and the unitary matrix (11.124) is often referred to as the KM matrix. The phenomenology of CP violation in gauge theories will be discussed in §12.2.

Neutrino mass degeneracy and the absence of leptonic mixings

We have seen that the standard theory gives a natural explanation for the presence of the Cabbibo angle and CP phases in quark charged currents. Similarly the same theory helps us to understand the absence of such features in the lepton sector; the masslessness of neutrinos implies that these mixings are physically unobservable. This can be seen as follows. Just as in eqn (11.121) we can write down the charged weak current for leptons

$$\begin{aligned} J^+_{l\mu} &= \bar{l}_{A\mathrm{L}}\gamma_\mu\tau^+ l_{A\mathrm{L}} \\ &= \bar{\nu}'_{A\mathrm{L}}\gamma_\mu \mathrm{e}'_{A\mathrm{L}} \\ &= \bar{\nu}_{A\mathrm{L}}[S^\dagger_{(\nu)}S_{(\mathrm{e})}]_{AB}\gamma_\mu \mathrm{e}_B \end{aligned} \qquad (11.131)$$

where

$$\begin{aligned} \nu'_\mathrm{L} &= S_{(\nu)}\nu_\mathrm{L} \\ \mathrm{e}'_\mathrm{L} &= S_{(\mathrm{e})}\mathrm{e}_\mathrm{L} \end{aligned} \qquad (11.132)$$

and the lepton doublets in terms of mass eigenstates should be

$$l_{A\mathrm{L}}\colon \quad \begin{pmatrix} \nu''_\mathrm{e} \\ \mathrm{e} \end{pmatrix}_\mathrm{L} \quad \begin{pmatrix} \nu''_\mu \\ \mu \end{pmatrix}_\mathrm{L} \quad \begin{pmatrix} \nu''_\tau \\ \tau \end{pmatrix}_\mathrm{L} \qquad (11.133)$$

where

$$\begin{pmatrix} \nu''_\mathrm{e} \\ \nu''_\mu \\ \nu''_\tau \end{pmatrix} = V \begin{pmatrix} \nu_\mathrm{e} \\ \nu_\mu \\ \nu_\tau \end{pmatrix} \qquad (11.134)$$

with $V = S^\dagger_{(\nu)}S_{(\mathrm{e})}$. However, neutrinos are massless, hence degenerate. Any unitary transformed ν-states can be taken as mass eigenstates; in particular $\nu_{A\mathrm{L}} = \nu''_{A\mathrm{L}}$ or V may be set to be the identity matrix. Thus, nontrivial lepton mixing angles can never show up in any physical processes if the νs are degenerate. In this connection we should remark that the conventionally referred to neutrino states ν_l are really weak-interaction (gauge) eigenstates. The ν_l is operationally defined to be the 'invisible' particle missing in the $\pi \to l$ decay and/or detected by l-production in ν-matter scattering. With $V = 1$, different lepton families are completely decoupled.

Remark: Here we assume that there is only one set of Higgs doublet. Thus the Yukawa couplings are directly proportional to the fermion mass matrix. Mass matrix diagonalization then also diagonalizes the Yukawa couplings. With more doublets the Yukawa couplings are usually flavour-changing (Bjorken and Weinberg 1977). However crude estimates would indicate that such a flavour-changing mechanism brings generally too large a rate for strangeness-changing neutral-current and muon-number violation effects.

The standard theory (with one doublet of Higgs scalar) possesses global symmetries corresponding to the separate conservation of e-, μ-, and τ-lepton numbers. Processes such as $\mu \to e\gamma$, $K^0 \to \mu e$, etc. are forbidden to all orders. Again we emphasize that these features are not *ad hoc* inputs but are natural consequences of the theory when neutrinos are degenerate. The possibility that νs may have small masses will be explored in §13.2. Even for massive neutrinos we would expect that the nontrivial lepton mixing angles would always appear multiplicatively with neutrino mass differences. Hence, muon-number nonconservation effects are invariably suppressed. These remarks will be illustrated with explicit calculations in Chapter 13.

Flavour conservation in neutral current interactions

The presence of the unitary transformation (11.124) means that hadronic charged currents are flavour nondiagonal and strangeness is not conserved in charged-current weak interactions. On the other hand, consistent with longstanding experimental observations, (11.124) does not bring about flavour-changing neutral currents. When we change in eqn (11.95)

$$\begin{aligned} J^0_\mu = {} & g^{\mathrm{p}}_{\mathrm{L}} \sum_A \bar{\mathrm{p}}'_{A\mathrm{L}} \gamma_\mu \mathrm{p}'_{A\mathrm{L}} + g^{\mathrm{p}}_{\mathrm{R}} \sum_A \bar{\mathrm{p}}'_{A\mathrm{R}} \gamma_\mu \mathrm{p}'_{A\mathrm{R}} \\ & + g^{\mathrm{n}}_{\mathrm{L}} \sum_A \bar{\mathrm{n}}'_{A\mathrm{L}} \gamma_\mu \mathrm{n}'_{A\mathrm{L}} + g^{\mathrm{n}}_{\mathrm{R}} \sum_A \bar{\mathrm{n}}'_{A\mathrm{R}} \gamma_\mu \mathrm{n}'_{A\mathrm{R}} \end{aligned} \tag{11.135a}$$

the gauge eigenstate bases to mass eigenstate bases through (11.122), the unitarity condition means that we can simply remove the primes on the above equation to obtain

$$J^0_\mu = \sum_f (g^f_{\mathrm{L}} \bar{f}_{\mathrm{L}} \gamma_\mu f_{\mathrm{L}} + g^f_{\mathrm{R}} \bar{f}_{\mathrm{R}} \gamma_\mu f_{\mathrm{R}}) \tag{11.135b}$$

for $f = \nu_e, \nu_\mu, \nu_\tau$, e, μ, τ, u, c, t, d, s, b, ... This is possible because there is fermion family replication, in the sense that fermions with the same charge and helicity have the same gauge group transformation property. In fact this was the original motivation which led Glashow, Iliopoulos, and Maiani (GIM) to introduce a charmed quark having the same charge and weak couplings as the up-quark. However the GIM mechanism means much more than the mere absence of flavour-changing neutral-current coupling in the Born approximation. Since the weak gauge coupling g is of order e (eqn (11.93)), one must be watchful that higher-order effects do not give rise to $O(G_{\mathrm{F}}\alpha)$ amplitudes for strangeness-changing neutral-current processes such as $K_{\mathrm{L}} \to \mu^+\mu^-$. (Experimentally they are observed to be at most $O(G_{\mathrm{F}}^2)$.) We shall study the GIM cancellation mechanism in the next chapter. It suffices to point out that the same relationship between mixing angles and fermion mass difference mentioned above in connection with muon-number nonconservation applies here: the induced amplitudes as expected, are of the magnitude $\propto g^2 G_F \, \Delta m_q^2 / M_W^2 \simeq G_F^2 \, \Delta m_q^2$, where Δm_q is the quark mass difference and we have used eqn (11.89).

Summary

Let us summarize the discussion on the fermion sector of the standard model. The fermion gauge eigenstates display a repetitive structure and they can be grouped into families. Each family consists of 15 two-component fields of leptons and quarks, with four LH doublets and seven RH singlets. The ABJ anomalies due to the lepton and quark loops precisely cancel. Thus the standard model with one family would be a completely self-consistent theory. With more than one family the principal new physical feature is the presence of mixing angles and complex phases in fermionic gauge couplings. This is because gauge invariance allows for nondiagonal mass matrices for each set of the same charged fermions. The mass eigenstates are introduced when we diagonalize these mass matrices. They are related to gauge eigenstates by unitary transformations. The procedure is irrelevant for leptons because of neutrino mass degeneracy. Thus in the standard theory with three families there are three mixing angles and one CP violation phase in the quark gauge couplings. Altogether we need to insert 17 parameters in such a theory (see the discussion at the end of §11.2). At present there is no deep understanding of this repetitive fermion structure. (The existence of family symmetry group? Lepton and quark substructure?) This is usually referred to as the *fermion family problem.*

12 Standard electroweak theory II: phenomenological implications

12.1 Flavour-conserving neutral-current processes

A basic requirement for any gauge model of electroweak interaction is that it should contain the well-established V–A charged weak currents in the low-energy limit. But the Glashow–Weinberg–Salam model also predicts a set of new low-energy phenomena associated with the neutral weak currents. Their experimental discovery and the subsequent confirmations in detail of the predicted structure have brought about the general acceptance of this model as the standard theory of the electroweak interaction. In this section we shall discuss processes corresponding to the tree-diagram neutral currents, which are diagonal in flavour space (see eqn (11.135)). Flavour-changing neutral-current effects as induced by higher-order loop diagrams will be studied, along with CP violation, in the next section.

For momentum transfer much less than the vector gauge boson masses, the effective Lagrangian for weak-current interaction may be written

$$\begin{aligned}\frac{1}{4}\mathscr{L}_{\text{eff}} &= \frac{1}{4}(\mathscr{L}_{\text{eff}}^{\text{CC}} + \mathscr{L}_{\text{eff}}^{\text{NC}}) \\ &= -\frac{G_{\text{F}}}{\sqrt{2}}(J_\mu^+ J^{-\mu} + \rho J_\mu^0 J^{0\mu}) \end{aligned} \tag{12.1}$$

where $J_\mu^\dagger$ and J_μ^0 are given in eqns (11.87) and (11.95) respectively, and ρ measures the relative strength of charged-current (CC) and neutral-current (NC) processes

$$\rho = \frac{M_{\text{W}}^2}{M_{\text{Z}}^2 \cos^2\theta_{\text{W}}}. \tag{12.2}$$

For the standard theory with doublet Higgs phenomenon only (eqn (11.82)) ρ is fixed at

$$\rho = 1. \tag{12.3}$$

The NC part in (12.1) reflects the basic coupling (eqn (11.92))

$$\mathscr{L}^{\text{NC}} = \frac{g}{\cos\theta_{\text{W}}} Z^\mu J_\mu^0 \tag{12.4}$$

with the neutral current being diagonal in quark–lepton flavours $f = l, \nu_l$, u, d, s, c, ...

$$\begin{aligned} J_\mu^0 &= g_{\text{L}}^f \bar{f}_{\text{L}}\gamma_\mu f_{\text{L}} + g_{\text{R}}^f \bar{f}_{\text{R}}\gamma_\mu f_{\text{R}} \\ &= \tfrac{1}{2}[g_{\text{L}}^f \bar{f}\gamma_\mu(1-\gamma_5)f + g_{\text{R}}^f \bar{f}\gamma_\mu(1+\gamma_5)f] \end{aligned} \tag{12.5}$$

where

$$g_L^f = T_3(f_L) - Q(f)\sin^2\theta_W \tag{12.6}$$

$$g_R^f = T_3(f_R) - Q(f)\sin^2\theta_W \tag{12.7}$$

where T_3 and Q are the third components of weak isospin and electric charge, respectively. We have deliberately parametrized the NC pieces in a sufficiently general form to include all models based on the group SU(2) × U(1), i.e. models having different fermion content and representation assignments. The standard model values $g_{L,R}^f$ for the first fermion family is given in eqn (11.95). Universality (and fermion replication) implies that, for i = L, R, we have $g_i^e = g_i^\mu = g_i^\tau$, $g_i^u = g_i^c = g_i^t$, $g_i^d = g_i^s = g_i^b$, and of course, all neutrinos have $g_L^\nu = 1/2$. Thus in the standard model prediction for neutral currents there is only one unknown parameter: the Weinberg angle θ_W which reflects the relative coupling strength of the SU(2) and U(1) gauge group factors.

The neutral currents were first discovered by the Gargamelle collaboration at CERN (Hasert *et al.* 1973). These and most of the subsequent data are on inclusive neutrino scattering off an (isoscalar) hadronic target. However we choose here to discuss first the theoretically simpler situation of leptonic neutral-current processes.

$\nu_\mu + e \to \nu_\mu + e$ and $\bar{\nu}_\mu + e \to \bar{\nu}_\mu + e$

These two processes are purely neutral-current processes, described by diagrams in Fig. 12.1. The amplitude for $\nu_\mu + e \to \nu_\mu + e$ is

$$T_\nu(\lambda, \lambda') = \frac{g^2}{4M_Z^2\cos^2\theta_W}\, g_L^\nu[\bar{u}(k')\gamma_\mu(1-\gamma_5)u(k)]$$
$$\times \{\bar{u}(p',\lambda')[g_L^e\gamma^\mu(1-\gamma_5) + g_R^e\gamma^\mu(1+\gamma_5)]\,u(p,\lambda)\}. \tag{12.8}$$

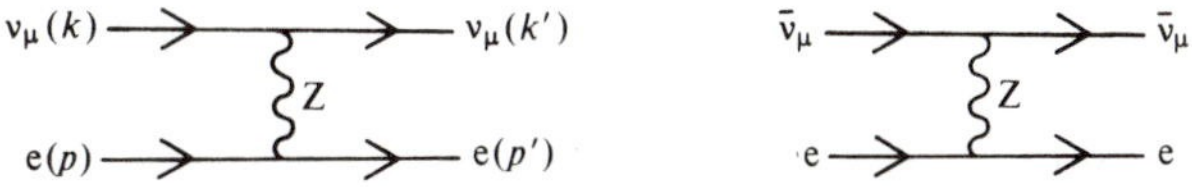

FIG. 12.1.

Summing over the final spin and averaging over initial electron spin (ν_μ is always left-handed),

$$\frac{1}{2}\sum_{\lambda,\lambda'}|T_\nu(\lambda,\lambda')|^2 = 2\left(\frac{G_F g_L^\nu}{\sqrt{2}}\right)^2 L_{\mu\nu}^-(k,k')[(g_L^e)^2 L^{-\mu\nu}(p,p')$$
$$+ (g_R^e)^2 L^{+\mu\nu}(p,p')] \tag{12.9}$$

where

$$L_{\mu\nu}^{\pm}(k,k') = \mathrm{tr}[\not{k}'\gamma_\mu(1\pm\gamma_5)\not{k}\gamma_\nu(1\pm\gamma_5)]$$
$$= 8(k'_\mu k_\nu + k'_\nu k_\mu - k\cdot k' g_{\mu\nu} \mp i\varepsilon_{\alpha\mu\beta\nu}k'^\alpha k^\beta). \tag{12.10}$$

We have neglected lepton masses; thus

$$(k\cdot p) = (k'\cdot p') \quad \text{and} \quad (k'\cdot p) = (k\cdot p'). \tag{12.11}$$

The leptonic tensor products in (12.9) are

$$L^{-}_{\mu\nu}(k,k')L^{\mp\mu\nu}(p,p') = 128\{[(k\cdot p)^2 + (k'\cdot p)^2] \pm [(k\cdot p)^2 - (k'\cdot p)^2]\} \tag{12.12}$$

and

$$\frac{1}{2}\sum |T_\nu|^2 = (16G_F)^2(g^\nu{}_L)^2[(g^e_L)^2(k\cdot p)^2 + (g^e_R)^2(k'\cdot p)^2]. \tag{12.13}$$

The cross-section is given by

$$\mathrm{d}\sigma = \frac{1}{2E_\nu}\frac{1}{2m_e}\frac{1}{|\mathbf{v}|}(2\pi)^4\,\delta^4(k+p-k'-p')\left(\frac{1}{2}\sum|T_\nu|^2\right) \times \frac{\mathrm{d}^3k'}{(2\pi)^3 2k'_0}\frac{\mathrm{d}^3p'}{(2\pi)^3 2p'_0}. \tag{12.14}$$

The phase space integral can be calculated easily

$$\int (2\pi)^4\,\delta^4(k+p-k'-p')\frac{\mathrm{d}^3k'}{(2\pi)^3 2E'_\nu}\frac{\mathrm{d}^3p'}{(2\pi)^3 2E_e} = \frac{1}{8\pi}\frac{\mathrm{d}E_e}{E_\nu} \tag{12.15}$$

where $E_\nu(E'_\nu)$ and E_e are the initial (final) neutrino and final electron energies in the laboratory reference frame. Then we have

$$\frac{\mathrm{d}\sigma(\nu_\mu e)}{\mathrm{d}E_e} = \frac{1}{32\pi m_e E_\nu^2}\left(\frac{1}{2}\sum|T_\nu|^2\right) = \frac{8m_e G_F^2}{\pi}(g^\nu_L)^2[(g^e_L)^2 + (g^e_R)^2(E'_\nu/E_\nu)^2]. \tag{12.16}$$

In terms of the conventional scaling variable (of fractional energy transfer),

$$y = E_e/E_\nu, \tag{12.17}$$

we have

$$\frac{\mathrm{d}\sigma(\nu_\mu e)}{\mathrm{d}y} = \frac{8G_F^2}{\pi}m_e E_\nu (g^\nu_L)^2[(g^e_L)^2 + (g^e_R)^2(1-y)^2] \tag{12.18}$$

and the total cross-section is

$$\sigma(\nu_\mu e) = \frac{8G_F^2}{\pi}m_e E_\nu (g^\nu_L)^2[(g^e_L)^2 + \tfrac{1}{3}(g^e_R)^2]. \tag{12.19}$$

It should be noted that the $(1-y)^2$ term in (12.18) and its corresponding 1/3 factor in (12.19) are associated with the term having the opposite helicity to that of the incoming neutrino. Thus the cross-section for the anti-neutrino process $\bar{\nu}_\mu + e \to \bar{\nu}_\mu + e$ can be inferred to be

$$\sigma(\bar{\nu}_\mu e) = \frac{8G_F^2}{\pi} m_e E_\nu (g_L^\nu)^2 [\tfrac{1}{3}(g_L^e)^2 + (g_R^e)^2] \tag{12.20}$$

since $\bar{\nu}_\mu$ has opposite helicity to ν_μ. From eqns (12.19) and (12.20) we see that measurements of the total cross-sections $\sigma(\nu_\mu e)$ and $\sigma(\bar{\nu}_\mu e)$ map out ellipses in the (g_L^e, g_R^e) plane (see Fig. 12.3) and these ellipses intersect at four allowed regions. The four solutions reflect the two sets of sign ambiguities: the cross-section results are insensitive to the substitutions (i) $g_L \leftrightarrow -g_L$, $g_R \leftrightarrow -g_R$, and (ii) $g_L \leftrightarrow g_L$, $g_R \leftrightarrow -g_R$. The latter substitution is simply a vector–axial-vector ambiguity. Thus we need two more independent measurements to resolve them.

$\nu_e + e \to \nu_e + e$ and $\bar{\nu}_e + e \to \bar{\nu}_e + e$

In these two processes both the charged and neutral currents contribute (see eqn (11.97) and Figs. 11.4 and 12.2). The cross-sections are computed to be

$$\sigma(\nu_e e) = \frac{8G_F^2}{\pi} m_e E_\nu (g_L^\nu)^2 [(1 + g_L^e)^2 + \tfrac{1}{3}(g_R^e)^2] \tag{12.21}$$

$$\sigma(\bar{\nu}_e e) = \frac{8G_F^2}{\pi} m_e E_\nu (g_L^\nu)^2 [\tfrac{1}{3}(1 + g_L^e)^2 + (g_R^e)^2]. \tag{12.22}$$

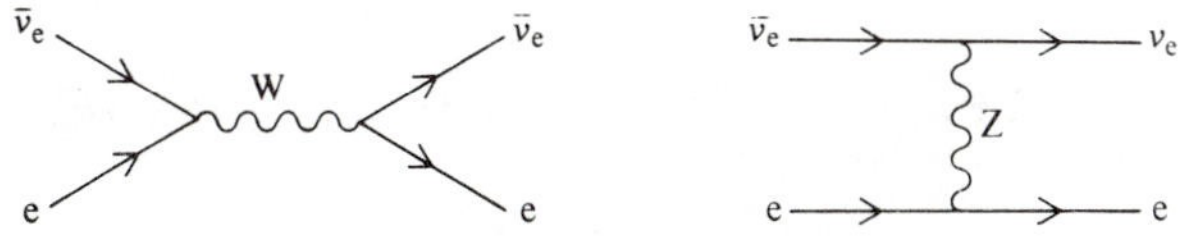

FIG. 12.2.

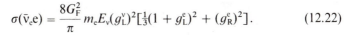

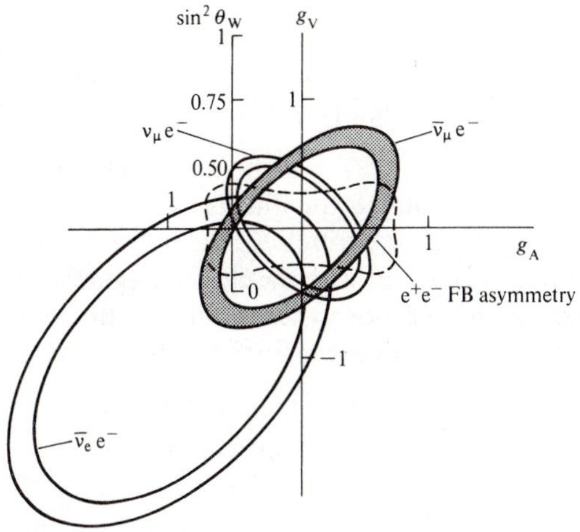

FIG. 12.3. νe scattering and e^+e^- annihilation experimental results for neutral-current couplings of the electron. $g_V = (g_L^e + g_R^e)$) and $g_A = (g_L^e - g_R^e)$. The only allowed region common to all four types of pure leptonic processes corresponds to that of the standard model with $\sin^2\theta_W \simeq 0.22$ (Barber *et al.* 1981).

The cross-section in (12.22) has been measured by using reactor neutrinos. This resolves the sign ambiguity (i) mentioned above as the sign of interference of the CC and NC amplitudes is now known through the linear g_L term in (12.22). This still leaves us with two possible solutions in Fig. 12.3 due to the vector and axial-vector ambiguity (ii).

The neutral-current effect in $e^+e^- \to \mu^+\mu^-$

The weak Z vector boson can contribute to Bhabha scattering $e^+e^- \to e^+e^-$ as well as to $e^+e^- \to \mu^+\mu^-, \tau^+\tau^-$. We shall concentrate on the simpler case of $e^+e^- \to \mu^+\mu^-$ (or $\tau^+\tau^-$) reactions with the diagrams shown in Fig. 12.4.

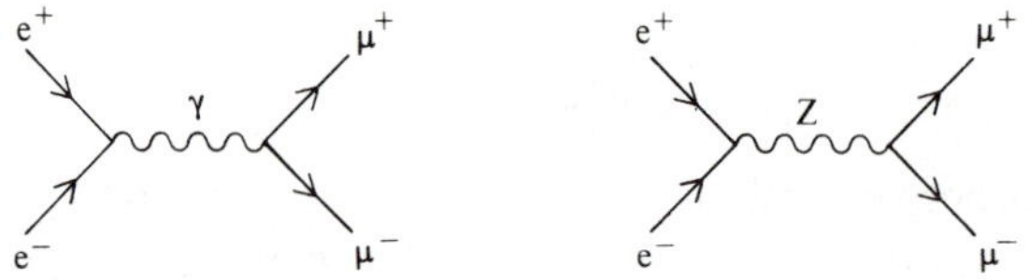

FIG. 12.4.

The cross-section can be worked out to be

$$\frac{d\sigma}{d(\cos\theta)} = \frac{\pi\alpha^2}{2s}[A(1+\cos^2\theta) + B\cos\theta] \tag{12.23}$$

with

$$A = 1 + 2\,\mathrm{Re}\,\chi g_V^e g_V^\mu + |\chi|^2[(g_V^e)^2 + (g_A^e)^2][(g_V^\mu)^2 + (g_A^\mu)^2] \tag{12.24}$$

$$B = 4\,\mathrm{Re}\,\chi g_A^e g_A^\mu + 8|\chi|^2 g_V^e g_V^\mu g_A^e g_A^\mu \tag{12.25}$$

where

$$\chi = \frac{G_F M_Z^2}{2\sqrt{2}\pi\alpha}\frac{s}{(s - M_Z^2 + iM_Z\Gamma)}. \tag{12.26}$$

$g_V = (g_L + g_R)$ and $g_A = (g_L - g_R)$. θ is the angle between the outgoing μ^- relative to the incoming e^- and Γ is the width of Z boson. The leading term in A of course is the pure photon contribution; the Re χ terms correspond to the interference of the neutral current with the electromagnetic current; the terms proportional to $|\chi|^2$ are pure Z-boson contributions. Because of the presence of the $\cos\theta$ term in (12.23) we have a forward–backward (FB) asymmetry

$$\Delta_{FB} = \frac{\int_0^1 d\cos\theta(d\sigma/d\cos\theta) - \int_{-1}^0 d\cos\theta(d\sigma/d\cos\theta)}{\int_{-1}^1 d\cos\theta(d\sigma/d\cos\theta)} = \frac{3}{8}\frac{B}{A}. \tag{12.27}$$

At low energies $s \ll M_Z^2$, we have

$$A \approx 1 - \frac{4sG_F}{\sqrt{2}\,e^2}g_V^2 \tag{12.28}$$

$$B \approx -\frac{8sG_F}{\sqrt{2}\,e^2} g_A^2 \tag{12.29}$$

where we have used universality to set $g_{V,A}^e = g_{V,A}^\mu$. We can see two weak effects.

(1) Modification of the total cross-section $\sigma(e^+e^- \to \mu^+\mu^-)$ from that of QED. At low energies this is proportional to g_V^2.

(2) The forward–backward asymmetry in the angular distribution, which at low energies measures g_A^2. Already at the highest PETRA energy, these weak effects can be measured to map out an allowed region in the (g_V, g_A) plane of Fig. 12.3. This singles out one of the remaining two solutions—that of the standard theory with $\sin^2\theta \approx 0.22$.

Neutrino–hadron neutral-current processes

While the pure leptonic processes considered above have the advantage of being free of strong interaction complications, their usefulness is diminished somewhat because the experimental data in this area generally have poor statistics. This is in contrast to the neutrino–nucleon scatterings where we have abundant and precise data, especially those of inclusive ν scatterings from isoscalar targets. To interpret the experimental results in the appropriate kinematic region one invokes the quark–parton model. In this way one can extract the first-generation quark weak couplings $g_{L,R}^{u,d}$ in an analogous manner to the ν–lepton scatterings considered above. For instance for the deep inelastic scattering $\nu + N \to \nu + X$ from an isoscalar target $N = \frac{1}{2}(n + p)$ (similar to eqn (12.18)), we have for $g_L^\nu = \frac{1}{2}$

$$\frac{d\sigma^{NC}(\nu N)}{dy} = \frac{G_F^2 M E_\nu Q}{\pi}\{[(g_L^u)^2 + (g_L^d)^2] + [(g_R^u)^2 + (g_R^d)^2](1-y)^2\} \tag{12.31}$$

where

$$Q = \int_0^1 x[u(x) + d(x)]\,dx \tag{12.32}$$

with $u(x)$, $d(x)$ being the quark distribution functions depending upon the Bjorken scaling variable x, and we have set the antiquark distributions $\bar{u}(x) = \bar{d}(x) = 0$ (see eqns (7.72) and (7.78)). We note that the coefficient in (12.31) is just the parton cross-section value for the charged-current process $\nu + N \to l^- + X$,

$$\frac{d\sigma^{CC}(\nu N)}{dy} = \frac{G_F^2 M E_\nu Q}{\pi}. \tag{12.33}$$

When the ratio $R_\nu \equiv \sigma^{NC}(\nu N)/\sigma^{CC}(\nu N)$ is taken, we are left with the weak couplings that we are seeking. Exactly similar results can be obtained for anti-neutrino scatterings, but keep in mind that here the $(1-y)^2$ factor is associated with the g_L^2s and $d\sigma^{CC}/dy$. Thus, $\sigma^{CC}(\bar{\nu}N) = \frac{1}{3}\sigma^{CC}(\nu N)$. One finds

$$\frac{\sigma^{NC}(\nu N) + \sigma^{NC}(\bar{\nu}N)}{\sigma^{CC}(\nu N) + \sigma^{CC}(\bar{\nu}N)} = (g_L^u)^2 + (g_L^d)^2 + (g_R^u)^2 + (g_R^d)^2 \qquad (12.34)$$

$$\frac{\sigma^{NC}(\nu N) - \sigma^{NC}(\bar{\nu}N)}{\sigma^{CC}(\nu N) - \sigma^{CC}(\bar{\nu}N)} = (g_L^u)^2 + (g_L^d)^2 - (g_R^u)^2 - (g_R^d)^2 \qquad (12.35)$$

Just as in the cases of $\nu_\mu e$ and $\bar{\nu}_\mu e$ scatterings, we need to supplement these νN, $\bar{\nu}N$ measurements with data from proton and neutron targets, or from semi-inclusive pion production off isoscalar targets, to resolve the sign and isospin ambiguities. We shall not give the details of this here. Suffice it to say that the results are again in accord with the standard theory. Here we will only mention a plot of the function corresponding to the standard model value for (12.35) of

$$\frac{\sigma^{NC}(\nu N) - \sigma^{NC}(\bar{\nu}N)}{\sigma^{CC}(\nu N) - \sigma^{CC}(\bar{\nu}N)} = \frac{1}{2}(1 - 2\sin^2\theta_W). \qquad (12.36)$$

This is the *Paschos–Wolfenstein relation* (1973). Thus measurements of R_ν and $R_{\bar{\nu}}$ immediately yield a value of $\sin^2\theta_W$. The most recent and accurate data are from CDHS (Geweniger *et al.* 1979) and CHARM (Jonker *et al.* 1981) collaborations.

It should be mentioned that in the actual phenomenological analysis much more detailed calculations of sea-quark contributions, QCD corrections, etc. are taken into account. Further discussion on the phenomenological values of the Weinberg angle and the ρ-parameter (12.2) will be presented in §12.3 when we discuss the W and Z intermediate vector bosons.

The electron–deuteron asymmetry measurement; a historical note

Although we have presented the electron weak coupling results $g_{L,R}^e$ as measured in the leptonic νe and e^+e^- reactions, historically the first elucidation of the electron weak couplings came from the semileptonic e–D scatterings in a beautiful experiment performed at SLAC by a SLAC–Yale collaboration (Prescott *et al.* 1978). At the time there was considerable confusion over the structure of electron weak couplings as a number of searches failed to detect atomic parity violation effects at the level predicted by the standard theory. There were also claims of experimental effects (an 'anomalous' y-distribution in νN scattering) that might be interpreted as indicating the presence of non-trivial $V + A$ charged currents, etc. Consequently a number of variants to the Weinberg–Salam model were proposed to account for these observations. The SLAC experiment measured parity violation asymmetry

$$A = \frac{\sigma(\lambda = 1/2) - \sigma(\lambda = -1/2)}{\sigma(\lambda = 1/2) + \sigma(\lambda = -1/2)}$$

where $\sigma(\lambda = \pm 1/2)$ is the double differential cross-section $d^2\sigma/d\Omega\, dE_e$ for the scattering of right- and left-handed electrons on deuterons. From the observed asymmetry parameter and the prior knowledge of $g^{u,d}_{L,R}$ one can then deduce $g^{e}_{L,R}$. The SLAC experiment was decisive in showing that the standard model was practically the only viable theory.

2.2 Weak mixing angles, the GIM mechanism, and CP violation

In this section we study the phenomenological consequences of the multi-family structure of the standard theory as presented in §11.3. In particular, we will concentrate on the strangeness-changing neutral current and the CP violation in the $\bar{K}^{\circ}$–K° system as examples illustrating the implications of the GIM mechanism.

We have shown in §11.3 that, because the Yukawa couplings generally involve fermions belonging to different particle-families (or generations), flavour-space fermion mass matrices are not diagonal. This means that mass eigenstates are different from weak eigenstates which have definite gauge transformation properties. For the simple case of two generations of fermions, this produces the Cabibbo mixing of the quarks in the charged current, and the quark weak eigenstates are

$$\begin{pmatrix} u \\ d' \end{pmatrix}_L \quad \begin{pmatrix} c \\ s' \end{pmatrix}_L \qquad u_R, \quad c_R, \quad d_R, \quad s_R$$

and

$$\begin{pmatrix} d' \\ s' \end{pmatrix}_L = \mathscr{R}_L \begin{pmatrix} d \\ s \end{pmatrix}_L = \begin{pmatrix} \cos\theta_c & \sin\theta_c \\ -\sin\theta_c & \cos\theta_c \end{pmatrix} \begin{pmatrix} d \\ s \end{pmatrix}_L,$$

i.e. the weak eigenstates d′, s′ are rotations of mass eigenstates d, s. Note that the Cabibbo angle θ_c is the difference of the rotation angles between the (u_L, c_L) and (d_L, s_L) sectors. This is why there is no mixing angle in the right-handed fermion sectors because (u_R, c_R)s do not couple to (d_R, s_R)s. The neutral current which is proportional to the operator $(Q \sin^2\theta_W - T_{3L})$ (see eqn (11.94)), has the important property that it is flavour-diagonal (or flavour-conserving). This follows from the fact that all fermions with the same charge and same helicity have the same transformation properties under the gauge group SU(2) × U(1) (Glashow and Weinberg 1977; Paschos 1977), so that the rotation matrices such as the above $\mathscr{R}_L$ commute with the neutral-current operator $(Q \sin^2\theta_W - T_{3L})$. For example, in the (d_L, s_L) sector, the $(1 - \gamma_5)$ part of the neutral current is

$$J^0_\mu(d, s) = (\bar{d}', \bar{s}')_L \left[-\frac{1}{3}\sin^2\theta_W + \frac{1}{2} \right] \begin{pmatrix} d' \\ s' \end{pmatrix}_L$$

$$= (\bar{d}, \bar{s})_L \left[-\frac{1}{3}\sin^2\theta_W + \frac{1}{2} \right] \begin{pmatrix} d \\ s \end{pmatrix}_L$$

which is flavour-conserving. Actually, this originally motivated Glashow, Iliopoulos, and Maiani (1970) to introduce the c quark coupled to s'_L so that s'_L has the same SU(2) × U(1) quantum number as d'_L to cancel the strangeness-changing neutral current (GIM mechanism). Otherwise, it would give rise to the order $G_F, \Delta S \neq 0$ neutral-current processes (e.g. $K_L \to \mu^+\mu^-$, $K^\pm \to \pi^\pm \nu\bar{\nu}$) which is phenomenologically unacceptable. It should be emphasized that the GIM mechanism achieves this suppression without any artificial adjustment of the parameters in the theory. In fact, as we have mentioned in §11.3, the GIM mechanism means much more than this tree-level cancellation; it also provides the additional suppression for the $\Delta S \neq 0$ neutral currents that are induced by higher-order loop diagrams. The need for this additional suppression is that without it these induced amplitudes would be of order $G_F\alpha$ while the experimental data on these processes are typically of order $G_F^2 m^2$ (with m being a few GeV). We shall illustrate this suppression mechanism with a calculation of the K_L–K_S mass difference: $\Delta m = 0.35 \times 10^{-14}$ GeV.

We mentioned in §11.3 that, for the case of three generations of fermions, one has the additional feature that there can be a CP-violating phase in the mixing matrix. We shall discuss the implication of this in the calculation of the CP-violating state-mixing parameter ε in the neutral kaon system.

Cabibbo–Kobayashi–Maskawa (CKM) mixing matrix

The basic charged-current (CC) interaction of eqn (11.86) with the mixing introduced by the need to diagonalize the quark mass matrices (eqn (11.124)) leads to the following CC couplings. In the simple two-family case, we have the familiar Cabibbo rotation, augmented by the GIM charmed quark,

$$\mathscr{L}_{CC} = \frac{g}{\sqrt{2}} (\bar{u}, \bar{c})_L \gamma^\mu \begin{pmatrix} \cos\theta_c & \sin\theta_c \\ -\sin\theta_c & \cos\theta_c \end{pmatrix} \begin{pmatrix} d \\ s \end{pmatrix}_L W_\mu^+ + \text{h.c.} \tag{12.37}$$

Since in this case the mass matrices can be taken to be real, the unitary transformations are just ordinary rotations. Also the amplitude for any strangeness-changing processes $d \leftrightarrow s$ must be proportional to $(m_c - m_u)\sin\theta_c$, as in the $m_c = m_u$ limit one can always choose one linear combination of c and u so that it does not couple to both s and d. A similar situation holds if we have d and s degeneracy, etc. In the three-family six-quark case, the mixing matrices are not just ordinary orthogonal matrices; we have

$$\mathscr{L}_{CC} = \frac{g}{\sqrt{2}} (\bar{u}, \bar{c}, \bar{t})_L \gamma^\mu U \begin{pmatrix} d \\ s \\ b \end{pmatrix}_L W_\mu^+ + \text{h.c.} \tag{12.38}$$

where the unitary matrix

$$U = \begin{pmatrix} U_{ud} & U_{us} & U_{ub} \\ U_{cd} & U_{cs} & U_{cb} \\ U_{td} & U_{ts} & U_{tb} \end{pmatrix} \tag{12.39}$$

can have one complex phase and can be parametrized in a form first introduced by Kobayashi and Maskawa (1973)

$$U = \begin{pmatrix} c_1 & s_1c_3 & s_1s_3 \\ -s_1c_2 & c_1c_2c_3 - s_2s_3\,e^{i\delta} & c_1c_2s_3 + s_2c_3\,e^{i\delta} \\ -s_1s_2 & c_1s_2c_3 + c_2s_3\,e^{i\delta} & c_1s_2s_3 - c_2c_3\,e^{i\delta} \end{pmatrix} \tag{12.40}$$

where we have used the abbreviations $c_i = \cos\theta_i$ and $s_i = \sin\theta_i$. By suitable choices of the signs of the quark fields, we can restrict the angles to the ranges $0 \leq \theta_i \leq \pi/2$ and $-\pi \leq \delta \leq \pi$.

In this KM parametrization θ_1 corresponds closely to the Cabibbo angle. It is also clear that θ_3 must be small because of the observed validity of approximate *Cabibbo universality*. To be more precise we recall that, from the muon lifetime and from the super allowed $0^+ \to 0^+$ nuclear β-decays, one can extract two values of the Fermi constant, which we indicate by G_μ^R and G_F^R respectively (R stands for renormalized parameters). Using (12.39), one has the theoretical prediction

$$[G_\mu^R/G_F^R] = U_{ud}^{-1}[1 + (\Delta_\mu - \Delta_F)] \tag{12.41}$$

where Δ_μ and Δ_F are, respectively, the radiative corrections to G_μ^R and G_F^R in which their finite part are expected to be different for μ- and for β-decay. This difference contributes a non-negligible correction ~ 2 per cent. One finds (Sirlin 1978, 1980).

$$|U_{ud}| = 0.9737 \pm 0.0025. \tag{12.42}$$

Also an overall fit to hyperon decays yields (Shrock and Wang 1978)

$$|U_{us}| = 0.219 \pm 0.003. \tag{12.43}$$

Combining the constraints (12.42) and (12.43), one finds that

$$|U_{ud}|^2 + |U_{us}|^2 = 0.996 \pm 0.004. \tag{12.44}$$

As the central value is less than one, this indicates a 'leakage' of u-quark coupling to the b-quark by

$$|U_{ub}| = 0.06 \pm 0.06 \tag{12.45}$$

which can be translated into the KM angle of

$$|s_3| < 0.28. \tag{12.46}$$

Thus θ_3 may be as small as (if not smaller than) the Cabibbo angle. At this stage of our knowledge, in order to obtain constraints on other KM parameters one needs to proceed by rather indirect routes, for example, by calculations of $K^\circ \leftrightarrow \bar{K}^\circ$ parameters (see calculation below). All indications are consistent with the qualitative feature that the diagonal elements of the CKM matrix (12.39) are the largest, and the magnitude of the matrix elements decreases as the element moves away from the diagonal

$$(U_{ud} \approx U_{cs} \approx U_{tb}) \gg (U_{cd} \approx U_{ts} \approx U_{us} \approx U_{cb}) \gg (U_{ub} \approx U_{td}). \tag{12.47}$$

Thus, all KM angles θ_i are small. This implies the dominant decay chain

$$t \to b \to c \to s.$$

Accordingly it is meaningful to talk about fermion families even in terms of the mass eigenstates (u, d), (c, s), (t, b), etc. as intrafamily couplings are the strongest.

In the KM parametrization (12.40) the CP violation phase δ appears only in the heavy quark sector. This is a convention. Clearly one can move δ to other sectors by redefining the phases of the quark fields. This means that one must involve more than one matrix element in the mixing matrix U to get CP violation. Physically this corresponds to the fact that CP violation comes from interference between amplitudes with different CP eigenvalues.

Phenomenology of K°–$\bar{K}^\circ$ mixing: some basic parameters

We will give a brief description of the K°–$\bar{K}^\circ$ system in order to set up the framework to study the CP violations in the standard model. (For details see Marshak, Riazuddin, and Ryan 1969; Kleinknecht 1976; Wolfenstein 1979.)

In a beautiful application of quantum mechanics Gell-Mann and Pais (1955) first discussed the decay of neutral K-mesons. The interesting feature is that, because the strong interaction conserves strangeness while the weak interaction does not, the neutral kaon eigenstates with respect to these interactions are different from each other. In particular, the strong-interaction eigenstates K° and $\bar{K}^\circ$ can mix through weak transitions such as $K^\circ \rightleftarrows 2\pi \rightleftarrows \bar{K}^\circ$. In this K°–$\bar{K}^\circ$ system, the $\bar{K}^\circ$ state is defined as the CP conjugate of K°,

$$|\bar{K}^\circ\rangle = \mathrm{CP}|K^\circ\rangle. \tag{12.48}$$

We will describe the weak-interaction-induced transition in the K°–$\bar{K}^\circ$ system by the S-matrix elements,

$$S_{\alpha'\alpha} = \langle\alpha'|T\exp\left(-\mathrm{i}\int H'_{\mathrm{w}}(t)\,\mathrm{d}t\right)|\alpha\rangle \tag{12.49}$$

where $\alpha,\ \alpha' = K^\circ$ or $\bar{K}^\circ$ and $H'_{\mathrm{w}}(t) = \mathrm{e}^{\mathrm{i}Ht}H_{\mathrm{w}}\,\mathrm{e}^{-\mathrm{i}Ht}$ where H_{w} is the weak Hamiltonian. To second order in H_{w}, we can write the transition matrix (T-matrix) element $T_{\alpha\alpha'}$ as

$$S_{\alpha'\alpha} - \delta_{\alpha\alpha'} = -2\pi\mathrm{i}T_{\alpha\alpha'}(E_\alpha) \tag{12.50}$$

where

$$\begin{aligned} T_{\alpha\alpha'}(E_\alpha) &= \langle\alpha'|H_{\mathrm{w}}|\alpha\rangle - \frac{\mathrm{i}}{2}\int \mathrm{d}t\langle\alpha'|T(H'_{\mathrm{w}}(t)H'_{\mathrm{w}}(0))|\alpha\rangle \\ &= \langle\alpha'|H_{\mathrm{w}}|\alpha\rangle + \frac{1}{2}\sum_\lambda\left[\frac{\langle\alpha'|H_{\mathrm{w}}|\lambda\rangle\langle\lambda|H_{\mathrm{w}}|\alpha\rangle}{E_{\alpha'} - E_\lambda + \mathrm{i}\varepsilon}\right. \\ &\qquad \left. + \frac{\langle\alpha'|H_{\mathrm{w}}|\lambda\rangle\langle\lambda|H_{\mathrm{w}}|\alpha\rangle}{E_\alpha - E_\lambda + \mathrm{i}\varepsilon}\right]. \end{aligned}$$

Since K° and $\bar{\mathrm{K}}$° are eigenstates of the strong interaction, we have for α, α' at rest

$$\langle\alpha'|H_{\mathrm{st}}|\alpha\rangle = \delta_{\alpha\alpha'}\mathrm{m}_\alpha \tag{12.51}$$

Thus combining eqns (12.50) and (12.51) we can represent the matrix element of the 'effective Hamiltonian' in the K°–$\bar{\mathrm{K}}$° system as

$$\langle\alpha'|H_{\mathrm{eff}}|\alpha\rangle \equiv H_{\alpha\alpha'} = m_{\mathrm{K}}\,\delta_{\alpha\alpha'} + T_{\alpha\alpha'}(m_{\mathrm{K}}) \tag{12.52}$$

For convenience, we will write the matrix element in eqn (12.52) as

$$H = \begin{pmatrix} H_{11} & H_{12} \\ H_{21} & H_{22} \end{pmatrix} \tag{12.53}$$

where the subscript 1 refers to the K° state and the subscript 2 refers to the $\bar{\mathrm{K}}$° state. Using the formula

$$\frac{1}{\chi - a + \mathrm{i}\varepsilon} = P\frac{1}{\chi - a} - \mathrm{i}\pi\,\delta(\chi - a), \tag{12.54}$$

we can decompose the matrix element of H_{eff} into real and imaginary parts,

$$H = M - \mathrm{i}\frac{\Gamma}{2} \tag{12.55}$$

or

$$H_{\alpha\alpha'} = M_{\alpha\alpha'} - \frac{\mathrm{i}}{2}\Gamma_{\alpha\alpha'}$$

where

$$M_{\alpha\alpha'} = m_{\mathrm{K}}\,\delta_{\alpha\alpha'} + \langle\alpha'|H_{\mathrm{w}}|\alpha\rangle + P\sum_\lambda \frac{1}{m_{\mathrm{K}} - E_\lambda}\langle\alpha'|H_{\mathrm{w}}|\lambda\rangle\langle\lambda|H_{\mathrm{w}}|\alpha\rangle \tag{12.56}$$

$$\Gamma_{\alpha\alpha'} = 2\pi\sum_\lambda \langle\alpha'|H_{\mathrm{w}}|\lambda\rangle\langle\lambda|H_{\mathrm{w}}|\alpha\rangle\,\delta(E_\lambda - m_{\mathrm{K}}). \tag{12.57}$$

Note that Γ, M are hermitian matrices

$$\Gamma^\dagger = \Gamma, \qquad M^\dagger = M \tag{12.58}$$

because of the hermiticity of H_{w} and H_{st}, while the 'effective Hamiltonian' H_{eff} is not hermitian. From CPT invariance, one can show that

$$H_{11} = H_{22}$$

or

$$\Gamma_{11} = \Gamma_{22} \quad \text{and} \quad M_{11} = M_{22}. \tag{12.59}$$

It is straightforward to diagonalize the Hamiltonian given in eqn (12.53) and the resulting eigenstates and eigenvalues are

$$H|K_{\mathrm{L,S}}\rangle = \lambda_{\mathrm{L,S}}|K_{\mathrm{L,S}}\rangle \tag{12.60}$$

where

$$\lambda_S = H_{11} + (H_{12}H_{21})^{\frac{1}{2}} \tag{12.61}$$

$$\lambda_L = H_{11} - (H_{12}H_{21})^{\frac{1}{2}}$$

$$|K_S\rangle = \frac{1}{\sqrt{2(1+|\varepsilon|^2)}}[(1+\varepsilon)|K_0\rangle + (1-\varepsilon)|\bar{K}_0\rangle] \tag{12.62}$$

$$|K_L\rangle = \frac{1}{\sqrt{2(1+|\varepsilon|^2)}}[(1+\varepsilon)|K_0\rangle - (1-\varepsilon)|\bar{K}_0\rangle]$$

with

$$\varepsilon = \frac{\sqrt{H_{12}} - \sqrt{H_{21}}}{\sqrt{H_{12}} + \sqrt{H_{21}}}. \tag{12.63}$$

The real and imaginary parts of $\lambda_{L,S}$ are, respectively the masses and decay rates of $K_{L,S}$

$$\lambda_{L,S} = m_{L,S} - \frac{i}{2}\gamma_{L,S}. \tag{12.64}$$

Note that CP invariance will imply that

$$M_{12} = M_{21} \quad \text{and} \quad \Gamma_{12} = \Gamma_{21} \tag{12.65}$$

or

$$H_{12} = H_{21}$$

which will give $\varepsilon = 0$ from eqn (12.63). Together with the hermiticity condition in eqn (12.58) this will imply that M_{ij} and Γ_{ij} are all real if CP is a good symmetry. Since the observed CP violations in the neutral kaon system is small, we will make the approximations that $\text{Im}\, M_{12} \ll \text{Re}\, M_{12}$, $\text{Im}\, \Gamma_{12} \ll \text{Re}\, \Gamma_{12}$, and ε is small. Then we can write

$$\begin{aligned}\varepsilon &= \frac{H_{12} - H_{21}}{4\sqrt{H_{12}H_{21}} + (\sqrt{H_{12}} - \sqrt{H_{21}})^2} \\ &\simeq \frac{H_{12} - H_{21}}{4\sqrt{H_{12}H_{21}}} \\ &\simeq \frac{(2\,\text{Im}\, M_{12} - i\,\text{Im}\,\Gamma_{12})}{2i(\lambda_L - \lambda_S)}.\end{aligned} \tag{12.66}$$

The parameter ε is a measure of the CP violation in the physical states of the K°–$\bar{K}^\circ$ system. From eqn (12.57) one can show, by examining the contributions coming from all possible intermediate states $\lambda = 2\pi$, 3π, $\pi e\nu \ldots$, that $|\text{Im}\,\Gamma_{12}| \ll |\text{Im}\, M_{12}|$. Then the difference in masses and decay rates can be approximated by

$$-\Delta\gamma \equiv \gamma_L - \gamma_S = 2|\Gamma_{12}| \tag{12.67}$$

$$\Delta m \equiv m_L - m_S = 2|M_{12}| \simeq 2\,\text{Re}\, M_{12} \tag{12.68}$$

as Im M_{12} is small. The ε parameter can then be written as

$$\varepsilon = \frac{-\text{Im}\, M_{12}}{\frac{1}{2}\Delta\gamma - \text{i}\Delta\text{m}}. \tag{12.69}$$

It is one of the magic properties of the K°–$\bar{K}^\circ$ system that with m_K being so close to the 3π threshold, and (reflecting the smallness of ε) with the dominant decay modes K_S and K_L being the CP-even 2π and -odd 3π, respectively, K_L and K_S have very different lifetimes

$$\gamma_S \simeq 500\gamma_L \simeq \Delta\gamma. \tag{12.70}$$

This allows for a clear separation of these two eigenmodes in the laboratory and made possible the eventual discovery by Cronin, Fitch and their collaborators (Christenson *et al.* 1964) of the very small decay mode $K_L \to 2\pi$. Detailed study of the interference between the K_L and K_S waves allows us to infer the extremely small mass difference $\Delta m = m_L - m_S$

$$\Delta m/m_K = 0.71 \times 10^{-14} \tag{12.71}$$

which corresponds to

$$\Delta m \simeq \tfrac{1}{2}\gamma_S. \tag{12.72}$$

We can then deduce the phase of ε (confirmed experimentally)

$$\arg \varepsilon \simeq \tan^{-1}(2\Delta m/\gamma_S) \simeq 45° \tag{12.73}$$

and write its magnitude in a simple form

$$|\varepsilon| \simeq \frac{1}{2}\left(\frac{\text{Im}\, M_{12}}{\text{Re}\, M_{12}}\right). \tag{12.74}$$

It is also clear from (12.72) that Δm is a $G_F^2 m^2$ effect and we shall show how such a result can be understood with the GIM cancellation mechanism.

The basic CP violation parameters that have been measured are the amplitude ratios

$$\eta_{+-} = \langle\pi^+\pi^-|H_w|K_L\rangle/\langle\pi^+\pi^-|H_w|K_S\rangle$$
$$\eta_{00} = \langle\pi^0\pi^0|H_w|K_L\rangle/\langle\pi^0\pi^0|H_w|K_S\rangle \tag{12.75a}$$

and the 'charge asymmetry'

$$\delta = \frac{\Gamma(K_L \to \pi^+ l^- \bar{\nu}) - \Gamma(K_L \to \pi^- l^+ \nu)}{\Gamma(K_L \to \pi^+ l^- \bar{\nu}) + \Gamma(K_L \to \pi^- l^+ \nu)}. \tag{12.75b}$$

By isospin decomposition of the final 2π state into $I = 0$ and $I = 2$ parts, one gets

$$\eta_{+-} = (\varepsilon_0 + \varepsilon_2)/(1 + 2^{-\frac{1}{2}}\omega)$$
$$\eta_{00} = (\varepsilon_0 - 2\varepsilon_2)/(1 - 2^{\frac{1}{2}}\omega) \tag{12.76}$$

where

$$\varepsilon_0 = \langle I=0|H_w|K_L\rangle/\langle I=0|H_w|K_S\rangle,$$

$$\varepsilon_2 = \frac{1}{\sqrt{2}}\langle I=2|H_w|K_L\rangle/\langle I=0|H_w|K_S\rangle,$$

and

$$\omega = \langle I=2|H_w|K_S\rangle/\langle I=0|H_w|K_S\rangle. \tag{12.77}$$

Because of the validity of the $\Delta I = 1/2$ rule for CP-conserving decays $\omega \ll 1$ and it can be neglected. Furthermore, we can parametrize the $K^\circ \to 2\pi$ amplitudes as

$$\langle I=n|H_w|K^\circ\rangle = A_n\, e^{i\delta_n} \tag{12.78}$$

where δ_n is the $\pi\pi$ phase shift in the $I = n$ channel coming from the final state interactions. A commonly adopted phase convention (Wu and Yang 1964) is to choose A_0 to be real,

$$\operatorname{Im} A_0 = 0. \tag{12.79}$$

In this case, it is easy to see from eqns (12.62), (12.77), and (12.78) that

$$\varepsilon_0 = \varepsilon$$

$$\varepsilon_2 = \frac{i}{\sqrt{2}}\, e^{i(\delta_0-\delta_2)}\, \frac{\operatorname{Im} A_2}{A_0} \equiv \varepsilon'. \tag{12.80}$$

Therefore,

$$\eta_{+-} = \varepsilon + \varepsilon'$$

$$\eta_{00} = \varepsilon - 2\varepsilon'$$

$$\delta = \operatorname{Re}\varepsilon. \tag{12.81}$$

Experimentally we have

$$|\varepsilon| \simeq 2\times 10^{-3} \tag{12.82}$$

$$|\varepsilon'/\varepsilon| < 1/50. \tag{12.83}$$

All these results are consistent with an early proposed theory of Wolfenstein (1964) which relegates all CP violations to a $\Delta S = 2$ superweak interaction. We shall demonstrate below that the standard electroweak theory with KM mixings can yield numbers mimicking the superweak theory results.

$\Delta S = 2$ effective Lagrangian for free quarks

The strategy, originally due to Gaillard and Lee (1974), is to construct first an effective $\Delta S = 2$ Lagrangian from the free quark model and then to sandwich this $\mathscr{L}_{\text{eff}}^{\Delta S=2}$ between the K° and $\bar{K}^\circ$ states to obtain Δm through (12.68) and $|\varepsilon|$ through (12.74).

One first computes the box diagrams of Fig. 12.5 with intermediate quarks

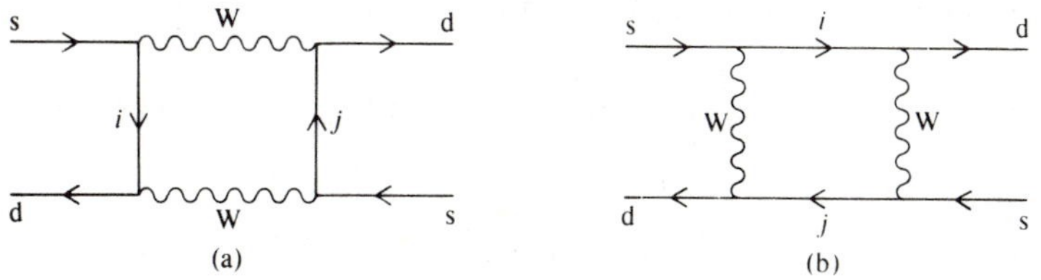

FIG. 12.5. Feynman diagrams for the $\Delta S = 2$ transition amplitude $\bar{s}d \to s\bar{d}$. The 'annihilation' term of (a) is equal to the 'scattering' term of (b). i, j = u, c, t.

$i, j =$ u, c, t in the approximation of taking all external momenta to be zero (as they are small compared to M_W and heavy quark masses). In the Feynman–'t Hooft gauge, we have

$$\mathrm{i}T(\bar{s}d \to s\bar{d}) = 2\left(\frac{g}{\sqrt{2}}\right)^4 \sum_{i,j} \xi_i \xi_j \int \frac{d^4k}{(2\pi)^4}\left(\frac{-\mathrm{i}}{k^2 - M_W^2}\right)^2$$
$$\times \left(\bar{d}_L \gamma^\mu \frac{\gamma\cdot k + m_i}{k^2 - m_i^2}\gamma^\nu s_L\right)\left(\bar{d}_L \gamma_\nu \frac{\gamma\cdot k + m_j}{k^2 - m_j^2}\gamma_\mu s_L\right) \tag{12.84}$$

where

$$\xi_i = U_{is} U_{id}^* \tag{12.85}$$

where the Us are elements of the CKM mixing matrix (12.39). First concentrate on the momentum integration given by

$$I_{\alpha\beta}(i,j) \equiv \int \frac{d^4k\, k_\alpha k_\beta}{(k^2 - M_W^2)^2 (k^2 - m_i^2)(k^2 - m_j^2)}$$
$$= \frac{-\mathrm{i}\pi^2}{4M_W^2} A(x_i, x_j) g_{\alpha\beta} \tag{12.86}$$

where

$$A(x_i, x_j) = \frac{J(x_i) - J(x_j)}{x_i - x_j}.$$

$$J(x_i) = \frac{1}{1 - x_i} + \frac{x_i^2 \ln x_i}{(1 - x_i)^2}, \tag{12.87}$$

and

$$x_i = \frac{m_i^2}{M_W^2}.$$

Using the identity

$$\gamma^\mu \gamma^\alpha \gamma^\nu = g^{\mu\alpha}\gamma^\nu + g^{\nu\alpha}\gamma^\mu - g^{\mu\nu}\gamma^\alpha - \mathrm{i}\varepsilon^{\mu\alpha\nu\beta}\gamma_5\gamma_\beta, \tag{12.88}$$

we can evaluate the Dirac matrices

$$[\gamma^\mu\gamma^\alpha\gamma^\nu(1-\gamma_5)/2]\ldots[\gamma_\nu\gamma_\alpha\gamma_\mu(1-\gamma_5)/2]$$
$$= 4[\gamma^\alpha(1-\gamma_5)/2]\ldots[\gamma_\alpha(1-\gamma_5)/2]. \tag{12.89}$$

Substituting the results of (12.86) and (12.89) into (12.84)

$$T(\bar{s}d \to s\bar{d}) = -\frac{G_F}{\sqrt{2}}\frac{\alpha}{\pi \sin^2\theta_W}(\bar{d}_L\gamma^\mu s_L)(\bar{d}_L\gamma_\mu s_L)\sum_{i,j}\xi_i\xi_j A(x_i, x_j). \tag{12.90}$$

After taking into account the (2!)(2!) Wick contractions, this is an amplitude that can be obtained directly from an effective Lagrangian

$$\mathscr{L}_{\text{eff}}^{\Delta S=2} = -\frac{G_F}{\sqrt{2}}\frac{\alpha}{16\pi}\left(\frac{1}{M_W \sin\theta_W}\right)^2 \lambda \mathcal{O}_{JJ} \tag{12.91}$$

where

$$\mathcal{O}_{JJ} = [\bar{d}\gamma^\mu(1-\gamma_5)s][\bar{d}\gamma_\mu(1-\gamma_5)s] \tag{12.91a}$$

and

$$\lambda = M_W^2 \sum_{i,j}\xi_i\xi_j A(x_i, x_j). \tag{12.91b}$$

QCD gluonic radiative corrections can, in principle, be taken into account. However, given all the other uncertainties, such corrections are not expected to change our conclusions materially and we shall ignore them.

The K_L–K_S mass difference and GIM cancellation mechanism

Given the effective Lagrangian (12.91), we can calculate Δm through eqn (12.68)

$$\Delta m = -2\,\text{Re}\,[\langle K| - \mathscr{L}_{\text{eff}}^{\Delta S=2}|\bar{K}\rangle]. \tag{12.92}$$

There are two aspects to this calculation. The first is the estimation of the matrix element $\langle K|\mathcal{O}_{JJ}|\bar{K}\rangle$; the second is the evaluation of the c-number (12.91b) corresponding to the sum of products of mixing angles and quark masses.

To get an order of magnitude of the size of $\langle K|\mathcal{O}_{JJ}|\bar{K}\rangle$, we make the 'vacuum saturation' approximation

$$\begin{aligned}\langle K|[\bar{d}\gamma^\mu(1-\gamma_5)s][\bar{d}\gamma_\mu(1-\gamma_5)s]|\bar{K}\rangle &= \frac{8}{3}\langle K|\bar{d}\gamma^\mu\gamma_5 s|0\rangle\langle 0|\bar{d}\gamma_\mu\gamma_5 s|\bar{K}\rangle \\ &= \frac{8}{3}\frac{f_K^2 m_K^2}{2m_K} \end{aligned}\tag{12.93}$$

where $f_K \simeq 1.23\, f_\pi$ is the kaon decay constant; the factor $(2m_K)^{-1}$ arises from the normalization of the state. The factor 8/3 corresponds to the four ways of Wick contraction times a colour factor 2/3. The hope in making such an approximation is that the simple vacuum intermediate state will give us a representative value of the four-fermion matrix element $\mathcal{O}_{JJ}$. A somewhat more realistic calculation of this matrix element in the MIT bag model yields a value about half the size (i.e. the same order of magnitude) as resulting from this vacuum insertion calculation (Shrock and Treiman 1979). Even so, there are still uncertainties as to whether these quark diagrams really dominate the low-energy parameters such as the $K_L - K_S$ mass difference (Wolfenstein 1979). Ignoring these complications, we combine eqns (12.91) and (12.93) to get

$$\Delta m = \frac{G_F}{\sqrt{2}}\frac{\alpha}{6\pi}\frac{f_K^2 m_K}{\sin^2\theta_W}\,\mathrm{Re}\left[\sum_{ij}\xi_i\xi_j A(x_i, x_j)\right]. \tag{12.94}$$

At first sight it appears that, as the right-hand side is $O(\alpha G_F)$, it will yield much too large a value for Δm. This is where the GIM cancellation mechanism comes in. Since $x_i \ll 1$ we can expand $A(x_i, x_j)$ of (12.87) in x_i and x_j. The leading constant terms cancel in the summation because of the unitarity condition

$$\sum_i \xi_i = \sum_i U_{is}U_{id}^* = 0. \tag{12.95}$$

In the remainder, the dominant terms are proportional to x_i

$$\sum_{i,j}\xi_i\xi_j A(x_i, x_j) = \sum_i \xi_i^2 x_i + \sum_{i\neq j}\xi_i\xi_j\frac{x_i x_j}{x_i - x_j}\ln\frac{x_i}{x_j}. \tag{12.96}$$

Thus we see that the factor x_i converts $G_F\alpha$ to $G_F^2 m_i^2$

$$G_F\alpha x_i\xi_i \simeq G_F^2 m_i^2\xi_i. \tag{12.97}$$

Taking $m_u = 0$, we may write

$$\frac{m_L - m_S}{m_K} = \frac{2}{3}\frac{G_F}{\sqrt{2}}f_K^2\frac{\alpha}{4\pi}\left(\frac{m_c}{37\ \mathrm{GeV}}\right)^2\sin^2\theta_c\cos^2\theta_c\,X \tag{12.98}$$

with

$$X = (\sin^2\theta_c\cos^2\theta_c)^{-1}\,\mathrm{Re}\left[(U_{cs}U_{cd}^*)^2 + (U_{ts}U_{td}^*)^2(m_t^2/m_c^2) + U_{cs}U_{cd}^*U_{ts}U_{td}^*\frac{2m_t^2}{m_t^2 - m_c^2}\ln\left(\frac{m_t^2}{m_c^2}\right)\right] \tag{12.99}$$

where we have used eqn (11.100). In the four-quark model, where we have a simple Cabibbo rotation (12.37) with $(U_{cs}U_{cd}^*)^2 = \sin^2\theta_c\cos^2\theta_c$, the factor $X = 1$ in (12.99). This is how Gaillard and Lee (1974) before the discovery of J/ψ first estimated that $m_c \simeq 1.5$ GeV. Even though m_t is expected to be very large ($\gtrsim 20$ GeV) still it is multiplied by a small mixing angle factor U_{td} (see eqns (12.45) and (12.47)) and we do not expect X to deviate significantly from unity. Thus the result in (12.98) must be regarded as a remarkable triumph of the GIM mechanism as embodied in the standard electroweak theory.

Before concluding this subsection on the GIM cancellation mechanism, two remarks are in order.

Remark (1). For a heuristic understanding of how the GIM cancellation comes about one may use the concept of *mass insertion* to view these GIM loop calculations (see for example Cheng and Li 1977). Mass insertion can be thought as either the simple expansion of the fermion propagator

$$\frac{1}{k\!\!\!/ - m} = \frac{1}{k\!\!\!/} + \frac{1}{k\!\!\!/} m \frac{1}{k\!\!\!/} + \dots \tag{12.100}$$

or as treating the fermion mass matrix ($\bar{\psi}_{iL} m_{ij} \psi_{jR}$ + h.c.) as a perturbation on the symmetric theory (where all fermions are massless). Such a mass insertion in an internal line flips the helicity of that particle and may also change the identity of the weak eigenstate if one of the off-diagonal elements in m_{ij} is used. For example the basic mechanism for s ↔ d transition is shown in Fig. 12.6. The couplings in these diagrams, by inspection, must be proportional to (see further comment in remark (2) below) $m_{ui} m_{ic} = \Sigma_i U_{di} U^*_{si} m_i^2$ which is just the leading term in a GIM cancellation. (Our example of the Δm calculation is slightly complicated by the fact that it is a $\Delta S = 2$ transition.)

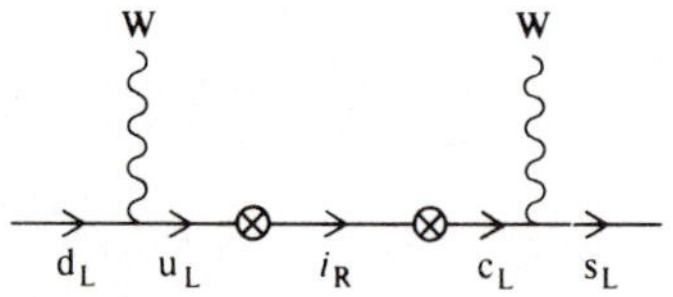

FIG. 12.6. The basic mechanism for the s ↔ d transition. All particles are weak-interaction eigenstates with i = u, c, t. Hence, the mass matrix has nondiagonal elements.

Remark (2). Whether the GIM mechanism produces a power suppression factor $\Sigma_i U_{di} U^*_{si} (m_i^2/M_W^2)$ which vanishes in the limit where all the m_is are equal depends on the convergence properties of the particular loop integration under consideration. There are cases where the GIM cancellation is much milder than this power suppression. It may take on the form $\Sigma_i U_{di} U^*_{si} \ln(m_i^2/M_W^2)$ which also vanishes in the equal-mass limit. An example of such a logarithmic GIM cancellation is the s ↔ d transition charge radius, which is physically relevant for example in the process $K \to \pi e^+ e^-$ (Gaillard and Lee 1974). In the 't Hooft–Feynman gauge the leading contribution comes from the diagram in Fig. 12.7(a). To lowest order in M_W^{-2} the W-propagator may be approximated by

$$\frac{-i}{k^2 - M_W^2} \simeq \frac{i}{M_W^2}. \tag{12.101}$$

One then makes a Fierz rearrangement (see Appendix A) in the resulting V–A four-fermion interaction

$$(\bar{s}_L \gamma^\mu q_{iL})(\bar{q}_{iL} \gamma_\mu d_L) = (\bar{s}_L \gamma^\mu d_L)(\bar{q}_{iL} \gamma_\mu q_{iL}). \tag{12.102}$$

Thus we can calculate the transition vertex directly from the diagram in Fig. 12.7(b)

$$\Gamma_\mu(p, p'; q) = \left(i \frac{2}{3} e\right)\left(-i \frac{G_F}{\sqrt{2}}\right)(\bar{s}_L \gamma^\nu d_L) \sum_i \xi I_{\mu\nu}(i) \tag{12.103}$$

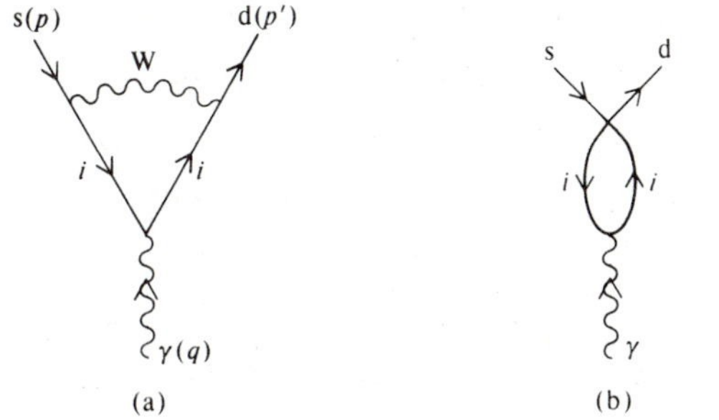

FIG. 12.7. The leading contribution to the s↔d transition charge radius. (b) is the Fierz-rearranged V–A four-fermion approximation to (a).

where

$$I_{\mu\nu}(i) = \mathrm{tr}\int \frac{\mathrm{d}^4k}{(2\pi)^4}\gamma_\nu \frac{i}{\gamma\cdot k - m_i}\gamma_\mu \frac{i}{\gamma\cdot(k-q) - m_i}. \tag{12.104}$$

$I_{\mu\nu}$ is just the familiar vacuum polarization tensor with the result that, for $m_i^2 \gg q^2$,

$$I_{\mu\nu}(i) \simeq \frac{1}{12\pi^2}(q_\mu q_\nu - g_{\mu\nu}q^2)[\ln\Lambda - \ln(m_i^2/q^2)]. \tag{12.105}$$

Again the (divergent) constant terms cancel because $\Sigma_i\, \xi_i = 0$ leaving behind a mild GIM suppression factor $\Sigma_i\, \xi_i \ln(m_i^2/q^2)$.

The kaon CP-violation state-mixing parameter and the CKM angles

The CP-violation parameter ε of eqn (12.74) is given by

$$|\varepsilon| \simeq \frac{1}{2}\frac{\mathrm{Im}\langle \mathrm{K}| -\mathscr{L}_{\mathrm{eff}}^{\Delta S=2}|\bar{\mathrm{K}}\rangle}{\mathrm{Re}\langle \mathrm{K}| -\mathscr{L}_{\mathrm{eff}}^{\Delta S=2}|\bar{\mathrm{K}}\rangle} \simeq \frac{1}{2}\left(\frac{\mathrm{Im}(\lambda)}{\mathrm{Re}(\lambda)}\right) \tag{12.106}$$

where the four-fermion matrix element $\langle \mathrm{K}|\mathcal{O}_{\mathrm{JJ}}|\bar{\mathrm{K}}\rangle$ is divided out and λ is given in (12.91b). Again in the limit $m_{\mathrm{u}} = 0$, we can use (12.95) to get

$$\mathrm{Im}(\lambda) = 2\left\{\xi_{\mathrm{cI}}\xi_{\mathrm{cR}}m_{\mathrm{c}}^2 + \xi_{\mathrm{tI}}\xi_{\mathrm{tR}}m_{\mathrm{t}}^2 + (\xi_{\mathrm{cI}}\xi_{\mathrm{tR}} + \xi_{\mathrm{tI}}\xi_{\mathrm{cR}})\frac{m_{\mathrm{c}}^2 m_{\mathrm{t}}^2}{m_{\mathrm{t}}^2 - m_{\mathrm{c}}^2}\ln\left(\frac{m_{\mathrm{t}}^2}{m_{\mathrm{c}}^2}\right)\right\} \tag{12.107}$$

$$\mathrm{Re}(\lambda) = \left\{(\xi_{\mathrm{cR}}^2 - \xi_{\mathrm{cI}}^2)m_{\mathrm{c}}^2 + (\xi_{\mathrm{tR}}^2 - \xi_{\mathrm{tI}}^2)m_{\mathrm{t}}^2 + (\xi_{\mathrm{cR}}\xi_{\mathrm{tR}} - \xi_{\mathrm{cI}}\xi_{\mathrm{tI}})\frac{2m_{\mathrm{c}}^2 m_{\mathrm{t}}^2}{m_{\mathrm{t}}^2 - m_{\mathrm{c}}^2}\ln\left(\frac{m_{\mathrm{t}}^2}{m_{\mathrm{c}}^2}\right)\right\}. \tag{12.108}$$

where we have used the notation $\xi_{i\mathrm{I}} \equiv \mathrm{Im}\,\xi_i$ and $\xi_{i\mathrm{R}} \equiv \mathrm{Re}\,\xi_i$. With the KM parametrization given in (12.40), we note that

$$\xi_{tI} = -\xi_{cI} = c_2 s_1 s_2 s_3 \sin\delta. \tag{12.109}$$

This allows us to factor out ξ_{cI} in (12.107). Thus Im(λ) and hence the CP parameter ε itself, is proportional to this combination of KM angles

$$\varepsilon \propto s_1 s_2 s_3 \sin\delta. \tag{12.110}$$

Then the CP-violation parameter is suppressed by all the KM angles. Since we have some grounds to expect all the θ_is to be small, this elegant theory of CP violation naturally gives us a small ε.

For definiteness, we shall write out eqns (12.106)–(12.108) for the case of small KM angles. Except for the overall ξ_{cI}, we shall drop all such terms and approximate

$$\xi_{cR} \simeq -s_1 c_1 c_2^2 c_3$$

$$\xi_{tR} \simeq -s_1 c_1 s_2^2 c_3.$$

The expression for ξ_{tR} is only valid for the special case

$$s_2 \gg s_3. \tag{12.111}$$

Such a situation is consistent with, but not demanded by, present experimental constraints. We have

$$\mathrm{Im}(\lambda) \simeq 2\xi_{cI}\left[\xi_{cR} m_c^2 - \xi_{tR} m_t^2 + (\xi_{tR} - \xi_{cR})\frac{m_c^2 m_t^2}{m_t^2 - m_c^2}\ln\left(\frac{m_t^2}{m_c^2}\right)\right] \tag{12.112a}$$

$$\mathrm{Re}(\lambda) \simeq \left[\xi_{cR}^2 m_c^2 + \xi_{tR}^2 m_t^2 + \frac{2\xi_{cR}\xi_{tR} m_c^2 m_t^2}{m_t^2 - m_c^2}\ln\left(\frac{m_t^2}{m_c^2}\right)\right]. \tag{12.112b}$$

Substituting (12.112) into (12.106), we have (Ellis *et al.* 1976*a*,*b*, 1977)

$$\varepsilon \simeq c_2 s_2 s_3 \sin\delta \left\{\frac{s_2^2(1 + \eta\ln\eta) - c_2^2\eta(1 + \ln\eta)}{s_2^4 + c_2^4\eta - 2s_2^2 c_2^2 \eta\ln\eta}\right\} \tag{12.113}$$

where $\eta = m_c^2/m_t^2$ and some higher-power terms in η have been dropped.

We conclude this discussion of CP violation with three brief comments.

(1) Calculating ε′. The calculation of ε', the CP-violation parameter in decays (as opposed to ε in the state), is less certain as it depends on our theoretical understanding of the nonleptonic weak decays. In most approaches ε' is found to be very small (much too small to be detectable).

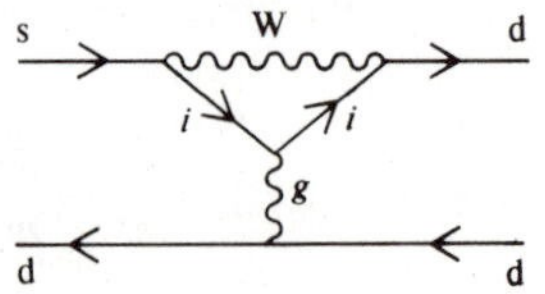

FIG. 12.8. Penguin diagram for $s\bar{d} \to d\bar{d}$ with $i = u, c, t$.

However if the $K \to 2\pi$ decay is dominated by the one-gluon exchange 'penguin diagram' (Gilman and Wise 1979) of Fig. 12.8, the calculated ratio $|\varepsilon'/\varepsilon|$ is fairly insensitive to the values of the mixing angles and is predicted to be 0.01 to 0.03. Such a value may be accessible to verification by the next generation of experiments.

(2) Neutron electric dipole moment. This is another important area where the standard theory with CP-violating complex CC couplings can simulate the superweak model. Both predict an electric dipole moment for neutron d_n/e much smaller than the already extraordinarily stringent limit of less than 10^{-26} cm (Ramsey 1978; Altarev *et al.* 1981). The basic reason for the smallness of this quantity in the standard electroweak theory is again related to GIM cancellation: the one-loop contribution such as Fig. 12.9(a) vanishes because it is 'self-conjugate' and there is no CP phase, $\Sigma_i U_{ui} U^*_{ui} = 1$. In such diagrams whatever the phase at one vertex it will be cancelled by the opposite phase coming from the other vertex. At the two-loop level we have the possibility of 'non-self-conjugate' diagrams, Fig. 12.9(b). This results in an electromagnetic vertex with 4-momentum transfer k as

$$e d_n(k^2) \bar{u} \sigma_{\mu\nu} k^\nu \gamma_5 u A^\mu . \tag{12.114}$$

The form factor d_n, being the electric dipole moment in the static limit, is suppressed by the usual GIM factor of $\Delta m^2/M_W^2 = \Sigma_i U_{qi} U^*_{q'i} m_i^2/M_W^2$. However Shabalin (1978, 1980) has shown that even this vanishes in the $k \to 0$ limit as signified by the cancellation between the two sets of diagrams shown in Fig. 12.9(c). The CP odd part is proportional to

$$\left[\frac{1}{\not p + \not k - m} \Gamma_{qq'} \frac{1}{\not p - m'} - \frac{1}{\not p + \not k - m'} \Gamma_{q'q} \frac{1}{\not p - m}\right]$$

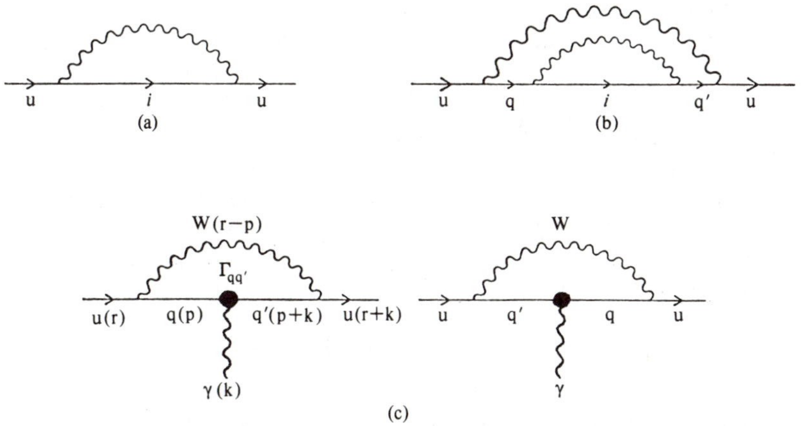

FIG. 12.9. Diagrams for induced u-quark dipole moment: the photon line is to be attached to all the charged lines in (a) one-loop and (b) two-loop diagrams. In (c) the black dots represent the quark e.m. transition vertices. Being at least one-loop, the graphs in (c) are actually two-loop diagrams, and they cancel in the $k \to 0$ limit.

where $\Gamma_{qq'}$s are the regularized quark electromagnetic transition moments, and it can be shown that up to one-loop

$$\Gamma_{qq'} = \Gamma_{q'q}.$$

In the static limit, $k \to 0$, we can drop the momentum k in the quark propagators and clearly the above contribution vanishes. Therefore, one must include higher order radiative corrections, the leading term being that due to QCD gluons. Thus the neutron dipole moment is expected to be of the order

$$\frac{d_{\mathrm{n}}}{e} \lesssim m_{\mathrm{u,d}}\left(\frac{\alpha_{\mathrm{s}}}{\pi}\right)\frac{g^4}{(2\pi)^4 M_{\mathrm{W}}^2}\left(\frac{m_q}{M_{\mathrm{W}}}\right)^4. \qquad (12.115)$$

The external factor of $m_{\mathrm{u,d}}$ reflects the helicity-flip nature of the dipole-moment operator. The numerical value generally quoted is

$$|d_{\mathrm{n}}/e| < 10^{-33}\,\mathrm{cm}. \qquad (12.116)$$

When one considers graphs with exchanges among different valence quarks in the neutrons, generally the same bound is obtained. However a possible large contribution may come from those involving Penguin diagrams; the GIM suppression being only logarithmic, they lead to a (d_{n}/e) value as large as 10^{-30} cm (see Gavela *et al.* 1982 and references cited therein).

(3) Hard vs. soft CP violations. In the standard theory there is only one Higgs doublet so that the vacuum expectation value $\langle\Phi\rangle_0$ has to be real because any possible phase had no physical significance and can be rotated away. The sources of the complex mass matrices are the Yukawa couplings themselves. This is usually called *hard CP violation* in the sense that it is an effect due to dimension-four operators. This is in contrast to another class of CP-violation theories where the violation arises from spontaneous symmetry breaking—hence a soft variety (T. D. Lee 1974). To get spontaneous CP violation, one has to extend the Higgs structure (e.g. two complex doublets) to get the complex vacuum expectation value. Physically the difference between hard and soft CP violation is that soft CP violation effects disappear at energies higher than the energy scale of the symmetry breaking, while hard CP violations will persist. One should also point out with more than one Higgs doublet the Yukawa coupling is naturally nondiagonal in the flavour space unless some discrete symmetries are imposed. These couplings generally tend to induce too large a rate for the strangeness-changing and muon-number nonconserving neutral-current effects (also see remarks in §11.3).

12.3 The W and Z intermediate vector bosons

The most basic and distinctive feature of a gauge theory with spontaneous symmetry breaking is the existence of a set of massive gauge bosons. In the standard model, there are three such intermediate vector bosons: W^+, W^-, and Z. Here we discuss in turn their masses, decays, and possible production mechanisms.

Masses

The W and Z masses can be expressed in terms of the Weinberg angle θ_W as in eqns (11.100) and (11.82).

$$M_W = \frac{1}{2}\left(\frac{e^2}{\sqrt{2}\,G_F}\right)^{1/2} \frac{1}{\sin\theta_W} = \frac{37.3\text{ GeV}}{\sin\theta_W} \tag{12.117}$$

$$M_Z = \left(\frac{e^2}{\sqrt{2}\,G_F}\right)^{1/2} \frac{1}{\sin 2\theta_W} = \frac{74.6\text{ GeV}}{\sin 2\theta_W}. \tag{12.118}$$

Experimental knowledge of θ_W will give precise predictions for these masses. Since these gauge bosons are rather heavy, at present we have only seen their virtual effects in the low-energy charged- and neutral-current phenomenology. In the standard model, the Higgs particles are in the SU(2) × U(1) doublet representation, M_W and M_Z are related (see eqn (11.82) and eqn (12.158) below)

$$\rho = \frac{M_W^2}{M_Z^2\cos^2\theta_W} = 1. \tag{12.119}$$

This fixes the relative strengths of the charged- and neutral-current reactions. From the data on the neutrino neutral-current processes a two-parameter fit to the standard model yields (Kim *et al.* 1981)

$$\rho = 0.998 \pm 0.050 \tag{12.120}$$

$$\sin^2\theta_W = 0.224 \pm 0.015. \tag{12.121}$$

The fact that ρ is very close to one lends support to the standard model. Of course this does not exclude the possibility that there can be more than one Higgs doublet or that there can be other representations of Higgs particles with small vacuum expectation values (except singlet Higgs scalars which do not couple to the W or Z).

If we restrict ourselves to $\rho = 1$, an average of the neutrino data yields

$$\sin^2\theta_W = 0.227 \pm 0.010. \tag{12.122}$$

Then the intermediate vector boson masses are

$$M_W = 78.5 \pm 1.7\text{ GeV}$$
$$M_Z = 89.3 \pm 2\text{ GeV}. \tag{12.123}$$

The fact that these masses are confined to a rather narrow range provides us with a clean experimental test of the standard model.

A more precise determination of $\sin^2\theta_W$ and the gauge boson masses requires the inclusion of radiative corrections of the weak processes involved and of the energy dependence of coupling constants (see §3.3). Since the Weinberg angle is defined through the coupling constant (eqn (11.80)), it will also depend on energy. After radiative corrections, the value of $\sin^2\theta_W$ at M_W is found to be

$$\sin^2\theta_W(M_W) = 0.215 \pm 0.10 \pm 0.004 \tag{12.124}$$

where ± 0.004 reflects the theoretical uncertainties in the radiative correction calculations (Marciano and Sirlin 1980; Sirlin and Marciano 1981; Llewellyn Smith and Wheater 1981). If one uses the fine structure constant at M_W

$$\frac{1}{\alpha(M_W)} = 127.49, \tag{12.125}$$

the gauge boson masses come out to be

$$M_W = \frac{38.5 \text{ GeV}}{\sin\theta_W(M_W)} = 83.0 \pm 2.4 \text{ GeV} \tag{12.126}$$

$$M_Z = M_W/\cos\theta_W = 93.8 \pm 2.0 \text{ GeV}. \tag{12.127}$$

Thus the higher-order effects increase the M_W and M_Z values by about 5 per cent. It is anticipated that M_Z will be measured to within 0.1 to 0.2 GeV in the high-energy e^+e^- machine, thereby probing the higher-order electroweak radiative corrections.

We now discuss W and Z decays. We shall see that the lifetime of these particles will be very short as they decay 'semi-weakly'.

W decays

The couplings of W to the fermions are given by eqns (11.86) and (11.121)

$$\mathscr{L}_W = \frac{g}{2\sqrt{2}} W_\mu^+ \left[(\bar{\nu}_e, \bar{\nu}_\mu, \bar{\nu}_\tau)\gamma^\mu(1-\gamma_5)\begin{pmatrix} e \\ \mu \\ \tau \end{pmatrix} + (\bar{u}, \bar{c}, \bar{t})\gamma^\mu(1-\gamma_5) U \begin{pmatrix} d \\ s \\ b \end{pmatrix} \right] + \text{h.c.} \tag{12.128}$$

where U is the Cabibbo–Kobayashi–Maskawa matrix, (12.39). Using (12.128) we can calculate the rate for various decay modes.

Consider the example

$$W^-(k) \to e(p) + \bar{\nu}(q). \tag{12.129}$$

The matrix element is given by

$$T_e = \frac{g}{2\sqrt{2}} \bar{u}_e(p)\gamma_\mu(1-\gamma_5)v_\nu(q)\varepsilon^\mu(k) \tag{12.130}$$

where $\varepsilon^\mu(k)$ is the polarization vector of the W. Summing over the spins of the fermions and averaging over the W polarizations, we have

$$\begin{aligned} \frac{1}{3}\sum_{\text{spin}} |T_e|^2 &= \frac{g^2}{3}[p_\mu q_\nu + p_\nu q_\mu - g_{\mu\nu}(p \cdot q)]\left(-g^{\mu\nu} + \frac{k^\mu k^\nu}{M_W^2}\right) \\ &= \frac{1}{3} g^2 M_W^2 \end{aligned} \tag{12.131}$$

where we have neglected the electron mass. The decay rate for W at rest is

$$\Gamma_e = \frac{1}{2M_W} \int (2\pi)^4\, \delta^4(k - p - q) \frac{1}{3} \sum |T_e|^2 \frac{d^3p}{(2\pi)^3 2p_0} \frac{d^3q}{(2\pi)^3 2q_0}$$

$$= \frac{g^2 M_W}{48\pi} = \frac{G_F}{\sqrt{2}} \frac{M_W^3}{6\pi}. \qquad (12.132)$$

This decay rate is proportional to the Fermi constant G_F, rather than to G_F^2 as in the usual weak decays; hence we say it is a semi-weak decay.

Clearly all other leptonic decays $W^- \to \mu\bar{\nu}_\mu$ and $W^- \to \tau\bar{\nu}_\tau$ will have the same decay rate: $\Gamma_\mu = \Gamma_\tau = \Gamma_e$. For the hadronic decays, we can calculate the rate of decay of W into a quark pair which then hadronizes with unit probability

$$\Gamma(W \to n_i \bar{p}_j) = 3|U_{ij}|^2 \Gamma_e \qquad (12.133)$$

where n_i = d, s, b and p_j = u, c, t, with Γ_e as given in eqn (12.132). The factor of 3 in front is due to the colour degree of freedom of quarks. $W \to n_i \bar{p}_j$ represents all decays into hadrons having the same quantum number as $n_i + \bar{p}_j$. For the light quarks decay products we expect them to have a two-jet structure, modified occasionally by the emission of a gluon jet. The total decay rate can be simplified by using the unitarity property of the U-matrix $\Sigma_i |U_{ij}|^2 = \Sigma_j |U_{ij}|^2 = 1$ (assuming $m_t \ll M_W$)

$$\Gamma_{tot} = 3\Gamma_e + \sum_{i,j} 3|U_{ij}|^2 \Gamma_e = 12\Gamma_e$$

$$= \frac{\sqrt{2}\, G_F M_W^3}{\pi} = 5.23(M_W \text{ in GeV})^3 \times 10^{-6}\,\text{GeV}. \qquad (12.134)$$

Thus for $M_W = 83$ GeV, the total width is

$$\Gamma = 2.99\ \text{GeV} \qquad (12.135)$$

with the leptonic branching ratios

$$B_e = \frac{\Gamma(W \to e\bar{\nu}_e)}{\Gamma(W \to \text{all})} = \frac{1}{12}, \quad B_e = B_\mu = B_\tau. \qquad (12.136)$$

The large width of (12.135) reflects both that W decays semiweakly and that M_W is large, so that there is plenty phase space.

Remarks

(1) In the standard model we have three families of leptons and quarks. If we generalize it to N families and if all of them are much lighter than M_W, the total width and leptonic branching ratio are given by

$$\Gamma_{tot} = 4N\Gamma_e = \frac{\sqrt{2}\, N G_F M_W^3}{3\pi}$$

$$B_e = \frac{1}{4N}. \qquad (12.137)$$

(2) In the calculation of the hadronic decays we assumed that quark masses could be neglected. This may be a poor approximation for decay modes involving the t-quark if m_t turns out to be substantially larger than the present experimental lower bound, ~20 GeV.

(3) Careful studies of the hadronic decay modes can yield useful information about the mixing matrix U_{ij}.

(4) It is clear from the Feynman rule that even if the Higgs mass $m_\phi < M_W$, W cannot decay directly into the Higgs ϕ^0 through the lowest-order Lagrangian. An interesting possibility of $W \to \phi^0 l\bar{\nu}_l$ involving a second-order diagram is shown in Fig. 12.10(a). At first one may expect that in

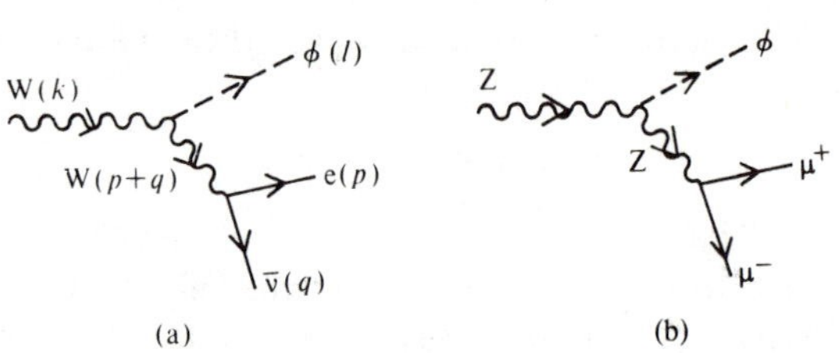

FIG. 12.10.

addition to the extra coupling g there will be further suppression coming from the W-propagator $\sim M_W^{-2}$. But the WWϕ coupling has a factor M_W; this and an additional factor coming from the three-body phase space will cancel the W-propagator suppression. Hence we expect this decay rate to be only down by an order of g^2 compared to $W \to e\nu_e$ and this could be useful in searching for the Higgs particle ϕ^0. The matrix element is given by

$$T_\phi = \frac{g^2 M_W}{2\sqrt{2}} \frac{1}{(p+q)^2 - M^2{}_W} \bar{u}_e(p)\gamma_\mu(1-\gamma_5)v_\nu(q)\varepsilon^\mu. \qquad (12.138)$$

Summing over fermion spins and averaging over the polarization of W, we get

$$\frac{1}{3}\sum |T_\phi|^2 = \frac{1}{3}\frac{g^4 M_W^2}{[(p+q)^2 - M_W^2]^2}\left[(p\cdot q) + \frac{2}{M_W^2}(p\cdot k)(q\cdot k)\right]. \qquad (12.139)$$

The decay rate is given by

$$\Gamma_\phi = \frac{1}{2M_W}\int (2\pi)^4\, \delta^4(k-l-p-q)$$
$$\times \frac{d^3 l}{(2\pi)^3 2l_0}\frac{d^3 p}{(2\pi)^3 2p_0}\frac{d^3 q}{(2\pi)^3 2q_0}\frac{1}{3}\sum |T_\phi|^2. \qquad (12.140)$$

Using the formula

$$I^{\alpha\beta} = \int \frac{d^3 p}{2p_0}\frac{d^3 q}{2q_0}\,\delta^4(Q-p-q)p^\alpha q^\beta$$
$$= \frac{\pi}{24}(Q^2 g^{\alpha\beta} + 2Q^\alpha Q^\beta) \qquad (12.141)$$

with $Q = k - l$, we get

$$\Gamma_{\phi} = \frac{G_F}{\sqrt{2}} \frac{M_W^3}{6\pi} \frac{g^2}{16\pi^2} \left[\ln \left| \frac{2m_{\phi}}{M_W} \right| + \frac{23}{24} \right] \tag{12.142}$$

where we have made the approximations $m_e = 0$ and $m_{\phi} \ll M_W$. Comparing this to $W \to e\bar{\nu}_e$, we have

$$\frac{\Gamma(W^- \to \phi^0 e\bar{\nu}_e)}{\Gamma(W^- \to e\bar{\nu}_e)} = \frac{\alpha}{4\pi \sin^2 \theta_W} \left[\ln \left| \frac{2m_{\phi}}{M_W} \right| + \frac{23}{24} \right]. \tag{12.143}$$

This indicates that if ϕ^0 is much lighter than W, the logarithm factor can give an enhancement. Experimentally this decay mode into ϕ^0 will be difficult to observe because of the missing neutrinos and the short lifetime of the ϕ^0.

Z decays

If we parametrize the coupling of Z^0 to any fermion f as

$$\mathscr{L}_{NC} = M_Z \left(\frac{G_F}{\sqrt{2}} \right)^{1/2} \bar{f} \gamma_{\mu} \frac{1}{\sqrt{2}} (g_V^f - g_A^f \gamma_5) f Z^{\mu} \tag{12.144}$$

then the width for each decay model is given by

$$\Gamma(Z \to \bar{f}f) = \frac{G_F M_Z^3}{24\pi\sqrt{2}} (|g_V^f|^2 + |g_A^f|^2). \tag{12.145}$$

In the standard model, we have

$$\begin{aligned} g_V^e = g_V^{\mu} = g_V^{\tau} &= -1 + 4\sin^2\theta_W, & g_A^e = g_A^{\mu} = g_A^{\tau} &= -1 \\ g_V^u = g_V^c = g_V^t &= 1 - \tfrac{8}{3}\sin^2\theta_W, & g_A^u = g_A^c = g_A^t &= 1 \\ g_V^d = g_V^s = g_V^b &= -1 + \tfrac{4}{3}\sin^2\theta_W, & g_A^d = g_A^s = g_A^b &= -1 \end{aligned} \tag{12.146}$$

and $g_V^{\nu} = g_A^{\nu} = 1$ for all neutrino flavours. Eqn (12.146) is obtained from eqn (11.95) with $g_V = 2(g_L + g_R)$ and $g_A = 2(g_L - g_R)$ and universality. We note that with the experimental value (eqn (12.124)) of $\sin^2\theta_W = 0.215$, which is close to 1/4, the vector coupling of the charged lepton $g_V^l = 0.14$ is small compared to the axial-vector coupling $g_A^l = 1$. The partial width for decay into a neutrino pair of a particular flavour f is

$$\begin{aligned} \Gamma_{\nu_f} = \Gamma(Z^0 \to \nu_f \bar{\nu}_f) &= \frac{G_F M_Z^3}{24\pi\sqrt{2}} \times 2 \\ &= 2.2(M_Z \text{ in GeV})^3 \times 10^{-7} \text{ GeV}. \end{aligned} \tag{12.147}$$

For the M_Z value in eqn (12.127), this gives

$$\Gamma_{\nu_f} = 0.18 \text{ GeV}. \tag{12.148}$$

The relative widths of $Z \to f\bar{f}$ for f in one given family can be just read off from (12.146)

$$\begin{aligned}\Gamma(Z \to \nu\bar{\nu}) &: \Gamma(Z \to l\bar{l}) : \Gamma(Z \to u\bar{u}) : \Gamma(Z \to d\bar{d}) \\ &= 2 : [1 + (1 - 4\sin^2\theta_W)^2] : 3[1 + (1 - \tfrac{8}{3}\sin^2\theta_W)^2] \\ &\quad : 3[1 + (1 - \tfrac{4}{3}\sin^2\theta_W)^2] \\ &= 2 : 1.02 : 3.54 : 4.53. \end{aligned} \tag{12.149}$$

The total width is

$$\begin{aligned}\Gamma_{tot} &= \Gamma(Z \to \text{all}) = 24(1 - 2\sin^2\theta_W + \tfrac{8}{3}\sin^4\theta_W)\Gamma_{\nu_f} \\ &= 3.0\ \text{GeV}. \end{aligned} \tag{12.150}$$

The branching ratio for decay into a $\mu^+\mu^-$ pair is

$$\begin{aligned} B(\mu^+\mu^-) &= \frac{\Gamma(Z \to \mu^+\mu^-)}{\Gamma(Z \to \text{all})} = \frac{1}{24}\left(\frac{1 - 4\sin^2\theta_W + 8\sin^4\theta_W}{1 - 2\sin^2\theta_W + \tfrac{8}{3}\sin^4\theta_W}\right) \\ &= 3.06 \times 10^{-2}. \end{aligned} \tag{12.151}$$

In eqns (12.149)–(12.151) we used the (12.124) value of $\sin^2\theta_W = 0.215$. The branching ratio of (12.151) is not very sensitive to $\sin^2\theta_W$ for $\sin^2\theta_W$ around 0.22. For example, as we vary $\sin^2\theta_W$ from 0.2 to 0.25, $B(\mu^+\mu^-)$ goes from 3.06 to 3.13 per cent. This will be a useful piece of information in our search for the hadronic-produced Zs.

Remarks

(1) If we generalize the standard model to N fermion families, the results in (12.150) and (12.151) become

$$\begin{aligned}\Gamma(Z \to \text{all}) &= 1.0 \times N\ \text{GeV} \\ B(\mu^+\mu^-) &= \frac{0.092}{N}. \end{aligned} \tag{12.152}$$

(2) While the decay $Z \to \phi^0\phi^0$ is forbidden by angular momentum and Bose statistics, we have the interesting decay model $Z \to \phi^0\mu^+\mu^-$ (Bjorken 1977). This is analogous to $W \to \phi^0 l\bar{\nu}$, as in Fig. 12.10(b). Similarly to eqn (12.143), we obtain

$$\frac{\Gamma(Z \to \mu^+\mu^-\phi^0)}{\Gamma(Z \to \mu^+\mu^-)} = \frac{\alpha}{4\pi\sin^2\theta_W\cos^2\theta_W}\left[\ln\left|\frac{2m_\phi}{M_W}\right| + \frac{23}{24}\right]. \tag{12.153}$$

Combining eqns (12.153) and (12.151) we get a branching ratio for this decay of order 10^{-4} for $m_\phi = 10$ GeV. The possibility of detecting this decay mode will be much more favourable when compared to $W \to \phi^0 e\nu$ because one can observe both μ^+ and μ^- and ϕ^0 should show up as a bump in the missing mass plot. Also in the future e^+e^- colliding machines it is expected that 10^6 Zs can be produced in a year, so that even decays with small branching ratios can be detected.

(3) Since the Zs will be produced in large quantities, one might hope to study the rare decay modes of Z which involve loop diagrams. Those

reactions with diagrams involving the trilinear gauge boson couplings will be of particular interest in checking the non-Abelian character of the theory. Most of the rare Z decays have been studied and are found to be still too small to be seen in the near-future (Albert *et al.* 1980). For example, suppose we wish to study a flavour-changing neutral-current decay such as $Z \to u\bar{c}$ in Fig. 12.11. Because of the GIM cancellation mechanism, the amplitude is of order

$$T \sim g^4 \left[\sum_i U_{ui} U_{ci}^* m_i^2 / M_W^2 \right]^2 \tag{12.154}$$

which gives the branching ratio

$$\frac{\Gamma(Z \to c\bar{u})}{\Gamma(Z \to c\bar{c})} \sim \left(\frac{\alpha}{\pi}\right)^2 \left(\frac{m_i^2}{M_W^2}\right)^2 \tag{12.155}$$

which is of order 10^{-9} even for m_i as large as 10 GeV.

FIG. 12.11. One-loop diagrams for $Z \to c\bar{u}$ with i = d, s, b.

Remarks on W and Z production

(1) We will just mention that the cleanest way of producing W and Z is through e^+e^- annihilations. The Z boson will show up as a sharp spike in an e^+e^- collision very much like the J/ψ or Υ particles. At the peak of Z resonance we have

$$R_Z = \frac{\sigma(e^+e^- \to Z^0 \to \text{all})}{\sigma(e^+e^- \to \gamma^* \to \mu^+\mu^-)} \simeq 5000 \tag{12.156}$$

which corresponds to five events per second if the luminosity is $10^{32}\ \text{cm}^{-2}\ \text{s}^{-1}$. We may also note that, reversing the situation encountered in the case of the J/ψ and Υ particles, the width of the Z boson ($\simeq$2–3 GeV) is much larger than the beam resolution ($\approx$100 MeV). So the shape of the resonance will allow us to deduce Γ_{tot} directly. From this one can 'count' the number of neutrino flavours N_ν using the relation

$$\Gamma_{\text{tot}} - \Gamma_{\text{visible}} = N_\nu \Gamma(Z \to \nu_f \bar{\nu}_f) \tag{12.157}$$

since we know the value of $\Gamma(Z \to \nu_f \bar{\nu}_f)$ (eqn (12.148)) and Γ_{visible} is determined by the observable cross-section at the resonance peak which, according to the Breit–Wigner resonance formula, should be proportional to $\Gamma(e^+e^-) \times \Gamma_{\text{visible}} \times \Gamma_{\text{tot}}^{-1}$.

(2) If the e^+e^- machine has $\sqrt{s} > 2M_W$, the Ws will be produced through the diagrams of Fig. 12.12. This $e^+e^- \to W^+W^-$ reaction is of particular interest because it involves the trilinear gauge boson couplings and will provide an important test of the non-Abelian nature of the underlying gauge

theories (Sushkov, Flambaum, and Khriplovich 1975; Alles, Boyer, and Buras 1977).

(3) In practical terms it is likely that the first production of W and Z will be in the proton–antiproton collisions through the Drell–Yan mechanism (Fig. 12.13). For a review see Quigg (1977).

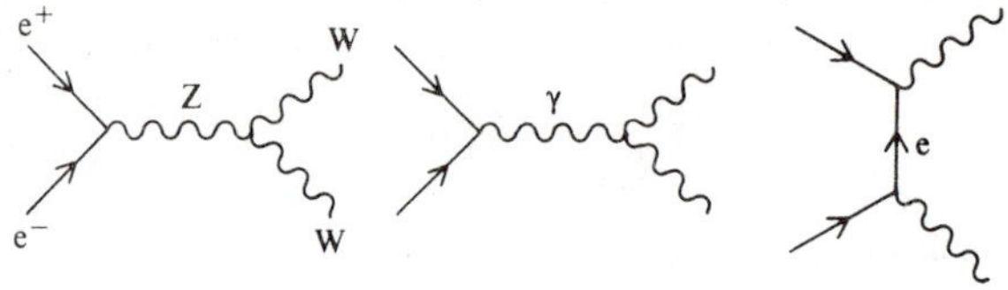

FIG. 12.12.

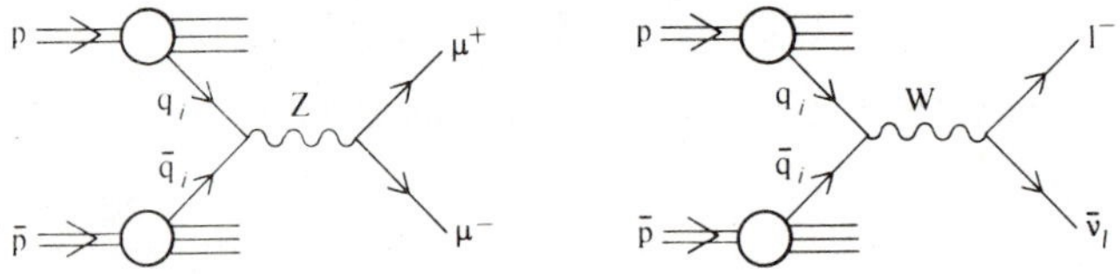

FIG. 12.13.

The Higgs particle

In the standard electroweak theory, we start out with a doublet of complex scalars. After spontaneous symmetry breaking three of the original four real scalar fields are eaten by the gauge particles and we are left with one neutral physical Higgs scalar ϕ^0. (For reviews of Higgs phenomenology see, for example, Ellis *et al.* 1976*a*,*b*; Li 1980).

General properties of the Higgs particle

The Higgs scalar of the minimal Weinberg–Salam model has the following basic properties.

$\boldsymbol{M_W = M_Z \cos\theta_W}$**.** As we have mentioned before (see, for example, eqns (11.82)–(11.84)) this relation follows from the doublet structure of Higgs particles and is well satisfied experimentally. Still, one cannot exclude the possibility that there can be more than one doublet in the theory. In fact there are other Higgs structures which also satisfy this relation (Tsao 1980). For a general SU(2) × U(1) multiplet of Higgs particle $\phi_{T,Y}$ with weak isospin T and weak hypercharge Y, the ρ parameter is given by (Lee 1972*c*)

$$\rho = \frac{M_W^2}{M_Z^2 \cos^2\theta_W} = \frac{\Sigma_{T,Y} |v_{T,Y}|^2 [T(T+1) - Y^2/4]}{2\,\Sigma_{T,Y} |v_{T,Y}|^2 Y^2/4} \tag{12.158}$$

where $v_{T,Y} = \langle 0|\phi_{T,Y}|0\rangle$ is the VEV of the Higgs particle. The requirement of $\rho = 1$ for arbitrary $v_{T,Y}$ means

$$T(T+1) = \tfrac{3}{4}Y^2. \tag{12.159}$$

Examples of solutions to (12.159) are $(T, Y) = (\frac{1}{2}, 1), (3, 4), (\frac{25}{2}, 15), \ldots$.

Higgs couplings to fermions. The Yukawa coupling of the Higgs particle ϕ^0 conserves parity and fermion flavour, and its strength is proportional to the fermion mass. This can be seen as follows. We have the Yukawa coupling

$$\mathscr{L}_{\mathrm{Y}} = f_{ij}\bar{\psi}'_{i\mathrm{L}}\phi\psi'_{j\mathrm{R}} + \text{h.c.} \tag{12.160}$$

where i, j are fermion flavour indices and the prime over the fermion fields indicates that they are weak eigenstates. The symmetry-breaking condition (eqns (11.63) and (11.90))

$$\langle\phi\rangle_0 = \frac{1}{\sqrt{2}}\begin{pmatrix}0\\ v\end{pmatrix}$$

with

$$v = (\mu^2/\lambda)^{\frac{1}{2}} = 2^{-1/4}G_{\mathrm{F}}^{-1/2} \simeq 250\ \text{GeV} \tag{12.161}$$

generates the fermion mass matrix

$$\mathscr{L}_{\mathrm{M}} = m_{ij}\bar{\psi}'_{i\mathrm{L}}\psi'_{j\mathrm{R}} + \text{h.c.} \qquad \text{with } m_{ij} = \frac{v}{\sqrt{2}}f_{ij}. \tag{12.162}$$

In this case, the fermion mass matrix is proportional to the Yukawa coupling matrix. Thus when we diagonalize the mass matrix

$$\mathscr{L}_{\mathrm{M}} = m_{ij}\bar{\psi}'_{i\mathrm{L}}\psi'_{j\mathrm{R}} + \text{h.c.} = m_i\bar{\psi}_{i\mathrm{L}}\psi_{i\mathrm{R}} + \text{h.c.} \tag{12.163}$$

with ψ_i being the mass eigenstates, we get diagonalized Yukawa couplings of the physical Higgs ϕ^0 to the fermion fields,

$$\mathscr{L}_{\mathrm{Y}} = 2^{3/4}\sqrt{G_{\mathrm{F}}}\, m_i\phi^0\bar{\psi}_{i\mathrm{L}}\psi_{i\mathrm{R}} + \text{h.c.} \tag{12.164}$$

which conserves fermion flavour and parity, and the strength of the coupling is proportional to the fermion mass. This property that mass and coupling matrices are proportional to each other is a consequence of the Higgs particles being in a single irreducible representation (Glashow and Weinberg 1977).

The experimental consequence of the proportionality of the coupling strength to the fermion mass is that the Higgs particle ϕ^0 can be produced more easily by heavy fermions and will decay predominantly into the heavy fermion channels that are allowed by kinematics. The factor $\sqrt{G_{\mathrm{F}}}$ in the Yukawa coupling (12.164) makes these coupling to the known fermions very small. Also the fact that these couplings conserve fermion flavours means that we cannot find their signature in the rare, but distinctive, flavour-changing processes.

Higgs couplings to gauge bosons. These couplings, being proportional to gM_{v} where M_{v} is the mass of the gauge bosons W or Z, are much larger than the Yukawa couplings studied above.

$$\mathscr{L}_{\phi\mathrm{VV}} = g\phi^0\left(M_{\mathrm{W}}W_\mu^+W^{-\mu} + \frac{1}{2\cos\theta_{\mathrm{W}}}M_ZZ_\mu Z^\mu\right) \tag{12.165}$$

The quartic ϕ^2VV couplings are listed in the Appendix.

Mass of the Higgs particle

In the standard model the mass of the physical Higgs scalar particle is given by eqn (11.72)

$$m_\phi = (2\mu^2)^{\frac{1}{2}} = (2\lambda)^{\frac{1}{2}}v.$$

While we know something about v, there is at present no information on the quartic coupling λ. This lack of a precise knowledge of the Higgs mass makes it very difficult to search for ϕ^0 experimentally. Theoretically, there are some prejudices as to the range of m_ϕ. If we require λ to be less than 1 so that perturbation theory remains valid, we get from (12.161) an upper bound

$$m_\phi < 350 \text{ GeV}. \tag{12.166}$$

On the other hand, if λ is too small, the symmetry-breaking vacuum will be unstable. This produces a lower bound as we will demonstrate below.

The Linde–Weinberg bound. The basic idea is that, if λ is too small, the one-loop contributions (particularly those from the gauge-boson loop) to the effective potential in §6.4 become relatively important; they cause $V(\langle\phi\rangle \neq 0)$ to be greater than $V(\langle\phi\rangle = 0)$ and SSB disappears (Linde 1976; Weinberg 1976*a*). In order to present the actual calculation we briefly summarize the results obtained in §6.4 and their generalization to the present case of SU(2) × U(1) gauge theory (Coleman and Weinberg 1973).

The basic calculation in §6.4 involves summations of an infinite number of one-loop diagrams (Fig. 6.12) with scalar, fermion, and gauge bosons running in the loops. It is clear that the non-Abelian nature of the theory is not particularly relevant as gauge-field trilinear and quartic self-couplings do not play a role. So we can simply take over the results of §6.4 (see in particular eqn (6.164)). In order to make sure that differences in definitions and normalizations of couplings and fields are taken into account, let us first concentrate on the SU(2) gauge-boson loop. If we decompose the complex scalars in terms of the real fields as

$$\phi = \frac{1}{\sqrt{2}}\begin{pmatrix}\phi_3 + i\phi_4\\ \phi_1 + i\phi_2\end{pmatrix}. \tag{12.167}$$

then

$$\phi^2 \equiv \phi^\dagger\phi = \frac{1}{2}\sum_{i=1}^{4}\phi_i^2 \tag{12.168}$$

with

$$\langle\phi^2\rangle_0 = \tfrac{1}{2}\langle\phi_1\rangle_0^2 = \tfrac{1}{2}v^2. \tag{12.169}$$

Since the effective potential can depend only on ϕ^2 it will be adequate for our purpose to explicitly calculate only loop diagrams with external ϕ_1s. Since the gauge boson–scalar coupling is of the form $\phi^\dagger(\mathbf{A}_\mu \cdot \mathbf{A}_\mu)\phi$, three gauge bosons will contribute equally. For each one, say A_μ^1, we can simply take over the scalar QED result of eqn (6.157). Keeping in mind that the $\phi_{1,2}$s in eqn

(6.154) are normalized differently from our ϕ_is and using the Feynman rules in Appendix B, we need to replace e^2 by $g^2/4$; we have

$$V(\phi_1^2) = \frac{3}{64\pi^2}\left(\frac{g^2}{4}\right)^2 \phi_1^4 \ln \phi_1^2/M^2 + \ldots$$

or

$$V(\phi^2) = \frac{3}{16\pi^2 v^4} M_W^4 \phi^4 \ln \phi^2/M^2 + \ldots \tag{12.170}$$

where we have used $M_W^2 = \frac{1}{4}g^2v^2$ and eqn (12.168). When the other gauge boson loops are included we have

$$V_g(\phi^2) = \frac{3}{16\pi^2 v^4} \sum_V m_V^4 \phi^4 \ln \phi^2/M^2 + \ldots. \tag{12.171}$$

The index V runs over the $W^\pm$ and Z vector bosons. Including the scalar and fermion loops, we finally obtain the one-loop effective potential similar to eqn (6.164)

$$V_1(\phi^2) = C\phi^4 \ln \phi^2/M^2 \tag{12.172}$$

with

$$C = \frac{1}{16\pi^2 v^4}\left(3\sum_V m_V^4 + m_\phi^4 - 4\sum_f m_f^4\right). \tag{12.173}$$

We should recall that the factor of 3 in the vector boson term comes from tracing the numerator of the gauge-boson propagator in the Landau gauge; the factor of 4 in the fermion term comes from tracing the Dirac matrices and the minus sign reflects the Fermi statistics. With

$$\begin{aligned} V(\phi) &= V_0(\phi) + V_1(\phi) \\ &= -\mu^2\phi^2 + \lambda\phi^4 + C\phi^4 \ln(\phi^2/M^2), \end{aligned} \tag{12.174}$$

we can determine v by

$$\left.\frac{\partial V}{\partial \phi}\right|_{\phi = v/\sqrt{2}} = 0 \tag{12.175}$$

$$-\mu^2 + \lambda v^2 + Cv^2(\ln v^2/2M^2 + \tfrac{1}{2}) = 0. \tag{12.176}$$

The mass of the Higgs particle is given by

$$m_\phi^2 = \left.\frac{1}{2}\frac{\partial^2 V}{\partial \phi^2}\right|_{\phi = v/\sqrt{2}} = 2v^2\left[\lambda + C\left(\ln\left(\frac{v^2}{2M^2}\right) + \frac{3}{2}\right)\right] \tag{12.177}$$

where we have used eqn (12.176) to eliminate the μ^2 term.

Now consider the case where λ is very small so that it can be neglected in eqns (12.174)–(12.177). The value of the potential at the minimum $\phi = v/\sqrt{2}$ is

$$V(v) = -\frac{1}{2}\mu^2 v^2 + \frac{Cv^4}{4}\ln(v^2/2M^2). \tag{12.178}$$

At $\phi = 0$, the value of the potential is

$$V(0) = 0, \tag{12.179}$$

when there is no SSB. Since we want (12.178) to be the absolute minimum, we must have

$$V(v) < V(0). \tag{12.180}$$

This implies, through eqns (12.176) and (12.178), that

$$\ln(v^2/2M^2) + 1 > 0. \tag{12.181}$$

When we substitute condition (12.181) into (12.177) we have

$$m_\phi^2 > Cv^2 = \frac{1}{16\pi^2 v^4} \sum_V 3m_V^4 = \frac{3\alpha^2(2 + \sec^4 \theta_W)}{16\sqrt{2}\, G_F \sin^4 \theta_W} \tag{12.182}$$

where we have neglected m_ϕ and m_f which are assumed to be small compared to m_V. For $\sin^2 \theta_W \simeq 0.215$, this gives

$$m_\phi > 7.9 \text{ GeV}. \tag{12.183}$$

Otherwise the radiative correction will make the asymmetric vacuum unstable.

The Coleman–Weinberg conjecture. An interesting suggestion is that $\mu^2 = 0$ in the standard model and SSB is driven completely by quantum radiative corrections (Coleman and Weinberg 1973; Weinberg 1979*b*). From eqns (12.176) and (12.177) we obtain

$$m_\phi = (2C)^{\frac{1}{2}} v \simeq 11 \text{ GeV}. \tag{12.184}$$

Even though this is a precise prediction about m_ϕ, one must keep in mind that at this stage there is no compelling physical ground to have $\mu^2 = 0$ although it is an intriguing proposition with its suggestive simplicity.

Production of the Higgs particle

It is clear from the discussions on Higgs couplings to fermions and gauge bosons above that most of the promising mechanisms for producing the Higgs particles make use of their couplings to

(1) heavy fermions;
(2) gauge bosons,

as they are less suppressed.

In category (1) perhaps the most promising reaction is

(a) $V_{Q\bar{Q}} \to \phi^0 + \gamma$

(Fig. 12.14) where $V_{Q\bar{Q}}$ is the 1^- $Q\bar{Q}$ bound state of heavy quarks (Wilczek 1977). Besides having a large fermion–Higgs coupling it has the additional advantage that the hadronic decay of quarkonium $V_{Q\bar{Q}}$ is suppressed by the

Zweig rule. The decay rate has been estimated to give

$$\frac{\Gamma(V_{Q\bar{Q}} \to \phi^0 + \gamma)}{\Gamma(V_{Q\bar{Q}} \to \gamma \to e^+e^-)} \cong \frac{G_F M_V^2}{4\sqrt{2}\,\pi\alpha}\left(1 - \frac{m_\phi^2}{M_V^2}\right) \qquad (12.185)$$

where M_V is the mass of $V_{Q\bar{Q}}$. For $M_V \simeq 30$ GeV and $m_\phi \ll M_V$ this ratio is about 8 per cent. One important feature of this mechanism is that the reaction produces a monochromatic photon and it may be a good experimental handle for its detection.

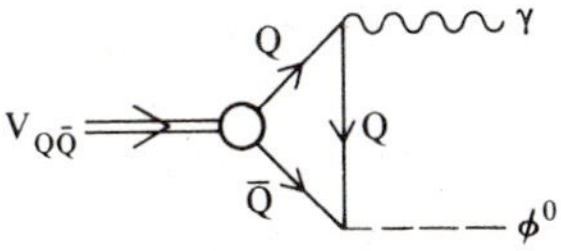

FIG. 12.14.

In category (2), there are several interesting processes.

(b) $Z^0 \to \phi^0 + \gamma$

(see Fig. 12.15), where the f in the second diagram is some heavy fermion which has a large coupling to ϕ^0 (Cahn, Chanowitz, and Fleishon 1979). It turns out that for $m_f < M_W$, the dominant contribution comes from the W-loop and the branching ratio is estimated to be 10^{-6} for $m_\phi \ll M_Z$. The photon coming from this decay is also monochromatic

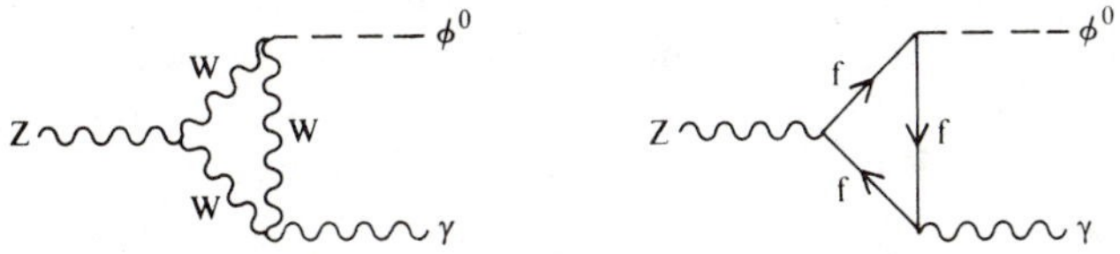

FIG. 12.15.

(c) $e^+ + e^- \to Z^0 + \phi^0$

(Fig. 12.16). At $\sqrt{s} \simeq 200$ GeV even a Higgs particle of mass close to 100 GeV could be produced with a cross-section larger than 10^{-37} cm^2. This corresponds to 1 event per day at a luminosity of 10^{32} cm^2 s^{-1} (Ellis *et al.* 1976*a*).

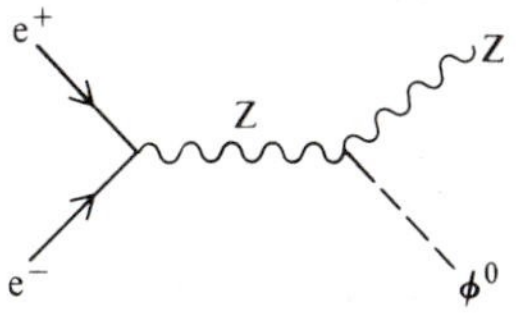

FIG. 12.16.

(d) $Z \to \phi^0 + e^+ + e^-$.

This decay has already been discussed in connection with Z decays in §12.3. The general feature is that ϕ^0 will show up as a mass bump in the recoil spectrum against the e^+e^- pair. The estimate gives $B(Z^0 \to \phi^0 e^+ e^-) \gtrsim 10^{-6}$ for $m_\phi < 40$ GeV. The related process $W^- \to \phi^0 e\bar{\nu}$ is also calculated in §12.3; however this W-decay mode is rather difficult to detect.

Charged Higgs particles

In the standard theory, there is no physical charged Higgs particle. But in many of the extensions of the minimal model there are charged Higgs particles, hence a richer phenomenology (Donoghue and Li 1979; Golowich and Yang 1979). Even though their masses and couplings are not very constrained by the theory, if they exist, they can be produced in e^+e^- annihilations through photon exchange (Fig. 12.17). For energy well above

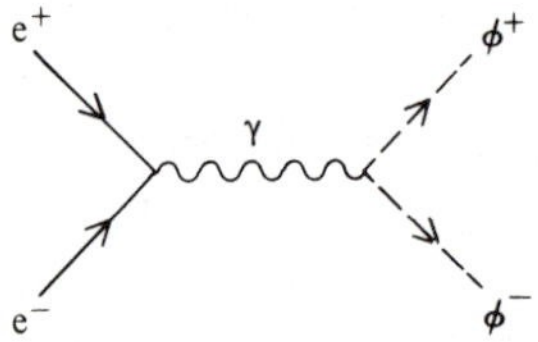

FIG. 12.17.

the threshold, the cross-section $\sigma(e^+e^- \to \phi^+\phi^-)$ is 1/4 of the standard point-like cross-section $\sigma(e^+e^- \to \mu^+\mu^-)$. Thus if there are several singly charged or doubly charged Higgs particles, their contributions to the e^+e^- total cross-section should be significant enough to be observable.

Also the charged Higgs particles will generate scalar or pseudo-scalar charged currents (McWilliams and Li 1981) which might contribute to low-energy charged-current interactions such as $\mu \to e\nu\bar{\nu}$, $\pi^+ \to \pi^0 e^+ \nu$, $n \to pe\bar{\nu}$, $\pi \to e\nu \ldots$, etc. Future high-precision measurements on these low-energy processes could shed some light on the properties of these charged Higgs particles. Any significant deviation from a V, A type of structure might indicate their existence.

3 Selected topics in quantum flavourdynamics

By the 'standard model' of strong and electroweak interactions one usually means the SU(3) × SU(2) × U(1) gauge theory with the SU(2) × U(1) electroweak group spontaneously broken down to $U(1)_{em}$ by one doublet of the elementary Higgs scalar fields. The neutrinos are massless. There are three families of fermions and CP violations result from a complex CKM mixing matrix in the weak charged-current couplings.

Several aspects of the standard electroweak theory may be revised and extended for various reasons. In §13.1 we discuss one particular realization of dynamic symmetry breaking based on an analogue to the QCD gauge theory of colour. Such schemes are referred to in the literature as *technicolour models* or *hypercolour models* and the purpose is to remove the elementary scalar field in the theory. In §13.2 the possibility of massive neutrinos is explored. We discuss the likely origin of neutrino mass terms and their phenomenological implications. We give details of a calculation (in §13.3) of $\mu \to e\gamma$ as a higher-order weak effect when there are massive neutrinos. The purpose is to illustrate the use of R_ξ-gauge Feynman rules in one nontrivial case where the unphysical 'would-be-Goldstone' bosons play an important role in maintaining the gauge invariance of the calculation.

3.1 Dynamical symmetry breaking and technicolour models

One attempts to replace the elementary Higgs scalars with composite ones (see, for example, Weinberg 1976*b*). A notable class of such models have the constituent fermions bound through gauge interactions that are modelled after the QCD theory of colour (Susskind 1979; Weinberg 1979*a*). Here one postulates a set of new gauge charges: the *technicolours*. One's aim is to have a spontaneous symmetry-breaking theory with gauge interactions alone: there is no elementary scalar with its self-couplings and Yukawa couplings. Such a successful theory has by no means yet been constructed. However progress has been made towards the realization of such a programme. (For reviews see Bég 1980; Farhi and Susskind 1981.) Our purpose here is to illustrate the possibility of dynamical models of spontaneous symmetry breaking. We follow the presentation by Sikivie (1982).

The motivation to replace the elementary Higgs scalars

Higgs scalars are used in gauge theories to cause spontaneous symmetry breakdown. In the electroweak theory they generate masses for W and Z gauge bosons. For one doublet of elementary Higgs particles, one gets the correct relative strength between neutral and charged currents: $M_W = M_Z \cos\theta_W$. In the standard model the leptons and quarks also acquire their masses during SSB through Yukawa couplings. The different sizes of the fermion masses can easily be accommodated by having different sizes of couplings and the complex Yukawa couplings can give rise to CP-violating charged-current couplings through the diagonalization of the fermion mass matrices (see §§11.3 and 12.2).

This 'versatility' of the elementary Higgs particle is related to the freedom one has in choosing the Higgs couplings: the scalar self-couplings and Yukawa couplings are quite unconstrained so long as they satisfy the requirements of gauge invariance. As a result a gauge theory with elementary Higgs scalars has many arbitrary parameters associated with the Higgs fields. This translates into the fact that in general masses and mixing angles cannot be calculated and must be introduced as parameters into the theory.

Furthermore as we shall see in Chapter 14 on grand unification that, when the standard $SU(3) \times SU(2) \times U(1)$ model is embedded into a simple gauge group, the above-mentioned arbitrariness is not much improved. On the other hand we have (see §14.2) the acute 'gauge hierarchy problem', which can be described as follows. Grand unified theories with group G require at least two stages of SSB corresponding to $G \to SU(3)_c \times SU(2)_L \times U(1)$ at energy scale M, and $SU(3)_c \times SU(2)_L \times U(1) \to SU(3)_c \times U(1)_{em}$ at scale μ. The scale μ is fixed by the weak interaction strength to be 250 GeV (see eqn (11.90)). The scale M is expected to be of order 10^{14} GeV both as a result of the lower bound on the proton lifetime and because of the successful prediction of $\sin^2\theta_W$. If these SSBs are due to elementary Higgs scalars, the ratio $\mu^2/M^2 \simeq 10^{-24}$ has to be introduced by hand and readjusted to 24 decimal places in each order of perturbation theory (see §14.2).

This motivates us to investigate the possibility of SSB without having to introduce elementary scalar fields. The notion of a composite Higgs scalar is really not a new one. We have already mentioned in §8.3 that the idea of the Higgs phenomena was first suggested by the theory of superconductivity. There the electromagnetic gauge symmetry is spontaneously broken by the condensate (i.e. non-vanishing ground-state expectation value) of the electron pairs (the Cooper pairs), which acts as an effective composite Higgs scalar. Thus SSB is brought on dynamically through the interactions of the electrons with the lattice phonons. (For an early work on dynamical symmetry breaking see Nambu and Jona-Lasinio 1961*a*,*b*.) Therefore the question one faces here in the electroweak case is that, if Higgs scalars are composite, what are their constituents? What interactions are responsible for binding them together? How can we have all the desired patterns of SSB without introducing many arbitrary parameters?

The basic technicolour idea

One naturally wonders whether the QCD strong force which binds the coloured quarks can be the interaction responsible for SSB in the electroweak interaction. We shall see that, although it fails to fulful this role, the way it fails suggests possible candidate theories.

Consider the standard $SU(3)_c \times SU(2)_L \times U(1)_Y$ model, but—this time—without the elementary Higgs scalar. Also, for simplicity, let us restrict ourselves to one family of fermions

$$q_L = \begin{pmatrix} u \\ d \end{pmatrix}_L \qquad l_L = \begin{pmatrix} \nu \\ e \end{pmatrix}_L . \tag{13.1}$$

The Lagrangian, for indices $\alpha = 1, \ldots, 8$; $i = 1, 2, 3$, is then

$$\begin{aligned} \mathscr{L} = &-\tfrac{1}{4}G^{\alpha\mu\nu}G^{\alpha}_{\mu\nu} - \tfrac{1}{4}A^{i\mu\nu}A^{i}_{\mu\nu} - \tfrac{1}{4}B^{\mu\nu}B_{\mu\nu} \\ &+ i(\bar{q}\gamma_\mu D^\mu q + \bar{l}\gamma_\mu D^\mu l)\,. \end{aligned} \tag{13.2}$$

Since there is no Higgs VEV to break the $SU(2) \times U(1)$ gauge symmetry, it would seem that all fermions and all gauge bosons, including W and Z, will remain massless. As we shall presently see, this is actually not the case.

Let us for the moment turn off the electroweak interaction and remember from Chapter 10 some of the basic features of the strong interaction as described by QCD. The fact that the u and d quarks are massless implies that we have the flavour symmetry $SU(2)_L \times SU(2)_R$. From the discussions in Chapter 5 on chiral symmetry and its breaking, all evidence is consistent with the picture that this symmetry is realized in the Goldstone mode. The symmetry is spontaneously broken with the vacuum being invariant only under the diagonal subgroup $SU(2)_{L+R}$

$$\langle \bar{u}u \rangle_0 = \langle \bar{d}d \rangle_0 \neq 0 \tag{13.3}$$

and there are three (exactly) massless Goldstone bosons $\pi^{\pm,0}$. Although this result has not been proven rigorously in QCD, all indications are compatible with this expectation.

To express the above in a more suggestive notation we define the effective scalar and pseudoscalar fields $(\sigma, \boldsymbol{\pi})$ having the quantum numbers of quark bilinears

$$\sigma \sim \bar{q}q \quad \text{and} \quad \boldsymbol{\pi} \sim i\bar{q}\boldsymbol{\tau}\gamma_5 q \tag{13.4}$$

where $q = \binom{u}{d}$. Thus we have just the $SU(2) \times SU(2)$ σ-model considered in §5.3. The SSB condition (eqn (5.169)) and a generalization of eqn (5.155) immediately imply that the magnitude of SSB in (13.3) is given by the pion decay constant f_π

$$\langle \sigma \rangle_0 = v \approx 95 \text{ MeV} \tag{13.5}$$

namely

$$\frac{1}{2v^2}\langle \bar{q}q \rangle_0 = f_\pi = 95 \text{ MeV}. \tag{13.6}$$

f_π in turn must be related to the QCD scale parameter $\Lambda_c \simeq 200$ MeV as it is the only scale in our theory.

Now we turn on the electroweak interaction. From (13.1) the quark bilinears transform as

$$\phi_{\text{eff}} = \begin{pmatrix} \pi_1 + \mathrm{i}\pi_2 \\ \sigma + \mathrm{i}\pi_3 \end{pmatrix} \tag{13.7}$$

The VEV $\langle\sigma\rangle = f_\pi$ then breaks the $\mathrm{SU(2)_L} \times \mathrm{U(1)}$ symmetry down to the electromagnetic U(1) with π being eaten by the three gauge bosons to become $\mathrm{W}^\pm$ and Z. From eqns (11.77) and (13.5) we have

$$M_\mathrm{W} = \tfrac{1}{2} g f_\pi \simeq 30 \text{ MeV} \tag{13.8}$$

which is about three orders of magnitude smaller than the value (80 GeV) required in the standard model. But this simple mechanism of dynamical symmetry breaking does obtain the correct relation between M_W and M_Z. This is because we have an $\mathrm{SU(2)_{L+R}}$ symmetry remaining (i.e. the isospin of the strong interaction) which will give $M_\mathrm{W} = M_\mathrm{Z} \cos\theta_\mathrm{W}$ (i.e. $\rho = 1$) as explained in §11.2 (see especially eqn (11.83)).

To see the above results more explicitly consider the vacuum polarization diagrams of Fig. 13.1; they are derivable from the Lagrangian in (13.2). From the AA diagram in Fig. 13.1 we have

$$\pi^{ij}_{\mu\nu}(k) = \delta^{ij} \left(\frac{g}{2}\right)^2 (g_{\mu\nu} - k_\mu k_\nu / k^2) k^2 \pi(k^2). \tag{13.9}$$

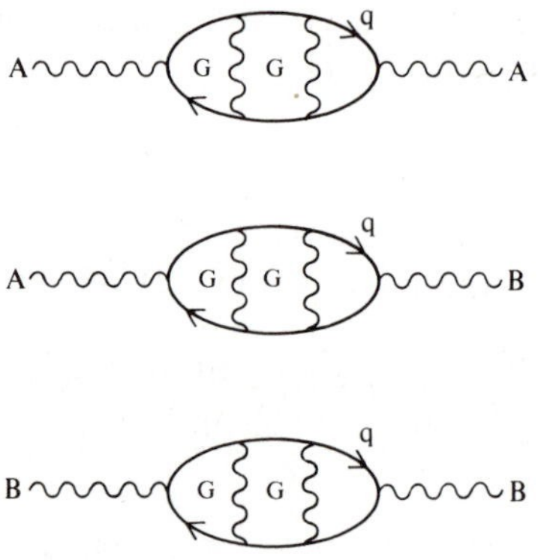

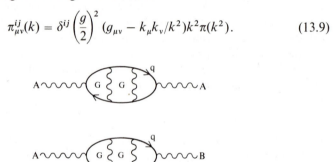

FIG. 13.1. Weak gauge boson vacuum polarization loops from the quark–gluon states.

Summing up the bubbles, the propagator of the gauge boson A_μ is modified from

$$\delta^{ij} \frac{(g_{\mu\nu} - k_\mu k_\nu / k^2)}{k^2}$$

to

$$\delta^{ij} \frac{(g_{\mu\nu} - k_\mu k_\nu / k^2)}{k^2[1 - g^2 \pi(k^2)/4]}. \tag{13.10}$$

Because chiral symmetry must be realized in the Goldstone mode QCD interaction implies a massless pole in the vacuum polarization bubble (see Fig.

13.2). This means that the vacuum polarization function in (13.9) is given by

$$\pi(k^2) = f_\pi^2/k^2. \tag{13.11}$$

Namely the contribution of the AA diagram in Fig. 13.1 to eqn (13.10) has a pole at

$$k^2 = M_{AA}^2 = g^2 f_\pi^2/4$$

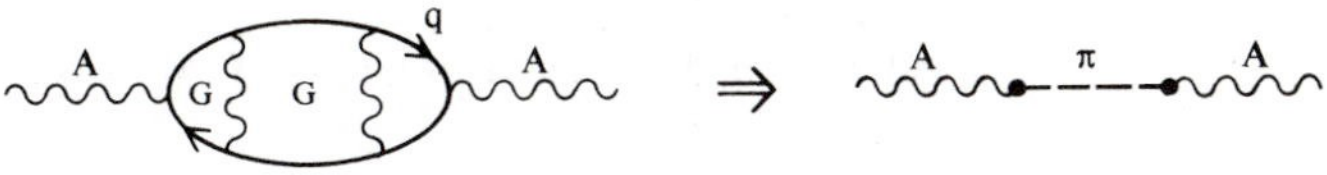

FIG. 13.2.

Similarly the AB diagram and the BB diagram in Fig. 13.1 have poles at

$$M_{AB}^2 = gg' f_\pi^2/4$$
$$M_{BB}^2 = g'^2 f_\pi^2/4.$$

Comparing this to eqn (11.78) we immediately obtain eqn (13.8) with $\rho = 1$ in (11.82). We should comment that if we had taken a three-family, six-quark flavour theory, the chiral symmetry would become SU(6) × SU(6) and there would be 35 Goldstone bosons. Three of them would become the longitudinal modes of $W^\pm$ and Z; the remaining 32 would acquire very small masses in higher orders of electroweak interactions i.e. they are pseudo-Goldstone bosons.

In any case we see that, in the right circumstances, QCD itself breaks down the electroweak gauge group in just the right pattern. However it falls short of being a realistic possibility because

(1) the scale is all wrong, we get $M_W \simeq 30$ MeV instead of 80 GeV as required;
(2) fermions remain massless.

It is relatively straightforward to overcome problem (1): we postulate the existence of another QCD-like interaction, called the *technicolour interaction* (TC), which has a scale parameter Λ_{TC} such that it produces the phenomenologically correct mass for W

$$M_W = \tfrac{1}{2} g_{TC} F_\pi \simeq 80 \text{ GeV} \tag{13.12}$$

Thus $F_\pi \simeq 250$ GeV and Λ_{TC} is of order 1 TeV. In other words the technicolour interaction, with a gauge group SU(3) for example, is in every way similar to QCD except that it produces condensate (or VEV) at energy three orders of magnitude larger than QCD. Thus there are fermions that carry technicolours (the techniquarks Q) with $SU(2)_L \times U(1) \times SU(3)_c \times SU(3)_{TC}$ transformation properties

$$\begin{pmatrix} U_L \\ D_L \end{pmatrix} \sim (\mathbf{2}, 1/3, \mathbf{1}, \mathbf{3})$$

$$U_R \sim (\mathbf{1}, 4/3, \mathbf{1}, \mathbf{3})$$

$$D_R \sim (\mathbf{1}, -2/3, \mathbf{1}, \mathbf{3}).$$

The familiar quarks and leptons are TC singlets. These technicolour quarks form bound states just like ordinary quarks under ordinary colour interactions. The TC chiral symmetry is also spontaneously broken with a magnitude given by the technipion decay constant

$$\frac{1}{2V^2}\langle\bar{Q}Q\rangle_0 = F_\pi \simeq 250\ \text{GeV}. \tag{13.13}$$

Thus we expect a rich spectrum of new particles in the TeV range.

Extended technicolour

The picture as developed so far still does not solve problem (2)—massless fermions. Quarks and leptons have separate chiral symmetries which remain unbroken. This situation is to be compared with that in the standard model where Yukawa couplings of elementary Higgs scalar to fermions produces fermion 'current masses' (see eqn (11.71)) as soon as $\langle\phi\rangle_0 \neq 0$. Thus we must find ways to produce effective Yukawa couplings between ordinary fermions and technimesons. One possible way to do this is to enlarge the technicolour group G_{TC} to an *extended technicolour* gauge group G_{ETC} by placing technifermions F (having, say, three technicolours) and ordinary fermions f (technicolour singlet) in a single irreducible representation of G_{ETC}.

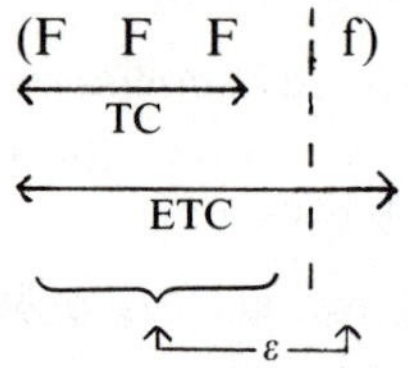

Extended technicolour (ETC) breaks down to technicolour at some energy scale μ. The $\mathcal{E}$ vector gauge boson being in ETC but not in TC acquires mass $M_\varepsilon \simeq g_{ETC}\mu$, and couples to currents of the form $\bar{F}\gamma_\mu f$. The effective four-fermion interactions mediated by $\mathcal{E}$ have the form

$$\frac{1}{2}\left(\frac{g_{ETC}}{M_\varepsilon}\right)^2(\bar{F}_L\gamma_\mu f_L)(\bar{f}_R\gamma^\mu F_R) \tag{13.14}$$

By a Fierz transformation we obtain

$$-\frac{1}{2\mu^2}[(\bar{F}F)(\bar{f}f) - (\bar{F}\gamma_5 F)(\bar{f}\gamma_5 f) + \ldots]$$

where we have used the relation $M_\varepsilon \simeq g_{ETC}\mu$. The condensation of technifermions $\langle\bar{F}F\rangle_0 \neq 0$ then produces a mass for the ordinary fermions,

$$m_f = \frac{1}{2\mu^2}\langle\bar{F}F\rangle_0. \tag{13.15}$$

Since $\langle\bar{F}F\rangle \sim (1\ \text{TeV})^3$, one needs $\mu \simeq 30$ TeV to produce $m_f \sim 1$ GeV.

As $\mathcal{E}$ vector bosons are $SU(3)_c \times SU(2)_L \times U(1)$ singlets, we need a set of technifermions for each ordinary fermion in order to give latter a mass. For one family of fermions, we need eight sets of technifermions,

$$\mathrm{U}_1,\ \mathrm{U}_2,\ \mathrm{U}_3,\ \mathrm{D}_1,\ \mathrm{D}_2,\ \mathrm{D}_3,\ \mathrm{N},\ \text{and}\ \mathrm{E},$$

where the subscripts are the QCD colour labels. The flavour symmetry of the TC interaction is then at least as large as $\mathrm{SU}(8)_{\mathrm{L}} \times \mathrm{SU}(8)_{\mathrm{R}}$. When this chiral symmetry is spontaneously broken, three of the Goldstone bosons combine with W^+, W^-, and Z and we are left with a large number of relatively light (on the TeV mass scale) pseudo-Goldstone bosons. This will allow for a possible early test of the TC approach to dynamical symmetry breaking.

Tumbling

To give masses to several families of ordinary fermions using one family of technifermions, we need to break the ETC gauge group down to TC in several successive stages. For example, a three-family model could be constructed by having the sequential breakdown

$$\mathrm{SU}(6)^{\mathrm{ETC}} \underset{\mu}{\rightarrow} \mathrm{SU}(5)^{\mathrm{E'TC}} \underset{\mu'}{\rightarrow} \mathrm{SU}(4)^{\mathrm{E''TC}} \underset{\mu''}{\rightarrow} \mathrm{SU}^{\mathrm{TC}}(3)$$

with fermions

$$\underbrace{\underbrace{(\mathrm{F}\quad \mathrm{F}\quad \mathrm{F}}_{\mathrm{TC}}\quad \mathrm{f}''\quad \mathrm{f}'\quad \mathrm{f})}_{\mathrm{ETC}}$$

transforming as a sextet of ETC. The first family $\mathrm{f} = \{\mathrm{e}, \nu, \mathrm{u}, \mathrm{d}\}$ would have mass $m_f \simeq (1/2\mu^2)\langle\bar{\mathrm{F}}\mathrm{F}\rangle$, the second family $\mathrm{f}' = \{\mu, \nu', \mathrm{c}, \mathrm{s}\}$ mass $m_{f'} \simeq (1/2\mu'^2)\langle\bar{\mathrm{F}}\mathrm{F}\rangle$, etc.

Can we have such a sequential SSB without an elementary Higgs scalar? Does one need to introduce a new TC interaction to perform each successive ETC symmetry breakdown? One possible way to avoid this proliferation of TC gauge groups is the idea of '*tumbling*'. This is a hypothesis as to the behaviour of unbroken asymptotically-free gauge groups with non-real fermion representation content. We adopt the convention under which all fermions are described by left-handed fields. For example QCD with n flavours has fermions $q_{\mathrm{L}i}$ and $(q^{\mathrm{c}})_{\mathrm{L}i} = i\gamma_2 q_{\mathrm{R}i}^{\dagger\mathrm{T}}$ $(i = 1, 2, \ldots, n)$ and we say its fermion content is $n(\mathbf{3} + \mathbf{3}^*)$ of $\mathrm{SU}(3)^{\mathrm{c}}$, which is real. It will be argued below that an asymptotically free gauge group with non-real fermion representation content breaks itself when the gauge coupling constants become large in the infrared region. Several successive breakings may occur before the fermion representation content become real under the unbroken subgroup, at which point the 'tumbling' stops. This allows us to establish a hierarchy of mass scales in an economical and natural way.

We illustrate the tumbling scheme by comparing QCD to an example of the SU(5) gauge group with fermions in the $\mathbf{5}^*$ and $\mathbf{10}$ representations. In the tensor notation a $\mathbf{5}^*$ is denoted by ψ^i and a $\mathbf{10}$ by the antisymmetric

$\psi_{ij} = -\psi_{ji}$ with $i, j = 1, 2, \ldots, 5$ (for more details, see §§4.3, 14.1, and 14.5). Possible scalar bound states composed of such fermion bilinears are

$$\begin{aligned} \mathbf{5}^* \times \mathbf{5}^* &= \mathbf{10}^*_{\mathrm{A}} + \mathbf{15}_{\mathrm{S}} \\ \mathbf{10} \times \mathbf{10} &= \mathbf{5}^* + \mathbf{45}^*_{\mathrm{A}} + \mathbf{50}_{\mathrm{S}} \\ \mathbf{5}^* \times \mathbf{10} &= \mathbf{5} + \mathbf{45}. \end{aligned} \tag{13.16}$$

This is to be compared to the QCD case

$$\begin{aligned} \mathbf{3} \times \mathbf{3} &= \mathbf{3}^* + \mathbf{6} \\ \mathbf{3}^* \times \mathbf{3}^* &= \mathbf{3} + \mathbf{6}^* \\ \mathbf{3} \times \mathbf{3}^* &= \mathbf{1} + \mathbf{8}. \end{aligned} \tag{13.17}$$

To have a Lorentz scalar the fermion bilinear must be a symmetric product. The bound state $\bar{\psi}_1\psi_2$ potential in the one-gauge-boson-exchange approximation is given by

$$V(r) = \frac{-\alpha(\mu)}{2r}(C_1 + C_2 - C) \tag{13.18}$$

where C_1 and C_2 are the Casimir operators of the constituents and C is that of the bound state. This corresponds to the familiar relation $-2\mathbf{T}_1 \cdot \mathbf{T}_2 = \mathbf{T}_1^2 + \mathbf{T}_2^2 - \mathbf{T}^2$ for $\mathbf{T} = \mathbf{T}_1 + \mathbf{T}_2$ in isospin symmetry. We then identify the *most attractive scalar channel* (MASC) for which $C_1 + C_2 - C$ is maximum. The basic proposition of the tumbling scheme states that, when we move from high energies to the low energy region where the running coupling becomes large, the MASC will condense first when

$$\alpha(\mu)(C_1 + C_2 - C) = O(1). \tag{13.19}$$

This is clearly consistent with QCD where the MASC is the singlet $\mathbf{1} \in \mathbf{3} \times \mathbf{3}^*$. For our SU(5) example, the MASC is the $\mathbf{5}^*$ contained in $\mathbf{10} \times \mathbf{10}$. We have

$$\langle 0|\psi_{ij}\psi_{kl}\varepsilon^{ijklm}|0\rangle \simeq \mu^3\delta_{m5}, \tag{13.20}$$

i.e. the difference is that here the condensate is not a singlet and it must single out a direction in the SU(5) space and hence breaks the SU(5) gauge symmetry: SU(5) → SU(4) at scale μ. The SU(5) gauge bosons that are not in SU(4) will pick up masses of order $g\mu$. The fermion multiplets split up according to

$$\begin{aligned} \mathbf{5}^* &= \mathbf{4}^* + \mathbf{1} \\ \mathbf{10} &= \mathbf{6} + \mathbf{4}. \end{aligned}$$

Just as in QCD, the fermions that participate in the condensate acquire a dynamical mass, closely related to 'constituent mass' of §5.5; $\langle \bar{\mathrm{q}}\mathrm{q} \rangle_0 = \mu^3$ implies a dynamical mass term $\sim \mu\bar{\mathrm{q}}\mathrm{q}$ for quarks. For the SU(5) case it is the sextet in $\mathbf{10}$ ($\psi_{ab} = -\psi_{ba}$, $a = 1, \ldots, 4$) that picks up masses. These dynamical masses are sharply energy-dependent and disappear at an energy scale above μ. Below μ the heavy particles decouple (Appelquist and Carazzone 1975); we have a theory based on the gauge group SU(4) with fermion representation content: $\mathbf{4} + \mathbf{4^*} + \mathbf{1}$. The gauge coupling constant now runs according to the SU(4) β-function. The new MASC is **4** and **4*** combining into a singlet. That channel condenses when

$$\alpha(\mu')(C_4 + C_{4^*} - C_1) = O(1).$$

The SU(4) gauge group remains unbroken and the tumbling stops because the condensate is an SU(4) singlet. The **4** and **4*** fermions acquire dynamical mass of order μ' which is the mass scale of the condensate of the SU(4) singlet. Only two mass scales are produced in this SU(5) example, but it is easy to construct examples that yield several more.

What is envisaged is a theory of low-energy (≤ 300 TeV) particle physics based on the gauge group

$$SU(2)_L \times U(1) \times SU(3)_c \times G_{ETC}.$$

The ETC fermion representation content is complex. From $\mu \sim 300$ TeV, ETC tumbles down to TC at μ' which is a mass scale of a few TeV. Under TC the fermion representation content is real. At $\mu'' \simeq 250$ GeV, the technifermion condensates break-down—$SU(2)_L \times U(1) \to U(1)_{em}$. The quarks and leptons are TC singlets but are ETC multiplets in common with the techniquarks and technileptons. They acquire masses through an effective Yukawa coupling as in eqn (13.15). In principle this scheme allows one to determine the quark and lepton masses, the Cabibbo angles, the $W^{\pm}$ and Z° masses, all in terms of a single parameter, the ETC gauge coupling constant at some given mass scale. In practice it is not easy to carry out this programme. Particularly one is concerned that some of the light pseudo-Goldstone bosons will induce quark (or lepton) flavour-changing neutral-current processes at much too large a rate. From our discussion it is also clear that in order to achieve the original goal of having a theory with very few adjustable parameters we must introduce a large number of particles. Thus an economical and elegant theory still eludes us.

13.2 Neutrino masses, mixings, and oscillations

We have already mentioned in §11.3 that the reason why there are no Cabibbo-like mixing angles in the lepton sector of the standard electroweak theory is neutrino mass degeneracy (i.e. all νs have the same mass—zero). This degeneracy means that there is no need to diagonalize the neutrino mass matrix (in fact no mass matrix to begin with). The absence of physically significant mixing angles brings about a set of conservation laws for the lepton flavours: the electron number, muon number, and the τ-lepton number. Processes such as $\mu \to e\gamma$ are forbidden. If neutrinos are not strictly massless, what are the phenomenological implications of small neutrino masses? Besides a nonzero rate for $\mu \to e\gamma$ (see calculation in the next section), we have the novel feature of neutrino oscillations.

Neutrino oscillations

This means that a beam of neutrinos (produced through weak interaction decays, corresponding to some definite flavour) can spontaneously change, or oscillate, into neutrinos of different flavour, e.g. $\nu_e \leftrightarrow \nu_\mu$, while travelling in vacuum. This property may explain the 'solar neutrino puzzle' (Davis, Harmer, and Hoffman 1968; Bahcall *et al.* 1980). While neutrino masses and mixings were discussed earlier by Sakata and his collaborators (see, for example, Maki *et al.* 1962; Nakagawa *et al.* 1963), these possibilities, particularly in connection with neutrino oscillation, have been studied most persistently by Pontecorvo (1958; 1968; Gribov and Pontecorvo 1969; Bilenky and Pontecorvo 1978).

If neutrinos are not massless, their mass matrix, just as in the case for quarks, will be nondiagonal and complex. One needs to transform it into a diagonal form by unitary rotations. Thus the mass eigenstates are different from gauge eigenstates

$$\nu_\alpha = \sum_i U_{\alpha i}\nu_i \tag{13.21}$$

where $\nu_\alpha = \nu_e, \nu_\mu, \nu_\tau$ are weak eigenstates and $\nu_i = \nu_1, \nu_2, \nu_3$ are mass eigenstates with mass eigenvalues m_1, m_2, and m_3. U is a unitary matrix which can be parametrized like the KM matrix for quark mixing angles (see eqn (12.40))

$$\begin{pmatrix} \nu_e \\ \nu_\mu \\ \nu_\tau \end{pmatrix} = \begin{pmatrix} c_1 & s_1c_3 & s_1s_3 \\ -s_1c_2 & c_1c_2c_3 - s_2s_3\,e^{i\delta} & c_1c_2s_3 + s_2c_3\,e^{i\delta} \\ -s_1s_2 & c_1s_2s_3 + c_2s_3\,e^{i\delta} & c_1s_2s_3 - c_2c_3\,e^{i\delta} \end{pmatrix} \begin{pmatrix} \nu_1 \\ \nu_2 \\ \nu_3 \end{pmatrix} \tag{13.22}$$

where $c_i \equiv \cos\theta_i$ and $s_i \equiv \sin\theta_i$. Of course there is no reason whatsoever to expect that these angles are in anyway similar to the Cabibbo–Kobayashi–Maskawa angles.

If at time $t = 0$, a beam of pure ν_e states is produced, say by $\pi^+ \to e^+\nu_e$ decays-in-flight, it is initially a superposition of mass eigenstates as

$$|\nu_e(0)\rangle = c_1|\nu_1\rangle + s_1c_3|\nu_2\rangle + s_1s_3|\nu_3\rangle. \tag{13.23}$$

The time evolution of a state is controlled by its energy eigenvalues. We assume that all neutrinos in the beam have a common fixed momentum p; then the mass eigenstates have energy eigenvalue

$$E_i^2 = p^2 + m_i^2 \tag{13.24}$$

and

$$|\nu_e(t)\rangle = c_1\,e^{-iE_1t}|\nu_1\rangle + s_1c_3\,e^{-iE_2t}|\nu_2\rangle + s_1s_3\,e^{-iE_3t}|\nu_3\rangle. \tag{13.25}$$

The probability of finding a ν_α at time t is given by $|\langle\nu_\alpha|\nu(t)\rangle|^2$. So for example the probability of finding a ν_e is

$$\begin{aligned} P_{\nu_e\to\nu_e}(t) &= 1 - 2c_1^2s_1^2c_3^2[1 - \cos(E_1 - E_2)t] \\ &\quad - 2c_1^2s_1^2s_3^2[1 - \cos(E_1 - E_3)t] - 2s_1^4s_3^2c_3^2[1 - \cos(E_2 - E_3)t]. \end{aligned} \tag{13.26}$$

For $p \gg m_i$, we have

$$E_i = (p^2 + m_i^2)^{\frac{1}{2}} \simeq p + (m_i^2/2p)$$

and

$$E_i - E_j = (m_i^2 - m_j^2)/2p. \tag{13.27}$$

It is convenient to define the *oscillation lengths*

$$l_{ij} = \frac{2\pi}{E_i - E_j} \simeq \frac{4\pi p}{|m_i^2 - m_j^2|} = 2.5\ \mathrm{m}\left[\frac{p\ (\mathrm{MeV})}{\Delta m^2\ (\mathrm{eV})^2}\right] \tag{13.28}$$

so that (13.26) may be stated as the probability of observing a ν_e at distance x from the source,

$$P_{\nu_e \to \nu_e}(x) = 1 - 2c_1^2 s_1^2 c_3^2 \left[1 - \cos\left(\frac{2\pi x}{l_{12}}\right)\right] - 2c_1^2 s_1^2 s_3^2 \left[1 - \cos\left(\frac{2\pi x}{l_{13}}\right)\right]$$
$$- 2s_1^4 s_3^2 c_3^2 \left[1 - \cos\left(\frac{2\pi x}{l_{23}}\right)\right]. \tag{13.29}$$

For $x \gg l_{ij}$ the harmonics are smoothed off and only the average intensity will be observable

$$\langle P_{\nu_e \to \nu_e}\rangle = 1 - 2c_1^2 s_1^2 - 2s_1^4 s_3^2 c_3^2. \tag{13.30}$$

The smallest average value possible is $\langle P_{\nu_e \to \nu_e}\rangle = 1/3$ corresponding to the special case $s_1^2 = 2/3$, $s_3^2 = 1/2$. Similarly,

$$\langle P_{\nu_e \to \nu_\mu}\rangle = 2c_1^2 s_1^2 c_2^2 + 2s_1^2 s_3^2 c_3^2 (s_2^2 - c_1^2 c_2^2) + 2s_1^2 s_2 s_3 c_1 c_2 c_3 \cos\delta (s_3^2 - c_3^2)$$
$$\langle P_{\nu_e \to \nu_\tau}\rangle = 2c_1^2 s_1^2 s_2^2 + 2s_1^2 s_3^2 c_3^2 (c_2^2 - c_1^2 s_2^2) + 2s_1^2 s_2 s_3 c_1 c_2 c_3 \cos\delta (s_3^2 - c_3^2). \tag{13.31}$$

Thus, one can in principle measure the leptonic CP angle δ through the transition rates of one neutrino to different neutrino flavours.

For the more general case

$$\nu_\alpha = \sum U_{\alpha i} \nu_i \qquad i = 1, 2, \ldots, N \tag{13.32}$$
$$|\nu_\alpha(t)\rangle = \sum U_{\alpha i} \nu_i\, e^{-iE_i t},$$

we can easily obtain the probability of finding flavour ν_β in a ν_α beam

$$\langle P_{\nu_\alpha \to \nu_\beta}\rangle = \sum_i |U_{\alpha i}|^2 |U_{\beta i}|^2 + \sum_{i \neq j} U_{\alpha i} U_{\beta i}^* U_{\alpha j}^* U_{\beta j} \cos\left(\frac{2\pi x}{l_{ij}}\right). \tag{13.33}$$

One can also show that the smallest possible value for the average probability $\langle P_{\nu_\alpha \to \nu_\beta}\rangle = 1/N$.

The magic of this oscillation phenomenon is of course intimately related to the quantum mechanical measurement theory. This possibility of one neutrino flavour ν_α spontaneously changing into another flavour reflects the uncertainty of energy–momentum measurement in these processes. For example, if we can pin down precisely which mass eigenstate ν_i is produced at the source, the oscillation pattern is destroyed, as the precision of momentum measurements required to do this will precisely prevent locating the source to an accuracy better than the oscillation length. Thus a more proper treatment should make use of wave packets, etc., but the same results are recovered (Kayser 1981).

The principal point one should keep in mind is that in order to have neutrino oscillation we must have nonzero (and non-degenerate) neutrino masses and mixing angles.

The question is then: can we have massive neutrinos in gauge theories? Before discussing the question of neutrino masses in SU(2) × U(1) electroweak theory we will first study briefly some of the special properties of neutrino mass terms.

Neutrino mass terms of Dirac and Majorana type

We adopt the conventions with respect to charge-conjugation (C) and helicity-projection operations

$$\psi^c = C\gamma^0\psi^* = i\gamma^2\psi^*, \qquad \bar{\psi}^c = \psi^T C$$
$$\psi_L = \tfrac{1}{2}(1-\gamma_5)\psi, \qquad \psi_R = \tfrac{1}{2}(1+\gamma_5)\psi. \tag{13.34}$$

We also use the notation

$$\psi_L^c \equiv (\psi_L)^c = \tfrac{1}{2}(1+\gamma_5)\psi^c = (\psi^c)_R. \tag{13.35}$$

The fermion mass terms connect left- and right-handed fields. A *Dirac-type mass* connects the L and R components of the same field,

$$\mathscr{L}_D = D(\bar{\psi}_L\psi_R + \bar{\psi}_R\psi_L) = D\bar{\psi}\psi. \tag{13.36}$$

Thus, the mass eigenstate is

$$\psi = \psi_L + \psi_R. \tag{13.37}$$

A *Majorana-type mass* connects the L and R components of conjugate fields. In the notation of (13.35), we can have

$$\mathscr{L}_{MA} = A(\bar{\psi}_L^c\psi_L + \bar{\psi}_L\psi_L^c) = A\bar{\chi}\chi \tag{13.38}$$

$$\mathscr{L}_{MB} = B(\bar{\psi}_R^c\psi_R + \bar{\psi}_R\psi_R^c) = B\bar{\omega}\omega. \tag{13.39}$$

The mass eigenstates are then self-conjugate fields

$$\chi = \psi_L + \psi_L^c, \qquad \chi^c = \chi$$
$$\omega = \psi_R + \psi_R^c, \qquad \omega^c = \omega. \tag{13.40}$$

These can be inverted to yield

$$\psi_L = \tfrac{1}{2}(1-\gamma_5)\chi; \qquad \psi_L^c = \tfrac{1}{2}(1+\gamma_5)\chi$$
$$\psi_R = \tfrac{1}{2}(1+\gamma_5)\omega; \qquad \psi_R^c = \tfrac{1}{2}(1-\gamma_5)\omega \tag{13.41}$$

when the γ_5 matrix is applied to the ψ, χ, and ω fields

$$\gamma_5\begin{pmatrix}\psi\\ \chi\\ \omega\end{pmatrix} = \begin{pmatrix}\psi'\\ \chi'\\ \omega'\end{pmatrix} = \begin{pmatrix}-\psi_L + \psi_R\\ -\psi_L + \psi_L^c\\ +\psi_R - \psi_R^c\end{pmatrix}. \tag{13.42}$$

Clearly this changes the sign of D, A, and B in eqns (13.36), (13.38), and (13.39). The fields ψ', χ', and ω' are interpreted as the correct mass eigenstates for the minus values of fermion masses.

When both Dirac and Majorana mass terms are simultaneously present we have

$$\begin{aligned}\mathscr{L}_{\rm DM} &= D\bar{\psi}_{\rm L}\psi_{\rm R} + A\bar{\psi}^{\rm c}_{\rm L}\psi_{\rm L} + B\bar{\psi}^{\rm c}_{\rm R}\psi_{\rm R} + \text{h.c.}\\ &= \tfrac{1}{2}D(\bar{\chi}\omega + \bar{\omega}\chi) + A\bar{\chi}\chi + B\bar{\omega}\omega\\ &= (\bar{\chi}, \bar{\omega})\begin{pmatrix} A & \tfrac{1}{2}D \\ \tfrac{1}{2}D & B\end{pmatrix}\begin{pmatrix}\chi\\ \omega\end{pmatrix}\end{aligned} \tag{13.43}$$

which can be diagonalized to yield two mass eigenvalues

$$M_{1,2} = \tfrac{1}{2}\{(A + B) \pm [(A - B)^2 + D^2]^{1/2}\} \tag{13.44}$$

corresponding to the Majorana mass eigenstates

$$\begin{aligned}\eta_1 &= (\cos\theta)\chi - (\sin\theta)\omega\\ \eta_2 &= (\sin\theta)\chi + (\cos\theta)\omega\end{aligned} \tag{13.45}$$

with

$$\tan 2\theta = D/(A - B). \tag{13.46}$$

We can easily invert (13.44) and (13.46) and obtain

$$\begin{aligned}D &= (M_1 - M_2)\sin 2\theta\\ A &= M_1\cos^2\theta + M_2\sin^2\theta\\ B &= M_1\sin^2\theta + M_2\cos^2\theta.\end{aligned} \tag{13.47}$$

Thus the most general mass term (13.43) for a four-component fermion field actually describes two Majorana particles with distinctive masses.

It is interesting to see that the usual four-component Dirac field formalism can be recovered in the limit of $A = B = 0$. When $\theta = \pi/4$, we have mass eigenstates $(\chi \pm \omega)/\sqrt{2}$ corresponding to eigenvalues $\pm D/2$. To flip the sign of the negative mass we need to apply a chiral transformation as in (13.42). Thus the fields

$$\frac{1}{\sqrt{2}}(\chi + \omega) = \frac{1}{\sqrt{2}}(\psi_{\rm L} + \psi^{\rm c}_{\rm L} + \psi_{\rm R} + \psi^{\rm c}_{\rm R}) \equiv \xi_1$$

and

$$\frac{1}{\sqrt{2}}(\chi' - \omega') = \frac{1}{\sqrt{2}}(-\psi_{\rm L} + \psi^{\rm c}_{\rm L} - \psi_{\rm R} + \psi^{\rm c}_{\rm R}) \equiv \xi_2 \tag{13.48}$$

have the same mass eigenvalue $\frac{1}{2}D$. Because of this degeneracy we are free to use any new combinations of fields so long they represent a rotation in the (ξ_1, ξ_2) plane. Thus,

$$\begin{aligned}\mathscr{L}_{\rm DM}(A = B = 0) &= \tfrac{1}{2}D(\bar{\xi}_1\xi_1 + \bar{\xi}_2\xi_2)\\ &= \tfrac{1}{2}D(\bar{\xi}'_1\xi'_1 + \bar{\xi}'_2\xi'_2).\end{aligned} \tag{13.49}$$

For the particular linear combination (i.e. another 45° rotation)

$$\xi_1' = \frac{1}{\sqrt{2}}(\xi_1 - \xi_2) = \psi_L + \psi_R$$

$$\xi_2' = \frac{1}{\sqrt{2}}(\xi_1 + \xi_2) = \psi_L^c + \psi_R^c, \tag{13.50}$$

it is obvious that (13.49) reduces to $D\bar{\psi}\psi$ with $\psi = \psi_L + \psi_R$.

Thus a Dirac fermion really corresponds to the degenerate limit of $A = B = 0$ in the more general case of two Majorana particles. Since the Majorana mass terms A (13.38) and B (13.39) violate the conservation of whatever additive quantum number that ψ carries, e.g. electric charge, all elementary fermions, except neutrinos, being charged must have $A = B = 0$. For neutrinos, Majorana mass terms violate lepton number by two units. The presence of such Majorana neutrino masses leads, for instance, to neutrinoless double β-decays: (Z − 1) → (Z + 1) + e + e, or kaon decays such as $K^- \to \pi^+ ee$. The quark diagram corresponding to these lepton-number nonconserving processes is shown in Fig. 13.3. (We do not consider theories with V + A charged currents coupled to heavy IVBs, W_R, where such an amplitude does not have to be proportional to a Majorana ν mass term.)

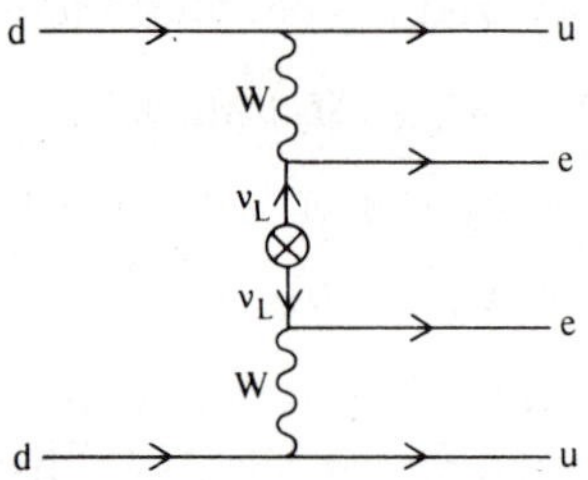

FIG. 13.3. Quark diagram for neutrinoless double β-decays. The symbol ⊗ corresponds to a Majorana-mass insertion $\nu_L \nu_L$ + h.c.

Possible neutrino mass sizes

At present the principal evidence for a possible non-zero neutrino mass comes from a single (not yet corroborated) experiment in tritium β-decay, $^3H \to {}^3He + e^- + \bar{\nu}_e$. It is found that the shape of the electron spectrum near the end-point can be interpreted as giving an m_{ν_e} ranging from 15 to 45 eV (Lyubimov *et al.* 1980). All other laboratory results only yield upper bounds on the m_νs

$$m_{\nu_e} < 60 \text{ eV},$$
$$m_{\nu_\mu} < 510 \text{ KeV},$$
$$m_{\nu_\tau} < 250 \text{ MeV}. \tag{13.51}$$

One can however obtain much more stringent bounds if one invokes cosmological theories. We shall give a brief account of this constraint.

Cosmological bound on neutrino masses. The standard model of *big-bang cosmology* is now generally accepted as being the correct theory of the universe. The principal ingredients are the cosmological principle and Einstein's field equations. It describes a universe which is spatially homogeneous and isotropic, and is expanding according to Hubble's law

$$V = HR \tag{13.52}$$

where V = velocity, R = distance, and H is the Hubble's constant ($\simeq 15$ km/s/million light years). Actually the basic features can be readily understood from Newtonian mechanics. Much like the notion of 'escape velocity', there is a *critical mass density* ρ_c corresponding to the precise cancellation of kinetic and (gravitational) potential energies

$$\rho_c = 3H^2/8\pi G \simeq 5 \times 10^{-30}\ \text{g/cm}^3 \tag{13.53}$$

where G is Newton's constant. If the mass density of the universe $\rho < \rho_c$ then it will continue to expand forever (the '*open universe*'); if $\rho > \rho_c$ then the expansion will slow down, eventually stop, and start contracting (the '*closed universe*'). The present bound on the total density of the universe, estimated from its age and deceleration, is

$$\rho^\circ \lesssim 4\rho_c\ (\simeq 2 \times 10^{-29}\ \text{g/cm}^3) \tag{13.54}$$

where the superscript zero denotes present time. On the other hand the observed galaxies and clusters can only account for a density ρ° that is at most a tenth of ρ_c. One can then speculate about the nonluminous masses in the universe.

The standard model of cosmology succeeds in providing a common meeting ground for a large variety of observational data. The matter of the universe is observed to reside primarily in the form of hydrogen atoms and a small portion in helium, and other light elements. If the galaxies are rushing apart from each other according to Hubble's law, they should have been closer in the past, making up a universe that was smaller and hotter. It is argued that such a hot universe would have 'cooked' all the hydrogen into heavier elements. However the small amounts of heavy elements observed in the universe is consistent with our picture of their being produced later in the galaxies. Namely they are not primordial in origin and the cosmological evolution should be such that no heavy elements are synthesized in the early universe. This can be the case if there was an intense field of radiation which would blast apart the heavy elements as soon as they were formed. Such an electromagnetic radiation, which was once in thermal equilibrium with matter, should still be present today, red-shifted by Hubble expansion to become a low-temperature black-body spectrum of background photons. This background radiation was indeed discovered and was measured to have an equivalent temperature of about 2.7 K. (This value is just compatible with a nucleosynthesis calculation of the above-mentioned cosmological helium abundance.) Using the standard black-body radiation formula relating number density n to temperature

$$n = T^3 \frac{\zeta(3)}{\pi^2} (g_B + \tfrac{3}{4} g_F) \tag{13.55}$$

with

$$\zeta(3) = \frac{1}{2} \int_0^\infty \frac{x^2\, dx}{e^x - 1} = 1 + \frac{1}{2^3} + \frac{1}{3^3} + \ldots \simeq 1.202$$

where g_B, g_F are the number of boson and fermion degrees of freedom, respectively (e.g. $g_B = 2$ for a photon gas (2 helicities), $g_F = 2$ for massless neutrinos, $g_F = 4$ for electrons, etc.), we have the present photon number density for $T = 2.7°K$

$$n_\gamma^0 \simeq 400\ \text{cm}^{-3}. \tag{13.56}$$

Similarly, the standard big-bang model suggests that the universe is immersed in a sea of primordial neutrinos. Because the neutrinos went out of thermal equilibrium before e^+e^- annihilation heated up the background radiation, the present cosmic black-body neutrinos should have a lower temperature (Peebles 1966)

$$\left(\frac{T_\nu^0}{T_\gamma^0}\right)^3 \simeq \frac{4}{11}. \tag{13.57}$$

or $T_\nu^0 \simeq 1.9$ K. Combining eqns (13.55)–(13.57), we have neutrino number density

$$n_\nu^0 \simeq \frac{3}{4}\left(\frac{T_\nu^0}{T_\gamma^0}\right)^3 n_\gamma^0 \simeq 110\ \text{cm}^{-3} \tag{13.58}$$

corresponding to a neutrino mass density,

$$\rho_\nu^0 \simeq \sum_i \frac{3}{11} n_\gamma^0 m_i = \sum_i 2m_i(\text{eV}) \times 10^{-31}\ \text{g/cm}^3 \tag{13.59}$$

where the sum is over neutrino mass eigenstates. The bound (13.54) can then be converted into a bound on the sum of neutrino masses (Gershtein and Zeldovich 1966; Cowsik and McClelland 1972; Szalay and Marx 1976)

$$\sum_i m_i < 100\ \text{eV}. \tag{13.60}$$

It is amusing to speculate that the universe is 'flat' with a mass density of precisely ρ_c and that the nonluminous masses all reside in the form of neutrino masses (i.e. ignoring all other possibilities such as magnetic monopoles, etc.); we then have

$$\sum_i m_i \simeq 25\ \text{eV}. \tag{13.61}$$

How can we have small neutrino masses? Fermion masses, whether coming from bare mass terms or Yukawa couplings (through Higgs mechanism), are arbitrary parameters in gauge theories due to infinite mass and coupling renormalization. Hence they are generally not calculable and have to be determined experimentally.

The neutrino masses, if nonzero, must be small when compared to all other mass scales. Theoretically one would like to have some way to understand their smallness. There are a few special cases where neutrino masses are not arbitrary and can be small.

(1) The most obvious possibility is that the neutrinos' mass terms are absent to the zeroth order and that higher-order radiative corrections give rise to masses (Georgi and Glashow 1973; Cheng and Li 1978). Consider the Dirac mass terms. If the particle content of the theory is such that there is neither the bare mass term $\bar{\psi}_L\psi_R$ (i.e. ψ_L and ψ_R transform differently) nor the Higgs-generated mass term $\bar{\psi}_L\Gamma^a\psi_R\langle\phi^a\rangle_0$ (because ϕ_a is absent or $\langle\phi_a\rangle_0 = 0$), then the diagram in Fig. 13.4 is finite as there are no possible counterterms to absorb the infinity. Such an m_ν in a theory where ν_R is assumed to transform nontrivially under the gauge group (and thus also to couple to the W-boson) should be calculable and is of the order $g^2 m_f$. A similar situation is also possible for Majorana-type mass terms.

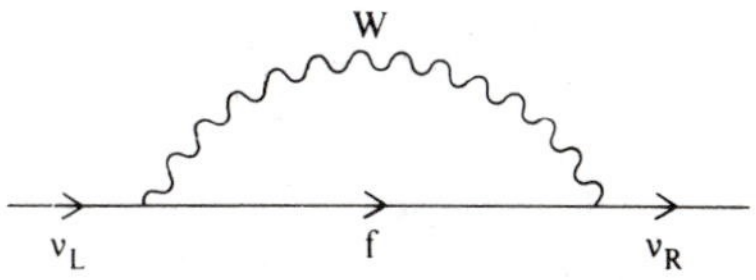

FIG. 13.4.

(2) As we shall discuss in the next chapter, grand unified theories (GUT) of the strong and electroweak interactions demand the existence in one theory of two vastly different mass–energy scales $\mu_{1,2}$ corresponding to two stages of symmetry breaking

$$G_{GUT} \underset{\mu_1}{\rightarrow} SU(3) \times SU(2) \times U(1) \underset{\mu_2}{\rightarrow} SU(3)_{colour} \times U(1)_{em}.$$

In certain situations this extremely small ratio $\mu_2/\mu_1 \lll 1$ (of order 10^{-13}) can be reflected in the fermion spectrum. The neutrino mass matrix of (13.43) may naturally have the form (Gell-Mann, Ramond, and Slansky 1979).

$$m_{ij} = \begin{pmatrix} 0 & m \\ m & M \end{pmatrix} \tag{13.62}$$

with $m \sim \mu_2$, $M \sim \mu_1$. The eigenvalues are $m_1 \simeq m^2/M$ and $m_2 \simeq M$; $(m_1/m_2) \simeq (\mu_2/\mu_1)^2 \lll 1$. Thus one ends up with one 'superheavy neutrino' and one extremely light particle which can be identified with the ordinary neutrino.

Neutrino masses in SU(2) × U(1) models

In the standard SU(2) × U(1) theory of electroweak interactions, neutrinos are massless because the simple Higgs structure of the theory leads to a global symmetry corresponding to lepton-number conservation which forbids the Majorana mass term $\bar{\nu}^c_L \nu_L$ *and* there are no ν_R that could combine with ν_L to form a Dirac mass term. In other words the masslessness of the neutrinos is related to the restricted particle content being considered in the standard model. When we consider a more complete unification it is inevitable that the number of fields will increase, with the consequential appearance of neutrino mass terms. (We shall see in §14.5 that the simplest grand unification scheme, the minimal SU(5) model, still has a global symmetry corresponding to the conservation of baryon number minus the lepton number $(B - L)$ which also forbids the Majorana mass terms.) However in GUT such as SO(10) $B - L$ is broken and neutrinos naturally acquire masses. Although recent interest in massive neutrinos is tied to the exploration of grand unification we shall study the $m_\nu \neq 0$ extensions of the minimal SU(2) × U(1) model directly (Cheng and Li 1980*a*) as all grand unified models must contain SU(2) × U(1) as a subtheory anyway and it is much simpler to work with the electroweak theory.

In the standard electroweak model, the lepton fields and Higgs scalar have the following SU(2) × U(1) transformation properties

$$l_L = \begin{pmatrix} \nu \\ l^- \end{pmatrix}_L \sim (\mathbf{2}, -1), \quad l_R^- \sim (\mathbf{1}, -2)$$
$$\Phi_i = \begin{pmatrix} \phi^+ \\ \phi^0 \end{pmatrix} \sim (\mathbf{2}, +1). \tag{13.63}$$

The first entries in parentheses on the right-hand sides of eqns (13.63) are the dimensions of SU(2) representations and the second entries are the U(1) hypercharge $Y = 2(Q - T_3)$. Lepton flavour indices are suppressed. Terms bilinear in lepton fields are

$$\bar{l}_L l_R \sim (\mathbf{2}, 1) \times (\mathbf{1}, -2) = (\mathbf{2}, -1)$$
$$\bar{l}^C_L l_L \sim (\mathbf{2}, -1) \times (\mathbf{2}, -1) = (\mathbf{1}, -2) + (\mathbf{3}, -2)$$
$$\bar{l}^C_R l_R \sim (\mathbf{1}, -2) \times (\mathbf{1}, -2) = (\mathbf{1}, -4). \tag{13.64}$$

With $\Phi \sim (\mathbf{2}, 1)$ only the Yukawa couplings $\bar{l}_R l_L \Phi$ + h.c. are present in the standard model and we have a global symmetry corresponding to lepton-number conservation.

There are many possible extensions of the standard model to give $m_\nu \neq 0$; they can be broadly categorized as

(1) Extension in the Higgs sector only;
(2) Extension in the lepton sector only;
(3) Extension in both Higgs and lepton sectors.

In case (1), other scalars, besides the doublet Φ, that can join the lepton bilinear in (13.64) to form SU(2) × U(1) gauge-invariant Yukawa couplings are triplet: $\mathbf{H} \sim (\mathbf{3}, 2)$, singly charged singlet: $h^+ \sim (\mathbf{1}, 2)$, and doubly charged singlet: $R^{++} \sim (\mathbf{1}, 4)$. For example the triplet $\mathbf{H}$ with $Y = 2$,

$$\boldsymbol{\tau}\cdot\mathbf{H} = \begin{pmatrix} H^{+} & \sqrt{2}\,H^{++} \\ \sqrt{2}\,H^{\circ} & -H^{+} \end{pmatrix} \tag{13.65}$$

gives rise to the additional Yukawa coupling and trilinear scalar coupling

$$f\bar{l}^{c}_{i\mathrm{L}} l_{j\mathrm{L}} \mathbf{H}(\varepsilon\boldsymbol{\tau})_{ij} + \mu\Phi_i\Phi_j\mathbf{H}^*(\varepsilon\boldsymbol{\tau})_{ij} + \text{h.c.} \tag{13.66}$$

where $\varepsilon = \mathrm{i}\tau_2$, with ε_{ij} and $(\varepsilon\boldsymbol{\tau})_{ij}$ being, respectively, antisymmetric and symmetric. When H develops a vacuum expectation value

$$\langle\boldsymbol{\tau}\cdot\mathbf{H}\rangle_0 = \begin{pmatrix} 0 & 0 \\ v_{\mathrm{H}} & 0 \end{pmatrix} \tag{13.67}$$

a Majorana mass term for the neutrino, $(v_{\mathrm{H}}f)\,\bar{\nu}^{c}_{\mathrm{L}}\nu_{\mathrm{L}}$, results. It should be noted that v_{H} also contributes to the W and Z masses with

$$\rho \equiv (M_{\mathrm{W}}/M_{\mathrm{Z}}\cos\theta_{\mathrm{W}})^2 = \frac{v_{\Phi}^2 + 2v_{\mathrm{H}}^2}{v_{\Phi}^2 + 4v_{\mathrm{H}}^2} \tag{13.68}$$

where v_{Φ} is the vacuum expectation value of the doublet. The phenomenological result (12.120), $\rho = 0.998 \pm 0.050$, restricts $(v_{\mathrm{H}}/v_{\Phi}) < 0.17$ if one standard deviation is allowed.

In case (2) the simplest scheme is obviously the addition of a neutral singlet, the right-handed neutrino ν_{R}. We then have the additional Lagrangian terms (for simplicity we first consider a one-flavour theory)

$$\mathscr{L}' = f\bar{l}_{i\mathrm{L}}\nu_{\mathrm{R}}\Phi_i^* + B\bar{\nu}^{c}_{\mathrm{R}}\nu_{\mathrm{R}} + \text{h.c.} \rightarrow D\bar{\nu}_{\mathrm{L}}\nu_{\mathrm{R}} + B\bar{\nu}^{c}_{\mathrm{R}}\nu_{\mathrm{R}} + \text{h.c.} \tag{13.69}$$

where $D = (1/\sqrt{2})v_{\phi}f$. A Majorana bare-mass term B is present because ν_{R} is totally neutral with respect to the SU(2) × U(1) group and we do not impose lepton-number conservation on the theory. Thus in this extension we are naturally led to consider neutrino mass terms of the Dirac and Majorana types (eqn (13.43)). Since the $A\bar{\nu}^{c}_{\mathrm{L}}\nu_{\mathrm{L}}$ term is absent (i.e. we do not introduce the triplet Higgs particles) the mass matrix is in fact of the form (13.62). If we make the plausible assumption that the Dirac mass term D is of the order of charge-2/3 quark masses and the Majorana mass term B is of the order of the energy scale when the GUT is broken into SU(3) × SU(2) × U(1), hence has a very large value. Then the weak eigenstates ν_{L} and ν_{R} are superpositions of two Majorana mass eigenstates with ν_{L} being predominantly a neutrino with a light mass $\simeq\frac{1}{4}D^2/B$ and a tiny admixture (D/B) of a superheavy neutrino with a mass $\simeq B$; the converse holds for ν_{R}.

One may also entertain the possibility that D and B are of comparable magnitude and, for whatever reason, are both small. In this case the mixing angle will not be small and the small mass eigenvalues allow for neutrino oscillations. But this is a type of oscillation different from those considered earlier (13.21) as the diagonalization of (13.43) (eqn (13.45)) means

$$\begin{pmatrix} \chi \\ \omega \end{pmatrix} = \begin{pmatrix} \cos\theta & -\sin\theta \\ \sin\theta & \cos\theta \end{pmatrix}\begin{pmatrix} \eta_1 \\ \eta_2 \end{pmatrix} \tag{13.70}$$

and the chiral projections of χ and ω (13.41) are just the usual weak eigenstate neutrino

$$\frac{1}{2}(1-\gamma_5)\begin{pmatrix}\chi\\ \omega\end{pmatrix}=\begin{pmatrix}\nu\\ \nu^c\end{pmatrix}_L \quad \text{or} \quad \begin{pmatrix}\nu_L\\ \nu_R\end{pmatrix}. \tag{13.71}$$

Thus we have neutrino–antineutrino oscillations $\nu_L \to (\nu^c)_L$ i.e. $\nu_L \to (\nu_R)^c$. Since the ν_Rs interact only through the superweak Yukawa couplings, such an oscillation with neutrinos turning into 'sterile particles' would be very different from the flavour oscillation discussed before (Barger *et al.* 1980; Cheng and Li 1980*a*). When the lepton-flavour family structure is taken into account, D and B become 3×3 matrices (for three lepton flavours). Each weak eigenstate neutrino will be a superposition of six Majorana mass eigenstates. One will then have both flavour-changing oscillations and particle–antiparticle oscillations: $\nu_{a_L} \leftrightarrow \nu_{b_L}$ and $\nu_{a_L} \leftrightarrow (\nu_b^c)_L$.

With the guiding principle that global symmetries should be determined by gauge invariance and renormalizability once the particle content is fixed, we find in this study of simple $m_\nu \neq 0$ extension of the minimal $SU(2) \times U(1)$ model such that, if $m_\nu \neq 0$, the physical neutrinos invariably turn out to be Majorana particles. This reflects mainly the point that if both Dirac and Majorana mass terms are present the mass eigenstates are still the self-conjugate Majorana fields. One can conclude that Majorana fields are natural representations of neutrino particles.

13.3 $\mu \to e\gamma$, an example of R_ξ-gauge loop calculations

Renormalizability is the principal feature of gauge theories of weak interaction. It will be instructive to go through the details of one non-trivial example of a higher-order weak radiative correction calculation. Here we study the neutrino-oscillation induced $\mu \to e\gamma$ as an illustration of R_ξ-gauge loop calculations. Thus we are working in the framework of the standard model (with one doublet of elementary Higgs scalars) modified by the presence of neutrino mass terms. We choose this example because, unlike the $s\bar{d} \to \bar{s}d$ box diagrams of §12.2, the would-be-Goldstone bosons are here not negligible in the leading order and must be included to obtain a gauge-invariant finite result. Also, the basic mechanism in Fig. 13.6 for $\mu \to e\gamma$ involves the distinctive non-Abelian features of the theory: the existence of trilinear couplings of gauge bosons. Of course it is a GIM-suppressed amplitude (by the neutrino mass difference). However, if observable (this unfortunately is not likely), its interpretation will not be complicated by the

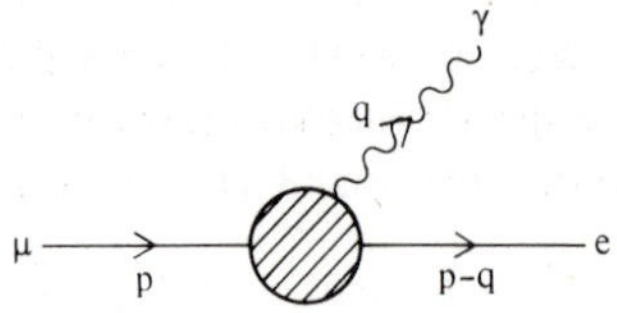

FIG. 13.5.

strong interaction effects as in the case of strangeness-changing neutral-current processes.

The $\mu \to e\gamma$ amplitude Fig. 13.5 can be written

$$T(\mu \to e\gamma) = \varepsilon^\lambda \langle e | J_\lambda^{em} | \mu \rangle \tag{13.72}$$

where $\varepsilon^\lambda(q)$ is the photon polarization. T has the Lorentz decomposition

$$\langle e | J_\lambda^{em} | \mu \rangle = \bar{u}_e(p-q)[iq^\nu \sigma_{\lambda\nu}(A + B\gamma_5) + \gamma_\lambda(C + D\gamma_5) + q_\lambda(E + F\gamma_5)]u_\mu(p) \tag{13.73}$$

where $A, B, \ldots, F$ are the invariant amplitudes. From electromagnetic gauge invariance

$$\partial^\lambda J_\lambda^{em} = 0, \tag{13.74}$$

we have the condition

$$-m_e(C + D\gamma_5) + m_\mu(C - D\gamma_5) + q^2(E + F\gamma_5) = 0$$

or

$$C = D = 0 \tag{13.75}$$

when the photon is on-shell $q^2 = 0$. And, since $\varepsilon^\lambda q_\lambda = 0$, the on-shell $\mu \to e\gamma$ amplitude is a magnetic transition

$$T(\mu \to e\gamma) = \varepsilon^\lambda \bar{u}_e(p-q)[iq^\nu \sigma_{\lambda\nu}(A + B\gamma_5)]u_\mu(p). \tag{13.76}$$

As (13.76) corresponds to a dimension-five operator, the on-shell $\mu \to e\gamma$ amplitude must be represented by a set of loop diagrams. They result in a finite amplitude since there can be no counterterm to absorb the infinities—the same reason why the $(g-2)$ anomalous magnetic moment of electron must be finite and calculable in QED.

As it is a lepton flavour-changing process it is strictly forbidden in the standard theory with $m_\nu = 0$ and muon number conservation. In this section we shall assume that neutrinos are not massless, and their mixings and oscillations mediate $\mu \to e\gamma$ (see Fig. 13.6). Initially we choose to work with, neutrinos having the more familiar (pure) Dirac mass terms,

$$\nu_\alpha = \sum_i U_{\alpha i} \nu_i \qquad \alpha = e, \mu, \tau;\ i = 1, 2, 3. \tag{13.77}$$

Extensions to the more general cases of Majorana neutrinos will be presented at the end of the section. The lowest-order diagrams contributing to the $\mu \to e\gamma$ amplitude in the R_ξ-gauge are displayed in Fig. 13.6. Since we know that the final amplitude must have the form of the magnetic transition (13.76), our strategy is to ignore all terms that cannot be reduced to the magnetic moment term. This means that there is no need to calculate the diagrams in Fig. 13.6(e) since they are all proportional to $\bar{u}_e \gamma^\lambda u_\mu$ and will be cancelled by terms of similar form coming from the diagrams in Fig. 13.6(a)–(d). As we shall make the approximation $m_e = 0$, the two $\mu \to e\gamma$ invariant amplitudes are equal,

$$A = B, \tag{13.78}$$

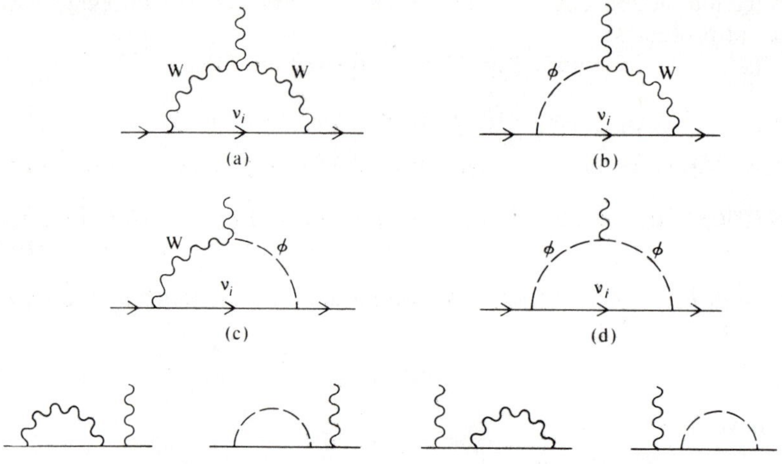

FIG. 13.6. One-loop diagrams for the $\mu \to e\gamma$ amplitude. The dashed lines represent the would-be-Goldstone bosons in renormalizable gauges. Thus, only (a) and the two graphs in (e) remain in the unitary gauge.

corresponding to the final-state electron being left-handed (it is e_L that couples to W and the helicity cannot be flipped in the $m_e = 0$ limit). Also, using Gordon decomposition (see Appendix A), we have

$$\begin{aligned} T &= A\bar{u}_e(p-q)(1+\gamma_5)\mathrm{i}\sigma_{\lambda\nu}q^\nu\varepsilon^\lambda u_\mu(p) \\ &= A\bar{u}_e(p-q)(1+\gamma_5)(2p\cdot\varepsilon - m_\mu\gamma\cdot\varepsilon)u_\mu(p). \end{aligned} \tag{13.79}$$

Thus in our calculation of the invariant amplitude A we need only to concentrate on the $p\cdot\varepsilon$ term. The momentum assignments of diagrams (a) to (d) in Fig. 3.6 are shown in Fig. 13.7.

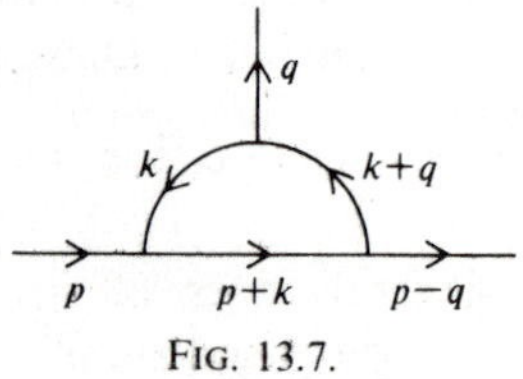

FIG. 13.7.

Diagram (a).

$$\begin{aligned} T_i(a) = -\mathrm{i}\int \frac{\mathrm{d}^4k}{(2\pi)^4}\Bigg[&\bar{u}_e(p-q)\left(\frac{\mathrm{i}g}{2\sqrt{2}}\right)U^*_{ei}\gamma_\mu(1-\gamma_5)\frac{\mathrm{i}}{\not p + \not k - m_i}\left(\frac{\mathrm{i}g}{2\sqrt{2}}\right) \\ &\times U_{\mu i}\gamma_\nu(1-\gamma_5)u_\mu(p)\Bigg][\mathrm{i}\Delta^{\nu\beta}(k)][\mathrm{i}\Delta^{\mu\alpha}(k+q)](-\mathrm{i}e)\Gamma_{\gamma\alpha\beta}\varepsilon^\gamma \end{aligned} \tag{13.80}$$

where $\Gamma_{\gamma\alpha\beta}\varepsilon^\gamma$ is the W-boson photon vertex: For $\Gamma_{\gamma\alpha\beta}(k_1, k_2, k_3) = [(k_3 - k_1)_\alpha g_{\gamma\beta} + (k_2 - k_3)_\gamma g_{\alpha\beta} + (k_1 - k_2)_\beta g_{\gamma\alpha}]$ with all the k_is flowing into the vertex, we have

$$\varepsilon^\gamma \Gamma_{\gamma\alpha\beta}(-q, k+q, -k) \equiv \Gamma_{\alpha\beta} = [(2k\cdot\varepsilon)g_{\alpha\beta} - (k+2q)_\beta \varepsilon_\alpha - (k-q)_\alpha \varepsilon_\beta] \tag{13.81}$$

where $\Delta^{\nu\beta}(k)$ and $\Delta^{\mu\alpha}(k+q)$ are the W-propagators in the R_ξ-gauge:

$$\Delta_{\mu\nu}(k) = -[g_{\mu\nu} - (1-\xi)k_\mu k_\nu/(k^2 - \xi M^2)]/(k^2 - M^2) \tag{13.82}$$

and M is the mass of W. U_{ei} and $U_{\mu i}$ are the mixing angles of (13.77). When we sum over all the diagrams corresponding to the three intermediate mass eigenstate v_is

$$\sum_i \left\{\frac{U^*_{ei}U_{\mu i}}{(p+k)^2 - m_i^2}\right\} = \sum_i U^*_{ei}U_{\mu i}\left\{\frac{1}{(p+k)^2} + \frac{m_i^2}{[(p+k)^2]^2} + \dots\right\}$$

$$= \sum_i \frac{U^*_{ei}U_{\mu i}m_i^2}{[(p+k)^2]^2} + \dots \tag{13.84}$$

the leading term vanishes, $\Sigma_i U^*_{ei}U_{\mu i} = 0$, reflecting the GIM cancellation mechanism. We then have

$$T(a) = \sum_i T_i(a) = \mathrm{i}c \int \frac{d^4k}{(2\pi)^4} \frac{R}{[(p+k)^2]^2} \tag{13.85}$$

where

$$c = \frac{g^2 e}{4} \sum_i U^*_{ei}U_{\mu i}m_i^2 \tag{13.86}$$

$$R = \Delta^{\nu\beta}(k)\Delta^{\mu\alpha}(k+q)N_{\mu\nu}\Gamma_{\alpha\beta} \tag{13.87}$$

with

$$N_{\mu\nu} = \bar{u}_e(p-q)\gamma_\mu(\not{p} + \not{k})\gamma_\nu(1-\gamma_5)u_\mu(p). \tag{13.88}$$

The W-boson propagator can be split up as

$$\Delta^{\mu\nu}(k) \equiv \Delta_1^{\mu\nu}(k) + \Delta_2^{\mu\nu}(k) \tag{13.89}$$

with

$$\Delta_1^{\mu\nu}(k) = -(g^{\mu\nu} - k^\mu k^\nu/M^2)/(k^2 - M^2)$$

$$\Delta_2^{\mu\nu}(k) = -(k^\mu k^\nu/M^2)/(k^2 - \xi M^2). \tag{13.90}$$

Substituting (13.89) into (13.87) we note that

$$\Delta_2^{\nu\beta}(k)\Delta_2^{\mu\alpha}(k+q)\Gamma_{\alpha\beta} = 0 \tag{13.91}$$

because

$$(k+q)^\alpha k^\beta \Gamma_{\alpha\beta} = 0. \tag{13.92}$$

We have for (13.85)

$$T(a) = \mathrm{i}c \int \frac{d^4k}{(2\pi)^4} \frac{1}{[(k+p)^2]^2}$$

$$\times \left\{\frac{S_1 - S_2 - S_3}{(k^2 - M^2)[(k+q)^2 - M^2]} + \frac{S_2}{(k^2 - \xi M^2)[(k+q)^2 - \xi M^2]}\right.$$

$$\left. + \frac{S_3}{(k^2 - M^2)[(k+q)^2 - \xi M^2]}\right\} \tag{13.93}$$

where

$$S_1 = \Gamma^{\mu\nu} N_{\mu\nu}$$
$$S_2 = (k^\lambda \Gamma^\mu_\lambda)(k^\nu N_{\mu\nu})/M^2$$
$$S_3 = [(k+q)^\lambda \Gamma^\mu_\lambda][(k+q)^\nu N_{\mu\nu}]/M^2. \tag{13.94}$$

Combining the denominators using Feynman parameters and shifting the integration variable, we get

$$T(a) = \mathrm{i}3!c \int \alpha_1\, \mathrm{d}\alpha_1\, \mathrm{d}\alpha_2 \left\{ \int \frac{\mathrm{d}^4k}{(2\pi)^4} \left[\frac{\tilde{S}_1 - \tilde{S}_2 - \tilde{S}_3}{(k^2-a^2)^4} + \frac{\tilde{S}_2}{(k^2-b^2)^4} + \frac{\tilde{S}_3}{(k^2-d^2)^4} \right\} \tag{13.95}$$

where

$$a^2 = (1-\alpha_1)M^2 + \dots$$
$$b^2 = [(1-\alpha_1-\alpha_2)\xi + \alpha_2]M^2 + \dots$$
$$d^2 = [(1-\alpha_1-\alpha_2) + \alpha_2\xi]M^2 + \dots. \tag{13.96}$$

Picking out only the $p\cdot\varepsilon$ terms,

$$S_1 \to \tilde{S}_1 = (p\cdot\varepsilon)[\bar{\mathrm{u}}_\mathrm{e}(1+\gamma_5)\mathrm{u}_\mu]2m_\mu[2(1-\alpha_1)^2 + (2\alpha_1 - 1)\alpha_2]$$
$$S_2 \to \tilde{S}_2 = -k^2(p\cdot\varepsilon)[\bar{\mathrm{u}}_\mathrm{e}(1+\gamma_5)\mathrm{u}_\mu](m_\mu/M^2) \times \{(3\alpha_2 - 1) + [2\alpha_1^2 - \alpha_1 + \alpha_2(2\alpha_1 - 1/2)]\}$$
$$S_3 \to \tilde{S}_3 = -k^2(p\cdot\varepsilon)[\bar{\mathrm{u}}_\mathrm{e}(1+\gamma_5)\mathrm{u}_\mu](m_\mu/M^2)[2\alpha_1^2 + \alpha_1 + (2\alpha_1 - 1/2)\alpha_2]. \tag{13.97}$$

After momentum integrations

$$\int \frac{\mathrm{d}^4k}{(2\pi)^4} \frac{1}{(k^2-a^2)^4} = \frac{\mathrm{i}}{96\pi^2 a^4}$$
$$\int \frac{\mathrm{d}^4k}{(2\pi)^4} \frac{k^2}{(k^2-a^2)^4} = \frac{-\mathrm{i}}{48\pi^2} \frac{1}{a^2} \tag{13.98}$$

and the integration over Feynman parameters α_1 and α_2, the contribution from diagram (a) to the invariant amplitude A is

$$A(a) = \frac{c}{64\pi^2} \frac{m_\mu}{M^4} \left[1 - \frac{1}{3} \frac{\ln \xi}{\xi - 1} + \left(\frac{1}{\xi - 1} \right) \left(\frac{\xi \ln \xi}{\xi - 1} - 1 \right) \right]. \tag{13.99}$$

Diagram (b).

$$T_i(b) = -\mathrm{i} \int \frac{\mathrm{d}^4k}{(2\pi)^4} \left[\bar{\mathrm{u}}_\mathrm{e}(p-q) \left(\frac{\mathrm{i}g}{2\sqrt{2}} \right) U^*_{\mathrm{e}i} \gamma_\lambda (1-\gamma_5) \frac{\mathrm{i}}{\not{p} + \not{k} - m_i} \times \left(\frac{-\mathrm{i}g}{\sqrt{2}M} \right) \frac{U_{\mu i}}{2} [m_i(1-\gamma_5) - m_\mu(1+\gamma_5)]\mathrm{u}_\mu(p) \right] \times [-\mathrm{i}\Delta^{\lambda\nu}(k+q)] \frac{\mathrm{i}}{k^2 - \xi M^2} (\mathrm{i}eM\varepsilon_\nu). \tag{13.100}$$

Splitting the W-propagator as in (13.89) and summing over all ν_i-diagrams using the approximation (13.84) and also

$$\sum_i \frac{U_{ei}^* U_{\mu i} m_i^2}{(p+k)^2 - m_i^2} \simeq \sum_i \frac{U_{ei}^* U_{\mu i} m_i^2}{(p+k)^2}, \tag{13.101}$$

we have

$$T(b) = \mathrm{i}c \int \frac{\mathrm{d}^4 k}{(2\pi)^4} \frac{N_\lambda}{[(p+k)^2]^2} \frac{1}{k^2 - \xi M^2} \times \left\{ \frac{[\varepsilon^\lambda - \varepsilon\cdot k(k+q)^\lambda/M^2]}{(k+q)^2 - M^2} + \frac{\varepsilon\cdot k(k+q)^\lambda/M^2}{(k+q)^2 - \xi M^2} \right\} \tag{13.102}$$

where c is the coupling constant of (13.86) and

$$N_\lambda = \bar{\mathrm{u}}_e(p-q)(1+\gamma_5)\gamma_\lambda[(p+k)^2 - m_\mu(\not{k} + m_\mu)]\mathrm{u}_\mu(p). \tag{13.103}$$

Combining denominators using Feynman parameters and shifting the integration variable, we have

$$T(b) = 6\mathrm{i}c \int \alpha_1\,\mathrm{d}\alpha_1\,\mathrm{d}\alpha_2 \int \frac{\mathrm{d}^4 k}{(2\pi)^4} \left[\frac{\tilde{N}_1 - \tilde{N}_2}{(k^2 - b^2)^4} + \frac{\tilde{N}_2}{(k^2 - a^2)^4} \right] \tag{13.104}$$

where a and b are given in (13.96). Again picking out $p\cdot\varepsilon$ terms in the numerator,

$$N_1 \equiv N_\lambda \varepsilon^\lambda \to \tilde{N}_1 = -2(p\cdot\varepsilon)[\bar{\mathrm{u}}_e(1+\gamma_5)\mathrm{u}_\mu]\alpha_2 m_\mu \tag{13.105}$$

$$N_2 \equiv N_\lambda (k+q)^\lambda k\cdot\varepsilon/M^2 \to \tilde{N}_2 = k^2(p\cdot\varepsilon)[\bar{\mathrm{u}}_e(1+\gamma_5)\mathrm{u}_\mu] \times \tfrac{1}{2}m_\mu[(1-4\alpha_1)(1-\alpha_1-\alpha_2) - (1-3\alpha_1)].$$

After the momentum integrations of (13.98) and Feynman parametric integrations we have the invariant amplitude from diagram (b)

$$A(b) = \frac{c}{64\pi^2}\frac{m_\mu}{M^4}\left[\frac{5}{6\xi} + \frac{4}{3}\frac{\ln\xi}{\xi - 1} - \frac{7}{3}\left(\frac{1}{\xi-1}\right)\left(\frac{\xi\ln\xi}{\xi-1} - 1\right)\right]. \tag{13.106}$$

Diagram (c). Following steps which are entirely similar to those followed in the calculation of diagram (b) we obtain

$$A(c) = \frac{c}{64\pi^2}\frac{m_\mu}{M^4}\left[\frac{5}{6\xi} - \frac{\ln\xi}{\xi - 1} + \frac{1}{3}\left(\frac{1}{\xi-1}\right)\left(\frac{\xi\ln\xi}{\xi-1} - 1\right)\right]. \tag{13.107}$$

Diagram (d).

$$T_i(d) = -\mathrm{i}\int \frac{\mathrm{d}^4 k}{(2\pi)^4} \left\{ \bar{\mathrm{u}}_e(p-q)\left(\frac{\mathrm{i}g}{\sqrt{2}M}\right)\frac{U_{ei}^*}{2}[m_i(1+\gamma_5) - m_e(1-\gamma_5)] \times \frac{\mathrm{i}}{\not{p} + \not{k} - m_i}\left(\frac{\mathrm{i}g}{\sqrt{2}M}\right)\frac{U_{\mu i}}{2}[m_i(1-\gamma_5) - m_\mu(1+\gamma_5)]\mathrm{u}_\mu(p)\right\} \times \frac{\mathrm{i}}{k^2 - \xi M^2}\frac{\mathrm{i}}{(k+q)^2 - \xi M^2}[\mathrm{i}e\varepsilon\cdot(2k+q)]. \tag{13.108}$$

Making the approximation $m_e = 0$ and using (13.101), we have

$$T(d) = \frac{-ic}{M^2}\int \frac{d^4k}{(2\pi)^4}[\bar{u}_e(p-q)(1+\gamma_5)ku_\mu(p)]$$
$$\times \frac{2k\cdot\varepsilon}{(p+k)^2}\frac{1}{k^2-\xi M^2}\frac{1}{(k+q)^2-\xi M^2}. \tag{13.109}$$

Combining denominators, we perform the momentum integration

$$T(d) = -4ic(p\cdot\varepsilon)[\bar{u}_e(1+\gamma_5)u_\mu]\int d\alpha_1\, d\alpha_2\, \alpha_1(\alpha_1+\alpha_2)\left(\frac{-i}{32\pi^2 a'^2}\right)\frac{m_\mu}{M^2} \tag{13.110}$$

where $a'^2 = (1-\alpha_1)\xi M^2$. After integrating over the parameters we have the invariant amplitude

$$A(d) = \frac{-c}{32\pi^2}\left(\frac{m_\mu}{M^4}\right)\frac{5}{6\xi}. \tag{13.111}$$

The contributions of diagrams (a) to (d) are summarized in Fig. 13.8.

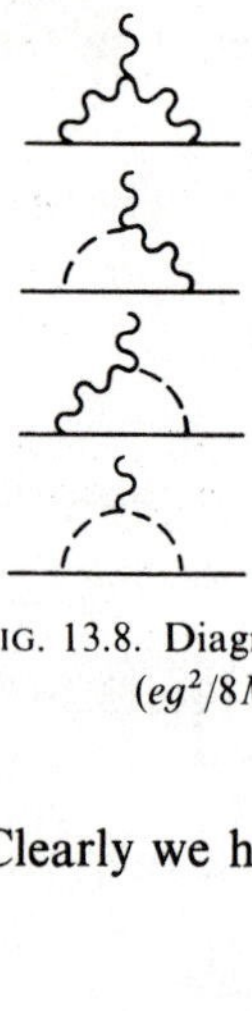

	R_ξ gauge	't Hooft gauge ($\xi = 1$)	unitary gauge ($\xi \to \infty$)
(a)	$1 - \frac{1}{3}f(\xi) + 2g(\xi)$	$\frac{5}{3}$	1
(b)	$\frac{5}{6\xi} + \frac{4}{3}f(\xi) - \frac{7}{3}g(\xi)$	1	0
(c)	$\frac{5}{6\xi} - f(\xi) + \frac{1}{3}g(\xi)$	0	0
(d)	$-\frac{5}{3\xi}$	$-\frac{5}{3}$	0

FIG. 13.8. Diagram contributions to the $\mu \to e\gamma$ invariant amplitude $A = B$ in units of $(eg^2/8M^2)(m_\mu/32\pi^2)\,\delta_\nu$ with $f(\xi) = \ln\xi/(\xi-1)$, $g(\xi) = [\xi f(\xi) - 1]/(\xi - 1)$.

Clearly we have for the total contribution of these diagrams,

$$A = B = e\frac{g^2}{8M^2}\frac{m_\mu}{32\pi^2}\delta_\nu \tag{13.112}$$

where δ_ν is the GIM suppression factor

$$\delta_\nu = \sum_i U^*_{ei}U_{\mu i}(m_i^2/M^2). \tag{13.113}$$

We note that in the final result the dependence on the gauge parameter ξ has been cancelled out. A straightforward calculation of the decay rate yields

$$\Gamma(\mu \to e\gamma) = \frac{m_\mu^3}{8\pi}(|A|^2 + |B|^2) \tag{13.114}$$

and using $\Gamma(\mu \to e\nu\bar{\nu}) = m_\mu^5 G_F^2/192\pi^3$ this can be converted into the branching ratio (Cheng and Li 1977; Petcov 1977; Marciano and Sandra 1977; Shrock and Lee 1977)

$$B(\mu \to e\gamma) \equiv \Gamma(\mu \to e\gamma)/\Gamma(\mu \to e\nu\bar{\nu}) \tag{13.115}$$
$$= \frac{3\alpha}{32\pi}\,\delta_\nu^2$$

where α is the fine structure constant and we have used $G_F/\sqrt{2} = g^2/8M^2$. Even if one takes a neutrino mass that saturates the cosmological bound 100 eV (eqn (13.60)), we still have a hopelessly small $B(\mu \to e\gamma) < 10^{-40}$.

We now comment very briefly on the situation when neutrinos are Majorana particles (Cheng and Li 1980*b*). Again we can distinguish two broad categories.

(1) The neutrino mass terms are pure Majorana $A\bar{\nu}_L^c\nu_L$. Each weak eigenstate is still a superposition of three (Majorana) mass eigenstates and the above $\mu \to e\gamma$ calculation should be carried through without modification;

(2) The neutrino mass terms have both Dirac and Majorana types. In particular when we have the situation of (13.69), $D\bar{\nu}_L\nu_R + B\bar{\nu}_R^c\nu_R$ + h.c. with $B \ggg D$. (Theoretically this is the most attractive possibility.) Each weak eigenstate is a superposition of six Majorana mass eigenstates: three are light with $m \sim D^2/B$; three are very heavy with $M \sim B(\gg M_W)$. This $\mu \to e\gamma$ calculation without the approximation of small intermediate neutrino masses has been carried out. As one would expect, the GIM cancellation is no longer complete because of the presence of superheavy neutrinos. However $B(\mu \to e\gamma)$ is still unobservably tiny as the admixture of heavy neutrinos in the ν_Ls is extremely small being $\theta \simeq D/B$ or $(m/M)^{\frac{1}{2}}$. One concludes that only in the case with the most general neutrino mass terms eqn (13.43) (i.e. having both Higgs triplet and ν_Rs) and with their magnitude coming out as $B \gg D \gg A$ is there any possibility of getting a neutrino-oscillation induced $B(\mu \to e\gamma)$ which is large enough, hopefully, to be detectable.

14 Grand unification

We have some confidence that elementary particle interactions down to distances as small as 10^{-16} cm are correctly described by SU(3) × SU(2) × U(1) gauge theory. This is the standard model of particle physics: quantum chromodynamics is the strong-interaction theory and the Glashow–Weinberg–Salam model provides the theory of weak and electromagnetic interactions.

Clearly it is desirable to have a more unified theory which can combine all these three interactions as components of a single force; a theory with only one gauge coupling. Georgi and Glashow (1974) showed that for the standard model, with the presently known quarks and leptons in each family, the simplest unification gauge group of colour and flavour is SU(5). However, because of the large differences in coupling strengths of strong and electroweak interactions, this unification would not become apparent until the energy scale of 10^{14} GeV was reached, corresponding to a distance scale of 10^{-28} cm.

Unification of colour and flavour was discussed earlier by Pati and Salam (1973). There are also other attractive models based on the gauge groups SO(10) (Georgi 1974; Fritzsch and Minkowski 1975) and E(6) (Gursey, Ramond, and Sikivie 1975). In this chapter we shall concentrate on the simplest grand unified model based on SU(5).

14.1 Introduction to the SU(5) model

We shall first give the basic structure of the SU(5) model. (For a detailed discussion see Buras, Ellis, Gaillard, and Nanopoulos 1978.) The original motivation of Georgi and Glashow for using the gauge group SU(5) will be presented at the end of this section.

A general representation under an SU(5) transformation may be expressed in tensor notation (see §4.3).

$$\psi^{ij\cdots}_{kl\ldots} \to U^{i}_{\ m} U^{j}_{\ n} U^{s}_{k} U^{t}_{l} \ldots \psi^{mn\ldots}_{st\ldots} \tag{14.1}$$

where all indices run from 1 to 5 and

$$[U]^{i}_{m} = [\exp(\mathrm{i}\alpha^{a}\lambda^{a}/2)]^{i}_{m}$$

is a 5 by 5 unitary matrix. $\{\lambda^{a}\}$, $a = 0, 1, \ldots, 23$, is a set of twenty-four 5 × 5 generalized Gell-Mann matrices, which are hermitian and traceless (so that the Us are unitary and have unit determinants), with normalization $\mathrm{tr}(\lambda^a\lambda^b) = 2\delta^{ab}$. For example,

$$\lambda^3 = \begin{bmatrix} 0 & & & & \\ & 0 & & & \\ & & 0 & & \\ & & & 1 & \\ & & & & -1 \end{bmatrix} \qquad \lambda^0 = \frac{1}{\sqrt{15}}\begin{bmatrix} 2 & & & & \\ & 2 & & & \\ & & 2 & & \\ & & & -3 & \\ & & & & -3 \end{bmatrix} \tag{14.2}$$

To obtain the SU(3) × SU(2) content of a representation we identify the first three of the SU(5) indices as the colour indices and the remaining two as $\mathrm{SU(2)_L}$ indices,

$$i = (\alpha, r) \quad \text{with} \quad \alpha = 1, 2, 3 \quad \text{and} \quad r = 4, 5. \tag{14.3}$$

Fermion content

In the standard SU(3) × SU(2) × U(1) model there are 15 left-handed (LH) two-component fermion fields in each family (generation). As we shall see, grand unification theories in general and the SU(5) model in particular do not shed light on the fermion replication problem (see §11.3). Thus for the sake of brevity, we shall initially write down the theory for the first (e) fermion family only

$$\begin{aligned} (\nu_e, e^-)_L &: (\mathbf{1}, \mathbf{2}) \\ e_L^+ &: (\mathbf{1}, \mathbf{1}) \\ (u_\alpha, d_\alpha)_L &: (\mathbf{3}, \mathbf{2}) \\ u_L^{c\alpha} &: (\mathbf{3^*}, \mathbf{1}) \\ d_L^{c\alpha} &: (\mathbf{3^*}, \mathbf{1}) \end{aligned} \tag{14.4}$$

where the SU(3) × SU(2) transformation properties of the family are displayed on the right-hand side of eqns (14.4). The superscript c stands for charge conjugate field (see eqns (13.34) and (13.35))

$$\begin{aligned} \psi^c &= C\gamma^o\psi^* = i\gamma^2\psi^* \\ (\psi_R)^c &= (\psi^c)_L \equiv \psi_L^c. \end{aligned} \tag{14.5}$$

The SU(3) × SU(2) contents of the simplest SU(5) representations are

$$\begin{aligned} &\text{The fundamental rep } \psi_i && \mathbf{5} = (\mathbf{3}, \mathbf{1}) + (\mathbf{1}, \mathbf{2}) \\ &\text{The fundamental conjugate rep } \psi^i && \mathbf{5^*} = (\mathbf{3^*}, \mathbf{1}) + (\mathbf{1}, \mathbf{2^*}) \\ &\text{The antisymmetric } \mathbf{5} \times \mathbf{5}\psi_{ij} = -\psi_{ji} && \mathbf{10} = (\mathbf{3^*}, \mathbf{1}) + (\mathbf{3}, \mathbf{2}) + (\mathbf{1}, \mathbf{1}). \end{aligned} \tag{14.6}$$

The less obvious points in the above decomposition are $\varepsilon^{\alpha\beta\gamma}\psi_{\alpha\beta} \sim (\mathbf{3^*}, \mathbf{1})$ and $\varepsilon_{rs}\psi^{rs} \sim (\mathbf{1}, \mathbf{1})$, where we have followed the index-labelling convention of

(14.3). A comparison of (14.4) and (14.6) shows that one family of fermions can be accommodated snugly in an SU(5) reducible representation of **5*** + **10**

$$\mathbf{5}^*: (\psi^i)_{\mathrm{L}} = (\mathrm{d}^{c1}\mathrm{d}^{c2}\mathrm{d}^{c3}\ \mathrm{e}^- \ -\nu_e)_{\mathrm{L}} \tag{14.7}$$

or

$$\mathbf{5}: (\psi_i)_{\mathrm{R}} = (\mathrm{d}_1\ \mathrm{d}_2\ \mathrm{d}_3\ \mathrm{e}^+ \ -\nu_e^c)_{\mathrm{R}} \tag{14.8}$$

and

$$\mathbf{10}: (\chi_{ij})_{\mathrm{L}} = \frac{1}{\sqrt{2}}\begin{bmatrix} 0 & \mathrm{u}^{c3} & -\mathrm{u}^{c2} & \mathrm{u}_1 & \mathrm{d}_1 \\ -\mathrm{u}^{c3} & 0 & \mathrm{u}^{c1} & \mathrm{u}_2 & \mathrm{d}_2 \\ \mathrm{u}^{c2} & -\mathrm{u}^{c1} & 0 & \mathrm{u}_3 & \mathrm{d}_3 \\ -\mathrm{u}_1 & -\mathrm{u}_2 & -\mathrm{u}_3 & 0 & \mathrm{e}^+ \\ -\mathrm{d}_1 & -\mathrm{d}_2 & -\mathrm{d}_3 & -\mathrm{e}^+ & 0 \end{bmatrix}_{\mathrm{L}} \tag{14.9}$$

It should be noted that in this one-family approximation gauge eigenstates are identical to mass eigenstates. We have thus labelled all the above gauge eigenstates by fields with definite masses. When the μ and τ fermion families are introduced (§14.5) they will also be assigned to **5*** and **10** representations. All the gauge eigenstates will then be some linear superpositions of mass eigenstates (see §11.3).

We have chosen the phase convention of having the neutrino field appear in **5*** (and **5**) with a minus sign. This conforms to our previous choice of $l^a = (\nu, \mathrm{e})_{\mathrm{L}}$ as a **2** under SU(2) and as being related to its conjugate $l_a = (\mathrm{e}, -\nu)_{\mathrm{L}}$ through the antisymmetric tensor $l^b = \varepsilon^{ab} l_a$. In the above we have only considered the SU(3) × SU(2) assignment. The correctness of the particle's U(1) charges will be shown below when we discuss 'charge quantization'.

Charge quantization

One immediate consequence of the SU(5) scheme is a very simple explanation for the experimentally observed charge quantization. In fact, whenever the unification gauge group is simple, the charge quantization will follow. This is because the eigenvalues of the generators of a simple non-Abelian group are discrete while those corresponding to the Abelian U(1) group are continuous. For example, in the SO(3) group of rotational symmetry the eigenvalues of the third component of the angular momentum can take only integer or half-integer values, while in the U(1) symmetry of translational invariance in time there is no restriction on the (energy) eigenvalues of the corresponding generator. Thus in SU(5) theory, since the electric charge Q is one of the generators, its eigenvalues are discrete and hence quantized.

Since the electric charge is an additive quantum number, Q must be some linear combination of the diagonal generators in SU(5). There are only four of these in SU(5) and, since Q commutes with all the $\mathrm{SU(3)}_{\mathrm{colour}}$ elements, we have

$$Q = T_3 + \frac{Y}{2} = T_3 + cT_0 \tag{14.10}$$

where T_3 and T_0 are the diagonal generators belonging to the SU(2) and U(1) subgroups, respectively; they are $\frac{1}{2}$ of eqn (14.2) for the fundamental representation. We see from eqn (14.10) that, for the SU(5) group, the formula for Q is not enlarged by terms beyond those already present in the Glashow–Weinberg–Salam model. The coefficient c which relates the operators Y and T_0 can be obtained by comparing in the fundamental representation the values of T_0 as given in (14.2) and the hypercharge values of the particles in (14.8), i.e. $Y(\mathbf{5}) = (-\frac{2}{3}, -\frac{2}{3}, -\frac{2}{3}, 1, 1)$,

$$c = -(5/3)^{1/2}. \tag{14.11}$$

The presence of the coefficient c signifies that the hypercharge Y is not properly normalized to be one of the SU(5) generators which have their scale fixed by the non-linear commutation relations

$$\left[\frac{\lambda^a}{2}, \frac{\lambda^b}{2}\right] = \mathrm{i}C^{abc}\frac{\lambda^c}{2}$$

with

$$\mathrm{tr}(\lambda^a\lambda^b) = 2\delta^{ab} \qquad a, b, c = 0, 1, \ldots, 23.$$

To check (14.11) we note that for the fundamental representation $T_a = \lambda_a/2$, eqns (14.10), (14.11), (14.2), and (14.8) yield

$$Q(\psi_i) = \begin{bmatrix} -1/3 & & & & \\ & -1/3 & & & \\ & & -1/3 & & \\ & & & 1 & \\ & & & & 0 \end{bmatrix} \equiv Q_i\,\delta_{ij}. \tag{14.12}$$

It follows that the fundamental conjugate representation **5*** has

$$Q(\psi^i) = -Q_i\,\delta_{ij} \tag{14.13}$$

From the transformation property shown in (14.1), a general tensor $\psi^{ij\cdots}_{k\ldots}$ has the same quantum number as $\psi^i\psi^j\psi_k\ldots$. Thus,

$$Q(\psi_{ij}) = Q_i + Q_j \tag{14.14}$$

$$Q(\psi_i^j) = Q_i - Q_j. \tag{14.15}$$

These quantities are, in a self-evident notation, the diagonal elements of Q for the **10** and **24** multiplets (also see eqn (4.141)).

The most interesting aspect of charge quantization as shown in (14.12) is the relation between colour SU(3) and charge. The traceless condition for the charge operator requires, for three colours,

$$3Q_{\mathrm{d}} + Q_{\mathrm{e}^+} = 0 \tag{14.16}$$

Quarks carry 1/3 of the lepton charge because they have three colours. Thus SU(5) theory provides a rational basis for understanding particle charges and the weak hypercharge assignment in the standard electroweak model is understood.

Anomaly cancellation

We should also check that the LH fermion assignment is free of anomaly. Of course, we already know from eqn (11.52) that each family is anomaly-free with respect to the $SU(3)_c \times SU(2)_L \times U(1)$ gauge bosons. Now we need to make certain that the fermionic couplings to all the remaining SU(5) gauge bosons do not introduce anomaly either.

In general the anomaly of any fermion representation R is proportional to the trace (see eqn (6.60); Georgi and Glashow 1972*b*; Banks and Georgi 1976; Okubo 1977)

$$\mathrm{tr}(\{T^a(R),\ T^b(R)\}T^c(R)) = \tfrac{1}{2}A(R)d^{abc} \tag{14.17}$$

where $T^a(R)$ is the representation matrix for R and the d^{abc}, are the totally symmetric constants appearing in the anticommutators (5.238)

$$\{\lambda^a, \lambda^b\} = 2d^{abc}\lambda^c. \tag{14.18}$$

We note that $A(R)$ in (14.17) which characterizes the anomaly of a given R is independent of the generators and is normalized to one for the fundamental representation. Thus we can use some simple generator to calculate $A(R)$ and to show that the anomalies cancel between the **5*** and **10** representations. Take, for example, $T^a = T^b = T^c = Q$; we immediately find

$$\frac{A(\mathbf{5}^*)}{A(\mathbf{10})} = \frac{\mathrm{tr}\, Q^3(\psi^i)}{\mathrm{tr}\, Q^3(\psi_{ij})}$$

$$= \frac{3(1/3)^3 + (-1)^3 + 0^3}{3(-2/3)^3 + 3(2/3)^3 + 3(-1/3)^3 + 1^3} = -1$$

and

$$A(\mathbf{5}^*) + A(\mathbf{10}) = 0. \tag{14.19}$$

Thus the combination **5*** and **10** of the fermion representation is anomaly-free.

Gauge bosons

The SU(5) adjoint representation A^i_j has dimension $5^2 - 1 = 24$ and an $SU(3) \times SU(2)$ decomposition

$$\mathbf{24} = (\mathbf{8}, \mathbf{1}) + (\mathbf{1}, \mathbf{3}) + (\mathbf{1}, \mathbf{1}) + (\mathbf{3}, \mathbf{2}) + (\mathbf{3}^*, \mathbf{2}). \tag{14.20}$$

Using the index-labelling convention of (14.3), we interpret this as

A^α_β; $(\mathbf{8}, \mathbf{1})$ are the $SU(3)_c$ gluons G^α_β of eqn (10.63)

A^r_s; $(\mathbf{1}, \mathbf{3})$ are the three SU(2) vector fields $\mathbf{W}$ with $W^\pm = (W^1 \mp W^2)/\sqrt{2}$ of eqns (11.38) and (11.76) with the notational change of $\mathbf{A}$ to $\mathbf{W}$.

$-\sqrt{\tfrac{1}{15}}A^\alpha_\alpha + \sqrt{\tfrac{3}{20}}A^r_r$; $(\mathbf{1}, \mathbf{1})$ is the U(1) B-field of eqn (11.39) corresponding to the diagonal element of A^i_j which does not belong to either SU(3) or SU(2).

The remaining 12 gauge fields have both SU(3) and SU(2) indices

$$A^r_\alpha: (\mathbf{3}, \mathbf{2}) \qquad A^\alpha_r: (\mathbf{3}^*, \mathbf{2}).$$

They are denoted as X, Y gauge bosons

$$A^r_\alpha = (\mathrm{X}_\alpha, \mathrm{Y}_\alpha), \qquad A^\alpha_r = \begin{pmatrix} \mathrm{X}^\alpha \\ \mathrm{Y}^\alpha \end{pmatrix}.$$

These vector particles, having non-zero triality with respect to the colour SU(3) group, carry fractional charges. According to eqn (14.15) we have

$$Q_\mathrm{X} = -4/3 \quad \text{and} \quad Q_\mathrm{Y} = -1/3. \tag{14.21}$$

If we put all the SU(5) gauge bosons in a 5×5 matrix form, $A = \Sigma^{23}_{a=0} A^a \lambda^a/2$,

$$A = \frac{1}{\sqrt{2}} \begin{bmatrix} & & & \mathrm{X}_1 & \mathrm{Y}_1 \\ & [\mathrm{G} - 2B/\sqrt{30}]^\alpha_\beta & & \mathrm{X}_2 & \mathrm{Y}_2 \\ & & & \mathrm{X}_3 & \mathrm{Y}_3 \\ \mathrm{X}^1 & \mathrm{X}^2 & \mathrm{X}^3 & \mathrm{W}^3/\sqrt{2} + 3\mathrm{B}/\sqrt{30} & \mathrm{W}^+ \\ \mathrm{Y}^1 & \mathrm{Y}^2 & \mathrm{Y}^3 & \mathrm{W}^- & -\mathrm{W}^3/\sqrt{2} + 3\mathrm{B}/\sqrt{30} \end{bmatrix}. \tag{14.22}$$

Spontaneous symmetry breaking is supposed to take place in two stages, characterized by two mass scales as given by the vacuum expectation values of two multiplets of Higgs fields $v_1 \gg v_2$

$$\mathrm{SU}(5) \xrightarrow{v_1} \mathrm{SU}(3) \times \mathrm{SU}(2) \times \mathrm{U}(1) \xrightarrow{v_2} \mathrm{SU}(3) \times \mathrm{U}(1).$$

This corresponds to the X, Y masses being superheavy $M_{\mathrm{X,Y}} \gg M_{\mathrm{W,Z}}$. The problem of spontaneous symmetry breaking in the SU(5) model will be studied in the next section.

Motivation for the SU(5) group

In constructing a grand unified theory of strong, weak, and electromagnetic interactions with one coupling, we seek a gauge group which is simple, or at most a product of identical simple groups (with the same coupling constants by some discrete symmetries). It should be large enough to contain the $\mathrm{SU}(3) \times \mathrm{SU}(2) \times \mathrm{U}(1)$ group of the standard model as a subgroup; thus it must be at least of rank 4. By this we mean that it must have at least four generators that can be simultaneously diagonalized, since it must contain the standard model which already has four mutually commuting generators: two from the colour SU(3) and two, the weak T_3 and the hypercharge Y, from $\mathrm{SU}(2) \times \mathrm{U}(1)$.

In fact one can make an exhaustive listing of Lie groups with a given rank (l)

$$A_l = \mathrm{SU}(l+1), \qquad B_l = \mathrm{O}(2l+1), \qquad C_l = \mathrm{Sp}(2l), \qquad D_l = \mathrm{O}(2l)$$

as well as (the exceptional groups) $E_{6,7,8}$, F_4, and G_2, the subscript indicating the rank. Thus the candidate $l = 4$ groups are SU(5), O(9), Sp(8), O(8), F_4, SU(3) × SU(3), and SU(2) × SU(2) × SU(2) × SU(2), etc.

However, all these possibilities, except SU(5) and SU(3) × SU(3), can be eliminated since they do not have complex representations. It is obvious from eqn (14.4) that we must have complex representations for fermions; in the standard model the fermion representations are not equivalent to their complex conjugates. The remaining possibility, SU(3) × SU(3), can quickly be eliminated since it cannot accommodate both the integrally and fractionally charged particles.

Thus SU(5), being rank-4, is the smallest group that can contain SU(3) × SU(2) × U(1) without introducing any new fermions. It has complex representations and can accommodate fractional charges. As we have seen, its anomaly-free reducible representation **5*** + **10** has just the right quantum numbers to fit one generation of leptons and quarks. Groups larger than SU(5) would necessarily involve particles other than the 15 two-component fermions with their familiar quantum numbers. In this sense SU(5) is the unique theory for the simplest grand unification scheme.

14.2 Spontaneous symmetry breaking and gauge hierarchy

The SU(5) model must have two mass scales: the X and W gauge boson masses. (For the remaining part of this chapter we shall often refer to the X and Y bosons collectively as X, and to the W and Z bosons collectively as W.) They characterize the spontaneous symmetry breakings (SSB) of SU(5) to SU(3) × SU(2) × U(1) and to SU(3) × U(1). Furthermore, as we shall see in the subsequent sections, we must have M_X larger than M_W by something like 12 orders of magnitude. Thus, there is a vast hierarchy of gauge symmetries.

As we discussed in Chapter 8, the Higgs phenomenon can provide masses for gauge bosons with elementary scalar fields developing a vacuum expectation value (VEV). This mechanism preserves the renormalizability of the theory. In order to have in one theory two mass scales, we must have two sets of scalars, and they develop vastly different VEVs which give rise to the desired gauge hierarchy. For the SU(5) model, this can be achieved with scalars in adjoint (H^i_j) and vector (ϕ_i) representations

$$\mathrm{SU}(5) \overset{\langle H\rangle}{\rightarrow} \mathrm{SU}(3) \times \mathrm{SU}(2) \times \mathrm{U}(1) \overset{\langle\phi\rangle}{\rightarrow} \mathrm{SU}(3) \times \mathrm{U}(1). \tag{14.23}$$

The general SU(5)-invariant fourth-order potential is

$$V(H, \phi) = V(H) + V(\phi) + \lambda_4(\mathrm{tr}\, H^2)(\phi^\dagger\phi) + \lambda_5(\phi^\dagger H^2 \phi). \tag{14.24}$$

with

$$V(H) = -m_1^2(\mathrm{tr}\, H^2) + \lambda_1(\mathrm{tr}\, H^2)^2 + \lambda_2(\mathrm{tr}\, H^4) \tag{14.25}$$

$$V(\phi) = -m_2^2(\phi^\dagger\phi) + \lambda_3(\phi^\dagger\phi)^2. \tag{14.26}$$

H is a traceless hermitian matrix. For simplicity we have imposed extra discrete symmetries: $H \to -H$ and $\phi \to -\phi$ to get rid of various cubic terms.

We first seek values of $H \neq 0$ that minimize the potential with $\phi = 0$. (This corresponds to the first stage of SSB.) Afterwards we look for deeper minima at small, but non-vanishing values of ϕ (the second stage of SSB).

It can be shown that, for $\lambda_2 > 0$ and $\lambda_1 > -7/30\ \lambda_2$, $V(H)$ has an extremum at $H = \langle H \rangle$ with

$$\langle H \rangle = v_1 \begin{bmatrix} 2 & & & & \\ & 2 & & & \\ & & 2 & & \\ & & & -3 & \\ & & & & -3 \end{bmatrix} \tag{14.27}$$

where

$$v_1^2 = m_1^2/(60\lambda_1 + 14\lambda_2). \tag{14.28}$$

We will only outline here the derivation given by Li (1974). First H can be diagonalized by a unitary transformation so that $H^i_j \to H_i\,\delta^i_j$, with $\Sigma_i H_i = 0$. The equations $\partial V/\partial H_i = 0$ are then cubic equations in the diagonal elements H_i, which can assume at most three different values. Detailed calculation then shows that, for the range of couplings indicated above, the potential V takes on its minimum when there are only two different values of H_i, which can be grouped as in eqn (14.27).

We shift the field to define a set of new scalars, which, in a matrix notation similar to that of (14.22), can be expressed

$$H' = H - \langle H \rangle = \left[\begin{array}{ccc|cc} & & & H_{X1} & H_{Y1} \\ & [H_8]^\alpha_\beta - 2H_0/\sqrt{30} & & H_{X2} & H_{Y2} \\ & & & H_{X3} & H_{Y3} \\ \hline H^\dagger_{X1} & H^\dagger_{X2} & H^\dagger_{X3} & \multicolumn{2}{c}{[H_3]^r_s + 3H_0/\sqrt{30}} \\ H^\dagger_{Y1} & H^\dagger_{Y2} & H^\dagger_{Y3} & & \end{array}\right] \tag{14.29}$$

Their mass spectrum can be obtained by a straightforward evaluation of the second derivative of V at $H = \langle H \rangle$. This is shown in Table 14.1.

TABLE 14.1

Scalar fields H: **24** *and* ϕ: **5** *in the minimal* $SU(5)$ *model. The massless* $(H_{X\alpha}, H_{Y\alpha})$ *are would-be-Goldstone bosons which are eaten by* (X^α, Y^α)

Scalar fields	SU(3) × SU(2) quantum numbers	[Mass]^2
$[H_8]^\alpha_\beta$	**(8, 1)**	$20\lambda_2 v_1^2$
$[H_3]^r_s$	**(1, 3)**	$80\lambda_2 v_1^2$
H_0	**(1, 1)**	$4m_1^2$
$(H_{X\alpha}, H_{Y\alpha})$	**(3, 2)**	0
$(H^\dagger_{X\alpha}, H^\dagger_{Y\alpha})$	**(3*, 2)**	0
$\phi_{t\alpha}$	**(3, 1)**	$-m_2^2 + (30\lambda_4 + 4\lambda_5)v_1^2$
ϕ_{dr}	**(1, 2)**	$-m_2^2 + (30\lambda_4 + 9\lambda_5)v_1^2$

Since H is in adjoint representation, the covariant derivative may be written

$$\begin{aligned} D_\mu H &= \partial_\mu H + ig[A_\mu, H] \\ &= D_\mu H' + ig[A_\mu, \langle H \rangle] \end{aligned} \tag{14.30}$$

where A^μ is the matrix gauge field of eqn (14.22). Thus the original 'kinetic energy' term $|D_\mu H|^2$ contains a factor of $g^2|[A_\mu, \langle H \rangle]|^2$. This is the mass term for gauge bosons.

From eqn (14.27) it is clear that $\langle H \rangle$ commutes with the generators of the subgroup SU(3) × SU(2) × U(1). Thus the mass terms for the G^α_β, W_r, B fields vanish. The X and Y gauge bosons acquire masses by combining with the would-be-Goldstone scalars H_X and H_Y. We have

$$M_X = M_Y = \sqrt{(25/2)}gv_1. \tag{14.31}$$

The fact H develops VEV also affects the ϕ system through the cross-couplings λ_4 and λ_5. The colour triplet ϕ_t: **(3, 1)** and the flavour doublet ϕ_d: **(1, 2)** components of $\phi = (\phi_{t\alpha}, \phi_{dr})$ acquire respective mass terms

$$\begin{aligned} m_t^2 &= -m_2^2 + (30\lambda_4 + 4\lambda_5)v_1^2 \\ m_d^2 &= -m_2^2 + (30\lambda_4 + 9\lambda_5)v_1^2. \end{aligned} \tag{14.32}$$

Thus after the first stage of SSB all non-zero values of particle masses are expected to be of the order v_1, namely M_X, which should be superheavy. For the second stage of SSB, SU(3) × SU(2) × U(1) → SU(3) × U(1) at 250 GeV, we need an SU(2) doublet scalars. We assume that 'somehow' m_d^2 of this doublet ϕ_d is vanishingly small (compared to v_1^2). Thus ϕ_d will survive to low energies (~250 GeV) as the superheavy particles (with masses of the order v_1) decouple. The relevant physics of the light particles is described by an SU(3) × SU(2) × U(1) invariant effective potential

$$V_{\text{eff}}(\phi_d) = -m_d^2\phi_d^\dagger\phi_d + \lambda_3(\phi_d^\dagger\phi_d)^2. \tag{14.33}$$

This is of course nothing but the Higgs potential for the Weinberg–Salam model (see Chapter 11). Gauge symmetry hierarchy means that ϕ_d develops a VEV much smaller than v_1,

$$\begin{aligned} \langle \phi_d \rangle &= \frac{1}{\sqrt{2}}\begin{pmatrix} 0 \\ v_2 \end{pmatrix} \\ v_2 &= (m_d^2/\lambda_3)^{\frac{1}{2}} \simeq 250 \text{ GeV}. \end{aligned} \tag{14.34}$$

We should emphasize that the above discussion has been carried out using the approximation of treating the two stages of SSB separately. A proper minimization of the combined system $V(H, \phi)$ would yield a VEV $\langle H \rangle$ that is slightly shifted from eqn (14.27). This would also break some of the mass degeneracies shown in Table 14.1 but all these corrections should be small, on the order of (v_2/v_1). Also the X and Y gauge bosons would have an $O(v_2)$ mass difference.

We shall see that the model requires $v_1 \gtrsim 10^{12}\, v_2$. It is challenging to understand the presence in a theory of such very different mass scales. A small m_d, as required by a small v_2 through eqn (14.34), appears to be unnatural since it receives a contribution from the large v_1 in eqn (14.32). Furthermore, there is the problem of a consistent implementation of such a large gauge hierarchy in the presence of radiative corrections (Gildener 1976), i.e., can we restrict light scalar particles to small masses to all orders in perturbation? The self-mass of an elementary scalar field is quadratically divergent; v_1 may be looked upon as the cut-off parameter for the low-energy effective theory of ϕ_d. One would expect large mass corrections from diagrams such as Fig. 14.1 with $\Delta m_d^2 \propto g^4 v_1^2$. With such corrections it appears that a small scalar mass can be obtained only by fine tuning of parameters at each order of perturbation. This is usually referred to as the *gauge-hierarchy problem*. For a review see, for example, Cheng and Li (1980*c*). The issue is whether one can find a model that avoids this unnatural feature. Such a model, if it exists, is likely to have additional symmetries. It is often suggested that they may be some form of supersymmetry which forbids quadratically divergent scalar masses. (Here the supersymmetric fermion partners to the gauge bosons and Higgs scalars must be included. The leading divergence in diagrams such as Fig. 14.1 is then cancelled by the corresponding fermion loops.) No realistic model has yet been constructed and the problem of a satisfactory implementation of gauge hierarchy is still an open one.

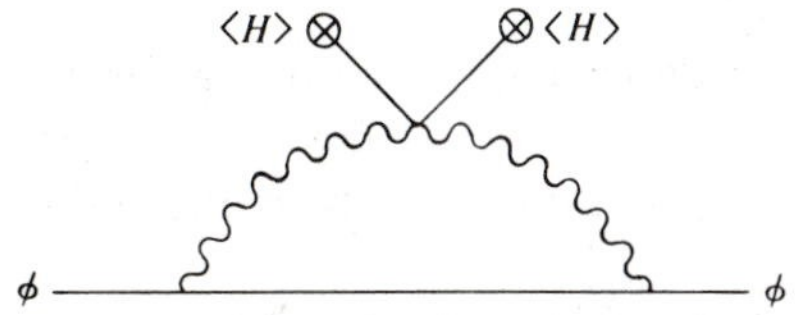

FIG. 14.1. A self-energy diagram for the ϕ scalar particle.

14.3 Coupling constant unification

The standard model describes the strong, weak, and electromagnetic interactions in the energy range $\lesssim 10^2$ GeV with the three different coupling constants: g_s, g, and g' for the gauge groups SU(3), SU(2), and U(1), respectively. Thus there is no real explanation of the different strengths displayed by the three interactions. One of the great virtues of GUT (in fact its original motivation) is that it provides such an understanding (Georgi, Quinn, and Weinberg 1974).

The grand unified theory, by definition, has only one coupling constant associated with the unified gauge group. This same coupling constant should apply to the subgroups as well. The possibility of different couplings for the various subgroups at low energies arises because of spontaneous symmetry breaking; the X, Y gauge bosons (or X bosons for short) of SU(5) acquire masses and decouple from coupling-constant renormalizations. This decoupling clearly will have different effects on the radiative corrections of different subgroup couplings, giving rise to different effective couplings at low energies through the energy dependence determined by the renormalization group equation illustrated in Chapter 3. The decoupling of heavy X bosons is reflected in the unequal renormalization group coefficients for subgroup couplings. Below the unification scale [M_X in the case of SU(5)] they evolve differently, giving rise to the observed disparities of interaction strengths.

Before proceeding with detailed analysis we shall first make some qualitative statements about the coupling constant unification, which is represented graphically in Fig. 14.2.

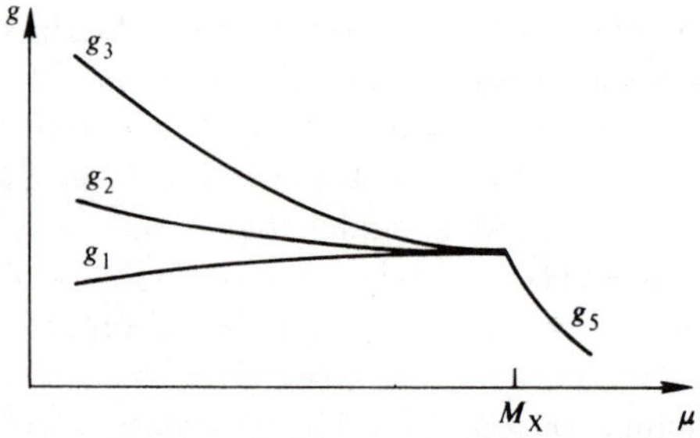

FIG. 14.2. Coupling constant unification. For the scale $\mu > M_X$, where the X boson mass can be neglected, the couplings corresponding to different subgroups remain unified. For the scale $\mu < M_X$, the couplings evolve in a pattern determined by the size of the gauge group.

(A) Since the energy dependence of coupling constants is only logarithmic and, since, in the energy region $\approx 10^2$ GeV, g_s, g, and g' are quite different, the unification scale M_X is expected to be many orders of magnitude larger than 10^2 GeV.

(B) From the analysis of the renormalization group equations, we have learned the following facts. For the non-Abelian gauge groups the coupling constant decreases with increasing energy and the rate of decrease is greater for larger groups. For the Abelian group the coupling constant increases with energy. Thus at energies less than M_X the ordering of coupling constants has to be $g_s > g > g'$. This pattern is in fact compatible with experimental observation.

(C) Furthermore, below M_X the three coupling trajectories in Fig. 14.2 must have just the correct relative strengths in order to all intersect at one point when reaching M_X. This implies a non-trivial consistency condition among g_s, g, and g'. This relation predicts the Weinberg angle θ_W, which relates g and g', in terms of the fine structure constant α and the QCD strong interaction coupling constant α_s (see eqn (14.57) below).

We now proceed with the detailed analysis. First we must study the relation between the coupling constants of SU(3) × SU(2) × U(1) and SU(5) at the unification scale M_X. Consider the covariant derivatives for these two groups; for definiteness we shall display those operating on the fundamental representations of the groups

$$D_\mu = \partial_\mu + ig_s \sum_{\alpha=1}^{8} G_\mu^\alpha \lambda^\alpha/2 + ig \sum_{r=1}^{3} W_\mu^r \tau^r/2 + ig' B_\mu Y/2 \qquad (14.35)$$

and

$$\mathrm{D}_\mu = \partial_\mu + \mathrm{i}g_5 \sum_{a=0}^{23} A_\mu^a \lambda^a/2. \tag{14.36}$$

The definition of coupling constants depends on the normalization of the generators. For the non-Abelian groups these normalizations are fixed by the nonlinear commutation relations of Lie algebra. Thus the Gell-Mann matrices $\{\lambda^\alpha\}$, along with their SU(5) generalized version $\{\lambda^a\}$, and the Pauli matrices $\{\tau^r\}$ are all similarly normalized. $\mathrm{tr}\{\lambda^a\lambda^b\} = 2\delta^{ab}$, etc. And we have

$$g_5 = g_3 = g_2 = g_1 \tag{14.37}$$

with

$$g_3 \equiv g_\mathrm{s}, \qquad g_2 \equiv g. \tag{14.38}$$

The coupling g_1 is that of the Abelian U(1) subgroup. Thus

$$\mathrm{i}g_1\lambda^\circ A_\mu^\circ = \mathrm{i}g' Y B_\mu$$

A_μ° is identified with the B_μ gauge field. One notes that the U(1) algebra does not provide any (nonlinear) restriction on its generator, and Y and λ° may be normalized differently. We can determine this difference in normalizations by noting that for the particle assignment of **5** in (14.7) we must have weak hypercharge

$$Y = \begin{bmatrix} -2/3 & & & & \\ & -2/3 & & & \\ & & -2/3 & & \\ & & & 1 & \\ & & & & 1 \end{bmatrix}. \tag{14.39}$$

Comparing (14.39) to the λ° of (14.2), we have, as in (14.11),

$$Y = -(5/3)^{\frac{1}{2}}\lambda^\circ, \qquad g' = -(3/5)^{\frac{1}{2}}g_1 \tag{14.40}$$

since $g'Y = g_1\lambda^\circ$. Eqns (11.80), (14.37), and (14.40) can be translated into the statement that

$$\sin^2\theta_\mathrm{W} \equiv g'^2/(g^2 + g'^2) = 3/8. \tag{14.41}$$

This and the equalities in (14.37) are valid in the SU(5) limit; i.e. for the energy scale $\mu > M_\mathrm{X}$. Now we need to study the regime $\mu < M_\mathrm{X}$. The evolution of the SU(n) gauge coupling constant is controlled by the renormalization group equation (see eqns (10.77)–(10.79))

$$\frac{\mathrm{d}g_n}{\mathrm{d}(\ln\mu)} = -\,b_n g_n^3 \tag{14.42}$$

where

$$b_n = (11n - 2N_\mathrm{F})/48\pi^2 \qquad \text{for } n \geq 2 \tag{14.43}$$

and

$$b_1 = -N_\mathrm{F}/24\pi^2. \tag{14.44}$$

Thus,

$$b_n - b_1 = 11n/48\pi^2. \tag{14.45}$$

We have ignored the contribution coming from the Higgs scalar. N_F is the number of quark flavours ($N_F = 6$ for a three-family theory) and thus the fermion effects on the relative rate of coupling constant evolution cancel out. Considering that only quarks couple to SU(3) gluons and that both quarks and leptons couple to SU(2) and U(1) gauge bosons, one may find it surprising that the fermion contributions should be equal in all gauge coupling renormalizations. We will now digress to explain this situation.

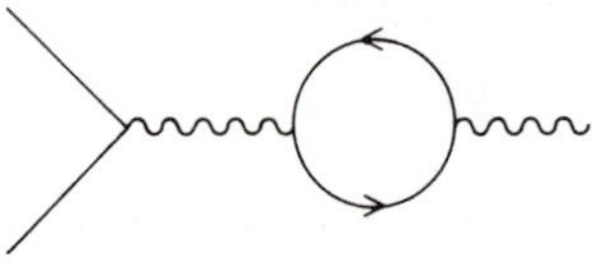

FIG. 14.3. Coupling-constant renormalization due to the fermion loop.

The fermions contribute through the loop diagram in Fig. 14.3 which is proportional to $F_n = N_m g_n^2 \operatorname{tr}(T_i T_j)$, where N_m is the number of multiplets of (two-component) fermions coupled to the gauge bosons and all representation matrices are similarly normalized with $\operatorname{tr}(T_i T_j) = \frac{1}{2}\delta_{ij}$. We shall first explicitly work out the F_n factor for each gauge group. For SU(3), N_m is the number of colour triplets. Since both quark and antiquark couple to gluons, we have $N_m = 2N_F$ and $F_3 = N_F g_3^2 \delta_{\alpha\beta}$. For SU(2), N_m is the number of doublets. Since for each lepton doublet there are three coloured quark doublets, we have $N_m = \frac{1}{2}(1+3)N_F$ and $F_2 = N_F g_2^2 \delta_{ab}$. For U(1), a straightforward summation of (squared) weak hypercharges with proper rescaling of the normalization according to (14.40) also yields $F_1 = N_F g_1^2$. From the SU(5) viewpoint, $F_3 = F_2 = F_1$ will no longer appear as a coincidental fact: all fermions form a complete (reducible) representation of the group, (**5*** + **10**), and all representation matrices for each subgroup factor are equally normalized; in addition to this, is the requirement that all members of this representation obtain comparable masses $\ll M_X$. This is to be contrasted with the situation for the gauge bosons, which end up with two very different mass scales: $M_W \ll M_X$ even though they form a complete SU(5) representation (the adjoint rep). And this split in the masses produces the different renormalization effects to the subgroup coupling constants.

We now return to the solutions of eqn (14.42). For g_1, g_2, and g_3, we have

$$g_1^{-2}(\mu) = g_1^{-2}(\mu_0) + 2b_1 \ln(\mu/\mu_0) \tag{14.46}$$

$$g_2^{-2}(\mu) = g_2^{-2}(\mu_0) + 2b_2 \ln(\mu/\mu_0) \tag{14.47}$$

$$g_3^{-2}(\mu) = g_3^{-2}(\mu_0) + 2b_3 \ln(\mu/\mu_0). \tag{14.48}$$

We can express the low-energy couplings in terms of more familiar parameters by using (11.80) $\tan\theta_W = g'/g$, (11.93) $e = g\sin\theta_W$, and (14.40)

$$g_1^2(\mu)/4\pi = (5/3)\alpha(\mu)/\cos^2\theta_W \tag{14.49}$$

$$g_2^2(\mu)/4\pi = \alpha(\mu)/\sin^2\theta_W \tag{14.50}$$

and

$$g_s^2(\mu)/4\pi \equiv \alpha_s(\mu). \tag{14.51}$$

We can recast eqns (14.46)–(14.48) in the form

$$\alpha_s^{-1}(\mu) = \alpha_5^{-1} + 8\pi b_3 \ln(\mu/M_X) \tag{14.52}$$

$$\alpha^{-1}(\mu)\sin^2\theta_W = \alpha_5^{-1} + 8\pi b_2 \ln(\mu/M_X) \tag{14.53}$$

$$(3/5)\alpha^{-1}(\mu)\cos^2\theta_W = \alpha_5^{-1} + 8\pi b_1 \ln(\mu/M_X) \tag{14.54}$$

where we have used eqn (14.37).

$$g_1(M_X) = g_2(M_X) = g_3(M_X) = g_5 \quad \text{and} \quad g_5^2/4\pi \equiv \alpha_5. \tag{14.55}$$

Taking the linear combination [2 × eqn (14.52) – 3 × eqn (14.53) + eqn (14.54)], we have

$$\begin{aligned} &2\alpha_s^{-1} - 3\alpha^{-1}\sin^2\theta_W + (3/5)\alpha^{-1}\cos^2\theta_W \\ &= 8\pi[2(b_3 - b_1) - 3(b_2 - b_1)]\ln(\mu/M_X) = 0. \end{aligned} \tag{14.56}$$

The right-hand side vanishes because of eqns (14.43)–(14.45). Thus

$$\sin^2\theta_W = 1/6 + 5\alpha(\mu)/9\alpha_s(\mu). \tag{14.57}$$

This is the consistency condition mentioned in Remark (C). The values of the coupling constants at $\mu = M_W$ where the Weinberg angle is deduced from neutral-current experiments are compatible with this prediction.

Taking the linear combination [(8/3) × eqn (14.52) – eqn (14.53) – (5/3) × eqn (14.54)] we have from eqns (14.43)–(14.45)

$$\ln(M_X/\mu) = (\pi/11)\left[\frac{1}{\alpha(\mu)} - \frac{8}{3\alpha_s(\mu)}\right]. \tag{14.58}$$

This determines the unification scale M_X. Also, combining eqns (14.57) and (14.58),

$$\sin^2\theta_W = 3/8 - (55/24\pi)\alpha(\mu)\ln(M_X/\mu). \tag{14.59}$$

We should note that in calculating b_n we have taken the simplest possible threshold behaviour: for an intermediate particle with mass $m < \mu$, the mass is taken to be zero; for $m > \mu$ the mass is taken to be infinite and the particle decouples. In particular for $\mu > M_X$, the SU(5) X, Y gauge bosons contribute and the coupling-constant equality of eqn (14.37) is maintained to all orders of perturbation theory. For $\mu < M_X$ they decouple and the b_ns become different for subgroups SU(3), SU(2), and U(1). With more careful treatment of thresholds we can actually identify M_X with the mass of the X boson. In eqns (14.43)–(14.45) we have also ignored the scalar contribution. By including the Higgs scalar of the Weinberg–Salam model and with more careful treatment of higher-order effects, one obtains a set of numerical results in the neighbourhood of

$$M_X \simeq 4 \times 10^{14}\,\text{GeV} \tag{14.60}$$

$$\sin^2\theta_W \simeq 0.21 \tag{14.61}$$

for a range of inputs of the QCD scale parameter $\Lambda \simeq 300$ MeV. (For reviews see Langacker 1981; Marciano and Sirlin 1981.)

14.4 Proton decay and baryon asymmetry in the universe

Proton decay

A prominent feature of GUTs is the nonconservation of baryon number. The SU(5) model has this property and the leading effective Lagrangian for proton decay arises from a set of tree-level X-boson exchange diagrams.

The gauge couplings of the **5* + 10** fermions (ψ^i, χ_{ij}) can be worked out, as usual, from their covariant derivatives. With gauge bosons in the matrix form A of (14.22) we have

$$g\bar{\psi}\gamma^\mu A_\mu^{\mathrm{T}}\psi + Tr\, g\bar{\chi}\gamma^\mu\{A_\mu, \chi\} = -\sqrt{\tfrac{1}{2}}\, g\mathrm{W}_\mu^{+}(\bar{\nu}\gamma^\mu \mathrm{e} + \bar{\mathrm{u}}_\alpha\gamma^\mu \mathrm{d}_\alpha)$$
$$+ \sqrt{\tfrac{1}{2}}g\mathrm{X}_{\mu\alpha}^{a}[\varepsilon^{\alpha\beta\gamma}\bar{\mathrm{u}}_\gamma^{\mathrm{c}}\gamma^\mu \mathrm{q}_{\beta a}$$
$$+ \varepsilon^{ab}(\bar{\mathrm{q}}_{\alpha b}\gamma^\mu \mathrm{e}^{+} - \bar{l}_b\gamma^\mu \mathrm{d}_\alpha^{\mathrm{c}})] + \ldots \quad (14.62)$$

where all the fermions are left-handed and g is the SU(5) gauge coupling constant. The SU(2) doublets are given the labels

$$\mathrm{X}_{\alpha a} = (\mathrm{X}_\alpha, \mathrm{Y}_\alpha) \quad (14.63)$$

$$\mathrm{q}_{\alpha a} = (\mathrm{u}_\alpha, \mathrm{d}_\alpha) \quad (14.64)$$

$$l_a = (\nu_{\mathrm{e}}, \mathrm{e}) \quad (14.65)$$

and $\varepsilon_{\alpha\beta\gamma}$ and ε_{ab} are the totally antisymmetric tensors. Once again it should be remembered that here we are working in the one-family approximation and when more families of fermions (all in the **5* + 10** representations) are incorporated, the fermion fields of eqns (14.62), (14.64), and (14.65) will all be replaced by suitable linear combinations of fermions having the same SU(3) × SU(2) × U(1) charges. A more detailed discussion of mixing angles and CP-violation phases will be postponed until the next section; for the present it suffices to note the expected presence of such mixing angles and phases.

It is important to notice that X-bosons couple to two-fermion channels with different baryon numbers (see Fig. 14.4). In one case they couple to quarks and leptons ($B_1 = 1/3$)—as such they are referred to as *leptoquarks*; in the other case they transform quarks to antiquarks ($B_2 = 2/3$)—they are then called *diquarks*. Consequently, through the mediation of X-bosons, a $B = -1/3$ channel can be converted into a $B = 2/3$ one; i.e. a

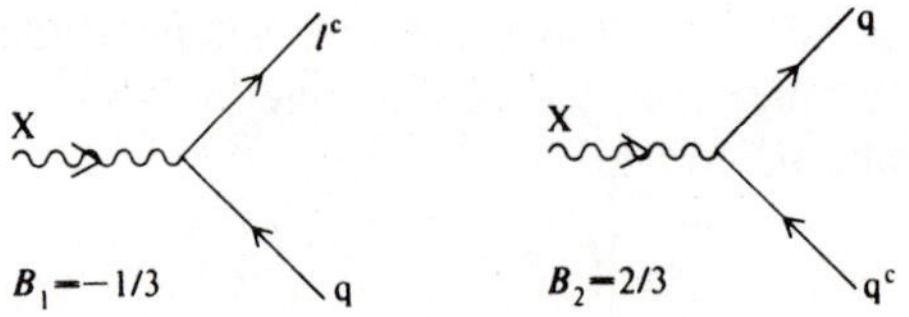

FIG. 14.4. X bosons as leptoquarks and diquarks.

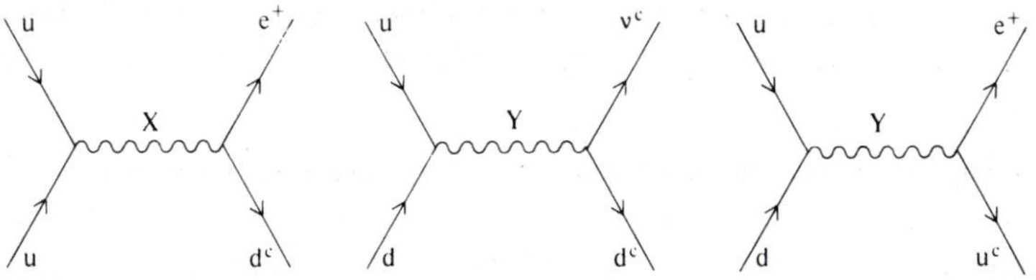

FIG. 14.5. Baryon-number violating processes in the lowest-order X- and Y-boson exchange diagrams.

baryon-number changing ($\Delta B = 1$) process at tree-level occurs (shown in Fig. 14.5).

Since M_X is large (compared to all the fermions), we can write down the effective four-fermion local interactions from (14.62)

$$\mathscr{L}_{\Delta B=1} = (g^2/2M_X^2)\varepsilon^{\alpha\beta\gamma}\varepsilon^{ab}(\bar{u}^c_\gamma \gamma^\mu q_{\beta a})(\bar{d}^c_\alpha \gamma_\mu l_b + \bar{e}^+ \gamma_\mu q_{\alpha b}). \qquad (14.66)$$

We note the following features of this $\Delta B = 1$ effective Lagrangian density.

(1) $\Delta(B - L) = 0$. Baryon minus lepton number is conserved; thus $p \to e^+\pi^\circ$ is allowed, but not $n \to e^-\pi^+$, etc.

(2) $SU(3) \times SU(2) \times U(1)$ *invariance*. All colour and flavour indices are contracted and electric charge is conserved in (14.66). This must be the case in view of our taking into account only one mass scale (i.e. M_X) and effectively treating all other particles including W and Z as massless.

This suggests that, independently of the validity of the SU(5) model, constructing the most general form of the dominant $\Delta B = 1$ amplitude involves listing all the lowest-dimensional SU(3) × SU(2) × U(1) symmetric operators (Weinberg 1979*d*; Wilczek and Zee 1979*a*). The lowest dimension is six as in (14.66) and a higher dimensional operator will necessarily be suppressed by extra powers of M_X in the denominator. Thus the procedure is just an extension of that used by Fermi for the β-decay effective amplitude where only a global U(1) charge conservation needed to be imposed. It turns out that, if we restrict ourselves to fermions with the familiar SU(3) × SU(2) × U(1) quantum numbers as in (14.4), the complete list of such dimension-six operators is rather small. (Many of them are related by Fierz rearrangement.) When such a list is examined, one finds that the selection rule $\Delta(B - L) = 0$ is still respected by this general amplitude. In fact, as we shall see in the next section, in the minimal version of SU(5) with a Higgs multiplet in the **5** + **24** representation, $B - L$ is an exact global symmetry of the model; hence it holds to all orders of perturbation.

We also make the parenthetical remark that the four-fermion effective Lagrangian in (14.66) contains a $\Delta S \leq 0$ rule for proton decay. When the second family of fermions is put in, the final d^c state in Fig. 14.5 will be replaced by Cabibbo-rotated d^c and s^c states. Thus $p \to e^+K^\circ$ is allowed, but not $p \to e^+\overline{K^\circ}$, etc. This selection rule is merely a consequence of the simple quark model because the lowest-dimension $\Delta B = 1$ operator has three quark fields and the strange quark has $S = -1$.

To obtain the proton life-time and branching ratios from $\mathscr{L}_{\Delta B=1}$ of (14.66) with the mixing angles of the minimal SU(5) we still need to perform two more sets of calculations.

(A) The effective Lagrangian is written down at the mass scale of M_X and it must be renormalized down to the typical hadron scale of $O(1\text{ GeV})$. The leading logarithmic radiative corrections to $\mathscr{L}_{\Delta B=1}$ due to the exchanges of SU(3) × SU(2) × U(1) gauge bosons can be estimated using the standard technique of the renormalization group discussed in Chapter 3. (An example of such calculations can be found in the next section.) This yields an amplitude enhancement factor of about 4 (Buras *et al.* 1978).

(B) Proton or bound-neutron decay can be viewed as arising from the four-fermion local interaction of (14.66) with the final antiquark combining with a spectator quark from the initial nucleon to form the final meson system. A variety of phenomenological hadron-physics techniques: SU(6), relativistic bag models, etc. have been used to evaluate the matrix element $\mathscr{L}_{\Delta B=1}$ between the hadron states. (For a review, see Langacker 1981.) Unfortunately there is considerable variation in the results so obtained—by a factor of about 5 in the amplitude. However most model calculations indicate that the minimal SU(5) would have substantial nucleon decay into two- or quasi-two-body channels, which should be relatively easy to detect

$$p \to e^+\pi^0,\ e^+\omega,\ \bar{\nu}\pi^+,\ \text{etc.}$$

$$n \to e^+\pi^-,\ \bar{\nu}\pi^0,\ e^+\rho^-,\ \text{etc.}$$

The prediction of proton lifetime τ_N is clearly very sensitive to uncertainties in our calculations of M_X since $\tau_N \propto M_X^4$. As M_X is directly related to the QCD scale parameter Λ, the value of which is still controversial, we should thus keep in mind that the oft-quoted SU(5) value $\tau_N \simeq 10^{30}$ years is probably uncertain by a multiplicative factor of $10^{\pm 2}$.

Baryon number asymmetry in the universe

In the remainder of this section we shall discuss briefly how a GUT, which predicts proton decay, may also explain why the universe does not seem to contain a large concentration of antimatter (Yoshimura 1978; Toussaint, Treiman, Wilczek, and Zee 1979; Dimopoulos and Susskind 1978; Weinberg 1979*c*). This cosmological asymmetry between matter (baryons) and antimatter (antibaryons) is a puzzle even at the SU(3) × SU(2) × U(1) level. No cosmology model can *generate* a net baryon number if all underlying physical processes conserve baryon number. Until the advent of GUT one had to impose on the standard model of cosmology an *ad hoc* asymmetric initial boundary condition (see eqn (14.68) below). This appears to many to be an unsatisfactory feature of the theory.

We have already mentioned in §13.2 in connection with our discussion on the cosmological bound on neutrino masses how the standard model of cosmology provides us with a very satisfactory picture that can account for a variety of observational data. In particular the observed 2.7 K degree of background black-body radiation is compatible with the nucleosynthesis calculation of the primordial helium abundance.

However the standard model with only baryon-number conserving interactions does not fix the ratio of the photon number density n_γ (corresponding to the value at 2.7 K) to the observed nucleon density n_N. We must put in by hand, as an initial condition, the value

$$\frac{n_N}{n_\gamma} \approx 10^{-9}. \tag{14.67}$$

When the universe was hot enough that baryons (quarks) and antibaryons (antiquarks) could be freely pair-created by radiation the above ratio implied the baryon-number asymmetry

$$\delta = \frac{n_q - n_{q^c}}{n_q + n_{q^c}} \approx 10^{-9}. \tag{14.68}$$

where n_q and n_{q^c} are the quark and anti-quark number densities, respectively.

Why should there be this asymmetry, with this particular value? It would be much more satisfying if starting with a symmetric state (or better, independent of initial conditions) such an asymmetry could be generated by the underlying physical interactions. To have such a situation we must postulate a new particle interaction beyond those given by the SU(3) × SU(2) × U(1) model. Besides being a baryon-number changing interaction, the new interaction must have the following general properties: (i) it must violate C and CP conservation; (ii) it must also have the feature that there was a certain period during the cosmological expansion when these B-, C-, and CP-violating processes were out of thermal equilibrium (Sakharov 1967). It is clear that charge conjugation and CP symmetries would automatically preclude a nonzero value for δ in (14.68) since these operations interchange n_q and n_{q^c}. The need for nonequilibrium is perhaps less obvious. Heuristically we can understand this by recalling that CPT invariance requires particle and antiparticle states to have the same mass, hence to be equally weighted in the Boltzmann distribution; thus no CPT-invariant interactions can generate a nonzero δ in thermal equilibrium.

GUTs, such as SU(5) models, have just the required properties to generate a nonzero δ. There is a set of B-, C-, and CP-violating processes, related to the X-boson couplings (and also to those to the Higgs particles), that can be forced out of equilibrium by the cosmological expansion.

To see how this can happen, we first need to calculate the various reaction rates as a function of energy (i.e. temperature). The criterion of thermal equilibrium is that the reaction rate be greater than the expansion rate of the universe. It turns out that the two-body collisions of Fig. 14.5 do not have the required equilibrium–nonequilibrium transition. However the decays and inverse decays of heavy X boson do have this transition because they have a threshold. For $kT > M_X$, the X bosons would exist in thermal equilibrium with an abundance comparable to ordinary particles (e.g. $N_X \simeq N_\gamma$). This mixture of X and X^c undergoes B and CP violating decays. This preferentially produces quarks relative to antiquarks (see discussions below). Normally such a net baryon number would have eventually been 'washed off' by inverse decays. However as the universe cools below M_X (i.e. $kT < M_X$) the number of X bosons (and the inverse decays) are suppressed by the Boltzmann factor $\exp(-M_X/kT)$. The baryon number production is essentially shut off and the net baryon number produced in the earlier stage is 'frozen in'.

As we have already pointed out, there are two classes of X decay channels with different baryon numbers: $B_1 = -1/3$ and $B_2 = 2/3$. Thus we have the following four decay rates for X and its antiparticle X^c.

$$\gamma_1 \equiv \Gamma(X \to l^c q^c) \quad \text{with} \quad B_1 = -1/3 \tag{14.69}$$

$$\gamma_2 \equiv \Gamma(X \to qq) \quad \text{with} \quad B_2 = 2/3 \tag{14.70}$$

and

$$\gamma_1^c \equiv \Gamma(X^c \to lq) \quad \text{with} \quad B_1' = 1/3 \tag{14.71}$$

$$\gamma_1^c \equiv \Gamma(X^c \to q^c q^c) \quad \text{with} \quad B_2' = -2/3. \tag{14.72}$$

CPT invariance requires that the total rates be the same for particles and antiparticles, i.e.

$$\gamma_1 + \gamma_2 = \gamma_1^c + \gamma_2^c. \tag{14.73}$$

However CPT demands that $\gamma_1 = \gamma_1^c$ and $\gamma_2 = \gamma_2^c$ only in the Born approximation. With C- and CP-violation couplings, higher-order interference terms such as those shown in Fig. 14.6 can lead to

$$\gamma_1 - \gamma_1^c = \gamma_2^c - \gamma_2 \neq 0 \tag{14.74}$$

Consequently, starting with an equal mixture of X and X^c their decay products can show a net baryon number in an nonequilibrium situation

$$\delta \propto \gamma_1 B_1 + \gamma_2 B_2 + \gamma_1^c B_1' + \gamma_2^c B_2' = (\gamma_1 - \gamma_1^c)(B_1 - B_2). \tag{14.75}$$

This explicitly displays the B, C, and CP nonconserving nature of δ, which should be small because $(\gamma_1 - \gamma_1^c)$ is necessarily a higher-order term (and multiplied possibly by some small CP phases). In fact in the minimal SU(5) model (see the next section) $\gamma_1 - \gamma_1^c$ receives its first nontrivial contribution at the 10th-order of perturbation leading to a δ many orders smaller than the observed asymmetry of 10^{-9}. A number of more complicated models have been proposed to remedy this. Further development in this direction is perhaps premature, since the quantitative calculation of δ at this stage is still uncertain. What is clear is that GUTs such as the SU(5) models have the qualitatively correct features that can, in the context of the standard model of cosmology, provide a natural explanation for the observed baryon-number asymmetry in the universe.

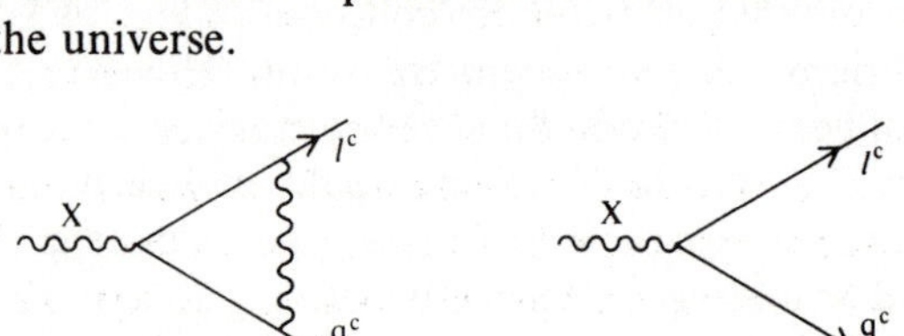

FIG. 14.6. $\gamma_1 - \gamma_1^c$ may receive a nonzero contribution from this fourth-order interference effect.

14.5 Fermion masses and mixing angles in the minimal SU(5) model

From our study of the Weinberg–Salam model in §11.3 we have learned that, when there are several particles of the same charge, the eigenstates of their mass matrix are generally different from those of fields having definite gauge-interaction quantum numbers. The mass eigenstates are related to the gauge-interaction eigenstates by some unitary transformation.

In this section all three **5*** + **10** fermion families will be included in the theory. The fermion fields used in previous sections of this chapter should be replaced by gauge eigenstates, which are vectors in a three-dimensional space of the family index $A = \mathrm{e}, \mu, \tau$.

$$\begin{aligned} \mathrm{e} &\to \mathrm{e}'_A = \delta_{AB}\,\mathrm{e}_B & \mathrm{e}_B &= (\mathrm{e}, \mu, \tau) \\ \nu_\mathrm{e} &\to \nu'_A = T^\dagger_{AB}\nu_B & \nu_B &= (\nu_1, \nu_2, \nu_3) \\ \mathrm{u} &\to \mathrm{p}'_A = U^\dagger_{AB}\mathrm{p}_B & \mathrm{p}_B &= (\mathrm{u}, \mathrm{c}, \mathrm{t}) \\ \mathrm{d} &\to \mathrm{n}'_A = V^\dagger_{AB}\mathrm{n}_B & \mathrm{n}_B &= (\mathrm{d}, \mathrm{s}, \mathrm{b}). \end{aligned} \tag{14.76}$$

We have chosen the basis such that charged lepton-gauge eigenstates e'_A are the same as their mass eigenstates e_A. For massless neutrinos any linear combination of the degenerate fields can be taken as their mass eigenstates; hence we can set the unitary matrix T_{AB} to be δ_{AB} also. Since the mass matrices are not necessarily symmetric, the unitary transformations U_{AB} and V_{AB} may be different for the left-handed (LH) and right-handed (RH) fermion fields. With our convention of always working with LH fermions in this chapter it means that we should distinguish between transformations for particle and antiparticles. Thus to (14.76) we should add

$$\mathrm{u}^\mathrm{c} \to \mathrm{p}'^\mathrm{c}_A = U^{\mathrm{c}\dagger}_{AB}\mathrm{p}^\mathrm{c}_B; \qquad \mathrm{p}^\mathrm{c}_B = (\mathrm{u}^\mathrm{c}, \mathrm{c}^\mathrm{c}, \mathrm{t}^\mathrm{c}) \tag{14.77}$$

$$\mathrm{d}^\mathrm{c} \to \mathrm{n}'^\mathrm{c}_A = V^{\mathrm{c}\dagger}_{AB}\mathrm{n}^\mathrm{c}_B \qquad \mathrm{n}^\mathrm{c}_B = (\mathrm{d}^\mathrm{c}, \mathrm{s}^\mathrm{c}, \mathrm{b}^\mathrm{c}) \tag{14.78}$$

and generally $U_{AB} \neq U^\mathrm{c}_{AB}$ and $V_{AB} \neq V^\mathrm{c}_{AB}$.

Kinship hypothesis

While the 'family' is a well-defined notion in terms of gauge eigenstates, historically we also have an intuitively sensible grouping of the fermion mass eigenstates

$$\begin{aligned} \mathrm{e} - \text{family:} &\quad (\mathrm{e}, \nu_1, \mathrm{d}, \mathrm{u}) \\ \mu - \text{family:} &\quad (\mu, \nu_2, \mathrm{s}, \mathrm{c}) \\ \tau - \text{family:} &\quad (\tau, \nu_3, \mathrm{b}, \mathrm{t}) \end{aligned} \tag{14.79}$$

with each family ascending on the mass scale. Furthermore this grouping scheme is supported by our experience with the weak interaction in the sense that charged-current transitions between different families are suppressed by small mixing angles (eqn (12.47)). The question naturally arises whether this feature can be generalized in a GUT in which there are new flavour-changing currents coupled to the X gauge bosons. Is the family structure (14.79) observed at low energies still valid in the SU(5) models? To focus the issue, this generalization is called the '*kinship hypothesis*'. Namely, it supposes that *all* interfamily transitions are suppressed by the appropriate small mixing angles. For example the kinship hypothesis says that the baryon-number changing couplings $\mathrm{u} \to \tau^+$ or $\mathrm{b^c}$ are suppressed relative to $\mathrm{u} \to \mathrm{e}^+$ or $\mathrm{d^c}$, etc. This of course would have important implications for the analysis of proton decay. A strong violation of this hypothesis would severely depress the decay rate and sharply alter its branching ratio pattern.

The SU(5) gauge couplings in eqn (14.62) can be transcribed in terms of gauge eigenstates according to eqns (14.76), (14.77) and (14.78)

$$\mathrm{W}(\bar{\mathrm{v}}'_A \mathrm{e}'_A + \bar{\mathrm{p}}'_A \mathrm{n}'_A) + \mathrm{X}(\mathrm{p}^{\mathrm{c}'}_A \mathrm{p}'_A + \bar{\mathrm{n}}'_A \mathrm{e}'^{+}_A + \bar{\mathrm{e}}'_A \mathrm{n}'^{\mathrm{c}}_A) \\ + \mathrm{Y}(\bar{\mathrm{p}}^{\mathrm{c}'}_A \mathrm{n}'_A + \bar{\mathrm{p}}'_A \mathrm{e}'^{+}_A + \bar{\mathrm{v}}'_A \mathrm{n}'^{\mathrm{c}}_A). \tag{14.80}$$

We have dropped all extraneous labels except the family index A. Expressing these couplings in terms of mass eigenstates we have

$$\mathrm{W}[\bar{\mathrm{v}}_A \mathrm{e}_A + \bar{\mathrm{p}}_A (UV^\dagger)_{AB} \mathrm{n}_B] + \mathrm{X}[\bar{\mathrm{p}}^{\mathrm{c}}_A (U^{\mathrm{c}} U^\dagger)_{AB} \mathrm{p}_B + \bar{\mathrm{n}}_A (V)_{AB} \mathrm{e}^+_B + \bar{\mathrm{e}}_A (V^{\mathrm{c}\dagger})_{AB} \mathrm{n}^{\mathrm{c}}_B] \\ + \mathrm{Y}[\bar{\mathrm{p}}^{\mathrm{c}}_A (U^{\mathrm{c}} V^\dagger)_{AB} \mathrm{n}_B + \bar{\mathrm{p}}_A (U)_{AB} \mathrm{e}^+_B + \bar{\mathrm{v}}_A (V^{\mathrm{c}\dagger})_{AB} \mathrm{n}^{\mathrm{c}}_B]. \tag{14.81}$$

The KM rotation matrix of (11.124) is just the combination $(UV^\dagger)$. It is clear that all the six other rotation matrices in the X, Y gauge boson couplings are not of this form. Hence in principle we may encounter very different mixing angles in these new interaction vertices; the kinship hypothesis may not be valid. However we shall display below that in the version of SU(5) with the simplest possible Higgs structure, the *minimal SU*(5) *model*, these new mixings all essentially collapse to the familiar KM rotation.

Given that the fermions are in the **5*** + **10** representation, the scalars that can form Yukawa couplings must be

$$\mathbf{5^*} \times \mathbf{5^*} = \mathbf{10^*} + \mathbf{15}$$
$$\mathbf{10} \times \mathbf{10} = \mathbf{5^*} + \mathbf{45^*} + \mathbf{50}$$
$$\mathbf{5^*} \times \mathbf{10} = \mathbf{5} + \mathbf{45}. \tag{14.82}$$

It is a straightforward exercise to check that only **5** and **45** have components transforming as **(1, 2)** under the subgroup SU(3) × SU(2). The minimal SU(5) model has Higgs scalars only in **24** and **5**, with **5** providing the breaking of SU(3) × SU(2) × U(1) → SU(3) × U(1) *and* giving masses to fermions.

We have the following Yukawa couplings

$$f^{(1)}_{AB}(\chi_{Aij})^{\mathrm{T}} C(\chi_{Bkl})\phi_m \varepsilon^{ijklm} + f^{(2)}_{AB}(\chi_{Aij})^{\mathrm{T}} C\psi^i_B \phi^{j\dagger} + \text{h.c.} \tag{14.83}$$

where C is the Dirac charge conjugation matrix. It then follows from the auticommutation of the fermion fields and the antisymmetric property of $C = \mathrm{i}\gamma^2\gamma^0$ that the Yukawa coupling matrix $f^{(1)}_{AB}$ is symmetric. When the **5** scalar ϕ develops VEV as in (14.34)

$$\langle\phi\rangle = (0, 0, 0, 0, v_2), \tag{14.84}$$

the couplings in (14.83) produce terms quadratic in fermion fields

$$v_2 f_{AB}^{(1)}(\bar{\mathrm{p}}_A \mathrm{p}_B) + v_2 f_{AB}^{(2)}(\bar{\mathrm{n}}_A \mathrm{n}_B + \bar{\mathrm{e}}_A \mathrm{e}_B). \tag{14.85}$$

The mass matrices for p, n, and e states then have the properties

$$\begin{aligned} v_2 f_{AB}^{(1)} &= M_{AB}^{(p)} = M_{BA}^{(p)} \\ v_2 f_{AB}^{(2)} &= M_{AB}^{(n)} = M_{AB}^{(e)}. \end{aligned} \tag{14.86}$$

The fact that the p-quark mass matrix is symmetric implies that $U^{\dagger}U^{c}$ is a diagonal unitary matrix (see (11.109)). The equality of $M^{(n)} = M^{(e)}$ is a consequence of the SU(4) symmetry of VEV in (14.84). This implies that the same biunitary transformations diagonalize both $M^{(n)}$ and $M^{(e)}$. Since we have chosen gauge eigenstates and mass eigenstates to be the same for charged leptons, it then follows that we can take V and V^{c} to be identity transformations also. Thus up to a complex unit matrix all the mixing angle matrices in (14.81) are of the form $(UV^{\dagger})$, which is just the KM rotation of weak interactions. For the minimal SU(5) model the kinship hypothesis is fully realized.

Lepton–quark mass relations

Is there any evidence in support of the minimal version of the SU(5) theory?

The equality of $M^{(n)} = M^{(e)}$ implies not only the equality of the diagonalization matrices but also of their eigenvalues. Hence we have (Georgi and Glashow 1974)

$$\begin{aligned} m_{\mathrm{e}} &= m_{\mathrm{d}} \\ m_{\mu} &= m_{\mathrm{s}} \\ m_{\tau} &= m_{\mathrm{b}}. \end{aligned} \tag{14.87}$$

Again, like the coupling-constant equality of (14.37), there are SU(5) symmetric relations, subject to significant renormalization corrections.

Perturbation calculation of fermion self-energy yields

$$m(\mu) = m - m g_n^2 b_m^{(n)} \ln(\Lambda/\mu). \tag{14.88}$$

Thus the renormalization group equation for the effective mass is

$$\frac{\mathrm{d}\ln m(\mu)}{\mathrm{d}\ln\mu} = b_m^{(n)} g_n^2(\mu). \tag{14.89}$$

This differential equation can be integrated since we know from eqn (10.80) the scale dependence of the coupling constant

$$g_n^2(\mu) = \frac{g_n^2(\mu_0)}{1 + 2b_n g_n^2(\mu_0)\ln(\mu/\mu_0)} \tag{14.90}$$

which is of course the solution to eqn (14.42)

$$\frac{\mathrm{d}\, g_n^2(\mu)}{\mathrm{d}\ln\mu} = -2b_n g_n^4(\mu). \tag{14.91}$$

We obtain the result

$$\frac{m(\mu)}{m(\mu_0)} = \left[\frac{g_n(\mu)}{g_n(\mu_0)}\right]^{-b_m^{(n)}/b_n} \tag{14.92}$$

where the b_ns are given in eqn (14.43). We now proceed to calculate $b_m^{(n)}$ using Fig. 14.7

$$b_m^{(n)} = -\frac{3}{8\pi^2}\sum_a (T^a T^a)_{ij} \tag{14.93}$$

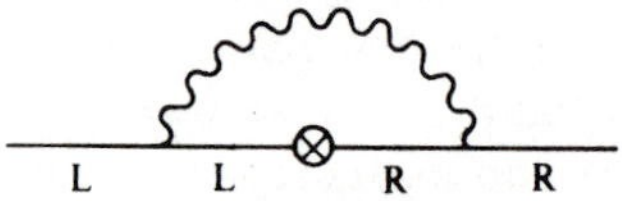

FIG. 14.7. Self-energy diagram for fermions.

where the T^as are the representation matrices appropriate for the fermions. For SU(n) with $n \geq 2$,

$$\sum_a (T^a T^a)_{ij} = \frac{n^2-1}{2n}\,\delta_{ij} \tag{14.94}$$

and, for the U(1) case (eqn (14.40)),

$$(T^0)^2 = \frac{3}{5}\left(\frac{Y}{2}\right)^2. \tag{14.95}$$

Using the expression for b_n in eqns (14.43) and (14.44), we obtain

$$\frac{m_p(\mu)}{m_p(\mu_0)} = \left[\frac{g_3(\mu)}{g_3(\mu_0)}\right]^{\frac{8}{11-2N_F/3}} \left[\frac{g_1(\mu)}{g_1(\mu_0)}\right]^{\frac{-6}{10N_F}} \tag{14.96}$$

$$\frac{m_n(\mu)}{m_n(\mu_0)} = \left[\frac{g_3(\mu)}{g_3(\mu_0)}\right]^{\frac{8}{11-2N_F/3}} \left[\frac{g_1(\mu)}{g_1(\mu_0)}\right]^{\frac{3}{10N_F}} \tag{14.97}$$

$$\frac{m_e(\mu)}{m_e(\mu_0)} = \left[\frac{g_1(\mu)}{g_1(\mu_0)}\right]^{\frac{-27}{10N_F}} \tag{14.98}$$

where N_F is the number of quark flavours. We note that there is no contribution from SU(2) gauge bosons because RH fermions are all singlet under SU(2) and the diagram in Fig. 14.7 involves one helicity flip. Dividing eqn (14.97) by eqn (14.98) and using eqn (14.87) for ($m_n(M_X) = m_e(M_X)$), we obtain (with $g_3(M_X) = g_1(M_X) = g_5(M_X)$)

$$\frac{m_n(\mu)}{m_e(\mu)} = \left[\frac{g_3(\mu)}{g_5(M_X)}\right]^{\frac{8}{11-2N_F/3}} \left[\frac{g_1(\mu)}{g_5(M_X)}\right]^{\frac{3}{N_F}} \tag{14.99}$$

If we take the effective current quark mass at the $q\bar{q}$ threshold

$$\mu = \mu_{th} \equiv 2\, m_q(\mu_{th}) \tag{14.100}$$

we then obtain for $n_3 = \mathrm{b}$, $e_3 = \tau$, and $\mu \simeq 10$ GeV

$$\frac{m_b}{m_\tau} \simeq 3 \tag{14.101}$$

which must be regarded as a successful prediction of the theory (Buras *et al.* 1978).

It is not clear how to properly evaluate the renormalization effects for lighter fermions since a smaller value of scale parameter μ must be involved. However, if we merely examine the renormalization-group invariant ratio as implied by eqn (14.87)

$$\frac{m_\mu}{m_e} = \frac{m_s}{m_d}, \tag{14.102}$$

we see that the two sides differ by something like a factor of 10, since current-algebra calculations indicate that $m_s/m_d \simeq 20$. Does this 'failure' definitely preclude the minimal SU(5) as a viable theory? Such a strong conclusion is perhaps unwarranted because, for the very light quark d ($m_d \simeq 7$ MeV), the renormalization group equation is not expected to hold. Thus the correctness of the minimal SU(5) model is still an open question.

B − L conservation

We close this section with a remark about another aspect of the minimal SU(5) model: it still possesses a global U(1) symmetry, corresponding to the conservation of $B - L$, baryon minus lepton number (Wilczek and Zee 1979*b*). It is clear that the X, Y gauge bosons conserve $B - L$ if we assign them the quantum number of $B - L = 2/3$. We need to check that the Yukawa couplings of (14.83) have this symmetry which is not spoiled by the spontaneous symmetry breaking of (14.84).

We can write (14.83) symbolically as

$$f^{(1)}(10_f)(10_f)(5_\phi) + f^{(2)}(10_f)(5_f^*)(5_\phi^*). \tag{14.103}$$

It is easy to check the conservation of a new charge (call it F). The first term requires

$$F(5_\phi) = -2F(10_f); \tag{14.104}$$

then the second term implies

$$F(5_f) = 3F(10_f). \tag{14.105}$$

This means that the conserved U(1) charge may be written in terms of number operators

$$F = (1/5)[N(10_f) + 3N(5_f) - 2N(5_\phi)]. \tag{14.106}$$

The normalization factor 1/5 is a convention. It is clear that $\langle\phi\rangle$ of (14.84) is not invariant under the U(1) generated by the F-charge of (14.106) nor under the U(1) generated by the weak hypercharge as neither

$$F(5_\phi) = -2/5\begin{bmatrix} 1 & & & & \\ & 1 & & & \\ & & 1 & & \\ & & & 1 & \\ & & & & 1 \end{bmatrix}$$

and

$$Y(5_\phi) = \begin{bmatrix} -2/3 & & & & \\ & -2/3 & & & \\ & & -2/3 & & \\ & & & 1 & \\ & & & & 1 \end{bmatrix} \qquad (14.107)$$

annihilate the vector in (14.84). However the linear combination $F' = F + \frac{2}{5}Y$ has the property that $F'(5_\phi)\langle\phi\rangle = 0$ and is thus still conserved after SSB. We note that F' is no other than $B - L$ since

$$F'(5_f) = \begin{bmatrix} 1/3 & & & & \\ & 1/3 & & & \\ & & 1/3 & & \\ & & & 1 & \\ & & & & 1 \end{bmatrix}. \qquad (14.108)$$

Just as B and L are accidental global symmetries in the standard SU(3) × SU(2) × U(1) model, the minimal SU(5) violates B and L conservation but still preserves the combination $B - L$.

15 Magnetic monopoles

The modern theory of magnetic charges was first formulated by Dirac over 50 years ago. Their existence has been under active experimental investigation ever since. An intriguing aspect of non-Abelian gauge theory is that it has objects with the properties of magnetic monopoles ('t Hooft 1974; Polyakov 1974). It is expected that monopoles associated with the spontaneous symmetry breakdown of grand unified gauge theories should be superheavy ($\sim 10^{16}$ GeV?) and such objects may well have escaped detection. The recent upsurge of an extensive search for magnetic monopoles has been prompted further by the possible evidence reported by Cabrera (1982). However, even if this first sighting is not confirmed by other experiments, the theoretical studies of monopoles should lead to a better understanding of the structure of non-Abelian gauge theories with spontaneous symmetry breaking.

In this chapter we shall present a brief introduction to the gauge theory of magnetic monopoles. In §15.1 we review the properties of the monopoles as originally proposed by Dirac (1931). We then discuss in §15.2 the general features of finite-energy solutions (solitons) to the equation of motion in field theory. In §15.3 we will illustrate some aspects of the 't Hooft–Polyakov monopole solution in non-Abelian gauge theory, which is a synthesis of the Dirac monopole and the soliton solution. It should be emphasized that our discussion will be kept at an elementary level and mathematical rigour is often sacrificed for simplicity of presentation. For excellent reviews on this subject the reader is referred to articles by Goddard and Olive (1978) and by Coleman (1975, 1981).

15.1 Dirac's theory of magnetic poles

In this section, we will review the properties of the magnetic monopole as originally introduced by Dirac. (For a more detailed discussion, see Goddard and Olive 1978.)

Classical electromagnetism and duality transformations

We start from classical electromagnetism which is well described by Maxwell's equations

$$\nabla \cdot \mathbf{E} = \rho \qquad \nabla \times \mathbf{B} - \partial_0 \mathbf{E} = \mathbf{j} \tag{15.1}$$

$$\nabla \cdot \mathbf{B} = 0 \qquad \nabla \times \mathbf{E} + \partial_0 \mathbf{B} = 0. \tag{15.2}$$

In terms of the electromagnetic field tensor $F^{\mu\nu}$, these equations can be combined into covariant form,

$$\partial_\nu F^{\mu\nu} = -j^\mu \tag{15.3}$$

$$\partial_\nu \tilde{F}^{\mu\nu} = 0 \tag{15.4}$$

where

$$j^\mu = (\rho, \mathbf{j}), \qquad F^{0i} = E^i, \qquad F^{ij} = -\varepsilon^{ijk}B^k \tag{15.5}$$

and the dual field tensor is defined as

$$\tilde{F}^{\mu\nu} = \tfrac{1}{2}\varepsilon^{\mu\nu\rho\sigma}F_{\rho\sigma}. \tag{15.6}$$

In vacuum, where $j_\mu = 0$, Maxwell's equations (15.3) and (15.4) are symmetric under the *duality transformation*

$$F^{\mu\nu} \to \tilde{F}^{\mu\nu}, \qquad \tilde{F}^{\mu\nu} \to -F^{\mu\nu} \tag{15.7}$$

which corresponds to the interchange of electricity and magnetism $\mathbf{E} \to \mathbf{B}$, $\mathbf{B} \to -\mathbf{E}$. This symmetry is broken by the presence of the electric current j_μ in eqn (15.3). One can introduce the magnetic current $k^\mu = (\sigma, \mathbf{k})$ on the right-hand side of (15.4) so that the modified Maxwell's equations read

$$\partial^\nu F_{\mu\nu} = -j_\mu \qquad \partial_\nu \tilde{F}^{\mu\nu} = -k^\mu. \tag{15.8}$$

They will be symmetric under the duality transformation (15.7), supplemented by the substitution

$$j^\mu \to k^\mu, \qquad k^\mu \to -j^\mu. \tag{15.9}$$

That is $(F, \tilde{F})$, (E, B) and (j, k) are 'duality vectors' and the transformation is a rotation (by 90°) in such two-dimensional planes. (These duality rotations will be useful in translating the familiar electromagnetic results involving electric charges into those involving magnetic charges.) The introduction of the magnetic current k^μ requires the existence of magnetically charged particles, the *magnetic monopoles*. In analogy to the electric current produced by point particles at x_i with charge q_i

$$j_\mu(x) = \sum_i q_i \int \mathrm{d}x_i^\mu\, \delta^4(x - x_i), \tag{15.10a}$$

the magnetic current due to point magnetic charges with strength g_i is given by

$$k_\mu(x) = \sum_i g_i \int \mathrm{d}x_i^\mu\, \delta^4(x - x_i) \tag{15.10b}$$

where the line integrals in (15.10a) and (15.10b) are taken along the particle world lines.

Consistency of the monopole with quantum theory

On the classical level, the magnetic monopole seems to be on the same footing as the electrically charged particle and it is mysterious that monopoles are not seen in the laboratory. It was first pointed out by Dirac that on the quantum level the existence of the monopole will lead to the condition

$$\frac{qg}{4\pi} = \frac{1}{2}n \qquad \text{where } n \text{ is an integer.} \tag{15.11}$$

This is the famous Dirac quantization condition which implies *charge quantization*, i.e. the possible value of the electric charge carried by any particle is an integral multiple of some basic unit (cf. discussion in §14.1). This can be seen as follows. Consider the case where particles may carry either electric or magnetic charge but not both. The possible values of the electric and magnetic charges are denoted by q_i and g_i, respectively. The Dirac quantization condition (15.11) implies that

$$\frac{q_i g_i}{4\pi} = \frac{1}{2}n_{ij}$$

where n_{ij} is an integer. Then for any fixed magnetic charge g_j, all electric charge q_i must be an integral multiple of $2\pi/g_j$. Similarly, for any fixed electric charge q_i, all magnetic charge must also be multiple of $2\pi/q_i$. If we assume that there exists a smallest electric charge q_0 and a smallest magnetic charge g_0, we have

$$q_i = n_i q_0, \tag{15.12}$$

$$g_i = n_i' g_0, \tag{15.13}$$

and

$$\frac{q_0 g_0}{4\pi} = \frac{1}{2}n_0 \tag{15.14}$$

where n_i, n_i', *and* n_0 are integers. Condition (15.14) also implies that the interaction between two magnetic monopoles which is proportional to

$$g_0^2 \sim q_0^2 \frac{n_0^2}{4}\left(\frac{4\pi}{q_0^2}\right)^2 \sim (n_0/2\alpha)^2 q_0^2 \tag{15.15}$$

is roughly $\alpha^{-2} \sim 10^4$ times stronger than the interactions between the electrically charged particles. In other words, the smallness of the electric charge coupling $q_0^2/4\pi \sim \alpha \sim 1/137$ implies a very strong interaction between monopoles, and it will be much more difficult to pair-produce the magnetic monopoles than the electrically charged particles.

Angular momentum and the Dirac quantization condition

One heuristic way to 'derive' the Dirac quantization condition (15.11) is to study the motion of a charged particle in the field of a monopole. The magnetic field due to a monopole of strength g fixed at the origin is given by

$$\mathbf{B} = \frac{g}{4\pi r^2}\hat{\mathbf{r}} \tag{15.16}$$

where $\hat{\mathbf{r}}$ is the unit vector in the radial direction. The motion of a particle with mass m and electric charge q in this field is given by

$$m\ddot{\mathbf{r}} = q\dot{\mathbf{r}} \times \mathbf{B} \tag{15.17}$$

We can calculate the rate of change of its orbital angular momentum,

$$\frac{\mathrm{d}}{\mathrm{d}t}(\mathbf{r} \times m\dot{\mathbf{r}}) = \mathbf{r} \times m\ddot{\mathbf{r}} = \frac{qg}{4\pi r^3}\mathbf{r} \times (\dot{\mathbf{r}} \times \mathbf{r}) = \frac{\mathrm{d}}{\mathrm{d}t}\left(\frac{qg}{4\pi}\hat{\mathbf{r}}\right).$$

This suggests that we can define the total angular momentum as

$$\mathbf{J} = \mathbf{r} \times m\dot{\mathbf{r}} - \frac{qg}{4\pi}\hat{\mathbf{r}} \tag{15.18}$$

so that it is conserved. The second term in eqn (15.18) can be interpreted as the angular momentum due to the electromagnetic field, because using (15.16) we have

$$\begin{aligned} \mathbf{L}_{\text{em}} &= \int \mathrm{d}^3x\mathbf{r} \times (\mathbf{E} \times \mathbf{B}) \\ &= \frac{g}{4\pi}\int \mathrm{d}^3x(\boldsymbol{\nabla}\cdot\mathbf{E})\hat{\mathbf{r}} = \frac{qg}{4\pi}\hat{\mathbf{r}}. \end{aligned}$$

Thus the conservation of the total angular momentum J means that angular momentum is passed back and forth between the particle and the field in the presence of electric and magnetic charges. When we quantize the theory we would expect the components of $\mathbf{J}$ to satisfy the usual angular momentum commutation relations. This would imply that the eigenvalues of J_i are half-integers. Since we expect the orbital angular momentum, the first term in (15.18), to have integral eigenvalues, we get

$$\frac{qg}{4\pi} = \frac{1}{2}(n = \text{integers})$$

which is just the Dirac quantization condition of eqn (15.11).

The Dirac string

A rigorous derivation of the Dirac quantization condition comes from considering the quantization of the motion of a particle in a given electromagnetic field. In the usual quantization of the electromagnetic field in the absence of the monopole, we write the electromagnetic field tensor $F_{\mu\nu}$ in terms of the four-vector potential $A_\mu = (\phi, \mathbf{A})$ as

$$F_{\mu\nu} = \partial_\mu A_\nu - \partial_\nu A_\mu$$

or

$$\mathbf{B} = \boldsymbol{\nabla} \times \mathbf{A}, \qquad \mathbf{E} = -\boldsymbol{\nabla}\phi - \frac{\partial \mathbf{A}}{\partial t} \tag{15.19}$$

so that the equation $\partial^\mu \tilde{F}_{\mu\nu} = 0$ is automatically satisfied. Then Schrödinger's equation for a particle moving in the electromagnetic field has the form

$$\left[\frac{1}{2m}(\mathbf{p}-e\mathbf{A})^2+e\phi\right]\psi=\mathrm{i}\frac{\partial\psi}{\partial t}, \tag{15.20}$$

which is invariant under the gauge transformation

$$\mathbf{A}(\mathbf{x})\to\mathbf{A}(\mathbf{x})+\frac{1}{e}\nabla\alpha(\mathbf{x})$$

$$\psi(\mathbf{x})\to \mathrm{e}^{\mathrm{i}\alpha(\mathbf{x})}\psi(\mathbf{x})$$

with $\alpha(\mathbf{x})$ being an arbitrary function. Thus the four-vector potential A_μ plays a crucial role as the basic dynamical variable in the quantization. But, if the monopole exists, the vector potential cannot exist everywhere because $\tilde{F}^{\mu\nu}$ satisfies eqn (15.8) rather than eqn (15.4). Does quantum mechanics preclude the existence of magnetic monopole? Dirac overcame this difficulty by introducing the concept of a string. To see this, consider the magnetic field of the monopole given in eqn (15.16). For any closed surface containing the origin, we have

$$g=\oint_S \mathbf{B}\cdot\mathrm{d}\mathbf{S} \tag{15.21}$$

It is clear that $\mathbf{B}$ cannot be written as $\nabla\times\mathrm{A}$ everywhere; otherwise the integral in (15.21) would be zero. However, we can define an $\mathbf{A}$ such that $\mathbf{B}$ is given by $\nabla\times\mathbf{A}$ everywhere except on a line joining the origin to infinity. To see that this is possible, consider the field due to an infinitely long and thin solenoid placed along the negative z-axis with its positive pole (with strength g) at the origin. Its magnetic field would be

$$\mathbf{B}_{\mathrm{sol}}=\frac{g}{4\pi r^2}\hat{\mathbf{r}}+g\theta(-z)\,\delta(x)\,\delta(y)\hat{\mathbf{z}} \tag{15.22}$$

where $\hat{\mathbf{z}}$ is a unit vector in the z-direction. This magnetic field differs from the field of the magnetic monopole (eqn (15.16)) by the singular magnetic flux along the solenoid, the second term in (15.22). Since the magnetic field given in (15.22) is source-free ($\nabla\cdot\mathbf{B}_{\mathrm{sol}}=0$), we can write

$$\mathbf{B}_{\mathrm{sol}}=\nabla\times\mathbf{A}. \tag{15.23}$$

Then from eqns (15.16), (15.22), and (15.23), the monopole field is given by

$$\mathbf{B}=\frac{g^2}{4\pi r^2}\hat{\mathbf{r}}=\nabla\times\mathbf{A}-g\theta(-z)\,\delta(x)\,\delta(y)\hat{\mathbf{z}} \tag{15.24}$$

with a pictorial representation shown in Fig. 15.1. The line occupied by the solenoid is called the *Dirac string*. It is not difficult to see that the vector potential $\mathbf{A}$ of the solenoid can be written, with the conventional definitions of polar and azimuthal angles, as

$$\mathbf{A}=\frac{g}{4\pi r}\left(\frac{1-\cos\theta}{\sin\theta}\right)\hat{\boldsymbol{\phi}} \tag{15.25}$$

which is singular on the negative z-axis. Eqn (15.24) means that the monopole field can be represented by a vector potential **A** together with a string.

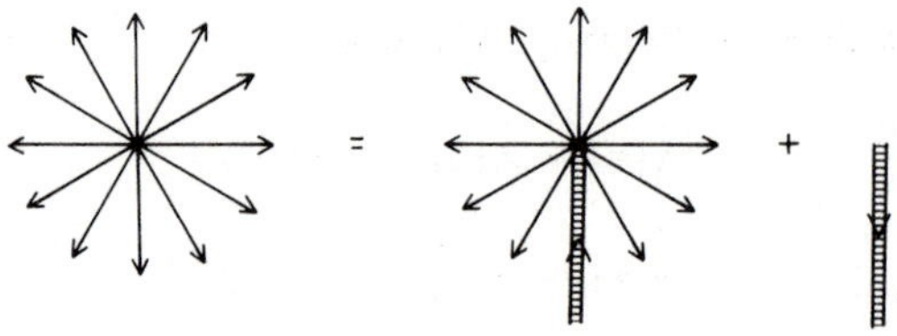

FIG. 15.1.

It should be emphasized that the Dirac string is not observable. For example, the Dirac string does not give rise to the Aharonov–Bohm effect (1959), which is the quantum-mechanical interference of charged particles in a region with $\mathbf{B} = 0$ but $\mathbf{A} \neq 0$. The condition for the *absence* of Aharonov–Bohm effect is that for a charged particle (q) in the two-slit experiment reaching the detector screen via two distinct paths labelled by 1 and 2 around the Dirac string is that

$$\left|\exp\left(iq\int_1 \mathbf{A}\cdot d\mathbf{l}\right)\psi_1 + \exp\left(iq\int_2 \mathbf{A}\cdot d\mathbf{l}\right)\psi_2\right|^2 = |\psi_1 + \psi_2|^2.$$

This is precisely the Dirac quantization condition (15.11):

$$qg = q\oint \mathbf{A}\cdot d\mathbf{l} = 2n\pi.$$

where the closed path consists of path 1 and (the reverse of) path 2.

Another way to see that the Dirac string is unobservable is to show that it can be moved around by a suitable gauge transformation. We demonstrate this as follows. The vector potential given in (15.25) is not unique for the solenoid field. If one makes a non-singular gauge transformation $\mathbf{A} \to \mathbf{A} + \nabla\chi$ where χ is a non-singular, single-valued function of position, then the $\nabla \times \mathbf{A}$ term in eqn (15.24) will remain unchanged and so must the Dirac string term. To move the string we need to extend the concept of the gauge transformation. Rewrite eqn (15.24) for the magnetic field of the monopole as

$$\mathbf{B}(\mathbf{r}) = \nabla \times \mathbf{A} + \mathbf{h}(\mathscr{C}, \mathbf{r}) \tag{15.26}$$

where $\mathbf{h}(\mathscr{C}, \mathbf{r})$ represents the contribution of the Dirac string along some curve $\mathscr{C}$ going from the origin to ∞ and has a flux of strength g,

$$\mathbf{h}(\mathscr{C}, \mathbf{r}) = g\int_{\mathscr{C}} d\mathbf{x}\, \delta^3(\mathbf{r} - \mathbf{x}). \tag{15.27}$$

Consider another string $\mathscr{C}'$ running from the origin to ∞ along curve $\mathscr{C}'$. Let Γ denote the curve $-\mathscr{C}'$ ($\mathscr{C}'$ taken in the reverse direction) followed by $\mathscr{C}$.

We may treat this as a closed path, either by making suitable assumptions about what happens at infinity or by assuming that $\mathscr{C}'$ differs from $\mathscr{C}$ only over a finite range (see Fig. 15.2). Let $\Omega(\mathbf{r})$ be the solid angle subtended at $\mathbf{r}$ by some particular surface spanning Γ. Various choices of the spanning surface

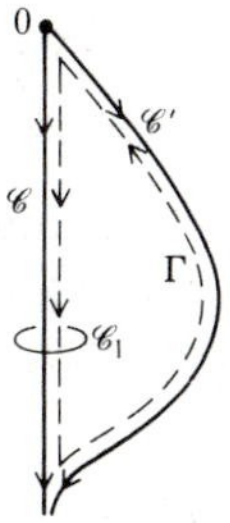

FIG. 15.2.

will lead to values of Ω differing by multiples of 4π, but will yield the same value for $\nabla\Omega$. Now consider the extended gauge transformation defined by

$$\mathbf{A} \to \mathbf{A}' = \mathbf{A} - \frac{g}{4\pi}\nabla\Omega. \tag{15.28}$$

Note that $\Omega(\mathbf{r})$ is a multi-valued function and is ill-defined for $\mathbf{r}$ on Γ. Then $\nabla \times \mathbf{A}' = \nabla \times \mathbf{A} = \mathbf{B}$ except on the two strings. Applying Stokes' theorem to a small loop encircling any point on Γ we see that the flux of $\nabla \times (\mathbf{A}' - \mathbf{A})$ along Γ is

$$\int_{\sigma_1} \nabla \times (\mathbf{A}' - \mathbf{A}) \cdot \mathrm{d}\boldsymbol{\sigma} = \oint_{\mathscr{C}_1} (\mathbf{A}' - \mathbf{A}) \cdot \mathrm{d}\mathbf{l} = \frac{g}{4\pi} \oint_{\mathscr{C}_1} \nabla\Omega \cdot \mathrm{d}\mathbf{l} = \frac{g}{4\pi} \oint_{\mathscr{C}_1} \mathrm{d}\Omega = g \tag{15.29}$$

where we have used eqn (15.28). From eqn (15.27), we can write eqn (15.29) as

$$\nabla \times (\mathbf{A}' - \mathbf{A}) = \mathbf{h}(\mathscr{C}, \mathbf{r}) - \mathbf{h}(\mathscr{C}'\mathbf{r})$$

or

$$\mathbf{B} = \nabla \times \mathbf{A} + \mathbf{h}(\mathscr{C}, \mathbf{r}) = \nabla \times \mathbf{A}' + \mathbf{h}(\mathscr{C}', \mathbf{r}). \tag{15.30}$$

This shows that the gauge transformation (15.28) moves the Dirac string around. The arbitrary position of the string shows that it is unphysical or, as one says, the Dirac string is a gauge artefact.

Derivation of the Dirac quantization condition

Eqn (15.11) follows from the crucial requirement that the generalized gauge transformation (15.28) acting on the wavefunction of the particle must be given an equivalent quantum-mechanical description. Thus, under the gauge transformation (15.28), the change in the wave function

$$\psi \to \psi' = \exp\left(-\mathrm{i}\frac{qg}{4\pi}\Omega\right)\psi \tag{15.31}$$

should not produce a multi-valued result. Since there is an ambiguity of 4π in $\Omega(\mathbf{r})$, we must have

$$qg = 2n\pi$$

which is just Dirac's quantization condition (15.11).

Comment 1 To make the above derivation more transparent it will be helpful to work out the example of vector potential $\mathbf{A}$ and $\mathbf{A}'$ for strings $\mathscr{C}$ and $\mathscr{C}'$ being the negative and positive z-axis respectively. We then have the explicit result from eqn (15.25):

$$\mathbf{A}\cdot\mathrm{d}\mathbf{r} = \frac{g}{4\pi}(1 - \cos\theta)\,\mathrm{d}\phi$$

$$\mathbf{A}'\cdot\mathrm{d}\mathbf{r} = \frac{-g}{4\pi}(1 + \cos\theta)\,\mathrm{d}\phi.$$

From their difference one can immediately verify the gauge function as

$$\chi = \frac{g}{4\pi}(2\phi) = \frac{g}{4\pi}\Omega.$$

Comment 2 We can also avoid the notion of singular potentials and gauge transformations by using different monopole potentials in different regions of space, namely $\mathbf{A}$ and $\mathbf{A}'$ defined over domains with $\mathscr{C}$ and $\mathscr{C}'$ (e.g. negative and positive z-axis) respectively excluded. The Dirac quantization condition then follows from the requirement that in the overlapping region of the two domains the potentials $\mathbf{A}$ and $\mathbf{A}'$ should be connected by a single-valued gauge transformation (Wu and Yang 1976). While this is an elegant formulation, in most practical calculations it is still more convenient to use the singular Dirac string potential.

15.2 Solitons in field theory

To prepare for discussion of the monopole in non-Abelian gauge theory, we will first present an elementary introduction to the classical finite-energy solutions of field theory, generally called *solitons*. (For further discussion, see Coleman 1975; Rajaraman 1975.) We will illustrate this by taking as an example $\lambda\phi^4$ theory in one space and one time dimension. The Lagrangian is given by

$$L = \int [\tfrac{1}{2}(\partial_0\phi)^2 - \tfrac{1}{2}(\partial_x\phi)^2 - V(\phi)]\,\mathrm{d}x \tag{15.32}$$

where

$$V(\phi) = \frac{\lambda}{2}(\phi^2 - a^2)^2 \tag{15.33}$$

and

$$a^2 = \mu^2/\lambda.$$

The Hamiltonian is given by

$$H = \int [\tfrac{1}{2}(\partial_0\phi)^2 + \tfrac{1}{2}(\partial_x\phi)^2 + V(\phi)]\,\mathrm{d}x. \tag{15.34}$$

As we discussed in §5.3, the classical ground-state configuration for the case $\mu^2 > 0$ is

$$\phi = \pm a = \pm\sqrt{\frac{\mu^2}{\lambda}} \tag{15.35}$$

and the ground-state energy is $E = 0$. An interesting feature of this model is that there exists a static (time-independent) finite-energy solution to the equation of motion (*solitons*). The time-independent solution to the equation of motion can be obtained from the Lagrangian L through the variational principle,

$$-\delta L = \delta \int \mathrm{d}x[\tfrac{1}{2}(\partial_x\phi)^2 + V(\phi)] = 0 \tag{15.36}$$

Mathematically, this is equivalent to the problem of motion of a particle of unit mass in a potential $-V(x)$, where the equation of motion is derived from

$$\delta \int \mathrm{d}t L' = \delta \int \mathrm{d}t \left[\frac{1}{2}\left(\frac{\mathrm{d}x}{\mathrm{d}t}\right)^2 + V(x)\right] = 0. \tag{15.37}$$

Every motion of the particle in the potential $-V(x)$ corresponds to a time-independent solution of the field equation. However, not all of these solutions are of finite energy. To get a finite-energy solution, we must require ϕ to go to a zero of $V(\phi)$ as $x \to \pm\infty$, so that the energy integral in eqn (15.34) is finite. In the particle problem, this corresponds to the condition that the particle must go to the zeros of the potential as $t \to \pm\infty$. Of course, the ground states where the particle sits at $x = a$ or $-a$ for all times will satisfy this requirement, but there are also non-trivial motions which also satisfy this requirement. Finiteness of energy requires the solution to take on the vacuum value ($\pm a$) at $t = \pm\infty$ but since we have a system of degenerate vacua the solution may take on different minima ($+a$ or $-a$) at different infinity points ($+\infty$ or $-\infty$). Thus for example there are motions where the particle starts on top of one hill and moves to the top of the other and has zero energy (Fig. 15.3). We can use this property of zero-energy motion to

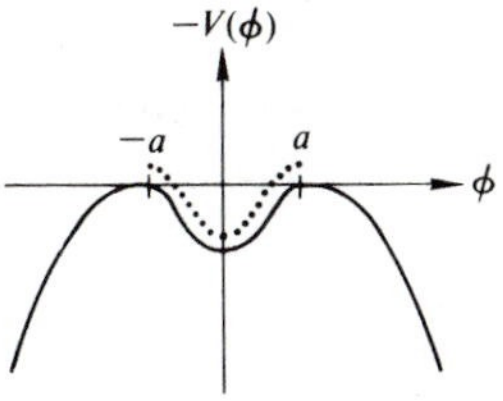

FIG. 15.3.

find the explicit form of the finite-energy solution to the field-theory case. From the energy conservation for the motion of the particle with zero total energy we have

$$\frac{1}{2}\left(\frac{dx}{dt}\right)^2 + [-V(x)] = 0$$

which corresponds to

$$\frac{1}{2}\left(\frac{d\phi}{dx}\right)^2 = V(\phi) \tag{15.38}$$

for the case of field theory. Eqn (15.38) can be solved easily by integration and the result is

$$x = \pm \int_{\phi_0}^{\phi} d\phi'[2V(\phi')]^{-1/2} \tag{15.39}$$

where ϕ_0 is the value of ϕ at $x = 0$ and can be any number between a and $-a$. The presence of this arbitrary parameter ϕ_0 is due to the translational invariance of eqn (15.38), i.e., if $\phi = f(x)$ is a solution, then $\phi = f(x - c)$ is also a solution where c is an arbitrary constant. For the case of $\lambda\phi^4$ theory, the potential is given in eqn (15.33) and the finite-energy solutions in eqn (15.39) can be written as

$$\phi_+(x) = a \tanh(\mu x) \tag{15.40}$$

$$\phi_-(x) = -a \tanh(\mu x). \tag{15.41}$$

The ϕ_+ is usually called the *kink* and ϕ_- the *anti-kink*. The energy of the kink (or anti-kink) can be calculated from eqns (15.40) and (15.34) to give

$$E = 4\mu^3/3\lambda \tag{15.42}$$

which is indeed finite. It is clear that as $x \to \pm\infty$, ϕ_+ (or ϕ_-) approaches the zeros of $V(\phi)$, i.e.

$$\phi_+ \to \pm a \quad \text{as} \quad x \to \pm\infty. \tag{15.43}$$

This behaviour is illustrated in Fig. 15.4.

These solutions can be shown to be stable with respect to small perturbations even though they are not the absolute minimum of the

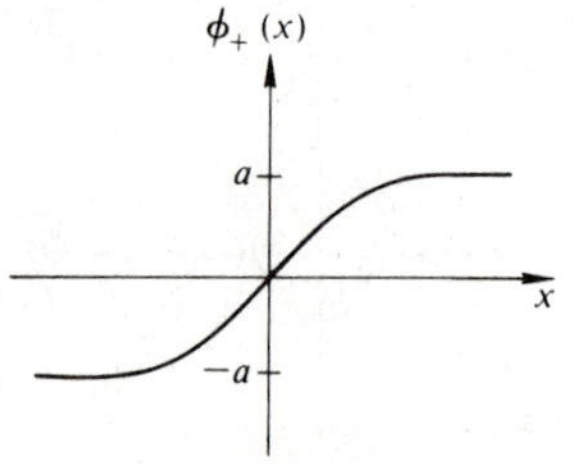

FIG. 15.4.

potential energy $V(\phi)$ (i.e. $\phi \neq \pm a$ for all x and t). The physical interest in these finite-energy solutions to the equation of motion comes from the fact that they resemble a particle, with structure, in the following respects.

(1) Its energy is concentrated in a finite region of space. This is because these solutions $\phi_\pm$ deviate from the ground-state configuration, $\phi = \pm a$ (zero energy) only in a small region around the origin.

(2) It can be made to move with any velocity less than unity. This is due to the fact that the equation of motion is Lorentz-covariant and we can apply a Lorentz boost to obtain a solution with non-zero velocity.

Topological conservation laws

The finite-energy solutions in $\lambda\phi^4$ theory in 1 + 1-dimensions have a rather interesting topological property which will make these solutions stable. This topological property can be easily generalized to a more complicated theory in higher dimensions and is very useful in finding stable finite-energy solutions.

The topological property of the kink (or anti-kink) solution in $\lambda\phi^4$ theory in two-dimensional space–time can be studied as follows. From the finite-energy requirement we have at spatial infinities

$$\phi(\infty) - \phi(-\infty) = n(2a) \tag{15.44}$$

where $n = 0$ corresponds to the ground state, $n = 1$ to the kink solution, and $n = -1$ to the anti-kink solution. Eqn (15.44) can be written as

$$\int_{-\infty}^{+\infty} (\partial_x\phi)\,\mathrm{d}x = n(2a). \tag{15.45}$$

Thus, if we define the current j_μ as

$$j_\mu(x) = \varepsilon_{\mu\nu}\,\partial^\nu\phi,$$

the current will be automatically conserved because $\varepsilon_{\mu\nu}$ is antisymmetric. The corresponding conserved charge is just eqn (15.45),

$$Q = \int_{-\infty}^{+\infty} j_0(x)\,\mathrm{d}x = \int_{-\infty}^{+\infty} \partial_x\phi\,\mathrm{d}x = n(2a). \tag{15.46}$$

This implies that the kink number n in (15.44) is a conserved quantum number. Thus there is no transition between kink (or anti-kink) solutions and ground states and they are stable. This conservation law, usually called the *topological conservation law* (Lubkin 1963), has a different origin from the usual Noether conservation laws (such as energy conservation) coming from the symmetry of the theory (see §5.1) in that it holds independently of the equations of motion. One intuitive way to understand this topological conservation law is that in order to convert the kink configuration (ϕ_+ or ϕ_-) to the ground-state configuration where $\phi = a$ or $-a$ for all x, we have to change the value of ϕ from $-a$ to a by penetrating the barrier around

$\phi = 0$ over an infinite range of x. This clearly will take an infinite amount of energy.

Thus the topological conservation law (eqn (15.46)) divides the finite-energy solutions into many distinct sectors; $n = 0$ (vacuum), $n = 1$ (kink), and $n = -1$ (anti-kink), etc. These different sectors can be characterized by their topological properties as follows. In two-dimensional space–time, the spatial infinities consist of two discrete points $\pm\infty$. Denote this set by S. For $\lambda\phi^4$ theory, the set of minima of the potential given in (15.35) also consists of two discrete points $\pm a$ and will be denoted by M_0,

$$M_0 = \{\phi\colon V(\phi) = 0\}. \tag{15.47}$$

The condition that the solution to the equation of motion must be of finite energy will imply that the asymptotic values of $\phi(x)$ must be zeros of $V(\phi)$, i.e.

$$\lim_{x \to \pm\infty} \phi(x) = \phi \in M_0. \tag{15.48}$$

This condition can be considered as a mapping from points in S to M_0. For example, in the ground-state configuration, both $\pm\infty$ are mapped to a (or $-a$), while the kink configuration ϕ_+ maps $+\infty$ to $+a$ and $-\infty$ to $-a$. These are topologically distinct mappings in the sense that it is impossible to continuously deform one mapping to the other. (For further discussion see §16.1.) This is the essence of the topological conservation laws. These topological properties will be very useful for more complicated theories in higher dimensions where explicit solutions are hard to come by.

To summarize, in $\lambda\phi^4$ theory in two dimensions, the finite-energy solutions to the equation of motion exist with non-trivial topological properties and these solutions are stable with respect to decaying into vacuum. It is clear that the existence of this type of topological, stable, finite-energy solution requires the theory to have degenerate vacua (spontaneous symmetry breaking) and non-trivial topological properties.

Solitons in four dimensions

So far we have only discussed the finite-energy solution in two-dimensional field theory. In the more realistic four-dimensional field theory we shall see that, in order to have topologically stable finite-energy solutions for the scalar field, long-range fields of magnetic type must be present. This will lead naturally to non-Abelian gauge theories.

First consider the finite-energy solution to a scalar field theory with some symmetry G in four dimensions. Write the Lagrangian density

$$\mathscr{L} = \tfrac{1}{2}(\partial_\mu\phi_i)^2 - V(\phi_i) \tag{15.49}$$

with

$$V(\phi_i) \geq 0. \tag{15.50}$$

Denote by M_0 the set of values of $\phi_i = \eta_i$ which minimize the potential $V(\phi_i)$,

$$M_0 = \{\phi_i = \eta_i\colon V(\eta_i) = 0\}. \tag{15.51}$$

Note that all the ϕ_is in M_0 are constant, independent of the space–time points x_μ. We will assume that these values are related by the symmetry group G. For example, the points $\phi = \pm a$ in the case of $\lambda\phi^4$ theory are related by the symmetry $\phi \to -\phi$. The possible directions in which the spatial coordinate $\mathbf{r}$ can go to infinity are labelled by a unit vector in the three-dimensional space,

$$S^2 = \{\hat{\mathbf{r}}: \hat{\mathbf{r}}^2 = 1\}. \tag{15.52}$$

It is clear that S^2 is a sphere in three-dimensional space. We say S^2 is a two-dimensional sphere, or two-sphere. Unlike the two-dimensional case where it consits of two discrete points ($\pm\infty$), S^2 is a connected set. This will make the topology of the spatial infinities in four-dimensional field theory very different from that of two-dimensional field theory. The energy for a given field configuration is given by

$$H = \int d^3x[\tfrac{1}{2}(\partial_0\phi_i)^2 + \tfrac{1}{2}(\nabla\phi_i)^2 + V(\phi_i)]. \tag{15.53}$$

The condition that the solution to the equation of motion has finite energy implies that as $r \to \infty$, ϕ_i approaches one of the zeros (minima) of $V(\phi_i)$, i.e.

$$\phi_i^\infty(\hat{\mathbf{r}}) = \lim_{R\to\infty} \phi(R\hat{\mathbf{r}}) \in M_0. \tag{15.54}$$

Note that for the ground-state configuration ϕ_i^∞ goes to the same value in all directions. S^2 being connected, ϕ_i^∞ would have to be constant (as it must be continuous) if M_0 is a discrete set (i.e. G is a discrete symmetry group). Then ϕ_i^∞ has the same topology as the vacuum configuration and is topologically trivial. To get a topologically non-trivial solution, M_0 must be a manifold with non-zero dimension. This requires the symmetry group G to be continuous. From eqn (15.53), the energy is bounded from below by

$$H \geq \int d^3x[\tfrac{1}{2}(\nabla\phi_i)^2 + V(\phi_i)]. \tag{15.55}$$

Write the gradient term $(\nabla\phi)^2$ as the sum of a radial and a transverse term

$$(\nabla\phi)^2 = \left(\frac{\partial\phi}{\partial r}\right)^2 + (\hat{\mathbf{r}} \times \nabla\phi)^2. \tag{15.56}$$

Since at infinity, ϕ_i^∞ is a function of direction $\hat{\mathbf{r}}$ only, if ϕ_i^∞ is not a constant, the second term in (15.56) is of order r^{-2} as $r \to \infty$. This will make the integral in H (15.55) divergent. Hence with scalar fields alone there are no topologically stable finite-energy solutions in four dimensions (Derrick 1964).

It turns out that this difficulty can be circumvented by adding gauge fields to the theory, i.e. by making the symmetry G a *local* symmetry. In this case the gradient term $\nabla_i\phi$ is replaced by the covariant derivative

$$D_i\phi = \nabla_i\phi + ig(\mathbf{A}_i\cdot\mathbf{T})\phi. \tag{15.57}$$

It is then possible (through subtle cancellation) to have $D_i\phi$ decrease like r^{-2}

while A_i^a and $\nabla_i\phi$ both decrease like r^{-1} such that the energy integral is convergent and the solution has non-trivial topological properties. Note that, for such time-independent solutions, the gauge field A_i^a decreases as r^{-1}; the field strength will decrease as r^{-2} which will correspond to a long-range magnetic field. (Cf. eqn (15.25).)

15.3 The 't Hooft–Polyakov monopole

As we mentioned in the last section, it is possible to find topologically non-trivial, finite-energy solutions in gauge theory with scalar fields. Since the electromagnetic interaction is described by an unbroken U(1) gauge symmetry and the finite-energy solution requires spontaneous symmetry breaking (the degenerate vacuum), we are led naturally to non-Abelian gauge theory in which the electromagnetic U(1) symmetry is embedded. Then the monopole will come out as a topologically non-trivial finite-energy solution. In this section, we will describe such a solution discovered by 't Hooft (1974) and Polyakov (1974), which has the properties of the magnetic monopole. The simplest example is the SO(3) model due to Georgi and Glashow (1972*a*). Even though this model is ruled out experimentally by the discovery of the neutral-current phenomena (see §11.2), it is the simplest example of a non-Abelian gauge theory having monopole solutions. And by studying the embedding of this SO(3) → SO(2) [i.e. SU(2) → U(1)] solution, one can also obtain monopoles occurring in theories of larger gauge groups—as we shall do so for the SU(5) grand unification model. Thus we concentrate first on this fundamental SO(3) case.

Soliton solution in the SO(3) model

The Georgi–Glashow model is based on the SU(2) gauge group with a triplet of Higgs scalars $\boldsymbol{\phi}$ (for the basic monopole solution we shall ignore the fermion fields). The Lagrangian density is

$$\mathscr{L} = -\tfrac{1}{4}F_a^{\mu\nu}F_{a\mu\nu} + \tfrac{1}{2}(\mathrm{D}^\mu\boldsymbol{\phi})\cdot(\mathrm{D}_\mu\boldsymbol{\phi}) - V(\boldsymbol{\phi}) \tag{15.58}$$

where

$$F_{\mu\nu}^a = \partial_\mu A_\nu^a - \partial_\nu A_\mu^a - e\varepsilon^{abc}A_\mu^b A_\nu^c \tag{15.59}$$

$$(\mathrm{D}_\mu\phi)^a = \partial_\mu\phi^a - e\varepsilon^{abc}A_\mu^b\phi^c \tag{15.60}$$

$$V(\boldsymbol{\phi}) = \frac{\lambda}{4}(\boldsymbol{\phi}\cdot\boldsymbol{\phi} - a^2)\,. \tag{15.61}$$

The equations of motion are

$$(\mathrm{D}_\nu F^{\mu\nu})_a = -e\varepsilon_{abc}\phi_b(\mathrm{D}^\mu\phi)_c \tag{15.62a}$$

$$(\mathrm{D}^\mu\mathrm{D}_\mu\phi)_a = -\lambda\phi_a(\boldsymbol{\phi}\cdot\boldsymbol{\phi} - a^2). \tag{15.62b}$$

In this model, the values of $\boldsymbol{\phi}$ which minimize the potential $V(\boldsymbol{\phi})$ in eqn (15.61) are given by

$$M_0 = \{\boldsymbol{\phi} = \boldsymbol{\eta};\, \eta^2 = a^2\}, \tag{15.63}$$

i.e. M_0 consists of points on a sphere in a three-dimensional internal symmetry space, and all points in M_0 are equivalent to each other by SO(3) transformations. For convenience, we choose

$$\boldsymbol{\phi} = (0, 0, a) \tag{15.64}$$

to be the ground-state configuration. The pattern of the spontaneous symmetry breaking is as follows (see discussion in §8.3)

$$\mathrm{SU}(2) \sim \mathrm{SO}(3) \rightarrow \mathrm{SO}(2) = \mathrm{U}(1)$$

because $\boldsymbol{\phi}$ given in eqn (15.64) is still invariant under the rotation around the 3-axis (SO(2) transformation). We will identify the unbroken U(1) symmetry as the electromagnetic interaction and the corresponding massless gauge bosons as the photon, $A_\mu^3 = A_\mu$. The electric and magnetic fields are then given by

$$F_3^{0i} = E^i, \qquad F_3^{ij} = \varepsilon^{ijk} B^k. \tag{15.65}$$

To get the finite-energy solution we require that as $\mathbf{r} \rightarrow \infty$, $\boldsymbol{\phi}(\mathbf{r})$ approaches values in M_0. Since the spatial infinities also form a two-sphere S^2 (see eqn (15.52)) which has the same topology as the set M_0, we can map each point in S^2 to the corresponding point in S^2 of M_0 to get a non-trivial topology

$$\phi_i^\infty = \eta_i = a\hat{\mathrm{r}}_i \tag{15.66}$$

It is not hard to see that this mapping cannot be deformed *continuously* into the mapping for the vacuum configuration where the whole S^2 is mapped to a point given in eqn (15.64). Thus, the mapping given in (15.66) is topologically stable. (For further discussion on mappings with nontrivial topology such as $S^1 \rightarrow S^1$ and $S^3 \rightarrow S^3$ see §16.1.)

Given (15.66) $\phi_b \propto \hat{r}_b$, the finite energy requirement that $D_\mu \phi_b = 0$ up to order r^{-2} (see discussion at the end of the last section) implies the gauge field asymptotic form of $A_i^b \propto \varepsilon_{bij} \hat{r}_j$. Believing that the lowest energy solutions correspond to those with the maximal symmetry, we can make the following *ansatz* for the explicit solution

$$\phi_b = \frac{r^b}{er^2} H(aer), \qquad A_b^i = -\varepsilon_{bij} \frac{r^j}{er^2}[1 - K(aer)], \qquad A_b^0 = 0 \tag{15.67}$$

where H and K are dimensionless functions to be determined by the equation of motion. A pictorial comparison of the $\boldsymbol{\phi}$ field configuration for the vacuum (15.64) and the monopole (15.67) is given in Fig. 15.5. For the ansatz given in eqn (15.67), which is time-independent, the energy of the system is given by

$$E = \frac{4\pi a}{e} \int_0^\infty \frac{\mathrm{d}\xi}{\xi^2} \left[\xi^2 \left(\frac{\mathrm{d}K}{\mathrm{d}\xi} \right)^2 + \frac{1}{2} \left(\xi \frac{\mathrm{d}H}{\mathrm{d}\xi} - H \right)^2 + \frac{1}{2} (K^2 - 1)^2 \right.$$
$$\left. + K^2 H^2 + \frac{\lambda}{4e^2} (H^2 - \xi^2)^2 \right] \tag{15.68}$$

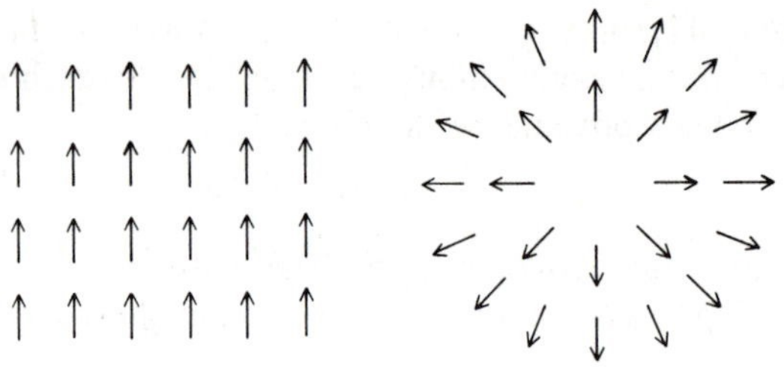

(a) (b)

FIG. 15.5. ϕ configurations (a) for the vacuum; (b) for the monopole.

where $\xi = aer$. The conditions for E to be stationary with respect to the variations of H and K are

$$\xi^2 \frac{d^2K}{d\xi^2} = KH^2 + K(K^2 - 1) \tag{15.69a}$$

$$\xi^2 \frac{d^2H}{d\xi^2} = 2K^2H + \frac{\lambda}{e^2} H(H^2 - \xi^2). \tag{15.69b}$$

These equations of motion for H and K can also be obtained from substituting the *ansatz* (15.67) into the equation of motion given in (15.62). The asymptotic condition (15.66) will imply that

$$H(\xi) \sim \xi \quad \text{as} \quad \xi \to \infty. \tag{15.70a}$$

Then in order to have a convergent integral in eqn (15.68), we require

$$K(\xi) \to 0 \quad \text{as} \quad \xi \to \infty \tag{15.70b}$$

and

$$H \leq O(\xi), \qquad K(\xi) - 1 \leq O(\xi) \quad \text{as} \quad \xi \to 0. \tag{15.70c}$$

It turns out that solutions to eqn (15.69) with boundary conditions (15.70) do exist and the functions H and K have the shapes shown in Fig. 15.6. The total energy of the solution, which will be interpreted as the classical mass, can be obtained from eqn (15.68)

$$\text{Mass} = \frac{4\pi a}{e} f(\lambda/e^2) \tag{15.71}$$

where $f(\lambda/e^2)$ is the value of the integral in (15.68) and has been calculated numerically to be of order unity for a wide range of values of λ/e^2. Thus the

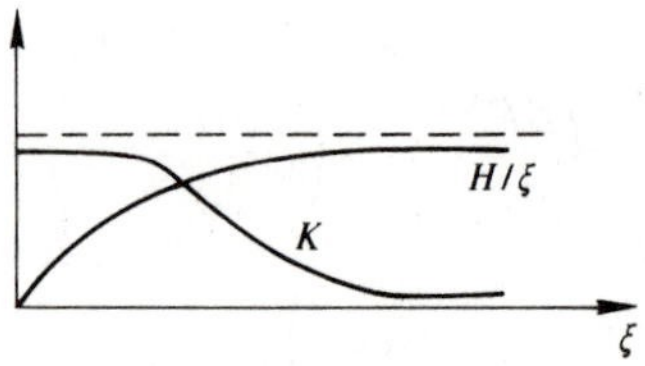

FIG. 15.6.

scale for the mass of this classical solution is set by the parameter a, which is the vacuum expectation value of the scalar field and is also the scale for the spontaneous symmetry breaking SO(3) → SO(2).

The 't Hooft–Polyakov soliton as a magnetic monopole

From the asymptotic condition (15.70b), we see that at large distances

$$F_a^{ij} \sim \frac{1}{er^4} \varepsilon^{ijk} r_a r_k \sim \frac{1}{aer^3} \varepsilon^{ijk} r^k \phi_a$$

which implies that the magnetic field in eqn (15.65) at large distance is

$$\mathbf{B} \sim \frac{-1}{e} \frac{\mathbf{r}}{r^3}. \tag{15.72}$$

Comparing this with eqn (15.16), we see that this field is due to a monopole which has a magnetic charge

$$g = -4\pi/e. \tag{15.73}$$

The constant e in (15.73) is the electromagnetic coupling constant which in this simple model is related to the electric charge operator by

$$Q = eT_3 \tag{15.74}$$

where T_3 is the third component of the weak isospin operators, which are the generators of SO(3) gauge symmetry. Since the smallest possible non-zero electric charge which might enter the theory is $q_0 = e/2$ corresponding to $T_3 = \frac{1}{2}$, we see that eqn (15.73) gives

$$\frac{q_0 g}{4\pi} = \frac{-1}{2}. \tag{15.75}$$

Thus the magnetic charge g of the monopole takes its lowest value when compared to the Dirac condition (15.11). This classical finite energy solution which is topologically non-trivial is called the 't Hooft–Polyakov monopole. Note that the quantization of the electric charge given in (15.74) is a consequence of the fact that the electromagnetic U(1) symmetry is embedded in the *simple* non-Abelian gauge group, in this case SO(3), and is independent of the existence of the monopole solution. Rather, here one discovers that both magnetic monopole and charge quantization are consequences of spontaneous symmetry breaking of a non-Abelian simple group down to the electromagnetic U(1).

This 't Hooft–Polyakov monopole differs from the Dirac monopole in two important aspects; the 't Hooft–Polyakov monopole has a finite core while the Dirac monopole is a point and there is no need for the Dirac string in the 't Hooft–Polyakov monopole. The finite size of the 't Hooft–Polyakov monopole core comes from the fact that for large ξ eqn (15.69a,b) becomes

$$\frac{d^2K}{d\xi^2} = K \qquad \frac{d^2h}{d\xi^2} - \frac{2\lambda}{e^2} h = 0$$

where $H = h + \xi$. Thus for large ξ, we have

$$K \sim \mathrm{e}^{-\xi} \cong \mathrm{e}^{-Mr}$$
$$H - \xi \sim \mathrm{e}^{-\mu\xi/M} \cong \mathrm{e}^{-\mu r}$$

where $\mu = (2\lambda)^{\frac{1}{2}}a$ and $M \simeq ea$ are the masses of the scalar and gauge bosons, respectively. This implies that the approach to the asymptotic form of each field is controlled by the masses of the corresponding particle. Hence we can think of the 't Hooft–Polyakov monopole as having a definite size determined by these masses. For distances smaller than this size, the massive fields play a role in providing a smooth structure and, for distances larger than this size, they vanish rapidly to give a field configuration indistinguishable from that of a Dirac monopole. From this one can also have a simple understanding of its mass value as given in (15.71). We divide the contribution to energy in (15.68) into two parts, coming from fields inside and outside the core respectively. Outside the core, $\mathrm{D}_\mu\phi = 0$ and the electric field $E = 0$, only the magnetic field survives:

$$\int \mathrm{d}^3x\tfrac{1}{2}B^2 = \frac{1}{2}\left(\frac{g}{4\pi}\right)^2 \int_{1/M}^{\infty} 4\pi r^2 \,\mathrm{d}r \frac{1}{r^4} = \frac{1}{2}\frac{4\pi}{e^2}M.$$

For the stationary solution the core contribution should be comparable, yielding $E = \mathrm{O}(M/\alpha)$. Thus the monopole is heavy because it has a small core and the Coulombic magnetic energy diverges as $r \to 0$.

As for the Dirac string, it is replaced in the 't Hooft–Polyakov monopole by the scalar field. To see this, write the asymptotic solution as

$$A_a^i = \varepsilon^{aij}\frac{r_j}{er^2}, \qquad \phi_b = \frac{ar_b}{r}. \tag{15.76}$$

We can write A_a^i in eqn (15.76) as

$$A_a^i = \frac{1}{a^2 e}\,\varepsilon^{abc}\phi^b\,\partial^i\phi^c \tag{15.77}$$

and the magnetic field tensor at large distance is given by

$$\begin{aligned} F_3^{ij} &= \partial^i A_3^j - \partial^j A_3^i - e(A_1^i A_2^j - A_2^j A_1^i) \\ &= \partial^i A_3^j - \partial^j A_3^i + \frac{1}{ea^3}\boldsymbol{\phi}\cdot(\partial^i\boldsymbol{\phi}\times\partial^j\boldsymbol{\phi}). \end{aligned} \tag{15.78}$$

Thus the magnetic field tensor has the form

$$F^{ij} = \partial^i A^j - \partial^j A^i + (\text{extra term}).$$

In the Dirac monopole the extra term is singular and involves the Dirac string, while in the 't Hooft–Polyakov monopole the extra term is smooth and involves the scalar fields. Thus, in some sense, in the 't Hooft–Polyakov monopole the singular Dirac string has been 'smoothed' into a scalar field.

To summarize, in the SO(3) model where the non-Abelian symmetry is spontaneously broken down to the electromagnetic U(1) symmetry, there exists a topologically non-trivial finite-energy solution, the 't Hooft–Polyakov monopole, with the following features.

(1) It has the same behaviour as the Dirac monopole at large distance;
(2) It has a finite core size determined by the masses of the gauge boson or scalar particle;
(3) The classical mass of the monopole is of order of the spontaneous symmetry-breaking scale–the vacuum expectation value of the scalar field;
(4) There is no need for the Dirac string.

Coupling of the spatial and internal symmetries

Finally let us take note of an important feature of the monopole in the non-Abelian gauge theory with spontaneous symmetry breaking: it mixes the spatial and the internal symmetries. In fact the 't Hooft–Polyakov ansatz (15.67) has the feature that it is symmetric with respect to rotations generated by

$$\mathbf{J} = \mathbf{L} + \mathbf{T} \tag{15.79}$$

where $\mathbf{L}$ is the spatial angular momentum (including the 'ordinary' spin) and $\mathbf{T}$ is the internal symmetry 'isospin' generator. Thus for example the form of $\boldsymbol{\phi}\cdot\mathbf{T} \propto \mathbf{r}\cdot\mathbf{T}$ in (15.67) shows clearly that it is invariant under $\mathbf{J}$. As a consistency check, we shall write out (15.79) more explicitly. The canonical momentum being $p_i = mv_i + e(\mathbf{A}_i\cdot\mathbf{T})$ with the gauge field outside the core being given by (15.76) we have

$$\mathbf{p} = m\dot{\mathbf{r}} + \frac{1}{r}\hat{\mathbf{r}} \times \mathbf{T} \tag{15.80}$$

and

$$\begin{aligned}\mathbf{J} &= \mathbf{r} \times \mathbf{p} + \mathbf{T}\\ &= \mathbf{r} \times m\dot{\mathbf{r}} + \hat{\mathbf{r}} \times \hat{\mathbf{r}} \times \mathbf{T} + \mathbf{T}\\ &= \mathbf{r} \times m\dot{\mathbf{r}} + (\hat{\mathbf{r}}\cdot\mathbf{T})\hat{\mathbf{r}}.\end{aligned} \tag{15.81}$$

Comparing to (15.18):

$$\mathbf{J} = \mathbf{r} \times m\dot{\mathbf{r}} - \frac{g}{4\pi} Q\hat{\mathbf{r}} \tag{15.82}$$

and noting that the magnetic charge of the 't Hooft–Polyakov monopole is given by (15.73), we have the consistency condition

$$Q = e\hat{\mathbf{r}}\cdot\mathbf{T}. \tag{15.83}$$

This is indeed the correct identification because outside the core we have $\hat{\boldsymbol{\phi}} = \hat{\mathbf{r}}$ and the unbroken electromagnetic U(1) symmetry corresponds to those SU(2) rotations that leave $\boldsymbol{\phi}$ invariant. For example, if $\hat{\mathbf{r}}$ is in the z-direction, $\hat{\boldsymbol{\phi}}$ will be in the 3-direction in the SO(3) space and $\hat{\mathbf{r}}\cdot\mathbf{T} = T_3$ is the generator for rotation around the 3-axis which will leave $\hat{\boldsymbol{\phi}}$ invariant.

This 'coupling' of the spatial and internal symmetries has some surprising consequences. For example, consider scalar particles (ϕ^+, ϕ^-) transforming as a doublet under **T**: with $Q = eT_3$ we have

$$Q\phi^{\pm} = \pm\tfrac{1}{2}e\phi^{\pm}.$$

Equation (15.79) informs us that the combined system of a monopole and this isodoublet can have half-interger angular momentum even though no fundamental fermion field has been introduced. This phenomenon of 'spin from isospin' has been discussed by Jackiw and Rebbi (1976), and by Hasenfratz and 't Hooft (1976). That this does not violate the usual connection between spin and statistics has been shown by Goldhaber (1976).

By the same reasoning we can see that the Dirac equation in the background monopole field $A_\mu(x)$ can have integral J partial wave solutions. In this connection there is another surprising result: fermions are found to change their nature when scattering off a monopole in the S-wave. As the angular momentum in (15.79) and (15.82) acquires a sign change when the fermion passes through the monopole core: $\hat{\mathbf{r}} \to -\hat{\mathbf{r}}$, $\mathbf{J}$ can be conserved only if (i) the charge, or (ii) the helicity of the fermion, make the corresponding change. This 'paradox' can be resolved only when one invokes results of the quantum monopole theory. (1) For the charge change, one discovers that a quantized monopole possesses a tower of excited states which have the charges of the gauge bosons coupling to the upper and lower members of an isodoublet. Such electrically and magnetically charged states are termed *dyons*. Thus overall charge conservation can be maintained in the fermion–monopole scattering as monopole turns into a dyon,

$$\mathcal{M} + \phi^+ \to \mathcal{D} + \phi^-. \tag{15.84}$$

(2) For the helicity change, this is possible in the presence of monopole fields because we have a nonvanishing axial anomaly: $\mathrm{F}\cdot\tilde{\mathrm{F}} = \mathbf{E}\cdot\mathbf{B} \neq 0$ and the possibility of chirality nonconservation (see §6.2), and for the massless fermions chirality and helicity changes are correlated.

Grand unified monopoles

The SU(2) → U(1) 't Hooft–Polyakov monopole solution may be regarded as the fundamental pattern when we consider the monopole solutions to larger gauge groups. Topological considerations lead to the general result that stable monopole solutions occur for any gauge theories in which a *simple* gauge group G is broken down to a smaller group $H = h \times \mathrm{U}(1)$ containing an explicit U(1) factor. For a review of the topological arguments see (Coleman 1975, 1981). Clearly this is compatible with our expectation that charge quantization and existence of monopole are related and that charge quantization follows from the spontaneous symmetry breaking of a *simple* gauge group. Thus in the grand unified theories where the symmetry is broken from some large *simple* group, e.g. SU(5), to $\mathrm{SU}(3)_c \times \mathrm{SU}(1)_{em}$, there are also monopole solutions of the 't Hooft–Polyakov type. The monopole mass is determined by the mass scale for the symmetry breaking, M_X, and is of order M_X/e^2. In the SU(5) model, we have $M_X \gtrsim 10^{14}$ GeV which implies a very heavy monopole mass $\sim 10^{16}$ GeV. This means that this type of monopole is out of reach for its production by accelerators. But it could be relevant for physics in the extreme early universe (see Preskill 1979; Guth and Tye 1980). In fact the attempts to suppress the monopole abundance in the conventional cosmology led originally to the '*inflationary universe scenario*' (Guth 1981) and its subsequent refinement (Linde 1982; Albrecht and Steinhardt 1982; Hawking and Moss 1982), which has the promise of solving several fundamental problems in cosmology.

It should be noted that the stable grand unified monopole having the smallest magnetic charge (hence the smallest mass) is expected to have both the 'ordinary' and the colour magnetic charges. (The colour magnetic fields are then supposed to be screened by the gluons.) Namely the SU(2) → U(1) embedding in G → $SU(3)_c \times U(1)_{em}$ is such that the final 'magnetic U(1)' factor sits in both $U(1)_{em}$ and $SU(3)_c$. To see this more explicitly consider the following SU(2) embedding in the SU(5) group of Chapter 14:

$$\mathbf{T} = \tfrac{1}{2}\begin{pmatrix} 0 & & & \\ & 0 & & \\ & & \boldsymbol{\tau} & \\ & & & 0 \end{pmatrix}, \qquad (15.85)$$

where $\boldsymbol{\tau}$ are the Pauli matrices. Thus they act both in the colour SU(3) (the first three components) and the electroweak SU(2) sectors. (The alternatives with the third colour replaced by either the first or the second colour are equally possible.) It has been shown that this embedding (15.85) leads to the smallest magnetic charges (Dokos and Tomaras 1980). Given the electromagnetic charge matrix of (14.12) in SU(5) model, the form of the monopole Higgs field $\hat{\boldsymbol{\phi}} \sim \hat{\mathbf{r}}$ suggests that the charge matrix can be written in the following spherically symmetric form

$$Q = \frac{e}{3}\begin{pmatrix} -1 & & & \\ & -1 & & \\ & & 1 - \frac{2}{3}\hat{\mathbf{r}}\cdot\boldsymbol{\tau} & \\ & & & 0 \end{pmatrix} \qquad (15.86)$$

which agrees with (14.12) when $\hat{\mathbf{r}} = \hat{\mathbf{z}}$. The monopole vector potential is then given as in (15.76):

$$A_i^a = \frac{1}{g_5 r}\varepsilon_{aij}\hat{r}_j \qquad (15.87)$$

where g_5 is the SU(5) gauge coupling constant and is related to the electromagnetic coupling e by (14.50) and the QCD $SU(3)_c$ coupling g_3 by (14.47):

$$e = \sqrt{(\tfrac{3}{8})}g_5, \qquad g_5 = g_3. \qquad (15.88)$$

Equation (15.87) yields a matrix-valued magnetic field as

$$\mathbb{B}_i = B_i^a T^a = \tfrac{1}{2}\varepsilon_{ijk}F_{jk}^a T^a = \frac{\hat{\mathbf{r}}\cdot\mathbf{T}}{g_5 r^2}\,\hat{r}_i. \tag{15.89}$$

One can project out the electromagnetic and colour components by taking the trace of the product of this magnetic field matrix $\mathbb{B}$ with the generator matrices of the $U(1)_{em}$ and $SU(3)_c$ gauge groups. For the normalization factors one should keep in mind the relations in (15.88) and the fact that for the fundamental representation $\mathrm{tr}(T^a T^b) = \frac{1}{2}\delta^{ab}$. Thus (15.89) has the electromagnetic component

$$B_i = 2\sqrt{\left(\frac{3}{8}\right)}\,\mathrm{tr}(\mathbb{B}_i Q) = \frac{1}{2e}\frac{\hat{r}_i}{r^2} \tag{15.90}$$

corresponding to a magnetic charge of one Dirac unit $g = 2\pi/e$. Similarly the colour magnetic fields are calculated to be

$$B_i^a = 2\,\mathrm{tr}(\mathbb{B}_i\lambda^a) = \frac{1}{\sqrt{3}}\,\delta^{a8}\,\frac{\hat{r}_i}{g_s r^2} \qquad a = 1, \ldots, 8. \tag{15.91}$$

Clearly (15.90) and (15.91) show that the 'magnetic charge matrix' has two components, represented in the unitary gauge as

$$T_3 = -\tfrac{1}{2}(Q + Q_c) \tag{15.92}$$

Q_c being the colour hypercharge $\lambda_8/\sqrt{3}$.

One can see this need for colour magnetic field more 'physically' by considering an Aharonov–Bohm experiment for a fractionally charged quark, say d_1 (i.e. a down-quark with the first colour). The null-effect condition is satisfied because of a cancellation due to the presence of the extra phase factor coming from the quark colour charge q_c multiplying the colour magnetic flux of the monopole

$$\exp\left\{-\mathrm{i}\frac{e}{3}\oint \mathrm{d}\mathbf{l}\cdot\mathbf{A}_{em} + \mathrm{i}q_c\oint \mathrm{d}\mathbf{l}\cdot\mathbf{A}_c\right\} = \mathrm{e}^{-\mathrm{i}2\pi/3}\,\mathrm{e}^{\mathrm{i}2\pi/3} = 1$$

where we have used $e = 2\pi/g$, etc.

Monopole catalysed proton decay. With respect to the SU(2) embedding (15.86) the fifteen left-handed members of the first SU(5) fermion family (14.7) and (14.9) have the following transformation properties:

$$\text{doublets:}\ \begin{pmatrix} e^+ \\ d_3 \end{pmatrix}_L \begin{pmatrix} d_3^c \\ e^- \end{pmatrix}_L \begin{pmatrix} u_2^c \\ u_1 \end{pmatrix}_L \begin{pmatrix} u_1^c \\ u_2 \end{pmatrix}_L$$

$$\text{singlets:}\ \nu_L, d_{1L}, d_{2L}, u_{3L}, d_{1L}^c, d_{2L}^c, u_{3L}^c. \tag{15.93}$$

With these doublets it is possible to have baryon-number-changing reactions such as

$$\mathscr{M} + u_1 + u_2 + d_3 \to \mathscr{M} + e^+. \tag{15.94}$$

This may be regarded as the inverse process of (15.84) with subsequent 'dyon decay'. Normally one would expect that such transitions be strongly suppressed by the super-heavy mass factor M_X. However studies of the fermion–monopole dynamics by Rubakov (1982), Callan (1982), and Wilczek (1982*a*,*b*) suggest that the fermion–monopole vacuum may be highly degenerate and condensates such as $\langle u_1 u_2 d_3 e^- \rangle_0$ exist in regions of space where the magnetic and chromomagnetic fields of the grand unified monopole coexist. We expect the colour magnetic field to extend as far as the confinement radius of order 1 fermi. Thus it is suggested that the baryon-number-nonconserving interaction (15.94) may have a typical strong interaction cross-section. This will manifest as proton decays strongly catalysed by monopoles. Although the theoretical understanding of this mechanism is still under investigation, it is clear that quantum theory of monopole–particle interactions will be highly nontrivial and should contain much interesting physics.

16 Instantons

In the previous chapter we encountered particle-like solutions of Yang–Mills field theory, the monopoles; they correspond to fields with nontrivial topological properties in ordinary three-dimensional space. We now study a class of solutions with topological structure in the Euclidean four-dimensional space–time, the instanton solutions. Like the soliton solutions of Chapter 15, they have finite spatial extension—thus the '-on' in its name—and unlike solitons, they are also structures in time (albeit imaginary time)—thus the 'instant-'. For the same reason, they are also called 'pseudo particles' in the literature.

Again, as in Chapter 15, the presentation will be given at an elementary level. The organization of this chapter is as follows. In §16.1, after a brief introduction to the topological notion of homotopy we show how the instanton solution can be obtained in the Euclidean Yang–Mills theory. The interpretation of instantons as tunnelling events between vacuum states with different topological quantum numbers is presented in §16.2. It shows that the vacuum of non-Abelian gauge theory in general, and QCD in particular, is not unique. A Yang–Mills Lagrangian actually represents a continuum of theories labelled by a parameter θ—just as each value of coupling constant describes a different theory. In §16.3 massless fermions are incorporated into the theory; their presence suppresses the vacuum tunnelling. One then sees how the nontrivial structure of QCD vacuum as revealed by the instanton solutions can help us solve the famous axial U(1) problem present in any quark theory with chiral symmetry. On the other hand the instantons themselves bring about strong P and CP violations. Possible ways out of this difficulty are briefly mentioned.

Throughout this chapter, especially in the first two sections, we work with the SU(2) Yang–Mills theory. For applications we consider mostly those for QCD which is SU(3). However all results are understood to be valid for the SU(2) subgroup of SU(3), or any other non-Abelian groups of higher rank.

16.1 The topology of gauge transformations

In Chapter 15 we encountered fields with nontrivial topology. They correspond to mappings from S^2 to S^2, where S^2 is a sphere in three-dimensional space and is a two-dimensional manifold, i.e. a two-sphere. In this chapter we need to consider mappings $S^3 \to S^3$, where S^3 is a sphere in four-dimensional Euclidean space, i.e. a three-sphere.

Homotopic classes

To study the topological properties of continuous functions, one can divide them into *homotopic classes*; each class is made up of functions that can be deformed continuously into each other. More precisely, let X and Y be two topological spaces and $f_0(x), f_1(x)$ two continuous functions from X to Y. Let I denote the unit interval on the real line $0 \leq t \leq 1$; f_0 and f_1 are said to be *homotopic* if and only if there is a *continuous* function $F(x, t)$ which maps the direct product of X and I to Y such that $F(x, 0) = f_0(x)$ and $F(x, 1) = f_1(x)$. The continuous function $F(x, t)$ which deforms the function $f_0(x)$ continuously into $f_1(x)$ is called the *homotopy*. We can then divide all functions from X to Y into homotopic classes such that two functions are in the same class if they are homotopic.

To illustrate the notion of homotopic classes, we consider the following examples.

$\mathbf{S}^1 \to \mathbf{S}^1$. Let X be the points on a unit circle labelled by the angle $\{\theta\}$, with θ and $\theta + 2\pi$ identified, and let Y be a set of unimodular complex numbers $u_1 = \{e^{i\sigma}\}$, which is topologically equivalent to a unit circle, a 'one-dimensional sphere'. We consider the mapping $\{\theta\} \to \{e^{i\sigma}\}$. The continuous functions given by

$$f(\theta) = \exp[i(n\theta + a)] \tag{16.1}$$

form a homotopic class for different values of a and a fixed integer n. This is because we can construct a homotopy

$$F(\theta, t) = \exp\{i[n\theta + (1 - t)\theta_0 + t\theta_1]\} \tag{16.2}$$

such that

$$f_0(\theta) = \exp[i(n\theta + \theta_0)]$$

and

$$f_1(\theta) = \exp[i(n\theta + \theta_1)] \tag{16.3}$$

are homotopic. One can visualize $f(\theta)$ of (16.1) as a mapping of a circle on to another circle. In this mapping, n points of the first circle are mapped into one point of the second circle and we can think of this as 'winding around it n times'. Thus, each homotopic class is characterized by the *winding number n*, also called the *Pontryargin index*. From (16.1), the winding number n for a given mapping $f(\theta)$ can be written

$$n = \int_0^{2\pi} \frac{d\theta}{2\pi} \left[\frac{-i}{f(\theta)} \frac{df(\theta)}{d\theta} \right]. \tag{16.4}$$

Of particular interest is the mapping with the lowest nontrivial winding number, $n = 1$,

$$f^{(1)}(\theta) = e^{i\theta}. \tag{16.5}$$

By taking powers of this mapping, we can get mappings of higher winding numbers. For instance, the mapping $[f^{(1)}(\theta)]^m$ will have winding number m.

We can also write the mapping in (16.5) in the Cartesian coordinate system as

$$f(x, y) = x + \mathrm{i}y \quad \text{with} \quad x^2 + y^2 = 1. \tag{16.6}$$

We can generalize the domain X of this mapping from the unit circle to the whole real line $-\infty \leq x \leq \infty$, by identifying the end-points $x = \infty$ and $x = -\infty$ to be the same point, i.e. the mappings are required to satisfy the property $f(x = \infty) = f(x = -\infty)$. Clearly this has the same topology as the unit circle. Examples of this type of mapping with winding number $n = 1$ are

$$f_1(x) = \exp\{\mathrm{i}\pi x/(x^2 + \lambda^2)^{\frac{1}{2}}\} \tag{16.7}$$

$$f'_1(x) = \exp\{\mathrm{i}2 \sin^{-1}[x/(x^2 + \lambda^2)^{\frac{1}{2}}]\} = \frac{(\lambda + \mathrm{i}x)^2}{\lambda^2 + x^2}. \tag{16.8}$$

where λ is an arbitrary number. In this case the topological winding number for a general mapping can be expressed as

$$n = \frac{1}{2\pi} \int_{-\infty}^{+\infty} \mathrm{d}x \left[\frac{-\mathrm{i}}{f(x)} \frac{\mathrm{d}f(x)}{\mathrm{d}x} \right] \tag{16.9}$$

which yields $n = 1$ for the functions $f_1(x)$ and $f'_1(x)$ given in (16.7) and (16.8).

$S^3 \to S^3$. We now consider mappings from a three-sphere to SU(2) space, i.e. mappings from the points on S^3, the sphere in four-dimensional Euclidean space labelled by three angles, to the elements of the SU(2) group, which are also characterized by three parameters. More explicitly, the manifold of the SU(2) group elements is topologically equivalent to a three-sphere S^3. This is because, any element in the SU(2) group can be written in terms of the Pauli matrices as $U = \exp\{\mathrm{i}\boldsymbol{\varepsilon} \cdot \boldsymbol{\tau}\}$. From the identities of the Pauli matrices, we can write $U = u_0 + \mathrm{i}\mathbf{u} \cdot \boldsymbol{\tau}$ with real u_0 and $\mathbf{u}$ satisfying

$$u_0^2 + \mathbf{u}^2 = 1$$

which follows from $UU^\dagger = U^\dagger U = 1$. This is clearly the equation for the sphere in four-dimensional Euclidean space, S^3. Mappings in this case are also characterized by the topological winding number n and the generalization of (16.6) with $n = 1$ is

$$f(x_0, \mathbf{x}) = x_0 + \mathrm{i}\mathbf{x} \cdot \boldsymbol{\tau} \quad \text{with} \quad x_0^2 + \mathbf{x}^2 = 1. \tag{16.12}$$

It can be shown that the topological winding number n can be expressed as (see, for example, Coleman 1977)

$$n = \frac{-1}{24\pi^2} \int \mathrm{d}\theta_1 \, \mathrm{d}\theta_2 \, \mathrm{d}\theta_3 \, \mathrm{tr}(\varepsilon_{ijk} A_i A_j A_k)$$

where

$$A_i = f^{-1}(x_0, \mathbf{x}) \, \partial_i f(x_0, \mathbf{x}) \tag{16.13}$$

and θ_1, θ_2, and θ_3 are the three angles that parametrize S^3.

One can also generalize to the case where the domain X is the whole three-dimensional space with all points at infinity identified. Examples of the mappings with $n = 1$ are

$$f_1(x) = \exp\{\mathrm{i}\pi\mathbf{x}\cdot\boldsymbol{\tau}/(x^2 + \lambda^2)^{\frac{1}{2}}\} \tag{16.14}$$

$$f'_1(x) = (\lambda\boldsymbol{\tau} + \mathrm{i}\mathbf{x})^2/(\mathbf{x}^2 + \lambda^2) \tag{16.15}$$

which are generalizations of the mappings given in (16.7) and (16.8). As for the winding number in (16.9), we now have the volume integral

$$n = \frac{-1}{24\pi^2}\int_V \mathrm{d}^3x \ \mathrm{tr}(\varepsilon_{ijk}A_iA_jA_k)$$

$$A_i = f^{-1}(x)\,\partial_i f(x). \tag{16.16}$$

As we shall see, SU(2) transformations of the form (16.12), (16.14), or (16.15) are very much of physical interest.

The instanton solution to Euclidean gauge theory

Here one seeks the (finite-action) solution to classical Yang–Mills theory in Euclidean space ($x^2 = x_0^2 + \mathbf{x}^2$). The SU(2) gauge fields

$$A_\mu = \frac{\tau^a}{2}A^a_\mu, \quad F_{\mu\nu} = \frac{\tau^a}{2}F^a_{\mu\nu} \quad a = 1, 2, 3$$

have the Lagrangian

$$\mathscr{L} = \frac{1}{2g^2}\,\mathrm{tr}\,F_{\mu\nu}F_{\mu\nu} \tag{16.17}$$

where for notational convenience we have scaled the gauge fields as

$$A_\mu \to \frac{\mathrm{i}}{g}A_\mu, \qquad F_{\mu\nu} \to \frac{\mathrm{i}}{g}F_{\mu\nu}$$

with

$$F_{\mu\nu} = \partial_\mu A_\nu - \partial_\nu A_\mu + [A_\mu, A_\nu]. \tag{16.18}$$

Under a gauge transformation U, we now have

$$A'_\mu = U^{-1}A_\mu U + U^{-1}\,\partial_\mu U. \tag{16.19}$$

We require the solution to satisfy the boundary condition that the Lagrange density vanishes, i.e. $F_{\mu\nu}F_{\mu\nu} = 0$, at infinity so that the Euclidean action,

$$S_\mathrm{E} = \int \mathrm{d}^4x \frac{1}{2g^2}\,\mathrm{tr}(F_{\mu\nu}F_{\mu\nu}),$$

is finite. This means that in Euclidean space

$$F_{\mu\nu}(x) \underset{|x|\to\infty}{\longrightarrow} 0. \tag{16.20}$$

Normally we take this to mean

$$A_\mu(x) \underset{|x|\to\infty}{\longrightarrow} 0. \tag{16.21}$$

From the viewpoint of gauge transformations (16.19), this is much too restrictive. Condition (16.20) only requires A_μ to approach the configuration

$$A_\mu(x) \underset{|x|\to\infty}{\longrightarrow} U^{-1}\,\partial_\mu U \tag{16.22}$$

which is obtained from $A_\mu(x) = 0$ by a gauge transformation (16.19) and thus also yields (16.20). We say such a $A_\mu(x)$ field is a '*pure gauge*'. One notes that points at infinity ($|x| \to \infty$) in four-dimensional Euclidean space are three-spheres and the gauge transformation U in (16.22) represents mappings from S^3 to SU(2) space. Thus the Us are just the $S^3 \to S^3$ type of functions discussed in connection with homotopic classes, with some topological winding number. The instanton solution discovered by Belavin, Polyakov, Schwartz, and Tyupkin (1975) corresponds to Us with a nontrivial winding number, i.e. $n = 1$ (with $n > 1$ solutions for instantons far apart in space-time obtainable by multiplication).

Let us first express the winding number in terms of the gauge fields. For this purpose we introduce an (unobservable) gauge-dependent current

$$K_\mu = 4\varepsilon_{\mu\nu\lambda\rho}\,\mathrm{tr}[A_\nu\,\partial_\lambda A_\rho + \tfrac{2}{3}A_\nu A_\lambda A_\rho]. \tag{16.23}$$

It is straightforward to check that

$$\partial_\mu K_\mu = 2\,\mathrm{tr}(F_{\mu\nu}\tilde{F}_{\mu\nu}) \tag{16.24}$$

where

$$\tilde{F}_{\mu\nu} = \tfrac{1}{2}\varepsilon_{\mu\nu\lambda\rho}F_{\lambda\rho} \tag{16.25}$$

is the dual of $F_{\mu\nu}$. Consider the volume integral

$$\int \mathrm{d}^4x\,\mathrm{tr}(F_{\mu\nu}\tilde{F}_{\mu\nu}) = \tfrac{1}{2}\int \mathrm{d}^4x\,\partial_\mu K_\mu = \tfrac{1}{2}\int_S \mathrm{d}\sigma_\mu K_\mu \tag{16.26}$$

where the surface integral is over the S^3 at infinity. In this region A_μ is given by (16.22) and using the antisymmetry properties of the indices and $U^\dagger U = 1$, we have

$$K_\mu = \tfrac{4}{3}\varepsilon_{\mu\nu\lambda\rho}\,\mathrm{tr}[(U^\dagger\,\partial_\nu U)(U^\dagger\,\partial_\lambda U)(U^\dagger\,\partial_\rho U)]. \tag{16.27}$$

Substituting (16.27) into (16.26) and comparing it to the expression for the $S^3 \to S^3$ winding number in (16.16) we obtain

$$n = \frac{1}{16\pi^2}\int \mathrm{d}^4x\,\mathrm{tr}(F_{\mu\nu}\tilde{F}_{\mu\nu}). \tag{16.28}$$

To find $A_\mu(x)$ satisfying the boundary condition (16.22) on the Euclidean three-sphere at infinity, we make use of the important positivity condition in Euclidean space

$$\mathrm{tr}\int (F_{\mu\nu} \pm \tilde{F}_{\mu\nu})^2\,\mathrm{d}^4x \geq 0. \tag{16.29}$$

Since

$$(F_{\mu\nu} \pm \tilde{F}_{\mu\nu})^2 = 2(F_{\mu\nu}F_{\mu\nu} \pm F_{\mu\nu}\tilde{F}_{\mu\nu}), \tag{16.30}$$

we have the inequality

$$\operatorname{tr}\int F_{\mu\nu}F_{\mu\nu}\,\mathrm{d}^4x \geq \left|\operatorname{tr}\int F_{\mu\nu}\tilde{F}_{\mu\nu}\,\mathrm{d}^4x\right| = 16\pi^2 n \tag{16.31}$$

where we have used (16.28). Thus, the Euclidean action satisfies the inequality

$$S_{\mathrm{E}}(A) \geq \frac{8\pi^2 n}{g^2}. \tag{16.32}$$

This implies from (16.29) that the action is minimized (i.e. equality achieved) when

$$F_{\mu\nu} = \pm\tilde{F}_{\mu\nu}, \tag{16.33}$$

i.e. the self-dual or antiself-dual fields are the (finite-action) solutions to the classical Euclidean Yang–Mills theory. We remark that the usual solution $A_\mu = 0$ which has trivial topological quantum number (i.e. $n = 0$) clearly satisfies condition (16.33).

To find nontrivial self-dual gauge-field solutions, Belavin *et al.* (1975) employed the strategy of first considering $F_{\mu\nu}$ of O(4) gauge theory, which is isomorphic to SU(2) × SU(2): one can identify one SU(2) with the SU(2) of the internal symmetry and the other SU(2) with the three-sphere at space–time infinity. Also in O(4), since $F_{\mu\nu}$ is really a matrix with four indices $F^{\alpha\beta}_{\mu\nu}$, where (α, β) are the internal O(4) space indices and (μ, ν) are the Euclidean O(4) indices, the self-dual condition of (16.33) can be translated into simple symmetry conditions on these charge and space–time indices. From this it is possible to construct the solution explicitly. We shall not provide the details here as the SU(2) gauge transformation has just the form discussed in eqn (16.12)

$$U(x) = \frac{x_0 + \mathrm{i}\mathbf{x}\cdot\boldsymbol{\tau}}{\rho} \tag{16.34}$$

where $\rho^2 = x_0^2 + \mathbf{x}^2$. This gives rise to a gauge field

$$A_\mu(x) = \left(\frac{\rho^2}{\rho^2 + \lambda^2}\right)U^{-1}\,\partial_\mu U \tag{16.35}$$

where λ is some arbitrary scale parameter, often referred to as *instanton size*. For $\rho \gg \lambda$, we have

$$A_\mu(x) \to U^{-1}\,\partial_\mu U \tag{16.36}$$

as required by the boundary condition. More explicitly we can write

$$A_0(x) = \frac{-\mathrm{i}\boldsymbol{\tau}\cdot\mathbf{x}}{\rho^2 + \lambda^2}, \qquad \mathbf{A}(x) = \frac{-\mathrm{i}(\boldsymbol{\tau}x_0 + \boldsymbol{\tau}\times\mathbf{x})}{\rho^2 + \lambda^2}. \tag{16.37}$$

One can further check that the corresponding action integral indeed has the value $8\pi^2/g^2$.

In the next section we shall discuss the physical interpretation of these nontrivial minima of the Euclidean action.

16.2 The instanton and vacuum tunnelling

The physical interpretation of the instanton solutions as quantum-mechanical events corresponding to tunnelling between vacuum states of different topological numbers was first advanced by 't Hooft (1976), and later substantiated in the work of Callan, Dashen, and Gross (1976) and of Jackiw and Rebbi (1976).

In Feynman path-integral formalism the basic vacuum-to-vacuum transition amplitude is expressed as a sum (a functional integral) over all possible paths between the initial and final states, weighted by the exponential of i times the action for the particular path. In previous discussions we have included in our sum of path history only the condition $A_\mu(x) \to 0$ on the boundary. We shall see that non-Abelian gauge-field theory has a vacuum with an unexpectedly rich structure. It corresponds to a superposition of vacuum states with different topological winding numbers. The instanton field configurations correspond to paths that connect initial and final vacuum states with different winding numbers. We shall examine the effect of including all such field configurations in our path-integral formalism.

Multiple vacuum states

To be more specific let us place our system inside a box. The vacuum condition

$$F_{\mu\nu} = 0 \tag{16.38}$$

is obtained for the region

$$t < -T/2, \qquad t > T/2, \quad \text{and} \quad |\mathbf{x}| > R \tag{16.39}$$

with T and R both very large. The 2 + 1 space–time version is depicted in Fig. 16.1. One then sums over all paths $A_\mu(x)$ that are consistent with the boundary condition that the vacuum state (16.38) be maintained outside the cylinder.

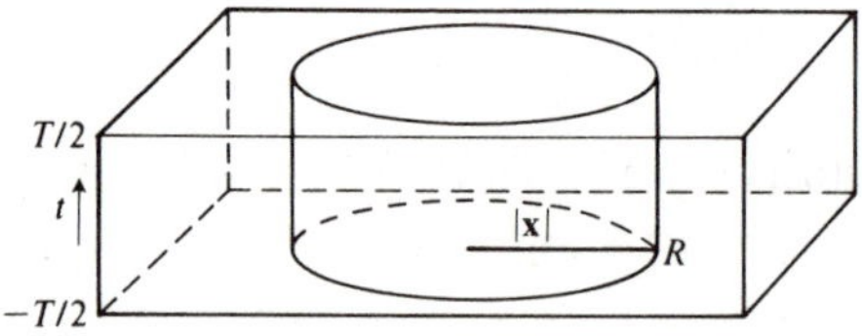

FIG. 16.1.

Throughout this section we shall be working with the gauge-fixing condition

$$A_0(x) = 0 \qquad \text{for all } x. \tag{16.40}$$

But we are still left with the freedom of time-independent gauge transformations

$$\partial_0 U(\mathbf{x}) = 0 \tag{16.41}$$

because

$$A_0(x) \to A'_0(x) = U^{-1}(\mathbf{x})A_0(x)U(\mathbf{x}) + U^{-1}(\mathbf{x})\,\partial_0 U(\mathbf{x}) = 0$$

where we have used both (16.40) and (16.41). Thus the vacuum will be described by a time-independent potential $A_i(x)$ which is a pure gauge potential

$$A_i(x) = A_i(\mathbf{x}) = U^{-1}(\mathbf{x})\,\partial_i U(\mathbf{x}). \tag{16.42}$$

At initial time $t = -T/2$, we can use this remaining gauge freedom to pick, for example,

$$A_i(\mathbf{x}) = 0 \tag{16.43}$$

by choosing

$$U(\mathbf{x}) = 1. \tag{16.44}$$

Then the vacuum state condition (16.38) implies that

$$F_{0i} = \partial_0 A_i = 0. \tag{16.45}$$

Thus (16.43) is maintained throughout the vacuum, i.e. throughout the space outside the cylinder in Fig. 16.1. In particular on the top-surface of the box ($t = T/2$) we have the situation of a disc with its edge identified, i.e. all points on the edge have the mapping (16.44). In realistic $3 + 1$ space–time dimensions, this corresponds to the $t = T/2$ vacuum being a pure gauge of mappings to the SU(2) gauge group manifold from three-dimensional space with infinities identified. As we have discussed in the previous section such gauge transformations, and hence the corresponding vacuum states (16.42), can be divided into inequivalent homotopic classes. An example of (16.42) with

$$U(\mathbf{x}) = \exp\{i\pi\boldsymbol{\tau}\cdot\mathbf{x}/(x^2 + \lambda^2)^{\frac{1}{2}}\} \tag{16.46}$$

as given by eqn (16.14) is an $n = 1$ vacuum state. Thus we conclude that there is a multiplicity of vacuum states $|n\rangle$, each characterized by its topological winding number.

The formal relation between multiple vacua and the instanton solution

The instanton solution, as we have shown in §16.1, is itself characterized by a winding number ν. We shall show that such a path in Euclidean space connects vacuum states that differ by that winding number

$$A_i^{\text{inst}}(\mathbf{x}, x_0 = -\infty) = A_i(\mathbf{x}) \text{ of vacuum state } |n\rangle$$

$$A_i^{\text{inst}}(\mathbf{x}, x_0 = \infty) = A_i(\mathbf{x}) \text{ of vacuum state } |n + \nu\rangle. \tag{16.47}$$

To see this formally let us recall the basic $\nu = 1$ instanton solution as given by eqns (16.34), (16.35), and (16.37). To cast it in the form of the $A_0 = 0$ gauge of (16.40) we make a gauge transformation on $A_\mu(x)$ of (16.37)

$$A_\mu(x) \to A'_\mu(x) = V^{-1}(x)A_\mu(x)V(x) + V^{-1}(x)\,\partial_\mu V(x). \tag{16.48}$$

The condition $A'_0(x) = 0$ implies that

$$\frac{\partial}{\partial x_0} V(x) = -A_0(x)V(x)$$

$$= \frac{\mathrm{i}\mathbf{x}\cdot\boldsymbol{\tau}}{x_0^2 + \mathbf{x}^2 + \lambda^2} V(x) \tag{16.49}$$

where we have written out the instanton solution (16.37). Eqn (16.49) can be integrated to yield

$$V(x) = \exp\left\{\frac{\mathrm{i}\mathbf{x}\cdot\boldsymbol{\tau}}{(\mathbf{x}^2 + \lambda^2)^{1/2}}\left(\tan^{-1}\frac{x_0}{(\mathbf{x}^2 + \lambda^2)^{1/2}} + \theta_0\right)\right\}. \tag{16.50}$$

We can set the integration constant to be

$$\theta_0 = (n + \tfrac{1}{2})\pi. \tag{16.51}$$

The result in (16.49) is obtained by requiring the boundary condition on the time-component $A_0(x) \to A'_0(x) = 0$. If we take the spatial component $A_i(x)$ to be zero at $x_0 = \pm\infty$, then

$$A'_i = V^{-1}(x)\,\partial_i V(x) \tag{16.52}$$

with

$$V(x_0 = -\infty) = \exp\left\{\mathrm{i}\pi\frac{\mathbf{x}\cdot\boldsymbol{\tau}}{(\mathbf{x}^2 + \lambda^2)^{1/2}}(n)\right\}$$

and

$$V(x_0 = \infty) = \exp\left\{\mathrm{i}\pi\frac{\mathbf{x}\cdot\boldsymbol{\tau}}{(\mathbf{x}^2 + \lambda^2)^{1/2}}(n+1)\right\}. \tag{16.53}$$

Thus the instanton solution of (16.37) indeed connects two vacuum states [cf. eqn (16.46)] that differ by one unit of winding number. This suggests that for a path-integral representation of the vacuum-to-vacuum transition amplitude, eqn (1.67) should be generalized to

$$\langle n|\mathrm{e}^{-\mathrm{i}Ht}|m\rangle_J = \int [\mathrm{d}A]_{\nu=n-m} \exp\left\{-\mathrm{i}\int(\mathscr{L} + JA)\,\mathrm{d}^4x\right\} \tag{16.54}$$

where $|n\rangle$ and $|m\rangle$ denote vacua with winding numbers n and m, respectively. Thus, we should sum over all gauge fields belonging to the same homotopic class with winding number $\nu = n - m$.

The instanton as a semi-classical tunnelling amplitude

The fact that the instanton is the minimum of the Euclidean action, i.e. it is the classical path for imaginary time, reminds one of the semi-classical (WKB) barrier-penetration amplitudes in non-relativistic quantum mechanics. This indeed turns out to be a fruitful analogy and we shall present a brief review of the elementary quantum-mechanical example of tunnelling between the two ground states of the double-well potential in Fig. 16.2(a)

$$V(q) = (q^2 - q_0^2)^2 \tag{16.55}$$

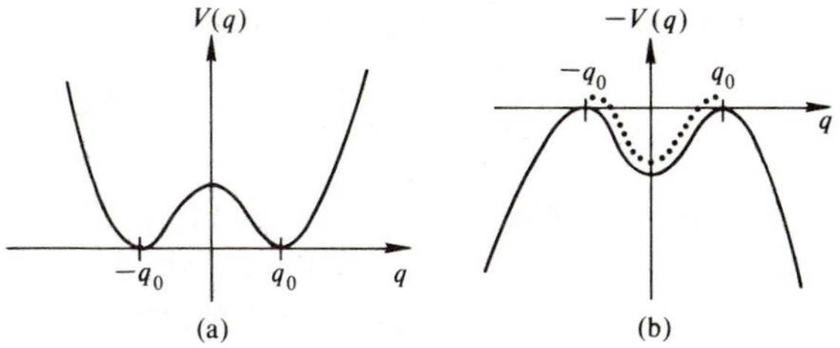

FIG. 16.2.

where $q(t)$ is some generalized coordinate. The energy of the system is then

$$E = \frac{1}{2}\left(\frac{\mathrm{d}q}{\mathrm{d}t}\right)^2 + V(q). \tag{16.56}$$

The classical ground state for this system is at $q = q_0$ or $q = -q_0$ with $E = 0$. In classical mechanics clearly there is no $E = 0$ path leading from q_0 to $-q_0$. However there can be quantum-mechanical tunnelling, so that the true ground state is neither $|q_0\rangle$ nor $|-q_0\rangle$ but their superposition

$$|\text{ground}\rangle = \frac{1}{\sqrt{2}}(|q_0\rangle + |-q_0\rangle). \tag{16.57}$$

The quantum-mechanical tunnelling amplitude can be calculated using the classical particle trajectory in the imaginary time system. This is because in the imaginary time system where $t = -\mathrm{i}\tau$ and $(\mathrm{d}q/\mathrm{d}t)^2 = -(\mathrm{d}q/\mathrm{d}\tau)^2$, the energy of the system is given by

$$-E = \left[\frac{1}{2}\left(\frac{\mathrm{d}q}{\mathrm{d}\tau}\right)^2 - V(q)\right] \tag{16.56a}$$

which is equivalent to a particle moving in the potential $-V(q)$ shown in Fig. 16.2(b). Thus in imaginary time there will be a path going between $-q_0$ and q_0 with $E = 0$. Setting $E = 0$ in eqn (16.56a) we can solve for this trajectory,

$$q(\tau) = q_0 \tanh(2^{\frac{1}{2}} q_0 \tau). \tag{16.58}$$

The action for this trajectory in the imaginary-time system is finite and can be calculated

$$S_t = \int_{-\infty}^{+\infty} \mathrm{d}\tau \left\{\frac{1}{2}\left(\frac{\mathrm{d}q}{\mathrm{d}\tau}\right)^2 - [-V(q)]\right\} = 2\int_{-\infty}^{+\infty} \mathrm{d}\tau\, V(q)$$

$$= 2\int_{-\infty}^{+\infty} [q^2(\tau) - q_0^2]^2\,\mathrm{d}\tau = \frac{4}{3}2^{\frac{1}{2}}q_0^3. \tag{16.59}$$

To get the tunnelling amplitude, we will use the path-integral formalism, in which the transition amplitude is given by

$$\langle q_\mathrm{f}|\mathrm{e}^{-\mathrm{i}Ht/h}|q_i\rangle = \int [\mathrm{d}q]\,\mathrm{e}^{\mathrm{i}S/h}. \tag{16.60}$$

In imaginary time (or Euclidean space), this becomes

$$\langle q_{\mathrm{f}}|\,\mathrm{e}^{-H\tau/\hbar}|q_i\rangle = \int [\mathrm{d}q]\,\mathrm{e}^{-S_{\mathrm{E}}/\hbar} \tag{16.61}$$

where $S_{\mathrm{E}} = -\mathrm{i}S$ is the Euclidean action. The right-hand side of eqn (16.61) can be visualized as the summation over all possible paths going from q_{i} to q_{f}. In the semi-classical approximation (expansion in powers of $\hbar$), the integral in (16.61) will be dominated by those paths for which S_{E} is stationary. It is not hard to see that the tunnelling amplitude for the case we are considering here will be of the form

$$T \sim \mathrm{e}^{-S_{\mathrm{i}}/\hbar}[1 + O(\hbar)] \simeq \exp(-4\sqrt{2}\, q_0^3/3\hbar). \tag{16.62}$$

This can be verified in this simple case by explicit calculations (for detail see Coleman 1977).

This example and eqn (16.53) suggest the interpretation that the instanton configuration corresponds to tunnelling between different vacuum states: $|n\rangle \to |n+1\rangle$. When we generalize the transition amplitude in (16.61) to field theory, we will find that not all field configuration (paths) give finite action because there are infinite degrees of freedom in field theory. But in the semi-classical approximation a configuration of infinite action will give zero to the path integration weighted by $\mathrm{e}^{-S_{\mathrm{E}}/\hbar}$. Thus the path integral will be dominated by configurations with finite action. Hence the tunnelling amplitude in the semi-classical approximation can be calculated in terms of the instanton configuration and has the form

$$T \sim \mathrm{e}^{-S_{\mathrm{E}}} \simeq \exp(-8\pi^2/g^2) \tag{16.63}$$

where we have used eqn (16.32). The form of eqn (16.63) also shows clearly that it is an effect that cannot be seen in ordinary perturbation theory.

θ-vacuum

As vacuum states $|n\rangle$ corresponding to different topological winding numbers are separated by finite-energy barriers and there are tunnellings between these states, we expect the true vacuum state to be a suitable superposition of these $|n\rangle$ states. We note that under a gauge transformation T_1 having a winding number 1 itself, we have

$$T_1|n\rangle = |n+1\rangle \tag{16.64}$$

and gauge invariance means that it commutes with the Hamiltonian

$$[T_1, H] = 0. \tag{16.65}$$

Thus we have a situation very similar to the familiar quantum-mechanical problem of periodic potential: with T being the translation operator and with the true ground state being the *Bloch wave* and there exists a conserved *pseudomomentum*. We can construct the true vacuum, the '*θ-vacuum*' as

$$|\theta\rangle = \sum_n \mathrm{e}^{-\mathrm{i}n\theta}|n\rangle \tag{16.66}$$

which is an eigenstate of the gauge transformation

$$T_1|\theta\rangle = \mathrm{e}^{\mathrm{i}\theta}|\theta\rangle. \tag{16.67}$$

Just like the pseudomomentum in the case of a periodic potential, θ labels the physically inequivalent sectors of the theory and within each sector we may study the propagation of gauge-invariant disturbances. Since different θ-worlds do not communicate with each other, there is no *a priori* method of determining the value of θ.

In terms of the θ-vacuum, the vacuum-to-vacuum transition amplitude in path-integral formalism should be

$$\langle\theta'|\mathrm{e}^{-\mathrm{i}Ht}|\theta\rangle_J = \delta(\theta - \theta')I_J(\theta). \tag{16.68}$$

Writing the left-hand side in terms of vacuum states with definite winding numbers we have

$$\begin{aligned}\langle\theta'|\mathrm{e}^{-\mathrm{i}Ht}|\theta\rangle_J &= \sum_{m,n} \mathrm{e}^{\mathrm{i}m\theta'}\,\mathrm{e}^{-\mathrm{i}n\theta}\langle m|\,\mathrm{e}^{-\mathrm{i}Ht}|n\rangle_J\\ &= \sum_{m,n} \mathrm{e}^{-\mathrm{i}(n-m)\theta}\,\mathrm{e}^{\mathrm{i}m(\theta'-\theta)}\int[\mathrm{d}A]_{n-m}\,\mathrm{e}^{\mathrm{i}\int(\mathscr{L}+JA)\mathrm{d}4x}\end{aligned} \tag{16.69}$$

where we have substituted in the expression (16.54). We can cast (16.69) in the standard form by relabelling $n - m \to v$; after doing the m-summation

$$\begin{aligned}I_J(\theta) &= \sum_v \mathrm{e}^{-\mathrm{i}v\theta}\int[\mathrm{d}A]_v \exp\left[-\mathrm{i}\int(\mathscr{L} + JA)\,\mathrm{d}^4x\right]\\ &= \sum_v \int[\mathrm{d}A]_v \exp\left[-\mathrm{i}\int(\mathscr{L}_{\text{eff}} + JA)\,\mathrm{d}^4x\right]\end{aligned} \tag{16.70}$$

and using eqn (16.29), we obtain

$$\mathscr{L}_{\text{eff}} = \mathscr{L} + \frac{\theta}{16\pi^2}\,\mathrm{tr}(F_{\mu\nu}\tilde{F}^{\mu\nu}). \tag{16.71}$$

Thus the rich structure of gauge theory vacuum corresponding to tunnelling between states with different topological winding numbers gives rise to an effective Lagrangian term which violates P and CP conservation. As we have already mentioned in §10.2, such a term is normally discarded because the $F\tilde{F}$ term can be expressed as the divergence of current (eqn (16.24)) and hence as a surface term in the action (eqn (16.26)). However because there is a nontrivial instanton gauge field configuration which does not vanish at infinity (eqn (16.22)), such an 'abnormal' term actually survives in non-Abelian gauge field theories.

16.3 Instantons and the U(1) problem

The U(1) problem

The ideas of chiral symmetry and quark–gluon interaction as described by QCD seem to lead to a contradiction. For simplicity we will illustrate this problem in the case where there are only two quark flavours in the theory, u and d quarks. In the limit $m_{u,d} \to 0$, the QCD Lagrangian has the symmetry $SU(2)_L \times SU(2)_R \times U(1)_V \times U(1)_A$ which is larger than the chiral symmetry $SU(2)_L \times SU(2)_R$ discussed in Chapter 5. The $U(1)_V$ symmetry generated by the transformation $q_i \to e^{i\alpha} q_i$ gives rise to the current

$$J^B_\mu = \bar{u}\gamma_\mu u + \bar{d}\gamma_\mu d.$$

This is just the baryon-number current for this two-flavour case and the $U(1)_V$ symmetry manifests itself in the baryon-number conservation. But the $U(1)_A$ symmetry generated by the transformation $q_i \to e^{i\beta\gamma_5} q_i$ with the current given by

$$J^5_\mu = \bar{u}\gamma_\mu\gamma_5 u + \bar{d}\gamma_\mu\gamma_5 d \tag{16.79}$$

does not seem to correspond to any observed symmetry in the hadron spectra (e.g. we do not observe a parity doubling of the baryon states). Thus we expect this $U(1)_A$ to be realized in the Goldstone mode and to give rise to an additional massless pseudoscalar besides the pion isotriplet which are the Goldstone bosons resulting from the spontaneous breakdown of the chiral $SU(2)_L \times SU(2)_R$ symmetry. As we turn on the quark mass, this new U(1) Goldstone boson is expected to have a mass comparable to that of the pion because they all have the same quark composition. (This statement will be made more quantitative later on.) Experimentally no such isoscalar pseudoscalar meson has been seen. The η-meson has the right quantum number, but is simply too heavy. This is usually referred to as the 'U(1) problem' or the 'η-mass problem' (Glashow 1967). As we shall see, the existence of instantons will solve the problem ('t Hooft 1976). There is another related U(1) problem that has to do with the decay $\eta \to 3\pi$. We shall comment on that problem and its resolution later on.

The η-mass problem. One may think that the presence of the ABJ anomaly in J^5_μ (which couples to gluons) will provide an escape from this paradox as the divergence of J^5_μ does not vanish in the $m_{u,d} \to 0$ limit

$$\partial^\mu J^5_\mu = 4\frac{g^2}{16\pi^2}\,\mathrm{tr}(G_{\mu\nu}\tilde{G}^{\mu\nu}) + 2\mathrm{i}m_u\bar{u}\gamma_5 u + 2\mathrm{i}m_d\bar{d}\gamma_5 d \tag{16.80}$$

where $G_{\mu\nu}$ is the gluon tensor matrix, $\tilde{G}$ its dual, and we have undone the scaling of the field normalization of (16.17). (The factor 4 in front of the first term on the right-hand side comes from the fact that there are two flavours $2N_F = 4$ in this theory.) However the matter is not so simple. We have already seen in eqn (16.24) that tr $G\tilde{G}$ itself is the divergence of a current,

$$\partial^\mu K_\mu = 4\frac{g^2}{16\pi^2}\,\mathrm{tr}\; G_{\mu\nu}\tilde{G}^{\mu\nu} \tag{16.81}$$

with

$$K_\mu = 4\frac{g^2}{16\pi^2}\,\varepsilon_{\mu\nu\lambda\rho}\,\mathrm{tr}\; G^{\nu\lambda}A^\rho \tag{16.82}$$

where $A_\mu = A_\mu^a \lambda^a/2$ is the gluon field. Thus we can define a new axial vector current

$$\tilde{J}_\mu^5 = J_\mu^5 - K_\mu \tag{16.83}$$

which is conserved in the $m_{u,d} \to 0$ limit,

$$\partial^\mu \tilde{J}_\mu^5 = 2im_u \bar{u}\gamma_5 u + 2im_d \bar{d}\gamma_5 d. \tag{16.84}$$

Of course the current K_μ, and hence also $\tilde{J}_\mu^5$, is unobservable, because it is not gauge-invariant. Nevertheless, because its charge

$$\tilde{Q}_5 = \int \tilde{J}_0^5 \, d^3x \tag{16.85}$$

is conserved, this symmetry when realized in the Goldstone mode would demand the existence of an $I = 0$ pseudoscalar meson with a mass m_0 (Weinberg 1975)

$$m_0 \leq (\sqrt{3}) m_\pi \tag{16.86}$$

To see this we can use the standard current-algebra technique to obtain a Ward identity entirely similar to eqn (5.228)

$$m_0^2 f_0^2 = i \frac{m_0^2 - k^2}{m_0^2} \left\{ ik^\nu \int d^4x \, e^{-ik\cdot x} \langle 0 | T(\partial^\mu \tilde{J}_\mu^5(0) \tilde{J}_\nu^5(x)) | 0 \rangle \right.$$

$$\left. + \int d^4x \, e^{-ik\cdot x} \langle 0 | \delta(x_0) [\partial^\mu \tilde{J}_\mu^5(0), \tilde{J}_0^5(x)] | 0 \rangle \right\} \tag{16.87}$$

where f_0 is the isoscalar meson decay constant. After taking the $k^\nu \to 0$ limit, since there is no zero mass pole in the first term on the right-hand side, we relate $m_0^2 f_0^2$ to a σ-term which is of identical form to that for $m_\pi^2 f_\pi^2$ encountered in §5.5. Thus,

$$m_0^2 f_0^2 = m_\pi^2 f_\pi^2. \tag{16.88}$$

This reflects the fact that $\tilde{J}_\mu^5$ has the same commutator with quark mass terms as J_μ^5. To proceed further we can write (16.79) as a sum of an SU(3) octet and singlet

$$J_\mu^5 = \frac{1}{\sqrt{3}} J_\mu^{(8)5} + \sqrt{\frac{2}{3}} J_\mu^{(0)5} \tag{16.89}$$

where

$$J_\mu^{(8)5} = \frac{1}{\sqrt{3}} (\bar{u}\gamma_\mu\gamma_5 u + \bar{d}\gamma_\mu\gamma_5 d - 2\bar{s}\gamma_\mu\gamma_5 s)$$

$$J_\mu^{(0)5} = \sqrt{\frac{2}{3}} (\bar{u}\gamma_\mu\gamma_5 u + \bar{d}\gamma_\mu\gamma_5 d + \bar{s}\gamma_\mu\gamma_5 s).$$

To the extent that SU(3) is a good symmetry and all pseudoscalar octet decay constants are equal, we obtain immediately

$$f_0 \geq \frac{1}{\sqrt{3}} f_\pi \tag{16.90}$$

and hence the bound in eqn (16.86) especially if we allow for the possibility that there may be more than one isoscalar pseudoscalar meson coupled to $\tilde{J}^5_\mu$.

It was originally pointed out by Kogut and Susskind (1975*b*) that one way to avoid the disastrous conclusion of (16.86) is for $\tilde{J}^5_\mu$ to be coupled to a 'particle' which is massless even if $m_{u,d} \neq 0$. Then the first term on the right-hand side of (16.87) does not drop out in the $k_v \to 0$ limit and the simple relation between $m_0^2 f_0^2$ and the σ-term is spoiled and there is no restriction on m_0. Since $\tilde{J}^5_\mu$ is gauge-variant, it is conceivable that this gauge-dependent massless pole does not generate poles in physical (gauge-invariant) quantities. As we shall see, in 't Hooft's resolution of the U(1) problem, the Kogut–Susskind mechanism is indeed realized in the new instanton θ-vacuum. The $U(1)_A$ symmetry is spontaneously broken without generating the $I = 0$ Goldstone boson.

The $\eta \to 3\pi$ problem. We should also remark that there is a second U(1) problem, related to $\eta \to 3\pi$ decay, which can also be solved by the Kogut–Susskind mechanism. We mentioned at the end of Chapter 5 that virtual photon exchange gives a vanishing amplitude for $\eta \to 3\pi$ in the chiral limit (Sutherland 1966). This decay must therefore be attributed to the isospin-violating quark-mass difference in the Lagrangian density

$$\mathscr{L}' = \tfrac{1}{2}(m_u - m_d)(\bar{u}u - \bar{d}d). \tag{16.91}$$

The leading term in chiral perturbation (Bell and Sutherland 1968) is then

$$\langle 3\pi|\mathscr{L}'|\eta\rangle \xrightarrow[q_\pi \to 0]{} (m_u - m_d)A/\sqrt{2}f_\pi^2 \tag{16.92}$$

with

$$A = \langle \pi\pi|(m_u\bar{u}\gamma_5 u + m_d\bar{d}\gamma_5 d)|\eta\rangle$$
$$= \frac{1}{2i}\langle \pi\pi|\partial^\mu \tilde{J}^5_\mu(0)|\eta\rangle \tag{16.93}$$

where the two-pion system and η have equal momenta and energy. Thus even with $m_u \neq m_d$ we still have a vanishing $\eta \to 3\pi$ amplitude because the amplitude A in (16.93) is a total divergence between states with the same four-momenta. Again we can see a possible resolution of this second U(1) problem with the existence of a massless pole coupled in $\tilde{J}^5_\mu$ giving a nonzero A in (16.93).

The θ-vacuum in the presence of massless fermions

We now show how the structure of the θ-vacuum implies the spontaneous breakdown of the $U(1)_A$ symmetry and how the associated massless particle decouples from physical quantities. Instead of following the original 't Hooft calculation in path-integral formalism, we shall only indicate the physical ideas with some heuristic arguments.

Since $\tilde{Q}_5$ is not gauge-invariant, under a gauge transformation characterized by topological quantum number n (see eqn (16.64)) it changes (see, for example, Callan *et al.* 1976) as

$$\tilde{Q}_5 \to \tilde{Q}'_5 = T_n^{-1}\tilde{Q}_5 T_n = \tilde{Q}_5 + 2nN_{\mathrm{F}} \tag{16.94}$$

where N_{F} is the number of massless quark flavours. A simple check of (16.94) is to take the original $\tilde{Q}_5$ to be zero; then from

$$\tilde{Q}'_5 = \int \mathrm{d}^3 x K'_0 \tag{16.95}$$

with

$$K'_0 = \frac{2N_{\mathrm{F}} g^2}{4\pi^2}\frac{1}{3}\varepsilon_{0ijk}\,\mathrm{tr}(A_i A_j A_k)$$

and

$$A_i = U^{-1}\,\partial_i U$$

we then have (16.94) when the above expression is compared with that for the winding number in eqn (16.16). Eqn (16.94) means that T_n acts like a 'raising operator' of chirality—instantons 'eat' massless quark pairs

$$[\tilde{Q}_5, T_n] = 2nN_{\mathrm{F}} T_n. \tag{16.96}$$

To be more explicit, T_n changes the winding number of the vacuum

$$T_n|0\rangle = |n\rangle. \tag{16.97}$$

But

$$\tilde{Q}_5|n\rangle = 2nN_{\mathrm{F}}|n\rangle \tag{16.98}$$

because of (16.96) and $\tilde{Q}_5|0\rangle = 0$. Thus the vacuum state with definite topological quantum number also has definite chirality (i.e. it is an eigenstate of $\tilde{Q}_5$). Since $\tilde{Q}_5$ is conserved

$$[\tilde{Q}_5, H] = 0, \tag{16.99}$$

the vacuum-to-vacuum transition amplitude vanishes unless the initial and final states have the same winding numbers,

$$\langle n|\mathrm{e}^{-\mathrm{i}Ht}|m\rangle \sim \delta_{nm} \tag{16.100}$$

or, more generally for an operator $P_\nu(x)$ with chirality $2N_{\mathrm{F}}\nu$,

$$\langle n|\mathrm{e}^{-\mathrm{i}Ht}P_\nu(x)|m\rangle \sim \delta_{n-m,\,\nu}. \tag{16.101}$$

Therefore, in the presence of massless fermions (i.e. $\tilde{Q}_5$ is conserved), tunnelling between vacua with different winding numbers as discussed in the previous section is suppressed.

Does this mean that the θ-vacuum is irrelevant in such a situation? Not so, as the $|n\rangle$ vacuum violates cluster decomposition. For example, in the vacuum state with $n = 0$, the expectation value for widely separated operators will not vanish

$$\langle 0|P_\nu^\dagger(x)P_\nu(y)|0\rangle \xrightarrow[|x-y|\to\infty]{} \sum_m \langle 0|P_\nu^\dagger(x)|m\rangle\langle m|P_\nu(y)|0\rangle$$
$$= \langle 0|P_\nu^\dagger(x)|\nu\rangle\langle \nu|P_\nu(y)|0\rangle \neq 0 \tag{16.102}$$

because of the presence of 'abnormal vacua' with winding number $n \neq 0$. Eqn (16.102) also indicates spontaneous symmetry breakdown as the VEV of operators carrying chirality is nonzero. Thus the θ-vacuum is still relevant.

What is the chiral property of the θ-vacuum? Under a chiral rotation by an angle α, we have

$$\begin{aligned} e^{-i\alpha \tilde{Q}_5}|\theta\rangle &= \sum_n e^{-in\theta - i\alpha 2nN_F}|n\rangle \\ &= |\theta + 2\alpha N_F\rangle, \end{aligned}$$

i.e. the θ-vacuum is changed to another one with $\theta' = \theta + 2\alpha N_F$. Thus the (gauge-variant) current conservation means that the theory is invariant under a rotation which changes its vacuum state $|\theta\rangle \to |\theta'\rangle$. This implies that in the presence of massless fermions θ has no physical meaning: one θ is as good as another!

Small θ-oscillations—the Kogut–Susskind pole

The situation described above is very much like that we encountered in the simple U(1) $\lambda\phi^4$ theory with SSB of eqn (5.138). There, under U(1) rotations, the vacuum states can be changed into each other as indicated in Fig. 5.3. Thus we can interpret the parameter θ in QCD as the parameter which characterizes the direction of symmetry breaking. However there is a crucial difference between these two spontaneously broken U(1) theories. In the U(1) theory of eqn (5.138) all the vacuum states connected by rotations are equivalent. In QCD the parameter θ is like the coupling constant; different values of θ correspond to different Hilbert spaces. In the U(1) theory of (5.138) small oscillations around the true vacuum in the angular direction (π) which do not cost any energy (i.e. zero energy excitations) are shown to be interpretable as physical massless particles (the Goldstone boson). On the other hand changes in the QCD θ-parameter are meaningless just as the pseudomomenta cannot be changed in the case of periodic potential. Hence small oscillations in the QCD θ correspond to an unphysical massless particle. But this is just the Kogut–Susskind pole which is required to solve the U(1) problem.

The strong CP problem

While the presence of instantons apparently resolves the U(1) problem, their presence also implies a θ-term in the effective Lagrangian (16.71) which violates P and conserves C (hence violates CP). In fact the stringent experimental upper limit on neutron dipole moment (see §12.2) can be translated into a bound on the QCD θ parameter, $\theta < 10^{-9}$. This is the strong CP problem. How to give a rationale for such a small value, considering that it is a strong interaction parameter and *a priori* we would expect it to be $O(1)$. One would think that the only plausible solution is that θ is effectively zero. Various ways to achieve this have been suggested; among them we list the following three approaches (see, for example, Wilczek 1978).

(The first two mechanisms both suggest that there may in fact be an exact global U(1) symmetry and the S-matrix, as we have explained above, is independent of θ so that it can be rotated to zero.)

(1) *Zero mass quark.* The exact U(1) symmetry comes about because one of the quark masses (presumably m_u) is zero. However all current-algebra calculations indicate that this is an extremely unlikely possibility.

(2) *The axion solution.* Increase the number of Higgs mesons so that even if $m_{u,d} \neq 0$, QCD together with the standard electroweak theory still has a global $U(1)_{PQ}$ symmetry (Peccei and Quinn 1977). However one of the scalars (the axion) plays the role of Goldstone boson for this $U(1)_{PQ}$. It is massless at the classical level and picks up a small mass (much like the η-meson) only through the axial anomaly and the instanton interaction (Weinberg 1978; Wilczek 1978). As the simplest version of this approach, constructed within the framework of the SU(2) × U(1) model, is ruled out by experiment, one is forced to postulate such a structure at the SU(5) GUT level producing an 'invisible axion' with infinitesimally small mass and couplings (Kim 1979; Dine, Fischler, and Srednicki 1981). However, this may pose problems for the standard theory of cosmology (see, for example, Sikivie 1982b).

(3) *Soft CP.* It has been suggested that perhaps one should set $\theta = 0$ as a symmetry requirement ('strong interaction conserves CP!'). However this in itself is not enough, since higher-order (CP-violating) weak interactions will generate γ_5-dependent quark mass terms. To eliminate them one has to apply chiral rotations which in turn induce a θ-term. To have a calculable and small ($<10^{-9}$) θ_{wk}, the CP violation in weak interactions must be soft (i.e. by operators with dimension less than four). So far no realistic model has been constructed.

Thus we can see that none of the suggested solutions are completely satisfactory and the value of θ remains a puzzle for QCD. For a recent review of QCD including topics such as instantons and the U(1) problem, etc. the reader is referred to Llewellyn Smith (1982).

Finally, we comment very briefly on the role of instantons. For a very small coupling constant g, corresponding to very small distances r, tunnelling is negligible as indicated by eqn (16.63) and QCD perturbation theory can be used in appropriate circumstances as shown in Chapter 10. For some intermediate values of g one can continue to use perturbation theory but with allowance made for vacuum tunnelling. This produces (Callan, Dashen, and Gross 1980) a sudden rapid increase in g as r is increased from zero to a distance comparable to the hadron size (~ 0.5 fm). This is compatible with the phenomenological quark models and with the results of lattice gauge theory calculations discussed at the end of §10.5. For large distances there is at present no reliable method for calculating the effect of instantons (in fact it is not even clear whether they are relevant at all) although they do remind us that the QCD vacuum must be very complicated.

Appendix A
Notations and conventions

Metric

Metric tensor $$g_{\mu\nu} = g^{\mu\nu} = \begin{pmatrix} 1 & 0 & 0 & 0 \\ 0 & -1 & 0 & 0 \\ 0 & 0 & -1 & 0 \\ 0 & 0 & 0 & -1 \end{pmatrix}. \quad \text{(A.1)}$$

Contravariant coordinate $$x^\mu = (x^0, x^1, x^2, x^3) = (t, x, y, z) = (t, \mathbf{x}). \quad \text{(A.2)}$$

Covariant coordinate $$x_\mu = g_{\mu\nu} x^\nu = (t, -\mathbf{x}).$$

Scalar product $$A \cdot B = A_\mu B^\mu = A_\mu g^{\mu\nu} B_\nu = A_0 B_0 - \mathbf{A} \cdot \mathbf{B}. \quad \text{(A.3)}$$

Derivatives $$\partial^\mu \equiv \frac{\partial}{\partial x_\mu} = \left(\frac{\partial}{\partial t}, -\nabla\right) \qquad \partial_\mu \equiv \frac{\partial}{\partial x^\mu} = \left(\frac{\partial}{\partial t}, \nabla\right) \quad \text{(A.4)}$$

where $$\nabla = \left(\frac{\partial}{\partial x}, \frac{\partial}{\partial y}, \frac{\partial}{\partial z}\right), \quad \text{(A.5)}$$

Four-divergence $$\partial^\mu A_\mu = \frac{\partial A_0}{\partial t} + \nabla \cdot \mathbf{A}. \quad \text{(A.6)}$$

Pauli matrices

$$\boldsymbol{\sigma} = (\sigma_1, \sigma_2, \sigma_3). \quad \text{(A.7)}$$

$$\sigma_1 = \begin{pmatrix} 0 & 1 \\ 1 & 0 \end{pmatrix} \qquad \sigma_2 = \begin{pmatrix} 0 & -\mathrm{i} \\ \mathrm{i} & 0 \end{pmatrix} \qquad \sigma_3 = \begin{pmatrix} 1 & 0 \\ 0 & -1 \end{pmatrix}. \quad \text{(A.8)}$$

$$[\sigma_i, \sigma_j] = 2\mathrm{i}\varepsilon^{ijk}\sigma_k \qquad \varepsilon^{ijk}\text{: totally antisymmetric} \qquad \varepsilon^{123} = 1. \quad \text{(A.9)}$$

$$\{\sigma_i, \sigma_j\} = 2\delta_{ij} \qquad \mathrm{tr}(\sigma_i \sigma_j) = 2\delta_{ij}. \quad \text{(A.10)}$$

Completeness: $$\sum_i (\sigma_i)_{ab}(\sigma_i)_{cd} = 2(\delta_{bc}\,\delta_{ad} - \tfrac{1}{2}\,\delta_{ab}\,\delta_{cd}). \quad \text{(A.11)}$$

Dirac matrices

$$\{\gamma^\mu, \gamma^\nu\} = 2g^{\mu\nu} \qquad \gamma_\mu = g_{\mu\nu}\gamma^\nu. \quad \text{(A.12)}$$

$$\gamma^\mu = (\gamma^0, \boldsymbol{\gamma}) \qquad \gamma_\mu = (\gamma_0, -\boldsymbol{\gamma}). \quad \text{(A.13)}$$

$$\gamma^5 = \gamma_5 = \mathrm{i}\gamma^0\gamma^1\gamma^2\gamma^3 \qquad \sigma_{\mu\nu} = \frac{\mathrm{i}}{2}[\gamma_\mu, \gamma_\nu]. \quad \text{(A.14)}$$

Spin matrix $s^i = \frac{1}{4}\varepsilon^{ijk}\sigma^{jk}, \qquad i = 1, 2, 3.$ (A.15)

Charge conjugation $\psi^c = C\psi^\dagger \qquad C\gamma_\mu C^\dagger = -\gamma_\mu^*.$ (A.16)

Identity $\gamma_\alpha\gamma_\beta\gamma_\lambda = g_{\alpha\beta}\gamma_\lambda + g_{\beta\lambda}\gamma_\alpha - g_{\alpha\lambda}\gamma_\beta + i\varepsilon_{\mu\alpha\beta\lambda}\gamma^\mu\gamma_5.$ (A.17)

Fierz transformation

Let $\Gamma_S = 1, \Gamma_V = \gamma_\mu, \Gamma_T = \sigma_{\mu\nu}, \Gamma_A = \gamma_\mu\gamma_5, \Gamma_P = \gamma_5.$ (A.18)

Then $\sum_i g_i(\Gamma_i)_{\alpha\beta}(\Gamma_i)_{\gamma\delta} = \sum_j \hat{g}_j(\Gamma_j)_{\alpha\delta}(\Gamma_j)_{\gamma\beta}$ (A.19)

where the indices i, j run over the set S, V, T, A, and P and the g_is are related to $\hat{g}_j$ by

$$\begin{bmatrix} \hat{g}_S \\ \hat{g}_V \\ \hat{g}_T \\ \hat{g}_A \\ \hat{g}_P \end{bmatrix} = \frac{1}{4}\begin{bmatrix} 1 & 4 & 12 & -4 & 1 \\ 1 & -2 & 0 & -2 & -1 \\ \frac{1}{2} & 0 & -2 & 0 & \frac{1}{2} \\ -1 & -2 & 0 & -2 & 1 \\ 1 & -4 & 12 & 4 & 1 \end{bmatrix}\begin{bmatrix} g_S \\ g_V \\ g_T \\ g_A \\ g_P \end{bmatrix}. \qquad \text{(A.20)}$$

Dirac representation

$$\gamma^0 = \begin{pmatrix} 1 & 0 \\ 0 & -1 \end{pmatrix} \qquad \boldsymbol{\gamma} = \begin{pmatrix} 0 & \boldsymbol{\sigma} \\ -\boldsymbol{\sigma} & 0 \end{pmatrix} \qquad \gamma_5 = \begin{pmatrix} 0 & 1 \\ 1 & 0 \end{pmatrix}. \qquad \text{(A.21)}$$

Hermitian conjugate

$$(\gamma^0)^\dagger = \gamma^0 \quad (\gamma^k)^\dagger = -\gamma^k, \quad (k = 1, 2, 3) \quad \sigma_{\mu\nu}^\dagger = \sigma^{\mu\nu} \quad \gamma_5^\dagger = \gamma_5. \quad \text{(A.22)}$$

Spin matrix $\mathbf{s} = \frac{1}{2}\begin{pmatrix} \boldsymbol{\sigma} & 0 \\ 0 & \boldsymbol{\sigma} \end{pmatrix}.$ (A.23)

Charge conjugate $C = i\gamma_2.$ (A.24)

Plane wave and Dirac spinor

Incoming plane wave $e^{-ik\cdot x} \equiv \exp[-i(\omega t - \mathbf{k}\cdot\mathbf{x})].$ (A.25)

Outgoing plane wave $e^{ik\cdot x} \equiv \exp[i(\omega t - \mathbf{k}\cdot\mathbf{x})]$ (A.26)

where $\omega = (\mathbf{k}^2 + m^2)^{1/2}.$ (A.27)

Space–time translation $A(x) = e^{ip\cdot x}A(0)\,e^{-ip\cdot x}$ (A.28)
where $p^\mu = (H, \mathbf{p})$ is the energy–momentum operator.

Klein–Gordon equation $(\partial^2 + \mu^2)\phi(x) = 0.$ (A.29)

Dirac equation $(i\gamma^\mu\,\partial_\mu - m)\psi(x) = 0$ (A.30)

In momentum space $(\not{p} - m)u(p, s) = 0, \; (\not{p} + m)v(p, s) = 0$ (A.31)

where $u(p, s)$, $v(p, s)$ are Dirac spinors.

Normalization

$$\bar{u}(p,s)u(p,s') = 2m\,\delta_{ss'} \qquad \bar{v}(p,s)v(p,s') = -2m\,\delta_{ss'} \tag{A.32}$$

Projection operators

$$\sum_s u_\alpha(p,s)\bar{u}_\beta(p,s) = (\not{p}+m)_{\alpha\beta}$$

$$\sum_s v_\alpha(p,s)\bar{v}_\beta(p,s) = (\not{p}-m)_{\alpha\beta}. \tag{A.33}$$

Gordon decomposition

$$\bar{u}(p,s)\gamma^\mu u(q,s) = \frac{1}{2m}\bar{u}(p,s)[(p+q)^\mu + \mathrm{i}\sigma^{\mu\nu}(p-q)_\nu]u(q,s) \tag{A.34}$$

$$\bar{u}(p,s)\gamma^\mu\gamma_5 u(q,s) = \frac{1}{2m}\bar{u}(p,s)[(p-q)^\mu\gamma_5 + \mathrm{i}\sigma^{\mu\nu}(p+q)_\nu\gamma_5]u(q,s).$$

In the Dirac representation (A.21), Dirac spinors are given by

$$u(p,s) = (E+m)^{\frac{1}{2}}\begin{pmatrix} 1 \\ \dfrac{\boldsymbol{\sigma}\cdot\mathbf{p}}{E+m}\end{pmatrix}\chi_s \quad v(p,s) = (E+m)^{\frac{1}{2}}\begin{pmatrix} \dfrac{\boldsymbol{\sigma}\cdot\mathbf{p}}{E+m} \\ 1\end{pmatrix}\chi_s \tag{A.35}$$

$s = 1, 2$

$$\text{where } \chi_1 = \begin{pmatrix}1\\0\end{pmatrix} \quad \chi_2 = \begin{pmatrix}0\\1\end{pmatrix} \qquad E = (\mathbf{p}^2+m^2)^{1/2}. \tag{A.36}$$

Normalization of states, cross sections and decay rates

One-particle states are normalized as

$$\langle \mathbf{p}, \alpha | \mathbf{p}', \alpha' \rangle = 2E(2\pi)^3\,\delta^3(\mathbf{p}-\mathbf{p}')\,\delta_{\alpha\alpha'} \tag{A.37}$$

where α, α' label the spin and/or internal symmetry indices. The projection for a one-particle state

$$\sum_{p,\alpha} |p,\alpha\rangle\langle p,\alpha| = \int \frac{\mathrm{d}^3 p}{(2\pi)^3 2E}\sum_\alpha |p,\alpha\rangle\langle p,\alpha|$$

$$= \int \frac{\mathrm{d}^4 p}{(2\pi)^3}\,\delta(p^2-m^2)\theta(p_0)\sum_\alpha |p,\alpha\rangle\langle p,\alpha|. \tag{A.38}$$

The transition probability per unit time per unit volume for the process $\mathrm{i}\to\mathrm{f}$ is

$$\omega(\mathrm{i}\to\mathrm{f}) = (2\pi)^4\,\delta^4(p_i - p_f)|\langle \mathrm{f}|T|\mathrm{i}\rangle|^2 \tag{A.39}$$

where $\langle \mathrm{f}|T|\mathrm{i}\rangle$ is the covariant T-matrix and is related to the S-matrix by

$$\langle \mathrm{f}|S|\mathrm{i}\rangle = \delta_{\mathrm{if}} + \mathrm{i}(2\pi)^4\,\delta^4(p_\mathrm{i}-p_\mathrm{f})\langle \mathrm{f}|T|\mathrm{i}\rangle. \tag{A.40}$$

Decay rate

The transition probability per unit time per unit volume to a specific final state f is given by

$$d\Gamma(\mathrm{i} \to \mathrm{f}) = \frac{\omega(\mathrm{i} \to \mathrm{f})}{\rho_\mathrm{i}} dN_\mathrm{f} \tag{A.41}$$

where $\rho_\mathrm{i} = 2E_\mathrm{i}$ is the density of the decaying state and

$dN_\mathrm{f} = \prod_{j=1}^{n} \frac{d^3 p_j}{(2\pi)^3 2E_j}$ is the density of the final state.

The total decay rate is

$$\Gamma(\mathrm{i} \to \mathrm{f}) = \frac{1}{2E_\mathrm{i}} \int \prod_{j=1}^{n} \frac{d^3 p_j}{(2\pi)^3 2E_j} (2\pi)^4 \, \delta^4\left(p_i - \sum_{j=1}^{n} p_j\right) |\langle \mathrm{f}|T|\mathrm{i}\rangle|^2 S \tag{A.42}$$

where S is the statistical factor which is obtained by including a factor $1/m!$ if there are m identical particles in the final state,

$$S = \prod_i \frac{1}{m_i!}. \tag{A.43}$$

Scattering cross-sections

The cross-section for $a_1 + a_2 \to f$ is given by

$$d\sigma(a_1 + a_2 \to f) = \frac{\omega(a_1 + a_2 \to f)}{J_i} dN_\mathrm{f} \tag{A.44}$$

where J_i is the flux of the incoming particles and is given by

$$J_i = \rho_1 \rho_2 v \tag{A.45}$$

where ρ_1, ρ_2 are the densities of the initial state and v is the relative velocity between two particles in the initial state. The flux factor J_i can be written in the centre-of-mass system as

$$\begin{aligned} J_i &= 2E_1 2E_2 \left| \frac{\mathbf{p}_1}{E_1} - \frac{\mathbf{p}_2}{E_2} \right| = 4|\mathbf{p}_1 E_2 - \mathbf{p}_2 E_1| = 4|\mathbf{p}_1|(E_1 + E_2) \\ &= 4[(p_1 \cdot p_2)^2 - m_1^2 m_2^2]^{1/2} \end{aligned} \tag{A.46}$$

Note that J_i in eqn (A.46) is expressed in terms of the Lorentz-invariant quantity. Hence it is valid in any other frame.

The cross-section is given by

$$d\sigma = \frac{(2\pi)^4 \, \delta^4(p_1 + p_2 - p_\mathrm{f})}{4[(p_1 \cdot p_2)^2 - m_1^2 m_2^2]^{1/2}} |\langle \mathrm{f}|T|\mathrm{i}\rangle|^2 \prod_{j=1}^{n} \frac{d^3 p_j}{(2\pi)^3 2E_j} S \tag{A.47}$$

Appendix B
Feynman rules

In field theory, the standard procedure for calculating any physically interesting quantities in perturbation theory is summarized in a set of Feynman rules. Efficient use of these rules will greatly simplify the calculation. The most important parts of the Feynman rules are the expressions for the propagators and vertices, which characterize the structure of the theory. In this appendix, we will outline a practical method for deriving the propagators and vertices for the most general situation. We will also summarize the results for the physically interesting cases at the end.

A practical guide to the derivation of Feynman rule propagators and vertices

The $\lambda\phi^4$ case

We will first discuss the simple case of $\lambda\phi^4$ theory for the spin-0 boson and then generalize the result to more complicated cases. When the $\lambda\phi^4$ theory Lagrangian is divided into the free part $\mathscr{L}_0$ and the interaction part $\mathscr{L}_I$

$$\mathscr{L} = \mathscr{L}_0 + \mathscr{L}_I \tag{B.1}$$

$$\mathscr{L}_0(\phi) = \tfrac{1}{2}(\partial_\mu\phi)^2 - \tfrac{1}{2}\mu^2\phi^2 \tag{B.2}$$

$$\mathscr{L}_I(\phi) = -\frac{\lambda}{4!}\phi^4, \tag{B.3}$$

the generating functional for the Green's functions is given by

$$\begin{aligned} W[J] &= \int [\mathrm{d}\phi] \exp\left\{ \mathrm{i} \int [\mathscr{L}_0(\phi) + \mathscr{L}_I(\phi) + J(x)\phi(x)]\, \mathrm{d}^4x \right\} \\ &= \exp\left\{ \mathrm{i} \int \left[\mathscr{L}_I\left(-\mathrm{i}\frac{\partial}{\partial J}\right)\right] \mathrm{d}^4x \right\} W_0[J] \end{aligned} \tag{B.4}$$

where

$$W_0[J] = \int [\mathrm{d}\phi] \exp\left\{ \mathrm{i} \int \mathrm{d}^4x [\mathscr{L}_0(x) + J(x)\phi(x)] \right\}. \tag{B.5}$$

Since the free Lagrangian $\mathscr{L}_0$ given in (B.2) is quadratic in the fields, the integral in eqn (B.5) can be evaluated as follows. Write (B.5) as

$$\begin{aligned} W_0[J] &= \int [\mathrm{d}\phi] \exp\left\{ \mathrm{i} \int [\tfrac{1}{2}(\partial_\mu\phi)^2 - \tfrac{1}{2}\mu^2\phi^2 + J(x)\phi(x)]\, \mathrm{d}^4x \right\} \\ &= \int [\mathrm{d}\phi] \exp\left\{ \mathrm{i} \int [\tfrac{1}{2}\phi(x)P(x)\phi(x) + J(x)\phi(x)]\, \mathrm{d}^4x \right\} \end{aligned} \tag{B.6}$$

where

$$P(x) = -(\partial^2 + \mu^2 - i\varepsilon) \tag{B.7}$$

is a Hermitian operator. In (B.6) we have inserted a factor $\exp\{-\frac{1}{2}\int \varepsilon\phi^2 \, d^4x\}$, with $\varepsilon > 0$, to make the path integral convergent. Let $\phi_c(x)$ be the solution of the classical free-field equation in the presence of the external source $J(x)$

$$P(x)\phi_c(x) = -J(x). \tag{B.8}$$

Using the usual Green's function technique, this equation can be inverted. Defining $\Delta_F(x)$ by

$$P(x)\,\Delta_F(x-y) = \delta^4(x-y), \tag{B.9}$$

we have

$$\phi_c(x) = -\int \Delta_F(x-y)J(y)\,d^4y. \tag{B.10}$$

It is easy to see that for the $P(x)$ given in (B.7), we have

$$\Delta_F(x) = \int \frac{d^4k}{(2\pi)^4}\, e^{-ik\cdot x}\,\Delta_F(k) \tag{B.11}$$

with

$$\Delta_F(k) = \frac{1}{k^2 - \mu^2 + i\varepsilon}. \tag{B.12}$$

We now change the integration variable in eqn (B.6) from $\phi(x)$ to $\phi'(x)$ defined by

$$\phi(x) = \phi_c(x) + \phi'(x). \tag{B.13}$$

Then (B.6) becomes, as in the case of usual Gaussian integrations (see eqns (1.49) and (1.81))

$$W_0[J] = N\exp\left\{-i\int [\tfrac{1}{2}J(x)\,\Delta_F(x-y)J(y)]\,d^4x\,d^4y\right\} \tag{B.14}$$

where the normalization factor

$$N = \int [d\phi']\exp\left\{i\int \tfrac{1}{2}\phi'(x)P(x)\phi'(x)\,d^4x\right\} \tag{B.15}$$

is independent of the source function $J(x)$ and is irrelevant for calculating the connected Green's function. The generating functional is then given by

$$W[J] = N\exp\left\{i\int d^4x\left[\mathscr{L}_I\left(-i\frac{\delta}{\delta J}\right)\right]\right\}\exp\left\{\frac{i}{2}\int d^4x_1\,d^4x_2\,J(x_1) \times \Delta_F(x_1 - x_2)J(x_2)\right\} \tag{B.16}$$

and the Green's function can be obtained from

$$\langle 0|T(\phi(x_1)\ldots\phi(x_n))|0\rangle = (-i)^n \left.\frac{\delta^n W[J]}{\delta J(x_1)\ldots\delta J(x_n)}\right|_{J=0}. \tag{B.17}$$

Using the rule for functional differentiation,

$$\frac{\delta J(y)}{\delta J(x)} = \frac{\delta}{\delta J(x)} \int \delta^4(y - x) J(x)\, \mathrm{d}^4 x = \delta^4(x - y) \tag{B.18}$$

it is not difficult to see that the functional differentiation in eqns (B.16) and (B.17) will reproduce Wick's theorem.

The connected Green's functions can be obtained by differentiating the $\ln W[J]$ functional (eqn (1.76))

$$\mathrm{i}^n \frac{\delta^n \ln W[J]}{\delta J(x_1) \ldots \delta J(x_n)}\bigg|_{J=0} = G^{(n)}(x_1 \ldots x_n) = \langle 0|T(\phi(x_1) \ldots \phi(x_n))|0\rangle_{\text{conn}}. \tag{B.19}$$

The free propagator, a two-point function, zeroth-order in the coupling λ, can be obtained from $\ln W_0[J]$ as

$$\langle 0|T(\phi(x)\phi(y))|0\rangle_{\text{free}} = \mathrm{i}^2 \frac{\delta^2 \ln W_0[J]}{\delta J(x)\, \delta J(y)}\bigg|_{J=0} = \mathrm{i}\Delta_{\mathrm{F}}(x - y). \tag{B.20}$$

To get the basic vertex in $\lambda\phi^4$ theory, we take the four-point function $G^{(4)}(x_1 \ldots x_4)$ and keep terms lowest order in λ. After some algebra we obtain

$$G^{(4)}(x_1 \ldots x_4) = -\mathrm{i}\lambda \int \mathrm{i}\Delta_{\mathrm{F}}(x_1 - x)\mathrm{i}\Delta_{\mathrm{F}}(x_2 - x)$$
$$\times\, \mathrm{i}\Delta_{\mathrm{F}}(x_3 - x)\mathrm{i}\Delta_{\mathrm{F}}(x_4 - x)\, \mathrm{d}^4 x \tag{B.21}$$

or, in momentum space,

$$G^{(4)}(k_1 \ldots k_4) = \mathrm{i}\Delta_{\mathrm{F}}(k_1)\mathrm{i}\Delta_{\mathrm{F}}(k_2)\mathrm{i}\Delta_{\mathrm{F}}(k_3)\mathrm{i}\Delta_{\mathrm{F}}(k_4)(-\mathrm{i}\lambda) \tag{B.22}$$

where

$$(2\pi)^4\, \delta^4(k_1 + k_2 + k_3 + k_4) G^{(4)}(k_1 \ldots k_4)$$
$$= \int \prod_{i=1}^{4} \mathrm{d}^4 x_i\, \mathrm{e}^{-\mathrm{i}k_i x_i} G^{(4)}(x_1 \ldots x_4). \tag{B.23}$$

After removing the propagators for the external lines, the four-point 1PI function (the vertex) is

$$\Gamma^{(4)}(k_1, k_2, k_3, k_4) = -\mathrm{i}\lambda. \tag{B.24}$$

To summarize, in $\lambda\phi^4$ theory, the propagator (B.12) and the vertex (B.24) are given by

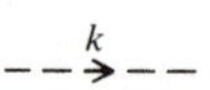

$$\frac{1}{k^2 - \mu^2 + i\varepsilon} \tag{B.25}$$

$$-i\lambda \tag{B.26}$$

Note that the propagator in momentum space is just the Fourier transform of the inverse of the operator $P(x)$ which appears in the quadratic term of the free Lagrangian (B.6). The vertex is just the coefficient of the interaction term multiplied by the factor i and by the number of permutations of identical particles in the interaction term.

Generalization

To generalize the above result to the case of more complicated interactions, consider the Lagrangian

$$\mathscr{L}(x) = \tfrac{1}{2}\phi_i(x)P_{ij}\phi_j(x) + \chi_i^*(x)V_{ij}(x)\chi_j(x) + \bar{\psi}_i(x)X_{ij}(x)\psi_j(x) + \mathscr{L}_{\text{I}}(\phi, \chi, \chi^*, \psi, \bar{\psi}) \tag{B.27}$$

where $\phi_i(\chi_i)$ denotes a set of real (complex) boson fields that may be scalar or vector fields and ψ_i the set of fermion fields. The index i stands for any spinor, Lorentz, isospin, etc. index. P, V, and X are matrix operators that may contain derivatives and must have an inverse. X and V are taken to be hermitian operators while P is taken to be a real symmetric operator

$$X^\dagger = X, \qquad V^\dagger = V, \qquad \text{and } P^{\text{T}} = P. \tag{B.28}$$

It is understood that they contain the iε factor so that the path integral will be convergent. A general term in $\mathscr{L}_{\text{I}}(x)$ has the explicit form:

$$\begin{aligned}\mathscr{L}_{\text{I}} = \int &\mathrm{d}^4x_1\,\mathrm{d}^4x_2 \ldots \\ &\alpha_{i_1 \ldots i_m \ldots i_n \ldots i_p \ldots i_q \ldots}(x; x_1, x_2 \ldots x_m \ldots x_n \ldots x_p \ldots x_q \ldots) \\ &\times \bar{\psi}_{i_1}(x_1) \ldots \psi_{i_m}(x_m) \ldots \phi_{i_n}(x_n) \ldots \chi_{i_p}^*(x_p) \ldots \chi_{i_q}(x_q) \ldots\end{aligned} \tag{B.29}$$

Define the inverses of P, V, X by

$$\begin{aligned}\sum_j P_{ij}(x)P_{jl}^{-1}(x-y) &= \delta_{il}\,\delta^4(x-y)\\ \sum_j V_{ij}(x)V_{jl}^{-1}(x-y) &= \delta_{il}\,\delta^4(x-y)\\ \sum_j X_{ij}(x)X_{jl}^{-1}(x-y) &= \delta_{il}\,\delta^4(x-y)\end{aligned} \tag{B.30}$$

and their Fourier transforms

$$\begin{aligned}P_{ij}^{-1}(x) &= \int \frac{\mathrm{d}^4k}{(2\pi)^4}\,\mathrm{e}^{-\mathrm{i}k\cdot x}\tilde{P}_{ij}^{-1}(k)\\ V_{ij}^{-1}(x) &= \int \frac{\mathrm{d}^4k}{(2\pi)^4}\,\mathrm{e}^{-\mathrm{i}k\cdot x}\tilde{V}_{ij}^{-1}(k)\\ X_{ij}^{-1}(x) &= \int \frac{\mathrm{d}^4k}{(2\pi)^4}\,\mathrm{e}^{-\mathrm{i}k\cdot x}\tilde{X}_{ij}^{-1}(k).\end{aligned} \tag{B.31}$$

Then the propagators are given by

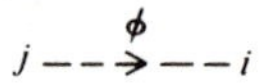

$$\Delta_{\mathrm{F}}^{\phi}(k)_{ij} = \int \mathrm{d}^4 x\, \mathrm{e}^{\mathrm{i}k\cdot x} \langle 0|T(\phi_i(x)\phi_j(0))|0\rangle = \mathrm{i}[\tilde{P}^{-1}(k)]_{ij}$$

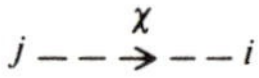

$$\Delta_{\mathrm{F}}^{\chi}(k)_{ij} = \int \mathrm{d}^4 x\, \mathrm{e}^{\mathrm{i}k\cdot x} \langle 0|T(\chi_i^*(x)\chi_j(0))|0\rangle = \mathrm{i}[\tilde{V}^{-1}(k)]_{ij}$$

$$S_{\mathrm{F}}(k)_{ij} = \int \mathrm{d}^4 x\, \mathrm{e}^{\mathrm{i}k\cdot x} \langle 0|T(\bar{\psi}_i(x)\psi_j(0))|0\rangle = \mathrm{i}[\tilde{X}^{-1}(k)]_{ij}. \tag{B.32}$$

For the vertex we define

$$\alpha_{i_1 i_2 \ldots}(x; x_1, x_2 \ldots) = \int \frac{\mathrm{d}^4 k_1}{(2\pi)^4} \frac{\mathrm{d}^4 k_2}{(2\pi)^4} \ldots \mathrm{e}^{\mathrm{i}k_1(x-x_1)+\mathrm{i}k_2(x-x_2)+\ldots} \times \tilde{\alpha}_{i_1 i_2 \ldots}(k_1, k_2 \ldots). \tag{B.33}$$

$\tilde{\alpha}$ contains a factor $\mathrm{i}k_{j_\mu}$ for every derivative $\partial/\partial x_{j_\mu}$ acting on a field with argument x_j. The vertex is then given by

$$I(k_1, k_2 \ldots) = \mathrm{i} \sum_{\{1\ldots m-1\}} \sum_{\{m\ldots n-1\}} \sum_{\{n\ldots p-1\}} \sum_{\{p\ldots q-1\}} \sum_{\{q\ldots\}} \times (-1)^P \tilde{\alpha}_{i_1, i_2}(k_1, k_2 \ldots). \tag{B.34}$$

The summations are over all permutations of indices and momenta as indicated in Fig. (B.1). The momenta are taken to flow inward. Any field $\bar{\psi}$

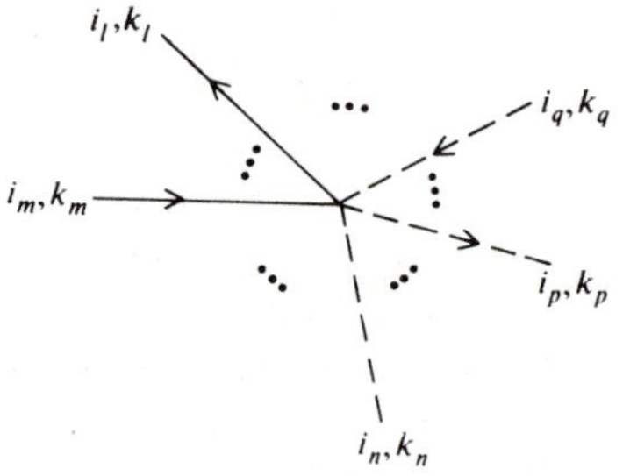

FIG. B.1.

(or χ^*) corresponds to a line with an arrow pointing out; a field ψ (or χ) has an oppositely directed arrow. The ϕ fields are represented by undirected lines. The factor $(-1)^P$ is only of importance if several fermion fields occur. There is a factor of (-1) for every permutation exchanging two fermion fields.

For example, in $\lambda\phi^4$ theory the interaction term can be written

$$-\frac{\lambda}{4!}\phi^4(x) = -\frac{\lambda}{4!}\int \mathrm{d}^4 x_1 \ldots \mathrm{d}^4 x_4\, \delta^4(x-x_1)\,\delta^4(x-x_2) \times \delta^4(x-x_3)\,\delta^4(x-x_4)\phi(x_1)\phi(x_2)\phi(x_3)\phi(x_4) \tag{B.35}$$

Then

$$\alpha(x; x_1 \ldots x_4) = -\frac{\lambda}{4!}\delta^4(x - x_1)\ldots\delta^4(x - x_4) \tag{B.36}$$

and

$$\tilde{\alpha}(k_1 \ldots k_4) = -\frac{\lambda}{4!}. \tag{B.37}$$

The vertex is given by

$$I(k_1 \ldots k_4) = \mathrm{i} \sum_{\{1,2,3,4\}} \tilde{\alpha}(k_1, k_2, k_3, k_4)$$

$$= \mathrm{i} \sum_{\{1,\ldots,4\}} (-\lambda/4!) = -\mathrm{i}\lambda. \tag{B.38}$$

This agrees with the result (B.24) obtained before.

General Feynman rules

Given the propagator and vertices we can now write down the Feynman rule for the computation of Green's functions.

(a) Draw all topologically distinctive, connected diagrams at desired order;

(b) In each diagram, attach a propagator to each internal line

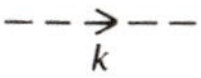

$\dfrac{\mathrm{i}}{k^2 - \mu^2 + \mathrm{i}\varepsilon}$ for a spin-0 boson

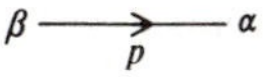

$\left(\dfrac{\mathrm{i}}{\not{p} - m - \mathrm{i}\varepsilon}\right)_{\alpha\beta}$ for a spin-1/2 fermion

Spin-1 boson propagators will depend on the theory;

(c) To each vertex, assign a vertex function given in (B.34), derived from the relevant term in the interaction Lagrangian;

(d) For each internal momentum k not fixed by the momentum conservation at vertices, give a factor $\displaystyle\int \frac{\mathrm{d}^4 k}{(2\pi)^4}$.

(e) Multiply the contribution for each diagram by

(i) a factor (-1) for each closed fermion loop:

(ii) a factor (-1) between graphs which differ from each other only by an interchange of two identical external fermion lines;

(iii) a symmetry factor S^{-1} with

$$S = g \prod_{n=2,3\ldots} 2^{\beta}(n!)^{\alpha_n}$$

where α_n is the number of *pairs* of vertices connected by n identical self-conjugate lines, β is the number of lines connecting a vertex with itself, and g is the number of permutations of vertices which leave the diagram unchanged with fixed external lines.

For example,

$g = 1, \alpha_2 = 1, \beta = 0$, and $S = 2!$

$g = 1, \alpha_3 = 1, \beta = 0$, and $S = 3!$

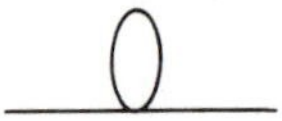

$g = 1, \alpha_n = 0, \beta = 1$, and $S = 2$

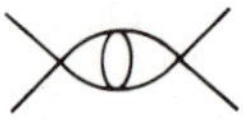

$g = 2, \alpha_2 = 1, \beta = 0$, and $S = 4$

(f) The proper (or one-particle irreducible) Green's function $\Gamma^{(n)}(p_1 \ldots p_n)$ comes from the one-particle irreducible (1PI) diagrams

(i) For the connected Green's function $G^{(n)}(p_1 \ldots p_n)$, attach propagators for the external lines;

(ii) For the scattering amplitude $T(p_1 \ldots p_n)$, put all the external lines on their mass-shell, i.e. $p_i^2 = m_i^2$ and provide external fermion lines with spinors: $u(p)$ [or $v(p)$] for fermions [or antifermions] entering with momentum p; $\bar{u}(p)$ [or $\bar{v}(p)$] for fermions [or antifermions] leaving with momentum p. Provide external vector bosons with polarization vectors: $\varepsilon_\mu(k, \lambda)$ [or $\varepsilon_\mu^*(k, \lambda)$] for the vector boson entering [or leaving] with momentum k.

Summary of $\lambda\phi^4$, Yukawa, QED and QCD propagators and vertices

(a) $\lambda\phi^4$ and Yukawa theory of scalar–fermion interactions.

$$\mathscr{L} = \frac{1}{2}(\partial_\mu \phi)^2 - \frac{\mu^2}{2}\phi^2 - \frac{\lambda}{4!}\phi^4 + \bar{\psi}(\mathrm{i}\gamma\cdot\partial - m)\psi + \mathrm{i}g\bar{\psi}\gamma_5\psi\phi$$

scalar propagator: 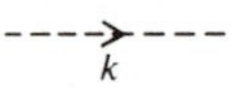

$$\mathrm{i}\Delta_\mathrm{F}(k) = \frac{\mathrm{i}}{k^2 - \mu^2 + \mathrm{i}\varepsilon}$$

fermion propagator: 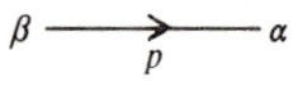

$$\mathrm{i}S_\mathrm{F}(p)_{\alpha\beta} = \left(\frac{\mathrm{i}}{\not{p} - m + \mathrm{i}\varepsilon}\right)_{\alpha\beta}$$

vertices: 

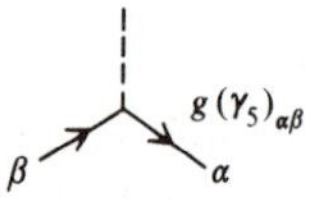

(b) QED.

$$\mathscr{L} = -\frac{1}{4}(\partial_\mu A_\nu - \partial_\nu A_\mu)^2 - \frac{1}{2\xi}(\partial_\mu A^\mu)^2 + \bar{\psi}(\mathrm{i}\gamma\cdot\partial - e\gamma\cdot A - m)\psi$$

Fermion propagator: 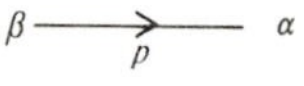

$$\mathrm{i}S_\mathrm{F}(p)_{\alpha\beta} = \left(\frac{\mathrm{i}}{\not{p} - m + \mathrm{i}\varepsilon}\right)_{\alpha\beta}$$

Photon propagator: 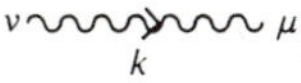

$$iD_F(k)_{\mu\nu} = \frac{-i}{k^2 + i\varepsilon}[g_{\mu\nu} + (\xi - 1)k_\mu k_\nu/k^2]$$

$\xi = 1$ Feynman gauge
$\xi = 0$ Landau gauge

Vertex: 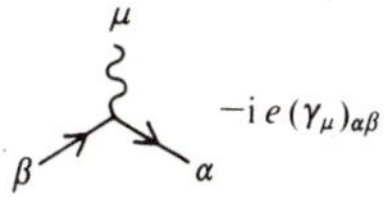

(c) Scalar QED.

$$\mathscr{L} = -\frac{1}{4}(\partial_\mu A_\nu - \partial_\nu A_\mu)^2 - \frac{1}{2\xi}(\partial_\mu A^\mu)^2$$
$$+ [(\partial_\mu + ieA_\mu)\phi]^\dagger[(\partial^\mu + ieA^\mu)\phi] - \mu^2\phi^\dagger\phi - \frac{\lambda}{4}(\phi^\dagger\phi)^2.$$

Scalar propagator: 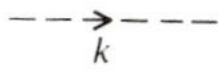

$$i\Delta_F(k) = \frac{i}{k^2 - \mu^2 + i\varepsilon}.$$

Photon propagator: 

$$iD_F(k)_{\mu\nu} = \frac{-i}{k^2 + i\varepsilon}[g_{\mu\nu} + (\xi - 1)k_\mu k_\nu/k^2].$$

$\xi = 1$ Feynman gauge
$\xi = 0$ Landau gauge

Vertices: 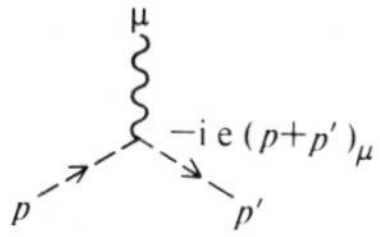

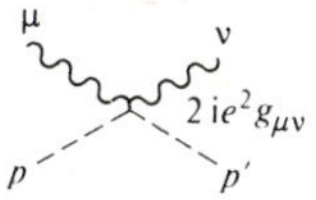

(d) Non-Abelian gauge theory.

$$\mathscr{L} = -\frac{1}{4}(\partial_\mu A^a_\nu - \partial_\nu A^a_\mu + gC^{abc}A^b_\mu A^c_\nu)(\partial^\mu A^{a\nu} - \partial^\nu A^{a\mu} + gC^{ade}A^{d\mu}A^{e\nu})$$
$$- \frac{1}{2\xi}(\partial_\mu A^{a\mu})^2 - \bar{\eta}^a\,\partial_\mu(\partial^\mu\,\delta^{ac} - gC^{abc}A^{b\mu})\eta^c$$
$$+ \bar{\psi}[i\gamma_\mu(\partial^\mu - igA^{a\mu}T^a) - m]\psi$$
$$+ [(\partial^\mu - igA^{a\mu}R^a)\phi]^\dagger[(\partial_\mu - igA^b_\mu R^b)\phi] - m^2_\phi\phi^\dagger\phi - \frac{\lambda}{4}(\phi^\dagger\phi)^2.$$

Vector meson propagator: 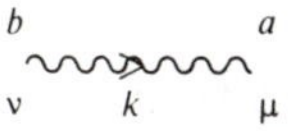

$$iD^{ab}_F(k)_{\mu\nu} = \frac{-i\,\delta^{ab}}{k^2 + i\varepsilon}[g_{\mu\nu} + (\xi - 1)k_\mu k_\nu/k^2]$$

$\xi = 1$ Feynman–'t Hooft gauge
$\xi = 0$ Landau gauge

Fermion propagator: 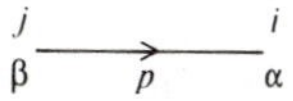

$$iS^{ij}_F(p)_{\alpha\beta} = \left(\frac{i\delta^{ij}}{\not{p} - m + i\varepsilon}\right)_{\alpha\beta}.$$

Scalar propagator:

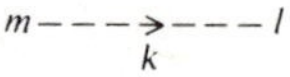

$$\mathrm{i}\Delta_{\mathrm{F}}^{lm}(k) = \frac{\mathrm{i}\delta^{lm}}{k^2 - m^2 + \mathrm{i}\varepsilon}.$$

Ghost propagator:

$$\mathrm{i}\Delta_{\mathrm{F}}^{ab}(k) = \frac{-\mathrm{i}\,\delta^{ab}}{k^2 + \mathrm{i}\varepsilon}.$$

Vertices:

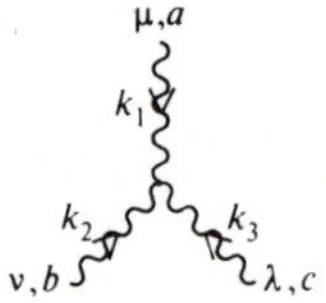

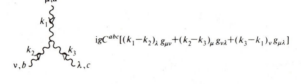

$$\mathrm{i}gC^{abc}[(k_1-k_2)_\lambda g_{\mu\nu}+(k_2-k_3)_\mu g_{\nu\lambda}+(k_3-k_1)_\nu g_{\mu\lambda}]$$

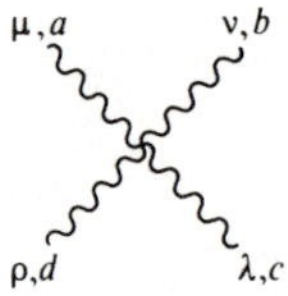

$$\begin{aligned}-\mathrm{i}g^2\,[&C^{eab}C^{ecd}(g_{\mu\lambda}g_{\nu\rho}-g_{\mu\rho}g_{\nu\lambda})\\ &+C^{eac}C^{edb}(g_{\mu\rho}g_{\lambda\nu}-g_{\mu\nu}g_{\lambda\rho})\\ &+C^{ead}C^{ebc}(g_{\mu\nu}g_{\rho\lambda}-g_{\mu\lambda}g_{\rho\nu})\end{aligned}$$

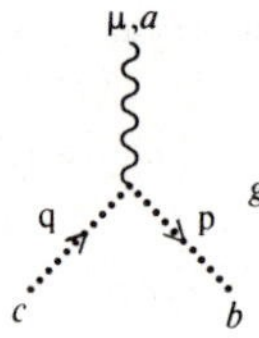

$$gC^{abc}p_\mu$$

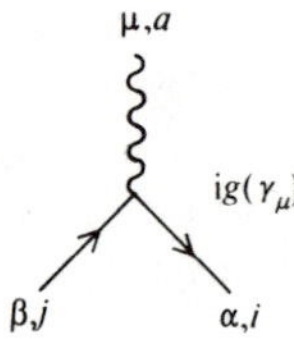

$$\mathrm{i}g(\gamma_\mu)_{\alpha\beta}(T^a)_{ij}$$

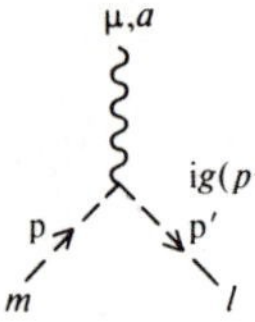

$$\mathrm{i}g(p+p')_\mu R^a_{lm}$$

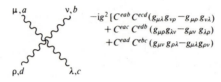

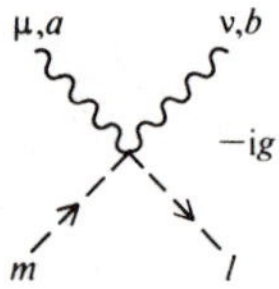

$$-\mathrm{i}g^2 g_{\mu\nu}\{R^a, R^b\}_{lm}$$

For QCD we have

$$C^{abc} = f^{abc} \qquad \text{with } a, b, c = 1, 2, \ldots, 8$$

and the representation matrices

$$(T^a)_{ij} = (\lambda^a/2)_{ij} \qquad \text{with } i, j = 1, 2, 3.$$

$$(R^a)_{lm} = 0$$

R_ξ gauge Feynman rules for the standard $SU(2) \times U(1)$ theory

Since these rules are slightly more complicated we provide some steps in their derivation.

The gauge-fixing and FP ghost terms in the effective Lagrangian

After spontaneous symmetry breaking the covariant derivatives of the scalar field give rise to mixing terms between vector bosons and scalar particles

$$(D_\mu\phi)^\dagger(D^\mu\phi) = \left[\left(\partial_\mu - ig\frac{\boldsymbol{\tau}}{2}\cdot\mathbf{A}_\mu - ig'\frac{B_\mu}{2}\right)(\phi' + \langle\phi\rangle_0)\right]^\dagger$$
$$\times\left[\left(\partial^\mu - ig\frac{\boldsymbol{\tau}}{2}\cdot\mathbf{A}^\mu - ig'\frac{B^\mu}{2}\right)(\phi' + \langle\phi\rangle_0)\right]$$
$$= ig\mathbf{A}^\mu\left[-\partial_\mu\phi'^\dagger\frac{\boldsymbol{\tau}}{2}\langle\phi\rangle_0 + \langle\phi^\dagger\rangle_0\frac{\boldsymbol{\tau}}{2}\partial_\mu\phi'\right]$$
$$+ ig'\frac{B^\mu}{2}[\langle\phi^\dagger\rangle_0\,\partial_\mu\phi' - \partial_\mu\phi'^\dagger\langle\phi\rangle_0] + \ldots. \qquad \text{(B.39)}$$

For practical calculations, it is more convenient to cancel such mixing terms. For this purpose we choose the gauge functions (9.27) to be

$$f_i = \partial_\mu A_i^\mu + ig\xi\left(\phi'^\dagger\frac{\tau_i}{2}\langle\phi\rangle_0 - \langle\phi^\dagger\rangle_0\frac{\tau_i}{2}\phi'\right) \qquad \text{for } SU(2)$$
$$f = \partial_\mu B^\mu + ig'\frac{\xi}{2}\left(\phi'^\dagger\langle\phi\rangle_0 - \langle\phi^\dagger\rangle_0\phi'\right) \qquad \text{for } U(1) \qquad \text{(B.40)}$$

so that the gauge-fixing term (9.68) is of the form

$$\mathscr{L}_{\text{gf}} = -\frac{1}{2\xi}\left[\partial_\mu\mathbf{A}^\mu + ig\xi\left(\phi'^\dagger\frac{\boldsymbol{\tau}}{2}\langle\phi\rangle_0 - \langle\phi^\dagger\rangle_0\frac{\boldsymbol{\tau}}{2}\phi'\right)\right]^2$$
$$-\frac{1}{2\xi}\left[\partial_\mu B^\mu + \frac{ig'\xi}{2}\left(\phi'^\dagger\langle\phi\rangle_0 - \langle\phi^\dagger\rangle_0\phi'\right)\right]^2. \qquad \text{(B.41)}$$

To calculate the Faddeev–Popov ghost from these gauge conditions, we make the infinitesimal SU(2) gauge transformation, with gauge function $\mathbf{u}(x)$,

$$\delta A_\mu^i(x) = \varepsilon^{ijk}u^j(x)A_\mu^k(x) - \frac{1}{g}\partial_\mu u^i(x)$$
$$\delta\phi(x) = -i\frac{\boldsymbol{\tau}}{2}\cdot\mathbf{u}(x)\phi(x)$$

or

$$\delta\phi' = -i\frac{\boldsymbol{\tau}}{2}\cdot\mathbf{u}(\phi' + \langle\phi\rangle_0). \qquad \text{(B.42)}$$

Then

$$\delta f_i = \partial_\mu\left[\varepsilon^{ijk}u^jA_\mu^k - \frac{1}{g}(\partial_\mu u^i)\right] + ig\xi\left[(\phi'^\dagger + \langle\phi^\dagger\rangle_0)i\frac{\boldsymbol{\tau}}{2}\cdot\mathbf{u}\frac{\tau_i}{2}\langle\phi\rangle_0\right.$$
$$\left. + \langle\phi^\dagger\rangle_0\frac{\tau_i}{2}i\frac{\boldsymbol{\tau}}{2}\cdot\mathbf{u}(\phi' + \langle\phi\rangle_0)\right]$$
$$\delta f = i\frac{g'}{2}\xi\left[(\phi'^\dagger + \langle\phi^\dagger\rangle_0)i\frac{\boldsymbol{\tau}}{2}\cdot\mathbf{u}\langle\phi\rangle_0 + \langle\phi^\dagger\rangle_0 i\frac{\boldsymbol{\tau}}{2}\cdot\mathbf{u}(\phi' + \langle\phi\rangle_0)\right]$$

or

$$\delta f_i = \left\{ -\frac{1}{g} \partial^\mu [\delta_{ij} \partial_\mu - g \varepsilon_{ijk} A^k_\mu] \right.$$

$$\left. - g\xi \left[\langle \phi^\dagger \rangle_0 \frac{1}{2} \langle \phi \rangle_0 \delta_{ij} + \phi'^\dagger \frac{\tau_j \tau_i}{4} \langle \phi \rangle_0 + \langle \phi^\dagger \rangle_0 \frac{\tau_i \tau_j}{4} \phi' \right] \right\} u_j$$

$$\delta f = -\frac{g'}{2} \xi \left[\langle \phi^\dagger \rangle_0 \tau_j \langle \phi \rangle_0 + \phi'^\dagger \frac{\tau_j}{2} \langle \phi \rangle_0 + \langle \phi^\dagger \rangle_0 \frac{\tau_j}{2} \phi' \right] u_j. \tag{B.43}$$

Similarly under the infinitesimal U(1) gauge transformation, with gauge function $\alpha(x)$,

$$\delta B_\mu = -\frac{1}{g'} \partial_\mu \alpha$$

$$\delta \phi' = -\mathrm{i} \frac{\alpha}{2} (\phi' + \langle \phi \rangle_0), \tag{B.44}$$

we have

$$\delta f_i = -\frac{g}{4} \xi [2 \langle \phi^\dagger \rangle_0 \tau_i \langle \phi \rangle_0 + \phi'^\dagger \tau_i \langle \phi \rangle_0$$

$$+ \langle \phi^\dagger \rangle_0 \tau_i \phi'] \alpha$$

$$\delta f = -\frac{1}{g'} \left\{ \partial^2 + \frac{g'}{4} \xi [2 \langle \phi^\dagger \rangle_0 \langle \phi \rangle_0 + \phi'^\dagger \langle \phi \rangle_0 \right.$$

$$\left. + \langle \phi^\dagger \rangle_0 \phi'] \right\} \alpha. \tag{B.45}$$

We can combine (B.43) and (B.45) in matrix notation

$$\delta \begin{pmatrix} f_i(x) \\ f(x) \end{pmatrix} = \int \mathrm{d}^4 y \begin{pmatrix} M_f(x, y)_{ij} & M_f(x, y)_i \\ M_f(x, y)_j & M_f(x, y) \end{pmatrix} \begin{pmatrix} u_j(y)/g \\ \alpha(y)/g' \end{pmatrix} \tag{B.46}$$

where

$$M_f(x, y)_{ij} = -\left\{ \partial^\mu [\delta_{ij} \partial_\mu - g \varepsilon_{ijk} A^k_\mu] + g^2 \xi \left[\frac{|\langle \phi \rangle_0|^2}{2} \delta_{ij} + \phi'^\dagger \frac{\tau_j \tau_i}{4} \langle \phi \rangle_0 \right. \right.$$

$$\left. \left. + \langle \phi'^\dagger \rangle_0 \frac{\tau_i \tau_j}{4} \phi' \right] \right\} \delta^4(x - y)$$

$$M_f(x, y)_j = \frac{-gg'}{2} \xi \left[\langle \phi^\dagger \rangle_0 \tau_j \langle \phi \rangle_0 + \phi'^\dagger \frac{\tau_j}{2} \langle \phi \rangle_0 + \langle \phi^\dagger \rangle_0 \frac{\tau_j}{2} \phi' \right]$$

$$\times \delta^4(x - y)$$

$$M_f(x, y) = -\left\{ \partial^2 + \frac{g' \xi}{4} [2 |\langle \phi \rangle_0|^2 + \phi'^\dagger \langle \phi \rangle_0 + \langle \phi^\dagger \rangle_0 \phi'] \right\} \delta^4(x - y). \tag{B.47}$$

The FP ghost-field Lagrangian (9.69) is then given by

$$\mathscr{L}_{\mathrm{FPG}} = \int \mathrm{d}^4x\, \mathrm{d}^4y(\omega_i^\dagger(x), \chi^\dagger(x)) \begin{pmatrix} M_f(x, y)_{ij} & M_f(x, y)_i \\ M_f(x, y)_j & M_f(x, y) \end{pmatrix} \begin{pmatrix} \omega_j(y) \\ \chi(y) \end{pmatrix}. \tag{B.48}$$

Propagators and vertices for bosons and FP ghosts

Define the physical vector bosons as

$$W_\mu^\pm = \frac{1}{\sqrt{2}}(A_\mu^1 \mp \mathrm{i}A_\mu^2)$$

$$Z_\mu = \cos\theta_\mathrm{W} A_\mu^3 - \sin\theta_\mathrm{W} B_\mu$$

$$A_\mu = \sin\theta_\mathrm{W} A_\mu^3 + \cos\theta_\mathrm{W} B_\mu$$

and write the scalar mesons as

$$\phi' = \begin{pmatrix} \phi^+ \\ \dfrac{\phi_1 + \mathrm{i}\phi_2}{\sqrt{2}} \end{pmatrix}.$$

For the ghost fields we can define similar combinations

$$\omega^\pm = \frac{1}{\sqrt{2}}(\omega_1 \mp \mathrm{i}\omega_2)$$

$$\omega_\mathrm{Z} = \cos\theta_\mathrm{W}\omega_3 - \sin\theta_\mathrm{W}\chi$$

$$\omega_\gamma = \sin\theta_\mathrm{W}\omega_3 + \cos\theta_\mathrm{W}\chi.$$

With these definitions, we can easily work out the propagators from the quadratic part of the Lagrangian $\mathscr{L}_1 + \mathscr{L}_3 + \mathscr{L}_{\mathrm{gf}} + \mathscr{L}_{\mathrm{FPG}}$ in eqns (11.37), (11.59), (B.41), and (B.48)

$\mathrm{W}^\pm$  $\dfrac{-\mathrm{i}}{k^2 - M_\mathrm{W}^2 + \mathrm{i}\varepsilon}[g_{\mu\nu} + (\xi - 1)k_\mu k_\nu/(k^2 - \xi M_\mathrm{W}^2)]$

Z  $\dfrac{-\mathrm{i}}{k^2 - M_\mathrm{Z}^2 + \mathrm{i}\varepsilon}[g_{\mu\nu} + (\xi - 1)k_\mu k_\nu/(k^2 - \xi M_\mathrm{Z}^2)]$

$\phi^\pm$ ------ $\dfrac{\mathrm{i}}{k^2 - \xi M_\mathrm{W}^2 + \mathrm{i}\varepsilon}$

ϕ_2 ------ $\dfrac{\mathrm{i}}{k^2 - \xi M_\mathrm{Z}^2 + \mathrm{i}\varepsilon}$

ϕ_1 ------ $\dfrac{\mathrm{i}}{k^2 - 2\mu^2 + \mathrm{i}\varepsilon}$

$\omega^\pm$ .... $\dfrac{-\mathrm{i}}{k^2 - \xi M_\mathrm{W}^2 + \mathrm{i}\varepsilon}$

ω_Z .... $\dfrac{-\mathrm{i}}{k^2 - \xi M_\mathrm{Z}^2 + \mathrm{i}\varepsilon}$

ω_γ .... $\dfrac{-\mathrm{i}}{k^2 + \mathrm{i}\varepsilon}$

where $\xi = 1$ 't Hooft–Feynman gauge, $\xi = 0$ Landau gauge, and $\xi = \infty$ unitary gauge.

The boson vertices are

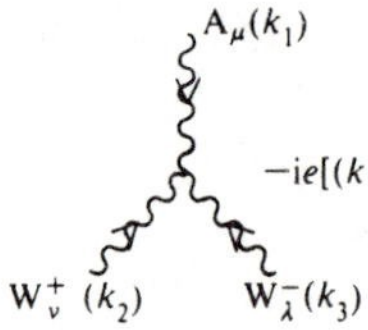

$$-ie[(k_1-k_2)_\lambda g_{\mu\nu}+(k_2-k_3)_\mu g_{\nu\lambda}+(k_3-k_1)_\nu g_{\lambda\mu}]$$

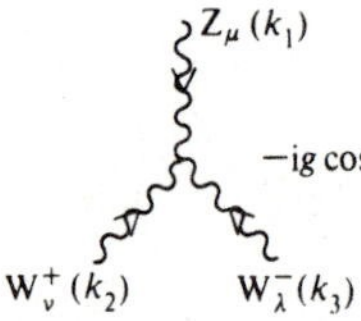

$$-ig\cos\theta_W[(k_1-k_2)_\lambda g_{\mu\nu}+(k_2-k_3)_\mu g_{\nu\lambda}+(k_3-k_1)_\nu g_{\lambda\mu}]$$

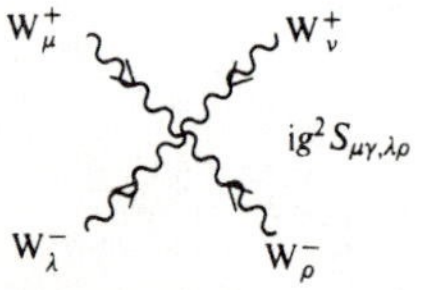

$$ig^2 S_{\mu\gamma,\lambda\rho}$$

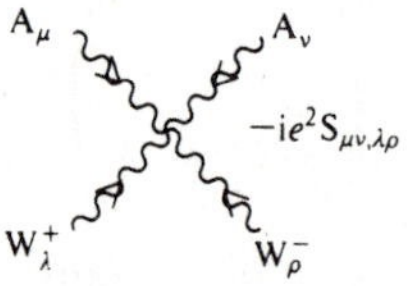

$$-ie^2 S_{\mu\nu,\lambda\rho}$$

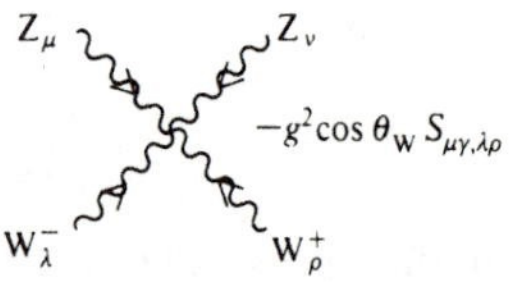

$$-g^2\cos\theta_W S_{\mu\gamma,\lambda\rho}$$

as well as the vertex $-ieg\cos\theta_W\ S_{\mu\nu,\lambda\rho}$ with $S_{\mu\nu,\lambda\rho} = 2g_{\mu\nu}g_{\lambda\rho} - g_{\mu\lambda}g_{\nu\rho} - g_{\mu\rho}g_{\nu\lambda}$. In graphs below *all* charged boson lines are taken to be entering *into* the vertices.

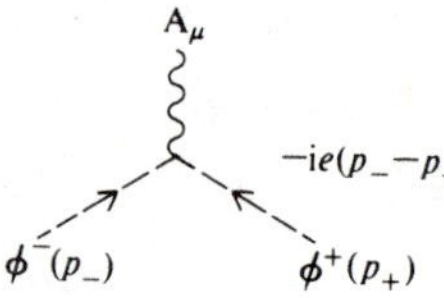

$$-ie(p_- - p_+)_\mu$$

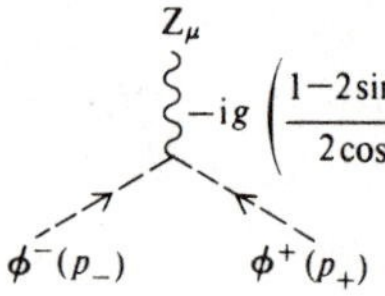

$$-ig\left(\frac{1-2\sin^2\theta_W}{2\cos\theta_W}\right)(p_- - p_+)_\mu$$

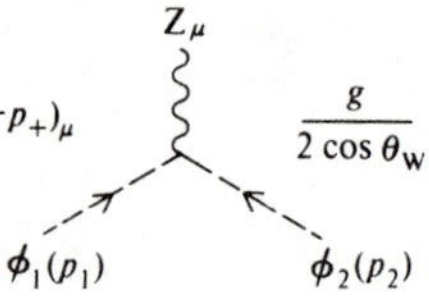

$$\frac{g}{2\cos\theta_W}(p_2-p_1)_\mu$$

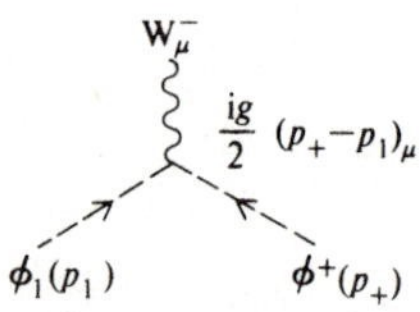

$$\frac{ig}{2}(p_+-p_1)_\mu$$

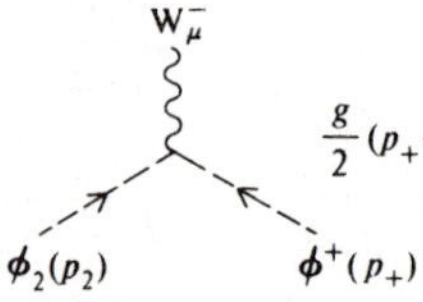

$$\frac{g}{2}(p_+-p_2)_\mu$$

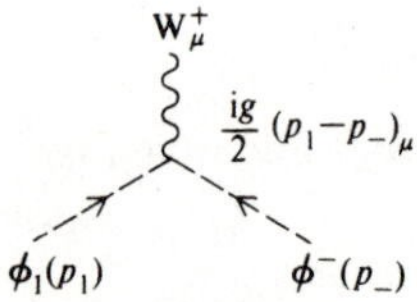

$$\frac{ig}{2}(p_1-p_-)_\mu$$

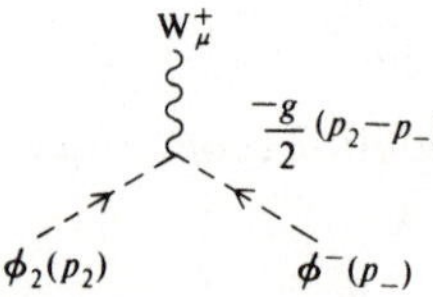

$$\frac{-g}{2}(p_2-p_-)_\mu$$

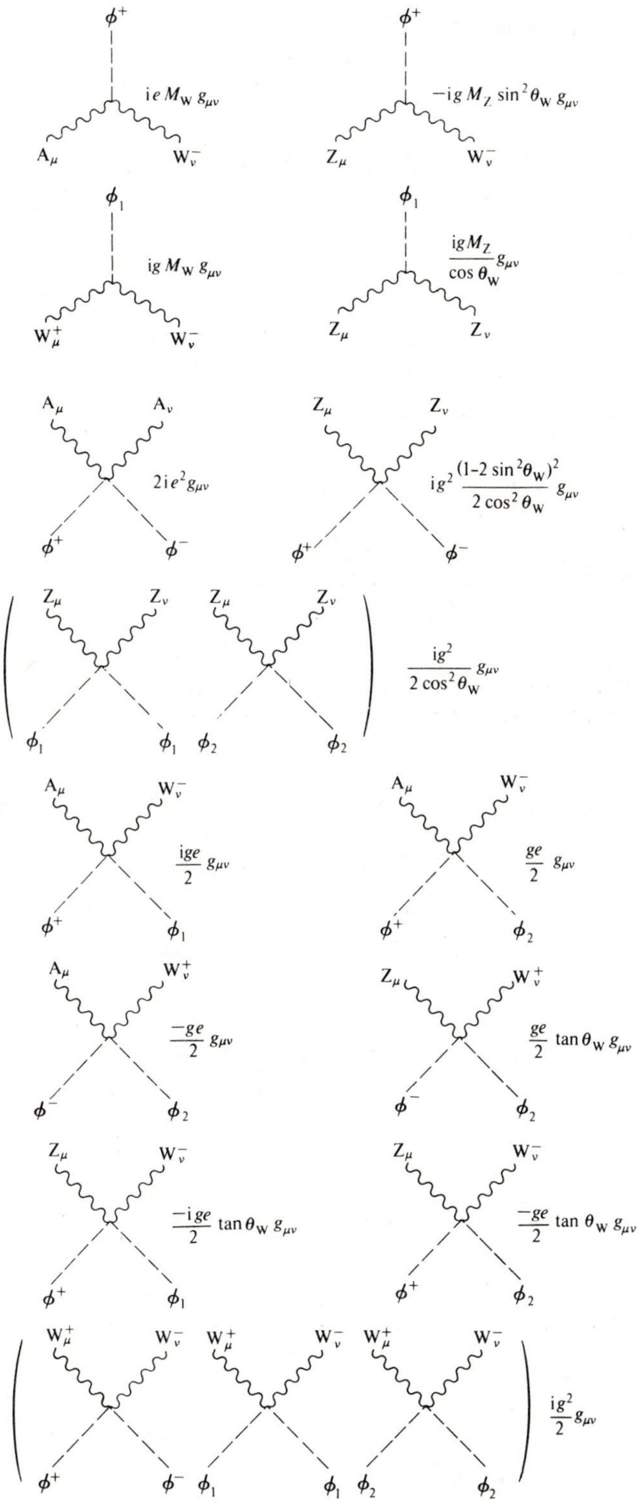

φ⁺ A_μ W_ν⁻ ie M_W g_μν
φ⁺ Z_μ W_ν⁻ −ig M_Z sin²θ_W g_μν
φ₁ W_μ⁺ W_ν⁻ ig M_W g_μν
φ₁ Z_μ Z_ν igM_Z/cos θ_W g_μν
A_μ A_ν φ⁺ φ⁻ 2ie² g_μν
Z_μ Z_ν φ⁺ φ⁻ ig² (1-2 sin²θ_W)²/2 cos²θ_W g_μν
Z_μ Z_ν φ₁ φ₁ Z_μ Z_ν φ₂ φ₂ ig²/2 cos²θ_W g_μν
A_μ W_ν⁻ φ⁺ φ₁ ige/2 g_μν
A_μ W_ν⁻ φ⁺ φ₂ ge/2 g_μν
A_μ W_ν⁺ φ⁻ φ₂ −ge/2 g_μν
Z_μ W_ν⁺ φ⁻ φ₂ ge/2 tan θ_W g_μν
Z_μ W_ν⁻ φ⁺ φ₁ −ige/2 tan θ_W g_μν
Z_μ W_ν⁻ φ⁺ φ₂ −ge/2 tan θ_W g_μν
W_μ⁺ W_ν⁻ φ⁺ φ⁻ W_μ⁺ W_ν⁻ φ₁ φ₁ W_μ⁺ W_ν⁻ φ₂ φ₂ ig²/2 g_μν

Inclusion of leptons and quarks

Propagator: $\xrightarrow{\quad p \quad}$ $\dfrac{i}{\not{p} - m_i + i\varepsilon}$.

Vertices for leptons: $l = (e, \mu, \tau)$, $\nu_l = (\nu_e, \nu_\mu, \nu_\tau)$

for quarks q: $p = (u, c, t)$, $n = (d, s, b)$ with the CKM mixing matrix U_{pn} of eqn (12.39).

A_μ ; l, l : $-ie\,\gamma_\mu$

A_μ ; q, q : $ie\,Q_q\,\gamma_\mu$

W_μ^- ; ν, l : $\dfrac{ig}{2\sqrt{2}}\,\gamma_\mu\,(1-\gamma_5)$

W_μ^- ; p, n : $\dfrac{ig}{2\sqrt{2}}\,\gamma_\mu\,(1-\gamma_5)\,U_{pn}$

Z_μ ; ν, ν : $\dfrac{ig}{4\cos\theta_W}\,\gamma_\mu\,(1-\gamma_5)$

Z_μ ; p, p : $\dfrac{ig}{4\cos\theta_W}\,\gamma_\mu\,[\,(1-\tfrac{8}{3}\sin^2\theta_W)+\gamma_5\,]$

Z_μ ; l, l : $\dfrac{ig}{4\cos\theta_W}\,\gamma_\mu\,[\,(-1+4\sin^2\theta_W)-\gamma_5\,]$

Z_μ ; n, n : $\dfrac{ig}{4\cos\theta_W}\,\gamma_\mu\,[\,(-1+\tfrac{4}{3}\sin^2\theta_W)-\gamma_5\,]$

ϕ^- ; ν, l : $\dfrac{-ig\,m_l}{2\sqrt{2}\,M_W}\,(1-\gamma_5)$

ϕ^- ; p, n : $\dfrac{-ig}{2\sqrt{2}\,M_W}\,[\,m_n(1-\gamma_5)-m_p(1+\gamma_5)\,]\,U_{pn}$

ϕ_1 ; l, l : $\dfrac{-ig\,m_l}{2M_W}$

ϕ_1 ; n, n : $\dfrac{-ig\,m_n}{2M_W}$

ϕ_1 ; p, p : $\dfrac{-ig\,m_p}{2M_W}$

ϕ_2 ; l, l : $\dfrac{g\,m_l}{2M_W}\,\gamma_5$

ϕ_2 ; n, n : $\dfrac{g\,m_n}{2M_W}\,\gamma_5$

ϕ_2 ; p, p : $\dfrac{-g\,m_p}{2M_W}\,\gamma_5$

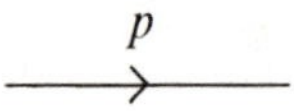

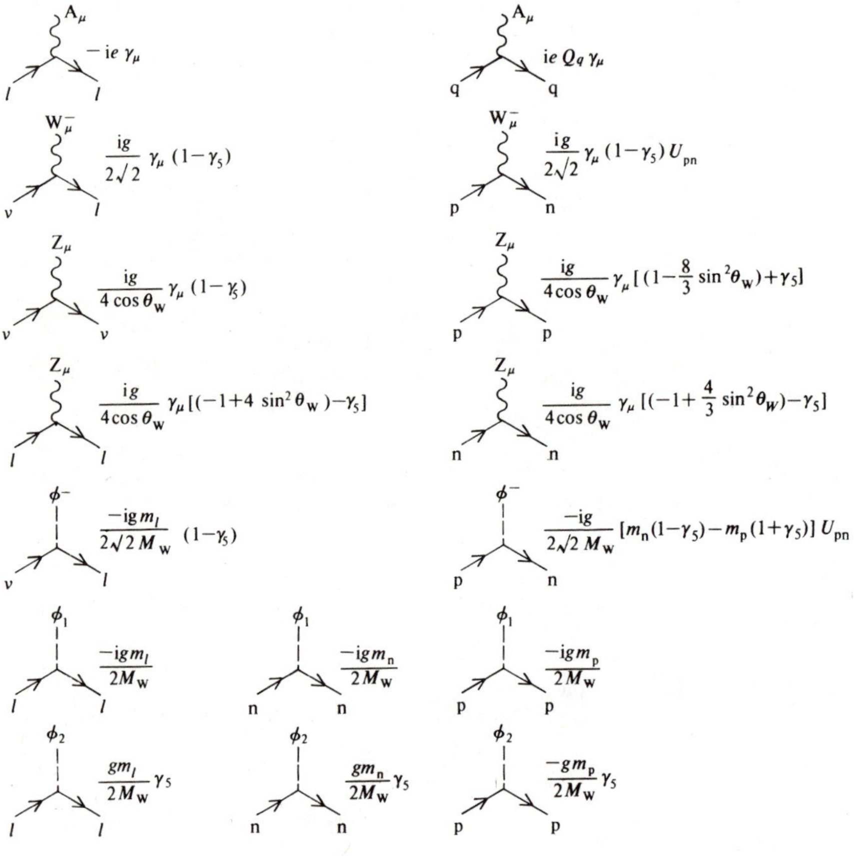

Bibliography

We have collected here a set of review articles and monographs pertaining to several of the general topics covered in this book. They are not a complete listing and inevitably some of the important contributions have been left out; we regret such omissions. Nevertheless it is hoped that a reader who wishes to go further will find this informal guide useful.

Particle physics

1. Källen, G. (1964). *Elementary particle physics.* Addison-Wesley, Reading, Massachusetts.
2. Gasiorowicz, S. (1966). *Elementary particle physics.* Wiley, New York.
3. Frazer, W. R. (1966). *Elementary particle physics.* Prentice-Hall, Englewood Cliffs, New Jersey.
4. Cheng, D. C. and O'Neill, G. K. (1979). *Elementary particle physics: an introduction.* Addison-Wesley, Reading, Massachusetts.
5. Lee, T. D. (1981). *Particle physics and introduction to field theory.* Harwood, New York.
6. Perkins, D. H. (1982). *Introduction to high energy physics* (2nd edn). Addison-Wesley, Reading, Massachusetts.
7. Leader, E. and Predazzi, E. (1982). *An introduction to gauge theories and the new physics.* Cambridge University Press, Cambridge.

Quantum field theory

1. Bjorken, J. D. and Drell, S. D. (1964). *Relativistic quantum mechanics.* McGraw-Hill, New York.
2. Bjorken, J. D. and Drell, S. D. (1965). *Relativistic quantum fields.* McGraw-Hill, New York.
3. Sakurai, J. J. (1967). *Advanced quantum mechanics.* Addison-Wesley, Reading, Massachusetts.
4. Beresteski, V. B., Lifshitz, E. M. and Pitaevskii, L. P. (1971). *Relativistic quantum theory,* Part I. Pergamon Press, New York.
5. Lifshitz, E. M., and Pitaevskii, L. P. (1971). *Relativistic quantum theory,* Part II. Pergamon Press, New York.
6. Nash, C. (1978). *Relativistic quantum fields.* Academic Press, New York.
7. Bogoliubov, N. N. and Shirkov, D. V. (1980). *Introduction to theories of quantized fields* (3rd edn). Wiley–Interscience, New York.
8. Itzykson, C. and Zuber, J. (1980). *Introduction to quantum field theory.* McGraw-Hill, New York.
9. Lee, T. D. (1981). *Particle physics and introduction to field theory.* Harwood, New York.

Feynman and path integral

1. Feynman, R. P. and Hibbs, A. R. (1965). *Quantum mechanics and path integrals.* McGraw-Hill, New York.
2. Fried, H. M. (1972). *Functional methods and models in quantum field theory.* MIT Press, Cambridge, Massachusetts.

3. Faddeev, L. and Slavnov, A. A. (1980). *Gauge fields, introduction to quantum theory*. Benjamin Cummings, Reading, Massachusetts.
4. Ramond, P. (1981). *Field theory; a modern primer*. Benjamin Cummings, Reading, Massachusetts.
5. Abers, E. and Lee, B. W. (1973). Gauge theories, *Phys. Rep.* **9C**, 1.
6. Marinov, M. S. (1980). Path integrals in quantum theory—an outlook of basic concepts. *Phys. Rep.* **60C**, 1.

Group theory

1. Hamermesh, M. (1962). *Group theory*. Addison-Wesley, Reading, Massachusetts.
2. Carruthers, P. (1966). *Introduction to unitary symmetry*. Interscience, New York.
3. Gilmore, R. (1974). *Lie groups, Lie algebra and some of their applications*. Wiley-Interscience, New York.
4. Wybourne, B. (1974). *Classical groups for physicists*. Wiley-Interscience, New York.
5. Lichtenberg, D. B. (1978). *Unitary symmetry and elementary particles* (2nd edn). Academic Press, New York.
6. Georgi, H. (1982). *Lie algebra in particle physics*. Benjamin Cummings, Reading, Massachusetts.
7. Slansky, R. (1981). Group theory for unified model building. *Phys. Rep.* **79C**, 1.

Quark model

1. Gell-Mann, M. and Ne'eman, Y. (1964). *The eightfold way*. Benjamin, New York.
2. Kokkedee, J. J. J. (1969). *The quark model*. Benjamin, New York.
3. Close, F. E. (1979). *An introduction to quarks and partons*. Academic Press, New York.
4. Lipkin, H. J. (1973). Quarks for pedestrians. *Phys. Rep.* **8C**, 173.

Current algebra

1. Adler, S. L. and Dashen, R. F. (1968). *Current algebra and application to particle physics*. Benjamin, New York.
2. Treiman, S. B., Jackiw, R. and Gross, D. J. (1972). *Lectures on current algebra and its applications*. Princeton University Press, Princeton, New Jersey.
3. de Alfaro, V., Fubini, S., Furlan, G. and Rossetti, C. (1973). *Current in hadron physics*. North-Holland, Amsterdam.

Parton model and light-cone

1. Roy, P. (1975). *Theory of lepton–hadron processes at high energies*, Oxford University Press, Oxford.
2. Close, F. E. (1979). *An introduction to quarks and partons*. Academic Press, New York.
3. Kogut, J. and Susskind, L. (1973). The parton picture of elementary particles. *Phys. Rep.* **8C**, 77.
4. Frishman, Y. (1974). Light cone and short distances. *Phys. Rep.* **13C**, 1.
5. Altarelli, G. (1982). Partons in quantum chromodynamics. *Phys. Rep.* **81C**, 1.

Gauge theory (general)

1. Taylor, J. C. (1976). *Gauge theory of weak interactions*. Cambridge University Press, Cambridge.
2. Itzykson, C. and Zuber, J.-B. (1980). *Quantum field theory*. McGraw-Hill, New York.

3. Faddeev, L. D. and Slavnov, A. A. (1980). *Gauge fields, introduction to quantum theory*. Benjamin Cummings, Reading, Massachusetts.
4. Konopleva, N. P. and Popov, V. N. (1981). *Gauge fields*. Harwood, New York.
5. Ramond, P. (1981). *Field theory: a modern primer*. Benjamin Cummings, Reading, Massachusetts.
6. Lopes, J. L. (1981). *Gauge field theories, an introduction*. Pergamon Press, New York.
7. Lee, T. D. (1981). *Particle physics and introduction to field theory*. Harwood, New York.
8. Aitchison, I. J. R. and Hey, A. J. G. (1982). *Gauge theories in particle physics*. Hilger, Bristol.
9. Abers, E. and Lee, B. W. (1973). Gauge theories. *Phys. Rep.* **9C**, 1.
10. Li, L.-F. (1981). Introduction to gauge theories of electromagnetic and weak interactions. (To be published in Proc. of 1981 Summer School on Particle Physics at Hefei, China.)

QCD

1. Politzer, H. D. (1974). Asymptotic freedom: an approach to strong interactions. *Phys. Rep.* **14C**, 129.
2. Marciano, W. and Pagels, H. (1978). Quantum chromodynamics. *Phys. Rep.* **36C**, 137.
3. Ellis, J. and Sachrajda, C. (1980). Quantum chromodynamics and its application. In *Quarks and leptons—Cargese* 1979 (ed. M. Levy *et al.*). Plenum Press, New York.
4. Buras, A. J. (1980). Asymptotic freedom in deep inelastic processes in the leading order and beyond. *Rev. mod. Phys.* **52**, 199.
5. Dokshitzer, Yu. L., Dyakonov, D. L. and Troyan, S. I. (1980). Hard processes in quantum chromodynamics. *Phys. Rep.* **58C**, 269.
6. Llewellyn Smith, C. H. (1980). Topics in quantum chromodynamics. In *Quantum flavordynamics, quantum chromodynamics and unified theories* (ed. K. T. Mahanthappa and J. Randa). Plenum Press, New York.
7. Reya, E. (1981). Perturbative quantum chromodynamics. *Phys. Rep.* **69C**, 195.
8. Muller, A. H. (1981). Perturbative QCD at high energies. *Phys. Rep.* **73C**, 237.
9. Altarelli, G. (1982). Partons in quantum chromodynamics. *Phys. Rep.* **81C**, 1.
10. Wilczek, F. (1982). Quantum chromodynamics: the modern theory of the strong interaction. *Ann. Rev. Nucl. and Part. Sci.* **32**, 177.
11. Close, F. E. (1982). First lap in QCD. *Phys. Scripta* **25**, 86.

Lattice QCD

1. Wilson, K. G. (1975). Quarks and strings on a lattice. In *New phenomena in subnuclear physics* (ed. A. Zichichi), Plenum Press, New York.
2. Drouffe, J. M. and Itzykson, C. (1978). Lattice gauge fields. *Phys. Rep.* **38C**, 134.
3. Kogut, J. B. (1979). An introduction to lattice gauge theory and spin system. *Rev. mod. Phys.* **51**, 659.

Weak interactions

1. Lee, T. D. and Wu, C. S. (1965). Weak interactions. *Ann. Rev. Nucl. Sci.* **15**, 381.
2. Marshak, R., Riazuddin, and Ryan, C. P. (1969). *Theory of weak interactions in particle Physics*. Wiley-Interscience, New York.
3. Commins, E. (1973). *Weak interactions*. McGraw-Hill, New York.

Standard model of electroweak interactions

1. Taylors, J. C. (1976). *Gauge theories of weak interactions*. Cambridge University Press, Cambridge.

2. Lopes, J. L. (1981). *Gauge field theories: an introduction*. Pergamon Press, New York.
3. Commins, E. and Bucksbaum, P. H. (1983). *The new weak interactions*. Cambridge University Press, Cambridge.
4. Abers, E. and Lee, B. W. (1973). Gauge Theories. *Phys. Rep.* **9C**, 1.
5. Bég, M. A. B. and Sirlin, A. (1974). Gauge theories of weak interactions. *Ann. Rev. Nucl. Sci.* **24**, 379.
6. Harari, H. (1978). Quarks and leptons. *Phys. Rep.* **42C**, 235.
7. Fritzsch, H. and Minkowski, P. (1981). Flavor dynamics of quarks and leptons. *Phys. Rep.* **73C**, 67.
8. Bég, M. A. B. and Sirlin, A. (1982). Gauge theories of weak interactions II. *Phys. Rep.* **88C**, 1.

GUT

1. Gaillard, M. K. and Maiani, L. (1980). New quarks and leptons. *Quarks and leptons—Cargese* 1979, p. 433 (ed. M. Levy *et al.*). Plenum Press, New York.
2. Langacker, P. (1981). Grand unified theories and proton decay. *Phys. Rep.* **72C**, 185.
3. Ellis, J. (1981). Phenomenology of unified gauge theories. Proc. 1981 Les Houches Summer School (to be published).
4. Zee, A. (1982). *The unity of forces in the universe*, Vol. I. World Science Press, Singapore.

Cosmology

1. Peebles, P. J. E. (1971). *Physical cosmology*. Princeton University Press, Princeton, New Jersey.
2. Sciama, D. W. (1972). *Modern cosmology*. Cambridge University Press, Cambridge.
3. Weinberg, S. (1972). *Gravitation and cosmology*. Wiley, New York.
4. Weinberg, S. (1977). *The first three minutes*. Basic Books, New York.
5. Wilczek, F. (1981). Erice lecture on cosmology. Proc. 1981 Int. Sch. of Subnucl. Phys. 'Ettore Majorana' (to be published).
6. Zee, A. (1982). *The unity of forces in the universe*, Vol. II. World Science Press, Singapore.

References

Abers, E. S. and Lee, B. W. (1973). *Phys. Rep.* **9C**, 1.
Adler, S. L. (1965*a*). *Phys. Rev.* **139**, B1638.
—— (1965*b*). *Phys. Rev.* **140**, B736.
—— (1966). *Phys. Rev.* **143**, 1144.
—— (1969). *Phys. Rev.* **177**, 2426.
—— (1970). In *Lectures on elementary particles and quantum field theory* Proc. 1970 Brandeis Summer Institute (ed. S. Deser *et al.*). MIT Press, Cambridge, Massachusetts.
—— Bardeen, W. A. (1969). *Phys. Rev.* **182**, 1517.
—— Dashen, R. F. (1968). *Current algebra and applications to particle physics.* Benjamin, New York.
Aharonov, Y. and Bohm, D. (1959). *Phys. Rev.* **115**, 485.
Aitchison, I. J. R. and Hey, A. J. G. (1982). *Gauge theories in particle physics.* Hilger, Bristol.
Albert, D. *et al.* (1980). *Nucl. Phys.* **B166**, 460.
Albrecht, A. and Steinhardt, P. J. (1982). *Phys. Rev. Lett.* **48**, 1220.
Alles, W., Boyer, C. and Buras, A. J. (1977). *Nucl. Phys.* **B119**, 125.
Altarelli, G. (1982). *Phys. Rep.* **81**, 1.
—— Parisi, G. (1977). *Nucl. Phys.* **B126**, 298.
Altarev, I. S. *et al.* (1981). *Phys. Lett.* **102B**, 13.
Anderson, P. W. (1958). *Phys. Rev.* **112**, 1900.
—— (1963). *Phys. Rev.* **130**, 439.
Appelquist, T. and Georgi, H. (1973). *Phys. Rev.* **D8**, 4000.
—— Carazzone, J. (1975). *Phys. Rev.* **D11**, 2856.
—— Politzer, H. D. (1975). *Phys. Rev. Lett.* **34**, 43.
—— Barnett, R. M. and Lane, K. D. (1978). *Ann. Rev. Nucl. Part. Sci.* **28**, 387.
Arnowitt, R. and Fickler, S. I. (1962). *Phys. Rev.* **127**, 1821.
Ashmore, J. F. (1972). *Nuovo Cimento Lett.* **4**, 289.
Aubert, J. J. *et al.* (1974). *Phys. Rev. Lett.* **33**, 1404.
Augustin, J. E. *et al.* (1974). *Phys. Rev. Lett.* **33**, 1406.
Bahcall, J. *et al.* (1980). *Phys. Rev. Lett.* **45**, 945.
Banks, T. and Georgi, H. (1976). *Phys. Rev.* **D14**, 1159.
Bander, M. (1981). *Phys. Rep.* **75C**, 205.
Barber, D. P. *et al.* (1979). *Phys. Rev. Lett.* **43**, 830.
—— *et al.* (1981). *Phys. Rev. Lett.* **46**, 1663.
Bardeen, W. A. (1969). *Phys. Rev.* **184**, 1848.
Barger, V. *et al.* (1980). *Phys. Rev. Lett.* **45**, 692.
Barnes, V. E. *et al.* (1964). *Phys. Rev. Lett.* **12**, 204.
Becchi, C., Rouet, A. and Stora, R. (1974). *Phys. Lett.* **52B**, 344.
Bég, M. A. B. (1980). In *Recent developments in high-energy physics* Proc. of Orbis Scientiae 1980 (ed. A. Perlmutter and L. F. Scott). Plenum Press, New York.
—— Sirlin, A. (1974). *Ann. Rev. Nucl. Sci.* **24**, 379.
—— —— (1982). *Phys. Rep.* **88C**, 1.
Belavin, A. A., Polyakov, A. M., Schwartz, A. S. and Tyupkin, Yu.S. (1975). *Phys. Lett.* **59B**, 85.
Bell, J. S. (1973). *Nucl. Phys.* **B60**, 427.
—— Jackiw, R. (1969). *Nuovo Cimento* **60A**, 47.
—— Sutherland, D. G. (1968). *Nucl. Phys.* **B4**, 315.

Berestetskii, V. B., Lifshitz, E. M. and Pitaevskii, L. (1971). *Relativistic quantum theory*, Part I. Pergamon Press, New York.
Berezin, F. A. (1966). *The method of second quantization*. Academic Press, London.
Berger, Ch. *et al.* (1979). *Phys. Lett.* **86B**, 418.
Bilenky, S. M. and Pontecorvo, B. (1978). *Phys. Rep.* **41C**, 225.
Bjorken, J. D. (1969). *Phys. Rev.* **179**, 1547.
—— (1976). In *Weak interactions at high energy and the production of new particles*. Proc. 1976 SLAC Summer Institute on Particle Physics (ed. M. C. Zipf). SLAC, Stanford, California.
—— Drell, S. D. (1964). *Relativistic quantum mechanics*. McGraw-Hill, New York.
—— —— (1965). *Relativistic quantum fields*. McGraw-Hill, New York.
—— Glashow, S. L. (1964). *Phys. Lett.* **11**, 255.
—— Paschos, E. A. (1969). *Phys. Rev.* **185**, 1975.
—— Weinberg, S. (1977). *Phys. Rev. Lett.* **38**, 622.
Bloch, F. and Nordsieck, A. (1937). *Phys. Rev.* **52**, 54.
Bogoliubov, N. N. and Parasiuk, O. S. (1957). *Acta. Math.* **97**, 227.
Bogoliubov, N. N. and Shirkov, D. V. (1959). *An introduction to theory of quantized fields* (1st edn). Wiley-Interscience, New York. [2nd edn 1980.]
Bollini, C. G. and Giambiagi, J. J. (1972). *Phys. Lett.* **40B**, 566.
Bosetti, P. C. *et al.* (1978). *Nucl. Phys.* **B142**, 1.
Bouchiat, C., Iliopoulos, J. and Meyer, Ph. (1972). *Phys. Lett.* **38B**, 519.
Brandelik, R. *et al.* (1979). *Phys. Lett.* **86B**, 243.
Brown, L. and Ellis, S. (1981). *Phys. Rev.* **D24**, 2383.
Buras, A. J. (1980). *Rev. mod. Phys.* **52**, 199.
—— Ellis, J., Gaillard, M. K. and Nanopoulos, D. V. (1978). *Nucl. Phys.* **B135**, 66.
Cabibbo, N. (1963). *Phys. Rev. Lett.* **10**, 531.
—— Parisi, G. and Testa, M. (1970). *Nuovo Cimento Lett* **4**, 35.
Cabrera, B. (1982). *Phys. Rev. Lett.* **48**, 1378.
Cahn, R., Chanowitz, M. and Fleishon, N. (1979). *Phys. Lett.* **82B**, 113.
Callan, C. G., Jr. (1970). *Phys. Rev.* **D2**, 1541.
—— (1982). *Phys. Rev. D.* **25**, 2141; **26**, 2058.
—— Dashen, R. F. and Gross, D. J. (1976). *Phys. Lett.* **63B**, 334.
—— —— —— (1980). *Phys. Rev. Lett.* **44**, 435.
—— Gross, D. (1969). *Phys. Rev. Lett.* **22**, 156.
Carruthers, P. (1966). *Introduction to unitary symmetry*. Interscience, New York.
Cheng, D. C. and O'Neill, G. K. (1979). *Elementary particle physics: an introduction*. Addison-Wesley, Reading, Massachusetts.
Cheng, T. P. (1976). *Phys. Rev.* **D13**, 2161.
—— Dashen, R. (1971). *Phys. Rev. Lett.* **26**, 594.
—— Eichten, E. and Li, L.-F. (1974). *Phys. Rev.* **D9**, 2259.
—— Li, L.-F. (1977). *Phys. Rev.* **D16**, 1425.
—— —— (1978). *Phys. Rev.* **D17**, 2375.
—— —— (1980*a*). *Phys. Rev.* **D22**, 2868.
—— —— (1980*b*). *Phys. Rev. Lett.* **45**, 1908.
—— —— (1980*c*). In *Proc. 1980 Guangzhou Conf. on Theor. Part. Phys.* p. 1469, Science Press, Beijing.
Chew, G. (1962). *S-matrix theory of strong interaction*. Benjamin, New York.
—— Frautschi, S. C. (1961). *Phys. Rev.* **123**, 1478.
Chou, K.-C. (1961). *Soviet Phys. JETP* **12**, 492.
Christ, N., Hasslacher, B. and Mueller, A. (1972). *Phys. Rev.* **D6**, 3543.
Christenson, J. H., Cronin, J. W., Fitch, V. L. and Turlay, R. (1964). *Phys. Rev. Lett.* **13**, 138.
Cicuta, G. M. and Montaldi, E. (1972). *Nuovo Cimento Lett.* **4**, 329.
Close, F. E. (1979). *An introduction to quarks and partons*. Academic Press, London.
Coleman, S. (1971*a*). In *Properties of the fundamental interactions*, Proc. of 1971 Int.

Sch. Subnucl. Phys. 'Ettore Majorana' (ed. A. Zichichi), p. 358. Editrice Compositori, Bologna.
—— (1971*b*). In *Properties of the fundamental interactions*. Proc. of 1971 Int. Sch. Subnucl. Phys. 'Ettore Majorana' (ed. A. Zichichi), p. 604. Editrice Compositori, Bologna.
—— (1973). In *Laws of hadronic matter*. Proc. 1973 Int. Sch. Subnucl. Phys. 'Ettore Majorana' (ed. A. Zichichi). Academic Press, London.
—— (1975). In *New phenomena in subnuclear physics*. Proc. 1975 Int. Sch. Subnucl. Phys. 'Ettore Majorana' (ed. A. Zichichi). Plenum Press, New York.
—— (1977). In *The whys of subnuclear physics*. Proc. 1977 Int. Sch. Subnucl. Phys. 'Ettore Majorana' (ed. A. Zichichi). Plenum Press, New York.
—— (1981). In Proc. 1981 Int. Sch. Subnucl. Phys. 'Ettore Majorana' (to be published).
—— Gross, D. J. (1973). *Phys. Rev. Lett.* **31**, 851.
—— Weinberg, E. (1973). *Phys. Rev.* **D7**, 1888.
Commins, E. (1973). *Weak interactions*. McGraw-Hill, New York.
—— Bucksbaum, P. H. (1983). *The new weak interactions*. Cambridge University Press, Cambridge.
Cornwall, J. M., Levin, D. N. and Tiktopoulos, G. (1974). *Phys. Rev.* **D10**, 1145.
Cowsik, R. and McClelland, J. (1972). *Phys. Rev. Lett.* **29**, 669.
Creutz, M. (1979). *Phys. Rev. Lett.* **43**, 553. *Erratum ibid.* (1979). **43**, 890.
Cutkosky, R. E. (1960). *J. Math. Phys.* **1**, 429.
Dashen, R. (1969). *Phys. Rev.* **183**, 1291.
Davis, R., Harmer, D. S. and Hoffman, K. C. (1968). *Phys. Rev. Lett.* **20**, 1205.
de Alfaro, V., Fubini, S., Furlan, G. and Rossetti, C. (1973). *Currents in hadron physics*. North-Holland, Amsterdam.
de Groot, J. G. H. *et al.* (1979). *Phys. Lett.* **82B**, 292.
Derrick, G. H. (1964). *J. Math. Phys.* **5**, 1252.
DeWitt, B. (1967). *Phys. Rev.* **162**, 1195, 1239.
Dimopoulos, S. and Susskind, L. (1978). *Phys. Rev.* **D18**, 4500.
Dine, M., Fischler, W. and Srednicki, M. (1981). *Phys. Lett.* **104B**, 199.
Dirac, P. A. M. (1931). *Proc. R. Soc.* **A133**, 60.
—— (1933). *Physikalische Zeitschrift der Sowjetunion*. **3**, 64.
Dokos, C. P. and Tomaras, T. N. (1980). *Phys. Rev. D.* **21**, 2940.
Dokshitzer, Yu. L., Dyakonov, D. I. and Troyan, S. I. (1980). *Phys. Rep.* **58C**, 269.
Donoghue, J. F. and Li, L.-F. (1979). *Phys. Rev.* **D9**, 945.
Drell, S. D. and Yan, T. M. (1971). *Ann. Phys.* (*NY*) **66**, 595.
Drouffe, J. M. and Itzykson, C. (1978). *Phys. Rep.* **38C**, 134.
Dyson, F. J. (1949). *Phys. Rev.* **75**, 486.
Eichten, E. *et al.* (1980). *Phys. Rev.* **D21**, 203.
Ellis, J. (1977). In *Weak and electromagnetic interactions at high energy*, Proc. of 1976 Les Houches Summer Sch. (ed. R. Balian and C. H. Llewellyn Smith). North-Holland, Amsterdam.
—— (1981). In Proc. 1981 Les Houches Summer Sch. (to be published).
—— Gaillard, M. K. and Nanopoulos, D. V. (1976*a*). *Nucl. Phys.* **B106**, 292.
—— —— —— (1976*b*). *Nucl. Phys.* **B109**, 216.
—— —— —— Rudaz, S. (1977). *Nucl. Phys.* **B131**, 285.
—— Sachrajda, C. (1980). In *Quarks and leptons—Cargese 1979* (ed. M. Levy *et al.*). Plenum Press, New York.
Englert, F. and Brout, R. (1964). *Phys. Rev. Lett.* **13**, 321.
Faddeev, L. D. and Popov, V. N. (1967). *Phys. Lett.* **25B**, 29.
—— Slavnov, A. A. (1980). *Gauge fields, introduction to quantum theory*. Benjamin Cummings, Reading, Massachusetts.
Farhi, E. and Susskind, L. (1981). *Phys. Rep.* **74C**, 277.
Felst, R. (1981). In *Proc. 1981 Int. Sym. on Lepton and Photon Interactions at High Energies*. (ed. W. Pfeil). University Press, Bonn.

Fermi, E. (1934*a*). *Nuovo Cimento* **11**, 1.
—— (1934*b*). *Z. Phys.* **88**, 161.
Feynman, R. P. (1948*a*). *Rev. mod. Phys.* **20**, 267.
—— (1948*b*). *Phys. Rev.* **74**, 939, 1430.
—— (1963). *Acta Phys. Polon.* **24**, 697.
—— (1972). *Photon–hadron interaction.* Benjamin, Reading, Massachusetts.
—— (1977). In *Weak and electromagnetic interactions at high energy*, Proc. of 1976 Les Houches Summer School (ed. R. Balian and C. H. Llewellyn Smith). North-Holland, Amsterdam.
—— Gell-Mann, M. (1958). *Phys. Rev.* **109**, 193.
—— Hibbs, A. R. (1965). *Quantum mechanics and path integrals.* McGraw-Hill, New York.
—— Speisman, G. (1954). *Phys. Rev.* **94**, 500.
Frazer, W. R. (1966). *Elementary particle physics.* Prentice-Hall, Englewood Cliffs, New Jersey.
Fried, H. M. (1972). *Functional methods and models in quantum field theory.* MIT Press, Cambridge, Massachusetts.
Frishman, Y. (1974). *Phys. Rep.* **13C**, 1.
Fritzsch, H., Gell-Mann, M. and Leutwyler, H. (1973). *Phys. Lett.* **47B**, 365.
—— Minkowski, P. (1975). *Ann. Phys. (NY)* **93**, 193.
—— —— (1981) *Phys. Rep.* **73C**, 67.
Fubini, S. and Furlan, G. (1965). *Physics* **1**, 229.
Fujikawa, K. (1979). *Phys. Rev. Lett.* **42**, 1195.
—— Lee, B. W. and Sanda, A. I. (1972). *Phys. Rev.* **D6**, 2923.
Gaillard, M. K. and Lee, B. W. (1974). *Phys. Rev.* **D10**, 897.
—— Maiani, L. (1980). In *Quarks and leptons—Cargese 1979* (ed. M. Levy *et al.*). Plenum Press, New York.
—— Lee, B. W. and Rosner, J. (1975). *Rev. mod. Phys.* **47**, 277.
Gasiorowicz, S. (1966). *Elementary particle physics.* Wiley, New York.
—— Rosner, J. L. (1981). *Am. J. Phys.* **49**, 954.
Gasser, A. and Leutwyler, H. (1975). *Nucl. Phys.* **B94**, 269.
Gavela, M. B. *et al.* (1982). *Phys. Lett.* **109B**, 83; 215.
Gell-Mann, M. (1953). *Phys. Rev.* **92**, 833.
—— (1961). California Institute of Technology Synchrotron Laboratory Report No. CTSL-20. (Reprinted in Gell-Mann, M. and Ne'eman, Y. (1964).)
—— (1962*a*). *Phys. Rev.* **125**, 1067.
—— (1962*b*). In *Proc. 11th Int. Conf. High Energy Phys. (CERN)* (ed. J. Prentki). CERN, Geneva.
——(1964*a*). *Physics* **1**, 63.
—— (1964*b*). *Phys. Lett.* **8**, 214.
—— Glashow, S. L. (1961). *Ann. Phys. (NY)* **15**, 437.
—— Levy, M. (1960). *Nuovo Cimento* **16**, 705.
—— Low, F. E. (1954). *Phys. Rev.* **95**, 1300.
—— Ne'eman, Y. (1964). *The eightfold way.* Benjamin, New York.
—— Oakes, R. J. and Renner, B. (1968). *Phys. Rev.* **175**, 2195.
—— Pais, A. (1955). *Phys. Rev.* **97**, 1387.
—— Ramond, P. and Slansky, R. (1979). In *Supergravity* (ed. D. Z. Freeman and P. van Nieuwenhuizen). North-Holland, Amsterdam.
Georgi, H. (1974). In *Particles and fields—1974* (ed. C. E. Carlson). American Institute of Physics, New York.
—— (1982). *Lie algebra in particle physics.* Benjamin Cummings, Reading, Massachusetts.
—— Glashow, S. L. (1972*a*). *Phys. Rev. Lett.* **28**, 1494.
—— —— (1972*b*). *Phys. Rev.* **D6**, 429.
—— —— (1973). *Phys. Rev.* **D7**, 2487.
—— —— (1974). *Phys. Rev. Lett.* **32**, 438.

Georgi, H. and Politzer, H. D. (1974). *Phys. Rev.* **D9**, 416.
—— Quinn, H. and Weinberg, S. (1974). *Phys. Rev. Lett.* **33**, 451.
Gershtein, S. and Zeldovich, Y. (1966). *Sov. Phys. JETP Lett.* **4**, 120.
Geweniger, C. (1979). In *Proc. Neutrino* 79. (Bergen) (ed. A. Haatuft and C. Jarlskog). European Phys. Society.
Gildener, E. (1976). *Phys. Rev.* **D14**, 1667.
Gilman, F. and Wise, M. B. (1979). *Phys. Lett.* **83B**, 83.
Gilmore, R. (1974). *Lie groups, Lie algebras and some of their applications.* Wiley-Interscience, New York.
Ginzburg, V. L. and Landau, L. D. (1950). *J. expl. theoret. Phys. USSR* **20**, 1064.
Glashow, S. L. (1958). Ph.D. Thesis. Harvard University.
—— (1961). *Nucl. Phys.* **22**, 579.
—— (1967). In *Hadrons and their interactions*, Proc. 1967 Int. School of Physics 'Ettore Majorana' (ed. A. Zichichi). Academic Press, New York.
—— Iliopoulos, J. and Maiani, L. (1970). *Phys. Rev.* **D2**, 1285.
—— Weinberg, S. (1968). *Phys. Rev. Lett.* **20**, 224.
—— —— (1977). *Phys. Rev.* **D15**, 1958.
Goddard, P. and Olive, D. I. (1978). *Rep. Prog. Phys.* **41**, 1357.
Goldberger, M. L. and Treiman, S. B. (1958). *Phys. Rev.* **110**, 1178.
Goldhaber, A. (1976). *Phys. Rev. Lett.* **36**, 1122.
Goldstone, J. (1961). *Nuovo Cimento* **19**, 154.
—— Salam, A. and Weinberg, S. (1962). *Phys. Rev.* **127**, 965.
Golowich, E. and Yang, T. C. (1979). *Phys. Lett.* **80B**, 245.
Greenberg, O. W. (1964). *Phys. Rev. Lett.* **13**, 598.
Gribov, V. N. and Lipatov, L. N. (1972). *Sov. J. Nucl. Phys.* **15**, 438, 675.
—— Pontecorvo, B. (1969). *Phys. Lett.* **28B**, 495.
Gross, D. (1976). In *Methods in field theory*, Proc. of 1975 Les Houches Summer School (ed. R. Balian and J. Zinn-Justin). North-Holland, Amsterdam.
—— Jackiw, R. (1972). *Phys. Rev.* **D6**, 477.
—— Llewellyn Smith, C. H. (1969). *Nucl. Phys.* **B14**, 337.
—— Wilczek, F. (1973*a*). *Phys. Rev. Lett.* **30**, 1343.
—— —— (1973*b*) *Phys. Rev.* **D8**, 3633.
—— —— (1974). *Phys. Rev.* **D9**, 980.
Guralnik, G. S., Hagen, C. R. and Kibble, T. W. B. (1964). *Phys. Rev. Lett.* **13**, 585.
Guralnik, G. S. *et al.* (1968). In *Advances in particle physics* (ed. R. Cool and R. Marshak), Vol. 2, p. 567. Interscience, New York.
Gursey, F., Ramond, P and Sikivie, P. (1975). *Phys. Lett.* **60B**, 177.
Guth, A. H. (1981). *Phys. Rev.* **23**, 347.
—— Tye, S. H. (1980). *Phys. Rev. Lett.* **44**, 631.
Hamermesh, M. (1962). *Group theory.* Addison-Wesley, Reading, Massachusetts.
Han, M. and Nambu, Y. (1965). *Phys. Rev.* **139B**, 1006.
Hanson, G. *et al.* (1975). *Phys. Rev. Lett.* **35**, 1609.
Harari, H. (1978). *Phys. Rep.* **42C**, 235.
Hasert, F. J. *et al.* (1973). *Phys. Lett.* **46B**, 138.
Hasenfratz, P. and 't Hooft, G. (1976). *Phys. Rev. Lett.* **36**, 1119.
Hawking, S. W. and Moss, I. G. (1982). *Phys. Lett. B* **110**, 35.
Hepp, K. (1966). *Comm. Math. Phys.* **2**, 301.
Herb, S. W. *et al.* (1977). *Phys. Rev. Lett.* **39**, 252.
Higgs, P. W. (1964*a*) *Phys. Rev. Lett.* **12**, 132.
—— (1964*b*). *Phys. Rev. Lett.* **13**, 509.
—— (1966). *Phys. Rev.* **145**, 1156.
Hughes, R. J. (1981). *Nucl. Phys.* **B186**, 376.
Iizuka, J. (1966). *Prog. theor. Phys. Suppl.* **37–8**, 21.
Itzykson, C. and Zuber, J.-B. (1980). *Quantum field theory.* McGraw-Hill, New York.
Jackiw, R. (1972). In *Lectures on current algebra and its applications.* Princeton University Press, Princeton, New Jersey.

Jackiw, R. and Rebbi, C. (1976). *Phys. Rev. Lett.* **36**, 1119; **37**, 172.
Jacob, M. and Wick, G. C. (1959). *Ann. Phys.* (*NY*) **7**, 404.
Jona-Lasinio, G. (1964). *Nuovo Cimento* **34**, 1790.
Jonker *et al.* (1981). *Phys. Lett.* **99B**, 265.
Källen, G. (1964). *Elementary particle physics*. Addison-Wesley, Reading, Massachusetts.
Kayser, B. (1981). *Phys. Rev.* **D24**, 110.
Kibble, T. W. B. (1967). *Phys. Rev.* **155**, 1554.
Kim, J. E. (1979). *Phys. Rev. Lett.* **43**, 103.
—— Langacker, P., Levine, M. and Williams, H. H. (1981). *Rev. mod. Phys.* **53**, 211.
Kinoshita, T. (1962). *J. Math. Phys.* **3**, 650.
Kleinknecht, K. (1976). *Ann. Rev. Nucl. Sci.* **26**, 1.
Kobayashi, M. and Maskawa, M. (1973). *Prog. theor. Phys.* **49**, 652.
Kogut, J. B. (1979). *Rev. mod. Phys.* **51**, 659.
—— Pearson, R. and Shigemitsu, J. (1979). *Phys. Rev. Lett.* **43**, 484.
—— —— —— (1981). *Phys. Lett.* **98B**, 63.
—— Susskind, L. (1973). *Phys. Rep.* **8C**, 77.
—— —— (1975*a*). *Phys. Rev.* **D11**, 395.
—— —— (1975*b*). *Phys. Rev.* **D11**, 1477.
—— —— (1975*c*). *Phys. Rev.* **D11**, 3594.
Kokkedee, J. J. J. (1979). *The quark model*. Benjamin, New York.
Konopleva, N. P. and Popov, V. N. (1981). *Gauge fields*. Harwood, New York.
Kuti, J. and Weisskopf, V. F. (1971). *Phys. Rev.* **D4**, 3418.
Landshoff, P. V. and Polkinghorne, J. C. (1971). *Nucl. Phys.* **B28**, 240.
Langacker, P. (1981). *Phys. Rep.* **72C**, 185.
LaRue, G. S., Fairbank, W. M. and Hebard, F. (1977). *Phys. Rev. Lett.* **38**, 1011.
Leader, E. and Predazzi, E. (1982). *An introduction to gauge theories and the new physics*. Cambridge University Press, Cambridge.
Lederman, L. M. (1978). In *Proc. 19th Int. Conf. High Energy Phys.* (*Tokyo*). (ed. G. Takeda). Physical Society of Japan, Tokyo.
Lee, B. W. (1972*a*). *Chiral dynamics*. Gordon & Breach, New York.
—— (1972*c*). In *Proc. 16th Int. Conf. on High Energy Phys.* (*Illinois*) (ed. J. D. Jackson and A. Roberts). Fermilab, Batavia, Illinois.
—— (1974). In *Renormalization and invariance in quantum field theory* (ed. E. R. Caianiello). Plenum Press, New York.
—— Shrock, R. (1977). *Phys. Rev.* **D16**, 1445.
—— Zinn-Justin, J. (1972). *Phys. Rev.* **D5**, 3121, 3137.
—— —— (1973). *Phys. Rev.* **D7**, 1049.
Lee, T. D. (1974). *Phys. Rep.* **9C**, 143.
—— (1981). *Particle physics and introduction to field theory*. Harwood, New York.
—— Nauenberg, M. (1964). *Phys. Rev.* **133**, 1549.
—— Wu, C. S. (1965). *Ann. Rev. Nucl. Sci.* **15**, 381.
—— Yang, C. N. (1956). *Phys. Rev.* **104**, 254.
Leibrandt, G. (1975). *Rev. mod. Phys.* **47**, 849.
Levin, D. and Tiktopoulos, G. (1975). *Phys. Rev.* **D12**, 420.
Li, L.-F. (1974). *Phys. Rev.* **D9**, 1723.
—— (1980). In *Proc. 1980 Guangzhou Conf. on Theor. Part. Phys.* Science Press, Beijing.
—— (1981). In Proc. 1981 Summer Sch. on Part. Phys. (Hefei).
—— Pagels, H. (1971). *Phys. Rev. Lett.* **26**, 1204.
Lichtenberg, D. B. (1978). *Unitary symmetry and elementary particles*. (2nd ed.) Academic Press, London.
Lifschitz, E. H. and Pitaevskii, L. P. (1971). *Relativistic quantum theory* Part II. Pergamon Press, New York.
Linde, A. D. (1976). *Sov. Phys. JETP Lett.* **23**, 64.
—— (1982). *Phys. Lett.* **B108**, 389.

Lipkin, H. J. (1973). *Phys. Rep.* **8C**, 173.
Llewellyn Smith, C. H. (1973). In *Proc. 5th Hawaii Topical Conf. Part. Phys.* (ed. P. N. Dobson *et al.*). Univ. of Hawaii Press, Honolulu, Hawaii.
—— (1974). *Phys. Rep.* **3C**, 264.
—— (1980). In *Quantum flavordynamics, quantum chromodynamics and unified theories* (ed. K. T. Mahanthappa and J. Randa). Plenum Press, New York.
—— (1982). *Phil. Trans. R. Soc. Lond.* **A304**, 5.
—— Wheater, J. F. (1981). *Phys. Lett.* **105B**, 486.
Lopes, J. L. (1981). *Gauge field theories, an introduction*. Pergamon Press, New York.
Low, F. E. (1954). *Phys. Rev.* **96**, 1428.
Lubkin, E. (1963). *Ann. Phys.* (*NY*) **23**, 233.
Lyubimov, V. A. *et al.* (1980). *Phys. Lett.* **94B**, 266.
Mandelstam, S. (1968). *Phys. Rev.* **94B**, 266.
—— (1979). In *Proc. 1979 Int. Sym. on Lepton and Photon Interactions at High Energies* (ed. T. B. W. Kirk and H. D. I. Abarbanel). Fermilab, Batavia, Illinois.
Maki, Z. *et al.* (1962). *Prog theor. Phys.* **28**, 870.
Marciano, W. and Pagels, H. (1978). *Phys. Rev.* **36C**, 137.
—— Sanda, H. (1977). *Phys. Lett.* **67B**, 303.
—— Sirlin, A. (1980). *Phys. Rev.* **D22**, 2695.
—— —— (1981). In *The Second Workshop on Grand Unification* (ed. J. Leveille, L. Sulak and D. Unger). Birkhauser, Boston.
Marinov, M. S. (1980). *Phys. Rep.* **60C**, 1.
Marshak, R., Riazuddin and Ryan, C. P. (1969). *Theory of weak interactions in particle physics.* Wiley-Interscience, New York.
McWilliams, B. and Li, L.-F. (1981). *Nucl. Phys.* **B179**, 62.
Muller, A. H. (1981). *Phys. Rep.* **73C**, 237.
Nachtmann, O. (1972). *Nucl. Phys.* **B38**, 397.
Nakagawa, M. *et al.* (1963). *Prog. theor. Phys.* **30**, 727.
Nambu, Y. (1960). *Phys. Rev. Lett.* **4**, 380.
—— (1966). In *Preludes in theoretical physics* (ed. de DeShalit). North-Holland, Amsterdam.
—— (1968). *Phys. Lett.* **26B**.
—— (1974). *Phys. Rev.* **D10**, 4262.
—— Jona-Lasinio, G. (1961*a*). *Phys. Rev.* **122**, 345.
—— —— (1961*b*) *Phys. Rev.* **124**, 246.
Nash, C. (1978). *Relativistic quantum fields*. Academic Press, London.
Ne'eman, Y. (1961). *Nucl. Phys.* **26**, 222.
Nielsen, N. K. (1981). *Am. J. Phys.* **49**, 1171.
Nishijima, K. and Nakano, T. (1953). *Prog. theor. Phys.* **10**, 581.
Noether, E. (1918). *Nachr. Kgl. Geo. Wiss Gottinger* **235**.
Okubo, S. (1962). *Prog. theor. Phys.* **27**, 949.
—— (1963). *Phys. Lett.* **5**, 163.
—— (1977). *Phys. Rev.* **D16**, 3528.
Panofsky, W. (1968). In *Proc. 14th Int. Conf. on High Energy Physics* (*Vienna*) (ed. J. Prentki and J. Steinberger). CERN, Geneva.
Paschos, E. (1977). *Phys. Rev.* **D15**, 1966.
—— Wolfenstein, L. (1973). *Phys. Rev.* **D7**, 91.
Pati, J. C. and Salam, A. (1973). *Phys. Rev.* **D8**, 1240.
Pauli, W. and Villars, F. (1949). *Rev. mod. Phys.* **21**, 434.
Peccei, R. D. and Quinn, H. R. (1977). *Phys. Rev.* **D16**, 1791.
Peebles, P. J. E. (1966). *Ast. J.* **146**, 542.
—— (1971). *Physical cosmology*. Princeton University Press, Princeton, New Jersey.
Perkins, D. H. (1982). *Introduction to high energy physics*, 2nd edn. Addison-Wesley, Reading, Massachusetts.
Perl, M. L. *et al.* (1975). *Phys. Rev. Lett.* **35**, 1489.
—— (1977). *Phys. Lett.* **70B**, 487.

Petcov, S. T. (1977). *Sov. J. Nucl. Phys.* **25**, 340.
Politzer, H. D. (1973). *Phys. Rev. Lett.* **30**, 1346.
—— (1974). *Phys. Rep.* **14C**, 129.
Polkinghorne, J. C. (1958). *Nuovo Cimento* **8**, 179.
Polyakov, A. M. (1974). *Sov. Phys. JETP Lett.* **20**, 194.
Pontecorvo, B. (1958). *Sov. Phys. JETP*, **6**, 429.
—— (1968). *Sov. Phys. JETP*, **26**, 984.
Popov, V. N. and Faddeev, L. D. (1967). Kiev. ITP report 67–36 (unpublished), available as Fermilab preprint NAL-THY-57 (1972).
Prescott, C. Y. *et al.* (1978). *Phys. Lett.* **77B**, 347.
Preskill, J. P. (1979). *Phys. Rev. Lett.* **43**, 1365.
Quigg, C. (1977). *Rev. mod. Phys.* **49**, 297.
—— Rosner, J. L. (1979). *Phys. Rev.* **56**, 169.
Rajaraman, R. (1975). *Phys. Rep.* **21C**, 227.
Ramond, P. (1981). *Field theory: a modern primer.* Benjamin Cummings, Reading, Massachusetts.
Ramsey, N. F. (1978). *Phys. Rep.* **43**, 409.
Regge, T. (1959). *Nuovo Cimento* **14**, 951.
Reya, E. (1981). *Phys. Rep.* **69C**, 195.
Roy, P. (1975). *Theory of lepton–hadron processes at high energies.* Oxford University Press, Oxford.
Rubakov, V. A. (1982). *Sov. Phys. JETP Lett.* **48**, 1148.
Sakharov, A. D. (1967). *Sov. Phys. JETP Lett.* **5**, 24.
Sakurai, J. J. (1958). *Nuovo Cimento* **7**, 649.
—— (1962). *Phys. Rev. Lett.* **9**, 472.
—— (1967). *Advanced quantum mechanics.* Addison-Wesley, Reading, Massachusetts.
Salam, A. (1968). In *Elementary particle physics* (*Nobel Symp No. 8*). (ed. N. Svartholm). Almqvist and Wilsell, Stockholm.
—— Ward, J. C. (1964). *Phys. Lett.* **13**, 168.
Schwinger, J. (1948). *Phys. Rev.* **73**, 416.
—— (1949). *Phys. Rev.* **75**, 898.
—— (1951*a*). *Phys. Rev.* **82**, 664.
—— (1951*b*). *Phys. Rev.* **82**, 914.
—— (1951*c*). *Proc. Natl. Acad. Sci.* **37**, 452.
—— (1957). *Ann. Phys.* (*NY*) **2**, 407.
—— (1959). *Phys. Rev. Lett.* **3**, 296.
Sciama, D. W. (1972). *Modern cosmology.* Cambridge University Press.
Shabalin, E. P. (1978). *Sov. J. Nucl. Phys.* **28**, 75.
—— (1980). *Sov. J. Nucl. Phys* **31**, 864; **32**, 228.
Shifman, M. A. (1981). In *Proc. 1981 Int. Sym. on Lepton and Photon Interactions at High Energies* (ed. W. Pfiel). University Press, Bonn.
Shrock, R. E. and Treiman, S. B. (1979). *Phys. Rev.* **D19**, 2148.
—— Wang, L.-L. (1978). *Phys. Rev. Lett.* **41**, 1692.
Sikivie, P. (1982*a*). In *Proc. 1980 Int. Sch. of Phys. 'Enrico Fermi'* (ed. G. Costa and R. Gatto). North-Holland, Amsterdam.
—— (1982*b*). *Phys. Rev. Lett.* **48**, 1156.
Sirlin, A. (1978). *Rev. mod. Phys.* **50**, 573.
—— (1980). *Phys. Rev.* **D22**, 971.
—— Marciano, W. (1981). *Nucl. Phys.* **B189**, 442.
Slansky, R. (1981). *Phys. Rep.* **79C**, 1.
Slavnov, A. A. (1972). *Theor. Math. Phys.* **10**, 99.
Sterman, G. and Weinberg, S. (1977). *Phys. Rev. Lett.* **39**, 1436.
Stueckelberg, E. C. G. and Peterman, A. (1953). *Helv. Phys. Acta.* **26**, 499.
Sudakov, V. V. (1956). *Sov. Phys. JETP* **3**, 65.
Sudarshan, E. C. G. and Marshak, R. E. (1958). *Phys. Rev.* **109**, 1860.

Sushkov, O. P., Flambaum, V. V. and Khriplovich, I. B. (1975). *Sov. J. Nucl. Phys.* **20**, 537.
Susskind, L. (1979). *Phys. Rev.* **D20**, 2619.
Sutherland, D. G. (1966). *Phys. Lett.* **B23**, 384.
—— (1967). *Nucl. Phys.* **B2**, 433.
Symanzik, K. (1970*a*). In *Fundamental interactions at high energies* (ed. A. Perlmutter). Gordan and Breach, New York.
—— (1970*b*). *Comm. Math. Phys.* **18**, 227.
Szalay, A. and Marx, G. (1976). *Astron. Astrophy.* **49**, 437.
Takahashi, Y. (1957). *Nuovo Cimento* **6**, 370.
Taylor, J. C. (1971). *Nucl. Phys.* **B33**, 436
—— (1976). *Gauge theory of weak interactions*. Cambridge University Press, Cambridge.
Taylor, R. E. (1975). In *Proc. 1975 Int. Symp. on Lepton and Photon Interactions at High Energies* (ed. W. T. Kirk), p. 679. Stanford, California.
't Hooft, G. (1971*a*). *Nucl. Phys.* **B33**, 173.
—— (1971*b*). *Nucl. Phys.* **B35**, 167.
—— (1972). Unpublished remarks at the 1972 Marseille Conference on Yang-Mills Fields as quoted, for example, in (Politzer 1974), p. 132.
—— (1973). *Nucl. Phys.* **B61**, 455.
—— (1974). *Nucl. Phys.* **B79**, 276.
—— (1976). *Phys. Rev. Lett.* **37**, 8.
—— Veltman, M. (1972). *Nucl. Phys.* **B44**, 189.
Tomonaga, S. (1948). *Phys. Rev.* **74**, 224.
Treiman, S. B., Jackiw, R. and Gross, D. J. (1972). *Lectures on current algebra and its applications*. Princeton University Press.
Tsao, H.-S. (1980). In *Proc. 1980 Guangzhou Conf. on Theor. Part. Phys.* Science Press, Beijing, China.
Toussaint, D., Treiman, S. B., Wilczek, F. and Zee, A. (1979). *Phys. Rev.* **D19**, 1036.
Utiyama, R. (1956). *Phys. Rev.* **101**, 1597.
Veltman, M. (1967). *Proc. R. Soc.* **A301**, 107.
—— (1970). *Nucl. Phys.* **B21**, 288.
Ward, J. C. (1950). *Phys. Rev.* **78**, 1824.
Weinberg, S. (1960). *Phys. Rev.* **118**, 838.
—— (1966). *Phys. Rev. Lett.* **17**, 616.
—— (1967). *Phys. Rev. Lett.* **19**, 1264.
—— (1972*a*). *Phys. Rev.* **D5**, 1412.
—— (1972*b*). *Phys. Rev. Lett.* **29**, 1698.
—— (1972*c*). *Gravitation and cosmology*. Wiley, New York.
—— (1973*a*). *Phys. Rev.* **D8**, 3497.
—— (1973*b*). *Phys. Rev. Lett.* **31**, 494.
—— (1975). *Phys. Rev.* **D11**, 3583.
—— (1976*a*). *Phys. Rev. Lett.* **36**, 294.
—— (1976*b*). *Phys. Rev.* **D13**, 974.
—— (1977). In *Festschrift for I.I. Rabi* (ed. L. Motz). New York Acad. Sci., New York.
—— (1978). *Phys. Rev. Lett.* **40**, 223.
—— (1979*a*). *Phys. Rev.* **D19**, 1277.
—— (1979*b*). *Phys. Lett.* **82B**, 387.
—— (1979*c*). *Phys. Rev. Lett.* **42**, 850.
—— (1979*d*). *Phys. Rev. Lett.* **43**, 1566.
Weisberger, W. I. (1966). *Phys. Rev.* **143**, 1302.
Weyl, H. (1919). *Ann. Physik.* **59.2**, 101.
—— (1921). *Space–time–matter* (translated by H. L. Brose). Dover, New York (1951).
—— (1929). *Z. f. Physik.* **56**, 330.

Wick, G. C. (1950). *Phys. Rev.* **80**, 268.
Wilczek, F. (1977). *Phys. Rev. Lett.* **39**, 1304.
—— (1978). *Phys. Rev. Lett.* **40**, 279.
—— (1981). In Proc. 1981 Int. Sch. Subnucl. Phys. 'Ettore Majorana' (to be published).
—— (1982*a*). *Phys. Rev. Lett.* **48**, 1148.
—— (1982*b*). *Ann. Rev. Nucl. and Part. Sci.* **32**, 177.
—— Zee, A. (1979*a*). *Phys. Rev. Lett.* **43**, 1571.
—— —— (1979*b*). *Phys. Lett.* **88B**, 311.
Wilson, K. G. (1969). *Phys. Rev.* **179**, 1499.
—— (1971). *Phys. Rev.* **D3**, 1818.
—— (1974). *Phys. Rev.* **D14**, 2455.
—— (1975). In *New phenomena in subnuclear physics*, Proc. 1975 Int. Sch. Subnucl. Phys. 'Ettore Majorana' (ed. A. Zichichi). Plenum Press, New York.
Wolfenstein, L. (1964). *Phys. Rev. Lett.* **13**, 352.
—— (1979). In *Neutrino 79* (Bergen) (ed. A. Haatuft and C. Jarlskog). European Phys. Society.
Wu, C. S. *et al.* (1957). *Phys. Rev.* **105**, 1413.
Wu, T. T. and Yang, C. N. (1964). *Phys. Rev. Lett.* **13**, 380.
—— —— (1976). *Nucl. Phys.* **B107**, 365.
Wybourne, B. (1974). *Classical groups for physicists*. Wiley-Interscience, New York.
Yang, C. N. (1975). In *Proc. 6th Hawaii Topical Conf. Part. Phys.* (ed. P. N. Dobson). University Press of Hawaii, Honolulu.
—— Mills, R. (1954). *Phys. Rev.* **96**, 191.
Yao, Y. P. (1973). *Phys. Rev.* **D7**, 1647.
Yoshimura, M. (1978). *Phys. Rev. Lett.* **41**, 281.
Zee, A. (1973*a*). *Phys. Rev.* **D7**, 3630.
—— (1973*b*). *Phys. Rev.* **D8**, 4038.
—— (1982). *The unity of forces in the universe*. I and II. World Science Press Singapore.
Zimmermann, W. (1970). In *Lectures on elementary particles and quantum field theory*, Proc. 1970 Brandeis Summer Institute (ed. S. Deser *et al.*). MIT Press, Cambridge, Massachusetts.
Zweig, G. (1964*a*). CERN report no. 8182/TH 401.
—— (1964*b*). CERN report no. 8419/TH 412.

Index